7th International Conference on Food and Agriculture (ICoFA 2024)

AA007126

IOP Conference Series: Earth and Environmental Science
Volume 1446

Banyuwangi, Indonesia
12 October 2024

ISBN: 979-8-3313-1422-4
ISSN: 1755-1307

Printed from e-media with permission by:

Curran Associates, Inc.
57 Morehouse Lane
Red Hook, NY 12571

Some format issues inherent in the e-media version may also appear in this print version.

Institute of Physics
Dirac House, Temple Back
Bristol BS1 6BE UK

Phone: 44 1 17 929 7481
Fax: 44 1 17 920 0979

techtracking@iop.org

Additional copies of this publication are available from:

Curran Associates, Inc.
57 Morehouse Lane
Red Hook, NY 12571 USA
Phone: 845-758-0400
Fax: 845-758-2633
Email: curran@proceedings.com
Web: www.proceedings.com

TABLE OF CONTENTS

The 7th International Conference on Food and Agriculture (ICoFA)
Introduction of ICoFA 2024

Dear colleagues,

We are glad to announce the release of the 7th International Conference of Food and Agriculture (ICOFA-2024). The ICOFA is a conference held by Politeknik Negeri Jember regularly every year. This year the theme of our conference is "Empowering Global Society in The Field of Agriculture, Food, Animal Husbandry, and Energy to Respond Climate Emergency for Sustainable Future"

Surely, we plan to conduct this conference physically and virtually (hybrid). However, due to the new normal working methods based on "Work From Anywhere (WFA)", the 7th ICOFA 2024 was carried out physically and virtually via the Zoom meeting platform. We took this option because this conference was already designated and funded. Keynote and invited speakers were also scheduled for this event. Many delegations and authors requested for this conference to be performed since they needed it to cover their publication and sharing knowledge requirements.

The key directions of the conference covered the discussion of advanced achievements and touched upon the main aspects of Agricultural Production and Agricultural Technology, Animal Husbandry and Fisheries, Agribusiness Management, Engineering, and Information Technology.

The conference was run as planned on 12th November 2024 and supported by a hybrid event organizer started from 9.00 am to 5.00 pm. The conference model are Plenary session with keynotes (40 minutes for each keynote speech) and participants' presentations (20 minutes for each presentation). A total of 135 participants from Universities, Research Institutes, and also Government Departments attended the conference in over four countries India, Indonesia, Malaysia, and Philippines.

Our honorable speakers were Prof. Benjamin C. McLellan from Graduate School of Energy Science, Kyoto University, Japan, Dr. Sumeth Piayura from Department of Food Technology, Sakon Nakhon Rajabhat University, Thailand, and Prof. Dr. Ir. Titi Candra Sunarti, M.Si, from Department of Agro-Industrial Technology Institut Pertanian Bogor, Indonesia.

An amount of 103 full papers was submitted to the 7th ICOFA 2024. After a pre-review on originality and language, the peer review process was arranged by the Editorial Committee, and 74 manuscripts were selected for publication in the IOP Conference Series: Earth and Environmental Science (EES).

We also wish to thank to keynote speaker, invite speakers, presenters and participants, and all authors for their contributions to the 7th ICOFA 2024 and to colleague members of the organizing committee, please accept my deep appreciation for your hard work in ensuring the success of the conference.

Editor-in-Chief, 7th ICOFA 2024
Dr. Yuana Susmiati, S.TP, MSi

IOP Conf. Series: Earth and Environmental Science **1446** (2025) 011001 doi:10.1088/1755-1315/1446/1/011001

Scientific Committee the 7th ICOFA 2024

Name	Institution
Prof. Benjamin C. McLellan	Graduate School of Energy Science, Kyoto University, Japan
Dr. Sumeth Piayura	Department of Food Technology, Sakon Nakhon Rajabhat University, Thailand
Prof. Dr. Ir. Titi Candra Sunarti, M.Si.	Department of Agricultural Technology, Institut Pertanian Bogor
Dr. Yuana Susmiati, S.TP., M.Si	Politeknik Negeri Jember, Indonesia
Dr. Silvia Oktavia Nur Yudiastuti, STP., MTP	Politeknik Negeri Jember, Indonesia
Dr. Ir. Budi Hariono, M.Si	Politeknik Negeri Jember, Indonesia
Dr. Elly Kurniawati, S.TP., MP	Politeknik Negeri Jember, Indonesia
Dr. Ir. Hariadi Subagja, SPt., M.P., IPM	Politeknik Negeri Jember, Indonesia
Dr. Nurul Zainal Fanani, S.T., M.T	Politeknik Negeri Jember, Indonesia
Dr. Denny Trias Utomo, S.Si, M.T	Politeknik Negeri Jember, Indonesia
Dr. Ir. Dadik Pantaya, M.Si., IPU	Politeknik Negeri Jember, Indonesia
Wendy Triadji Nugroho, ST., MT., M.Sc.Eng., Ph.D	Politeknik Negeri Jember, Indonesia
Dr. Ir. Nantil Bambang Eko Sulistyono	Politeknik Negeri Jember, Indonesia
Netty Ermawati, SP., Ph.D	Politeknik Negeri Jember, Indonesia

Technical Committee the 7th ICOFA 2024

Name	Position
Saiful Anwar, STP., MP	Chairman of International Advisory Board
Surateno, S.Kom., M.Kom	Advisor
Dr. Yossi Wibisono, STP., MP	Advisor
Wahyu Kurnia Dewanto, S.Kom., M.T	Advisor
Agung Wahyono, SP., MSi., Ph.D	Advisor
Wendy Triadji Nugroho, ST., MT., M.Sc.Eng., Ph.D	Chairman of Organizing Committee
Muhammad Yunus, S.Kom., M.Kom.	Secretary
Dr. Yuana Susmiati, S.TP., M.Si	Editor in Chief ICOFA
Dr. Ir. Budi Hariono, M.Si	Editor
Dr. Elly Kurniawati, S.TP., MP	Editor
Dr. Ir. Hariadi Subagja, SPt., M.P., IPM	Editor
Dr. Nurul Zainal Fanani, S.T., M.T	Editor
Dr. Denny Trias Utomo, S.Si, M.T	Editor
Dr. Ir. Dadik Pantaya, M.Si., IPU	Editor
Dr. Ir. Nantil Bambang Eko Sulistyono	Editor
Netty Ermawati, SP., Ph.D	Editor
Dr. Silvia Oktavia Nur Yudiastuti, STP., MTP	Financial Treasury Coordinator
Mery Hadiyah Wati, S.Kom	Financial Treasury Support
Khafidurrohman Agustianto, S.Pd, M.Eng	IT Publication System Coordinator
David Juli Ariyadi, S.Kom., M.Kom	IT Publication System Support

IOP Conf. Series: Earth and Environmental Science **1446** (2025) 011001 doi:10.1088/1755-1315/1446/1/011001

Evan Hendra Lukito, A.Md	IT Publication System Support
Ahmad Nuril Firdaus, SE	Publication and Book Program
Yahya Ade Wahyu Prasetya, A.Md	Publication and Book Program
Irene Margaret, A.Md	Publication and Book Program
Beni Widiawan, S.ST, M.T	Event Coordinator
Rizki Amalia Nurfitriani, S.Pt., M.Si.	Event Coordinator Support
Resti Pranata Putri, S.Si., M.Sc.	Event Coordinator Support
Risse Entikaria Rachmanita, S.Pd., M.Si.	Event Coordinator Support
Anugrah Trial Akbar, A.Md	Supporting Staff
Syafiratul Hikmah, A. Md	Supporting Staff

IOP Conf. Series: Earth and Environmental Science **1446** (2025) 011002 doi:10.1088/1755-1315/1446/1/011002

Peer Review Statement

All papers published in this volume have been reviewed through processes administered by the Editors. Reviews were conducted by expert referees to the professional and scientific standards expected of a proceedings journal published by IOP Publishing.

- **Type of peer review:** Single Anonymous
- **Conference submission management system:** Morressier
- **Number of submissions received:** 73
- **Number of submissions sent for review:** 73
- **Number of submissions accepted:** 62
- **Acceptance Rate (Submissions Accepted / Submissions Received × 100):** 84.93
- **Average number of reviews per paper:** 1
- **Total number of reviewers involved:** 7
- **Contact person for queries:**
  Name: Muhammad Yunus
  Email: m.yunus@polije.ac.id
  Affiliation: Politeknik Negeri Jember

IOP Conf. Series: Earth and Environmental Science **1446** (2025) 012001 doi:10.1088/1755-1315/1446/1/012001

Urban farming analysis to enhances sustainable development through the utilization of existing land in urban areas

N Istiqomah[1], L Asnan Qodri[2], M P Priambodo[1], N A Yunikawati[1]

[1]Department of Development Economics, Faculty of Economics and Business, Universitas Negeri Malang
[2]Department of Development Economics, Faculty of Economics and Business, Universitas Merdeka Malang

Corresponding author: nimatul.istiqomah.fe@um.ac.id, asnan.lutfi@unmer.ac.id

Abstract. Urban areas are one of the centers of growth in a region. There are government centers, offices, and dense settlements. This study uses an exploratory qualitative method. Qualitative data is used to find out and find information in actual conditions in real life. The qualitative stage is carried out by in-depth interviews with the snowball sampling method. The study was located at the Gelintung Go Green Urban Farming Village, Purwantoro Blimbing Village, Malang City. The results of this study are that the application of urban farming in the Purwantoro area creates various benefits that are seen from economic, social, and ecological aspects. The implementation of Gelintung Go Green Urban Farming is realized through the development of the younger generation, the development of the community's socio-economic, the Green Business Movement, and the Water Saving Movement. This study concludes that from an economic perspective, Social, and Ecology. The society of Kampung Gelintung Go Green feels that their economy has improved, especially with the implementation of greening until it is realized as Kampung Wisata Gelintung Go Green, the most basic social aspect of its benefits is the increase in interaction between residents in various places.

1. Introduction

Protection of agricultural land is an inseparable part of spatial planning. [1] [2] [3] A city will depend on rural areas around the city, especially in providing food. [4] [5] Therefore, protection of agricultural land needs to be done by determining agricultural areas that need to be protected. [6] [7] [8] Not only in rural areas, but in reality agricultural land located in urban areas also needs protection. [9] [10] Protection of agricultural areas and agricultural land includes planning and determination, development, research, utilization and guidance, control, supervision, development of information systems, protection and empowerment of farmers, community participation, the role of academics and financing. Protection of agricultural areas and land is carried out by respecting local cultural wisdom and customary communal rights. [11] [12] [13]

One solution in striving for a food security system is with an urban agricultural policy or commonly called "Urban Farming". [14] [15] The urgency of implementing the Urban Farming concept due to the increasing population growth rate, high urbanization rate, [16] [17] increasing poverty in urban areas threatens the food security situation of Bandung City itself. Therefore, Urban Farming is a very important alternative solution. [18] [19] In relation to urban dynamics, one solution to this problem is to increase urban productivity with the Urban Farming concept. This Urban Farming concept can be applied through the utilization of narrow land, abandoned land that has not been or is not used at all such as yards, marginal land or remaining land on riverbanks, roadsides, alleys. Basically, urban farming uses polybag media, pots, yards and technological innovations such as green houses, organic farming, vertical farming.

Urban Farming can increase food security by improving the availability and reducing household food expenditure. [20] As in the Malang City area there is an "Urban Farming" Village, namely on Jl Citepus 2, RW 04, Padjajaran, Cicendo, Bandung City. This is one form of positive innovation implementation, besides being a solution in utilizing land in an urban area in realizing food independence, especially for local residents of the gardening village RW 04 Cicendo, Bandung City. The geographical conditions of the village adjacent to the river make this village sometimes isolated from flooding.

Gelintung Go Green Village or what is called 3G Village is a village with the concept of Urban Farming Village, where the village is located in the middle of the city and only has minimal land. Gelintung Go Green Village which is located in RW 23 consists of 4 Neighborhood Associations. The total population consists of 320 Heads of Families or around 1,154 People. However, the proportion of productive age in RW 23 is around 800 people. This can certainly be an opportunity to build Gelintung Go Green Village to develop. The majority of the livelihoods of the residents of RW 23, Gelintung Village, are private employees, factory workers and small entrepreneurs. The boundaries of Gelintung Go Green Village RW 23 to the north are RW 06, Blimbing Village/Carrefour, to the south is RW 05, Purwantoro Village/Tax Office, to the west is the main road into Malang City, while to the east is the Railway Line.

Problems that occur in narrow urban areas often become local problems for residents to get access to land for planting. [21] [22] The narrow location of agriculture and land around the house encourages residents to find a solution to be able to utilize the surrounding environment into a productive area and can produce various social and economic benefits for the community. [23] [24] One alternative is to open an urban farming area. [25] This article will discuss how the community applies urban farming as one of the main forces in realizing productive areas that have social and economic value.

2. Method

This study uses an exploratory qualitative method. While the data used is qualitative data. Qualitative data is used to find out and seek information in its actual conditions in real life. The qualitative research process involves important efforts, such as asking questions and procedures, collecting specific and real data from participants, analyzing data inductively from specific themes to general themes, and interpreting the meaning of the data. In this method, the researcher uses the Explanatory Survey approach, namely making observations that occur, and looking for factors that may be the cause through certain data. The qualitative research stages were carried out by means of in-depth interviews with the main source, namely the Head of RW 23, Purwantoro Village, Blimbing District, Malang City. This is because the Head of RW, namely Bambang Irianto, is the initial initiator of the Gelintung Go Green Urban Farming Village.

In addition to Bambang Irianto as the head of RW, interviews were also conducted with residents of the Gelintung Go Green community who were selected randomly. The results of the interviews were then analyzed using thematic categories related to the processing process. This is certainly done to see the extent of participation carried out by the surrounding community in realizing the Urban Farming Gelintung Go Green village. In this study, the data sources used are primary data and secondary data. In the primary data, the researcher conducted observation techniques, interviews with the community of RW 23, Purwantoro Village in depth and continued with documentation. The researcher took secondary data from related agencies such as RW, the Department of Agriculture and Food Security of Malang City.

Data processing in this study is divided into four stages, namely:

2.1 Data collection

Data collection in this study is collecting data that has been obtained from direct field observations.

2.2 *Data reduction*
Data reduction is the combination of data that has been obtained and made into a single written form (script) that will be analyzed according to each format.
2.3 *Data display (data presentation)*
This data presentation can be done in the form of a brief description. Can use tables or graphs. Data presentation is intended so that researchers can easily understand the data as a whole and can also help determine the activities that will be carried out to analyze its contents.
2.4 *Drawing conclusions*
Concluding is a process where the initial conclusions made by researchers are still unclear so that they become detailed conclusions until finally the conclusion of a study is obtained.

3. Results and discussion

Results

3.1 Analysis of the Vision and Mission of Gelintung Go Green Village

3.1.1 Development of the young generation. Based on the researcher's findings in the field, the researcher has seen the activities of the Qu'ran Education Park which are carried out every day after the Asr prayer, in addition, a routine activity of developing the environment for young people has been carried out through education and education given to the children of Kampung Gelintung Go Green which also collaborates with universities that happen to be conducting KKN in Kampung Gelintung Go Green. This is very important where TPQ is formed to form a good character and morals in the future because regeneration is very much needed to realize the sustainability of Kampung Gelintung Go Green. While the intellectual level is built in synergy with universities around Malang City. With this, sustainability or regeneration in the future will run well.

3.1.2 Community social economic development. Economic development in the community in Kampung Gelintung Go Green in the form of socialization of education on how to make various crafts, delegation of responsibility for each spot whose purpose is to be a guide for tourists when there is a visit, as well as empowerment given to the community when there is a visit, especially related to snacks, accommodation, services and assistance. In addition, the promotion on the website of the role of the 3G Cooperative is very important to provide stimulus to residents who want to innovate. Related to the promotion of sales of hydroponic planting media, it is also very helpful in increasing community income. Of course, everything is obtained from comparative studies that have been held by Kampung Gelintung Go Green itself or in collaboration with the Malang City Government.

3.1.3 Green business movement. Green business movement is based on business activities or buying and selling in the form of hydroponic planting media, plant seeds, and educational tourism. Greening can indirectly provide a multiplier effect on the community, in addition to the increasingly beautiful scenery and increasing air quality, holding an educational tour that brings tourists from outside also provides great benefits to the community's economy. In this educational tour is a business activity that brings benefits to the community without ignoring environmental sustainability. In the midst of conventional businesses that have existed in the modern era, this business is one of the good solutions and has a sustainable development character. However, the focus of human resource capacity must also be considered because it is to overcome the high level of demand someday.

3.1.4 Water saving movement. The implementation of the Water Saving Movement was initiated in 2017 and has been inaugurated by the Ministry of Home Affairs. Even this effort has yielded results, namely as the 15th best village development in the world according to the International Award for Urban Innovation in Guangzhou, China. Based on the findings of researchers, Gelintung Village has injection wells and Biopores that are very good for overcoming flooding which is the main problem of the village before the greening was realized at that time. The Water Saving Movement itself can be utilized in realizing Micro Hydro Electricity, which can later provide independent electricity for Gelintung Village RW 23, Purwantoro Village.

3.2 Road map for the development of gelintung go green village
Kampung Gelintung Go Green also has its own motto which is already known worldwide, namely "Eat Fruits and Vegetables Before Medicine Becomes Your Food". This motto is intended to motivate villagers to plant vegetables that are beneficial because vegetables are one source of vitamins for our health. In the implementation of realizing a village with a greening concept, it is not easy, of course, sacrifices are needed from the beginning of planning to implementation. It is also certainly inseparable from problems and obstacles in the implementation of Kampung Gelintung Go Green. Therefore, through an organization led by the Regional Manager who is also the Head of RW 23, a road map was built for the realization of Kampung Gelintung Go Green from the initial phase to implementation and monitoring.

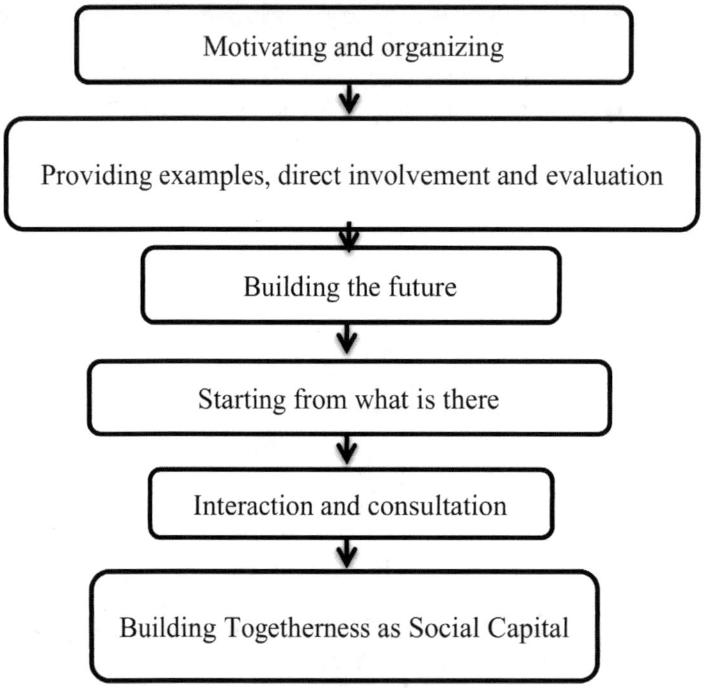

Figure 1. Road Map for the Development of Gelintung Go Green Village
This road map is needed to unite thoughts and future goals from the beginning of the idea of Kampung Gelintung Go Green, now, and in the future.
3.2.1 Motivating and organizing. In the beginning of the idea, motivation was very much needed, this motivation came from Mr. Bambang Irianto as the initial initiator of the establishment of Kampung Gelintung Go Green. This motivation is related to the beauty of the village that will be obtained from the realization of a village with a greening concept. In addition, the main problem of the village regarding flooding will also be resolved along with the implementation of the greening concept, which later the flood water can be absorbed into the ground properly.

I told my people at that time that in the next five years there will be one to two buses that will visit here to see our beautiful and green village. Motivation

This motivation is very much needed when the condition and trust of the community in realizing the greening is low, besides the RW cash fund which is only Rp3,030,000 is still not enough to realize the greening. When the motivation and hope are given, the community response can only rely on funds from the local government through Musrenbang. However, the RW as the community leader thinks that this will actually take a long time to implement it. From there, the community is required to work together to build independence in realizing Kampung Gelintung Go Green through deliberation efforts between RTs for one year, community awareness then begins to build.

3.2.2 Providing examples, direct involvement and evaluation. In realizing a goal, there must be someone who initiates it, especially in the context of society, because a society with such a large population has different characters, so a leader is needed who is able to move and provide an example in terms of implementation. Mr. Bambang Irianto as the Head of RW and the initiator of the realization of Kampung Gelintung Go Green is always directly involved in every deliberation event or joint work service held every week. Not only that, every community that has awareness and is oriented towards success must also provide a real example in the implementation of realizing Kampung Gelintung Go Green. In practice, this is one of the capitals where the community is given an example to realize their dreams in the future. In this case, the community has the belief that the culture, mentality and mindset that have been embedded in the Gelintung Go Green community can be used as capital in realizing the ideals of building a village.

The Gelintung Go Green RW 23 community is trying to dismantle the old paradigm where the development process is still monotonous, in the sense that it still depends on other parties, namely the local government. Therefore, every community that seems to have the same thoughts at the beginning of the Gelintung Go Green Village idea is required to play an active role in every activity.

As time goes by, there are certainly some people who do not have the same thoughts and goals in its implementation, for example, not participating in community service, rarely participating in village development discussions and not wanting to plant plants in front of their houses. Therefore, as the Head of the RW and the initiator of the Gelintung Go Green Village, he determined that every resident who is reluctant to plant plants in front of their house will not be served for their administrative needs. This is quite phenomenal where many people need the simplest RW stamp for various permits. This policy also made the community react strongly to the point of judging that their RW is no longer healthy. However, from this, the community was moved to participate by at least planting plants in front of their houses and it would be even better if they were made from used items such as used cans, used pans and so on.

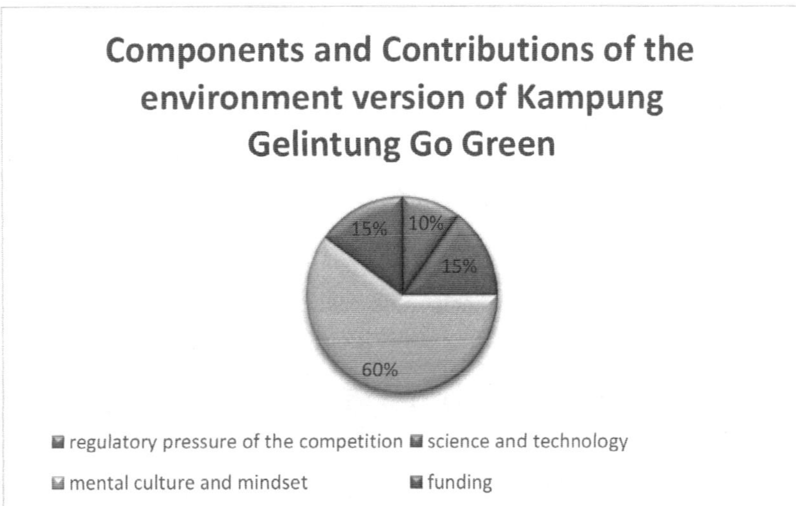

Figure 2. Components and Contributions of the environment version of Kampung Gelintung Go Green

According to the diagram above, it is a concept of thought of the initiator of Kampung Gelintung Go Green, Culture, Mentality and Mindset are the main capital for the formation of a common goal, namely realizing Kampung Gelintung Go Green. The initial doctrine in realizing the greening concept is "starting from ourselves" because it has been described by the Founding Fathers in the song Indonesia Raya, namely "Build your soul" and "Build your body" because a strong body will grow in a healthy soul.

The Gelintung Go Green Community RW 23 is trying to dismantle the old paradigm where the development process is still monotonous, in the sense that it still depends on other parties, namely the local government. Therefore, every community who has the same thoughts at the beginning of the idea of Kampung Gelintung Go Green is required to play an active role in all its activities.

Over time, there must have been some people who did not have the same thoughts and did not have the same goals in its implementation, for example not participating in community service, rarely participating in village development discussions and not wanting to plant plants in front of their houses. Therefore, as the Head of RW and the initiator of Kampung Gelintung Go Green, he determined that every resident who is reluctant to plant plants in front of their house will not be served their administrative needs. This is quite phenomenal where many people need the simplest RW stamp for various licensing needs. This policy also made the community react strongly to the point of judging that their RW was no longer healthy. However, from that, the community was moved to participate at least by planting plants in front of their house and it would be even better if it was from used goods such as used cans, used pans and others.

3.2.3 Building the future. Efforts to build the future have always been an encouragement for every community effort to build Kampung Gelintung Go Green. From a hope to build the future, a cheerful atmosphere will be created for every resident because they always remember their beautiful dreams. As time goes by, the cheerful atmosphere will create an optimistic soul, always enthusiastic, especially if every effort is done together. The atmosphere is always happy, because the community collectively has a dream of a beautiful, green, and comfortable village, as time goes by, even slowly the dream has shifted, as if the atmosphere is already in their grasp, always present, in everyday life.

Implementation of an action to realize a green, clean, comfortable, and shady village according to the initial agreement in a simple way by involving experts in their fields. One example of this small action according to the understanding of the Gellintung community is about the breath that we breathe every day, namely with the assumption that every day we breathe 2.88 liters of air / day. The image below explains the calculation of the total oxygen we need assuming we live to the age of 50 years multiplied by the unit price of oxygen according to WHO (World Health Organization).

3.2.4 *Starting from what is there*. In the initial planning to realize Kampung Gelintung Go Green, of course, it requires a large capital, but the RW cash at that time which was only Rp 3,030,000 was certainly not enough to finance the needs of equipment and materials. Therefore, residents are required to start from existing materials for planting media such as used pots, used pans, used cans, etc. Community service is also often encouraged to create a vertical garden concept that is placed on the side of the alley entrance made of bamboo requested from the neighboring village.

This movement is a social capital to build a village that starts from what is there and is also required to act practically. Practical action as an option, because through practical action it requires little energy with a short time. On the other hand, impractical actions tend to be burdensome, will drain energy and take time to realize it. Therefore, the action that must be taken is to start from oneself, from now on, and from what is most likely to be done. The habit of finishing work as early as possible by the Gelintung Go Green community is a manifestation of practical action, because by postponing tomorrow it will actually have its own difficulties and be difficult to do. In this way, it can reduce the reasons for delaying, so that practical action is more efficient and effective in realizing the collective dream of Kampung Gelintung Go Green. As an illustration, the residents of RW 23 are of productive age and on average work. Therefore, when the collective dream of a beautiful, comfortable, and shady Kampung Gelintung Go Green is carried out through mutual cooperation and cooperation to work together which is only done during the day, it tends to still take a long time. As a result, it will certainly affect the achievement of collective dreams. Therefore, residents began to make an innovation or breakthrough in terms of the schedule of community service activities, because the spirit of practical action was so strong, residents made an innovation of community service at night, when residents finished working during the day to earn a living. Community service at night to realize a shared dream by working together is what was later given the term "Suku Dalu". In implementing practical actions, Bambang Irianto as the initiator of Kampung Gelintung Go Green has the slogan "four pillars". The slogan is certainly simple, but very effective in dealing with all unexpected possibilities that will affect the achievement of results. The four pillars are time, reason, garbage, and friends.

3.2.5 Interaction and consultation. As the Head of RW 23 Gelintung who is also the Regional Manager, of course he also has limitations related to realizing Kampung Gelintung Go Green, therefore this consultation is needed to help in terms of implementing Kampung Gelintung, especially regarding each discipline. The interaction and consultation have been achieved with various agencies and media such as Malang Post, Government, Brawijaya University Malang, and Earth Hour Indonesia.

This Interaction and Consultation Phase is related to technical matters such as workshops, attending seminars aimed at residents of Kampung Gelintung Go Green and competitions, etc. The involvement of experts from the local government such as DKP (Department of Cleanliness and Parks in accordance with their fields certainly also provides real contributions to the Kampung Gelintung Go Green program which can directly provide confidence that the development of Kampung Gelintung Go Green which is based on needs from below (Bottom up) can also be the target of community service programs from universities such as Brawijaya University. Likewise with the reciprocity given by the Brawijaya University Campus, namely by contributing concepts about making injection wells that can be implemented in the Kampung Gelintung environment which is famous for its flood problems. This is also in accordance with the implementation of the Vision and Mission related to the procurement of 1000 Biopores, Injection Wells, and Micro Hydro Electricity in Kampung Gelintung Go Green.

3.2.6 *Building Togetherness as Social Capital.* The process of education and awareness for the community about the importance of plants is not instant. It requires commitment from leaders as role models and the seriousness of the residents of Kampung Gelintung Go Green in realizing togetherness to realize their collective dreams. To realize awareness of development that is jointly needed, continuous and consistent efforts need to be made, developing various methods of awareness for the community, especially for people who are pessimistic about the ideals of Kampung Gelintung Go Green in the future. This is a challenge for the management of Kampung Gelintung Go Green.

Realizing the weight of the challenge, Mr. Bambang Irianto as the Regional Manager took the initiative to build togetherness in realizing a beautiful, comfortable, and shady village with the motto Gelintung Go Green. One of the methods used is to use the RW stamp as a means of education, namely that it is required for residents who are taking care of administrative letters, the person concerned must be able to show the RW administrator that the person concerned has planted plants in front of his house. The education is that planting is not by buying and must be done by yourself without the help of others. For residents who will take care of letters for administrative purposes and happen to have no plants at home, the letter will not be signed by Bambang Irianto as the RW Head. This method turns out to be quite effective in encouraging the participation of village residents to realize Kampung Gelintung Go Green.

Discussion

Urban Farming activities in Kampung Gelintung Go Green by utilizing narrow urban land which is known for its hustle and bustle of urban situations that cause pollution and low air quality is one of the renewable solutions, especially in Indonesia, [26] [27] this is also intended to create sustainable development that is environmentally conscious. The initiator of Kampung Gelintung Go Green saw that the worsening level of environmental quality caused various problems for residents, especially RW 23, Purwantoro Village, Blimbing District. Flooding, which is the main problem of Kampung Gelintung, is also a classic problem experienced, especially in narrow villages in urban areas. This problem has a complex effect on people's lives, one of which is degenerative diseases, ulcers, nausea, skin diseases, heart disease, asthma and others. In addition, of course, it also has an impact on the economic sector of residents, especially residents of RW 23, Purwantoro Village. In terms of human resources, most of the community's level of concern for environmental preservation is low and tends to be indifferent. Based on these problems, the RW and the community work together to find solutions to the problems of Kampung Gelintung to become a green, beautiful, and comfortable village.

Urban Farming activities in Kampung Gelintung Go Green have the aim of developing environmentally friendly (organic) and sustainable urban agriculture. This is expected to improve the quality of the environment and the problems in Kampung Gelintung Go Green, especially flooding and community welfare. In addition, an ideal from the residents of Kampung Gelintung Go Green can be realized, namely not only realizing greening but also creating an Educational Tourism Village using Green Business elements. [28] [29]

By involving various elements of society and the active leadership of the RW head, it is the main capital in the development of a village that was once known to be often hit by floods. Although the initial implementation often encountered various obstacles, these efforts did not discourage Bambang Irianto and the community who wanted change in their village. Although it had to go through a little coercion and threats not to be served with all administrative matters, it succeeded in making the community who were initially passive towards the development and implementation of Kampung Gelintung Go Green become involved, planting plants at least in front of their respective houses. Not only the role in terms of human resources, but greening also requires infrastructure support, especially in the village development stage. Meanwhile, developing environmentally friendly green infrastructure requires the development of harmony between humans and nature. Building a green environment is a long-term effort. For that, all efforts need to be made from a long-term perspective. Once stakeholders behave badly towards the local environment, it will have a bad impact and it will take a long time to restore public trust. Kampung Gelintung Go Green has formed an effort by involving elements of harmony between the environment and all its aspects. The concept of green infrastructure development is one of the efforts for sustainable development that is environmentally friendly according to the Green Economy concept. In addition, the Gelintung Go Green concept has been socialized in various meetings at the RW and RT levels for a period of one year, and in the end the Gelintung Go Green concept became the collective goal of the majority of Kampung Gelintung residents. This kind of change in the mindset of citizens is an important foundation for the sustainability of sustainable development.

The concept of Green Economy itself is a new paradigm in economic development that has replaced environmental policies that in the past often focused on short-term solutions. [30] [31] Green Economy can be simply defined as an economic concept that does not harm the environment. [32] Green Economy itself harmonizes economic growth, social welfare, and environmental sustainability. [33] Based on the above explanation, we can see that urban farming is one of the alternatives in food security in various regions. This is in accordance with research from Zezza and Tascioti [34]. In addition, the role of the community is very important in supporting urban farming programs in an area, especially in densely populated areas, in line with research from Surls, et al. [35]

4. Conclusion

From the problems that have been discussed, it can be concluded that when viewed from an economic aspect, the people of Gelintung Go Green Village feel that their economy has improved, especially with the implementation of greening until it was realized as Gelintung Go Green Tourism Village. Starting from income from the sale of souvenirs in the form of 3G t-shirts, 3G sandals, etc. made by the community itself, mentoring, selling small shops to renting accommodation for visitors can provide separate income for the community.

From the social aspect, the most basic benefit is the increase in interaction between residents in various places, to the continuation of a joint effort to work together to work together in realizing the ideals of greening, not only ending there, even after the realization of Gelintung Go Green Tourism

Village, in terms of cooperation to welcome visitors, the community does not hesitate to help each other, especially as tour guides in order to present the best quality of service for the good name of Gelintung Go Green Village.

From the ecological aspect, namely the creation of a plant planting medium that has so far been thought to only be implemented in rural areas that have large areas of land, but Kampung Gelintung Go Green proves that planting can also be done in urban areas. This is a form of preserving nature and a healthy environment.

References

[1] Handayani SW, Sujadi S, Sumantri AS. A critical review of the discrepancy between the spatial planning map and the spatial planning regulation of agricultural lands in Indonesia: Kulon Progo Regency case study. World Review of Science, Technology and Sustainable Development. 2020;16(4):343.

[2] Munir S, Siti Afiyah, Munir A. FORMS OF LEGAL PROTECTION DUE TO THE CONVERSION OF AGRICULTURAL LAND INTO RESIDENTIAL AND INDUSTRIAL LAND. EQUALEGUM International Law Journal [Internet]. 2023 [cited 2024 Oct 6];14–31. Available from: http://syntificpublisher.com/index.php/equalegum/article/view/2

[3] Suwari Akhmaddhian, Budiman H, Erga Yuhandra, Septian Apriditiya. Local Government Policy: Converting Agricultural Land to Housing. UNIFIKASI Jurnal Ilmu Hukum. 2021 Jun 29;8(1):28–33.

[4] van Berkum S. How Urban Growth in the Global South Affects Agricultural Dynamics and Food Systems Outcomes in Rural Areas: A Review and Research Agenda. Sustainability [Internet]. 2023 Jan 1;15(3):2591. Available from: https://www.mdpi.com/2071-1050/15/3/2591

[5] de Bruin S, Dengerink J, van Vliet J. Urbanisation as driver of food system transformation and opportunities for rural livelihoods. Food Security. 2021 Jun 28;13(4):781–98.

[6] Chabibah Kusuma Dewi, Harvini Wulansari. Agricultural Land Protection for Food Security: Policy Integration of Protected Rice Field Map in Spatial Planning in Demak Regency. Marcapada Jurnal Kebijakan Pertanahan [Internet]. 2024 Aug 20 [cited 2024 Oct 6];3(2):81–94. Available from: https://mail.jurnalmarcapada.stpn.ac.id/index.php/JM/article/view/52

[7] Mariane I, Listyowati Sumanto, Ignatius Pradipa Probondaru, Idris. Problems of Converting Agricultural Land and the Need to Anticipate Its Control After the Enactment of the Job Creation Law. Volksgeist Jurnal Ilmu Hukum dan Konstitusi [Internet]. 2024 Jun 30 [cited 2024 Oct 6];177–98. Available from: https://openurl.ebsco.com/EPDB%3Agcd%3A3%3A15004307/detailv2?sid=ebsco%3Aplink%3Ascholar&id=ebsco%3Agcd%3A178823125&crl=c

[8] Moran D, MacLeod M, Wall E, Eory V, McVittie A, Barnes A, et al. Developing carbon budgets for UK agriculture, land-use, land-use change and forestry out to 2022. Climatic Change. 2010 Aug 21;105(3-4):529–53.

[9] Nicolau R, Condessa B. Monitoring Net Land Take in a Metropolitan Region—Portugal. Urban Science [Internet]. 2024 Jan 18 [cited 2024 Oct 6];8(1):8–8. Available from: https://www.mdpi.com/2413-8851/8/1/8

[10] Alğın Demir K. Spatial transformation of agriculture in urban-rural relations: Torbalı district (İzmir). MEGARON / Yıldız Technical University, Faculty of Architecture E-Journal [Internet]. 2024 [cited 2024 Oct 6];231–44. Available from: https://jag.journalagent.com/megaron/pdfs/MEGARON_19_2_231_244.pdf

[11] Kartika Winkar Setya, Abdul Aziz Nasihuddin, Izawati Wook. Fulfilling Communal Rights through the Implementation of the Second Principle of Pancasila towards the Regulation on Agrarian Reform. Volksgeist. 2023 Jun 30;89–102.

[12] Wijaya, Made SI, Ni S, Nazarrudin T, Sulaiman. Patterns of Indigenous Land Tenure and Utilization: Comparison of Customary Law in Balinese and Acehnese People - Scholar Guardian. Sdpublishersorg [Internet]. 2024 Mar 18 [cited 2024 Oct 6]; Available from:

http://science.sdpublishers.org/id/eprint/2629/
[13] Y. Sinaga R, B. Barus L. LEGAL PLURALISM IN RURAL LAND RESILIENCE [Internet]. Nuruz Zaman M, editor. https://scientium.co.id/journals/index.php/jsdi/article/view/496. Journal of Sustainable Development Issues; 2023 [cited 2024 Oct 6]. Available from: https://scientium.co.id/journals/index.php/jsdi/index
[14] Gibson M. Food Security—A Commentary: What Is It and Why Is It So Complicated? Foods [Internet]. 2012 Dec 3;1(1):18–27. Available from: https://www.ncbi.nlm.nih.gov/pmc/articles/PMC5302220/
[15] Saed. Urban Farming: The Right to What Sort of City? Capitalism Nature Socialism. 2012 Dec;23(4):1–9.
[16] Akaeze O, Nandwani D. Urban agriculture in Asia to meet the food production challenges of urbanization: A review. Urban Agriculture & Regional Food Systems. 2020 Jan;5(1).
[17] Yan D, Liu L, Liu X, Zhang M. Global Trends in Urban Agriculture Research: A Pathway toward Urban Resilience and Sustainability. Land. 2022 Jan 11;11(1):117.
[18] Yan D, Liu L, Liu X, Zhang M. Global Trends in Urban Agriculture Research: A Pathway toward Urban Resilience and Sustainability. Land. 2022 Jan 11;11(1):117.
[19] Ritu Bharat Kukde, Negi P, Vikram Kad. Urban Agriculture: A New Dimension in Alternative Farming Systems for Achieving the Target of Food for All. Advances in Science, Technology & Innovation/Advances in science, technology & innovation. 2024 Jan 1;39–51.
[20] Ni`matul Istiqomah, None Wahjoedi, Lutfi Asnan Qodri, Mohd. Urban Farming in Pandemic Covid-19 and How the Economic Impact Analysis for Communities Kauman Village, Malang City. E3S web of conferences. 2023 Jan 1;444:01002–2.
[21] Lovell ST. Multifunctional Urban Agriculture for Sustainable Land Use Planning in the United States. Sustainability. 2010 Aug 4;2(8):2499–522.
[22] Nicholls E, Ely A, Birkin L, Basu P, Goulson D. The contribution of small-scale food production in urban areas to the sustainable development goals: a review and case study. Sustainability Science. 2020 Mar 28;15.
[23] Wästfelt A, Zhang Q. Keeping agriculture alive next to the city – The functions of the land tenure regime nearby Gothenburg, Sweden. Land Use Policy. 2018 Nov;78:447–59.
[24] Duan J, Ren C, Wang S, Zhang X, Reis S, Xu J, et al. Consolidation of agricultural land can contribute to agricultural sustainability in China. Nature Food [Internet]. 2021 Dec 1;2(12):1014–22. Available from: https://www.nature.com/articles/s43016-021-00415-5
[25] Nowysz A, Mazur Ł, Vaverková MD, Koda E, Winkler J. Urban Agriculture as an Alternative Source of Food and Water Security in Today's Sustainable Cities. International Journal of Environmental Research and Public Health. 2022 Nov 24;19(23):15597.
[26] Haryanto B. Climate Change and Urban Air Pollution Health Impacts in Indonesia. Climate Change and Air Pollution. 2017 Oct 6;215–39.
[27] Surya B, Muhibuddin A, Suriani S, Rasyidi ES, Baharuddin B, Fitriyah AT, et al. Economic Evaluation, Use of Renewable Energy, and Sustainable Urban Development Mamminasata Metropolitan, Indonesia. Sustainability. 2021 Jan 22;13(3):1165.
[28] Roman M, Iryna Kudinova, Viktoriia Samsonova, Kawęcki N. Innovative Development of Rural Green Tourism in Ukraine. Tourism and Hospitality [Internet]. 2024 Jun 24 [cited 2024 Oct 6];5(3):537–58. Available from: https://www.mdpi.com/2673-5768/5/3/33
[29] Wulandari LD, D Asikin, Firsandy BA, Sari KE, A F I Johanda. Edu-Tourism Concept in Eco-Friendly Farm Development Design in Ngronggot Village, Nganjuk, East Jawa. IOP conference series Earth and environmental science. 2024 Mar 1;1310(1):012005–5.
[30] Mert Mentes. Sustainable development economy and the development of green economy in the European Union. Energy, Sustainability and Society. 2023 Aug 22;13(1).
[31] Endang Sungkawati. Opportunities and Challenges: Adopting "Blue-Green Economy" Terms to Achieve SDGs. Revenue Journal. 2024 Mar 14;2(1):01-13.
[32] Financial and Trade Globalization, Greener Technologies and Energy Transition [Internet].

Google Books. 2023 [cited 2024 Oct 6]. Available from: https://books.google.co.id/books?hl=en&lr=&id=nQitEAAAQBAJ&oi=fnd&pg=PA133&dq=%5B32%5D%09Zhang

[33] Radoslava Kanianska BANSKÁ BYSTRICA 2017 MATEJ BEL UNIVERSITY BANSKÁ BYSTRICA FACULTY OF NATURAL SCIENCES [Internet]. [cited 2024 Aug 2]. Available from: https://www.enviroportal.sk/dokument/f/green-growth-and-green-economy.pdf

[34] Zezza A, Tasciotti L. Urban agriculture, poverty, and food security: Empirical evidence from a sample of developing countries. Food Policy. 2010 Aug;35(4):265–73.

[35] Surls R, Feenstra G, Golden S, Galt R, Hardesty S, Napawan C, et al. Gearing up to support urban farming in California: Preliminary results of a needs assessment. Renewable Agriculture and Food Systems. 2014 Mar 5;30(1):33–42.

Effect of Refugia Plant Combination with Biopesticide Application on Population and Attack Intensity of *Spedoptera exigua* (Hubner) (Lepidoptera; Noctuidae)

A A Rezki[1], S Sjam[1*] and Tamrin[1]

[1] Department of Plant Pest and Disease, Faculty of Agriculture, Hasanuddin University, Makassar, Indonesia 90245

*Email: sylviasjam@yahoo.com

Abstract. *Spodoptera exigua* is an insect that is an essential pest in shallot plants. Spedoptera exigua is generally controlled with chemical insecticides, which can impact the environment and natural enemy populations. Using a combination of flowering plants and applying plant extracts is one of the environmentally friendly alternatives to reduce population and damage. This study was conducted in Enrekang Regency in South Sulawesi using an RCBD with five treatments and three replications. The treatments are flower, plant extract, flower and plant extract, farmer's method, and control. Flowering plants were planted at the edge of the bed. Observations were made visually and using a sweepnet. The application of plant extracts and observations were made seven times weekly. Observations were made on the parameters of egg groups, larval populations, attack intensity, and natural enemy populations. The results showed that the treatment refugia plant and plant extract reduced the egg population by 76.64% and the larval population by 92.59%. It increased the population of natural enemies by 72.26%. The farmer treatment reduced the intensity of the attack by 65.03%, which was not significantly different from the flower treatment with plant extracts, with a percentage reduction of 59.38%.

1. Introduction

Shallots are a horticultural crop commodity with tremendous business opportunities in the agricultural world. Farmers widely cultivate them in the low and highlands. [1]. This pest attack can cause heavy damage to the leaves and decrease production or yield loss [2]. The problem in the cultivation process of farmers is the high use of pesticides used to control S. exiqua, which is a crucial pest and attacks throughout the year, both dry and rainy seasons [3].

Flowering plant management systems in shallot plantations can reject the presence of pests for oviposition on the main plant. Still, it can attract beneficial insects, such as pollinators and natural enemies, to the shallot planting area [4].

This strategy uses a combination of behavior-altering stimuli to manipulate the distribution and abundance of pests and beneficial insects, such as natural enemies, for pest management [5]. These natural enemies will suppress the pest population so that there is no outbreak or explosion of pests that can harm shallot farmers [6].

Farmers can utilize Mary Gold or sunflower plants as attractant plants for natural enemies in the design arrangement system of flowering plants in shallot plants. Still, they can be repellent to *S. exigua* oviposition on shallot plants. So, there needs to be an alternative control by other means, including using flowering plants and vegetable pesticides against the onion caterpillar pest *S. exigua*.

2. *Material and methods*

2.1. Planting refugia plants

The refugia plants used were Mary Gold and Sun Refugia plants. Before planting, seedlings were first sown and then planted one week before the main plant. The layout of the refugia plants can be seen in Figure 1, which are red marigold and blue sun refugia plants placed on the edge of the bed. Figure 2 is the concept of the Refugia Plant trap, which aims to invite pests and natural enemies not to land on the central plants.

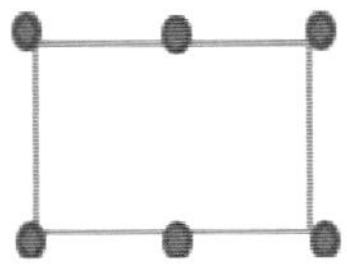

Figure 1. Location of refugia plants

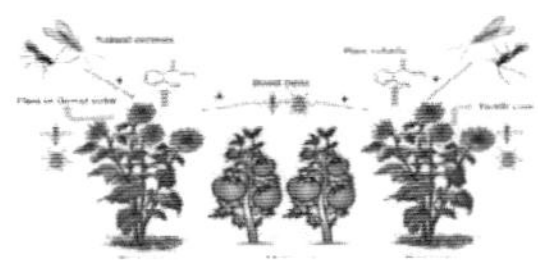

Figure 2. It is a natural enemy trap concept

2.2 Planting shallot plants

The main crop was planted using Super Philips seeds, which have a spicy flavor and are more extensive.

2.3 Application of extract

The application is done once every week, starting when the onion plants are 10 DAP. The concentration was 5%, with a total volume of 1 liter per plot and a sprayer size of 15 liters. The biopesticide is an extract formulated from a mixture of maja fruit and bidi from the laboratory of Hasanuddin University natural ingredients.

2.4 Observation Method

Initial observations were made one day before application (First observation at 10 DAP). Each plot was observed on ten clumps of onion plants diagonally and moving. There are two techniques for population observation: direct observation (visual) and observation with insect nets (sweeping net) sampling with a line transect pattern. Sweeping net observations were made by swinging the insect net three times with the direction of swinging left and right quickly.

2.5 Insect Identification

Insects obtained in the field were then sprayed into collection bottles with 70% alcohol. The insects observed were distinguished between those that were pests and natural enemies. Insects obtained in the field were then photographed and labeled based on the type of treatment.

2.6 Experiment design

This research method uses a Randomised Group Design (RGD) consisting of refugia plants, biopesticides, refugia plants and biopesticides, farmer methods, and controls with three replications. Observation parameters include egg groups, larval population, intensity of Spedoptera exigua, and attack by natural enemies. The formula for calculating the population and intensity of Spedoptera exigua attack is as follows [7].

Larval population level $P = \frac{n}{N}$

Description:
P = Pest Population
n = Number of larvae found on plants/clump
N = Number of clumps observed

Level of attack intensity $I = \frac{\Sigma(ni.vi)}{N.Z} \times 100\%$

Description:
I = intensity of attack (%)
ni = Number of leaves attacked with a particular attack scale
vi = Scale value for each attack category
Z = Scale of the highest attack category
N = Number of leaves observed[8].

Then, the results of the observations were analyzed using the ANOVA method. If the treatment affected the analysis, further tests were carried out using the Smallest Real Difference test (BNT) with a level of 5%.

3. Results and Discussion

3.1 Observation of *Spedoptera exigua* Egg Group

The results of the observation of the average group of S. exigua eggs with the results of the analysis of variance showed that using a combination of refugia plants with Biopesticides had a significant effect on the observation of 5 MST and 7 MST results that were significantly different from the group of Spedoptera exigua eggs in the BNT further test.

Table 1. Average Population of *Spodoptera Exigua* Egg Groups Per Plant Clump

Treatment	Observations						
	1 WAP	2 WAP	3 WAP	4 WAP	5 WAP	6 WAP	7 WAP
P0	0,13a	0,03a	0,10a	0,13a	0,11 a	0,10a	0,10 a
P1	0,00a	0,07a	0,00a	0,14a	0,07 ab	0,03a	0,10 a
P2	0,03a	0,10a	0,10a	0,10a	0,00 b	0,03a	0,07 a
P3	0,00a	0,17a	0,00a	0,00a	0,03 b	0,03a	0,07 a
P4	0,00aa	0,07a	0,03a	0,07a	0,00 b	0,00a	0,00 b
NP BNT 5%	0,126tn	0,162tn	0,170tn	0,095tn	0,065	0,094tn	0,065

Numbers followed by the same letter in the same column indicate results that are not significantly different at the 5% confidence level BNT further test. (P0: Control; P1: Farmer's method; P2: Refugia Plants; P3: Biopesticide; P4: Refugia + Biopesticide).

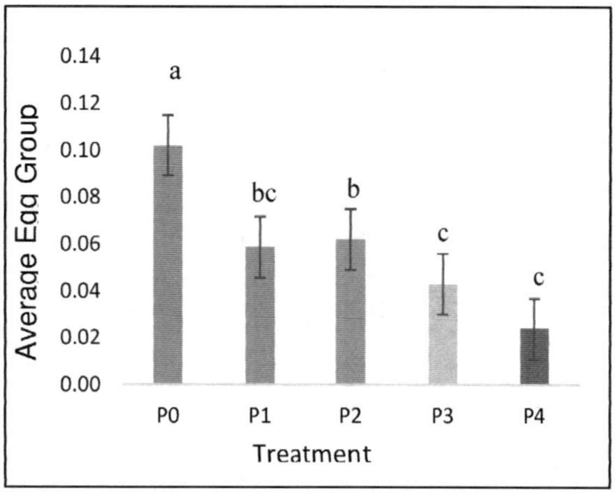

Figure 3. Average of S. exigua egg clusters during seven week observations

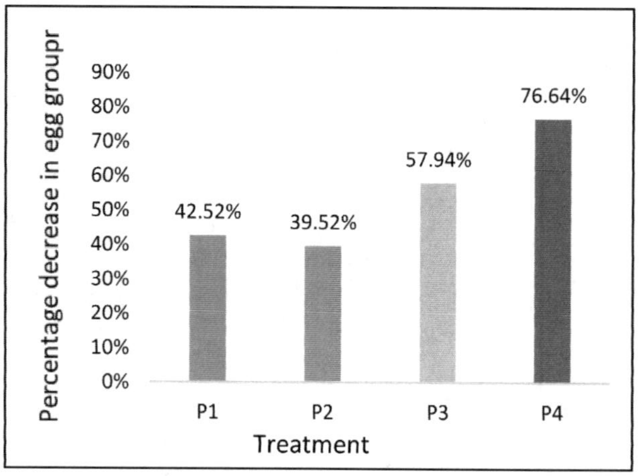

Figure 4. Diagram of Percentage Reduction of S. exigua Egg Clusters.

Based on Table. 1, the Farmer Method factor significantly affected the observation five and seven weeks after planting. In the observation of 5 weeks after planting, P2 and P4 showed a lower number of egg groups of 0.00 and significantly different from the treatment P0, as much as 0.11 / clump of plants. At the observation of 7 weeks after planting, the farmer method showed a low number of egg groups at 0.00 / clump of plants, significantly different from all treatments P0, P1, P2, P3, and P4. This is due to the fifth week of Biopesticide application affecting the behavior of S.exigua imago in laying eggs. Figure 3 shows that the average Number of egg groups is the lowest in the P4 treatment due to the effect of the combination of refugia plants; it can significantly suppress the population of egg groups that can attract and stimulate S. exigua to lay their eggs [9]. Figure 4. shows the presentation diagram of the highest egg group reduction is P4. Coupled with the application of biopesticides that contain secondary metabolites from biduri in the form of terpenoid compounds that are repellent so that they can repel insect pests. It is coupled with the application of biopesticides that contain secondary metabolites from C. gigantea in the form of terpenoid compounds that are repellent so that they can repel insect pests. This is in line with the results of research that volatile terpenoid compounds can cause repellent effects for pests because they provide a distinctive aroma and have proven effective in inhibiting the process of laying and hatching eggs from the pest Paraeucosmetus pallicornis [10]. The presence of refugia plants that provoke the presence of natural enemy arthropods to come to the ecosystem so that their performance as predators or parasitoids will be maximized [11].

3.2 *Observation of Spedoptera exigua Larvae Population*

Analysis of variance showed that the treatment of Refugia Plants, Biopesticides, and Refugia Plants + Biopesticides significantly affected the average parameter of S. exigua larval population.

Table 2. Average Population of Spodoptera exigua Larvae Per Plant Clump

Treatment	Observations						
	1 WAP	2 WAP	3 WAP	4 WAP	5 WAP	6 WAP	7 WAP
P0	0,13a	1,79 a	0,89 a	0,62 a	0,49a	0,33a	0,26a
P1	0,00a	0,17 b	0,82 ab	0,23 b	0,44a	0,46a	0,30a
P2	0,03a	0,13 b	0,64 b	0,25 b	0,40a	0,32a	0,18a
P3	0,00a	0,07 b	0,07 c	0,07 b	0,30a	0,29a	0,17a
P4	0,00a	0,07 b	0,07 c	0,03 b	0,17a	0,07a	0,07a
NP BNT 5%	0,126tn	0,652	0,238	0,257	0,546tn	0,292tn	0,150tn

Numbers followed by the same letter in the same column indicate results that are not significantly different at the 5% confidence level BNT further test. (P0: Control; P1: Farmer's method; P2: Refugia Plants; P3: Biopesticide; P4: Refugia + Biopesticide).

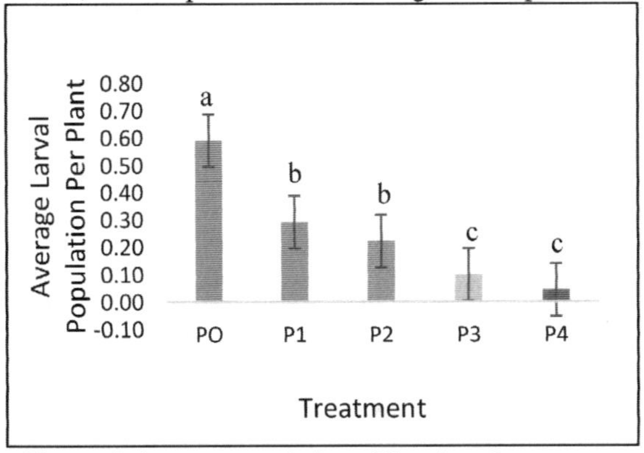

Figure 5. Average population of S. exigua larvae during seven week observations.

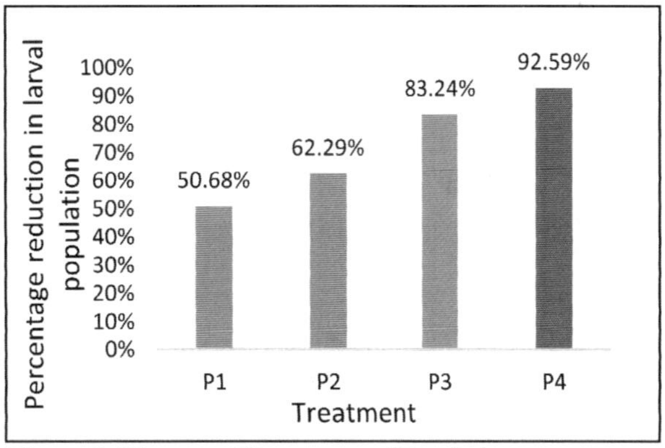

Figure 6. Diagram of Percentage of Population Decrease of S. exigua Larvae

The results of the observation of the average larval population are presented in Table 2. Analysis of variance showed that the treatment significantly affected the parameters of the average population of S. exigua larvae. Observations of treatments that had a natural effect occurred in the observations two weeks after planting, three weeks after planting, and four weeks after planting. Observations of the 2 WAP treatment showed that the lowest average larval population was found in the P4 and P3 treatments, which was 0.07/plant, significantly different from P0 but not significantly different from P1 and P2. The combination of these two treatments can reduce S. exigua larvae. This is supported by the results of research [12] showing that the presence of S. exigua eggs began to be monitored at 14 DAP. Figure 5 shows that the average Number of larval populations during the 7 observations was the lowest in the P4 treatment at 0.04%. Figure 6 shows that the highest percentage of larval population reduction was P4 (Plant Refugia + biopesticide) at 92.59%. The combination of refugia plants and biopesticides effectively reduced the population of S. exigua larvae. The incidence of damage was evenly distributed by early instar larval attacks starting at 14 DAP, while larvae outside the leaves were observed at 18 HST [13]. Biduri plants and maja fruit contain various chemical compounds that are useful as the main ingredients for medicines. Because they contain secondary metabolites that are toxic to insects that interfere with feeding activities [14]. The presence of larvae outside the leaves was observed at the age of 2 weeks [15]. Invasion is characterized by the presence of S. exigua imago on cultivated plants, further stating that there is competition with refugia plants that act as alternative hosts for the pest [16].

3.3 *Observation of Spedoptera exigua Attack Intensity*

Table 3. Average Attack Intensity of *S. exigua* larvae per clump Plants

Treatment	Observations						
	1 WAP	2 WAP	3 WAP	4 WAP	5 WAP	6 WAP	7 WAP
P0	0,00a	1,33a	9,26 ab	11,93a	24,37 a	32,44 a	42,74 a
P1	0,00a	1,67a	2,96 b	7,11a	7,70 c	10,74 c	12,44 c
P2	0,00a	1,00a	11,33 a	13,40a	17,13 b	23,60 b	29,53 b
P3	0,00a	1,00a	6,80 b	8,80a	11,00 c	13,07 c	15,53 c
P4	0,00a	0,67a	5,67 b	8,27a	10,40 c	11,27 c	13,27 c
NP BNT 5%	0,000tn	1,275tn	4,354	2,949tn	5,986	6,258	7,788

Numbers followed by the same letter in the same column indicate results that are not significantly different at the 5% confidence level BNT further test. (P0: Control; P1: Farmer's method; P2: Refugia Plants; P3: Biopesticide; P4: Refugia + Biopesticide).

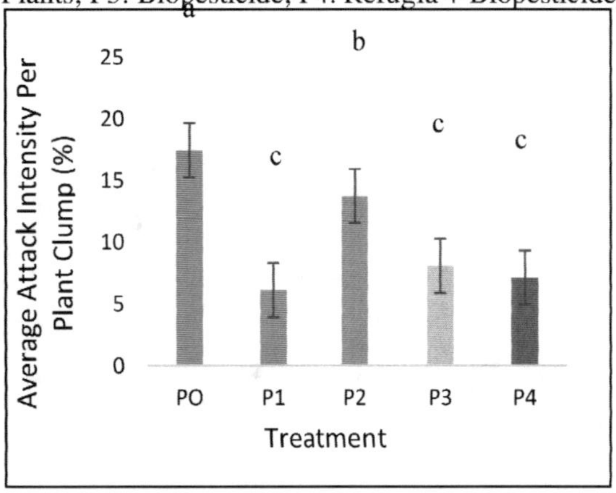

Figure 7. Average attack intensity of S. exigua larvae during seven week observations

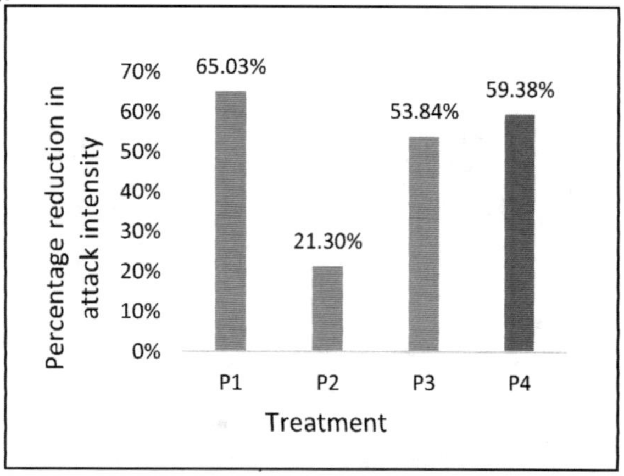

Figure 8. Diagram of Percentage Reduction of S. exigua Larvae Attack Intensity

The results of the average observation of attack intensity are presented in Table 3, which shows that the treatment of P3 and P4 in the observation of 3 weeks after planting, five weeks after planting, and seven weeks after planting had a significant effect. P1 showed the lowest average intensity of S.exigua attack, namely 10.74 / clump of plants. Based on the field under P1, the materials and methods of control still use chemical insecticides whose doses are too high and the volume of spraying up to two or three times a week so that the level of intensity of S. exigua attack is negligible. In line with the excessive use of chemical insecticides, they can cause environmental damage and resistance.

Figure 7 shows that the lowest average attack intensity was in treatment P1. Figure 8 shows the presentation of the highest reduction in attack intensity in P1 than in P4. Regarding this, the lowest intensity of attack is in the farmer's method because it uses a higher dose of chemical insecticides. The morphology and structure of soft and delicate shallot plants also affect the intensity of S. exigua attacks [17] and the larval population. The population and intensity of S. exigua attacks are not yet present at the beginning of the vegetative phase, precisely when the age of the plant enters the age of 1 - 2 MST, and only appears when entering the age of 3 WAP; this happens because the pest S. exigua has just invaded and colonized or the new imago lays eggs The intensity of the attack is closely related to the population and feeding ability of the larvae [18]. Combining refugia plants with biopesticides shows a lower population and intensity of attack than those without refugia plants and their applications. Refugia plants can provide food sources, shelter, or other resources for natural enemies [19]. The high intensity of S. exigua attack on shallot plants without refugia plant treatment and the Farmer Method is higher than the other treatments. [20]. Food suitability is closely related to the dynamics of insect life in choosing suitable food sources for population growth and offspring development [21]. The unwillingness of larvae to eat shallot leaves can be caused by the morphology of leaves that are difficult to consume or the presence of secondary metabolite compounds that act as repellent allelochemicals that affect larval growth and survival [22]. Insects that are given other types of food, or food that is not usually eaten, will go through a long adaptation stage [23]. Another conjecture is that larvae experience stress due to environmental changes [24].

3.4 Observation of *Spedoptera exigua* natural enemy population

Table 4. Total Population of Natural Enemies at Each Observation

Treatment	Observations						
	1 WAP	2 WAP	3 WAP	4 WAP	5 WAP	6 WAP	7 WAP
P0	0,33a	1,67a	3,33 b	11,67 b	7 c	13,33 b	13a
P1	1a	1,33a	3 b	17 b	13,33 c	8 b	6a
P2	8,33a	9,67a	18 a	35,67 ab	33,67 a	28,67 a	17a
P3	1,33a	1,33a	3,67 b	23,33 b	12,33 c	13,00 b	18,33a
P4	7,33a	8,33a	23 a	38,67 a	21,67 b	31,67 a	12,33a
NP BNT 5%	6,43tn	9,24tn	6,66	13,71	10,56	10,14	8,56tn

Numbers followed by the same letter in the same column indicate results that are not significantly different at the 5% confidence level BNT further test. (P0: Control; P1: Farmer's method; P2: Refugia Plants; P3: Biopesticide; P4: Refugia + Biopesticide).

Table 5. Natural Enemy Population in Each Observation Plot

Ordo	Family	Common Name	Treatment					Total	Average
			P0	P1	P2				
Araneae	araneidae	Spiders	13	7	19	13	17	69	14
Lepidoptera	Peridae	Butterflies	1	13	60	16	74	164	33
	Formicidae	Red Ants	30	66	78	61	73	308	62
	Formicidae	Black Ants	18	16	84	46	76	240	48
Hymenoptera	Ichneumonidae	Eriborus sinicus	4	5	33	17	26	85	17
		Bees	15	4	64	7	72	162	32
		Wasps	9	13	50	13	37	122	24
Coleoptera	Coccinelidae	Koksi beetle	16	11	22	23	26	98	20
Dermaptera	Chelisochidae	Cocopet	9	9	22	11	17	68	14

Ordo	Family	Common Name	Treatment					Total	Average
			P0	P1	P2				
Scolopendromorpha	Scolopendridae	Centipede	4	5	21	13	11	54	11
Total			119	149	453	220	429		

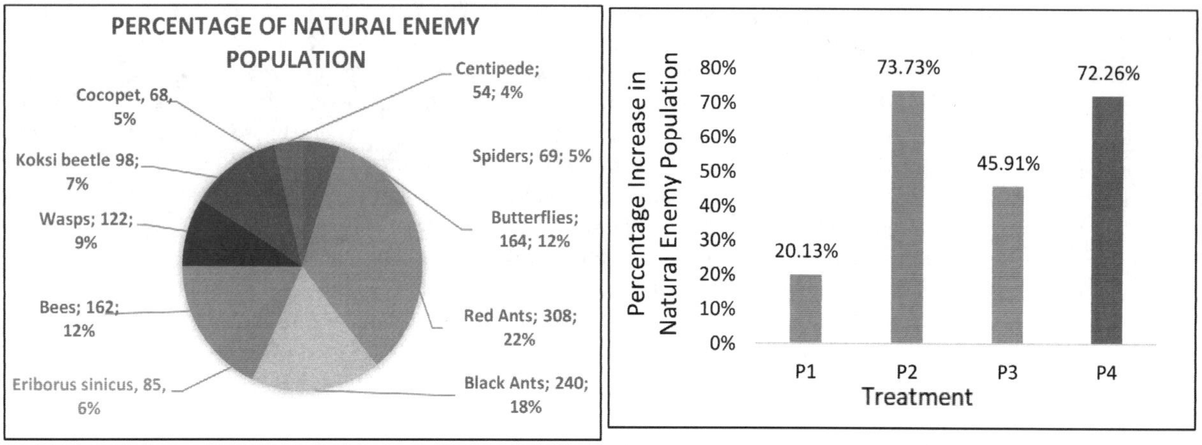

Figure 9. Percentage diagram of each type of natural enemy during seven week observations. Diagram of Percentage Increase in Natural Enemy Population during 7 Observations.

Table 4. Analysis of variance showed that the observations of 3 WAP, 4 WAP, 5 WAP, and 6 WAP influenced the average parameters of natural enemy populations in several observations. It showed the average population of natural enemies seen at 1 WAP and increased every week until the observation of 7 MST. The highest population was found in the observation of 4 WAP, with the highest in the P4 treatment (Refugia + plant extracts) with an average population of 13.71 heads/m2. The lowest population was found in the control treatment at observation 1 WAP with an average population of 6.43 tails/m2. Table. 5 shows that the population of natural enemies with the highest species is red ants from the Formicidae family, with 62 individuals/m2. Then, the lowest population of natural enemies was centipedes from the Scolopendridae family, totaling 11 individuals/m2. They are seen in Figure 9. The diagram of the percentage of each type of natural enemy during the seven observations in the presentation of every kind of natural enemy is highest in red ants with a percentage of 308;22%. The lowest presentation of natural enemies is centipedes, with a percentage of 54.4%. They are seen in Figure 10. In the diagram of the presentation, the highest increase in natural enemies is in the P2 treatment, with a percentage value of 73.73%. The lowest increase in natural enemies is presented in the P1 treatment, with a percentage value of 20.13%. This is also due to the large population of natural enemies found during the early vegetative phase of plants that can attack S. exigua larvae. [25] stated that the presence of Refugia Plants in an agricultural field makes the population of natural enemies very high [26]. Which can suppress the activities of insect pests; Refugia Plants from marigold plants and sun Refugia Plants, which have many types of colors, make insects easily provoked to come [27]. A decrease will follow the decreased pest population in the intensity of the attack of shallot plants that use refugia plants can reduce the intensity of pest attacks to increase production [28].

4. Conclusion

The combination of refugia plants and the application of biopesticides S. exigua attack can be suppressed with the lowest average population in the treatment of refugia plants and biopesticides with a value of 0.2%. The highest, with a value of 6.10%, is in the farmer's method, which Can provoke the arrival of natural enemies.

5. Acknowledgments

BIMA by the Ministry of Education, Culture, Research, and Technology and the Department of Plant Pests and Diseases, Faculty of Agriculture, Hasanuddin University, Makassar, Indonesia, for providing funding for this research from sampling to the publication of this paper; all parties involved are greatly appreciated.

References

[1] Saptorini, Supandji, and Taufik. (2019). Testing of ZA Fertilizer Administration on the Growth and Production of Bauji Variety of Red Onion Plants. AGRINIKA Journal. 3(2): 134-148A reference

[2] Moekasan, TK, Basuki, RS & Prabaningrum, L. 2012. Application of Threshold for Controlling Plant Pest Organisms in Shallot Cultivation to Reduce Pesticide Use. J Hort., Vol. 22 (1): 47-65.

[3] Mahmoud, Y., Henaish, H. & Neven, RR 2020. Identification and Taxonomic Notes of Spodoptera species (Lepidoptera : Noctuidae) Known to Occur in Egypt. J. Biologist. Sci., Vol 13 (2): 161-175.

[4] Nurariaty A., ID. Daud, Nur amin and S. Nuraminah, 2016. Development of pellet formulation as a supplement of Coccinella sp. predator for controlling rice aphids/planthoppers. Research report, LP2M. Hasanuddin University. Makassar.

[5] Rismayani. 2013, "Benefits of Maja Fruit as Botanical Pesticide for Cocoa Fruit Borer Pests (Conopomorpha Cramerella) Research News.

[6] Sjam S., Rosmana A., Sartika VD Ekawati DS, Surapati U., Herawati A. 2017. Oviposition Deterrent and Ovicidal Properties of Calatropis gigantea (L.) Leaf Extract to Paraucosmetus pallicornis (Dallas) in Rice. Journal of Plant Protection Research. Vol. 57 (3) : 243-24.

[7] Paparang, M. 2016. Population and Percentage of Spodoptera exigua Hubner Larvae Attacks on Leeks and Shallots in Ampreng Village, Langowan Bara District. Vol 3.

[8] Hanafiah, KA 2010. Experimental Design. Sriwijaya University. Palembang

[9] Oszmianski, J., JK, Ostek, Biernat, A. 2015. The content of phenolic compounds in leaf tissues of Aesculus glabra and Aesculus parviflora Walt. Molecules, Vol. 20 : 2176-2189.

[10] Sjam, S. and Sari, DE 2017. Repellent Effect of Calotropis gigantea R. Extract Against Paraeucosmetus pallicornis Dallas. Jurnal Agrominansia Vol. 3: 103-109

[11] Nurariaty, A., D. Itji Diana and A. Tamrinn, 2018. Conservation of natural enemies by providing pellets and refugia plants for controlling major pests in rice plantations. Report on PSN- INSTITUTION (DIKTI- UNHAS).

[12] Moekasan, Basuki RS and Prabaningrum, L. 2012. Application of Plant Pest Control Threshold in Shallot Cultivation in an Effort to Reduce Pesticide Use. J. Hort. Vol. 22. No. 1 Pp. 47-56.

[13] Nigam, V., and VS Nambiar. 2015. Therapeutic Potential Of Aegle marmelous (L) Correa Leaves As An Antioxidant And Antidiabetic Agent: A Review. International Journal of Pharma Sciences and research. Vol. 6(3): 611-621

[14] Palayukan, PA, Sjam, S., Melina, Sulaeha, Rosmana, A., Dewi, VS, & Budirman. 2021. Application of the combination of Calontropis gigantea L. and Crescentia cujete L. against Schirpophaga innotata and Leptocorisa acuta Thunb. and predators in paddy plants. IC-FSSAT : IOP Conf. Series: Earth and Environmental Science 807

[15] Op. Cit. "Determination of Saponin Levels in Snake Plant Leaf Extract by Gravimetry" Jurnal UNSRAT. Vol. 2. No. 2 (March 2015).

[16] Fariantika, A., Rahayu, SE, Dharmawan, A. 2019. Test of Maja Fruit Extract (Aegle marmelos) on Mortality and Inhibition of Feeding of Spodoptera litura. Journal of Life Sciences, Vol. 3(1): 31-37

[17] Setyadin, Y., SH Abida, H. Azzamuddin, SF Rahmah, and AS Leksono. 2017. "The Effect of Refugia on Corn Plants (Zea mays) and Long Bean Plants (Vigna cylindrica) on Insect Visit Patterns in Rice Fields (Oryza sativa) in Balong Hamlet, Karanglo, Malang." Biotropica5 (2): 54–58

[18] Capinera, JL Original Published 1999 Revised 2017. Beet Armyworm, Spodoptera exigua (Hübner)

(Insecta : Lepidoptera : Noctuidae). Entomology and Nematology Department. University of Florida.

[19] S Zheng, XL, XP Cong., XP Wang. & CL Lei. 2011. A Review of geographic distribution, overwintering and migration in Spodoptera exigua Hübner. (Lepidoptera: Noctuidae). J. Entomol. Res. Soc. 13 (3) : 39-48.

[20] Suroto, A., Haryani, AL, & Minarni, EW 2020. Cannibalistic Behavior of Spodoptera Frugiperda JE Smith (Noctuidae: Lepidoptera) on Various Types of Plant Leaf Feed. National Seminar on Integrated Animal Husbandry Agriculture. 2(3) : 199-203.

[21] Putra, ILI, & Wulanda, A. (2021). Life Cycle of Spodoptera frugiperda JE Smith with Green Spinach Leaves and Green Prickly Spinach Leaves in the Laboratory. Biome: Scientific Journal of Biology, 10(2), 201-216.

[22] Sjam, S. 2006. Utilization of Maja Fruit Extract with EM4 Against Cocoa Fruit Borer Conopomorpha cramerella Snellen (Lepidoptera: Gracillaridae). Research Bulletin 9 (1): 18-23.

[23] Letourneau, D.K., I. Ambrecth, B.S. Rivera, J.M. Lerna, & E.J. Carmona. 2011. Does Plant Diversity Benefit Agroecosystems? A Synthetic Review. Ecological Applications 21: 9–21

[24] Alifuddin, 2022. Effectiveness test of suren leaf extract, maja fruit and Macbio against red onion caterpillars in Enrekang Regency. Faculty of Agriculture. Hasanuddin University.

[25] M Kurniawati N and Martono E. 2015. The Role of Refugia Plants as a Conservation Media for Natural Enemies of Arthropods. Indonesian Plant Protection Journal 19(2), pp.53–59

[26] Wahyuni R. 2013. Increasing the diversity of plants with Refugia Plants as an attraction for rice pest predators. Sebelas Maret University.

[27] Nonci, N, J. Tandiabang, Masmawati, A Muis, 2000. Inventory of natural enemies of corn stem borer (Ostrinia furnacalis) in corn production centers in South Sulawesi. Agricultural Research. Food Crops Research Center. Agricultural Research and Development Agency. Pages 38-49.

[28] Waterhouse DF, 1993. The major arthropod pests and weeds of agriculture in Southeast Asia. The major arthropod pests and weeds of agriculture in Southeast Asia., v + 141 pp.; [ACIAR Monograph No. 21].

IOP Conf. Series: Earth and Environmental Science **1446** (2025) 012003 doi:10.1088/1755-1315/1446/1/012003

Workload analysis of noodle sheeting slitting machine operators using the heart rate variability method

T Rahman[1], N K I Mayasti[1]. A Indriati[1*], F Novianti[1], N Rahman[1], D Sumaryadi[1], N Komalasari[1], R C E Andriansyah[2] and I N Afiah[3]

[1]Research Center for Appropriate Technology, National Research and Innovation Agency, Subang, Jawa Barat, Indonesia
[2]Research Center for Food Technology and Processing, National Research and Innovation Agency, Yogyakarta, Indonesia
[3]Department of Industrial Engineering, Faculty of Industrial Technology, Universitas Muslim Indonesia, Makassar, Indonesia

*ashriindriati@gmail.com

Abstract. One of the most most popular foods in the community is noodles. In the noodle processing process, poor design and work systems can increase the workload of machine operators. This study uses heart rate variability (HRV) monitoring to examine the workload of noodle machine operators. The operator's heart rate was measured using a Garmin Pro device during each stage of the noodle moulding process: preparation, operation, and finishing. Software Kubios HRV Standard 3.5.0 was used to process the data, and the analysis parameters that were used were SDNN, RMSSD, LF, HF, and LF/HF. The data processing results show that the preparation stage bears the heaviest workload, as evidenced by the lowest average values of SDNN, RMSSD, and HF. Conversely, the operation stage shows the lightest workload, reflected by the highest average values of SDNN, RMSSD, and HF.

1. Introduction

Workload is an important and fundamental concept for investigating people's interactions with technological devices [1]. Human activities can be divided into physical work and mental work. Mental work is relatively less energy-consuming than physical work, but the two are interrelated [2]. Workload is not only a function of task demands but also of how the task is perceived by the individual [3].

Each individual has a unique capacity and set of limitations when it comes to work. This capacity is influenced by both their physical and mental condition. To ensure productivity and efficiency, the workload assigned should be aligned with a person's physical and mental capabilities, ensuring balance and appropriateness [4]. The negative impact of excessive workload is a decrease in work quality due to physical and mental fatigue, so work results are less good. This can cause complaints from customers because the work results are not as expected [5]. The workload analysis can provide accurate measurements to determine the right workload for each operator, identify operator needs, and act as a benchmark for employee performance [6] so that productivity can be increased.

Assessment of workload levels can be measured using subjective and objective approaches. Subjective assessments of mental workload levels include The Cooper-Harper Rating Scale, NASA - TLX, Subjective Workload Assessment Technique (SWAT), and Rating Scale Mental Effort (RSME). Objective measurements based on physiological characteristics can provide honest and unbiased data.

One objective method that can be used to analyze workload is the Heart Rate Variability (HRV) method. This study employs the HRV method because it provides objective data on how the body responds to both physical and mental workloads, complementing subjective evaluations from individuals. This approach allows for a more comprehensive analysis of the effects of workload on both physical and mental health [7]. Heart rate (HR) refers to the number of heartbeats within a specific period, while heart rate variability (HRV) is the variation in time intervals between successive heartbeats. This variation is obtained by observing the time interval between beats or rate-to-rate (R-R) intervals. HRV reflects the activity of the central nervous system and cognitive performance in a job [8]. Increased heart rate and decreased HRV are associated with higher workload and are influenced by various factors [9]. The results of Saleem et al.'s research [10] show that the design of a usable information system can affect the mental burden of its users.

Noodle products are a popular food ingredient in Indonesia. The process of making noodles generally uses the sheeting sliting method, namely thinning and slicing the dough sheets into long strands [11]. Indriati et al. [12] reported that the working posture of noodle machine operators has a high level of ergonomic risk. Pain complaints experienced during production include activities such as placing dough into the machine, mixing ingredients, flipping the dough, cutting and storing the noodles. The current design of noodle machines has not implemented ergonomic principles, namely poor working postures such as bending over for a long time and repeated movements for a certain period while doing the work. Furthermore, an uncomfortable work environment and a fast-paced work phase can also affect the mental workload of machine operators. Therefore, the aim of this study is to analyze the workload of noodle machine operators during their tasks by examining heart rate variability (HRV). This study uses the HRV method, an objective approach based on physiological measurements that has never been applied to food processing machine operators before. Previous studies have predominantly used subjective or empirical methods to analyze workload. This study could serve as an initial step toward using the HRV method to measure workload in other food processing machine operators.

2. Research methodology

2.1. A subsection

The data collection process was carried out for 4 months, from June - September 2023. The study was carried out at the Equipment Testing Laboratory of the Research Center for Appropriate Technology - BRIN.

2.2. Participant

Participants of this study consisted of 22 participants (27% female, 73% male). Participants were recruited voluntarily, and they consisted of technicians, operators, and users of agricultural processing machines. Participants were aged between 16 and 58 years (Mean = 36.72, SD = 12.0) and had experience using agricultural processing machines, varying from 1 year to 30 years. All participants experimented with healthy conditions. Before starting the study, participants were asked to fill in their biodata and sign an informed consent.

2.3. Data collection

Data collection was carried out using the experimental method, namely direct data collection from participants. In the experiment, participants were instructed to complete a series of tasks. The task scenario resembled a case in real conditions. The machine used in this study is shown in Figure 1.
The tasks given can be divided into three scenarios (tasks), namely:
Task 1. Installing the tool (set-up)
Task 2. Setting/operating the tool (operation)
Task 3. Tidying up and cleaning the tool (Finish)

A more detailed task scenario is presented in Table 1.

IOP Conf. Series: Earth and Environmental Science **1446** (2025) 012003 doi:10.1088/1755-1315/1446/1/012003

Table 1. Stages of the sheeting slitting task scenario

No	Task
Set-up	
1	Preparing raw materials
2	Turn on the machine and adjust the distance of the flattening roll
Operation	
3	Put the noodle dough into the flattening roll
4	The flattening roll spacing adjuster becomes narrower
5	Turn off the machine and install the noodle cutter
6	Turn the machine back on and cut the noodle dough
Finish	
7	Turn off the engine and remove the cutting tool
8	Cleaning the remaining material stuck to the flattening roll and conveyor

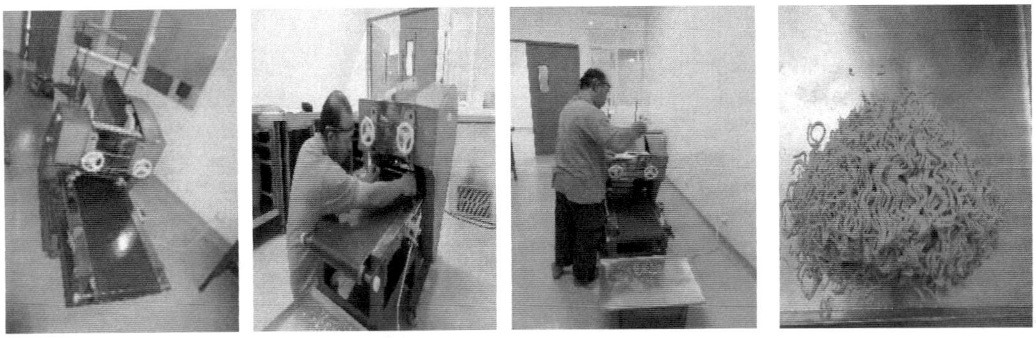

(a) (b) (c) (d)

Figure 1. Sheeting slitting noodle making machine. Figure 1(a) Sheeting slitting machine, Figure 1(b) process of installation of noodle cutter, Figure 1(c) process of inserting noodle dough into flattening roll, Figure 1(d) noodle products

The specifications of the sheeting slitting noodle printing machine used in this study can be seen in Table 2.

Table 2. Sheeting slitting machine specifications

Parameter	Information
Type	KT 955
Motor	1 HP – 1 Phase
Power	750 watt – 220 volts
Dimensions	160 x 60 x 131 cm
Netto	190 kg

Before the study began, participants received a brief explanation of the procedure and tasks. Heart rate data was collected using the Garmin HRM Pro device. This device is a belt looped around the chest equipped with a sensor that can detect heart rate. Heart rate data is connected to a mobile phone via Bluetooth. The recorded data is then downloaded for further processing.

2.4. Data processing

Heart rate variability data were analyzed using Kubios HRV Standard 3.5.0 software. The analysis included Time Domain Analysis, with parameters SDNN and RMSSD, and Frequency Domain Analysis, which involved LF, HF, and LF/HF. The results were then analyzed statistically using repeated measures ANOVA, with a significance level set at a p-value < 0.05.

3. Results and discussion

Heart Rate Variability (HRV) or RR interval is the time elapsed between two consecutive R waves. There are two methods of HRV analysis: time-based analysis (time domain analysis) and frequency-based analysis (frequency domain analysis), which is a wave transformation from a time function to a frequency function using the Fast Fourier Transform (FFT) method using KUBIOS software.

3.1. Time Domain Analysis Data Processing

Time domain analysis is a method of HRV analysis based on time. It involves several parameters, including NN (the total sum of R-R intervals), NN50 (the sum of NN intervals longer than 50 ms), pNN50 (the ratio of NN50 to the total number of NN intervals), SDNN (the standard deviation of all NN intervals), SDANN (the standard deviation of the average NN interval over 5-minute intervals throughout a 24-hour period), and RMSSD (the square root of the average squared differences between consecutive NN intervals).

In this study, the time-domain analysis parameters used were SDNN and RMSSD. SDNN reflects the overall power of spectral analysis and the variability within the measurement period, while RMSSD assesses short-term components of HRV. A decrease in RMSSD and SDNN values suggests an increase in mental workload [13]. The results of the SDNN and RMSSD tests are presented in Table 3.

Table 3. SDNN and RMSSD values at each stage of the process

Participant	SDNN (ms)			RMSSD (ms)		
	Set-up	*Operation*	*Finish*	*Set-up*	*Operation*	*Finish*
1	12.5872	7.8177	5.9334	8.395	6.4514	5.1707
2	11.943	14.4521	15.661	7.1533	8.9853	5.7237
3	14.9304	23.2913	25.824	14.0536	25.046	19.4586
4	7.5268	7.98	7.6587	7.7055	8.1496	6.1876
5	16.2259	16.2367	16.7273	11.0524	13.1657	11.9801
6	15.6942	25.5214	39.0512	11.8004	26.7573	32.0508
7	18.1664	20.0541	23.206	18.8201	25.0952	25.3156
8	20.6998	53.2264	37.2442	12.1024	62.3184	34.1583
9	15.4012	15.2298	20.4097	7.3018	11.9231	13.8736
10	27.9587	75.5751	55.3612	27.1544	88.415	69.1601
11	20.1937	33.1362	33.4138	19.4377	42.2711	36.4573
12	12.2187	20.301	17.9228	11.957	20.8903	17.8184
13	15.2774	24.3578	20.2145	12.092	17.2116	10.0767
14	22.5204	21.1322	32.2964	27.7316	18.3726	37.3006
15	9.1393	13.7393	21.4499	5.4292	8.7586	24.5643

Table 3. SDNN and RMSSD values at each stage of the process

Participant	SDNN (ms)			RMSSD (ms)		
	Set-up	*Operation*	*Finish*	*Set-up*	*Operation*	*Finish*
16	12.9821	50.4273	14.1116	17.6483	61.6134	5.9358
17	10.8867	12.2439	11.7471	7.3364	13.478	8.3034
18	47.2765	39.6245	22.4842	59.5422	40.7498	20.6317
19	7.4627	9.0499	7.177	5.2544	10.8301	6.3472
20	19.2806	15.9502	13.1476	9.8415	12.8739	8.0599
21	11.1575	11.1319	7.3311	8.0736	7.9952	5.9638
22	15.8402	13.6268	15.0909	16.0792	13.5458	13.011

Table 1 shows that the highest SDNN values are at the set-up, operation and finish stages, namely 47.2765 ms, 75.5751 ms and 55.3612 ms. The lowest SDNN values at the set-up, operation and finish stages are, respectively, 7.4627 ms, 7.8177 ms and 5.2544 ms. The highest RMSSD value of the HRV test results is at the finish stage with a value of 69.1601 ms with the 10th participants, while the lowest RMSSD value is the first participants at the finish stage with a value of 5.1707 ms.

3.2. Frequency Domain Analysis Data Processing

Frequency domain analysis is a type of HRV analysis based on frequency. In the frequency domain, HRV is divided into three components: Very Low Frequency (VLF) within the range of ≤ 0.04 Hz, Low Frequency (LF) between 0.04 and 0.15 Hz, and High Frequency (HF) between 0.15 and 0.4 Hz. In this study, operator workload was assessed based on the LF, HF, and LF/HF ratio values. The results for LF, HF, and the LF/HF ratio are presented in Table 4.

Table 4. LF, HF and LF/HF values at each process stage

Participant	LF (ms^2)			HF (ms^2)			LF/HF (ms^2)		
	Set-up	*Operation*	*Finish*	*Set-up*	*Operation*	*Finish*	*Set-up*	*Operation*	*Finish*
1	28.2098	33.3148	10.0538	9.0805	9.5539	5.6818	3.1066	3.487	1.7695
2	39.8729	135.3956	119.9441	4.8032	12.3656	12.0961	8.3014	10.9493	9.916
3	163.1013	237.427	441.2947	34.8365	104.7241	120.8434	4.6819	2.2672	3.6518
4	24.4121	32.9665	22.2019	11.2337	9.5658	5.5336	2.1731	3.4463	4.0122
5	154.7176	125.8347	100.18	45.2866	42.1865	19.3594	3.4164	2.9828	5.1748
6	70.8032	335.9521	806.9111	13.2019	165.5993	165.6766	5.3631	2.0287	4.8704
7	39.9382	108.8605	56.5758	26.3059	82.3425	69.7761	1.5182	1.322	0.8108
8	147.9256	2005.194	818.5405	13.4251	2008.652	113.0708	11.0186	0.9983	7.2392
9	57.8318	122.5266	82.9467	5.8153	24.2291	22.6458	9.9449	5.057	3.6628
10	377.5918	9050.415	373.3366	81.2112	4184.657	802.6124	4.6495	2.1628	0.4652
11	251.9839	992.3646	306.0969	154.0243	1283.73	161.4051	1.636	0.773	1.8965
12	66.0377	190.106	254.3746	19.5743	76.1024	63.7486	3.3737	2.498	3.9903
13	78.3317	423.5697	271.4028	2.2947	38.0775	19.864	34.1352	11.1239	13.663
14	100.7218	259.481	201.3073	242.8666	71.1876	138.4442	0.4147	3.645	1.4541
15	38.9962	121.8263	95.8217	16.4369	39.3823	31.6541	2.3725	3.0934	3.0271
16	12.1046	4622.545	121.857	11.1046	4800.228	6.6525	1.09	0.963	18.3175

Table 4. LF, HF and LF/HF values at each process stage

Participant	LF (ms^2)			HF (ms^2)			LF/HF (ms^2)		
	Set-up	*Operation*	*Finish*	*Set-up*	*Operation*	*Finish*	*Set-up*	*Operation*	*Finish*
17	22.1912	57.1484	115.9016	4.4187	32.0853	26.8035	5.0221	1.7811	4.3241
18	286.8102	688.5973	212.7965	75.9134	421.8351	33.1489	3.7781	1.6324	6.4194
19	15.4118	35.272	28.7256	4.8181	15.7333	7.8718	3.1987	2.2419	3.6492
20	104.0224	159.0164	204.3365	11.5021	27.9248	10.9754	9.0438	5.6945	18.6177
21	100.1112	89.469	47.7129	9.6548	16.0423	5.7922	10.369	5.5771	8.2374
22	61.3309	83.5555	122.404	29.6653	44.6923	119.0756	2.0674	1.8696	1.028

LF values are affected by the activation of both the parasympathetic nervous system (PNS) and the sympathetic nervous system (SNS). The SNS prepares the body to react to potentially threatening situations, while the PNS is more active in less demanding situations and helps return the body to a resting state. Generally, LF values are higher under conditions of increased cognitive load. The test results indicate that the highest LF value occurs during the operation stage, while the lowest LF value is observed at the finishing stage.

PNS activation will affect the HF value. The test results show that the highest HF value occurs when the operator is at the operation stage, and the lowest HF value occurs at the setup stage. HF values tend to increase when cognitive load conditions are light [14]. The highest LF/HF ratio is recorded at the setup stage, measuring 34.1352 ms^2, while the lowest LF/HF ratio, 0.4652 ms^2, occurs at the finish stage. The LF/HF ratio serves as an indicator of the balance between the PNS and SNS. When cognitive load is high, the LF/HF ratio typically increases [15].

3.3. Statistical test

The results of the repeated measures ANOVA statistical test are shown in Table 5 and Table 6. The test reveals that three parameters—SDNN, RMSSD, and HF—have a significance value of less than 0.05. For these three parameters, the null hypothesis (H0) is rejected, and the alternative hypothesis (H1) is accepted. Therefore, it can be concluded that SDNN, RMSSD, and HF exhibit significant differences across the stages of the process: setup, operation, and finish.

Table 5. ANOVA test results

Parameter	p-value	Information
SDNN	0.027*	significantly different
RMSSD	0.025*	significantly different
LF	0.062	Not significantly different
HF	0.027*	significantly different
LF/HF	0.079	Not significantly different

Nb: * = significant $p < 0.05$

Table 6. Average values of SDNN, RMSSD, LF, HF and LF/HF ratio

	SDNN	RMSSD	LF	HF	LF/HF
Setup	16.61±8.5^a	14.82±11.8^a	101.93±95.8	37.61±57.9^a	5.94±7.0
Operation	23.82±17.1^b	24.77±21.6^b	905.04±2085.2	614.13±1348.4^b	3.44±2.8
Finish	21.07±12.3^{ab}	18.98±15.6^{ab}	218.85±223.8	89.22±168.6^{ab}	5.74±5.2

The process stage with the heaviest workload from the test results is the set-up stage, which is indicated by the lowest average SDNN, RMSSD and HF values compared to the other two stages, namely the operation and finish stages. On the other hand, the process stage with the lightest workload is the operation stage, which is indicated by the highest average SDNN, RMSSD and HF values. Further test results on the SDNN, RMSSD and HF parameters show a significant difference between the set-up and operation process stages. However, there is no significant difference between the set-up and finish stages or the operation and finish stages.

4. Conclusion

From the research conducted, each process stage has a different workload. The process stage with the heaviest workload is the set-up stage, and the process stage with the lightest workload is the operation stage. The test results show that the SDNN, RMSSD and HF parameters have significant differences at the process stage. Further test results on the SDNN, RMSSD and HF parameters show significant differences between the set-up and operation process stages. However, there is no significant difference in workload between the set-up and finish stages or the operation and finish stages.

Acknowledgements

The authors wish to express their gratitude to ORPP-BRIN for funding the research through the Appropriate Technology and Process program and the technicians who have provided much assistance in data collection.

References

[1] Longo L and Dondio P 2015 *Proc IEEE/WIC/ACM International Conference on Web Intelligence and Intelligent Agent Technology* pp 345-352.
[2] Sutalaksana I Z, Anggawisastra R and Tjakraatmadja J H 2006 *Work System design techniques* (Bandung: ITB Press).
[3] Megaw T 2005 The definition and measurement of mental workload *Evaluation of Human Work 3rd* ed J R Wilson and N Corlett (New York: Taylor dan Francis Group LLC).
[4] Tarwaka 2004 *Ergonomics for Safety, Occupational Health and Productivity* (Surakarta: UNIBA Press).
[5] Hakiim A, Suhendar W and Sari D A 2018 *Barometer* **3(2)** 142-146.
[6] Arifin H 2020 *Teknoin* **26(2)** 83-95.
[7] Mulder L J M 1992 *Biological Psychology* **34** 205–236.
[8] Hansen A, Johnsen B and Thayer J 2003 *International Journal of Psychophysiology* **48** 263-274
[9] Hancock P A and Meshkati N 1988 *Human mental workload: advances in psychology* vol 52 (Amsterdam: Elsevier Science Publisher BV).
[10] Saleem J J, Patterson E S, Militello L, Anders S, Falciglia M and Wissman J A 2007 *Journal of the American Medical Informatics Association* **14(5)** 632–640.
[11] Kusuma P T W W and Mayasti N K I 2014 *Agritech* **34(2)** 194-202.
[12] Indriati A, Mayasti N K i, Rahman T, Litaay C, Anggara C E W, Astro H M, Surahman D N, Novianti F and Andriana Y 2022 *Proc. IOP Conf. Series: Earth and Environmental Science* **980** (2022) 012048.
[13] Heine T, Lenis G, Reichensperger P, Beran T, Doessel O and Deml B 2017 *Applied Ergonomics* **61** 31–43.
[14] Novani N F, Arief M L and Anjasmara R 2019 **3(2)** 90-95.
[15] Abbasi A M, Darvishi E, Rodrigues M A and Sayehmiri K 2022 *Applied Acoustics* **189** 108602.

IOP Conf. Series: Earth and Environmental Science **1446** (2025) 012004 doi:10.1088/1755-1315/1446/1/012004

IoT- Enabled soil moisture management for precision plant sprinkler system for organic farming

M S Maghirang[1], R Maaño[1*], H S Santonil[1], D N Canela[1], M J De la Roca[1], D J Buban[1] and R A Maaño[2]

[1]College of Engineering, Manuel S. Enverga University Foundation, Lucena City Philippines,
[2]College of Computing and Multimedia Studies, Manuel S. Enverga University Foundation, Lucena City Philippines

[*]Email : ronaldo.maano@mseuf.edu.ph

Abstract. In daily operations related to farming, watering stands out as crucial yet labour-intensive task regardless of weather conditions. The traditional way of watering crops involves manual irrigation, which poses challenges such as inefficiency, labour intensity, and uneven water distribution that often results in overwatering in some areas leading to potential crop yield losses. It requires two important aspects: when and how much to water. With that, this project study aimed to provide an alternative way of irrigating crops and embrace automation to increase crop productivity, less water consumption, and reduce labour intensity. Farmers can benefit from automated watering systems to optimize irrigation practices. To address farming challenges, a microcontroller-based irrigation system with Arduino microcontroller, soil moisture sensors, relay modules, and water sprinkler was developed. This system aids in transitioning from traditional farming to digital farming, boosting productivity and allowing real-time monitoring via a mobile app. Additionally, it is integrated into a railing system to automate irrigation based on the crops' watering needs. It offers a solution to labor-intensive manual watering and farming and is claimed to be effective based on the evaluation results, which yielded strong agreement in functionality, effectiveness and efficiency, with scores of, 4.24, 4.17, and 4.1 respectively. It has the potential to be an innovation in the development of more eco-friendly and sustainable agricultural practices.

1. Introduction

Agriculture has existed in various forms since the beginning of time. People depend on farming every day to provide us with fruits, vegetables, plants, and rice-making wheat. The type of farming that is most advantageous to both farmers and consumers is organic agriculture. The awareness of hazards caused by the use of synthetic chemicals in agriculture makes organic farming attract the attention of both the producer and consumer levels. The irrigation sector, however, has continuously failed to live up to expectations despite receiving over half of the overall budget for agriculture in recent years. In order to prevent drought, a routine schedule in watering crops is necessary to maximize yields and crop growth [1].

Throughout the years, farmers implement the traditional way of watering plants by manual irrigation, prompting the need for effective solutions like smart farming due to the effects of the El Niño phenomenon which caused significant agricultural damage and incorporating organic agriculture [2].

There are a lot of opportunities for innovation in agriculture, enabling cost-effective management such as resource and crop management through wireless sensor networks that shows how the Internet for Things (IoT) capabilities. These IoT-powered systems offer benefits such as reduced operational costs, enhanced resource utilization, and environmental conservation [3][4]. Through projects like the one employing Arduino technology, which continuously monitors and adjusts water levels, modern technology becomes accessible to small-scale farmers, improving agricultural productivity and sustainability [5] [6]. However, challenges such as underdevelopment and poor management in the irrigation sector require medium to long-term solutions for sustainable agricultural development [7]. Implementing IoT-based automated irrigation systems offers remote control capabilities, allowing farmers to adjust watering schedules and conserve water resources efficiently [8]. Precision agriculture (PA) stands out as a significant approach that maximizes farming methods by utilizing information, communication, and control technologies. IoT and cloud computing integration improves connectivity and data exchange, enabling more accurate and customized farming operations. PA helps farmers to apply water and nutrients more effectively, reducing waste and fostering healthier crops in organic agriculture, where every input counts [9]. Overall, integrating IoT technology into agriculture presents numerous advantages in terms of resource management, productivity enhancement, and environmental protection, laying the foundation for a more sustainable and efficient farming industry.

2. System Design and Methodology

The system design consists of two parts which are the hardware design and the software design which is the program of the control system. The flowchart is used to describe the mechanism of the control system. This flowchart outlines the smart irrigation system's operation sown in Figure 1. It shows the process of how the system works starting from the reading of soil moisture sensor until it finishes watering the crops.

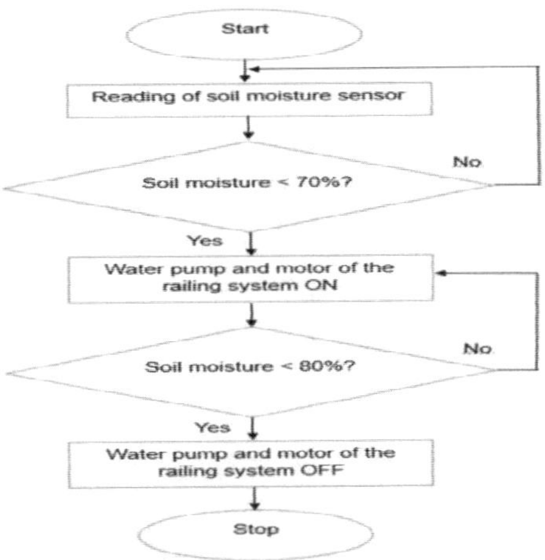

Figure 1. Flowchart of the control system mechanism

Requirements planning involved data gathering from previous studies and the expertise of farmers and engineers. The agile methodology is a project management technique. The fabrication of the prototype focused on the development, design and testing phases. The concept of the Agile methodology revolves around iterative and incremental development [10]. Table 1 shows cited research studies on automated watering systems. The components of the system include the Arduino Uno microcontroller, soil moisture sensors and relay modules.

Table 1. Technologies Used for Previous Automated Watering Systems

Title	Components	Author
Design of Automatic Sprinkler based on Arduino Uno Microcontroller	Arduino Uno R3; ultrasonic sensors; soil moisture sensor	[8]
Arduino-based smart water irrigation using water flow sensor, soil moisture sensor, temperature sensor and ESP8266 Wi-Fi module	Arduino Uno R3; water flow sensor; soil moisture sensor; temperature sensors; ESP8266 Wi-Fi Module	[11]

A set of carpentry tools, including grinder, electric drill, and welding tool are needed to build the plant watering prototype's wailing system. Materials used were angle bar (200 cm x 600cm), flat bar (500mm x 3mm), hex bolts (14mm, 15mm), sprinkler nozzle (4.5mm), plywood, galvanized iron pipe, water pump, chain and sprocket (14 teeth).

Whereas materials and components used to construct the designed irrigation control system were Arduino Uno R3, microcontroller, 20x4 LCD, soil moisture sensors, relay modules, I2C backpack module, 24V power supply, and switches. An IP65 junction box was used as the panel box and all components are integrated inside it. Arduino IDE was used to programmed and compile the program of the whole system.

2.1. Block Diagram

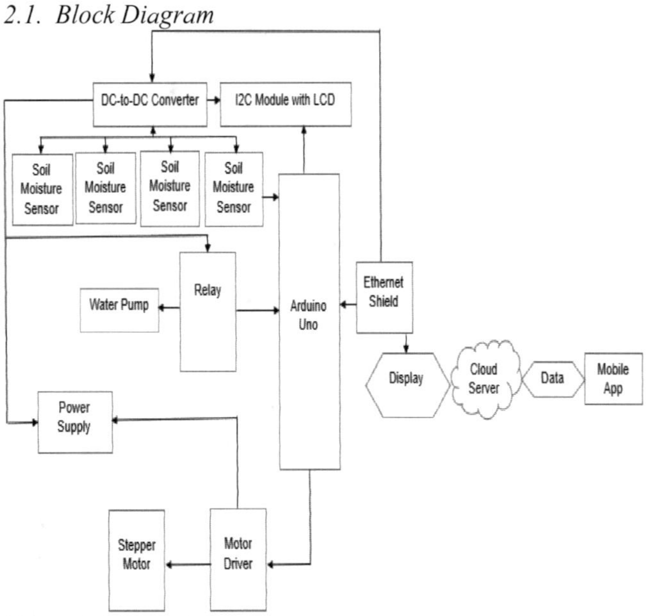

Figure 2. The block diagram

The block diagram shown in figure 2 shown outlines the system architecture, illustrating the interconnected components of the smart irrigation system. Key modules include the soil moisture sensor, relay module, motor driver, ethernet shield, interface with the Arduino Uno board, powered by a 24V power supply. This supply also powers the railing system, enabling comprehensive crop irrigation management [12].

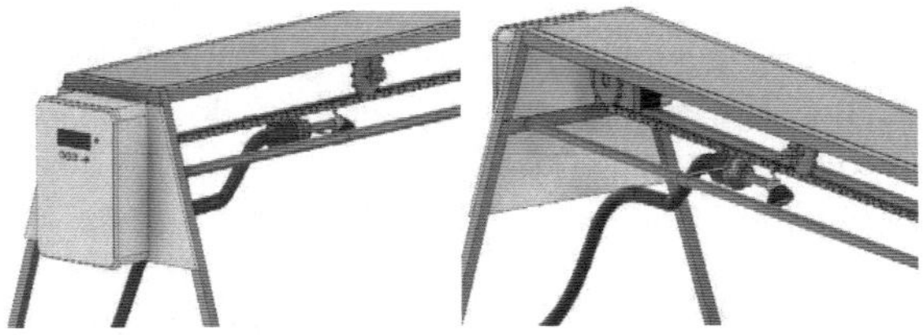

Figure 3. CAD mockup of the prototype

The figure shows the CAD mockup of the prototype. The frame, which supports the railing system, is seen in the illustration as holding and supporting the chain and sprocket. The stepper motor is also connected to the chain and sprocket. To water the crops, a water sprinkler that oscillates back and forth is attached to a chain and sprocket. The junction box on the side of the entire prototype contains the control system. Inside the junction box at the side of the entire prototype are components such as the Arduino Uno microcontroller, Ethernet shield, relay, etc. For power, the control system and water pump have an AC cord that will be plugged into a 220V outlet. The automated water sprinkler will cover an area of 8x1 meters, equivalent to one crop plot on the farm. However, due to the prototype's small scale, it will only cover 2.5x1 meters of crops.

3. Results and Discussion

3.1. Assembly of the prototype

The researchers gathered data on how to start the initial phase of the development by facilitating site visits, interviews, and questionnaire from the farm. The parameters that contribute to the overall effectiveness and efficiency of the automated watering system, are the soil moisture content, watering duration, water flow rate, weather conditions enabling precise and tailored irrigation management for optimal plant health and growth.

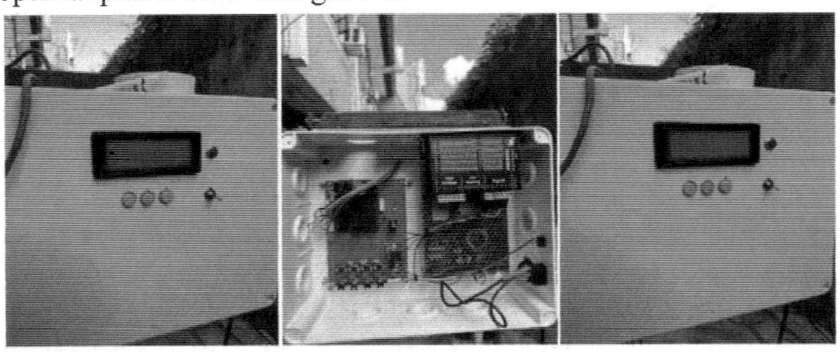

Figure 4. Automated Water Sprinkler System

Figure 5. Integration of the control system and the railing system

The figure 5 shows the integration of the control system and the railing system. The final prototype was tested using a variety of vegetable crops, including *pechay* and *mustasa.* The data gathered on the initial phase of the development were found useful. The parameters that contribute to the overall effectiveness and efficiency of the automated watering system, are the soil moisture content, watering duration, water flow rate, weather conditions enabling precise and tailored irrigation management for optimal plant health and growth. The 80% setting point value is determined based on Hariadi's statement that the required absolute moisture content is approximately 80% [8]. The watering duration depends on the soil moisture of each crop since it is automated, it must reach a minimum of 80% and a maximum of 100%. Since the system is automated, the watering schedules will not depend on the weather conditions but instead on the soil moisture readings.

3.2. Mobile app dashboard

Figure 6. The dashboard using Blynk app for the vegetable crop monitoring system of the vegetable crops

The dashboard using Blynk app for the vegetable crop monitoring system of the vegetable crops is shown in figure 6. The four quadrants displayed on the dashboard correspond to the four soil moisture sensors that have been placed in the lot. The first quadrant's soil reading of 59% indicates that it is already wet. The second quadrant's readings of 61%, the third quadrant's readings of 60%, and the fourth quadrant's reading of 65% all indicate that the soil is already wet and is being watered until it reaches 80%, which is the minimum percentage that will trigger the water pump to turn off as well as the stepper motor of the railing system. This indicates that the soil moisture readings is connected to the dashboard.

3.3. Testing and evaluation of prototype

Prototype testing aims to gather insightful input from real users in order to lower launch related risks. Integration testing is done on the power supply unit and control system. To acquire reliable input, testing prototypes with target users attempts to deliver an accurate innovation experience. Projects move forward, innovators require this kind of feedback to make precise design decisions [13]. Table 2 displays a test case for the prototype. Integration testing is the process of assembling individual components and evaluating them to determine whether their interactions lead to any issues. The evaluation process involved gathering feedback from a total of 24 respondents, comprising seven respondents in the field of irrigation systems and farming, twelve respondents in the field of planting, and five respondents in the field of agriculture technology.

Table 2. Test cases used for the prototype

Title	Test Case	Expected Result
Arduino-based automated watering system	Soil Moisture Sensor reading	The result must show that indicators all work once the soil moisture is detected and will automatically turn on the water pump and the use of Ethernet Shield will utilize IoT in the mobile app
IoT Cloud	Data Reflected on IoT Dashboard	The soil moisture sensors displayed real-time data of the crops

3.4. *Calibration of the prototype*

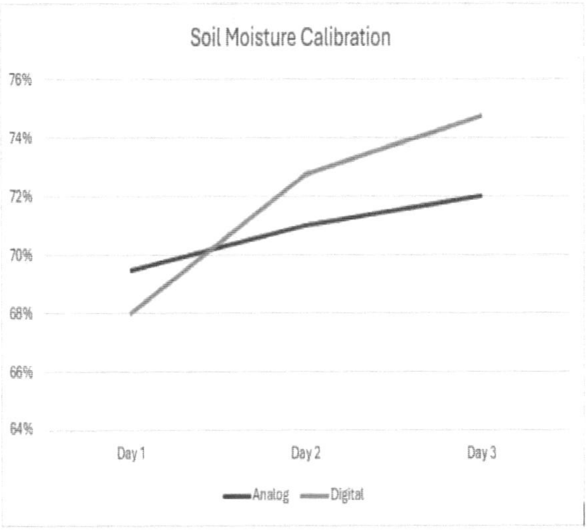

Figure 7. Calibrations process

The Figure 7 shows calibration process involves comparing the sensor's output with known moisture levels in the soil to establish a correlation and adjust the sensor's readings accordingly. By calibrating the sensor, you can improve its accuracy and consistency, enabling more precise irrigation control and better decision-making in agriculture. Comparing it to a digital soil moisture sensor to analog soil moisture meter allows for validation of the sensor's performance and helps determine any discrepancies or inaccuracies between the two measurement methods.

4. Conclusions and Future Works

The prototype effectively detects soil moisture levels and automatically waters the crops when it reaches 70%. The system is turned off when the soil moisture reaches a minimum of 80% and a maximum of 100%. Soil moisture sensor readings are displayed on the LCD, and users can monitor their vegetable crops via the mobile app developed by the researchers. This innovation not only reduces labor costs and enhances irrigation methods but also promotes sustainable farming practices and increased crop yields. Engaging various stakeholders, the study highlighted how IoT technology can optimize water usage and improve overall farm profitability.

For recommendation, additional sensors like temperature and humidity sensors to collect more environmental data that could assist in developing irrigation schedules that are more adaptable to weather conditions.

References

[1] S. Chu and A. Majumdar, "Opportunities and challenges for a sustainable energy future," *Nature*, vol. 488, no. 7411, pp. 294–303, Aug. 2012, doi: 10.1038/nature11475.

[2] Y. Yu, Y. He, and X. Zhao, "Impact of demand information sharing on organic farming adoption: An evolutionary game approach," *Technological Forecasting and Social Change*, vol. 172, p. 121001, Nov. 2021, doi: 10.1016/j.techfore.2021.121001.

[3] T. M. Bandara, W. Mudiyanselage and M. Raza, "Smart farm and monitoring system for measuring the Environmental condition using wireless sensor network - IOT Technology in farming," *2020 5th International Conference on Innovative Technologies in Intelligent Systems and Industrial Applications (CITISIA),* pp. 17, Nov. 2020, doi:10.1109/CITISIA50690.2020.9371830.

[4] A. Ali, T. Hussain, N. Tantashutikun, N. Hussain, and G. Cocetta, "Application of smart techniques, internet of things and data mining for resource use efficient and sustainable crop production," *Agriculture 2023*, vol. 13, no. 2, p. 397, Feb. 8, 2023, doi:10.3390/agriculture13020397.

[5] K. Sekaran, M. Meqdad, P. Kumar, S. Rajan, and S. Kadry, "Smart agriculture management system using internet of things", *TELKOMNIKA Telecommunication, Computing, Electronics and Control*, vol. 18, no. 3, pp. 1275-1284, Jun. 2020, doi: 10.12928/TELKOMNIKA.v18i3.14029.

[6] R. C. Maaño, R. A. Maaño, P. J. De Castro, E. P. Chavez, S. C. De Castro and C. D. Maligalig, "SmartHatch: An Internet of Things–Based Temperature and Humidity Monitoring System for Poultry Egg Incubation and Hatchability," *2023 11th International Conference on Information and Communication Technology (ICoICT)*, Melaka, Malaysia, 2023, pp. 178-183, doi: 10.1109/ICoICT58202.2023.10262810.

[7] A. Anand, N. Trivedi, V. Gautam, R. Tiwari, D. Witarsyah, and A. Misra, "Applications of Internet of Things (IoT) in agriculture: the need and implementation," *IEEE Conference Publication | IEEE Xplore*, Nov. 23, 2022, doi: 10.1109/ICADEIS56544.2022.10037505.

[8] A. Priyati, M. Sahbandi, G. M. D. Putra, and D. A. Setiawati, "The Design of Automatic Sprinkler based on Arduino Uno Microcontroller," *IOP Conference Series*, vol. 355, no. 1, p. 012088, Nov. 2019, doi: 10.1088/1755-1315/355/1/012088.

[9] M. A. Zamora-Izquierdo, J. Santa, J. A. Martínez, V. Martínez, and A. F. Skarmeta, "Smart farming IoT platform based on edge and cloud computing," *Biosystems Engineering*, vol. 177, pp. 4–17, Nov. 2018, doi: 10.1016/j.biosystemseng.2018.10.014.

[10] A. Rasnacis and S. Bērziša, "Method for adaptation and implementation of Agile project management methodology," *Procedia Computer Science*, vol. 104, pp. 43–50, Jan. 2017, doi: 10.1016/j.procs.2017.01.055.

[11] P. Singh and S. Saikia, "Arduino-based smart irrigation using water flow sensor, soil moisture sensor, temperature sensor and ESP8266 WiFi module," *IEEE Conference Publication | IEEE Xplore*, Dec. 01, 2016, doi: 10.1109/R10- HTC.2016.7906792.

[12] N. Putjaika, S. Phusae, A. Chen-Im, P. Phunchongharn, K. Akkarajitsakul, "A control system in an intelligent farming by using arduino technology," *IEEE Conference Publication | IEEE Xplore*, May 2016, doi: 10.1109/ICTISPC.2016.7519234.

[13] K. Mandula, R. Parupalli, C.H.A.S. Murty, E. Magesh, R. Lunagariya, "Mobile based home automation using Internet of Things (IoT)," *IEEE Conference Publication | IEEE Xplore*, Dec. 01, 2015, doi: 10.1109/ICCICCT.2015.7475301. 2019.

iPond: An Internet of Things–Based Water Quality Monitoring System for Fishpond

Ahlihya Bianca E. Gulifardo, Buen Jasper D. Balane, Mark T. Ebora, Melissa R. Serrano, Erwin P. Ellazar, Ariel V. Nicolas, Ronaldo C. Maaño

College of Engineering, Manuel S. Enverga university Foundation, Lucena City, Philippines

ronaldo.maano@mseuf.edu.ph

Abstract. The cultivation of aquatic species faces a significant challenge due to water quality degradation, which impacts farming operations and poses a threat to livelihoods. To contribute to advancing aquaculture informatics and enabling more precise and convenient monitoring of aquaculture ponds, this study introduces an IoT-based Water Quality Monitoring System for Fishponds. This system is designed to measure pH and salinity values in real-time, aiming to address the pressing need for continuous monitoring and management of water quality parameters for cultivating milkfish (Chanos chanos). Users can remotely access and control measurements by integrating the prototype with the Blynk IoT system, ensuring that water quality remains within optimal thresholds or 6.5 to 8.5 pH for acidity and 18 to 29 ppt for salinity. In addition, the system is equipped with an alert system to notify end-users when the parameters exceed predetermined limits promptly. The prototype received an average functionality score of 4.325 and an efficiency score of 4.13. The error percentages recorded were 0.027% for acidity, 0.187% for salinity, and 4.36% for overall effectiveness. This system enhances cultivation security, enabling farmers to make informed decisions and take timely actions to maintain optimal water quality conditions in fishponds.

1. Introduction

The Food Agriculture Organization stated that "Fish is a vital source of food for people. It is man's most important source of high-quality protein, providing ~16% of the animal protein consumed by the world's population." It is one of the most consumed foods in the world that provides vital nutrients for our health [1]. From 2000-2022, 184.6 metric tons in 2022 and 178.1 metric tons in 2021 are rapidly increasing and popular in most countries. There are different ways of producing aquatic resources in the Philippines, including Aquaculture Production [2][3]. However, culturing marine resources such as fish is susceptible to threat. There are several factors to consider in developing and sustaining the production of marine resources for both soil and water for fish farmers.[4][5].

The Philippines belongs to tropical countries where the acidity and salinity of water are typical. Acid rain can also cause acid water in most water, especially in developing countries, due to modernization, development, and urbanization, thus decreasing the water pH, and affecting the fish species unfavorably [6].

The location is one of the primary fish sources and is known for having fishponds in aqua culturing and exporting milkfish to other markets and communities. However, their home is undergoing road development, low waste management, and water contamination, affecting the water quality. The researchers interviewed the fishpond owner, with an area of 480 m2 hectares and 1.5ft – 3ft deep, one of the fish exporters in the city, and conducted water testing using a temperature meter, pH meter, and electroconductivity meter every morning to afternoon every hour for a total of 4 days. The gathered data shows that the temperatures are within the acceptable range for milkfish culturing which is 20-43 °C, and the data resulted in 29.8 °C as the lowest and 32.12 °C as the highest [7]. However, the acidity and salinity of the water show the instability of the water quality having 5.57 pH as the lowest and 9.178 pH as the highest average per day while the salinity has the lowest average of 21 ppt and 34 as the highest. The acidity level exhibits irregular fluctuations, while the salinity remains below the optimal threshold for cultivating milkfish. The water's inherent instability and unpredictability make it hard for the fish to nurture and grow, as the salinity levels fluctuate irregularly, and the acidity remains too low for milkfish cultivation. Specifically, the salinity levels range between 16 and 29 parts per thousand (ppt), and the pH levels oscillate between 6.5 and 8.5. These parameters, falling outside the requisite ranges, undermine the viability of the water body for the intended culturing activities [7].

IoT monitoring with an alert system will help secure better fishpond management, providing a safer habitat for Milkfish (Chanos chanos). Implementing a real-time data monitoring system with an alert mechanism will enable prompt action by owners or caretakers when acidity and salinity levels deviate from the necessary thresholds, ensuring the well-being of the aquatic environment [8]. Modern technology benefits both people and marine life, creating an advantage for local fishpond farmers. With automation and interconnectivity, vulnerability to threats is addressed and prevented, maintaining a secured fishpond, and supplying healthy fish for the community [9].

2. Methodology

The main objective of this study is to develop an IoT-based system that monitors the water quality of a fishpond. The research and development method were adopted by the researchers in developing, fabricating and designing the IoT monitoring system. The researchers employed this research to establish and acquire the relevant data that provided the basis for the conceptualized design in this project study. The data was useful in realizing the concept, process as well as the procedures of the study about the preferred method and the prototype. Systematic analysis of designing, developing, and evaluating instructional operations, processes, and policies, including a selection of materials, was used to meet desired consistency and effectiveness. Furthermore, the R&D method consists of activities that help the researchers attain the most efficient, economical, and ecological solution to resolve a particular issue encountered in the study and the different processes undertaken for consideration. The process allowed the researchers to express their abilities and capabilities in designing alternatives and improvements.

The researchers employed various research methodologies including research, assessment, and interviews to acquire necessary information to manage the fishpond including the process, methodologies, and the existing solution to the challenge faced. The researchers were able to acquire average data samples using some of the available equipment like temperature gauges, pH meters and Electrical Conductivity meters. This approach gave a clue on the important factors for milkfish production especially noting the importance of pH and salinity on the performance of milkfish and milkfish production respectively.[10]. The study aimed to elucidate which parameters predominantly impact milkfish growth and contribute to developing effective management strategies for enhancing aquaculture practices in the selected locale.

2.1 System Requirements and Architecture

Figure 1 illustrates an IoT-based Water Quality Monitoring System for fishponds. The system uses sensors to collect Analog data, which is processed by a microcontroller that converts it to digital form. This digital data is sent to an IoT server and displayed on a user interface. If water quality parameters, such as pH (< 6.5 or > 8.5) and salinity (16–29 ppt), fall outside acceptable ranges, the microcontroller sends alert notifications via the Blynk app and GSM module to inform the owner about the water conditions [3][4].

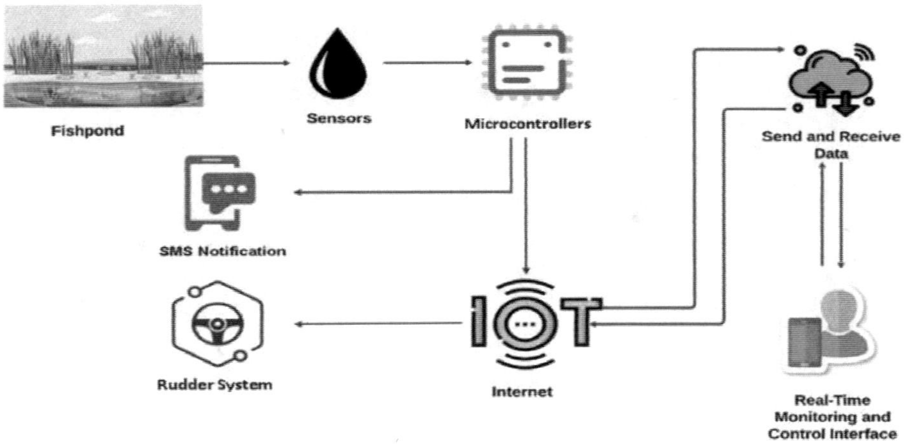

Figure 1. System Architecture of the IoT-based Monitoring System

2.2 Design and Development

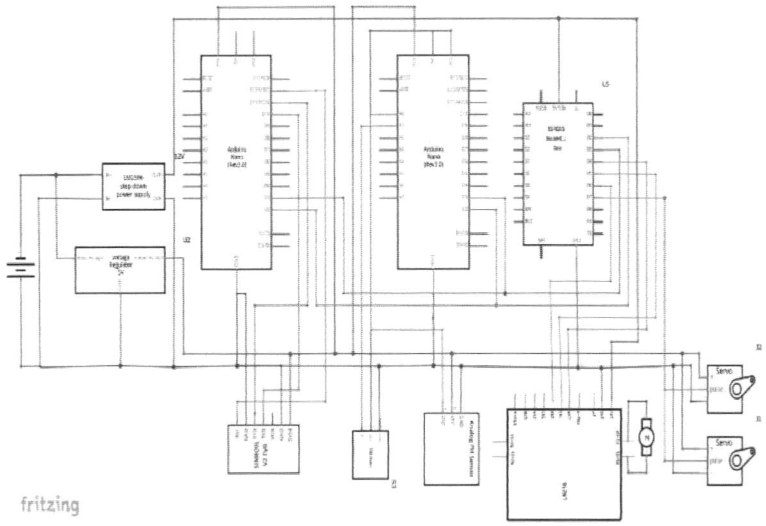

Figure 2. Schematic diagram

Figure 2 provides a comprehensive schematic diagram of the study's system circuitry, illustrating the precise connections and configurations among its components. This visual representation showcases the layout and interdependencies between sensors, controllers, power supplies, and communication modules. The program flow begins with one microcontroller processing data from the salinity and pH sensors, while another microcontroller manages additional system controls. The pH and TDS sensors continuously monitor the fishpond's acidity and salinity levels. The data collected is displayed in real-time on the Blynk app dashboard shown in Figure 3. When the pH levels surpass the established range of 6.5 to 8.5pH, the IoT and GSM modules trigger an alert to notify the user. Similarly, notifications are sent if the salinity values exceed the threshold of 16 to 29 ppt.

The Internet of Things (IoT) has emerged as a transformative technology, offering practical support to agricultural and aquacultural operations worldwide by enabling real-time monitoring of water quality parameters essential for maintaining stable fishpond production [11] When the system is deployed, it can increase the yield from the fishponds by giving precise pond monitoring and instantaneous responses to any changes that would have harmed the fish.[12]. The interface allowed simple monitoring of salinity and acidity levels. The dashboard displays the salinity and acidity of the water in real-time in the fishpond, which can be viewed on a laptop or mobile device.

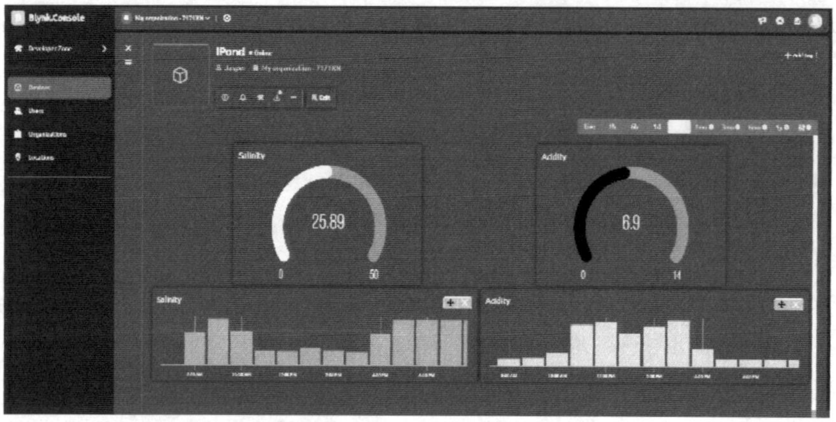

Figure 3. Blynk App IoT Cloud Dashboard

2.3 Testing and Evaluation of prototype

Prototype testing is a critical phase in product development, involving the evaluation of preliminary versions to gather feedback and refine designs before finalization. This process validates functionality, usability, and performance early on, minimizing risks and aligning the final product with user expectations. The study employs black box testing, focusing on end-user perspectives to assess functionality and uncover usability issues and functional gaps. Integration testing ran experiments by integrating the components to assess the system's performance by comparing the output response to the expected result. These contain tests to determine how the prototype will measure the data and respond to the control command through the Blynk. System acceptance testing follows integration testing and aims to evaluate the functionality of the finished IoT Water Quality Monitoring System. The main goal is to check that all the integrated functions in the prototype meet the acceptance criteria. It is an automatic system that tends to regulate the salinity and acidity of fishponds and warn the users if the values surpass certain standards while also sending information about the actual conditions of fishponds to the cloud for further control.

Salinity and acidity were measured in the fishpond from 8:00 am to 5:00 pm, during which the prototype was deployed to monitor water quality in real-time. This proved that acidity increases in the afternoon due to eh consumption of CO2 by aquatic plants and algae, and other environmental factors then decreases in the evening until sunrise [13]. The measured water quality data is displayed on the Blynk app dashboard, ensuring the parameters remain within predetermined thresholds. The prototype evaluation involves guidelines developed in the usability and human-centred design standards stipulating ISO 9241 and Nielsen's Usability Heuristics for establishing basic principles that enable users to achieve planned goals efficiently, effectively, and with minimal errors within a predetermined context of use. Using such standards and heuristics, it is hoped that this type of evaluation will help determine the usability, experience and layout of the prototype leading to improvements in its efficiency and effectiveness for the user. [14].

3. Results and Discussion

This study involves experienced fishpond's owners and pond caretakers with vast experience in the practices of aquaculture. Their ideas are valuable for assessing the efficiency and utility of an IoT water quality monitoring system specifically for fishponds. It uses real- time control of such parameters as pH, salinity, and dissolved oxygen with remote access. Issues in water quality can be noted instantly. This approach hinges on the use of technology which enables efficient collection of data that will enable better decision making as well as improvement of resource management, use and conservation of natural resources when engaging in aquaculture hence improving productivity and sustainability in fishponds.

Table 1 shows the comparison of the salinity readings from the prototype and a traditional TDS measuring instrument. The measurement time, prototype and actual TDS readings, error values, and accuracy percentages are listed in the columns. With a total average of 99.083431% accuracy and 0.187% error percentage.

Table 1. The IoT-based Water quality Monitoring Accuracy for Salinity

Time	T S Measuring Device	Prototype's Measurement	Error (%)	Accuracy (%)	Pass	Fail
8:00	24.87	24.8	0.07	99.718536	✓	
8:00	25.59	25.51	0.08	99.687378	✓	
8:00	22.4	22.1	0.3	98.660714	✓	
8:00	17.81	17.65	0.16	99.101628	✓	
8:00	17.93	17.57	0.36	97.992192	✓	
8:00	17.9	17.74	0.16	99.106145	✓	
8:00	18.12	18.03	0.09	99.503311	✓	
8:00	17.69	17.43	0.26	98.530243	✓	
8:00	25.14	25.12	0.02	99.920446	✓	
8:00	26.69	26.32	0.37	98.613713	✓	
		Total Average:	**0.187%**	**99.083431%**	✓	

Table 2 present a comparison of pH measurements obtained from an actual pH measuring device and the prototype measurement. The columns list the measurement time, actual pH readings, prototype pH readings, error values, and accuracy percentages. With a total average of 99.6408% accuracy percentage, the prototype's readings exhibit exceptional accuracy with a 0.027% error percentage.

Table 2. The IoT-based Water quality Monitoring Accuracy for Acidity

Time	pH Measuring Device	Prototype's Measurement	Error (%)	Accuracy (%)	Pass	Fail
8:00	6.76	6.74	0.02	99.70414	✓	
8:00	6.92	6.91	0.01	99.85549	✓	
8:00	6.81	6.8	0.01	99.85316	✓	
8:00	8.78	8.73	0.05	99.43052	✓	
8:00	8	7.97	0.03	99.625	✓	
8:00	8.43	8.39	0.04	99.5255	✓	
8:00	8.64	8.63	0.01	99.88426	✓	
8:00	8.8	8.77	0.03	99.65909	✓	
8:00	6.2	6.13	0.07	98.87097	✓	
8:00	6.5	6.5	0	100	✓	
		Total Average:	**0.027%**	**99.6408%**	✓	

The integration of DC motors with the Blynk IoT cloud-enabled seamless connectivity facilitates continuous data transfer and allows the motors to respond to various control signals, including start, stop, direction change, and speed adjustment. Additionally, the Blynk interface displayed precise, real-time data on measured salinity (ppt) and acidity (pH), ensuring responsiveness. The system incorporated an alert mechanism that informed the administrator when collected fishpond data exceeded preset limits, using the Blynk IoT platform and a GSM module. As a result, the prototype demonstrated effective functionality under all operating conditions.

For the evaluation of the prototype, 20 fishpond owners/fishpond caretakers were given a survey form. The weighted mean value for each question ranged from 3.9 as the lowest and 4.85 as the highest. The result of the evaluation in terms of functionality has an average of 4.325. Regarding efficiency and effectiveness, the average values are 4.13 and 4.36, respectively.

4. Conclusions and Recommendations

The researchers conducted experiments to assess the water quality of an inland fishpond, noting unstable acidity and salinity levels. Recognizing the need for continuous monitoring due to consistent fluctuations, they designed an IoT-based Water Quality Monitoring System using pH and TDS sensors (SEN0161-V2 and SEN0244). The system's circuit, comprising Arduino Nano microcontrollers and NodeMCU ESP8266 for Wi-Fi connectivity, enables real-time data transmission to the cloud for remote monitoring and control. The prototype underwent a rigorous fabrication process, including design, coding, testing, implementation, and support. The low error percentages recorded for accuracy—0.027% for acidity and 0.187% for salinity—alongside an overall rate of 4.27%, indicate that respondents agree the prototype meets its intended functions effectively. To enhance the study, the researchers suggest incorporating an Analog Electroconductivity sensor for precise salinity measurement and utilizing Arduino Uno R4 for hardware flexibility. They also recommend expanding the parameters beyond pH and salinity and adding a microSD card for extended data storage and analysis. The prototype's functionality, effectiveness, and efficiency were evaluated through unit testing, integration testing, system acceptance testing, and user feedback, highlighting its potential for improving fishpond management and environmental monitoring.

Acknowledgment

The authors would like to express gratitude for Manuel S. Enverga University Foundation and Kalawakan Fishpond's assistance in providing us with the data required to conduct this investigation. Our endeavour would not have been a success without their support.

5. References

[1] J. H. Tidwell and G. L. Allan, "Fish as food: aquaculture's contribution: Ecological and economic impacts and contributions of fish farming and capture fisheries," *EMBO Rep*, vol. 2, no. 11, p. 958, Nov. 2001, doi: 10.1093/EMBO-REPORTS/KVE236.

[2] "World fish production 2022 | Statista." Accessed: Nov. 05, 2023. [Online]. Available: https://www.statista.com/statistics/264577/total-world-fish-production-since-2002/

[3] A. B. Tahiluddin and E. Terzi, "An Overview of Fisheries and Aquaculture in the Philippines," *Journal of Anatolian Environmental and Animal Sciences*, vol. 6, no. 4, pp. 475–486, Dec. 2021, doi: 10.35229/JAES.944292.

[4] R. Ismail, K. Shafinah, and K. Latif, "A Proposed Model of Fishpond Water Quality Measurement and Monitoring System based on Internet of Things (IoT)," *IOP Conf Ser Earth Environ Sci*, vol. 494, no. 1, Aug. 2020, doi: 10.1088/1755-1315/494/1/012016.

[5] "pH in Fish Farming | Yokogawa Philippines." Accessed: Oct. 30, 2023. [Online]. Available: https://www.yokogawa.com/ph/library/resources/application-notes/ph-in-fish-farming/

[6] Islam Mannan, Rimi Ruksana, and Meghla Nowara Tamanna, "(PDF) Impact of Water Quality on Fish Growth and Production in Semi-Intensively Managed Aquaculture Farm." Accessed: Oct. 30, 2023. [Online]. Available: https://www.researchgate.net/publication/236007029_Impact_of_Water_Quality_on_Fish_Growth_and_Production_in_Semi-Intensively_Managed_Aquaculture_Farm

[7] G. Ganesh Research Scholar *et al.*, "Evaluation of water quality parameters in grow out phase of bracksh water fish Chanos chanos (Milk fish) in floating net cages," *~ 460 ~ Journal of*

Entomology and Zoology Studies, vol. 8, no. 5, pp. 460–464, 2020, Accessed: May 01, 2024. [Online]. Available: http://www.entomoljournal.com

[8] H. Herawati and I. Rostini, "WATER QUALITY MANAGEMENT OF MILK FISH (Chanos chanos) AT THE SOUTH AREA AND SEA FISHERIES CENTER PANGANDARAN," vol. 8, p. 2022, Accessed: Mar. 06, 2024. [Online]. Available: www.ijariie.com

[9] M. J. Burnett, G. C. O'Brien, F. J. Jacobs, F. Botha, G. Jewitt, and C. T. Downs, "The southern African inland fish tracking programme (FISHTRAC): An evaluation of the approach for monitoring ecological consequences of multiple water resource stressors, remotely and in real-time," *Ecol Indic*, vol. 111, Apr. 2020, doi: 10.1016/j.ecolind.2019.106001.

[10] A. H. Sambu *et al.*, "Modular system pond cultivation model for the survival of milkfish (Chanos chanos): case study of Pangkajene Kepulauan Regency, Indonesia," Jan. 2024, doi: 10.2103/RS.3.RS-38738/V1.

[11] N. Ya'acob, N. N. S. N. Dzulkefli, A. L. Yusof, M. Kassim, N. F. Naim, and S. S. M. Aris, "Water Quality Monitoring System for Fisheries using Internet of Things (IoT)," *IOP Conf Sr Mater Sci Eng*, vol. 1176, no. 1, p. 012016, Aug. 2021, doi: 10.108/1757-899X/1176/1/012016.

[12] A. N. Natividad, C. Miranda, J. C. Valdoria, and D. Balubal, "An IoT Based pH Level Monitoring Mobile Application on Fishponds using pH Sensor and Waterproof Temperature Sensor," *Journal for Educators, Teachers and Trainers*, vol. 14, no. 3, p. 4, Jun. 2023, doi: 10.47750/JETT.2023.14.03.091.

[13] Environmental Protection Agency (USA Government), "Factsheet on Water Quality Parameter", Accessed: May 02, 2024. [Online]. Available: https://www.epa.gov/system/files/documents/2021-07/parameter-factsheet_ph.pdf

[14] "ISO 9241-210:2019 - Ergonomics of human-system interaction — Part 210: Human-centred design for interactive systems." Accessed: May 02, 2024. [Online]. Available: https://www.iso.org/standard/77520.html

IOP Conf. Series: Earth and Environmental Science **1446** (2025) 012006 doi:10.1088/1755-1315/1446/1/012006

Effect of starch and chitosan on the physical and mechanical properties of durian mesocarp cellulose biodegradable foam

W Suryaningsih [*1)] , T Budiati [1)], M P D Utami [2)] , A C Dewi [2)] , O Arddiarisca [3)]

[1)] Food Engineering Technology Study Program, Departement of Agricultural Technology, State Polytechnic of Jember, Indonesia

[2)] Poultry Production Management Study Program, Department of Animal Husbandry, State Polytechnic of Jember

[3)] Public Sector Accounting Study Program, Department of Business, State Polytechnic of Jember

Email: wahyu_surya@polije.ac.id

Abstract. Durian mesocarp cellulose biodegradable foam is an environmentally friendly alternative food packaging material to replace styrofoam. The research aims to determine the effect of starch and chitosan types on the physical and mechanical properties of biodegradable foam. The research method was a completely randomized factorial design with variations in the types of starch (corn, arrowroot) and chitosan (0%, 1%, 2%, 3%, 4%, 5%, and 6%), using analysis of variance and the Duncan Multiple Range mean test. The results show corn starch type has a compressive strength value of durian mesocarp cellulose foam biodegradable products 0.076 N/mm2, lower than arrowroot starch 0.96 M/mm2. The chitosan concentration of 6% produced the best physical and mechanical properties having a water absorption value of 13.55%, thickness of 4.47 mm, density of 0.76 Gram/mm3, compressive strength of 0.79 N/mm2, but had a low biodegradation rate of 9.86%. The combination of arrowroot starch with 4% chitosan produced the best physical, chemical, and mechanical properties of biodegradable foam, having a water absorption value of 12.90%, a thickness of 4.67 mm, a density of 0.66 grams/cm3, a compressive strength of 0.91 N/mm2, and a biodegradability level of 12.90% with an estimated complete degradation time of 76.58 days.

1. Introduction

The use of single-use food packaging materials is increasing frequently with the development of the fast food industry. Foam packaging materials are widely used because they are lightweight, low-density, non-porous, malleable, waterproof, and strong against physical stress [1]. Most of the commercially produced foam food packaging is styrofoam. Styrofoam made from polystyrene polymers is difficult to degrade and has a bad impact on the environment. If burned, it will produce dioxin compounds that can interfere with breathing and trigger the growth of cancer cells [2]. For this reason, it is necessary to find alternative materials to replace Styrofoam with natural materials that are easily degraded, such as bio-foam. Biodegradable foam is a starch polymer product from renewable natural materials. Biofoam must be easy to form, waterproof, lightweight, can withstand cold and hot temperatures, and has a low production price [3].

The main raw materials for bio foam are cellulose and starch. Cellulose is used as a filler to increase the compactness of bio-foam. Durian skin mesocarp cellulose is an excellent raw material for making commercial bio-foam. It is abundant, completely degradable, biocompatible, and easy to modify. The cellulose content of durian skin is between 50 and 60% [4] Cabbage cellulose 34% [5] Durian skin cellulose is an excellent raw material for making bioplastics. It improves mechanical properties, water resistance, and biodegradation.

Starch is a structural material to replace polystyrene. Corn starch and arrowroot starch, can be used as raw materials for bio-foam. Corn starch contains 74-76% amylopectin and 24-26% amylose. Corn starch has hygroscopic properties with a moisture content of about 11% . Meanwhile, arrowroot starch contains 24.64% amylose and 75.36% amylopectin. Arrowroot starch tends to be highly hygroscopic due to its protein content. Arrowroot starch is bright white, odorless, and widely available. Arrowroot starch contains 19.4% amylose, 59.35% amylopectin, and 0.72% protein [7]. Starch biofoam has weaknesses in that it is fragile and hygroscopic. If it is to be used as food packaging, it is necessary to improve these properties by adding other materials that can increase strength and water resistance [8]. Materials that can improve the quality of starch-based biofoam are plasticizers, synthetic polymers and chitosan.

Chitosan is a natural biopolymer consisting of d-glucosamine and N-acetyl-d-glucosamine which is a β (1-4) glycosidic bond [9]. The function of chitosan in making biofoam is as a mixing biopolymer by forming hydrogen bonds between amylopectin chains and amylose in starch, thereby improving mechanical properties. Chitosan can increase the tensile strength and water resistance of bio-foam [8].

This study aims to determine the effect of starch type, chitosan, and the interaction of starch type and chitosan on the physical and mechanical properties of durian mesocarp cellulose biodegradable foam. Find starch type and chitosan concentration, and determine the combination between starch type and chitosan concentration to produce durian mesocarp cellulose biodegradable foam with the best physical and mechanical properties.

2. Materials and Methods

2.1. *Materials*

The materials used in making biodegradable foam cellulose from mesocarp durian (*Jambe arum variety*), 3% sodium hydroxide (NaOH), 5% hydrogen peroxide (H_2O_2), corn starch, arrowroot starch, chitosan with a deacetylation degree of 97.03% , acetic acid 1%, distilled water, polyvinyl alcohol (PVA), magnesium stearate, glycerol and compost soil.

2.2. *Equipments*

The equipment used is a food dehydrator (Harvest), 60 mesh sieve, pH meter, grinder, oven (Venticell), hot plate, magnetic stirrer, mixer, scrub micrometer, color reader (Konika Minolta Cr-10 Plus), and texture analyzer (LLOYD, UK).

2.3. *Metodology*

The research used a factorial completely randomized design (CRD) method, consisting of 2 factors, namely chitosan concentration (K) and starch type (P). The concentration of chitosan (K) consists of 7 levels, namely 0% (K1), 1% (K2), 2% (K3), 3% (K4), 4% (K5), 5% (K6), and 6% (K7) from the starch weight. The starch type P factor consists of 2 levels, namely corn starch (P1) and arrowroot starch (P2), with four replications. Data analysis using analysis of variance (ANOVA) at 95% confidence level

($p<0.05$) processed with SPSS v27. If there is a significant difference, it is followed by Duncan Multiple Range Test (DMRT) at the level of ($p<0.05$).

2.4. Production of Biodegradable Foam from Durian Mesocarp Cellulose

The stages of the process of making biodegradable foam from durian mesocarp cellulose consist of dissolving chitosan, dissolving polyvinyl alcohol (PVA) and glycerol, mixing and shaking, molding, drying, and cooling. Dissolving chitosan by weighing chitosan according to the specified treatment in a glass jar. Add 15 ml of 1% acetic acid solution, heating with a hot plate at 70 °C and stirring until homogeneous. The mixing was carried out in a basin. Chitosan solution, PVA solution, glycerol, 10 grams of starch, 1 gram of magnesium stearate, and 0.3 grams of cellulose were mixed and shaken with a mixer at a speed of 2 for 3 minutes until the dough expanded and formed foam. The molding process uses a Petri dish that has been coated with aluminum foil. The dough-drying process is carried out in an oven at 105°C for 4 hours. The dried biofoam was then cooled in a desiccator for 1 -2 days and stored in an airtight container.

2.5 Parameter

The observation parameters for durian mesocarp cellulose biodegradable foam are water absorption capacity referring to SNI 1969:2008, thickness, density, sensory test, and biodegradability referring to ASTM G-21-70 with the Soil Burial Test method and SNI 7188:2016, compressive strength with texture analyzer (LLOYD, UK)

3. Results And Discussions

3.1. Sensory Properties of Biodegradable Foam Durian Mesocarp Cellulose

Biodegradable foam cellulose mesocarp durian from various concentrations of chitosan and starch types using the baking process method produces different product appearances. The higher concentration of chitosan resulted in a rough appearance and texture and a cloudy white colour. The use of corn starch produced a coarser appearance than arrowroot starch, while the colour was the same, which was cloudy white. Chitosan influenced the sensory properties of the appearance and colour of durian mesocarp biodegradable foam. The sensory properties *of* biodegradable foam durian mesocarp cellulose can be seen in Figure 1.

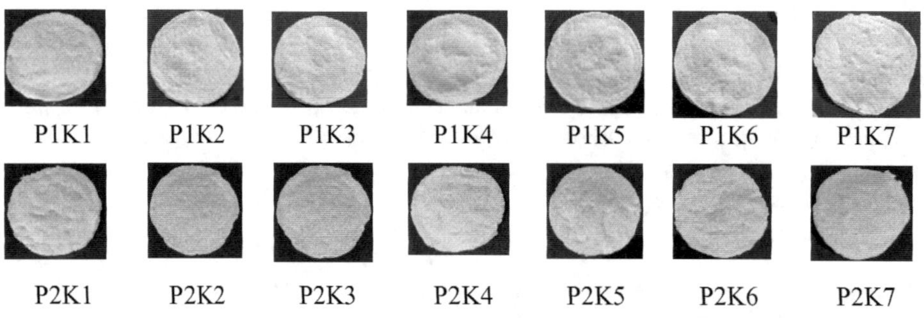

Description: P1K1 (corn starch and chitosan 0%), P1K2 (corn starch and chitosan 1%), P1K3 (corn starch and chitosan 2%), P1K4 (corn starch and chitosan 3%), P1K5 (corn starch and chitosan 4%), P1K6 (corn starch and chitosan 5%), P1K7 (corn starch and chitosan 6%), P2K1 (arrowroot starch and chitosan 0%), P2K2 (arrowroot starch and chitosan 1%), P2K3 (arrowroot starch and chitosan 2%), P2K4 (arrowroot starch and chitosan 3%), P2K5 (arrowroot starch and chitosan 4%), P2K6 (arrowroot starch and chitosan 5%), P2K7 (arrowroot starch and chitosan 6%).

Figure 1. The biodegradable foam of durian mesocarp cellulose foam profile

3.2. *Physical and Mechanical Properties of Biodegradable Foam Durian Mesocarp Cellulose*

The physical properties of durian mesocarp cellulose biodegradable foam products are determined by the parameters of water absorption, thickness, and density. Chemical properties are determined by biodegradability parameters, and mechanical properties are compressive strength. The results of data processing with ANOVA ($P>0.05$) showed that the treatment factor of starch kind, and the interaction between starch type and chitosan did not significantly affect the parameters of physical and chemical properties, but significantly affected the mechanical properties, namely compressive strength. Parameter data on physical, chemical and mechanical properties of biodegradable foam cellulose mesocarp durian treated with chitosan concentration and starch type can be seen in Table 1.

Table 1. Physical, chemical, and mechanical characteristics of durian mesocarp cellulose biodegradable foam treated with chitosan concentration and starch type.

Treatment	water absorption (%)	Thickness (mm)	Density (g/cm^3)	Compressive Strength (N/mm^2)	Biodegradability	
					Biodegradation (%)	Estimated degradation time 100% (day)
P1K1	47,70±2,49 ^{a}	5,58±0,79 ^{a}	0,60±0,01 ^{a}	0.74±0.02^c	46,32±6,90 ^{a}	30,22
P1K2	40,49±8,58 ^{a}	5,47±0,53 ^{a}	0,55±0,02 ^{a}	0.33±0.01^d	41,80±2,85 ^{a}	33,50
P1K3	44,92±4,04 ^{a}	5,79±0,03 ^{a}	0,61±0,02 ^{a}	0.65±0.02^c	23,28±8,68 ^{a}	60,14
P1K4	30,35±5,08 ^{a}	5,07±0,53 ^{a}	0,59±0,08 ^{a}	0.64±0.08^c	13,73±4,83 ^{a}	102,00
P1K5	28,70±7,81 ^{a}	4,89±0,80 ^{a}	0,64±0,12 ^{a}	1.11±0.00^b	19,31±5,44 ^{a}	72,50
P1K6	17,02±6,36 ^{a}	4,79±0,17 ^{a}	0,66±0,01 ^{a}	1.15±0.34^b	14,12±1.32 ^{a}	99,15
P1K7	12,17±3,94 ^{a}	4,27±0,32 ^{a}	0,75±0,01 ^{a}	0.67±0.03^c	6,82±3,75 ^{a}	205,27
P2K1	55,71±1,17 ^{a}	6,40±0,14 ^{a}	0,70±0,01 ^{a}	1.10±0.16^b	48,62±1,36 ^{a}	28,79
P2K2	45,92±0,83 ^{a}	6,09±0,12 ^{a}	0,57±0,12 ^{a}	0.86±0.07^c	43,22±3,75 ^{a}	32,39
P2K3	46,45±1,85 ^{a}	5,48±0,07 ^{a}	0,59±0,04 ^{a}	0.97±0.05^b	39,43±1,69 ^{a}	35,51
P2K4	34,68±4,04 ^{a}	5,13±0,37 ^{a}	0,61±0,02 ^{a}	1.36±0.11^a	18,28±5,09 ^{a}	76,60
P2K5	35,01±0,87 ^{a}	5,33±0,05 ^{a}	0,67±0,05 ^{a}	0.95±0.05^b	15,10±5,32 ^{a}	92,69
P2K6	26,10±4,94 ^{a}	5,08±0,64 ^{a}	0,68±0,04 ^{a}	0,56±0,05^d	13,54±2,82 ^{a}	103,43
P2K7	14,94±1,46 ^{a}	4,66±0,75 ^{a}	0,77±0,15 ^{a}	0.91±0.09^c	12,90±2,52 ^{a}	108,57

Description: P1K1 (corn starch and chitosan 0%), P1K2 (corn starch and chitosan 1%), P1K3 (corn starch and chitosan 2%), P1K4 (corn starch and chitosan 3%), P1K5 (corn starch and chitosan 4%), P1K6 (corn starch and chitosan 5%), P1K7 (corn starch and chitosan 6%), P2K1 (corn starch and chitosan 0%), P2K2 (corn starch and chitosan 1%), P2K3 (corn starch and chitosan 2%), P2K4 (corn starch and chitosan 3%, P2K5 (corn starch and chitosan 4%), P2K6 (corn starch and chitosan 5%), P2K7 (corn starch and chitosan 6%). The same letter notation in the treatment indicates that the treatment is not significantly different based on the Duncan test at the 5% level ($P>0.05$)

3.2.1. Water Absorption (%). The water absorption capacity of durian mesocarp cellulose biodegradable foam indicates its water resistance. A good biodegradable foam must be able to retain water, so the water absorption value is low. Table 1. shows that the water absorption of chitosan concentration treatment is 9.38%-56.54%. The concentration of chitosan has a very significant effect ($P<0.01$) on the water absorption of the biodegradable foam made from durian peel cellulose. The chitosan concentration of 6% has the lowest water absorption value. Chitosan is hydrophobic, so it is difficult to bind water. The structure of chitosan is composed of intramolecular and intermolecular hydrogen bonds, which makes it unable to effectively interact with water molecules [10]. In addition, chitosan has amine groups (NH2-) in the C2 position and also due to primary and secondary hydroxyl groups (OH-) in the C3 and C positions. These amino acids are positively charged so they can bind to negative surface charges through hydrogen or ionic bonds.

3.2.2 Compressive Strength Test (N/mm2). Compressive strength is the maximum ability of biodegradable foam to withstand external forces before being damaged or broken. Broken biodegradable is caused by excessive pressure and structural deformation. The compressive strength shows the resistance of biodegradable foam material to pressure or stress [9]. Based on the ANOVA ($P<0.05$) showed the treatment of chitosan concentration, starch type, and the interaction of chitosan concentration with starch type significantly affected the compressive strength value of biodegradable foam. Based on the DMRT test ($P<0.05$), the interaction of chitosan concentration with the type of starch is significantly different.

Biodegradable foam in arrowroot starch treatment and 3% chitosan concentration (P2K4) has the highest compressive strength value of 1.36 N/mm2, and this is significantly different from corn starch treatment and 1% chitosan concentration (P1K2) which has the lowest compressive strength of 0.33 N/mm2. The same results show that chitosan concentration 6.5 % has the highest tensile strength value of bio foam, namely 1.11 Mpa, while chitosan concentration 0 has a low tensile strength value, namely 0.44 MPa. The addition of high chitosan concentration can improve the mechanical properties of the compressive strength of biodegradable foam [6]. However, after reaching the maximum point, the compressive strength value shows a decrease. Biodegradable foam using starch with high amylopectin content has smaller pores and lower density, compared to bio foam using high amylose [11]. The water content of the starch solution can influence the melting temperature, foaming process, crystallinity, cell structure, and mechanical properties of the final starch foam. Water can also function as a plasticizer and blowing agent during the starch foam manufacturing process [12]

3.2.3 Biodegradable. Biodegradability is a parameter that does not affect the quality of the use of foam but is an important element in a foam product. However, biodegradable foam has poor biodegradability, which is potentially harmful to the environment [9]. The biodegradation test of Durian peel cellulose biofoam was conducted for 14 days. The biodegradation test results of durian peel cellulose biofoam with chitosan concentration treatment and starch type ranged from 4.17% to 51.20%. Based on the results of ANOVA ($p<0.05$) showed that the concentration of chitosan had a very real effect. Based on the DMRT test results ($P<0.05$) showed that the biodegradation of durian peel cellulose biodegradable foam is significantly different from 2% concentration with 3%, 4%, and 5% chitosan concentrations.

The highest biodegradation value at 0% chitosan concentration was 48.62 ± 1.36 with a predicted time of complete degradation of about 28.79 days, while the low value at 6% chitosan concentration was $6.82 \pm 3.75\%$ with a predicted time of degradation of about 205.27 days. Biodegradable foam with a high concentration of chitosan will be difficult to degrade. A high chitosan concentration (7%) results in a low degradation value and requires a long time (Figure 2).

The decrease in degradation rate can be influenced by chitosan, which is hydrophobic and antimicrobial [10]. The biodegradation rate of biodegradable foam is influenced by environmental conditions and soil type. The biodegradability is caused by chemical, microbiological, enzymatic, mechanical, and light decomposition [13]. Starch-based biodegradables absorb moisture from the environment, easily interact with water and microorganisms and are easily exposed to physicochemical influences. The degradation rate of durian peel mesocarp cellulose foam biodegradable products with the addition of chitosan has a higher value with a short time than the Standard (ASTM 5336) for commercial plastic films takes 60 days to be completely degraded [14]. According to SNI, bio-foam can be 100% degraded for 6-9 months. The high degradation rate of 48.62% with a time of 28.79 days shows that the addition of chitosan can improve the mechanical properties of biodegradable foam and can contribute to environmental sustainability, given its nature that does not contain hazardous materials and is easily recycled.

Description: - P1K1 (corn starch and chitosan 0%), P1K2 (corn starch and chitosan 1%), P1K3 (corn starch and chitosan 2%), P1K4 (corn starch and chitosan 3%), P1K5 (corn starch and chitosan 4%), P1K6 (corn starch and chitosan 5%), P1K7 (corn starch and chitosan 6%)

Figure 2. Degradation rate of durian mesocarp cellulose foam biodegradable products of different concentrations of chitosan and arrowroot starch

3.2.4 .Thickness (mm). The thickness of biodegradable foam is influenced by shaking and formulation of ingredients. The process of foam formation occurs when the interaction of durian cellulose with starch in bio foam dough acts to bind water molecules to form hydrogen bonds. The more hydrogen bonds formed affect the thickness of bio foam. The thickness of durian peel cellulose biodegradable foam in the treatment of chitosan concentration and starch type ranged from 4.04 mm - 6.51 mm. DMRT test results ($P < 0.05$) showed that chitosan concentrations of 5% and 6% were thinner than chitosan concentrations of 1%, 2%, 3%, 4%. The dissolution process between high-concentration chitosan and acetic acid produces a thick dough and at the time of molding there are still many residues left in the container.

3.2.5 Density (g/cm³) . The highest biodegradable foam density is produced by a concentration of 6% (0.76 g/cm³), while the lowest density is achieved with a chitosan concentration of 1% (0.56 g/cm³). A chitosan solution has a high viscosity, which produces a small and tight cell structure in the biodegradable foam and increases the density. states that smaller cell-sized foam has a higher density, which is evident in the high density of biodegradable foam. This results in a bio-foam that is denser and has a large compressive strength. Conversely, bio foam with low-density makes bio foam brittle and easily absorbent of water, which results in increased water absorption values [15].

4. Conclusion

Chitosan concentration treatment significantly affects the physical, chemical, and mechanical properties. The type of starch, and the interaction between starch type and chitosan concentration did not significantly affect the physical and chemical properties, except for the mechanical properties, namely compressive strength. The treatment of corn starch has a compressive strength value of durian mesocarp cellulose foam biodegradable products 0.076 N/mm2, lower than arrowroot starch 0.96 M/mm2. The chitosan concentration of 6% produced the best physical and mechanical properties having a water absorption value of 13.55%, thickness of 4.47 mm, density of 0.76 Gram/mm3, compressive strength of 0.79 N/mm2, but had a low biodegradation rate of 9.86%. The combination treatment of arrowroot starch with 4% chitosan produced the best physical, chemical, and mechanical properties of biodegradable foam, having a water absorption value of 12.90%, a thickness of 4.67 mm, a density of 0.66 grams/cm3, a compressive strength of 0.91 N/mm2, and a biodegradability level of 12.90% with an estimated complete degradation time of 76.58 days.

Acknowledgement

This research was funded by PNBP of Jember State Polytechnic 2024.

REFERENCES

[1] Liao J, Hou Y, Zhang X, Li J, Zhang M, Dong Y, Zheng Q and Chen X 2024 Bird-nest inspired cellulose-based biofoam with excellent performance enabled by gas–liquid interface co-assembly and thermal welding for plastic replacement *Chem. Eng. J.* **495** 153412

[2] Sutiarno S, Muryani M, Sucipto A, Rahmawati R, Fegiliani F and Riyanto A 2022 Pengaruh Penambahan Tepung Umbi Porang dan Variasi Konsentrasi Flavonoid Kulit Nanas dalam Pembuatan Biofoam *J. Fis. Flux J. Ilm. Fis. FMIPA Univ. Lambung Mangkurat* **19** 101

[3] Kaisangsri N, Kerdchoechuen O and Laohakunjit N 2012 Biodegradable foam tray from cassava starch blended with natural fiber and chitosan *Ind. Crops Prod.* **37** 542–6

[4] Nugraheni H M, Mulyanti T A and Badriyah L 2018 Pemanfaatan Limbah Kulit Buah Durian Mentega sebagai Carboxymethyl Cellulose(CMC) *Semin. Nas. Sains* 115–22

[5] Wang T and Ma Z 2017 A novel insoluble dietary fiber-based edible paper from Chinese cabbage *Cellulose* **24** 3411–9

[6] Yun S-M, Kang M, Park S Y, Eun J-B and Chun H H 2023 Characterization of biodegradable corn starch-based foam container incorporating kimchi cabbage (Brassica rapa L. pekinensis) by-product *LWT* **188** 115432

[7] Malki M K S, Wijesinghe J A A C, Ratnayake R H M K and Thilakarathna G C 2023 Characterization of arrowroot (Maranta arundinacea) starch as a potential starch source for the food industry *Heliyon* **9** e20033

[8] Marlina R, Kusumah S S, Sumantri Y, Syarbini A, Cahyaningtyas A A and Ismadi I 2021 Karakterisasi Komposit Biodegradable Foam Dari Limbah Serat Kertas dan kulit Jeruk Untuk Aplikasi Kemasan Pangan. *J. Kim. dan Kemasan* **43** 1

[9] Muharram F I 2020 Penambahan Kitosan Pada Biofoam Berbahan Dasar Pati *Edufortech* **5**

[10] Vlacha M, Giannakas A, Katapodis P, Stamatis H, Ladavos A and Barkoula N-M 2016 On the efficiency of oleic acid as plasticizer of chitosan/clay nanocomposites and its role on thermo-mechanical, barrier and antimicrobial properties – Comparison with glycerol *Food Hydrocoll.* **57** 10–9

[11] Ikhsan M H, Dewata I and Umar Kalmar Nizar M A 2021 Pengaruh Penambahan Karagenan Terhadap Kuat Tarik Dan Biodegradasi Edible Film Dari Pati Kulit Singkong *J. Kependud. ...* **2** 44–50

[12] Trongnit J, Mayakun J and Kaewtatip K 2024 The effect of agar from the seaweed Gracilaria fisheri on properties of biodegradable starch foam *Int. J. Biol. Macromol.* **273** 132952

[13] Akmala A and Supriyo E 2020 Optimasi Konsentrasi Selulosa pada Pembuatan Biodegradable Foam dari Selulosa dan Tepung Singkong *Pentana J. Penelit. Terap.* **1** 27–40

[14] Hendrawati N, Lestari Y I and Wulansari P A 2017 Pengaruh Penambahan Kitosan dalam Pembuatan Biodegradable Foam Berbahan Baku Pati *J. Rekayasa Kim. Lingkung.* **12** 1

[15] Sefrin Speroni C, Rigo Guerra D, Beutinger Bender A B, Stiebe J, Ballus C A, Picolli da Silva L, Lozano-Sánchez J and Emanuelli T 2021 Micronization increases the bioaccessibility of polyphenols from granulometrically separated olive pomace fractions *Food Chem.* **344** 128689

IOP Conf. Series: Earth and Environmental Science **1446** (2025) 012007 doi:10.1088/1755-1315/1446/1/012007

Occupational Health and Safety Risk Control at X Teaching Factory

M. Adhyatma[1], Gandu Eko Julianto Suyoso[1*], Angga Rahagiyanto[1], Faruq Avero Azhar[1], Veronika Vestine[1]

[1]Politeknik Negeri Jember

*Email : gandu.eko.js@polije.ac.id

Abstract. UHT (Ultra High Temperature) milk factories are complex environments with numerous inherent risks due to the high-temperature processes, sophisticated machinery, and extensive use of chemicals. These factors create a hazardous workplace where managing occupational safety and health is crucial to prevent accidents, protect worker health, and maintain production efficiency. The primary objective of this study is to identify and control the hazard in the UHT milk facility of X Teaching Factory (Tefa). The data collected by walkthrough survey, observation and interview. The team then conducted a literature review to map out the types of hazards, sources of hazards, potential impacts, analyze the risk and provide risk control recommendations. The study identified five types of workplace hazards: explosion and fire hazards, physical hazards, chemical hazards, ergonomic hazards, and slips, falls and trips hazards. Among these, The highest risks present at the X Tefa were fall, slip, and trip hazards as well as fire and explosion hazards. The recommended controls focused on substitution, engineering controls, administrative controls, and personal protective equipment, as it was not feasible to eliminate the machines/materials/processes used at X Tefa. Although X Tefa is not yet operational, it is crucial to implement risk control measures as early as possible to prevent workplace accidents and occupational diseases once X Tefa becomes fully operational.

1. Introduction

UHT milk facility converts perishable milk into a long-shelf-life product.[1] The milk is partially sterilized by heating it for 1-2 seconds at a temperature exceeding 135°C, preserving nutrients and allowing it to be stored for long periods of time at ambient temperature.[2] Those operation involves machinery and production processes that pose occupational health and safety risks to its workers.

A study indicated that dairy workers experience a two-fold higher rate of occupational injuries, with rates of 3.3 to 6.6 injuries per 100 full-time workers [3]. A population-based study of 600 dairy workers found an injury rate of 16.6% per year, equating to approximately 166 injuries per 1,000 workers annually. While specific statistics on occupational diseases in the dairy industry are less frequently reported, the nature of work exposes employees to various health risks, including respiratory issues from dust and aerosols, skin irritations from chemicals, and musculoskeletal disorders from repetitive tasks. The physically demanding nature of dairy work can lead to chronic

IOP Conf. Series: Earth and Environmental Science 1446 (2025) 012007 doi:10.1088/1755-1315/1446/1/012007

conditions such as back pain and repetitive strain injuries, which are common among workers engaged in manual handling and machine operation.

X Tefa have a new milk processing facility that has not yet begun operations. In the future, X Tefa will serve as a practical field for both lecturers and students to apply their knowledge. Based on the initial assessment conducted by the Occupational Health and Safety (P2K3) team at Y Polytechnic in collaboration with external auditors, it was found that UHT milk processing facility at X Tefa has not yet reached the maximum score across 64 criteria of the Occupational Health and Safety Management System (SMK3) evaluation. The report also highlighted several findings that need to be addressed promptly by X Tefa. Considering the report and the existing findings, a thorough study of potential hazards and risk controls must be conducted, referencing both literature and best practices from the field, with the ultimate goal of improving occupational health and safety as well as worker productivity.

2. Method

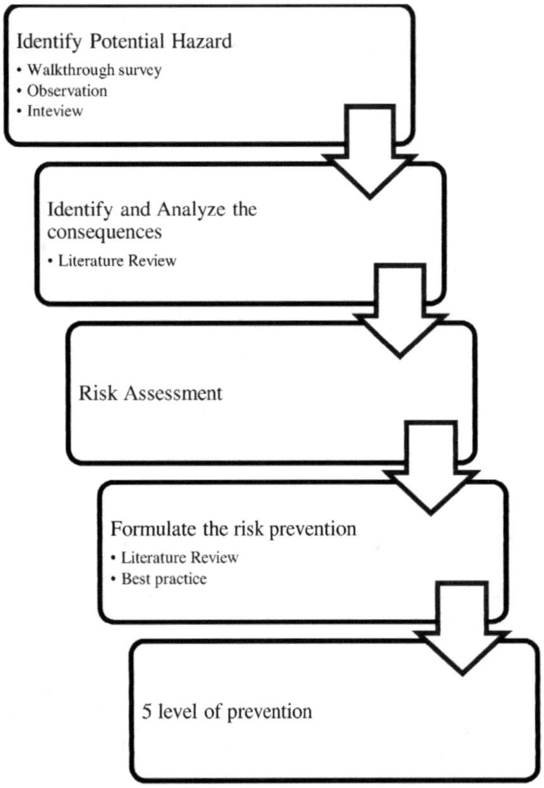

Figure 1. Research flow-chart

Based on figure 1, the study began with data collection through a walkthrough survey and observation at X Tefa to identify potential hazards using a checklist. This was followed by interviews with the person in charge of X Tefa. Subsequently, the author conducted a literature review to explore the potential impacts of these hazards on workers. After identifying the hazards and their associated impacts, a risk assessment was carried out using the risk level framework adapted from the AS/NZ 4360 Standard Risk Matrix (Table 1). The risk evaluation process was followed by formulating risk control recommendations based on the 5 levels of prevention concept: elimination, substitution, engineering controls, administrative controls, and personal protective equipment (PPE).

IOP Conf. Series: Earth and Environmental Science **1446** (2025) 012007 doi:10.1088/1755-1315/1446/1/012007

Table 1. Risk Level, adapted from the AS/NZ 4360 Standard Risk Matrix and NHS QIS Risk Matrix

		Consequences (C)				
		Insignificant (1)	Minor (2)	Moderate (3)	Major (4)	Catastrophic (5)
Likelihood (L)	Almost Certain (5)	Medium	High	High	Extreme	Extreme
	Likely (4)	Medium	Medium	High	High	Extreme
	Possible (3)	Low	Medium	Medium	High	High
	Unlikely (2)	Low	Low	Medium	Medium	High
	Rare (1)	Low	Low	Low	Medium	Medium

3. Result and Discussion

Based on data collected, the following is a summary of the identified workplace hazards, their sources, and the potential impacts associated with these hazards, risk level along with recommended risk control measures:

Table 2. Identification Results of Work Hazard Types, Hazard Sources, Impacts, Risk Level and Risk Control Recommendations

Work Hazard type	Hazard source	Potential Impact	L	C	Risk Value (Risk Level)	Control
Fire and Explosion	Boiler tank and retort machine	Injuries, burn wounds, and fatalities	2	5	10 (High)	Engineering control: a. Use of an integrated process monitoring system in the control room b. Installation and testing of active and or passive fire protection systems. Administrative control: a. A thorough testing inspection of the production machines should be conducted by certified personnel. b. Training and certification for personnel operating the boiler
	Use of LPG gas cylinders as a fuel source for heating the boiler tank that located close to	Injuries, burn wounds, and fatalities	2	5	10 (High)	Engineering control: a. Moving the LPG gas cylinder to a safer position (not too close to the boiler)

Work Hazard type	Hazard source	Potential Impact	L	C	Risk Value (Risk Level)	Control
	the boiler					b. Installation of active and passive fire protection systems.
	Electrical system of production machine	Electric shock, burn injuries, and fatalities	1	5	5 (Medium)	Engineering control: a. Installation of active and passive fire protection systems Administrative control: a. Periodic monitoring of the electrical system by certified electrical safety personnel
Physic	Production machine noise	Increased hearing threshold up to hearing loss	3	3	9 (Medium)	Administrative control: a. Noise measurement and monitoring in the production area PPE: a. If the noise measurement results exceed the permissible exposure limit, workers must wear earmuffs or ear plugs
	Heat from the sterilization process	Burn injuries	2	3	6 (Medium)	Engineering control: a. Equip all hot surfaces and pipes with thermal insulation Administrative control: b. Installation of hazard signs c. Installation of floor marking
Ergonomy	Repetitive movements and standing working posture during packing	musculoskeletal disorders in the areas of the hands, arms, neck, shoulder, back and legs	4	2	8 (Medium)	Administrative control: a. Selection of workers whose body anthropometry matches the dimensions of the workstation. b. Warming up before work and stretching during

Work Hazard type	Hazard source	Potential Impact	L	C	Risk Value (Risk Level)	Control
						work breaks
	Work posture when lifting raw materials and lifting packed milk	Low back pain	4	2	8 (Medium)	Engineering control: a. Use of trolleys for transporting raw materials/packed milk Administrative control: a. If manual lifting is unavoidable, implement signage indicating the weight limit that can be safely lifted by workers. Additionally, provide socialization and posters illustrating ergonomic methods for lifting items to ensure proper technique and prevent injury. PPE: a. Use of back support
Fall, slip and trip	Piping systems and parts of the production process equipment that cross the workers' mobility area	Minor injuries to fractures	4	3	12 (High)	Administrative control: a. Use of floor marking and hazard sign
	Liquid spills	Minor injuries to fractures;	2	3	6 (Medium)	Administrative control: a. Should be dried immediately, with hazard signs placed first. b. Regular inspections in high-risk spill areas. c. Inspections of the drainage system. PPE: a. Use of anti-slip safety shoes untuk keseharian

Work Hazard type	Hazard source	Potential Impact	L	C	Risk Value (Risk Level)	Control
						pekerja. b. Use of anti-slip safety shoes for workers' daily activities. c. Safety gloves when cleaning up liquid spills.
	Falling from a ladder	Minor injuries to fractures	3	3	9 (Medium)	Engineering control: a. Installation of anti-slip rubber on handrails and steps of the ladder b. Installation of poster or signs about the three-point contact rule near stairways. PPE: a. Use of anti-slip safety shoes and hand glove
Chemical	Chemical substance	Irritation on workers' skin, eyes and respiratory tract disorder	2	3	6 (Medium)	Subtitution: Subtitute chemical agent with another less dangerous agent Engineering control: a. Install local exhaust ventilation Administrative Control: a. Chemical handling SOP b. Training of chemical handling c. Ensure material safety data sheet d. Regular health check e. Spills must be promptly addressed using a spill kit PPE: a. Safety goggle b. Chemical-resistant glove and apron

Based on Table 2, it is evident that the sources of hazards classified as high risk include the boiler tank, LPG cylinders located near the boiler tank, and piping systems crossing worker mobility areas. Meanwhile, medium-risk hazards include the electrical systems of production machinery, noise from production machines, heat generated during the sterilization process, repetitive movements and standing working postures during packaging, work postures when lifting raw materials and packed milk, liquid spills, falling from ladders, and exposure to chemical substances.

The Hierarchy of Controls is a framework used to minimize or eliminate exposure to workplace hazards, organized by effectiveness[4]. The most effective method is elimination, which involves completely removing the hazard. When elimination is not feasible, substitution replaces the hazard with a less dangerous alternative. Engineering controls isolate workers from hazards by implementing physical changes, such as ventilation or machine guards. Administrative controls reduce exposure by altering work practices, policies, or training. Finally, personal protective equipment (PPE) serves as the last line of defense, providing protection such as gloves, masks, or goggles when other controls cannot fully eliminate the risks.[5]

Based on Table 2, there are five types of hazards present at X Tefa. Including physical hazards, fire and explosion hazards, ergonomic hazards, chemical hazards, and fall and trip hazards. Among these, the hazard with the most severe impact is the fire and explosion hazard. If a malfunction or safety failure occurs with the boiler, it could trigger a catastrophic event affecting everyone in the vicinity. Based on the table, it is also noted that only four levels of prevention (substitution, engineering control, administrative control, and personal protective equipment) are utilized, as it is not feasible to eliminate the objects/processes that pose potential hazards.

Figure 2. The LPG cylinder is located in the same room as the boiler

The risk of fire and explosion is very potential due to the use of boilers in the production process[6], along with the boiler's energy source (Liquid Petroleum Gas (LPG)) and the electrical systems of the UHT milk production machinery itself. Considering the significant impact that could occur if there is a malfunction, and the fact that X Tefa is located within the same building as classrooms and laboratories, an integrated production process monitoring system in a single control room is necessary. This control room would monitor all production process indicators, including safety indicators such as vessel pressure, vessel temperature, and the electrical system. If a malfunction occurs, the control room operator can halt the entire production process from the control room without needing to go directly to the production area. Additionally, we discovered that the LPG cylinders were placed in the same room as the boiler, which poses a significant explosion risk (figure 2). We recommend that the LPG cylinders be located in a protected and separate area from the boiler to ensure that, in the event of a safety failure in the boiler, the LPG cylinders are not affected by any potential explosion.

Figure 3. Cracks in one part of the retort machine.

Figure 4. The emergency exit is located behind the retort machine

The process monitoring system must be supported by several other measures. First, inspection and testing of the retort machine and UHT milk production machinery by certified personnel are necessary to determine whether the equipment can function within its capacity[7], or if there are manufacturing defects that could cause malfunctions, as we found cracks in one part of the retort machine (Figure 3). Second, there needs to be both active fire protection systems (smoke detectors, alarms, fire suppression systems, smoke control, portable fire extinguishers) and passive fire protection systems (compartment systems, fire-resistant doors and windows, fire-resistant walls, evacuation facilities, and evacuation aids)[8], those must have been tested by the relevant authorities. Third, an emergency response plan (ERP) should be developed so that workers know how to respond if a fire or explosion occurs. The Emergency Response Plan (ERP) should also include the management of evacuation routes, as we found that the emergency exit can only be accessed after passing through the boiler area, and there are still holes and pipes just before reaching the door (Figure 4). This ERP should be regularly simulated to achieve the ERP's objective, which is workers to act quickly and effectively in the event of a disaster with minimal risk of injuries or casualties[9]. At Last, there should be capacity-building initiatives, including certified training for boiler operators and electrical system monitors, to ensure that they better understand the machinery and can take corrective action if any abnormalities are detected.

Physical hazards at the research location include noise hazards and high temperatures. Noise can be generated by the machines involved in the homogenization process and filling milk into bottles. When simulating the production process, noise measurement must be conducted in the production area of X Tefa. Likewise, when production is underway, there must be a system in place to monitor the noise level in the production area, or at the very least, periodic measurements should be taken. If production is conducted for 8 hours a day, the noise level in the area should not exceed 85 dB[10]. If the noise exceeds this threshold and there is no risk control, it can potentially increase workers' hearing thresholds or even lead to hearing loss if not treated promptly. Using personal protective equipment like earmuffs or earplugs can protect workers from noise exposure[11]. The choice between the two depends on the noise measurement results. For frequencies ranging from 800 Hz to 125 Hz, earplugs have shown reduced efficiency, while earmuffs show a rapid decline in efficiency from 1000 Hz to 300 Hz, nearly 25 dB. It appears that at higher sound levels, earmuffs better protect the hearing system against high-frequency noise, whereas earplugs lose their ability to provide protection against low-frequency noise[12].

Figure 5. Heat hazard may arise at the surface of equipment or piping system

Based on the observations at the site, heat hazards arise from the use of high temperatures as a key element in the UHT milk production process. The most likely sources of heat hazards for workers are the UHT sterilization equipment systems such as the plate heat exchanger, steam heating system, retort machine and piping system, all of which generate or are exposed to high temperatures. These heat hazards can result in burns if workers come into direct contact with hot surfaces[13]. Engineering controls that can be implemented include installing thermal insulation on the surfaces of equipment and piping that may come into contact with workers. Administrative controls may involve placing heat hazard signs, applying floor markings, and regularly monitoring workplace temperatures. Personal protective equipment (PPE) for workers should include heat-resistant gloves, aprons, and protective clothing to safeguard against burns.

Figure 6. Packaging workstation may be too high for some individuals

Based on the data obtained, it was found that ergonomic hazards may arise during the product packaging process and the manual lifting and transportation of raw materials or packaged products. During packaging, workers are likely to perform repetitive movements and maintain standing postures for extended periods. Prolonged repetitive movements can potentially cause musculoskeletal disorders [14][15], in this case the symptoms will arise in the hands, arms, neck, and back, while a static awkward standing posture increases the risk of musculoskeletal issues in the lower back, waist, and lower limbs. We also found that the packaging workstation may be too high for some individuals (Figure 6). To address these risks, if the packaging workstations are not designed to be adjustable, it is recommended that the selection of workers for packaging roles be based on their anthropometric measurements. Matching the workers' body dimensions with the workstation design can enhance comfort for the involved body parts, thereby minimizing the risk of musculoskeletal disorders[16].

Warming up before work and stretching during breaks can also help relax the muscles before and during work.

Manual lifting and transporting of raw materials or packaged products, if not done properly, can lead to musculoskeletal disorders in the lower back, especially if workers lift excessive weights. The way workers lift can impact the strain placed on the lower back[17]. If the back is not kept straight while lifting, the force generated may cause excessive tension or even tearing in the annulus fibrosus structure[18], leading to low back pain or even weakness[19]. To prevent such incidents, workers should be provided with assistive equipment such as trolleys to lift, transport, or move raw materials or packaged UHT milk. Additionally, posters and training sessions on the maximum allowable weight for manual lifting and the correct lifting techniques should be implemented. The use of personal protective equipment, such as back support, is also necessary to prevent excessive flexion of the back while lifting.

Figure 7. The floor of UHT milk processing facility at X Tefa has been painted green, but there is no hazard marking yet

The hazards of falls, slips and trips at X Tefa arise from the piping system, liquid spills, and several staircases in various points within the production area. In some areas, the piping system and production equipment components at X Tefa are installed across the floor or in worker mobility pathways (figure 7). Visually, these obstacles are quite apparent to the workers and can easily be avoided. However, this may not be the case when a worker's attention is diverted during work, walking, or even running during emergency situations. In such instances, a worker may stumble, trip, and fall. The potential consequences range from minor injuries to serious injuries such as bone fractures. The floor of X Tefa has been painted green; however, there are no hazard indicators on the floor, such as for pipes crossing or the presence of drainage gaps (Figure 7). We recommend the use of floor marking in the X Tefa production area to reduce these risks. The use of distinct colors in the floor markings is expected to improve the visibility of which areas are safe, which areas require caution, and which areas are prohibited from being accessed[20][21].

UHT milk production involves various liquid substances, including production materials, cleaning agents, and finished products, which can pose hazards for falls, slips, and trips. It is possible that liquid spills may occur at some point due to pipe leaks, steam or condensation buildup, cleaning processes, or maintenance of production machinery. These spills pose a risk of workers slipping, which could result in minor to severe injuries. We recommend implementing a spill response procedure that all workers must be trained in and required to follow whenever they notice a liquid spill. If a spill cannot be immediately addressed, warning signs should be placed, and the incident should be reported to the supervisor. In addition, regular inspections should be conducted in areas prone to leaks, spills, and liquid accumulation. Inspections of the drainage system are also crucial to ensure it is functioning properly, preventing hazardous liquid pooling in the workplace. Workers should be equipped with personal protective equipment (PPE), including anti-slip safety shoes for daily use and safety gloves when cleaning up spills. An emergency spill kit is also essential to have on hand to address spills quickly, which should include absorbent materials, mops, and/or wet vacuums.

Figure 8. Insufficient grip on the handrail and steps of the stair

The presence of stairways in the production area can also contribute to worker injuries. We found that several stairways lacked sufficient grip, both on the handrails and the steps themselves (figure 8). It is crucial for workers to have a firm grip on both when ascending and descending, as inadequate grip can cause them to slip and fall. We recommend adding rubber coating to the handrails and steps to improve the grip for hands and feet. In addition, placing signs or posters regarding the three-point contact rule (maintain three points of contact with the ladder—two hands and one foot, or two feet and one hand[22]) in visible areas near the stairways is essential. Regular inspections of stair cleanliness are also necessary. In line with the risks associated with liquid spills, workers should be equipped with anti-slip safety shoes and gloves with good grip for better handling of tasks.

The work process in the UHT milk factory involves the use of various chemicals for cleaning, sanitation, and product formulation, it is also stated on guidelines that released by IFC[23]. Some of these chemicals can be hazardous to workers if not handled properly. The chemicals involved include cleaning and sanitizing agents, sterilization-related chemicals, and additives. If workers come into direct contact with these substances without proper protection, it can cause irritation to their skin, eyes, and respiratory system[24]. Several risk control steps can be taken to reduce chemical hazards, including substituting harmful chemicals with less dangerous ones, installing local exhaust ventilation systems to remove harmful chemical fumes in areas where these substances are used, and implementing standard operating procedures (SOPs) for chemical handling. Furthermore, workers should receive chemical management training, and chemical substances must be labeled appropriately, accompanied by their respective material safety data sheets (MSDS). Regular health check-ups are crucial[25], particularly for workers managing these chemicals. The final line of defense for workers includes wearing chemical-resistant safety gloves, safety goggles or face shields to protect their eyes from chemical splashes, and protective clothing or aprons to safeguard their skin from irritating chemicals.

Based on the observations, the OHS (Occupational Health and Safety) team should be one of the most fundamental aspects in X Tefa, but it has not yet been established. The OHS team is not only an operational requirement in the UHT milk facility but also a strategic asset that protects workers, ensures regulatory compliance, identifies hazards and assesses risks, prevents workplace accidents and occupational diseases, monitors workers' health status, provides training and education, and handles emergency response and preparedness. By addressing these areas, the OHS team can reduce legal and financial risks, support the smooth and safe operation of the UHT milk processing plant, and enhance worker productivity.

4. Conclusion

X Tefa faces multiple occupational hazards that necessitate immediate and comprehensive risk management strategies to ensure worker safety. The hazards identified include fire and explosion, ergonomic challenges, noise, heat exposure, fall, slip, trip and chemical hazards. The highest risk was fall, slip and trip hazard as well as fire and explosion hazard. To mitigate these risks, a robust integrated monitoring system is essential, allowing for centralized control of production processes and prompt response to malfunctions. Additionally, implementing active and passive fire protection systems, regular equipment inspections, and an emergency response plan will further enhance safety. Physical issues need to be follow up by noise measurement, hazard sign and the use of appropriate and comfortable personal protective equipment (PPE). Ergonomic issues related to repetitive movements and manual lifting must be addressed through workstation design tailored to workers' anthropometric measurements, along with proper lifting techniques and training. Addressing slips, trips, and falls by improving floor markings, spill response protocols, and stairway safety is critical. The management of chemical hazards requires substituting harmful substances, providing appropriate PPE, and ensuring proper chemical handling training. Establishing an Occupational Health and Safety (OHS) team is fundamental to ensuring regulatory compliance, reducing workplace incidents, and promoting a culture of safety. Through these integrated measures, X Tefa can significantly enhance workplace safety, safeguard employee health, and optimize operational efficiency. X Tefa is not yet operational, but risk control processes must be implemented as early as possible to prevent workplace accidents and occupational diseases when X Tefa eventually becomes operational.

References

[1] H. Deeth, "13 - Improving UHT processing and UHT milk products," in *Improving the Safety and Quality of Milk*, M. W. Griffiths, Ed. Woodhead Publishing, 2010, pp. 302–329.
[2] K. H. Gedam, R. Prasad, and V. K. Vijay, "The study on UHT processing of milk: a versatile option for rural sector.," *World J. Dairy \& Food Sci.*, vol. 2, pp. 49–53, 2007.
[3] L. M. Menger-Ogle, F. Pezzutti, A. D. Menger-Ogle, L. Stallones, J. Rosecrance, and I. N. Roman-Muniz, "Occupational Safety and Health of Foreign-Born, Latinx Dairy Workers in Colorado.," *J. Occup. Environ. Med.*, vol. 61, no. 1, pp. 61–68, Jan. 2019, doi: 10.1097/JOM.0000000000001472.
[4] G. A. Morris and R. Cannady, "Proper Use of the Hierarchy of Controls," *Prof. Saf.*, vol. 64, no. 08, pp. 37–40, Aug. 2019.
[5] CDC, "Hierarchy of Controls." 2023.
[6] T. A. Azad, I. Hasan, M. K. Saha, R. Ahmmed, S. J. Moni, and M. H. Kabir, "Risk of fire disaster: consequences on industry sectors in bangladesh," *Int. J. Energy Sustain. Dev.*, vol. 3, no. 3, pp. 52–63, 2018.
[7] M. S. Ali and H. Habibullah, "A review on the current status of boiler inspection and safety issues in Bangladesh," *Energy Procedia*, vol. 160, pp. 614–620, 2019, doi: https://doi.org/10.1016/j.egypro.2019.02.213.
[8] R. Burke, *Fire Protection: Systems and Response*. CRC Press, 2007.
[9] K. K. Hall, "Emergency Response Planning," in *Planning and Managing the Safety System*, London: Bernan Press, 2017.
[10] OSHA, "Occupational Noise Exposure - Standards | Occupational Safety and Health Administration." .
[11] V. A. R. Silva *et al.*, "Are hearing protection devices used in the workplace really efficient? A systematic review," *Work*, vol. 74, pp. 477–484, 2023, doi: 10.3233/WOR-205290.
[12] F. Forouharmajd, K. Nazaryan, A. Fuente, S. Pourabdian, and H. Asady, "The Efficiency of Hearing Protective Devices against Occupational Low Frequency Noise in Comparison to the New Subjective Suggested Method.," *Int. J. Prev. Med.*, vol. 13, p. 143, 2022, doi: 10.4103/ijpvm.IJPVM_700_20.
[13] T. N. McManus, *Management of Hazardous Energy: Deactivation, De-Energization, Isolation, and Lockout*. CRC Press, 2016.
[14] T. C. Aw, K. Gardiner, and J. M. Harrington, *Occupational Health: Pocket Consultant*. Wiley, 2006.
[15] A. Luttmann, M. Jager, B. Griefahn, G. Caffier, F. Liebers, and W. H. Organization, "Preventing musculoskeletal disorders in the workplace," 2003.
[16] H. K. Z. R. Waldemar Karwowski, *Social and Occupational Ergonomics*. AHFE INTERNATIONAL, 2022.
[17] A. Mital, *Guide to Manual Materials Handling*. CRC Press, 2017.
[18] M. S. Sommers, *Davis's Diseases & Disorders: A Nursing Therapeutics Manual*. F.A. Davis, 2022.
[19] M. Y. Rusmayanti and S. N. Kurniawan, "HNP LUMBALIS," *J. Pain, Headache Vertigo*, vol. 4, no. 1, pp. 7–11, 2023, doi: 10.21776/ub.jphv.2023.004.01.2.
[20] D. F. de Korne, J. D. H. van Wijngaarden, J. van Rooij, L. S. G. L. Wauben, U. F. Hiddema, and N. S.

Klazinga, "Safety by design: effects of operating room floor marking on the position of surgical devices to promote clean air flow compliance and minimise infection risks.," *BMJ Qual. Saf.*, vol. 21, no. 9, pp. 746–752, Sep. 2012, doi: 10.1136/bmjqs-2011-000138.

[21] K. Rzepecki, "How Floor Markings Can Improve Safety and Productivity in Your Facility," *Ehstoday.com*. EHS Today, 2019.

[22] A. K. Das, *Principles of Industrial Safety Management: Understanding The WS of Safety at Work*. PHI Learning Pvt. Ltd., 2020.

[23] IFC, "Environmental, Health, and Safety Guidelines for Dairy Processing," 2007.

[24] A. Akbarzadeh, E. Abasi, M. Ghanei, A. Hasanzadeh, and Y. Panahi, "The Effects of Various Chemicals on Lung, Skin and Eye: a Review," *Toxin Rev.*, vol. 35, no. 3–4, pp. 187–195, Oct. 2016, doi: 10.1080/15569543.2016.1187174.

[25] P. Chaturvedi, *Occupational Safety, Health & Environment And Sustainable Economic Development*. Concept Publishing, 2007.

Effect of microwave-assisted extraction (MAE) method on the characteristics of protein extract from edamame (*Glycine max (L.) Merrill*)

R P Putri[1*], A Choirun[1], N A Indis[2], Rusdiarti[3], A G Rakhmadevi[2]

[1)] Food Engineering Technology Study Program, Department of Agricultural Technology, Politeknik Negeri Jember
[2)] Food Industry Technology Study Program, Department of Agricultural Technology, Politeknik Negeri Jember
[3)] Clinical Nutrition Study Program, Department of Health, Politeknik Negeri Jember

[*]Email: resti.pranata@polije.ac.id

Abstract. Edamame, a type of soybean, is a promising source of plant-based protein. Protein extraction from edamame was conducted using the microwave-assisted extraction (MAE) method, which utilizes thermal damage to the tissue to facilitate protein extraction with a solvent. Food-grade distilled water was chosen as the solvent, as the protein extract is intended to enhance the nutritional value of meatballs and sausages. The study employed a Completely Randomized Design (CRD) with variations in extraction time (0, 20, 40, and 60 minutes) and pH levels (7, 9, and 11), with each condition tested three times. In the conducted study, it was found that microwave-assisted extraction methods significantly enhanced the quality of protein content from edamame. The thermal effects of microwaves effectively disrupted tissue, facilitating the breakdown of cells and the extraction of valuable components. Our findings showed that protein yield improved with increasing extraction pH, making alkaline conditions—particularly at pH 9—beneficial for maximizing protein extraction. The optimum protein yield of 77.76 ± 0.20% was achieved at pH 9 with a 60-minute extraction time. However, higher alkaline pH levels were associated with decreased protein purity. Consequently, the highest protein content, at 94.12 ± 0.21%, was achieved at pH 7 with a 60-minute extraction time.

1. Introduction

Sufficient protein consumption is essential for maintaining health and preventing diseases caused by protein deficiency, such as kwashiorkor[1]. Beyond individual health, protein deficiency can contribute to stunting, a serious health issue in Indonesia that results in physical and brain growth disorders in toddlers[2]. Data showed that the stunting incidence rate in East Java Province in 2023 is 17.7%, while the stunting incidence rate in Jember Regency is 7.43%, or 10,414 children[3]. Addressing this issue requires exploring alternative protein sources, and plant-based proteins are gaining attention for their health benefits and lower environmental impact compared to animal-based proteins[4]. Furthermore, utilizing plant-based proteins is essential to meet the growing global demand for protein and to combat stunting effectively.

Edamame, a type of soybean, are unique among plants because they offer all the essential amino acids needed by the body, such as tryptophan, leucine, phenylalanine, threonine, valine, methionine, lysine, histidine, and isoleucine, though their protein quality is somewhat lower than animal protein. In Jember, edamame is a key regional product, with export-quality shipped to Japan, while about 15% of the lower-grade produce is sold locally at discounted prices[5]. Edamame can be eaten directly, processed into multigrain rice[6], or further extracted to make a protein concentrate[5]. To enhance the protein content, edamame protein concentrate is produced by processing edamame flour to remove roughly half of its carbohydrates and some minerals. This concentrate must contain at least 70% protein by dry weight and is highly valued for its functional properties in food products[7].

The Microwave Assisted Extraction (MAE) method has been shown to be highly effective in isolating protein, offering higher extraction rates than the traditional alkaline method. This technique utilizes high power to heat the plant materials evenly, creating pressure that breaks down cell walls and facilitates the release of proteins. Additionally, electromagnetic microwave waves help to significantly reduce treatment time [8]. Although still a relatively new technique, various studies have reported the successful use of microwave technology in protein extraction from peanuts [8], sesame bran [9], and coffee silverskin [10]. Research indicated that this method yields more protein from various materials compared to the standard alkaline procedure[11].

Edamame protein concentrate from lower-grade edamame has a broad range of applications in food and beverage products. Protein extracts play a key role in enhancing the functional properties of various foods, such as meatballs and sausages. Although [6] examined the extraction of edamame using the alkaline method, a study on the effects of the Microwave Assisted Extraction (MAE) method had not yet been conducted. Therefore, this study aims to explore how microwave-assisted extraction affects the protein extraction process from lower-grade edamame, evaluating its potential as a source of protein concentrate.

2. Materials and Methods

2.1 Materials

The materials utilized in this study included edamame sourced from Tanjung market, Jember regency, along with NaOH and HCl obtained from Aneka Kimia, Jember regency.

2.2 Equipments

The equipments required for this research included a modified microwave, Soxhlet apparatus, centrifuge, chromatometer, Kjeldahl flasks, porcelain crucibles, Erlenmeyer flasks, measuring cylinders, pipettes, funnels, filter paper, analytical balance, oven, desiccator, and dehydrator.

2.3 Research Design

Pretreatment of edamame modified from the research conducted by [9] as outlined here : The edamame was initially cleaned of any foreign materials, then dried using a tray dryer set at 60°C for 12 hours. After drying, it was sieved to a 20 mesh size. The processed edamame was free of impurities, with an average total protein content of 31.5% [12]. The samples were then stored in aluminum foil packaging until the extraction process.

To investigate the effects of microwave-assisted protein extraction from edamame, process time and pH were chosen as independent variables. Their influence on protein content (%), protein yield (%), and water content was analyzed with two-way ANOVA. Statistical differences in the parameters were determined using Duncan test ($p < 0.05$) The experimental setup included 12 runs, each conducted three times. Edamame was mixed with deionized water at a 1:10 (w/v) ratio. Then, it was placed in a microwave extractor set at 60°C for varying times (0, 20, 40, and 60 minutes) and pH levels (7, 9, and 11). The pH was adjusted using a 1 N sodium hydroxide solution, and the mixture was then incubated in a water bath at 45°C for 30 minutes. After incubation, the supernatant was collected following centrifugation at 3000 rpm for 30 minutes, and the pH was adjusted to 4.5 with 1 N hydrogen chloride to precipitate the protein. The resulting mixture was centrifuged again at 3000 rpm for 30 minutes to obtain the filtrate, which was then dried using a tray dryer at 50°C for 3 hours and prepared for further analysis.

2.4 Analysis
The protein concentrate was further analyzed to measure water content using the oven-drying method. The Kjeldahl method was used to determine total protein content, while protein yield was subsequently calculated using Equation (1).

$$Protein\ yield\ (\%) = \frac{(m_{extract})(\%_{extract})}{(m_{edamame})(\%_{edamame})} \times 100 \quad (1)$$

3. Results And Discussions
The chemical characteristics of protein extracted from edamame using MAE method, as presented in Table 1, revealed significant variations across different treatments in terms of water content, protein content, and protein yield. Two-way ANOVA (results presented in Table 2) showed that pH and extraction time both had significant effects on water content, protein content, and yield parameters of protein extracted from edamame ($p < 0.001$). The interaction between pH and extraction time on these parameters was also significant ($p < 0.001$).

Table 1. Chemical characteristic of protein extracted from edamame

pH	Time (minutes)	Water content (%)	Protein content (%)	Protein yield (%)	Category
7	0	4.58 ± 0.07^{g}	73.34 ± 0.11^{g}	47.52 ± 0.45^{k}	EPC
	20	6.51 ± 0.16^{f}	88.67 ± 0.04^{b}	57.91 ± 0.39^{i}	EPC
	40	6.15 ± 0.14^{f}	85.08 ± 0.08^{c}	56.34 ± 0.12^{j}	EPC
	60	6.88 ± 0.03^{e}	94.12 ± 0.21^{a}	59.93 ± 0.05^{h}	EPI
9	0	7.68 ± 0.14^{c}	72.88 ± 0.30^{h}	60.56 ± 0.16^{g}	EPC
	20	7.11 ± 0.19^{d}	75.85 ± 0.22^{f}	70.51 ± 0.09^{e}	EPC
	40	8.21 ± 0.26^{b}	84.82 ± 0.06^{c}	74.63 ± 0.14^{c}	EPC
	60	6.78 ± 0.09^{e}	81.26 ± 0.17^{d}	77.76 ± 0.20^{a}	EPC
11	0	7.74 ± 0.29^{c}	62.49 ± 0.34^{k}	60.40 ± 0.62^{h}	-
	20	9.36 ± 0.22^{a}	63.11 ± 0.07^{j}	69.55 ± 0.06^{f}	-
	40	9.13 ± 0.08^{a}	68.59 ± 0.05^{i}	75.28 ± 0.31^{b}	-
	60	6.38 ± 0.24^{f}	76.85 ± 0.18^{e}	71.23 ± 0.12^{d}	EPC

Note: * [a-k] Means in a row without a common superscript letter differ ($P < 0.05$) as analyzed by two-way ANOVA and the DUNCAN test.
** EPC referred to Edamame Protein Concentrate, EPI referred to Edamame Protein Isolate, while - referred to samples with protein content lower than 70%.

Table 2. Two-way ANOVA results for the effects of pH and extraction on water content, protein content, and protein yield parameters of protein extracted from edamame

Parameter	*P*-value		
	pH	Time	P×T
Water content	<0.001	<0.001	<0.001
Protein content	<0.001	<0.001	<0.001
Protein yield	<0.001	<0.001	<0.001

Note: P × T = pH × Time interaction effect.

3.1 Water content
Water content is a key determinant of the quality and shelf life of protein extracts, as it influences the conditions for microbial growth and potential spoilage. As shown in Table 1, the water content of edamame protein extract falls between 4.58% and 9.13%, significantly enhancing its shelf life in

comparison to fresh edamame beans, which contain up to 70.8% of water [12]. According to the SNI (Indonesian National Standard) for soybean powder, the water content should not exceed 10% [13]. This indicates that all of the extracted edamame protein meets the standard. The water content of a product is heavily affected by both the drying duration and temperature. Extended drying at higher temperatures generally reduces water content, as it allows for greater moisture evaporation. Conversely, inadequate drying time can result in a subpar protein extract, with higher moisture levels that may compromise its quality [12].

3.2 Protein content

As indicated in Table 2, the lowest average protein content in the extract, approximately 62.49%, was observed at pH 11 without the use of microwave-assisted extraction. In contrast, the extract at pH 7 and with a 60-minute extraction time exhibited the highest average protein content, reaching 94.12%. The protein content in extracts obtained through microwave-assisted extraction ranged from 63.11% to 94.12%, as shown in Table 2. This range highlights the variability in extraction efficiency, which is influenced by both pH and extraction time.

Figure 1(a) shows the effect of pH on protein content. The figure compares the conventional method (control) with the MAE method at a 60-minute extraction time. The lowest protein content for both methods was observed at pH 11, and protein content decreased significantly with increasing pH. This finding aligns with the results of [14], which reported that soy proteins extracted under alkaline conditions are associated with lower overall protein content.

Figure 1(b) illustrates the effect of extraction time on protein content at pH 11. Protein content increased significantly with longer extraction times, particularly between 20 and 60 minutes. The results demonstrate that the MAE method significantly increased protein content compared to the control with longer extraction times. This improvement is attributed to the use of high power in the MAE method, which heats the plant materials evenly, creates pressure that breaks down cell walls, and facilitates the release of proteins [8].

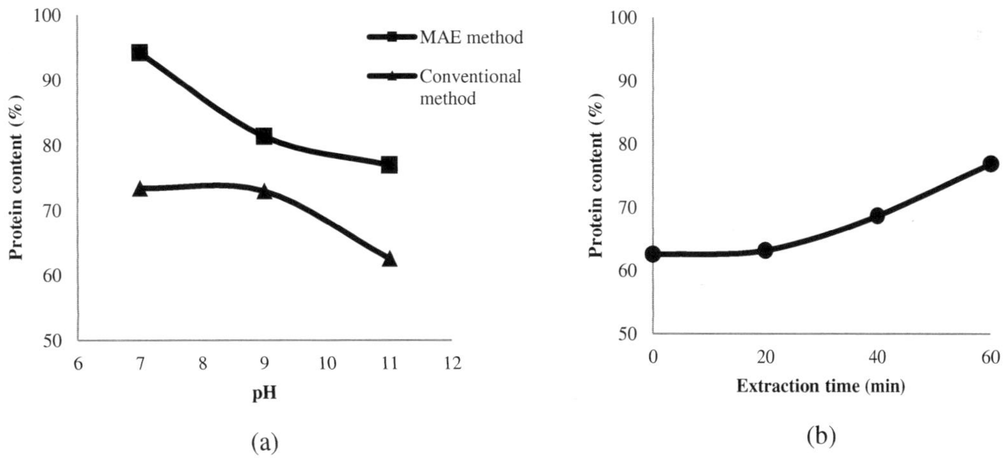

Figure 1. Effect of (a) pH and (b) extraction time on protein content

Proteins derived from edamame can be separated into two primary forms, protein concentrates and protein isolates. Concentrates contain at least 70% protein by dry weight, while isolates must have more than 90% protein [15]. By this definition, we can categorize protein extracted from edamame using microwave-assisted methods into three groups: EPI (Edamame Protein Isolate), EPC (Edamame Protein Concentrate), and samples that do not fall into either category due to a protein content lower than 70%.

Table 2 demonstrates these classifications. Protein content categorized as EPI was achieved only at pH 7 with a 60-minute extraction time, where the highest protein content (94.12 ± 0.21%) was observed. Most of the samples fall under the EPC category, with protein content ranging between 73.34 ± 0.11% and 88.67 ± 0.04%. Three samples, however, did not meet the criteria for categorization. These samples had protein contents lower than 70%, specifically 62.49 ± 0.34%, 63.11 ± 0.07%, and 68.59 ± 0.05%. These values indicated the least effective extraction conditions, which occurred at pH 11 with extraction times of 20–40 minutes.

Microwave-assisted methods significantly affect the quality of protein content. The thermal effects of microwaves align well with the needs of tissue disruption, making them suitable for breaking down cells to extract valuable components from plant tissues. Moreover, a study by [16] showed that microwave treatment can damage the microstructures of edamame. Prolonged exposure leads to the extraction of protein bodies from within the cell walls, resulting in increased cell damage. This corresponds with the rise in protein content as extraction time extends.

3.3 Protein yield

Protein yields obtained from edamame using the microwave-assisted extraction method, as shown in Table 2, varied from 47.52% to 77.76%, showing an increase with longer extraction times. This increase is attributed to the detrimental effects of microwaves on the cellular structure of edamame [16]. Figure 2 showed the effect of pH to protein yield. The protein yield from edamame increases with extraction pH, rising from 7 to 9, but shows no significant improvement when the pH is further increased from 9 to 11. Therefore, alkaline conditions are advantageous for extraction. However, as the extraction pH becomes more alkaline, the purity of the soy protein decreases, as shown as protein content in Table 2. According to [14], soy proteins extracted at alkaline condition demonstrate the high solubility and emulsification properties, which are associated with high protein yield but lower overall protein content.

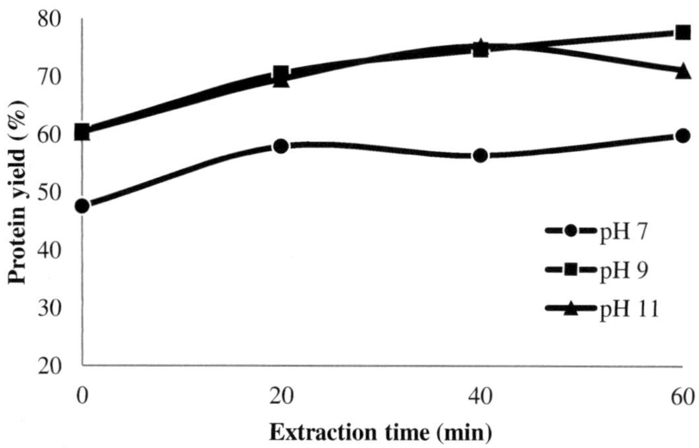

Figure 2. Effects of pH and extraction time on protein yield

4. Conclusion

The study concluded that the use of microwave-assisted extraction methods led to a considerable improvement in the protein content extracted from edamame. The thermal effects of microwaves effectively disrupted tissue, facilitating the breakdown of cells and the extraction of valuable components. Our findings showed that protein yield improved with increasing extraction pH, making alkaline conditions—particularly at pH 9—beneficial for maximizing protein extraction. The optimum protein yield of 77.76 ± 0.20% was achieved at pH 9 with a 60-minute extraction time. However, higher alkaline pH levels were associated with decreased protein purity. Consequently, the highest protein content of 94.12 ± 0.21% in the extracted protein was obtained at pH 7 with an extraction time of 60 minutes.

Acknowledgement
The authors gratefully acknowledge the financial support from P3M Politeknik Negeri Jember through the 2024 PNBP funding, which made this research possible.

References

[1] Katz K A, Mahlberg M H, Honig P J and Yan A C 2005 *J. Am. Acad. Dermatol.* **52** S69–72
[2] Haryani S, Astuti A P and Sari K 2021 *J. Pengabdi. Kesehat.* **4** 30–9
[3] Kementerian Kesehatan 2023 *Laporan Tematik Survei Kesehatan Indonesia 2023* (Jakarta: Kementerian Kesehatan)
[4] González A D, Frostell B and Carlsson-Kanyama A 2011 *Food Policy* **36** 562–70
[5] Hertamawati R T, Nurkholis and Rahmasari R 2021 *IOP Conf. Ser. Earth Environ. Sci.* **888** 012069
[6] Kurniawati E, Putri R P and Wahyono A 2024 *IOP Conference Series: Earth and Environmental Science* vol 1338 p 012049
[7] Liener I E 1994 *Crit. Rev. Food Sci. Nutr.* **34** 31–67
[8] Ochoa-Rivas A, Nava-Valdez Y, Serna-Saldívar S O and Chuck-Hernández C 2017 *Food Bioprocess Technol.* **10** 543–55
[9] Görgüç A, Özer P and Yılmaz F M 2020 *J. Food Process. Preserv.* **44**
[10] Wen L, Álvarez C, Zhang Z, Poojary M M, Lund M N, Sun D-W and Tiwari B K 2021 *Biomass Convers. Biorefinery* **11** 1575–85
[11] Phongthai S, Lim S-T and Rawdkuen S 2016 *J. Cereal Sci.* **70** 146–54
[12] Elvizahro L and Majid V M 2021 *Acad. Hosp. J.* **3** 10–7
[13] BSN 2011 SNI 7612:2011 Bubuk minuman kedelai
[14] Song H, Zhong M, Sun Y, Li Y and Qi B 2023 *LWT* **174** 114455
[15] Wang H, Johnson L A and Wang T 2004 *J. Am. Oil Chem. Soc.* **81** 713–7
[16] Choi I, Choi S J, Chun J K and Moon T W 2006 *J. Food Process. Preserv.* **30** 407–19

Test of the superiority of several soybean (*Glycine max* (L.) Merril) pure lines

N Sjamsijah[1], L Kurniasari[1*], S Rahayu[1], R Firgiyanto[1], DA Saputra[1] and DF Sabiku[1]

[1]Department of Agriculture Production, State Polytechnic of Jember, Mastrip Street PO BOX 164 Jember, Indonesia.
*kurniasari@polije.ac.id

Abstract. Soybeans are a strategic food commodity with increasing demand but production is still low. The aim of this research is to see the superiority of production stability of each Harapan Jember lines. The object of research used genotype of pure lines soybean produced by the main researcher, which is a candidate for a new superior variety with a potential yield. The study used a Non-Factorial Randomized Block Design which only used one factor, namely the soybean pure lines (GHJ1, GHJ2, GHJ3, GHJ4, GHJ5), Malabar and Anjasmoro Variety. Based on the analysis of all tested parameters (number of pods per plant, seed yield per plant, seed yield per plot, seed yield per hectare, and 100 seeds weight), GHJ4 is a stable and superior line. Although not always the highest in every category, the combination of good results in all aspects makes it a balanced and consistent variety. In addition, its 100 seeds weight is still in the good category, although not the highest. GHJ4 also shows consistency in results across various parameters, including yield per plot and per hectare. These results indicate decent seed quality and good yield potential.

1. Introduction

Soybeans are legumes or beans and are classified as food crops which are used as a source of vegetable protein, however, the need for soybeans has not been met due to a decline in soybean production. Soybeans are a strategic food commodity with increasing demand along with the growth of Indonesia's population. Based on BPS data, domestic soybean production is around 555,000 tons, while national demand reaches 2.7 million tons [1]. This low soybean production causes Indonesia to still import up to 2.2 million tons in 2023 [2]. This imbalance is a problem that needs to be solved.

Efforts to increase production and increase the value of soybeans continue to be carried out, considering that soybeans are a strategic product to meet the vegetable protein needs of the Indonesian population [3]. One of the efforts made to increase soybean production through plant breeding activities is through the use of superior varieties [4]. Soybean breeding to produce superior varieties is adjusted to consumer and farmer preferences for soybeans such as large seeds, yellowish color, if planted resistance to lodging, early maturity, and high productivity.

One of the efforts to address this problem is through the use of certified and quality seeds which are believed to be able to support productivity problems in the soybean cultivation system. The use of superior and quality soybean varieties with a cultivation system and using the right technology is one form of effort to increase national soybean production [5]. Superior varieties can be obtained conventionally through plant crossing using superior parents to obtain a new plant that has superior characteristics from both parents, obtained by selection from plants resulting from repeated crossings to obtain a promising line to improve plant character. This promising line is obtained from yield testing until a stable plant is obtained so that the line can be released as a superior variety.

To increase production, the development and use of superior varieties need to be attempted, the use of superior lines with high productivity that are resistant to pests and diseases, one of which is the Galur Harapan Jember 1,2,3,4,5 which is expected to have a big impact on farmers. Based on the description above, it is necessary to conduct research on testing several soybeans lines of Galur Harapan Jember 1,2,3,4,5, in order to determine the level of superiority of the candidate varieties and it is necessary to use comparative varieties, namely Malabar and Anjasmoro soybean varieties, so that the productivity of Harapan Jember soybeans can be known.

In this study, there is the use of previous research results from Porwoko & Sjamsijah [6], soybean has produced crosses with the results of four crosses, namely Unej-1 × Malabar, Malabar × Unej-1, Unej-2 × Malabar and Malabar × Unej-2, and then further testing was carried out using F2 plant selection which obtained soybean genotypes GHJ 1 (Unej 1 × Malabar), GHJ 2 (Malabar × Unej 1), GHJ 3 (Unej 2 × Malabar), GHJ 4 (Malabar × Unej 2) and GHJ 5 (Malabar × Unej 2). The results of preliminary testing based on the results of the tested lines produced characteristics at a short harvest age of 74-76 days with a potential productivity of more than 3 tons/ha with an early maturity age referto Sjamsijah [7] and produced a weight of 100 seeds weighing 15.5 - 16.3 grams.

2. Methods

The study was conducted from May to September 2024 at experimental garden of State Polytechnic of Jember, Jember, East Java, Indonesia at an altitude of 103 m asl. The research design used was a non-factorial randomized block design which only used one factor with four replication. The factor was several soybean varieties with 7 levels consisting of; K0: Soybean pure lines namely Galur Harapan Jember 1 (GHJ1) K1: Soybean pure lines namely Galur Harapan Jember 2 (GHJ2) K2: Soybean pure lines namely Galur Harapan Jember 3 (GHJ3) K3: Soybean pure lines namely Galur Harapan Jember 4 (GHJ4) K4: Soybean pure lines namely Galur Harapan Jember 5 (GHJ5), K5: Malabar variety soybeans K6: Anjasmoro variety soybeans.

2.1.1 Procedure

The implementation of the experiment was carried out starting from land processing to harvest. Land processing was carried out using a tractor. Planting was carried out with 2 seeds per hole. if the soil condition is acidic (ph <5) then it is necessary to give dolomite first. After plowing is complete, then make a plot or bed with a height of 0.3 m and a width of 2 x 2 m and a drainage channel with a width of 0.5 m. Planting uses a planting distance of 40 x 20 cm. Planting is done by inserting soybean seeds into the planting hole as many as 3 grains per planting hole and given furadan then covered with soil.

Plant maintenance includes replanting, irrigation, fertilization, weeding, rouguing, hilling, and pest and disease control. Fertilization is carried out twice when the soybean plants are 12 HST and 30 HST using Urea [8], TSP and KCL fertilizers. According to Retno [9] the recommended dose of fertilization for 1 soybean plant is Urea 50 kg/ha, TSP 75 kg/ha, and KCL 50 kg/ha, fertilization is done by digging. The recommended dose of 2nd fertilization is Urea 50 kg/ha, TSP 75 kg/ha, and KCl 100 kg/ha [10].

Soybean harvesting is carried out at the age of 80-92 HST when the seeds reach the ripe phase or which is marked by 95% of the pods having turned blackish brown and most of the leaves on the plant have fallen off, harvesting is carried out in the morning by cutting the base of the stem. After all the plants have been harvested, the soybean stalks are immediately dried in direct sunlight when the stalks are dry, the seeds are threshed, drying of the soybean seeds is carried out until the water content reaches 14%.

2.1.2 Data Analyzed

The data obtained were analyzed by analysis of variance and if result of F-test have significant effect, it will continue with BNT test at α=5%. Parameter observed were number of pods, yield per plant, yield per plot, yield per hectar, and weight of 1000 seeds.

3. Result

This study aims to determine the production advantages of each pure lines and comparative variety. The results show that pure lines originating from GHJ 1 produces the same number of pods as the comparison variety Anjasmoro, but is not significantly different from genotype GHJ4 and the comparison variety of Malabar. The data presented in Figure. 1 shows that several pure lines of Galur Harapan Jember have advantages that are not significantly different to the comparison varieties. The number of pods per plant produced by GHJ1 and GHJ4 were not significantly different from the Malabar variety and the Anjasmoro variety.

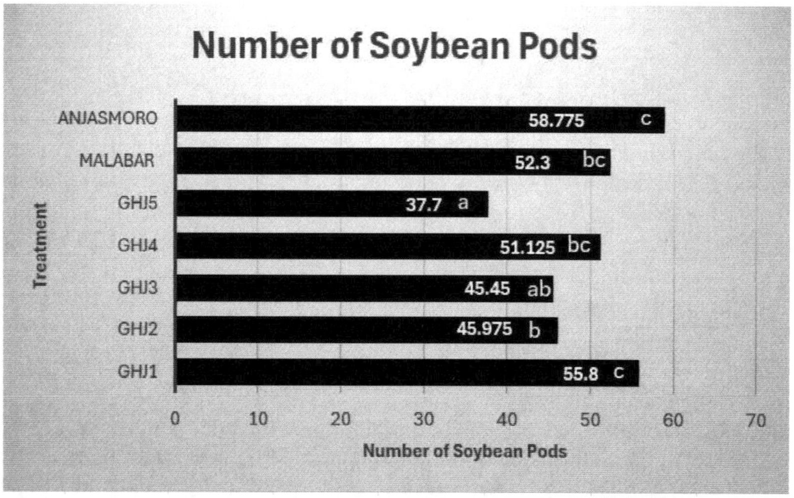

Figure 1. Number of Soybean Pods from several pure lines of Galur Harapan Jember and comparison varieties.
Notes: Numbers followed by the same letter *were not signicantly different using BNT α = 0,05*

The Anjasmoro variety produced the highest number of pods per plant (58,775 pods) although it was not significantly different from GHJ1 (55.8 pods), Malabar (52,300 pods) and GHJ4 (51,125 pods). If we look at all the soybean pure lines tested, GHJ1 and GHJ 4 have the potential to produce the best number of pods per plant while GHJ5 produces the lowest. The number of pods per plant is an important parameter in soybean seed production because it is an indicator of production, yield stability and part of the selection activities in soybean plant breeding. According to Xu [11] increasing the number of pods per plant can significantly increase soybean seed yield. Meanwhile Milioli [12] state that genotypes that have high and stable yields are the main goal of plant breeding programs.

In the parameter of seed yield per plant (Figure 2), GHJ1 (16.980 g) and GHJ4 (17.735 g) were the lines with the highest seed yield among all tested lines and with two comparison varieties that Malabar variety which produced 6,991 g and the Anjasmoro variety which produced 13,385 g. In fact, in terms of the number of pods, Malabar and Anjasmoro Variety did not differ significantly in producing the number of pods per plant. This inconsistent result is suspected because GHJ1 and GHJ4 produced more seeds than the comparison varieties and other lines. But this indicates that GHJ1 and GHJ4 may have superior genetic quality and are suitable for existing growing environmental conditions. Among all the test lines compared with the reference varieties, GHJ 4 has the greatest potential for seed production because it produces the highest number of seeds per plant. Seeds yield per plant is a very important parameter in soybean seed breeding and production. In addition, seed yield per plant also contributes to the quality of the yield. Previous study from Sjamsijah [13] shows that the number of pods and the number of seeds effect soybean seed yield. Meanwhile, according to Adi & Krisnawati [14], the increase in soybean yield is greatly influenced by the number of filled pods per plant, the more filled pods the higher the seed yield.

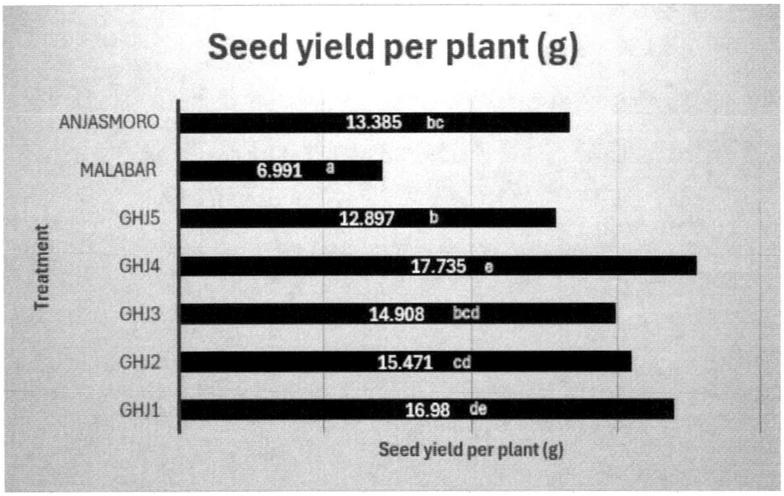

Figure 2. Seed yield per plant from several pure lines of Galur Harapan Jember and comparison varieties.
Notes: Numbers followed by the same letter *were not signicantly different using BNT α = 0,05*

Seed yield per plot is influenced by the potential yield of each lines, so that each lines has different results (Figure 3.). In the parameter of seed weight per plot, the tested soybean lines had significant differences with the comparison varieties. GHJ3 produced the highest seed weight per plot but was not significantly different from GHJ1, GHJ2, GHJ4 and GHJ5. Seed yield parameters per plot showed that GHJ3 lines produced the highest seed yield (1.275 kg) among all the lines and comparison varieties. The seed yield per plot is used to determine how much the line/variety can produce [15]. Therefore, the yield per plot can be used to determine the pure lines that have high yields

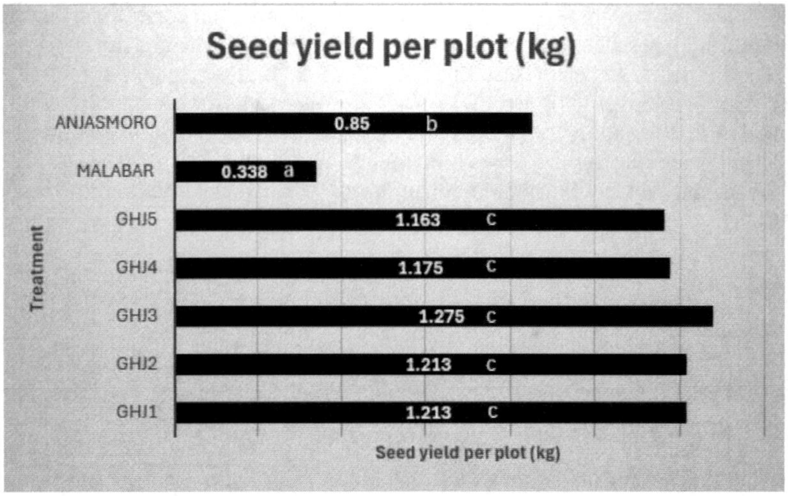

Figure 3. Seed yield per plot from several pure lines of Galur Harapan Jember and comparison varieties. Notes: Numbers followed by the same letter *were not signicantly different using BNT α = 0,05*

Soybean lines tested with comparison varieties showed that the seed weight per hectare produced by all lines statistically showed significant differences (Figure 4.). However, there was no significant difference between lines (GHJ1, GHJ2, GHJ3, GHJ4, GHJ5). This high yield indicates that all lines are very productive and have good genetics and are suitable for growing conditions in the field. This indicates that all of lines have excellent potential in terms of productivity, and genetic factors or adaptation to environmental conditions may contribute to the better yield. The low yield in Malabar suggests the need for further evaluation of the genetic resistance and suitability of these varieties to local conditions. These data suggest that the GHJ variety is superior in terms of seed yield per hectare and is more suitable for cultivation under the conditions tested.

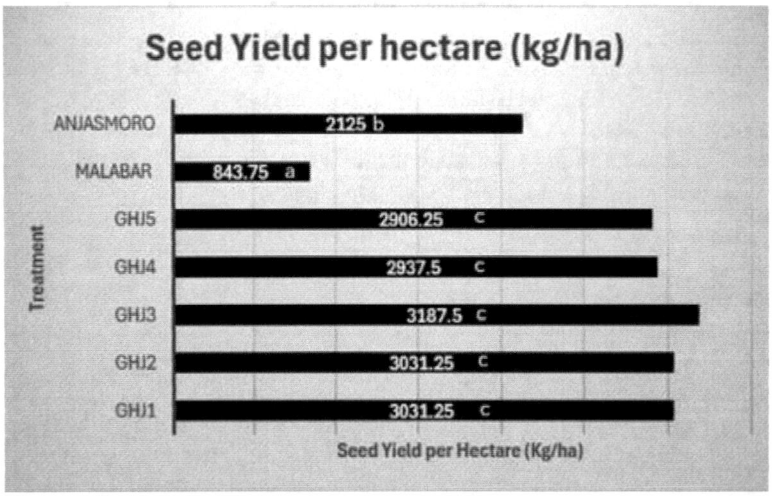

Figure 4. Seed yield per hectare (Kg/ha) from several pure lines of Galur Harapan Jember and comparison varieties. Notes: Numbers followed by the same letter *were not signicantly different using BNT α = 0,05*

In the parameter of 100 seed weight, GHJ 2 has the highest 100 seed weight (13.515 g) among all varieties (Figure 5.). This higher weight usually indicates better seed quality, high yield potential, and better resistance. Meanwhile, GHJ 3 and GHJ4 lines had almost similar weights (GHJ 3=13.121 g and GHJ4=13.070 g) and slightly lower than GHJ 2. These results indicate that both GHJ 3 and GHJ 4 also have good potential in terms of seed size. The parameter of the weight of 100 seeds is important in lines selection activities. Refer to Sjamsijah [13] the weight of 100 seeds can be used as a criterion in the selection of high-yielding soybean genotypes. [16] stated that the size of soybean seeds is categorized into small seeds (<10 g/100 seeds), medium seeds (10 - 12 g/100 seeds), and large seeds (13 -18 g/100 seeds). In fact, according to the variety description, Anjasmoro is a national superior soybean variety with a potential yield of 2.25-2.03 tons/ha with a weight of 100 grains of 14.8-15.3 grams and Malabar soybean variety has large seeds (12 g/100 grains). Malabar soybean variety is one of the superior varieties widely used by soybean farmers, the Malabar variety has a high yield of 1.27 t/ha of dry beans for paddy fields and 0.79 t/ha of dry beans for dry land. Thus, based on the weight of 100 seeds, GHJ2, GHJ3, GHJ4 and GHJ5 were included in the category of large seeds while GHJ 1 is categorized as medium (Figure 6.). Furthermore, according to Arifiana & Sjamsijah[17], the size and weight of soybean seeds can be maximized if the needs for nutrients and water can be met optimally.

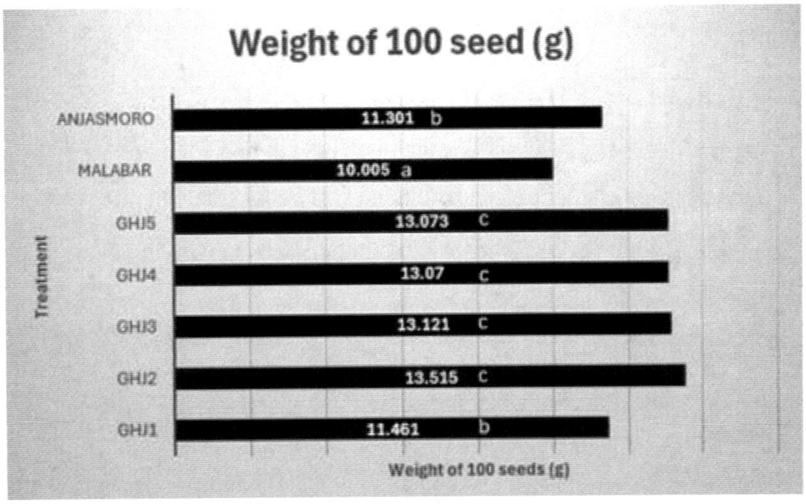

Figure 5. Weight of 100 seeds (g) from several pure lines of Galur Harapan Jember and comparison varieties. Notes: Numbers followed by the same letter *were not signicantly different using BNT α = 0,05*

Figure 6. Seeds from the tested line with the reference variety

4. Conclusion

Based on the analysis of all tested parameters (number of pods per plant, seed yield per plant, seed yield per plot, seed yield per hectare, and 100 seeds weight), GHJ 4 is a stable and superior line. Although not always the highest in every category, the combination of good results in all aspects makes it a balanced and consistent variety. In addition, its 100 seeds weight is still in the good category, although not the highest. GHJ 4 also shows consistency in results across various parameters, including yield per plot and per hectare. These results indicate decent seed quality and good yield potential.

5. Acknowledgements

The author would like to thank the Kementerian Pendidikan, Riset, dan Pendidikan Tinggi for the distribution of PNBP research funds in 2024 through the Center for Research and Community Service of the Jember State Polytechnic.

IOP Conf. Series: Earth and Environmental Science **1446** (2025) 012009 doi:10.1088/1755-1315/1446/1/012009

6. References

[1] Badan Pusat Statistik 2024 *Statistik Indonesia* (Jakarta: Indonesia) 852 p

[2] Balai Pengujian Standar Instrumen Tanaman Aneka Kacang 2023 *Aneka Kacang Siap Dukung Peningkatan Produktivitas Kedelai di Jawa Timur* (Malang: Indonesia)

[3] Kementrian Pertanian 2023 *Info Teknologi: Mengenal Ragam Manfaat Kedelai* (Jakarta:Indonesia)

[4] MSA Harahap, RS Handayani, K Kunci 2022 *J Ilmiah Mahasiswa Agroekoteknologi*, **1** 73–76.

[5] Kementerian Pertanian 2016 *Kementan Mendukung Meningkatkan Produksi Kedelai Nasional dengan Penggunaan Varietas Unggul Bermutu* (Jakarta:Indonesia)

[6] MS Poerwoko and N Sjamsijah 2019 *Bioscience Research*, **16** 2139-2149

[7] N Sjamsijah, Kuswanto, B Guritno, N Basuki 2016 *Agriculture and Agricultural Science Procedia*, **9** 510–517.

[8] Z Arifin and LE Sulistyowati 2020 *Jurnal Masyarakat Mandiri*, **4** 4-10.

[9] W Retno 2021 *Dosis Pupuk N, P, K Untuk Tanaman Jagung Pada Lahan* Sawah

[10] RH Nurhafsah and IA Fitriawaty 2021 *Petunjuk Teknis Budidaya Kedelai*. 1st edn, *BPTP Balitbangtan Suawesi Barat*

[11] C Xu , R Li, W Song, T Wu, S Sun, S Hu, T Han and C Wu, 2021 *Agriculture* **11** 1-12.

[12] A S. Milioli, AD. Zdziarski1 LG. Woyann, R Santos, AC. Rosa, A Madureira1 and G Benin 2018 *Chilean Journal Of Agricultural Research* **78** 299-309.

[13] N Sjamsijah, N Varisa, Suwardi 2018 *J of Applied Agricultural Sciences* **2** 106-116

[14] MM Adie, A Krisnawati 2015 In Prosiding Seminar Nasional Masyarakat Biodiversitas Indonesia 787–791

[15] A Kustera 2013. *Keragaman Genotipe dan Fenotipe Galur-Galur Padi Hibrida di Desa Kahuman, Polanharjo, Klaten* (Solo: Universitas Sebelas Maret)

[16] T Adisarwanto, 2008 *Budidaya Kedelai Tropika* (Jakarta:Penebar Swadaya) 76 p

[17] NB Arifiana & N Sjamsijah 2017 *Agriprima, Journal of Applied Agricultural Sciences* **1** 50–58.

IOP Conf. Series: Earth and Environmental Science **1446** (2025) 012010 doi:10.1088/1755-1315/1446/1/012010

Enhancing Hydroponic Systems with ESP32: An IoT Approach to Real-Time Monitoring and Automation

Arvita Agus Kurniasari[1], Pramuditha Shinta Dewi Puspitasari[2], Lukie Perdanasari[3], Dia Bitari Mei Yuana[4], Jumiatun[5]

Department of Information Technology, Politeknik Negeri Jember[1,2,3,4]
Department of Agriculture, Politeknik Negeri Jember[5]

*Email: arvita@polije.ac.id

Abstract. Indonesia's limited arable land, exacerbated by urbanization and industrialization, poses a significant challenge to sustainable agriculture. With only 7.1 million hectares of paddy fields and 30% of the land deemed critical or unproductive, innovative methods like hydroponics are essential. Hydroponics, a cultivation technique relying on water and nutrient solutions instead of soil, requires precise environmental management. This research introduces an IoT-based system to optimize hydroponic farming in Indonesia. The system utilizes an ESP32 microcontroller to monitor and control environmental factors such as temperature, humidity, pH, nutrient concentration, and light intensity. By automating adjustments, the system aims to enhance plant growth and yield. A prototyping method was adopted to ensure effective system development, resulting in a final product that meets desired objectives. Through testing, the system components demonstrated their expected performance, accurately collecting data and responding automatically to environmental changes. The 10 tests conducted under varying conditions revealed that the IoT device accurately measured temperature, humidity, and soil moisture. These results validate the successful integration of IoT technology into hydroponic systems, contributing to sustainable and efficient hydroponic agriculture in Indonesia.

1. Introduction

Indonesia, the fourth most populous country in the world, is facing a pressing issue of food security. With a population of 270.6 million in 2020, expected to rise to 297.5 million by 2045, the demand for food is escalating. However, the available agricultural land is shrinking due to urbanization, industrialization, and infrastructure development. In 2019, the remaining area of rice fields in Indonesia was only around 7.1 million hectares, with 30% of the country's land area classified as degraded or unproductive[1]. This situation intensifies the challenges of sustainably meeting the nation's food needs, necessitating immediate and innovative solutions.

The limitation of agricultural land has driven both the government and the community to seek innovative solutions to maximize food production. One increasingly popular approach is the adoption of hydroponic systems. Hydroponics is a method of growing plants without soil, where plants are cultivated using a mineral nutrient solution dissolved in water. This system is environmentally friendly and efficient in land use, making it highly suitable for urban areas or regions with limited agricultural land[2]. Hydroponics also allows for better control over nutrients and environmental conditions, which can significantly enhance plant productivity.

IOP Conf. Series: Earth and Environmental Science **1446** (2025) 012010 doi:10.1088/1755-1315/1446/1/012010

Despite the advantages of hydroponics, this system requires precise and careful management. The nutrient solutions must be maintained at optimal levels for plant growth. Water pH, humidity, temperature, light intensity, and nutrient concentration must be monitored regularly and adjusted when necessary. Manual control of these factors can be time-consuming, labor-intensive, and prone to inaccuracies or human error, impacting plant quality. To address these challenges, an automated system is needed to monitor and control environmental conditions in real-time, enhancing the efficiency and productivity of hydroponic systems. The Internet of Things (IoT) presents a promising solution. IoT enables hardware and software integration to automate system monitoring and control, reducing the need for manual intervention and minimizing the risk of human error[3][4].

This study proposes using the ESP32 microcontroller as the core of an automation system for hydroponics. The ESP32 is a microcontroller equipped with Wi-Fi and Bluetooth connectivity, capable of connecting to various sensors to collect real-time environmental data. The sensors used in this study include the DHT11 for measuring temperature and humidity, a water pH sensor, a TDS meter from DFRobot for measuring nutrient concentration, and the BH1750 for measuring light intensity. The data collected by these sensors allows the ESP32 to control various devices responsible for regulating the hydroponic environment, such as water pumps, lights, and fans. The system is designed to make automatic adjustments based on the data received, maintaining optimal conditions for plant growth. For example, if the sensor detects that the water pH is too high or too low, the system automatically adds the necessary ingredients to balance the pH. Similarly, the system will add nutrients to the solution if the nutrient concentration is too low[5][6].

The implementation of this IoT-based system offers several key advantages. First, automation enables more efficient management, as the system can operate without continuous human intervention[7]. This saves time and effort that would otherwise be spent manually monitoring and regulating conditions. Second, automation reduces the risk of human error, which is often a factor in manually managed systems. With real-time data collection and automatic adjustments, stability and consistency in plant growth can be maintained. Third, continuous monitoring allows for a rapid response to changes in environmental conditions that could affect plant health, thereby reducing the risk of losses due to diminished plant quality[8][9].

Furthermore, using IoT technology in hydroponics facilitates more effortless system scalability. IoT-enabled hydroponic systems can be operated in multiple locations with minimal intervention, making them suitable for commercial and home applications. This technological advancement positions hydroponics as a more efficient and effective solution to address the challenges of limited land availability in Indonesia, contributing to national food security. This study explores the potential of IoT integration in hydroponic systems, focusing on developing an ESP32-based system that can monitor and control environmental conditions in real-time. The findings of this study will significantly contribute to increasing agricultural productivity in Indonesia and serve as a step toward more sustainable agriculture.

2. Research Method

This system automatically collects temperature and humidity data from a DHT sensor and soil moisture levels, displaying them in real-time on an LCD screen. The collected data is then periodically sent to the Firebase cloud platform for storage and further analysis. This enables continuous monitoring of environmental conditions and allows for data access from anywhere with an internet connection.

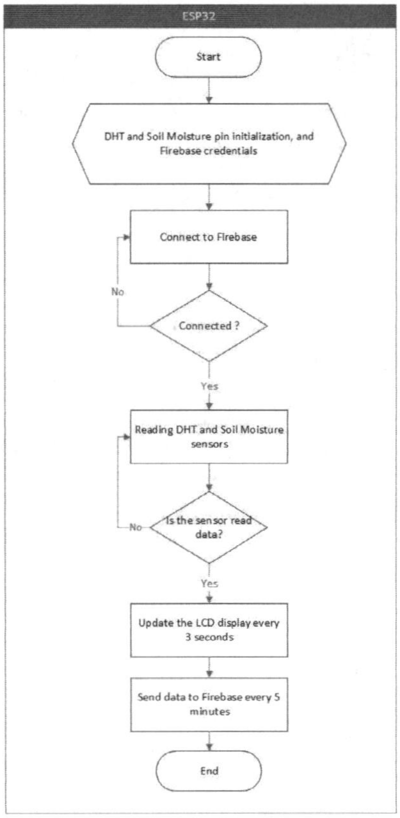

Figure 1. Flowchart IoT System

This IoT system development adopts an effective approach known as the prototyping method.The prototyping methodology is particularly beneficial for IoT development, as it enables developers to receive direct user feedback. This iterative process, involving prototyping, testing, and feedback collection, allows for early identification and resolution of issues, design refinements, and ensuring the final system aligns with user requirements [4].

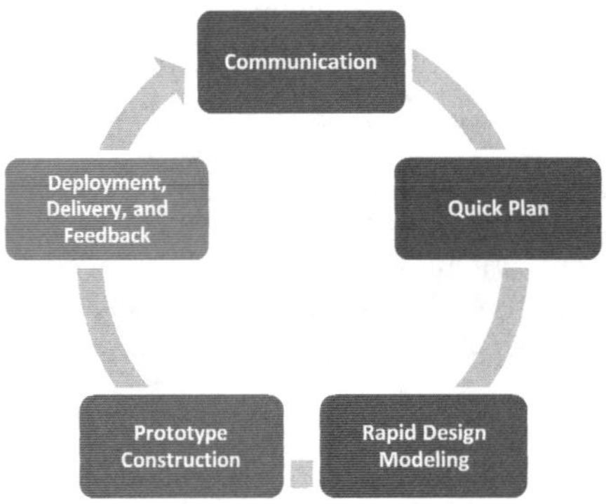

Figure 2. Protoype Method [4]

IOP Conf. Series: Earth and Environmental Science **1446** (2025) 012010 doi:10.1088/1755-1315/1446/1/012010

The prototyping method involves several stages [4]: First, developers and stakeholders communicate to identify the project goals, gather requirements, and understand limitations. Then, a quick plan is created based on the gathered information. Next, a quick design is developed to visualize the system's interface and functionality. A prototype is then built to provide users with a tangible experience and gather feedback. Finally, the prototype is evaluated by stakeholders, and their feedback is used to refine the requirements and design for subsequent iterations. This iterative process allows developers to create a product that meets user needs and expectations.

3. Proposed architecture

The proposed architecture leverages the ESP32 microcontroller to enhance hydroponic systems through real-time monitoring and automation, creating an efficient IoT-based solution for plant growth management. By integrating sensors like the capacitive soil moisture sensor, DHT11 temperature and humidity sensor, pH sensor, and TDS meter, the system continuously monitors critical environmental factors such as moisture levels, air humidity, temperature, water pH, and nutrient concentration. This data is processed by the ESP32, which triggers automatic actions such as activating the cooling fan to regulate temperature or powering the water pump when soil moisture drops below a set threshold. A connected LCD provides real-time feedback for users, while the ESP32's Wi-Fi capability enables remote monitoring and control through a cloud platform or mobile app. This setup ensures optimal plant growth conditions with minimal manual intervention, promoting efficient water use, maintaining nutrient balance, and automating climate control. Scalable and adaptable, this architecture is ideal for both small-scale and commercial hydroponic systems, providing a modern, data-driven approach to agriculture.

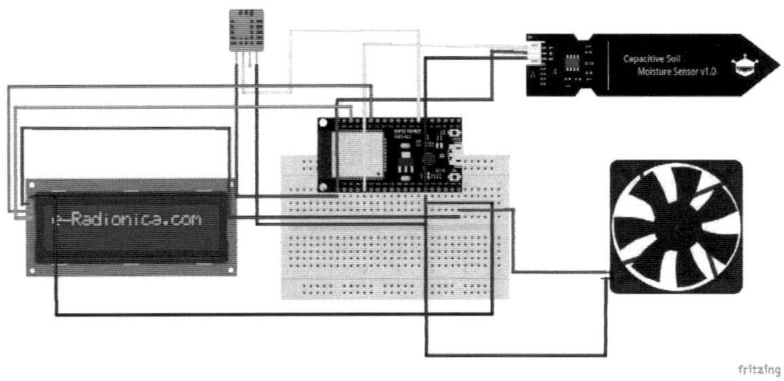

(a)

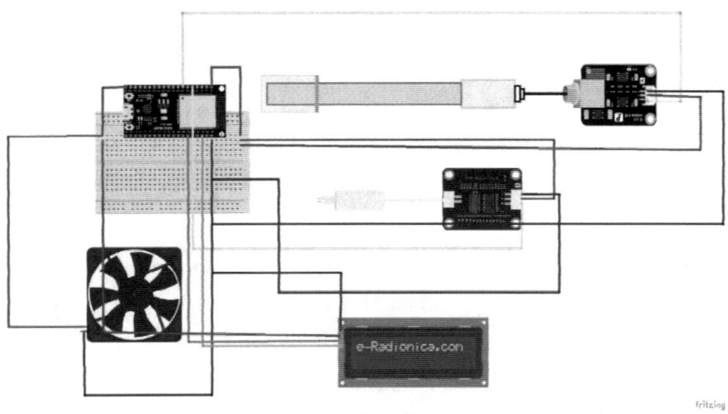

(b)

Figure 3. (a) environmental conditions architecture, (b) nutrient concentration architecture

The fundamental components of the proposed system are as follows:

3.1. ESP32

The ESP32 is a microcontroller developed by Espressif Systems, the successor to the ESP8266 microcontroller. This microcontroller is designed to meet the needs of Internet of Things (IoT) applications by providing integrated Wi-Fi and Bluetooth connectivity and higher processing capabilities than its predecessor. With specifications that include two Xtensa LX6 processor cores that can operate at up to 240 MHz, the ESP32 offers efficient performance for various applications, from network sensors to complex control systems[10][11].

One of the standout features of the ESP32 is its ability to support a variety of communication protocols, including Wi-Fi and Bluetooth, allowing developers to create wirelessly connected applications easily. This makes it an excellent choice for IoT projects, where connectivity and energy savings are critical. With support for multiple GPIO (General Purpose Input/Output) pins, the ESP32 also provides flexibility in hardware development[10][12].

Figure 4. ESP32.

3.2. Soil Moisture Sensor

Soil Moisture Sensor is a device designed to measure soil moisture content, an essential factor in agricultural and irrigation management. This sensor utilizes various physical principles to detect changes in soil water content, which can affect plant growth and water use efficiency. In agriculture, soil moisture monitoring is essential to optimize water use, increase crop yields, and prevent damage due to drought or excess water[13][14].

Figure 5. Soil Moisture sensor measures soil moisture.

3.3. DHT Sensor

DHT Sensor, which stands for Digital Humidity and Temperature Sensor, is a device used to measure air temperature and humidity digitally. This sensor is very popular in home automation applications, environmental monitoring systems, and Internet of Things (IoT) projects due to its ease of use and relatively low cost. DHT Sensor is available in several variants, including DHT11 and DHT22, each with different specifications and accuracy[15][16].

Figure 6. DHT11 sensor measures air temperature and humidity.

3.4. *TDS*

A TDS meter, or Total Dissolved Solids meter, measures the total concentration of dissolved solids in water. These solids include various ions, salts, minerals, and organic compounds. TDS measurement is essential in multiple applications, including water quality management, agriculture, aquaculture, and industry, because TDS concentrations can affect the health of plants, aquatic organisms, and industrial processes[17].

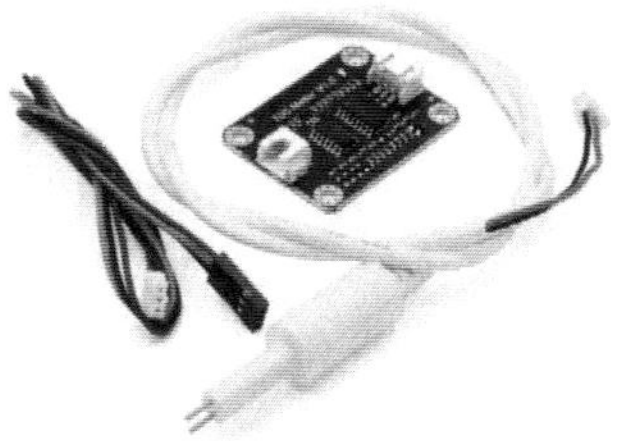

Figure 7. TDS meter

3.5. *BH1750*

The BH1750 is a digital light sensor that measures light intensity in lux units. It uses an I2C interface for communication, which makes it easy to integrate with various microcontrollers such as Arduino and Raspberry Pi. The BH1750 can provide accurate and fast readings on ambient light levels, making it a popular choice in different applications, including home automation, smart farming, and automated lighting systems[18].

Figure 8. BH1750 Light Sensor.

3.6. *Water PH Sensor*

A water pH sensor is a device used to measure the degree of acidity or alkalinity of a solution expressed in the pH scale. The pH scale ranges from 0 to 14, where pH values below 7 indicate acidic properties, pH 7 indicates neutral properties, and pH above 7 indicates basic properties [19][20]. These sensors are critical in various applications, including water quality management, agriculture, aquaculture, and environmental research, because pH is a crucial indicator of the chemical conditions of water that can affect the lives of organisms and biochemical processes[21].

Figure 9. Water PH Sensor.

4. Experiment and Resul

4.1. Hardware Implementation

In this research, as illustrated in Figure 8, we implemented a hardware setup within a controlled screen house, employing the ESP32 microcontroller as the central component of our Internet of Things (IoT) approach. The configuration incorporated various sensors, including the DHT11 for temperature and humidity, a pH water sensor, the TDS Meter from DFRobot for nutrient concentration, and the BH1750 for light intensity, which collectively monitored critical environmental parameters. Additionally, actuators were integrated to manage water pumps and nutrient delivery systems, facilitating optimal conditions for plant growth.

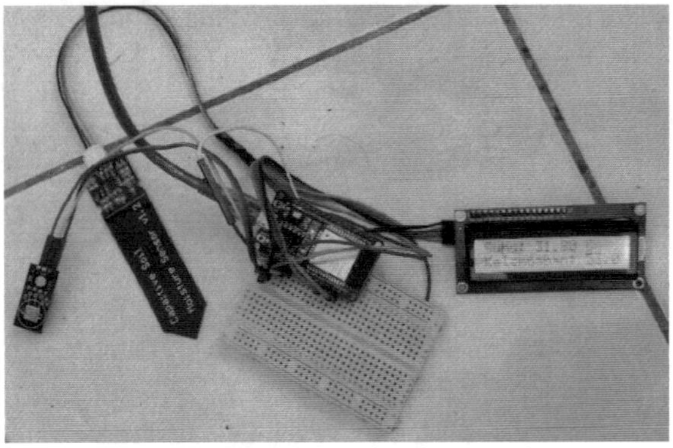

Figure 10. Hardware implementation.

4.2. Display on Real-Time Monitoring

The sensors continuously transmit information through a microcontroller connected to the internet. Once the ESP32 microcontroller establishes this connection, it enables simultaneous data transmission to and from the Firebase database, facilitating the automatic adjustment of nutrient settings and enhancing operational efficiency. Users can access and monitor all relevant data conveniently displayed on website. The firebase database will display the amount of TDS, PH, Humidity, Soil Moisture, and Temperature.

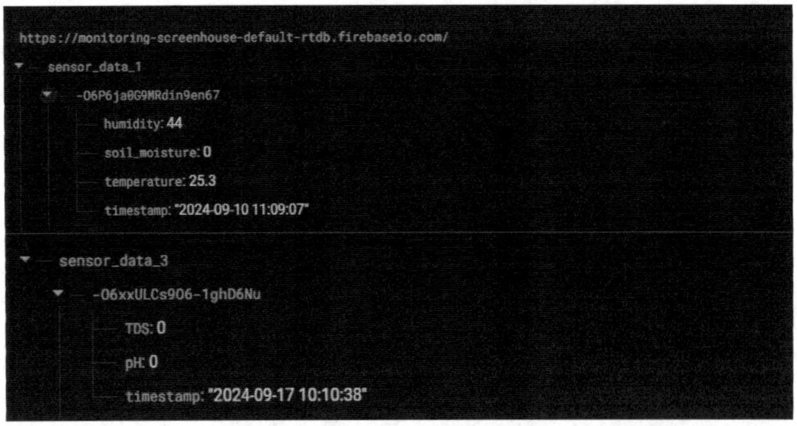

Figure 11. Firebase database.

4.3. Result of IoT Technology Deployment

The deployment of our IoT technology within the controlled screen house has yielded significant results in optimizing hydroponic systems. Utilizing the ESP32 microcontroller as the core component, the system successfully integrated various sensors, including the DHT11 for temperature and humidity, a pH water sensor, the TDS Meter from DFRobot for nutrient concentration, and the BH1750 for light intensity.

As illustrated in Figure 10, the comprehensive monitoring setup provided real-time data on critical environmental parameters, enabling precise adjustments to maintain optimal conditions for plant growth. Integrating actuators to control water pumps and nutrient delivery systems facilitated automated responses to fluctuations in environmental conditions, ensuring that plants received necessary resources promptly.

Figure 12. Hardware check-in screen house.

Data collected from the sensors was transmitted to the Firebase database, allowing for remote monitoring and analysis. The accompanying images show the real-time results displayed on the Firebase dashboard, showcasing metrics such as nutrient levels, light intensity, and water temperature. The integrated LCD screen enhances accessibility, providing quick visual feedback for immediate decision-making.

4.4. Hardware Testing

Tests are conducted to assess the sensor's operational efficacy as designed. The outcomes are presented in Table 1.

Table 1. Hardware Component.

No	Hardware Component	Quantity	Status
1	ESP 32 can send data to Firebase and reads all sensors	2	√
2	The TDS sensor can read the nutritional value	1	√
3	The Soil sensor can read the humidity of the soil	2	√
4	DHT11 can read temperature and humidity	2	√
5	BH1750 can read light around plants	1	√
6	pH water sensor can read the pH around hydroponic plants	1	√
7	The LCD's temperature and humidity.	1	√

Note :
√ = successful
X = not working

The ESP32 microcontroller successfully sends data to Firebase and reads all connected sensors, acting as a reliable central hub. According to Table X, the TDS sensor accurately measures nutritional value, while two soil sensors monitor humidity levels for optimal plant growth. DHT11 sensors provide temperature and humidity readings, and the BH1750 sensor measures light intensity around the plants. The pH water sensor checks the pH levels of the nutrient solution, ensuring it remains balanced. Additionally, the integrated LCD shows real-time temperature and humidity data, making critical information easily accessible. Overall, the successful testing of these components is detailed in Table 1.

4.5. Testing the soil with air temperature

Before analyzing the relationship between soil moisture and environmental conditions, precisely air temperature and humidity, it is essential to examine the similarities of the soil to these atmospheric factors. By recording data on air temperature, humidity, and soil moisture multiple times, we can better understand how air temperature and moisture levels affect the soil's ability to retain air content. Table 2 presents the data collected, showing air temperature, humidity, and soil moisture. This data provides insight into the interactions between these variables and how changing weather patterns can affect soil conditions over time. The following data was obtained:

Table 2. Resulting in temperature, humidity, and soil moisture.

No	Timestamps	Temperature	Humidity	Soil Moisture
1	2024-09-10T15:05:05.000	38	44	37
2	2024-09-10T15:06:05.000	33.3	56	37
3	2024-09-10T15:07:05.000	30	68	36
4	2024-09-17T08:02:55.000	28.5	62	55
5	2024-09-17T08:03:55.000	29.3	61	37
6	2024-09-17T08:43:59.000	25.8	51	38
7	2024-09-17T08:44:59.000	25.8	50	38
8	2024-09-17T08:45:59.000	25.8	50	37
9	2024-09-17T11:13:41.000	33.8	44	51
10	2024-09-17T11:19:21.000	32.8	47	38

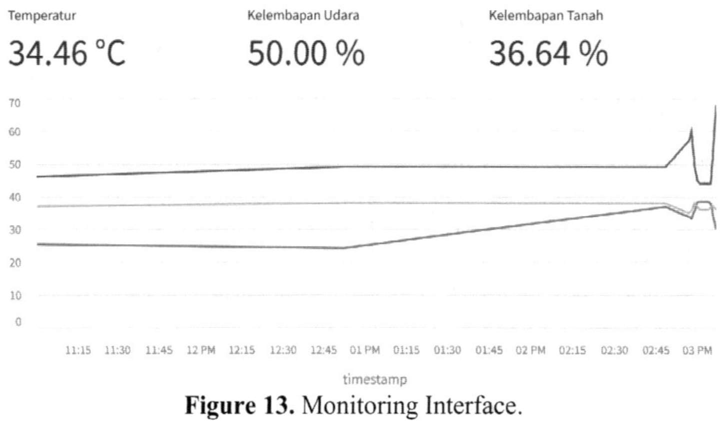

Figure 13. Monitoring Interface.

Based on 10 tests conducted at different times, the following conclusions can be drawn:

- Temperature: The measured temperature ranged from 25.8°C to 38°C. There was a significant fluctuation in temperature, likely influenced by the measurement time and ambient conditions.

IOP Conf. Series: Earth and Environmental Science **1446** (2025) 012010 doi:10.1088/1755-1315/1446/1/012010

- Humidity: The relative humidity varied between 44% and 68%. This significant fluctuation indicates the influence of weather conditions on the surrounding environment.
- Soil Moisture: Soil moisture was relatively stable, ranging from 36% to 55%. Although there was some variation, soil moisture generally remained consistent throughout the testing period.

Table 3. Resulting in temperature, humidity, and soil moisture in a week

No	Timestamps	Temperature	Humidity	Soil Moisture
1	2024-09-18T07:00:00	24.5	75	40
2	2024-09-18T12:00:00	31.2	60	38
3	2024-09-18T17:00:00	27.8	65	39
4	2024-09-19T07:00:00	24	76	41
5	2024-09-19T12:00:00	30.8	59	37
6	2024-09-19T17:00:00	27.5	64	38
7	2024-09-20T07:00:00	24.2	74	42
8	2024-09-20T12:00:00	31.5	58	37
9	2024-09-20T17:00:00	28	63	39
10	2024-09-21T07:00:00	23.8	75	43
11	2024-09-21T12:00:00	31	60	38
12	2024-09-21T17:00:00	27.9	66	40
13	2024-09-22T07:00:00	24.1	76	44
14	2024-09-22T12:00:00	31.8	59	39
15	2024-09-22T17:00:00	28.2	64	41
16	2024-09-23T07:00:00	24.3	73	42
17	2024-09-23T12:00:00	30.7	58	36
18	2024-09-23T17:00:00	27.6	63	38

Based on 18 tests conducted at different times, the following conclusions can be drawn:

- Temperature: The measured temperatures ranged from 23.8°C to 31.8°C. This range reflects significant fluctuations, largely influenced by the time of day, with the highest temperatures occurring around midday and the lowest during early morning.
- Humidity: The relative humidity varied between 58% and 76%. This wide range suggests environmental factors such as morning moisture and afternoon dryness play a significant role in the readings.
- Soil Moisture: Soil moisture levels ranged from 36% to 44%. Despite minor variations, the data indicates relatively stable soil conditions throughout the testing period.

5. Conclusion

The system successfully reduces manual intervention, minimizes the risk of human error, and ensures more consistent growing conditions for plants. By automating the monitoring and regulation of factors such as water pH, nutrient levels, humidity, temperature, and light intensity, the ESP32-based solution enhances overall productivity and plant health, leading to higher yields. Furthermore, remote monitoring capabilities via Firebase and the LCD interface allow users to control the system anywhere, increasing convenience and operational efficiency.

The system components performed as expected through testing, with accurate data collection and automated responses to environmental changes. The 10 tests conducted under varying conditions revealed that the IoT device accurately measured temperature, humidity, and soil moisture. The measured temperature ranged from 25.8°C to 38°C, and humidity varied between 44% and 68%, demonstrating sensitivity to ambient conditions. Soil moisture, however, remained relatively stable between 36% and 55%. The results confirm the effectiveness of IoT-based automation in optimizing hydroponic systems, making it a promising approach to address the challenges of limited agricultural land in Indonesia. As a scalable technology, this system can be implemented in small-scale urban farms and more extensive commercial hydroponic operations, contributing to the country's sustainable agriculture and food security.

Future work could explore further advancements, such as integrating machine learning algorithms for predictive analysis and improving energy efficiency, to make the system more adaptive and robust. This research offers a foundational framework for applying IoT in agriculture, paving the way for more sophisticated solutions to modern farming challenges.

6. Acknowledgment

The authors would like to express their sincere gratitude to the Director and the Center for Research and Community Service at Jember State Polytechnic for their invaluable support in facilitating international seminars, providing sponsorship for our research endeavors, and continuing to organize international conferences on food and agriculture. Without their commitment and contributions, this research would not have been possible.

References

[1] H. V. Rahmi, S. D. N. Perwitasari, and L. H. Pratopo, "Business Development Strategy for Smart Watering Product of PT. Hidroponik Padjadjaran Indonesia," *J. Manaj. Dan Agribisnis*, 2023, doi: 10.17358/jma.20.3.477.

[2] Y.-H. Hsieh, S.-P. Li, and T.-I. Lee, "The Application of Organic Hydroponics on Homegrown Urban Agriculture in Taiwan," 2018, doi: 10.2495/sdp180831.

[3] Y. V. Bhargava, P. K. Chittoor, and C. Bharatiraja, "IoT Powered Smart Hydroponic System for Autonomous Irrigation and Crop Monitoring," *Ecs Trans.*, vol. 107, no. 1, pp. 11937–11944, 2022, doi: 10.1149/10701.11937ecst.

[4] L. Perdanasari *et al.*, "Pengukuran Karakteristik Lahan Berbasis Internet of Things," *JTIM J. Teknol. Inf. Dan Multimed.*, vol. 3, no. 3, pp. 169–175, 2021.

[5] N. Ramsari and T. Hidayat, "Monitoring System and Hydroponic Plant Automation Using Microcontroller Internet of Things Based (IoT)," *Compiler*, vol. 11, no. 2, 2022, doi: 10.28989/compiler.v11i2.1365.

[6] H. J. Christanto *et al.*, "Design and Development of the Mobile-Based Hydroponic Planting Machine Application MyHydro," *J. Inf. Syst. Informatics*, vol. 5, no. 4, pp. 1321–1339, 2023, doi: 10.51519/journalisi.v5i4.581.

[7] H. Y. Riskiawan *et al.*, "Artificial Intelligence Enabled Smart Monitoring and Controlling of IoT-Green House," *Arab. J. Sci. Eng.*, 2023, doi: 10.1007/s13369-023-07887-6.

[8] M. Martin and E. Molin, "Environmental Assessment of an Urban Vertical Hydroponic Farming System in Sweden," *Sustainability*, vol. 11, no. 15, p. 4124, 2019, doi: 10.3390/su11154124.

[9] I. K. Wibowo *et al.*, "SLiCE: Implementation of Automation Technology and Internet of Things in the Greenhouse," *Abdimas J. Pengabdi. Masy. Univ. Merdeka Malang*, vol. 8, no. 2, pp. 315–325, 2023, doi: 10.26905/abdimas.v8i2.9816.

[10] A. Ramelan *et al.*, "Design and Implementation RESTful API for IoT Based Smart Home Systems," *E3s Web Conf.*, vol. 465, p. 2061, 2023, doi: 10.1051/e3sconf/202346502061.

[11] H. I. Shahadi, M. K. Kadhim, N. M. Almeyali, and A. T. Hadi, "Design and Implementation of a Smart, Interactive and Portable System for Monitoring of Human Vital Signs," *J. Ilm. Tek. Elektro Komput. Dan Inform.*, vol. 7, no. 1, p. 30, 2021, doi: 10.26555/jiteki.v7i1.20209.

[12] J. Kim, "Analysis and Optimization of DC Supply Range for the ESP32 Development Board," 2020, doi: 10.36227/techrxiv.12798410.v3.

[13] M. Pramanik *et al.*, "Evaluation of Capacitance-Based Soil Moisture Sensors in IoT Based

Automatic Basin Irrigation System," 2023, doi: 10.21203/rs.3.rs-3043138/v1.

[14] M. S. Kukal, S. Irmak, and K. Sharma, "Development and Application of a Performance and Operational Feasibility Guide to Facilitate Adoption of Soil Moisture Sensors," *Sustainability*, vol. 12, no. 1, p. 321, 2019, doi: 10.3390/su12010321.

[15] "Development of IoT Based Meteorological Measurement System," *Int. J. Innov. Technol. Explor. Eng.*, vol. 9, no. 2S3, pp. 68–71, 2019, doi: 10.35940/ijitee.b1017.1292s319.

[16] A. Pangestu, M. Yusro, W. Djatmiko, and A. Jaenul, "The Monitoring System of Indoor Air Quality Based on Internet of Things," *Spektra J. Fis. Dan Apl.*, vol. 5, no. 2, 2020, doi: 10.21009/spektra.052.06.

[17] M. M. Akram, M. Ramezannezhad, A. Nikfarjam, S. Kabiri, and S. Ehyaei, "A Strip-based Total Dissolved Solids Sensor for Water Quality Analysis," *Iet Sci. Meas. Technol.*, vol. 16, no. 3, pp. 208–218, 2022, doi: 10.1049/smt2.12098.

[18] M. Suhairi and Halimah Tuzsakdiah, "Sistem Kontrol Dan Monitoring Intensitas Cahaya dan Suhu Pada Greenhouse BerBasis IoT," 2023. doi: 10.35143/elementer.v9i1.5888.

[19] I. P. Y. pramesia Pratama, K. S. Wibawa, and I. M. A. D. Suarjaya, "Perancangan PH Meter Dengan Sensor PH Air Berbasis Arduino," *Jitter J. Ilm. Teknol. Dan Komput.*, vol. 3, no. 2, p. 1034, 2022, doi: 10.24843/jtrti.2022.v03.i02.p02.

[20] A. Rahman and A. N. Salim, "Sistem Kendali pH Dan Kekeruhan Air Pada Aquascape Menggunakan Wemos D1 Mini Esp8266 Berbasis IoT," *J. Teknol. Terpadu*, vol. 8, no. 1, pp. 22–30, 2022, doi: 10.54914/jtt.v8i1.526.

[21] I. Irawan and J. C. Chandra, "Sistem Pemantauan Dan Kendali Aquaponic Menggunakan Arduino Berbasis Web," *Kresna J. Ris. Dan Pengabdi. Masy.*, vol. 3, no. 1, pp. 77–85, 2023, doi: 10.36080/kresna.v3i1.62.

IOP Conf. Series: Earth and Environmental Science **1446** (2025) 012011 doi:10.1088/1755-1315/1446/1/012011

Effect of Low Temperature on Physical Characteristics of Chicken (*Gallus domesticus*) Meat During Storage

H Risnawati[1], R Triani[1], P B Purwandoko[2*], N K I Mayasti[2], N D Susanti[2], F Novianti[2], I F Aprianto[2], C Litaay[2], I S Kuala[2], A Indriati[2], E J Pristianto[3], A N Rahman[3] and E D Kurniawan[3]

[1]Department of Food Technology, Faculty of Engineering, Pasundan University, Bandung 40153, West Java, Indonesia

[2]Research Center for Appropriate Technology, National Research and Innovation Agency, Subang 41213, West Java, Indonesia

[3]Research Center for Telecommunication, National Research and Innovation Agency, Bandung 40135, West Java, Indonesia

[*]prad011@brin.go.id

Abstract. Chicken meat is highly perishable and requires special handling because it spoils quickly during the metabolic processes after slaughter. Poor handling and storage methods can lead to a decline in meat quality. Storing chicken at low temperatures, such as in a refrigerator at 4°C, helps extend its shelf life and preserve its qualities. This study aimed to examine the changes in the physical properties of raw chicken meat stored at 4°C. The research was conducted using a Completely Randomized Design (CRD) with five storage periods: day 0, day 2, day 4, day 6, and day 8. The physical characteristics observed were pH, color, aroma, texture, and drip loss. The length of storage at 4°C significantly affected the chicken's pH, color, texture, water loss, and aroma quality ($P<0.05$). The color analysis showed that the lightness (L*) and greenness (a*) values increased, while the yellowness (b*) values also increased. The texture, including hardness, springiness, cohesiveness, gumminess, chewiness, and resilience, decreased over time. Prolonged storage at low temperatures (4°C) reduced pH, increased water loss, and changes in color, aroma, and texture. Overall, storage time significantly impacted the physical characteristics of raw chicken meat stored at 4°C.

1. Introduction

Chicken meat is an animal-based food ingredient widely consumed by the Indonesian people [1]. According to National Food Agency Indonesia data, the average Indonesian person consumed 7.46 kilograms/capita/year of broiler chicken meat in 2023. Chicken meat contains high protein, low fat, and essential fatty acids [2]. It is a perishable food that requires special treatment because it spoils quickly [3].

The public widely consumes chicken meat because it has an affordable purchase value and is considered an alternative source of protein that is healthier than red meat because it contains more protein and lower fat [4]. In addition, chicken meat is also rich in nutrients such as calcium, magnesium, phosphorus, sodium, vitamin B3, vitamin A, and vitamin B6. However, this nutrient content can also be a medium for microorganisms to grow [5]. As a result, issues regarding product quality and safety can arise at different stages of the production process [6,7]. Microorganisms can grow in chicken meat due to contamination that occurs during the handling, storage, and distribution processes, which decreases the quality of chicken meat [8].

The quality of chicken meat can degrade due to factors such as storage temperature, packaging type, and the condition of storage environments and equipment used [3]. Storing chicken at low temperatures does help slow microorganism growth, limit chemical reactions, and reduce enzyme activity, but some level of quality deterioration is still inevitable during storage [8]. Sustainable practices in the broiler chicken supply chain play a key role in maintaining meat quality [9,10]. Effective supply chain management helps ensure that chicken remains fresh throughout storage and distribution, minimizing waste and enhancing operational efficiency [11,12].

This study aims to investigate the deterioration in quality and the physical property changes of chicken meat stored at low temperatures. The research used a Completely Randomized Design (CRD) to ensure accurate results by reducing errors. This method was chosen because it is effective for studying the effects of one factor, such as storage time, on multiple physical properties like pH, texture, and aroma. This study provides insight into the impact of low-temperature storage on the physical properties of chicken meat, offering a comprehensive analysis of changes in pH, texture, and aroma during storage.

2. Research Method

In this study, we observed changes in the characteristics of raw chicken meat, such as the pH value, color, texture, drip loss value, and odor. The chicken was six weeks old, and only the breast was selected. The chicken breast meat was packaged using Wayang brand polypropylene plastic measuring 25 cm × 35 cm and stored in a refrigerator at 4°C. This storage temperature was chosen because it is a standard refrigeration temperature that can slow microbial growth and enzymatic activity in meat, thereby reducing spoilage [13]. Storage durations of 0, 2, 4, 6, and 8 days were selected based on previous research by Sangadji et al. [14], which evaluated the effects of different storage periods on the quality of broiler chicken meat, particularly its protein content and bacterial load. Before testing, the meat was removed from the refrigerator and allowed to thaw for 30 minutes.

2.1. Acidity

Acidity is related to pH measurements, which are crucial for assessing the quality of food products, including chicken meat. The pH measurement of the chicken meat was conducted using a Mettler Toledo TM FE 20 pH meter. Chicken breast samples were ground, mixed with 100 mL of distilled water, and homogenized. Before measurement, the pH meter was calibrated with standard buffer solutions at pH 4 and 7. Once calibrated, the electrode was immersed in the sample solution to obtain the pH reading.

2.2. Color

The color of chicken meat was analyzed using a chromameter (digital colorimeter, 3NH-NH300), which operates based on the CIELAB color system. In this system, L represents lightness, while a and b* represent the red-green and yellow-blue axes, respectively. Measurements were taken on three different surfaces of the chicken breast to ensure accuracy and consistency. This method allows for an objective assessment of color changes during storage.

2.3. Texture

The texture of chicken meat was analyzed using texture profile analysis (TPA) with a texture analyzer. A chicken breast meat sample cut measuring 2 × 2 × 2 cm3 was pressed using a probe with a diameter of 6 mm, with the probe speed set at 5 mm/s to 30% of the initial height of the sample [15,16]. The sample pressing was carried out two times. The data emerge from the system: hardness, springiness, cohesiveness, gumminess, chewiness, and resilience.

2.4. Drip loss

Water or drip loss in raw chicken meat was measured by weighing the sample and recording it as the initial weight. The sample was then neatly wrapped in a refrigerator at 4°C for the test duration. During the test period, remove the sample from the refrigerator in the open air for 30 minutes, weigh it, and record it as the final weight. Fluid loss or drip loss was calculated using the following formula [17]:

$$Drip\ loss = \frac{W1-W2}{W1}\ x\ 100\%$$

W1 = weight of the sample before storage (day 0)
W2 = weight of the sample after the storage period

2.5. Odor
Hedonic quality testing was performed on 20 trained panelists. During testing, panelists will be given samples to assess their quality, namely, odor. The assessment used a scale written in the questionnaire provided.

3. Results and Discussion

3.1. Acidity
The pH value of chicken meat stored at 4°C decreased gradually from day 0 to day 8 ($P<0.05$). On day 0, the pH was 6.33, but it dropped to 6.22 on day 2, 6.14 on day 4, 6.08 on day 6, and 5.99 on day 8. This pH reduction can be attributed to biochemical processes occurring after the chicken is slaughtered. During slaughter, the chicken loses a significant amount of blood, causing oxygen levels to decrease until depleted. This condition triggers a shift to anaerobic metabolism, specifically anaerobic glycolysis, which produces lactic acid. The accumulation of lactic acid in the meat leads to a decrease in pH. Also, pH may drop due to protein denaturation during storage [4].

Table 1. pH value of raw chicken meat during storage at 4°C

Storage time (day)	pH
0	$6{,}33 \pm 0{,}07^{c}$
2	$6{,}22 \pm 0{,}06^{bc}$
4	$6{,}14 \pm 0{,}12^{ab}$
6	$6{,}08 \pm 0{,}11^{ab}$
8	$5{,}99 \pm 0{,}05^{a}$

Note: Values with different letter superscripts represent a statistically significant difference ($P<0.05$)

3.2. Color
The results of this study for the color of chicken meat show that storage time at 4°C significantly affects the color characteristics of raw chicken meat ($P<0{,}05$), as shown in Table 2. Color measurements were conducted using a chromameter based on the CIELAB color scale, which evaluates three parameters: L (lightness), a* (red-green), and b* (yellow-blue) [18]. The color of the chicken meat stored at 4°C changed from day 0 to day 8. The results showed that the L value (brightness) and a value (redness) decreased from day 0 to day 8, while the b value (yellowish) increased. Color changes during storage are related to lipid oxidation and drip loss [19]. The number of myoglobin levels, the chemical structure of the meat, genetics, feed, environment during rearing, conditions during slaughter, post-slaughter conditions, and pH of the meat also influence the occurrence of color changes in chicken meat [20].

The L value (brightness) read by the chromameter decreased, causing the color of raw chicken meat to become darker during storage at 4°C. The color of chicken meat stored for 8 days becomes darker because a lipid oxidation process occurs during storage, which produces metmyoglobin [8]. On day 0, the average value of L was 55.02, which decreased from day 2 to 54.22. The L value on days 4, 6, and 8 also decreased with an average L value sequentially of 52,61, 48,04, and 42,09.

The a value (reddishness) decreases and causes the color of the chicken meat to change to green as it remains at a temperature of 4°C. On day 0, raw chicken meat had an average value of -0.007 and decreased to -3.133 on day 8. A decrease in the value of a can occur because the myoglobin contained in chicken meat reacts with oxygen in the environment and undergoes an oxidation reaction [21]. During storage, chicken meat can change color to greenish because the myoglobin in chicken meat reacts with hydrogen sulfide to form sulphmyoglobin [22]. This reaction can occur because of the growth of microorganisms in the meat. Microorganisms can grow on chicken meat owing to external contamination.

Table 2. L, a, and b value of raw chicken meat during storage at 4°C

Storage time (day)	L	a	b
0	55,02 ± 0,03^{e}	-0,007 ± 0,24^{d}	2,76±0,14^{a}
2	54,22 ± 0,61^{d}	-0,51 ± 0,27^{b}	4,12±0,11^{b}
4	52,61 ± 0,59^{c}	-1,83 ± 0,15^{a}	5,78±0,06^{c}
6	48,04 ± 0,07^{b}	-2,257 ± 0,13^{a}	6,11±0,07^{d}
8	42,09 ± 0,08^{a}	-3,133 ± 1,64^{a}	7,37±0,03^{e}

Note: Values with different letter superscripts represent a statistically significant difference (P<0.05)

The b value (yellowish) in chicken meat stored at a temperature of 4°C changes with an increase in the value read by the chromameter. Raw chicken meat on day 0 had a b value of 2.76, increasing to 4.12 on storage on day 2 and continued to expand on days 4, 6, and 8 with b values of 5.78, 6.11, and 7.37, respectively. This condition can happen because meat pigment cells or myoglobin experience oxidation during storage [21].

3.3. Texture

The findings indicate that storage time significantly affects the texture of raw chicken meat stored at 4°C, including hardness, springiness, cohesiveness, gumminess, chewiness, and resilience (Table 3). This decline is primarily due to structural changes in the proteins within the meat during storage. Factors such as decreasing pH levels, protein damage, and degradation contribute to these changes [23]. During cold storage, the chicken meat enters the post-rigor mortis phase, weakening the muscle structure and affecting its texture. Maintaining optimal storage conditions can help mitigate these effects but cannot entirely prevent the natural degradation process over time [21].

Table 3. The texture of raw chicken meat during storage at 4°C

Storage time (day)	Variable					
	Hardness	Springiness	Cohesiveness	Gumminess	Chewiness	Resilience
0	7732,56 ± 83,651^{e}	0,87 ± 0,009^{e}	0,64 ± 0,001^{e}	4972,09 ± 61,455^{e}	4316,16 ± 100,833^{e}	0,38 ± 0,003^{e}
2	7259,73 ± 339,381^{d}	0,70 ± 0,008^{d}	0,62 ± 0,002^{d}	4479,67 ± 223,300^{d}	3135,41 ± 191,943^{d}	0,36 ± 0,003^{d}
4	5108,27 ± 158,619^{c}	0,62 ± 0,006^{c}	0,51 ± 0,007$_{c}$	2614,27 ± 111,162^{c}	1630,88 ± 85,458^{c}	0,30 ± 0,004^{c}
6	3662,52 ± 19,298^{b}	0,50 ± 0,006^{b}	0,49 ± 0,001^{b}	1806,86 ± 15,096^{b}	895,66 ± 18,331^{b}	0,26 ± 0,004^{b}
8	2753,61 ± 46,185^{a}	0,48 ± 0,004^{a}	0,44 ± 0,001^{a}	1217,13 ± 23,153^{a}	582,66 ± 15,925^{a}	0,24 ± 0,003^{a}

Note: Values with different letter superscripts represent a statistically significant difference (P<0.05)

The reduction in the hardness of chicken meat during storage is primarily due to decreased water-holding capacity over time. The hardness of meat is directly influenced by the structure of the protein matrix and its ability to retain water [24]. As the storage period lengthens, the meat's ability to hold water diminishes, leading to a softer texture. Similarly, the values for springiness and cohesiveness are also affected by the water content in the meat. Generally, higher hardness, springiness, and cohesiveness indicate higher water content in the chicken meat [16]. However, water loss increases when the meat is stored from day 0 to day 8, resulting in a decline in these texture properties. The moisture loss weakens the protein bonds and reduces the meat's structural integrity, making it less elastic (springy) and less cohesive. This overall degradation in texture contributes to a decline in the meat's quality, as the firmness and chewiness desirable in fresh meat are lost during prolonged storage. Proper storage methods that minimize moisture loss are critical to maintaining these essential textural qualities.

The protein structure in meat plays a crucial role in determining its gumminess. The gumminess value tends to increase when the spaces within the protein matrix are filled, making the structure denser and requiring more energy to break down [25]. However, during storage, the gumminess value decreases as the proteins in chicken meat become damaged. This condition reduces the proteins' ability to bind water, causing moisture loss and creating empty spaces in the protein matrix. The meat's pH decreases when storage continues and activates the cathepsin enzyme. This enzyme breaks down the muscle fibers, weakening their bonds and reducing the meat's cohesiveness and gumminess. The weakening of the protein structure not only affects the texture but also impacts the overall quality and chewiness of the meat. Managing these biochemical changes during storage is essential to preserving the texture and ensuring the meat retains its desirable qualities for extended periods.

Chewiness is influenced by gumminess. The greater the gumminess value or the denser the matrix structure of the meat, the greater the chewiness value or chewing power required to crush the meat [25]. The chewiness value decreased because the gumminess content also decreased. Furthermore, resilience is the ability of a tissue structure to form layers of hydration and protein to maintain tissue structure when pressure is applied. The decrease in resilience value was caused by protein decomposition due to the activity of microorganisms, which damaged the tissue structure of the meat.

3.4. Drip loss

The results revealed that storage time significantly impacted the drip loss of chicken meat stored at 4°C ($P<0.05$). As shown in Table 4, the drip loss value increased steadily from day 0 to day 8. This rise in drip loss indicates a loss of water that was initially bound within the meat's structure. During storage, water and water-soluble substances are released, leading to increased fluid loss from the meat. This phenomenon occurs because drip loss is closely tied to the water-holding capacity of chicken meat. As storage progresses, the meat's ability to retain water diminishes, possibly due to changes in the protein structure and a reduction in the pH, which negatively affects its water-holding properties. Managing drip loss is crucial in preserving the quality and juiciness of the meat, as excessive loss can lead to drier texture and reduced product weight, impacting consumer satisfaction and overall product value [26].

Table 4. Drip loss value of raw chicken meat during storage at 4°C

Storage time (day)	Drip loss (%)
0	$0 \pm 0,00^{a}$
2	$1,9 \pm 1,06^{ab}$
4	$2,79 \pm 1,29^{bc}$
6	$3,82 \pm 1,61^{bc}$
8	$4,41 \pm 1,65^{c}$

Note: Values with different letter superscripts represent a statistically significant difference ($P<0.05$)

Drip loss refers to the reddish liquid that leaks from the surface of meat after it has thawed from a frozen state. This liquid loss is influenced by various factors, including the composition of the muscle, pH level, the sex and age of the animal, storage duration, and thawing process [20]. The increase in drip loss is primarily caused by a reduction in the meat's water-holding capacity, which is closely related to its pH level. As the meat's pH decreases, its protein structure loosens, causing it to lose its ability to retain water. This decrease in water-holding capacity becomes more pronounced as the pH value drops and approaches the meat's isoelectric point, where proteins hold the least amount of water. When this happens, the muscle fibers can no longer retain moisture effectively, leading to increased drip loss. This loss not only affects the texture and juiciness of the meat but can also lead to a decrease in weight and quality.

3.5. Odor

The odor of chicken meat, evaluated using the hedonic quality test by 20 panelists, showed a significant decrease during storage at 4°C from day 0 to day 8 ($P<0.05$). This decline in aroma is attributed to microbial activity and chemical reactions. Microorganisms can contaminate chicken meat during handling and storage, growing on the surface and within the meat. Under anaerobic conditions, these microorganisms break down proteins, forming volatile compounds such as H_2S from sulfur-containing amino acids, mercaptans, indole, ammonia, and amines [21]. Bhawana et al. [27] explained that psychrotrophic bacteria produce large amounts of volatile compounds, including amines, mercaptans, and H_2S, during storage at cold temperatures, directly impacting the decline in meat aroma quality.

Table 5. Odor hedonic quality value of raw chicken meat during storage at 4°C

Storage time (day)	Hedonic quality value
0	$4,32 \pm 0,19^{d}$
2	$3,68 \pm 0,33^{d}$
4	$2,87 \pm 0,33^{c}$
6	$1,78 \pm 0,37^{b}$
8	$1,23 \pm 0,25^{a}$

Note: Values with different letter superscripts represent a statistically significant difference ($P<0.05$)

Lipid oxidation in chicken meat during storage contributes to the decline in aroma quality. This process involves the oxidation of triglycerides and phospholipids, producing aldehydes, ketones, and free fatty acids, which give the meat an undesirable rancid odor. Bhawana et al. [27] noted that lipid oxidation, triggered by oxygen and lipase enzymes, exacerbates the loss of sensory quality. Additionally, Katiyo et al. [22] highlighted that changes in fatty acid composition due to phospholipid hydrolysis accelerate lipid oxidation, further impacting aroma quality

4. Conclusion

The storage duration significantly affects the characteristics of raw chicken meat stored at a low temperature (4°C). Prolonged storage can decrease pH value, water content loss, and changes in color, odor, and texture. According to research findings, raw chicken meat stored at 4°C maintains its quality for less than four days.

5. References

[1] Central Statistical Agency 2024 *Average Per Capita Consumption of Several Important Food Ingredients Per Week* (Jakarta: BPS).
[2] Purwandoko PB, Kuala SI, Susanti ND, Apriyanto IF, Novianti F and Tribowo RI 2023 *BIO Web Conf.* **69** 03018.
[3] Cortez-Vega WR, Pizato S and Prentice C 2012 *J Food Saf* **32** 360–8.

[4] Zhang M, Zhu L, Zhang Y, Mao Y, Zhang M, Dong P, Niu L, Luo X and Liang R 2019 *Asian-Australas J Anim Sci* **32** 701–10.
[5] Hailemariam A, Esatu W, Abegaz S, Urge M, Assefa G and Dessie T 2022 *Applied Food Research* **2** 100233.
[6] Purwandoko PB, Seminar KB, Sutrisno and Sugiyanta 2019 *Information* **10**(10) 288.
[7] Purwandoko PB, Seminar KB, Sutrisno and Sugiyanta 2018 *IOP Conf Ser Earth Environ Sci* **147** 012044.
[8] Ab Aziz MF, Hayat MN, Kaka U, Kamarulzaman NH and Sazili AQ 2020 *Foods* **9**(6) 741.
[9] Purwandoko PB, Solahudin M, Novianti F, Mayasti NKI, Rahman N, Susanti ND, Indriati A, Gultom N, Rahmawati L and Andriansyah RCE 2024 *IOP Conf Ser Earth Environ Sci* **1338** 012053.
[10] Susanty A, Sutrimo WHWM and Saptadi S 2021 *2021 IEEE 8th International Conference on Industrial Engineering and Applications (ICIEA)* pp 226–30.
[11] Purwandoko PB, Seminar K, Sutrisno and Sugiyanta 2022 *Pangan* **31** 1–12.
[12] Purwandoko PB, Seminar K B, Sutrisno, Sugiyanta, Litaay C, Sriharti, Triyono A and Mayasti NKI 2023 *IOP Conf Ser Earth Environ Sci* **1168** 012050.
[13] Dourou D, Spyrelli ED, Doulgeraki AI, Argyri AA, Grounta A, Nychas GJE, Chorianopoulos N G and Tassou C C 2021 *Foods* **10** (4) 765.
[14] Sangadji I, Jurianto and Rijal M 2019 *Jurnal Biology Science & Education* **8** 47–58.
[15] Aguirre ME, Owens CM, Miller RK and Alvarado CZ 2018 *Poult Sci* **97** 1456–61.
[16] Indiarto R, Nurhadi B and Subroto E 2012 *Jurnal Teknologi Hasil Pertanian* **5** 106–16.
[17] Yunanda AW, Razali, Gani FA, Iskandar CD, Ismail and Akmal M 2020 *Jurnal Ilmiah Mahasiswa Veteriner (JIMVET)* **4** 87–95.
[18] Sagita D, Mardjan SS, Suparlan, Purwandoko PB and Widodo S 2024 *Journal of Food Composition and Analysis* **133** 106478.
[19] Kaewthong P, Pomponio L, Carrascal JR, Knøchel S, Wattanachant S and Karlsson AH 2019 *Journal of Poultry Science* **56** 308–17.
[20] Mir N A, Rafiq A, Kumar F, Singh V and Shukla V *J Food Sci Technol* **54** 2997–3009.
[21] Erkmen O and Faruk Bozoglu T 2016 *Spoilage of Meat and Meat Products* vol 1 (New York: John Wiley & Sons) pp 279–95.
[22] Katiyo W, de Kock HL, Coorey R and Buys EM 2020 *LWT* **128** 109468.
[23] Irfan, Lubis YM, Ryan M, Yunita D and Lahmer RA 2023 *Indonesian Journal of Halal Research* **5** 1–11.
[24] Miwada INS, Sukada IK and Ariana INT 2022 *Majalah Ilmiah Peternakan* **25** 135–40.
[25] Prijambodo OM, Trisnawati CY and Sutedja AM 2014 *Jurnal Teknologi Pangan dan Gizi* **13** 6–11.
[26] Shen S, Jiang Y, Liu X, Luo Y and Gao L 2015 *J Food Sci Technol* **52** 5204–11.
[27] Bhawana I, Malik A, Raposo A, Singh S, Yadav S, Zandonadi RP, Lho LH, Han H and Thakur N 2023 *Front Nutr* **10** 1184005.

An innovation of high protein biscuit incorporation of jackfruit seed flour, millet seed, and eel for stunting prevention

P R Ratri [1*], Y Yuanta [1], S D Puspita [1], Rusdiarti [1]
[1] Health Departement, Politeknik Negeri Jember, Jalan Mastrip PO BOX 164, Jember 68121, Indonesia.

* email: putri_ratri@polije.ac.id

Abstract. Stunting is still a focus of attention for the Indonesian government. Indonesia is rich in local food which can be developed into healthy supplementary food (PMT). This study aimed to develop a high protein biscuit as a supplementary feeding formula (PMT) with higher quality and highly nutritious. The method of this research is an experimental laboratory with five distinct formulations for comparative analysis. This research analyzed the organoleptic, and chemical quality of biscuits. The chemical analysis was carried out using the proximate test and the results are compared to SNI 01-7111.2-2005 in a quantitative descriptive manner. Five levels of treatment consisting of the main ingredients: jackfruit seed flour, millet seed, and eel with the composition were graded. The additional ingredients are sugar, butter, egg yolks, milk powder, and vanilla. The organoleptic test showed the best formulation is F4 with the highest recapitulation value in aroma (90%), flavor (90%), and texture (95%). The proximate analysis reveals the F4 formulation also showed a higher value of protein content (12,44% ± 0,01) than other formulas. The conclusion of this study is the biscuits made from jackfruit seed flour, millet seed, and eel have the potential to be developed into high-protein biscuits.

1. Introduction

Stunting is still a focus and attention of the Indonesian government because the problems caused by stunting cases that cannot be overcome are related to the emergence of various diseases and are related to children's intelligence so that they can reduce the quality of the nation's future human resources. The government still uses stunting reduction rates as an indicator of health development targets contained in the 2020 to 2024 RPJMN with a target reduction rate of 14%. According to data from the 2021 Indonesian Toddler Nutrition Status Survey (SSGI), stunting in Indonesia reached 24.4%. This number exceeds the limit set by WHO, which is 20%. [1][2]

There are multifactorial risk factor for stunting, with chronic malnutrition in the first 1000 days of life being the major cause. Stunting in children can be influenced by the child's unresponsive feeding. Providing additional food greatly influences the growth of toddlers [3]. To meet the nutritional status of toddlers, one solution is to develop a formula for giving supplementary food (PMT) to toddlers that is of higher quality and high nutrition to solve the problem of stunting. Supplementary food (PMT) that can be developed into a functional food product is healthy biscuits which are made using local food ingredients that are rich in nutrients that are beneficial for health.ss

Biscuits are food that comes from soft dough, has a high fat content, is crunchy, and when it's broken the cross-section of the piece has a less dense texture. The characteristic of biscuits is that they have a high sugar and fat content and low water content (less than 5%) so they have a crunchy texture [4]. Biscuits are a type of dry food product that is popular on the market so they can be chosen to be used as a functional food product because the product selection pattern for PMT needs to pay attention to aspects of taste, practicality, shelf life, ease of serving, ease of obtaining and is well known in society [5].

The main ingredient for making biscuits, namely wheat flour, can be modified with other ingredients that have more nutritional value. This research developed a formulation for making functional biscuits by utilizing local food sources of jackfruit seed flour, millet seed, and eel as a source of fulfillment of macro and micronutrients. Seeds of jackfruit are abundant in macronutrients such as carbohydrates, proteins, and dietary fiber. They also contain essential micronutrients including vitamins (B-complex vitamins, vitamin A, and vitamin C) and minerals (potassium, magnesium, and zinc). Jackfruit seed flour has many advantages compared to other flour, such as the protein content is 4.2%. The high protein content in the seeds makes them a valuable plant-based protein source, contributing to our health. [6]

Apart from jackfruit seeds, the biscuits formulated in this research also contain millet seeds. Millet is a source of carbohydrates and can provide a high energy. The nutritional content of millet (g/100g) is, protein 11,6-11,8 grams, fat 4,8-5 grams, and carbohydrates 67 grams. Furthermore, millet also contains minerals, B vitamins, especially niacin, B6, and folacin, and essential amino acids such as isoleucine, leucine, phenylalanine, and threonine. [7] The source of animal protein in the biscuits that we formulate is eel. The results of research by Sandita et al. (2015) reported that the fat contained in eel meat is entirely hexadecanoic acid or palmitic acid [8]. According to Supriyanti et al. (2015), the fat content of eels is a source of polyunsaturated fatty acids (PUFA) besides salmon fat. The EPA and DHA content in rice eels is high (7.16%). Based on the explanation above, this research aims to create innovation in the form of biscuits that have rich nutritional content originating from local foods such as jackfruit seeds, millet seeds, and eels [9].

2. Methods

2.1. Materials

The materials are ingredient scales, blenders, flour grinders, containers, filters, bottles, spoons, knives, baking paper, and ovens.

2.2. Methods

This type of research is experimental laboratory research with five distinct formulation for comparative analysis. The formula development was then analyzed for nutritional content and compared according to SNI 01-7111.2-2005 in a quantitative descriptive manner. The study used five levels of treatment with the main ingredients: oat flour, jackfruit seed flour, millet seed and eel. Meanwhile, other additional ingredients are sugar, butter, egg yolks, milk powder and vanilla.

The making of the biscuit refers to a modification of the method of Sinuraya et al. (2024). [10] Powdered milk, sugar, butter, vanilla, and egg yolks are mixed with a mixer at high speed for 15 minutes. Jackfruit seed flour, milled seed flour, and eel flour are added according to the formulation. Wheat flour is added gradually as much as the dough while stirring until smooth. The dough is flattened, molded with a mold, placed on a baking sheet lined with baking paper, then baked in the oven at 100°C for 60 minutes. Biscuits are placed at room temperature for 15 minutes. The ratio of main ingredients for each biscuit formulation can be seen in Table 1. In the other hand, the additional ingredients is showed in Table 2.

Table 1. The main ingredients for each formulation

Ingredients	F1	F2	F3	F4	F5
Jackfruit seed flour (g)	0	30	50	55	100
Millet seed flour (g)	0	5	10	15	0
Eel (g)	0	15	20	25	0
Oat Flour (g)	100	50	20	5	0
Total Ratio	100	100	100	100	100

F1 = Formula 1; F2 = Formula 2; F3 = Formula 3; F4 = Formula 4; F5 = Formula 5

Table 2. The additional composition biscuit for each formulation

Ingredients	F1	F2	F3	F4	F5
Sugar (g)	25	25	25	25	25
Butter (g)	40	40	40	40	40
Egg Yolk (g)	30	30	30	30	30
Milk powder (g)	7	7	7	7	7
Vanilla essence (g)	2	2	2	2	2

F1 = Formula 1; F2 = Formula 2; F3 = Formula 3; F4 = Formula 4; F5 = Formula 5

This research analyzed the organoleptic, and chemical quality of biscuits. The five biscuit formulas were tested organoleptically to find the most liked formula. The organoleptic assessment involved hedonic tests where 20 semi-trained panelists evaluatedthe color, aroma, taste, and texture of the biscuits. The organoleptic test was conducted at Laboratorium Pendidikan Gizi, Politeknik Negeri Jember. Moreover, the five products were chemically tested for their nutritional content in the laboratory through a proximate analysis. The nutritional content tests analyzed are the carbohydrate content, fat content, protein content, water content, ash content, and crude fiber content. The proximate test was carried out at the food analysis laboratory of Jember State Polytechnic and the Integrated Laboratory of Jember University.

The carbohydrate analysis uses the SNI 01-2891-1992 method, protein analysis uses the IK 2.7.2.10 test method, fat analysis uses the SNI 01-2891-1992 point 8.1 test method, water content analysis uses the IK 2.7.2.3 method, content analysis ash using the SNI 3751:2009 point A6 test, crude fiber analysis using the IK test. 2.7.2.10. The nutritional content value data was analyzed descriptively by comparing results with the complementary food standard for toddler (SNI MPASI) SNI 01-7111.2-2005.

3. Results and Discussion

Organoleptic Test

In this research, five formulations have been successfully produced with different compositions of each ingredient (jackfruit seed flour, oat flour, millet deed flour, and eel). The composition is shown in Table 1. Biscuits are a mixture of flour and water added to dough with other ingredients and baked to a crispy texture [11]. Jackfruit seed flour contains high starch (40-50%), so it can replace the basic ingredient of flour for making biscuits [12]. Based on visual observations, the biscuits produced with the P5 formula showed the brightest color of other biscuit formulas. The effect of proteins that combine with sugar/starch in a hot atmosphere will cause the color to darken [13]. The color of cookies was also influenced by baking in the oven due to high temperatures and the Maillard reaction when protein and sugar meet, resulting in a characteristic brownish color [14]. A visualization of the five biscuit formulations is shown in Figure 1.

Figure 1. Five formulation biscuit (Left to Right: P1 (Formula 1), P2 (Formula 2), P3 (Formula 3), P4 (Formula 4), and P5 (Formula 5))

The selected formula was determined based on the recapitulation result of the hedonic test of biscuit. The hedonic test is describing as total percentage of hedonic value of the product for kinda like, like, and really like. The hedonic result is described in table 3 below.

Table 3. Percentage of Hedonic Value for Each Group Formulation

Attribute	Score	Percentage Of Hedonic Value (%)				
		F1	F2	F3	F4	F5
Color	Kinda Like	15	15	45	20	15
	Like	10	15	30	35	25
	Really Like	0	5	0	10	35
	Total	25	35	**75**	65	**75**
Aroma	Kinda Like	35	45	35	15	5
	Like	30	30	50	45	25
	Really Like	0	0	0	30	40
	Total	65	75	85	**90**	70
Flavor	Kinda Like	20	30	35	30	15
	Like	10	20	45	35	30
	Really Like	5	5	5	25	20
	Total	35	55	85	**90**	65
Texture	Kinda Like	25	25	50	10	20
	Like	0	10	30	60	35
	Really Like	0	0	5	25	30
	Total	25	35	85	**95**	85

F1 = Formula 1; F2 = Formula 2; F3 = Formula 3; F4 = Formula 4; F5 = Formula 5

Table 3 shows organoleptic attribute for each group formulation. The result showed the highest percentage of hedonic values obtained by F4 (formula 4) for aroma (90%), flavor (90%), and texture (95%) attributes. The F4 formula contains 55 % jackfruit seed flour, 15% millet seed, 25% eel, and 5% oat flour. On the other hand, the highest value of color attribute was performed in F3 and F5 (75%). This aligns with several studies that make food products by substituting jackfruit seed flour. The panellists preferred food products that used jackfruit seed flour as a substitute. The more jackfruit seed flour was added, the more the panellists' assessment of the organoleptic value increased. [15] [16]

Formula F4 contains the most eels and millet seeds. This affects the balance of aroma, taste, and texture. The F4 group with the highest eel composition contributes to the protein content in the biscuits. This can strengthen the aroma that appears. The higher protein content of the material results in a stronger aroma from the Maillard reaction [17]. In addition, the Maillard reaction produces volatile compounds with the potential to form aromas [18]. Apart from that, the protein content can affect the taste, making the biscuits savorier.

Chemical Analysis

Supplementary food (PMT) for toddlers is food besides the main meal for toddlers that is made from local food ingredients with appropriate nutritional content to be given to toddlers aged 6 – 59 months to meet nutritional needs. Providing additional food for toddlers is expected and can be applied to providing daily food on the family level. [19] The nutritional analysis for the carbohydrate, protein, fat, water content, ash content, and crude fiber content are as follows:

Table 4. Nutritional content of biscuits in each group formulation

No.	Component	F1	F2	F3	F4	F5	SNI standard
1	Carbohydrate (%)	61,11 ± 0,08	62,93 ± 0,03	64,78 ± 0,04	66,66 ± 0,08	69,12 ± 0,04	Max 30
2	Protein (%)	10, 75 ± 0,01	11,26 ± 0,06	12,19 ± 0,03	12,44 ± 0,01	8,58 ± 0,02	Min 6*
3	Fat (%)	22,62 ± 0,22	21,94 ± 0,10	25,11 ± 0,33	22,75 ± 0,37	28,53 ± 0,34	Max 18
4	Water content (%)	7,49 ± 0,02	8,32 ± 0,03	6,89 ± 0,00	9,88 ± 0,05	7,30 ± 0,02	Max 5
5	Ash content (%)	1,90 ± 0,02	2,03 ± 0,02	2,19 ± 0,01	2,18 ± 0,03	2,14 ± 0,01	Max 3,5*
6	Crude Fiber (%)	5,81 ± 0,02	5,72 ± 0,17	6,11 ± 0,14	5,37 ± 0,08	4,7 ± 0,05	Max 5

Source: SNI 01-7111.2-2005
F1 = Formula 1; F2 = Formula 2; F3 = Formula 3; F4 = Formula 4; F5 = Formula 5
*Meet the SNI standard

Based on laboratory tests, it is known that the carbohydrate content of the five formulations is still higher than the SNI standard for PMT biscuits which is 30 g/100g (30%). Carbohydrates are the main source of calories for the body. Carbohydrates also have an important role to the food ingredients' characteristic taste, color, and texture [13]. Based on laboratory test results, it is known that the carbohydrate content of the F5 formulation has a higher carbohydrate content than the other groups. This result is related to jackfruit seed composition. F5 contains 100% jackfruit seed flour. The jackfruit seeds have a rich carbohydrate content, reaching 76.1% [6].

On the other hand, the protein content of the five formulations meets SNI standards, with the highest protein content in the F4 group (12,44 ± 0,01). This is because group P4 has more eel composition compared with other groups. The protein content in eels is sufficient to meet the Recommended Dietary Allowances (RDA) of Indonesian people (57 grams/day). Candra & Rahmawati (2018), state that the protein content of fresh eel is 18.49 grams/100 grams, its higher than egg protein [20]. Insufficient protein intake plays a role in children's growth so that the risk of stunting increases. Several studies show that the protein intake of stunted children is 13 grams lower than that of normal children [21].

Similar to the carbohydrate content, the fat content in the five formulas also does not meet the SNI standards yet. The fat content in each group was still relatively high (> 6%). The composition of additional ingredients containing fat such as butter, milk, and eggs can be recalculated for a more precise formulation so that the fat content can meet standards according to SNI.

Besides macronutrient analysis, proximate analysis (water content, ash content, and crude fiber) of five biscuit formulations was analyzed. Table 4 show the ash content of the five groups meets SNI standards, while the water and crude fiber content is still above SNI standards. The water and fiber content in oat flour and jackfruit seed flour are quite high. Therefore, certain cooking process needs to be reconsidered to reduce their value. Food processing at high temperatures and for a certain time can reduce the water and fiber content in food. The cooking process using heat can cause physicochemical, organoleptic, and nutritional changes in food ingredients. [22] [23]. The baking process was 60 minutes at a temperature of 100°C.

Even though the chemical analysis of the five product formulations does not meet the SNI standards yet in carbohydrate, fat, and water content, the biscuits made from jackfruit seed flour, millet seed, and eel have the potential to be developed into high-protein biscuits. Therefore, further research is needed to formulate PMT biscuits made of jackfruit seed flour, millet deed, and eel with the right ingredient composition.

4. Conclusion

Based on the organoleptic test, the best formulation in this research is F4 in aroma (90%), flavor (90%), and. Texture (95%). The F4 formulation also showed the highest value of protein content (12,44% ± 0,01). Even though the chemical analysis of the five product formulations does not meet the SNI standards yet in carbohydrate, fat, and water content; the biscuits made from jackfruit seed flour, millet seed, and eel have the potential to be developed into high-protein biscuits. Therefore, further research is needed to formulate PMT biscuits with the right ingredient composition.

Acknowledgement

This study is supported by a PNBP grant 2024 with the contract number SP DIPA-023.18.2.677607/2024 Politeknik Negeri Jember.

References

[1] Ministry of Health of the Republic of Indonesia. 2018. Laporan Nasional Riskesdas (2018) Jakarta: Kementerian Kesehatan RI.
[2] Yuliani F,& A. Sadad. 2024. Factors Inhibiting Implementation of Disaster Management Policies (Case Study of Stunting Management in Rokan Hilir Regency). in IAPA 2023 Annual International Conference, KnE Social Sciences, pages 333–341
[3] Irwan, T. Mery, Kadir S. Al. 2020. Efektivitas Pemberian Pmt Modif Berbasis Kearifan Lokal Terhadap Peningkatan Status Gizi Balita Gizi Kurang Dan Stunting. J Heal Sci.;4: 59–67.
[4] Mayasari R. 2015. Kajian Karakteristik Biskuit Yang Dipengaruhi Perbandingan Tepung Ubi Jalar (*Ipomea Batatas* L.) Dan Tepung Kacang Merah (*Phaseolus Vulgaris* L.). Universitas Pasundan.
[5] Setyowati Wt, Nisa Fc. 2014. Formulasi Biskuit Tinggi Serat (Kajian Proporsi Bekatul Jagung: Tepung Terigu Dan Penambahan Baking Powder). J Pangan Dan Agroindustri. 2(3):224–31.
[6] Shajahan M.S., Susmita D, Arti S, Ab W.W., Neha R., Sanjeev K., and Meraj A. 2024. Jackfruit seed: A Fuctional Food Companion. *African Journal of Biological Science.* 6(12): 2330 - 2351.
[7] Amadou I, Gounga ME, Guo-Wei L. 2013. Millets: Nutritional composition, some health benefits and processing—a Review. *Emir J Food Agric.* 25(7):501–508. doi: 10.9755/ejfa.v25i7.12045.
[8] Sandita, A., Maulana, I.T., & Syafnir, L. 2015. Perbandingan komposisi asam lemak antara minyak belut (*Monopterus albus*) dan minyak sidat (*Anguilla sp.*) dengan metode KG-SM. Prosiding Penelitian Sivitas Akademika Unisba: 388-396.
[9] Supriyanti, F.M.T., Herwiandani D.P., & Kusrijadi, A. 2015. Mikroenkapsulat minyak belut (*Monopterus albus*) beromega-3 sebagai fortifikan keju cottage. Chimica et Natura Acta3(2): 70-75.
[10] Sinuraya, T.U., Pranata, F.S., Swasti, Y.R. 2024. Kualitas Biskuit Tepung Uwi Ungu (*Dioscorea alata*) dan Tempe Kacang Merah (*Phaseolus vulgaris L*). AGRITEKNO: Jurnal Teknologi Pertanian.
[11] Hindah, M. 2007. Healthy Cooking Biscuit. Gramedia Pustaka Utama: Jakarta.
[12] Winarti, S., & Purnomo, Y. 2006. Olahan Biji Buah. Trubus Agrisarana: Surabaya.
[13] Winarno FG. 2004. Food and Nutrition Chemistry. Jakarta: Gramedia.

[14] Hendrayati, Suaib F, Tamrin A, Yani N. 2022. Daya Terima dan Kandungan Protein serta Zat Besi Pada Cookies dengan Substitusi Tepung Jewawut (*Setaria italica*) dan Tepung Ikan Kembung (Rastrelliger kanagurtaL.). Media Gizi Pangan. 29(2):1-9.

[15] Yosua F, S., Evawany Y, A., dan Ernawati, N. 20015. Pemamfaatan Tepung Biji Nangka Untuk Pembuatan Kue Nastar Dan Nilai Gizinya, Skripsi, Fakultas Kesehatan Masyarakat, Universitas Sumatera Utara, Medan

[16] Sofian, D.P.F., Ansharullah, and Mariani L. 2023. Pengaruh subtitusi tepung biji nangka (*Artocarpus heterophyllus*) terhadap karakteristik organoleptic dan nilai gizi brownies panggang. Jurnal Riset Pangan, 1(2): 107 – 116.

[17] Oktaviana AS, Hersoelistyorini W, Nurhidajah. Kadar Protein, Daya Kembang, dan Organoleptik Cookies dengan Substitusi Tepung Mocaf dan Tepung Pisang Kepok. Jurnal Pangan dan Gizi. 2017;7(2):72-81. https://doi.org/10.26714/jpg.7.2.2017.72-81.

[18] Estiasih T, Harijono, Waziiroh E, Fibrianto K. 2016. Kimia dan Fisik Pangan. Jakarta: Bumi Aksara.

[19] Ministry of Health of the Republic of Indonesia. 2011. Panduan Penyelenggaraan Pemberian Makanan Tambahan: Pemulihan Bagi Balita Gizi Kurang. Kemenkes RI.

[20] Candra, C., & Rahmawati, H. 2018. Peningkatan Kandungan Protein Mie Basah Dengan Penambahan Daging Ikan Belut (Monopterus albus Zuieuw). Jukung (Jurnal Teknik Lingkungan), 4(1), 82–86. https://doi.org/10.20527/jukung.v4i1.4665

[21] Fikawati S, Syafiq A, Ririyanti RK, Gemily SC. 2021. Energy and proteinintakes are associated with stunting among preschool children in Central Jakarta, Indonesia: a case-control study. Malays J Nutr. 2021;27(1).

[22] Muchtadi, D. 2001. Sayuran sebagai sumber serat pangan untuk mencegah timbulnya penyakit degeneratif. Jurnal Teknologi dan Industri Pangan, XII(1): 61-71.

[23] Nilasari, O.W., Susanto W.H., and Maligan J.M. 2017. Pengaruh suhu dan lama pemasakan terhadap karateristik lempok labu kuning (waluh). Jurnal Pangan dan Agroindustri. 5(3): 15 – 26.

IOP Conf. Series: Earth and Environmental Science **1446** (2025) 012013 doi:10.1088/1755-1315/1446/1/012013

Ergonomic Evaluation of a Drum Dryer Machine Using the RULA (Rapid Upper Limb Assessment) Method: Case Study Instant Sweet Potato Flour Production

N K I Mayasti[1], F Novianti[1*], H Septiyansyah[2], D D Hidayat[1], A Indriati[1], M Furqon[1], A Herminiati[1], T Rahman[1], R C E Andriansyah[3], and A Rahayuningtyas[1]

[1]Research Center for Appropriate Technology, National Research and Innovation Agency, Subang, West Java, Indonesia
[2]Industrial Engineering Study Program, Bandung University of Technology, Bandung, West Java, Indonesia
[3]Research Center for Food Technology and Processing, National Research and Innovation Agency, Yogyakarta, Indonesia

[*]Email : fith002@brin.go.id

Abstract. The instant sweet potato flour production process using a drum dryer machine consists of 3 operators with different tasks and unnatural working postures. The working principle of the drum dryer machine is to dry the instant flour dough with a double cylinder equipped with a heater so that dry dough sheets are produced. Operator 1 is tasked with inserting liquid dough into the top of the cylinder, and operators 2 and 3 take dry dough sheets from the front and back. This research evaluates ergonomics by analyzing working posture when operating a drum dryer machine to make instant flour. The method used is RULA (Rapid Upper Limb Assessment). The research results show that the working posture of operators 1 and 2 has a final score of 7 with an ergonomic level of 4. Operator 3 has a score of 6 with an ergonomic level of 3. Ergonomic risk level 4 indicates that the working posture of operators 1 and 2 has a high risk and requires immediate improvement to minimize the risk of injury to MSDs.

1. Introduction

Developments in people's lifestyles prefer to consume food that is healthy, nutritious, and practical in presentation. Instant sweet potato flour is one of the raw materials that can be processed into various processed food products. Even just adding hot water or milk can become instant porridge that is practical to serve and has a soft texture that is easy to digest [1]. The technology for making instant flour can use several drying methods, including vacuum, cabinet, oven, and drum dryers. Rif'an et al. (2017) reported that drying pumpkin soup was best using a drum dryer compared to other drying methods based on the product's physicochemical characteristics [2].

The material dried with a drum dryer is in the form of liquid, slurry, or puree. The slurry is poured onto the surface of a hot rotating drum so that it dries and forms a thin, dry layer. The knife will erode this thin layer, forming product fragments [3]. Pua et al. (2010) researched optimizing jackfruit flour production's drum dryer drying process [4]. Rodiawati et al. (2017) researched the manufacture of instant flour and porridge from *gayong* using a double drum dryer. Research with drum dryers focuses more on product formulation and optimization [5]. Research regarding operator work comfort during the drum dryer production process has yet to be widely studied. Work comfort is closely related to the working posture of the operator. Working postures that are not ergonomic will cause operators to get tired more quickly and, in the long term, will cause complaints of pain in body parts, product defects, and even physical disabilities. Work fatigue affects workers' productivity [6].

The instant flour production process with a drum dryer requires the operator to put the liquid material onto the layer of the drying drum and the operator to take the thin layer that has dried on top of the drum cylinder. Based on data collection on work postures during the production of instant sweet potato flour using a drum dyer, there are work postures that are considered less ergonomic, including standing and bending work postures when picking up dry flour, the position of the spine and hands rotating for long periods when putting the material into the cylinder.

This research analyses the operator's working posture when operating a drum dryer machine. Repetitive work done in less effective time can increase the risk of musculoskeletal disorders (MSDs), especially if the work is done with an inappropriate workload and posture [7,8]. Several types of MSDs complaints felt by workers in their body parts can be identified using the Nordic Body Map (NBM) questionnaire. Through this questionnaire, it can be seen which muscle parts are experiencing complaints [9]. According to Santoso et al. (2014), the Nordic Body Map method can be used to find out more details about the parts of the body that experience disorders or pain while working. However, it is subjective. This questionnaire has been standardized and is valid for use [10]. Priangkoso et al. (2024) stated that the measurement of operator musculoskeletal complaints could use the NBM method and analyze the risk of work posture [11].

Assessment of work posture can use several methods, including Rapid Upper Limb Assessment (RULA) [11,12, 13], Rapid Entire Body Assessment (REBA) [13,14,15,16,17], Ovako Working Posture Analysis System (OWAS) [18,19]. Rapid Upper Limb Assessment (RULA) is used in this research. This method was chosen because assessing the working posture for operating a drum dryer machine is dominant in the upper body posture. Using this method, one can determine the musculoskeletal load associated with a profession that poses dangers to the body area, including the upper limbs, neck, and stomach. The neck, back, and upper arm postures can be evaluated without requiring specialized equipment. Every move has a preset score assigned to it. RULA was created to identify risk factors in the work posture. The technique is intended to evaluate the working operator and identify the musculoskeletal burden that may result in issues with the upper limbs. This research aims to analyze the musculoskeletal complaints of drum dryer operators in producing instant sweet potato flour using the NBM and RULA methods. The results of this work procedures analysis will be used to evaluate drum dryer machine design using an ergonomics approach.

2. Methods

This quantitative and descriptive research uses direct observation techniques to examine the pilot-scale instant flour production process using a drum dryer. This study began with an assessment of work complaints (Nordic Body Map) with a predetermined Likert scale, starting from no pain (1), somewhat pain (2), pain (3) and very pain (4) [20]. The next stage is an ergonomic evaluation, which analyzes the work posture (RULA method) of the drum dryer's work activities carried out by the three operators. The research stages of work posture research with RULA are shown in Figure 1.

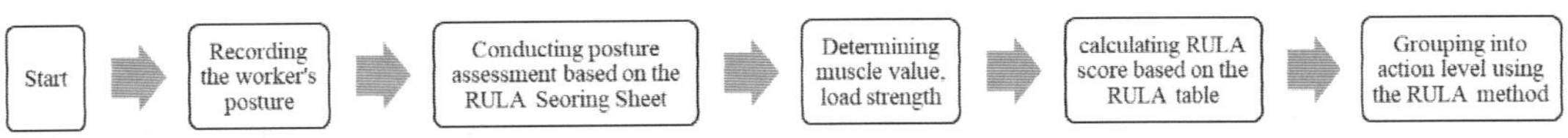

Figure 1. Production Process Flow Chart [15] (Modified)

Data is collected by recording operator activities and determining the angle of the operator's body parts. The data analysis method used in this research is the RULA (Rapid Upper Limb Assessment) method. This method was chosen because it can systematically assess the operator's body posture in static and dynamic work. The RULA method analyses the posture and movement of an activity, investigating the level of risk generated by the activity [13].

3. Results and discussion

3.1. Instant sweet potato flour production process with drum dryer on a pilot scale
The production process of instant sweet potato flour begins with washing to clean the remaining dirt and dirt that sticks until the thin skin peels off. Before steaming, the sweet potato is cut into smaller pieces and steamed for approximately 40 minutes with a steamer pan. The next stage is grinding with a grinding machine. The last stage is drying the sweet potato with a double drum dryer (temperature 120-124° C, pressure 2 bar, and rotation speed 13.6 rpm).

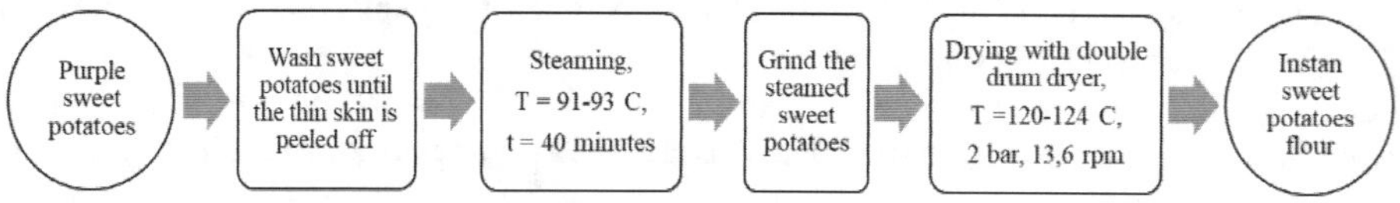

Figure 2. Stages of instant sweet potato flour production using a drum dryer

3.2. Drum dryer specifications for drying instant sweet potato flour

Figure 3. Drum dryer description: (1) Scrapper; (2) Steam pipe; (3) Drum; (4) Drop chute; (5) Panel box; and (6) Stairs.

The drying machine used in this study is a double-drum atmospheric dryer, as shown in Figure 3. The drum is made of Ferro Casting Ductile 450 coated with hard chrome, 400 mm long, 500 mm outer diameter, and 20 mm thick. The surface area of the drum available for drying is 1.26 m2. The drum is driven by 3.75 kW of power. The capacity of the drying drum can reach 10 kg/batch, depending on the raw materials used. Herminiati et al. (2023) researched the manufacture of sweet potato flour with a double-drum dryer. The research results show that the optimal machine setting is a temperature of 120 C and a rotation speed of 0.7 rpm [21].

When the material is fed into the drum/cylinder, both drums rotate, and the incoming material will thin and dry to form a thin-dry layer. The distance between the two drums/cylinders is 0.5 mm. The drum is heated internally with steam at 3-4 bar pressure. The steam to heat the inside of the drum comes from a 9000 W steam boiler. A thin layer of dry instant flour is then scraped off with a scraper knife and falls into the material dispensing hopper.

3.3. Respondent/operator profile
Operator 1 is a male who is younger (23 years) than operators 2 and 3, who are female, namely 56 and 47 years old. Operator 1's job requires more power than operators 2 and 3. Operator 1 must climb up the ladder and sit at the top to pour the material until the material feeding process is complete.

Table 1. Number of respondents

Operator	Gender	Age (years)
Operator 1	L	23
Operator 2	P	56
Operator 3	P	47

IOP Conf. Series: Earth and Environmental Science **1446** (2025) 012013 doi:10.1088/1755-1315/1446/1/012013

3.4. The operator's working posture data

The number of operators working with the drum dryer is three people. Operator 1 inserts the sweet potato paste into the top; operators 2 and 3 take thin sheets of dried instant sweet potato. The working postures of the three operators are not expected. Operator 1 works in a sitting position with his shoulders turned towards the drum (45^0) to insert material. Operators 2 and 3 work in a sitting position with their backs bent to pick up a thin layer of dried sweet potato that has been dried as the final material.

Table 2. Body posture and length of work

	Working posture	Job description
Operator 1		Pour the material onto the drum dryer
Operator 2	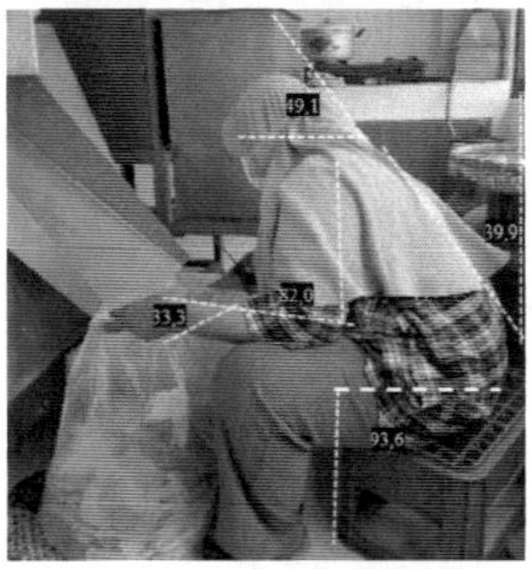	Take a thin layer that has dried (from the rear cylinder side) and pack it in a package
Operator 3		Wrap the dried thin layer (from the front side of the cylinder) in a package

3.5. Assessment of operator 1,2,3 work complaints with Nordic Body Map

Respondents were asked to assess of the part of their body that felt painful during work activities.

Table 3. NBM assessment for operator 1,2,3

No	Body part	Complaint assessment		
		Operator 1	Operator 2	Operator 3
1	Upper neck	3	1	1
2	lower neck	3	1	1
3	left shoulder	1	1	1
4	right shoulder	3	3	3
5	left upper arm	1	1	1
6	back	4	1	1
7	right upper arm	4	3	3
8	waist	4	1	1
9	buttock	4	3	3

No	Body part	Complaint assessment		
		Operator 1	Operator 2	Operator 3
10	bottom	4	1	1
11	left elbow	1	1	1
12	right elbow	1	1	1
13	left lower arm	1	2	2
14	right lower arm	3	1	1
15	left wrist	1	1	1
16	right wrist	3	3	3
17	left hand	1	1	1
18	right hand	1	1	1
19	left thigh	1	1	1
20	right thigh	1	1	1
21	left knee	3	3	2
22	right knee	3	3	2
23	left calf	1	1	1
24	right calf	1	1	1
25	left ankle	1	1	1
26	right ankle	1	1	1
27	left foot	1	1	3
28	right foot	1	1	3
		57	41	43

Based on the total score of the complaint assessment experienced by operator 1,2,3, it can be seen that the highest complaint score for operator 1 is 57. A score of 57 is classified as a moderate risk level (50 – 70), which means corrective action is needed later [22]. The highest NBM score for operator 1 indicates that operator 1 has the highest level of complaints compared to operators 2 and 3. Operator 1 reported that the work activity when feeding the drum dryer material caused painful complaints (score 4) in the right upper arm, back, waist, buttock, and bottom body parts. The NBM assessment scores of operator 2 and operator 3 were 41 and 43, indicating that operator 3's work activity complaints were greater than those of operator 2. Operator 3's work activities were standing, taking a thin layer of dried instant sweet potatoes, and cleaning the drum dryer. Operator 2's work activities were almost the same as operator 3's; the only difference was that his work position was sitting.

The complaints felt by operators 2 and 3 were pain in the right shoulder, right upper arm, buttock, right wrist, left foot, and right foot. Unergonomic work postures such as bending forward, bending to the side, rotating the back and spine, and repetitive standing result in stiffness and acute or chronic back pain [23]. They were working with a continuous work system with little rest time, which resulted in fatigue, aches, and back pain, impacting work comfort and productivity [24]. Baihaqi et al. (2024) also reported similar things and examined the influence of non-ideal work posture and work duration on back pain complaints [25]. The working posture of sweeping on the highway with a working time of 8 hours/day is positively correlated and has a significant impact on causing back pain complaints. Operators 1, 2, and 3's non-ideal working posture with a working duration of 7 hours and a rest period of 1 hour could pose a risk of musculoskeletal injury in the coming period.

3.6. The assessment of working posture with RULA on operator 1

3.6.1 Group A's body posture. Group A scores are obtained based on the assessment results of the sum of the body posture scores for the upper arm, forearm, and wrist. The calculation of group A scores from operator 1 will be shown in Table 4. Group A's work posture score based on the table is 4, with the following value details:

a. The posture of the upper arm, the angle formed is 65.6, so the value = 3
b. Lower arm body posture, the angle formed is 33.8, so the value = 2
c. Wrist twist, the angle formed is 28.4, so the value = 3
d. Wrist twist = 1

The muscle activity score of Group A is 1, based on static posture lasting more than 10 minutes. Because the load is less than 2 kg, the load value is 0. So, the total value of group A is 4 + 1 + 0 = 5.

3.6.2 Group B's body posture. Group B scores are obtained based on the assessment results of the sum of body posture scores for the neck, back, and legs. Table 5 shows the Group B score from operator 1. Group B's work posture score based on the table is 5, with the following value details:

a. Neck body posture, the angle formed is 55.9, so the value = 3
b. Back body posture, the angle formed is 31.1, so the value = 4
c. Body posture of the legs = 1

The muscle activity score of group B is 1. This value is based on static posture lasting more than 10 minutes. Because the load is less than 2 kg, the load value is 0. So, the total value of group B is 5 + 1 + 0 = 6.

Table 4. The working posture of Group A operator 1

Upper arm	Forearm	Wrist 1 PP 1	Wrist 1 PP 2	Wrist 2 PP 1	Wrist 2 PP 2	Wrist 3 PP 1	Wrist 3 PP 2	Wrist 4 PP 1	Wrist 4 PP 2
1	1	1	2	2	2	2	3	3	3
	2	2	2	2	2	3	3	3	3
	3	2	3	3	3	3	3	4	4
2	1	2	3	3	3	3	4	4	4
	2	3	3	3	3	3	4	4	4
	3	3	4	4	4	4	4	5	5
3	1	3	3	4	4	4	4	5	5
	2	3	4	4	4	4	4	5	5
	3	4	4	4	4	4	5	5	5
4	1	4	4	4	4	4	5	5	5
	2	4	4	4	4	4	5	5	5
	3	4	4	4	5	5	5	6	6
5	1	5	5	5	5	5	6	6	7
	2	5	6	6	6	6	7	7	7
	3	6	6	6	7	7	7	7	8
6	1	7	7	7	7	7	8	8	9
	2	8	8	8	8	8	9	9	9
	3	9	9	9	9	9	9	9	9

Table 5. The working posture of Group B operator 1

Neck	Back 1 Foot 1	Back 1 Foot 2	Back 2 Foot 1	Back 2 Foot 2	Back 3 Foot 1	Back 3 Foot 2	Back 4 Foot 1	Back 4 Foot 2	Back 5 Foot 1	Back 5 Foot 2	Back 6 Foot 1	Back 6 Foot 2
1	1	3	2	3	3	4	5	5	6	6	7	7
2	2	3	2	3	4	5	5	5	6	7	7	7
3	3	3	3	4	4	5	5	6	6	7	7	7
4	5	5	5	6	6	7	7	7	7	7	8	8
5	7	7	7	7	7	8	8	8	8	8	8	8
6	8	8	8	8	8	8	8	9	9	9	9	9

3.6.3 Group C operator 1 final score and risk level. The final score is obtained from the combined results of group A (Table 4) and group B (Table 5) score calculations, which are combined into group C (Table 6). The results of group C in Table 6 are added to those from additional activities.

Group C's total score is 7. After the assessment of operator 1's body posture is completed, the assessment of operators 2 and 3 continues. As explained in the previous discussion, the assessment of operators 2 and 3 uses the RULA method. Based on Wijaya and Muhsin (2018), a score of 1-2 has a low level, a score of 3-4 has a medium level, a score of 5-6 has a high-risk level, a score > 7 has a very high-risk level and an action level that requires immediate possible changes [14]. Table 7 summarizes the results of the body posture assessments of operators 1, 2, and 3.

Table 6. The final score operator 1

Group score		B						
		1	2	3	4	5	6	7+
	1	1	2	3	3	4	5	5
A	2	2	2	3	4	4	5	6
	3	3	3	3	4	4	5	6
	4	3	3	3	4	5	6	6
	5	4	4	4	5	6	7	7
	6	4	4	5	6	6	7	7
	7	5	5	6	6	7	7	7
	8	5	5	6	7	7	7	7

Table 7. Recapitulation of work posture assessment results for 3 operators using the RULA method

Operator	Final score	Risk level
Operator 1	7	4
Operator 2	6	3
Operator 3	6	3

Based on the analysis of operator 1's body posture, the calculation results for Group A are 5, with a score of 4 plus the muscle usage and load scores. The calculation for Group B's operator 1's body posture obtained a score of 5 plus the muscle usage and load scores, making the total score 6. So, the total score for Group C operator 1 is 7.

Operator 1 works in the position of the spinal frame of the arm, which rotates when the material is inserted into the cylinder. This work posture has a final score of 7, with a risk level of 4 or a high level of risk, meaning that a condition like this is very dangerous if left for too long. Therefore, it is necessary to check and improve the work system shortly to avoid increasing the risk of musculoskeletal injury. Operator 1 must climb a ladder to get to the top and pour the material to be dried. The ladder is used because the height of the cylinder/drum is 150 cm. Mayasti et al. (2023) reported that the recommended height of the drying machine is 123.55 cm based on the height of the shoulder of the hand above the floor so that the hand can easily reach/put materials into the drying cupboard [26]. The height of the drum dryer machine, which is higher than the recommended dimensions, can cause the working posture to be non-ideal; this is in line with Pegiardi et al. (2017), who said that non-ideal working posture is often caused by the location of facilities that are not following the operator's anthropometry, thus affecting the operator's performance [27]. Work factors, including workload, work duration, work posture, and repetition, influence muscle pain complaints in certain body parts. Age, gender, body mass index (BMI), and length of service are several individual factors. Then, work environment and psychosocial factors include vibration factors [28].

3.7. The assessment of working posture with RULA on operator 2

The results of the assessment of operator 2's work posture using the RULA method obtained a score for Group A in Tabel 3, which has a score of 4 plus 1 for the muscle usage score and 0 for load scores, so the total score for section A is 5. Operator 2's group B body posture scored 4 plus 1 for muscle use and 0 for load scores, making the total score 5. Then, the total score for Group C operator 2 is 6 (see Table 6). The working posture of operator 2, who sits with a chair height too short and has the back and hands reach forward, has a score of 6 with a level risk of 3 (see Table 7). Operator 2's non-ideal working posture is sitting and bending for a long time; this is in line with what Rohman (2014) stated: unnatural work postures, for example, bending long, can cause discomfort and pain in one of the body parts.

3.8. The assessment of working posture with RULA on operator 3

The results of the assessment of operator 3's work posture using the RULA method obtained a score for Group A in Tabel 3, which has a score of 4 plus 1 for muscle usage and 0 for load scores, so the total score for section A is 5. For the body posture of operator 3's Group B, a score of 4 was obtained, plus the score for muscle usage and load, making the total score 5 (see Table 5). Moreover, the final total score for Group C operator 3 based on Table 6 is 6. Operator 3's working posture is standing and bending with the position of the hands reaching forward when picking up the dry instant flour. Based on Table 7, this work posture has a score of 6 with a risk level of 3, which means the risk level is medium; in this condition, the work system needs to be improved immediately.

Operator 3's non-ideal working posture is standing and bending during the production process. According to Pegiardi et al. (2017), working postures that lack safety, such as standing, bending over, or lifting weights without the help of tools or other material handling for a long period of time, can result in complaints of pain in one of the body parts [27].

3.9. Improvement of ergonomic evaluation result on drum dryer machine

From the complaint assessment result with NBM, it is known that the complaint of work activity and work risk score (RULA) of operator 1 is the highest compared to the work activity of operators 2 and 3. Therefore, the improvement made is adding a hopper above the drum to reduce the fatigue of operator 1. Improvement of the drum dryer machine is by adding a hopper or material container equipped with a control system so that it can flow material automatically so that the operator does not need to pour material manually, which takes quite a long time; the container is placed above the heating drum equipped with automatic valve controlled by the control system.

Figure 4. Image of material container: (1) Material container; and (2) Automatic valve.

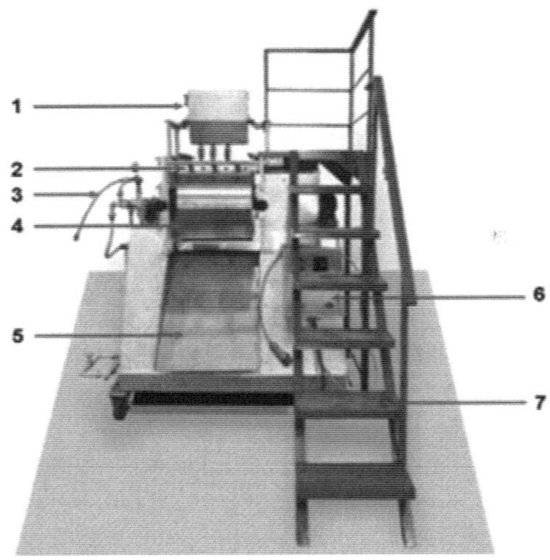

Figure 5. Drum dryer repair description: (1) Scrapper; (2) Steam pipe; (3) Drum; (4) Drop chute; (5) Panel box; and (6) Stairs.

4. Conclusion

Based on the assessment of working complaints and working posture of operators 1, 2, and 3 when operating the drum dryer machine, the results of the ergonomic risk evaluation were obtained. The study results showed that the work posture of operator 1 was the highest, namely 57 (moderate risk level and requires improvement), compared to the work activities of operators 2 and 3. The results of the work posture analysis with RULA showed that operator 1 has a score of 7, which indicates that the working posture of operators 1 has a very high risk and requires immediate improvement. Operators 2 and 3 also have a score of 6, which indicates that operators 2 and 3's work posture has a high risk. Improvement of the drum dryer machine is by adding a hopper or material container equipped with a control system so that it can flow material automatically so that the operator does not need to pour material manually, which takes quite a long time.

5. References

[1] Palijama S, Breemer R, and Topurmera M 2020 *J.Agritekno.* **9(1)** 20–27.
[2] Rifan, Nurrahman, and Aminah S 2017 *J. Pangan dan Gizi.* **7(2)** 104–116.
[3] Utami SR 2022 *Pengaruh Suhu dan Kecepatan Putar Silinder Drum Dryer Terhadap Karakteristik Bubur Instant Berbasis Millet Merah (Eleusine coracana) dan Kacang Merah (Phaseolus vulgaris L.)* (Bandung: Universitas Pasundan)
[4] Pua CK, Hamid NSA, Tan CP, Mirhosseini H, Rahman RBA, and Rusul G 2010 *J Food Science and Technology.* **43(2)** 343–349.
[5] Rodiahwati W, Mikhratunnisa, and Putri YE 2017 *J Tambora.* **2(1)** 1–12.
[6] Erliana CI, and Amri K 2020 *J Industrial Engineering.* **9(1)** 37–44.
[7] Indriati A, Mayasti NKI, Rahman T, Litaay C, Anggara CEW, Astro HM, Surahman DN, Novianti F, and Andriana, Y 2022 *4th International Conference on Food and Agriculture IOP Conf. Series: Earth and Environmental Science* vol 980 (England: IOP Publishing) pp. 1–9.
[8] Rini FA, Katili PB and Ummi N 2015 *J Teknologi Industri Untirta* **3(2)** 1–6.
[9] Anggraini DTC, Herwanto D and nugroho RE 2022 *J Sains, Tek. dan Industri.* **20(1)** 147–155.
[10] Santoso S, Yasra R and Purbasari A 2014 *J Profesiensi* **2(2)** 155–164.
[11] Priangkoso RTA, Puti NPS, Wardhana DH, Andriansyah DF, Setyandi AMF, and Rosyidiin AF 2024 *J Profisiensi.* **12(1)** 71-80.
[12] Rezki AS, Maksum AH, Herwanto D, and Rachmat MT 2023 *J Media Teknik & Sistem Industri.* **7(2)** 86-94.
[13] Hunusalela ZF, Perdana S, and Dewanti GK 2022 *J IKRAITH-Teknologi.* **6(1)** 1–10.
[14] Wijaya ISA, and Muhsin A 2018 *J Optimasi Sistem Industri.* **11(1)** 49–57.
[15] Oesman TI, Irawan E, and Wisnubroto P 2019 *J Ergonomi Indonesia.* **5(1)** 39–46.
[16] Ahmad NP, Hidayat R, and Hamdani R 2020 *Bulletin of Applied Industrial Engineering Theory.* **2(1)** 59–63.
[17] Rozi F, and Kn H 2021 *J Baut dan Manufaktur.* **3(2)** 41–47.
[18] Fitri M, Laila W, and Fendi 2017 *J Sains dan Teknologi.* **17(2)** 138–146.
[19] Bastuti S, and Zulziar M 2020 *J Ilmiah Teknik dan Manajemen Industri.* **2(2)** 116–125.
[20] Pratama P, Tannady H, Nurprihatin F, Ariyono HB, Sari SM 2017 *J PASTI.* **11(1)** 13–21.
[21] Herminiati A, Andriansyah RCE, Hidayat DD, Indriati A, Mayasti NKI, Komalasari N, Santoso T, Sutrisna and Sukwati 2024 *1st International Conference on Food and Agricultural Sciences AIP Conf. Proc.* **2957(1)** 050038.
[22] Tarwaka 2010 *Ergonomi Industri: Dasar-Dasar Pengetahuan Ergonomi dan Aplikasi di Tempat Kerja* (Solo: Harapan Press).
[23] World Health Organization 2006 *Neurological disorders: public health challenges* (World Health Organization).
[24] Susanti L, Zadry H, and Yuliandra B 2015 *Pengantar Ergonomi Industri* 1st ed (Padang: Andalas University Press).
[25] Baihaqi KT, Dewi NS, and Pratama SB 2024 *J Ilmu Kedokterean dan Kesehatan.* **11(3)** 550–556.
[26] Mayasti NKI, Purcawadikesit E, Indriati A, HaryantoA, Hanifah U, Karim MA, Siregar YH, Anggara CEW, and Purwandoko PB 2023 *5th International Conference on Food &Agriculture IOP Conf. Proc.* **1168 (2023)** 012035.
[27] Pegiardi I, Handika FS, and Supriyadi 2017 *J INTECH.* **3(2)** 73–77.
[28] Diana M and Saftarina F 2016 *J Ergonomi.* **1(2)** 369–379.

Biochemical and sensory characteristics of the fermented golden apple snail extract with the addition of kaffir lime (*Citrus hystrix*) leaves at several ratios

S L Asmono[1*]; Rahmawati[1]; N B Arifiana[1]; DP Andini[1]; M Syarief[1]

[1]Politeknik Negeri Jember, Po BoX 164 Jember, East Java, Indonesia

*sepdian@polije.ac.id

Abstract. Golden Apple Snail (GAS) (*Pomacea canaliculate* L.) is a source of macro nutrient which have the potential to be the main raw material for biostimulants for plant growth. Biostimulants are made through a fermentation process and produce an unpleasant aroma. The addition of organic ingredients which have strong aromatic compounds such as Kaffir Lime peel leaves is expected to cover up the unpleasant aroma of the GAS extract produced. Apart from that, users of this biostimulant will not be disturbed by the aroma of GAS extract when applied in the field. Therefore, the purpose of this research is to determine the characteristics of the fermentation results of GAS extract with the addition of Kaffir Lime leaves in four GAS extract formulations: lime leaves, including a ratio (ml:gr) of 100:0 (GAS extract without lime leaves); ratio 100:1; 100:5; 100:10. Kaffir Lime leaf slices measuring 1-2 cm^2 are mixed in the fermentation process which is carried out for 21 days. Anova analysis and Duncan's advanced test showed that there were significant differences in aroma parameters, but there were no differences in C-Organic, N-Total, P_2O_5, K_2O, and C/N Ratio, color and consistency. The aroma produced from fermentation in the 100:10 ratio formulation is the aroma of Kaffir Lime leaves with a strong quality and can cover the fishy aroma of GAS. Therefore, the addition of Kaffir Lime leaves can have the potential to improve the aroma quality of the GAS Biostimulant product, but does not reduce the nutrient content in it.

1. Introduction

GAS (*Pomacea canaliculate* L.) is a potential source of amino acids as the main raw material for biostimulants for plant growth. Several important roles of biostimulants for plants include stimulating the formation and growth of plant roots and shoots, thus helping in the formation of strong plant structures and optimal vegetative development [1], [2]. As in previous research results, it has been proven that in the fermentation of GAS extract there are 12 amino acid compounds including Arginine, Aspartic acid, Tycosine, Valine, Glutamic-acid, Alanine, Glycine, Leucinei-soleucine, Cysteine, Methionine, Proline Threonine ; the auxin hormone is 0.838% [3], [4].

The potential of the GAS biostimulant is certainly very good as an alternative to chemical fertilizers. However, the Biostimulant results in an unpleasant aroma. Of course, this becomes a problem during application in the field. This is supported by the results of previous research that MOL (Local Micro Organism) products from Keong Mas produce an unpleasant aroma [5]. Therefore, the main idea of this research is to modify the fermentation formulation of GAS extract by adding organic ingredients that have strong aromatic compounds such as Orange aroma [6], [7]. According to Kaswindiarti & Khotimah (2020) sweet Orange peel can be used as an aromatherapy ingredient. The research results of Dewi et al., (2022), show that the eco-enzyme from orange peel has a pungent and fresh sour orange peel aroma.

Therefore, this study aims to test several ratios of Kaffir lime leaves added to GAS fermentation extract, so that the biostimulant has a fragrant aroma or at least can be accepted by humans.

2. Method

This research was carried out in July to November 2024 at the "Rumah Organik" Teaching Factory and Field Laboratory, Politeknik Negeri Jember. The snail was collected from rice fields in Tegal Gede Village, Sumbersari, Jember, East Java, Indonesia, at coordinates 8°09'01"S 113°43'48"E with a height of 110 meters above sea level. Fermented ingredients include whole GAS complete with shell and flesh, auqadest, brown sugar, Kaffir Lime leaves, EM4. As well as fermentation equipments including knives, meat grinders, fermentation biorectors, measuring cups, stationery, pH meters, scales.

2.1. Fermentation of GAS and kaffir lime leaves

Fermentation of the basic Biostimulant ingredients for each treatment is carried out by mixing ingredients such as 1500 ml of GAS extract (flesh and shell), 375 ml of Aquadest, 25 ml of Em4, 300 ml of brown sugar, so the total volume of GAS Biostimulant for each treatment is 2000 ml [3].

In this study there were 4 treatments. Each biostimulant treatment is mixed with chopped Kaffir Lime leaves measuring 1-2 cm^2 at several treatment formulation ratios including:
a) Ratio 1 =100:0 (2000 ml GAS extract without adding Kaffir Lime leaves);
b) Ratio 2 = 100:1 (2000 ml GAS extract + 20 gr Kaffir Lime leaves);
c) Ratio 3= 100:5 (2000 ml GAS extract + 100 gr Kaffir Lime leaves);
d) Ratio 4 = 100:10 (2000 ml GAS extract + 200 gr Kaffir Lime leaves).
Anaerobic fermentation at room temperature (25°-30°C) for 21 days, and stirring every 2 days.

2.2. Observation parameters

The research method used was an organoleptic test involving 30 semi-trained panelists. The selected panelists are lay people aged >18 years. The parameters assessed are the biochemical quality observed, namely, Total Nitrogen using the Kjeldahl method, P_2O_5 and K_2O using the SNI 7763:2018 method, C-Organic using the Walkley and Black method, C/N Ratio and pH.

Apart from that, the sensory character parameter is aroma with the rating scale for the aroma parameter used being 1 = Very unpleasant smell; 2 = Bad smell; 3 = Unpleasant odor mixed with weak Kaffir Lime aroma; 4 = Medium Kaffir Lime aroma; 5 = Strong Kaffir Lime aroma; 6 = Very Strong Kaffir Lime aroma. Apart from that, there is a color parameter with a score of 1 = Brown; 2=Dark Brown. The consistency parameter uses a score of 1 = Liquid; 2=Quite liquid, 3=Thick.

2.3. Data analysis

The research data were analyzed using ANOVA and Duncan's advanced test at the 5% level.

3. Results and discussion

3.1. Biochemical content

The results of ANOVA analysis on the biochemical content of several GAS Biostimulant ratio formulations: Kaffir lime leaves showed significantly different results in the C-Organic test, but not significantly different in the N-Total, P_2O_5, K_2O and C/N Ratio. Microorganisms need carbon as an energy source for the formation and growth of new cells during the decomposition process [10]. Based on the analysis results in **Table 1**, the highest levels of C-Organic were shown in the 100:10 ratio formulation, namely 4.994%, but there was also no significant difference in the GAS extract added with kaffir lime leaves at either the 100:5 or 100:1 ratio. Meanwhile, at a ratio of 100:0 or without the mixture, kaffir lime leaves have the lowest C-Organic content, namely 4.389%.

IOP Conf. Series: Earth and Environmental Science **1446** (2025) 012014 doi:10.1088/1755-1315/1446/1/012014

Tabel 1. Nitrogen, Phosphorus, Potassium content in several ratios of Biostimulants and Kaffir Lime Leaves

Ratio Biostimulant : Kaffir Lime Leaves	C-Organik (%)		N-Total (%)		P_2O_5 (%)		K_2O (%)		C/N Ratio (%)	
100:0	4.389	±0.840	0.920	±0.032	0.096	±0.032	0.319	±0.009	4.790	±0.994
100:1	4.465	±0.604	0.873	±0.048	0.094	±0.022	0.308	±0.006	5.154	±0.945
100:5	4.894	±0.810	0.917	±0.060	0.084	±0.036	0.324	±0.005	5.319	±0.661
100:10	4.994	±0.148	0.957	±0.102	0.099	±0.014	0.328	±0.006	5.278	±0.732

From these results it is known that the addition of chopped Kaffir Lime leaves is able to increase the C-Organic content so that microorganisms get an additional supply compared to just pure GAS extract, which is thought to continue to utilize the existing carbon elements for the decomposition process so that the resulting C-Organic content decreases. This is in accordance with research on the addition of chopped salak leaves which can also increase the C-Organic content [11].

The results of the nitrogen content in the ratio formulation of GAS extract and Kaffir Lime leaves showed relatively similar results or were not significantly different **(Table 1).** The average amount of N-Total produced is 0.8 to 0.9%. However, the average results show that the addition of Kaffir Lime leaves increases the N content produced, it is proven that a ratio of 100:10 has a total N content of 0.957%. Based on these results, the nitrogen content of the GAS extract has met the minimum technical requirements for N-Total in liquid organic fertilizer which is 0.5% [12].

Apart from that, in the phosphorus content test results **(Table 1.),** the fermentation results of several ratio formulations also showed that the results were not significantly different with the average P_2O_5 produced being 0.08% to 0.09%. These results are in accordance with previous research which obtained a P2O5 content of 0.08% [13] and 0.04% [14]. In the Potassium (K_2O) test results in the formulation, the ratio of GAS extract and kaffir lime leaves had an average of 0.3%and the overall treatment showed no significant difference.

Based on the results of variance analysis, it can be seen that the addition of Kaffir Lime leaves also has no effect on the C/N Ratio results. In **Table 1**, the average C/N ratio is highest in treatments with ratios of 100:5 and 100:10. Thus, even though the results were not significantly different, the addition of Kaffir Lime leaves was able to increase the C/N ratio. The results of the C/N ratio which were not significantly different could be due to the fact that during the fermentation process the kaffir lime leaves were still large and difficult to dissolve within 21 days of fermentation. The results of previous research show that pulverizing leaves by soaking in water and mud alone can take months [15]. However, in this case the purpose of mixing kaffir lime leaves is just to get the fragrant aroma.

Several ratios of GAS extract to Kaffir Lime leaves produce relatively similar pH values during the fermentation period, namely an average pH of 5.2 on the first day and a pH of 6.3 on the 14th day, as in **Figure 1** below.

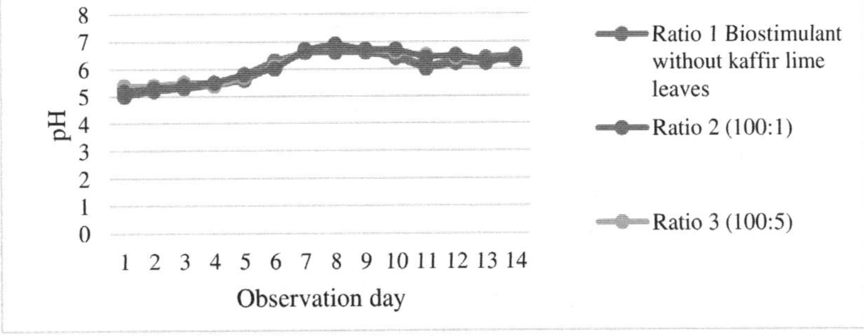

Figure 1. pH during the fermentation process

It is known that the acidic pH (5.2) on the first day of fermentation is caused by the activity of microorganisms which begin to decompose organic matter [16]. The pH value during the fermentation process process is in accordance with the minimum standard technical requirements for liquid organic fertilizer with a pH value between 4-9 based on minimum technical requirements for organic and biological fertilizers [12].

3.2. Sensory test

The results of the variance analysis of the sensory test showed that the aroma parameters gave significant results, while the color and consistency were not significantly different. Therefore, further tests were carried out on the aroma parameters as shown in the following table.

Tabel 2. The results of adding kaffir lime leaves on the aroma, color and consistency of the biostimulant produced

Ratio	Aroma		Color		Consistency	
	Average score	Description	Average score	Description	Average score	Description
100:0	1.6 [a] ±0.5	Bad smell 4;	1.8 ±0.4	Dark brown	2.1 ±0.5	Quite liquid
100:1	3.1 [b] ±0.8	Unpleasant odor mixed with weak Kaffir Lime aroma;	1.8 ±0.4	Dark brown	2.1 ±0.6	Quite liquid
100:5	4.0 [c] ±0.6	Medium Kaffir Lime aroma	1.7 ±0.4	Dark brown	2.0 ±0.6	Quite liquid
100:10	5.0 [d] ±0.8	Strong Kaffir Lime aroma	1.7 ±0.4	Dark brown	2.1 ±0.6	Quite liquid

Note: Numbers followed by the same letter in the same column show no significant difference at the DMRT Test 5%

The treatment ratio of GAS biostimulant: Kaffir Lime leaves of 100:5 and 100:10 showed a distinctive fermented aroma of kaffir lime leaves which was significantly different compared to the pure biostimulant from GAS extract without the addition of Kaffir Lime leaves. Based on initial observations, the resulting fermentation aromas all resemble the pungent aroma of ammonia, mud and fishy snails. However, after 7 to 21 days of fermentation, the aroma of kaffir lime and sourness begins to be strong and the aroma of ammonia begins to decrease. These results show that the decomposition process is starting to occur and is mixed with the aroma of acid and ammonia. In accordance with previous research, the decomposition process of GAS takes quite a long time, namely 21 days [17].

Figure 2. The color of the 21 day fermentation results from several ratios of GAS Biostimulant: Kaffir Lime leaves a) Ratio 100:0; b) 100:1; c) 100:5; d) 100:10

GAS biostimulant with the addition of chopped kaffir Lime leaves in several ratios shows the same color as pure biostimulant, namely dark brown (**Figure 2**). The same color was also obtained in previous research using golden snail material to make bioactivators [18]. This color indicates that the material components inside have begun to decompose and rot is occurring [19]. This color also does not come from dissolved materials, but also suspended materials[20].

Based on observations, the texture of GAS Biostimulant is quite liquid. This is because the material produced from grinding Mas Snail shells cannot dissolve and leaves residue, apart from that, the kaffir lime leaves that are mixed are not ground but are in chopped form so they cannot dissolve until the 21st day. As in making liquid organic fertilizer which uses egg shells. Chicken also produces a fairly liquid texture because there are still insoluble residues [21].

4. Conclusion

Based on the results of this research, the characteristics of the fermentation results of GAS extract with the addition of Kaffir Lime leaves based on biochemical parameters show that there is a significant difference in the C-Organic content but there is no significant difference in the content of N-Total, P_2O_5, K_2O, and the C/N-Organic Ratio. Apart from that, in terms of sensory characteristics, there are significant differences regarding aroma, but color and consistency are not significantly different. The aroma produced from fermentation in the 100:10 ratio formulation is the good aroma and is able to cover the fishy aroma of GAS. Therefore, the addition of Kaffir Lime leaves has the potential to improve the aroma quality of the GAS Biostimulan product, but does not reduce the nutritional content in it.

Acknowledgments

This research was funded by the Politeknik Negeri Jember PNBP Fund in accordance with Director's Decree Number 816/PL17.4/PG/2024

References

[1] O. I. Yakhin, A. A. Lubyanov, I. A. Yakhin, and P. H. Brown, "Biostimulants in plant science: a global perspective," *Front. Plant Sci.*, vol. 7, p. 238366, 2017.

[2] P. Du Jardin, "Plant biostimulants: Definition, concept, main categories and regulation," *Sci. Hortic. (Amsterdam).*, vol. 196, pp. 3–14, 2015.

[3] S. L. Asmono, Rahmawati, N. B. Arifiana, and Suharjono, "Phytohormone and amino acid content in golden apple snail (Pomaceae canaliculata L.) extract which fermented at several effective microorganism 4 (EM4) concentrations," *IOP Conf. Ser. Earth Environ. Sci.*, vol. 1338, no. 1, p. 12008, May 2024, doi: 10.1088/1755-1315/1338/1/012008.

[4] S. L. Asmono, S. Suharjono, N. B. Arifiana, and others, "Kandungan NPK Ekstrak Keong Mas (Pomacea canaliculata L.) yang Difermentasi Pada Beberapa Konsentrasi EM4 (Effective Microorganism 4): NPK Content of Golden Snail Extract Fermented at Some EM4 (Effective Microorganism 4) Concentrations," *J. Ilm. Inov.*, vol. 24, no. 2, pp. 127–130, 2024.

[5] R. Subula, W. D. Uno, and A. Abdul, "Kajian Tentang Kualitas Kompos Yang Menggunakan Bioaktivator Em4 (Effective Microorganism) Dan Mol (Mikroorganisme Lokal) Dari Keong Mas," *Jambura Edu Biosf. J.*, vol. 4, no. 2, pp. 54–64, 2022.

[6] A. S. Aryani, H. Taher, and K. Subandi, "Desain Model Identifikasi Dini Pemanfaatan Tanaman Aromatik Indonesia," *J. Teknoinfo*, vol. 17, no. 2, pp. 538–547, 2023.

[7] M. C. González-Mas, J. L. Rambla, M. P. López-Gresa, M. A. Blázquez, and A. Granell, "Volatile compounds in citrus essential oils: A comprehensive review," *Front. Plant Sci.*, vol. 10, p. 433929, 2019.

[8] S. Kaswindiarti and T. H. Khotimah, "Pengaruh Aromaterapi Jeruk Manis (Citrus Sinensis) Terhadap Penurunan Tekanan Darah Dan Denyut Nadi Anak Usia 6-9 Tahun Pada Kunjungan Pertama Ke Dokter Gigi," *JIKG (Jurnal Ilmu Kedokt. Gigi)*, vol. 3, no. 2, 2020.

[9] S. P. Dewi, S. Devi, and S. Ambarwati, "Pembuatan dan uji organoleptik eco-enzyme dari kulit buah jeruk," in *Prosiding Seminar Nasional Hukum, Bisnis, Sains dan Teknologi*, 2022, vol. 2, no. 1, p. 649.

[10] S. Jahan *et al.*, "Role of Nitrogen for Plant Growth and Development: A review," *Adv. Environ. Biol.*, vol. 10, pp. 209–218, 2016.

[11] R. Amnah and N. Friska, "Pengaruh aktivator terhadap kadar unsur C, N, P dan K kompos pelepah daun salak sidimpuan," *J. Pertan. Trop.*, vol. 6, no. 3, pp. 342–347, 2019.

[12] Kementerian Pertanian Republik Indonesia, "Persyaratan Teknis Minimal Pupuk Organik, Pupuk Hayati dan Pembenah Tanah. Nomor: 261/KPTS/SR.310/M/4/2019," 2019.

[13] A. Setiawan, S. Safruddin, and R. Mawarni, "Pengaruh Pemberian Pupuk Mikoriza dan Pupuk Organik Cair (POC) Keong Mas Terhadap Pertumbuhan dan Produksi Tanaman Mentimun (Cucumis sativus L.)," *Bernas J. Penelit. Pertan.*, vol. 16, no. 1, pp. 71–80, 2020.

[14] S. Madusari, G. Lilian, and R. Rahhutami, "Karakterisasi pupuk organik cair keong mas (Pomaceae canaliculata L.) dan aplikasinya pada bibit kelapa sawit (Elaeis guineensis Jacq.)," *J. Teknol.*, vol. 13, no. 2, pp. 141–152, 2021.

[15] Y. Astuti, D. Anugrah, and H. Faruq, "Efektivitas Fermentasi Bahan Organik dalam Pengelupasan Jaringan Mesofil Daun Kupu-Kupu (Bauhinia purpurea L.). JBIO: JURNAL BIOSAINS (The Journal of Biosciences), 6 (2), 32--36." 2020.

[16] A. Ismayana, N. S. Indrasti, A. M. Suprihatin, and A. F. Tip, "Faktor rasio C/N awal dan laju aerasi pada proses co-composting bagasse dan blotong," *J. Teknol. Ind. Pertan.*, vol. 22, no. 3, 2012.

[17] Rusmini, R. Manullang, Daryono, and A. Sadikin, "Peningkatan Kualitas Bioaktivator Keong Mas Dengan Penambahan Buah Maja (Qualiy Improvement Of Golden Apple Snaol Bioaktivator With Addition Of Maja Fruit).," 2019.

[18] R. Rusmini, D. Daryono, L. Mudi, R. Anwar, and A. Sadikin, "PENINGKATAN KUALITAS BIOAKTIVATOR KEONG MAS DENGAN PENAMBAHAN RUMEN KAMBING YANG BERBEDA," *J. Hutan Trop.*, vol. 11, no. 2, pp. 151–159, 2023.

[19] S. Hadisuwito, *Membuat pupuk organik cair*. AgroMedia, 2012.

[20] H. Effendi, "Telaah kualitas air, bagi pengelolaan sumber daya dan lingkungan perairan. Kanisius," *Yogyakarta. ISBN*, pp. 972–978, 2003.

[21] F. I. Putri and A. Asngad, "Pemanfaatan Azolla microphylla dan Cangkang Telur Ayam sebagai Pupuk Organik Cair dengan Bioaktivator Rebung Bambu," in *Prosiding SNPBS (Seminar Nasional Pendidikan Biologi dan Saintek)*, 2022, pp. 255–261.

Effect of Seed Biopriming with *Pseudomonas fluorescens* for Increasing Seed Quality of Kenaf (*Hibiscus cannabinus* L.) Cultivar KR 14

M Azizah*, DG Cahyaningrum, CD Utami, F Rohman, GF Dinata

Department of Agricultural Production, Politeknik Negeri Jember, Mastrip Street PO BOX 164, Jember, Indonesia

maria_azizah@polije.ac.id

Abstract. Kenaf is a seasonal plant that produces fiber with biomass that has the potential to be developed as animal feed at the Animal Feed Tefa of Jember State Polytechnic. The quality of seed that planted can affect crop yield and quality. Seed priming can stimulate germination to overcome the problem of low physiological quality of kenaf seeds due to storage. Trichoderma is one of the biological agents that can be used as a seed priming material. This study aims to determine the effect of biological agents for seed priming on increasing the quality of kenaf seeds. The research was conducted from June to August 2024 at Seed Technology Laboratory of Jember State Polytechnic. The research was arranged using a nonfactorial completely randomized design. The treatment applied was control (without soaking agent), 1 hour seed soaking with *Pseudomonas fluorescens* isolate, 2 hours seed soaking with *P. fluorescens* isolate, and 3 hours seed soaking with *P. fluorescens* isolate. The observations made were Seed Germination (%), Growth Rate (%), Vigor Index (%), sprout shoot length (cm) and root length (cm). The results showed that seed biopriming using *P. fluorescens* increasing seed germination and seed vigor percentage. The recommended seed soaking treatment for seed priming kenaf seeds with low initial viability is soaking for 1 hour.

Introduction

Kenaf is a seasonal plant that produces fiber as raw material for the paper, textile and automotive parts industries. Currently, kenaf biomass is developed as animal feed. Kenaf can replace alfalfa straw or crude protein from soybean meal up to around 66% in dairy cattle feed [1]. Research on several kenaf lines in Malaysia showed that the leaves crude protein content ranged from 13.6% to 22.3% [2]. Kenaf flour has the potensial for replacing soybean and fish meal as a protein supplement for livestock [3]. Kenaf also has the potential as feed because of its ability to grow on marginal, dry, acidic land [4] and ex-mining land [5].

Jember State Polytechnic as a vocational college has various Teaching Factories (Tefa) that support student learning with conditions that are adjusted to small-scale industries. Dairy Cattle Tefa is one of the Tefa managed by the Integrated Agricultural Production UPA. The problems in managing Dairy Cattle Tefa are increasing dairy cattle productivity and developing types of green fodder. The development of kenaf plants also has the potential to be developed as feed in Tefa Animal Feed.

Developing kenaf plant as feed material can be started with the development of kenaf plant cultivation, especially the availability of kenaf seeds. The simultaneity of seed growth is the quality of seeds that needs to be considered because it is related to the need for planting materials. Seeds with low physiological quality can be caused by suboptimal seed filling, seed harvest age and seed deterioration during storage. Kenaf seeds can still survive for up to 5 years at 0°C temperature storage, while viability decreases when seeds are stored using 12% Relative Humidity and temperature higher than 10°C [6]. Seed priming is carried out to overcome the problem of low physiological quality of kenaf seeds because storage uses materials that can stimulate germination. Seed priming of kenaf using GA3 50 ppm for 3 hours can improve seed physiological quality [7]. In addition to the use of synthetic gibberellin, research is needed on the application of natural PGPR from Trichoderma sp and *Pseudomonas fluorescens* which are environmentally friendly and can suppress pathogen attacks.

Bio-priming is a priming activity that uses beneficial microorganisms to protect seeds to control diseases and improve physiological quality such as hydrating seeds. The Biopriming method is generally carried out by moistening the seeds and then mixing the formulated bioagent product with the previously imbibed seeds [8]. Seed soaking using *Pseudomonas fluorescens* reduced the number of nematode eggs the most (86.39 %) and suppressed disease intensity by 71.95 % when compared to the control. [9]. Fungal and bacterial bioagents whether they are biofertilizer or biopesticide such as Azotobacter, Rhizobium, Arthrobacter, Agrobacterium, Azospirillum, Enterobacter, Streptomyces, Bacillus, Burkholderia, Klebsiella, PSB, Pseudomonas fluorescence, Trichoderma viride, Trichoderma harzianum and Vesicular Arbuscular Mycorhiza may be useful as biopriming agents. Seed biopriming is useful in almost all crops and ecologically friendly for chemical fungicides substitute [10].The application of biological agents in seed priming is expected to improve the physiological quality and simultaneity of seeds that play a good role in crop production.

Currently, the Sweetener and Fiber Crop Instrument Standard Testing Center is one of the centers that provides certified kenaf seeds and has not met Indonesian market demand. This research needs to be done to improve the quality of kenaf seeds through seed priming and kenaf seed production with fertilization so that seeds with high production and quality are obtained. The quality seeds produced will later be used for research in the following year on kenaf cultivation as animal feed that can be developed in Tefa Animal Feed.

Method

The research was conducted from June to August 2024, at the Seed Technology Laboratory of Jember State Polytechnic. The materials used in the study included Kenaf seeds of Foundation Seed Class KR 14 from BSIP-TAS, germination paper, plastic bags, *Pseudomonas fluorescens* isolate, distilled water, labels, and 70% alcohol. The tools used in the study included hand sprayers, scales, measuring glasses, beaker glasses, digital cameras, laminar air flow, autoclaves, petri dishes, micropipettes, pipette tips, Erlenmeyers, and stationery

Seed priming is a method of increasing germination and seed growth simultaneity using materials that can stimulate seed germination. The research was designed with a non-factorial completely randomized design. The treatments applied were control (without a soaking agent), 1 hour seed soaking with *Pseudomonas fluorescens* isolate, 2 hours seed soaking with *Pseudomonas fluorescens* isolate, and 3 hours seed soaking with *Pseudomonas fluorescens* isolate. The treatments were repeated 5 times to obtain 20 experimental combinations. The observations made were Seed Germination (%), Growth Rate (%), and Vigor Index (%), sprout shoot length (cm) and root length (cm). The data were analyzed using ANOVA, followed by a Tukey test at the 5% significance level if significant effects were found.

The propagation of *Pseudomonas fluorescens* isolates was carried out by growing pure isolates in King`s B liquid media (contain of 20 gram protease peptone, 1.5 gram K_2HPO_4, 1.5 gram $MgSO_4.7H_2O$, Glycerol 15 ml, and 1 liter of distilled water) and incubated at room temperature for 24 hours. The purity of the bacterial isolates was indicated by colonies that produced yellowish green pigments as an indication of antibiotic and siderophore compound production. The application of *Pseudomonas fluorescens* isolates for seed soaking was carried out at a concentration of 10% (100 ml isolate/1L water).

Seed quality was assessed using the between-paper method. Seeds that had been given soaking treatment with *Pseudomonas fluorescens* were planted on germination paper. In each replication, 50 seeds were planted. The seeds that had been arranged were covered and rolled in plastic and placed on the germinator. The germinator was set with lighting and was in humid conditions. Germination observations were carried out 7 days after planting. Observation of vigor index (first count vigor index) was conducted 5 days after planting, and observation of growth rate was conducted 6 days after planting. Observation of sprout shoot length (cm) and root length (cm) was conducted on sprouts aged 7 days after planting.

Result and Discussion

Research shows that seed soaking treatment using *Pseudomonas fluorescens* isolates has an effect on kenaf seed germination with initial viability having decreased. Commercial kenaf seeds typically have a germination rate of more than 90%. The seeds used in the test had undergone initial testing with an average viability result of 84% and an average vigor index (first count vigor index) of 70%.

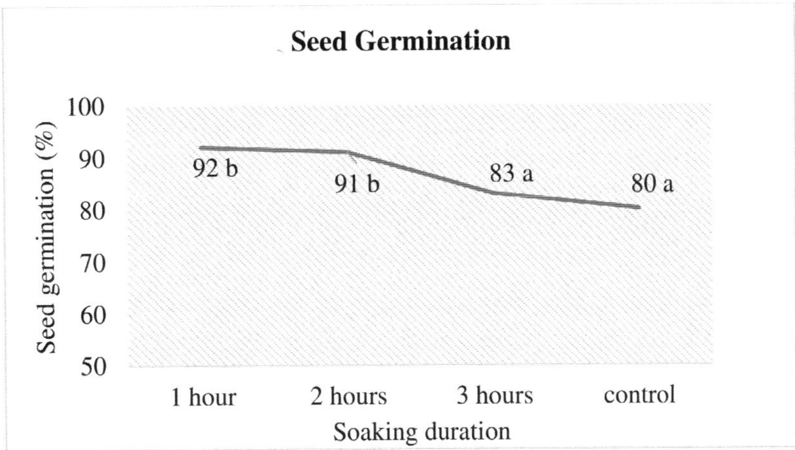

Figure 1. Effect of seed soaking using *P. fluorescens* isolate on seed germination

Seed germination is an observation parameter to determine the quality of seeds to be planted. The results of the study in **Figure 1** show that soaking for 1 and 2 hours can increase seed germination compared to the control. Priming can stimulate enzyme activity, metabolism, and promote plant grow faster. Biopriming using *P. fluorescens* enhances seed germination and seedling vigor in chili by improving the germination rate, percentage, mean root length, mean shoot length, and biomass production when compared to seeds that have not been prime [11]. Biopriming with P. fluorescens ensures strong bacterial establishment and adhesion to the seed before planting, making it a recommended treatment to enhance the seed index and improve the growth of sunflower seedlings [12]. Seed priming is an effective, cost-efficient method to boost germination, growth, and the productivity of crops. It involves using various priming agents such as water, inorganic salts, sugars, solid media with water and nutrients, beneficial microbes, micronutrients, hormones, rhizobacteria, and organic substances. Seed priming not only enhances plant growth and yield but also boosts tolerance to abiotic stress [13]. Biopriming seed is a promising and environmentally friendly method to prevent the transmission of bacterial fruit blotch from seeds to seedlings. Liquid or solid matrix priming using *Pseudomonas fluorescens* improved the germination energy for watermelon and germination rate of melon compared with unprimed seeds [14].

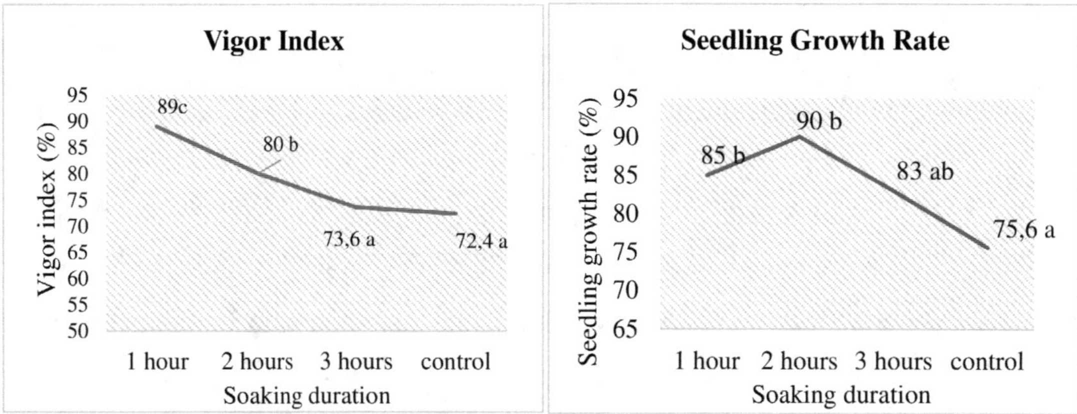

Figure 2. Effect of seed soaking using *Pseudomonas fluorescens* isolates on vigor and seedling growth index

Seed vigor refers to the seeds' capacity to thrive in less-than-ideal conditions. There are multiple methods to assess seed vigor, generally using the calculation of the first count vigor index and seedling growth rate. The results of the study in **Figure 2** show that the vigor index calculated at the first count in the seed soaking treatment using *Pseudomonas fluorescens* isolates showed the best vigor index number among the other soaking treatments. However, the seedling growth rate parameter for soaking treatments of 1, 2 and 3 hours did not show a significant result on seedling growth. Priming treatment can help produce uniform sprouts. By stimulating uniform germination, primed seeds produce seedlings with uniform size and growth. This provides an adaptive advantage for plants in dealing with adverse environmental conditions. [15]. Biopriming using *P. fluorescens* is recommended for control of seed-borne diseases. Biopriming chili seeds with *P. fluorescens* has shown effective suppression of diseases [16].

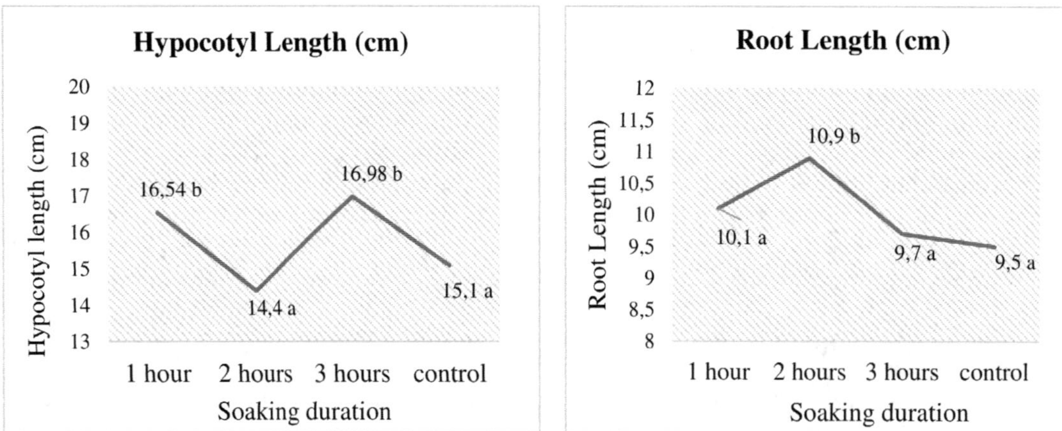

Figure 3. Effect of seed soaking using *Pseudomonas fluorescens* isolate on seedling hypocotyl length and seedling root length

Hypocotyl and root lengths are among the parameters used to assess the vigor of seeds being tested. Seeds with high vigor will grow faster than seeds with low vigor. The results of the study in **Figure 3.** show that the 1 and 3 hour soaking treatments affect the average hypocotyl length compared to the

IOP Conf. Series: Earth and Environmental Science **1446** (2025) 012015 doi:10.1088/1755-1315/1446/1/012015

control. The root length parameter shows that seeds soaked for 2 hours have better root lengths than other treatments, but the hypocotyl length is not better than the control.

Overall research results show a tendency that the longer the seed is soaked, the more the seed viability and vigor decreases. This occurs because the seed coat of long-stored seeds starts to deteriorate. Damaged seed coats when soaked in liquid will absorb water too quickly. The imbibition process that is too fast can actually injure the seeds and can actually inhibit germination. The imbibition process, which involves adding water to the seed to increase its water content to a certain level, is the first stage of germination. The imbibition process with the presence of water causes the outermost layer of the seed to rupture. After this procedure, the seed cells will divide and go through a number of biochemical processes until they finally develop into plants. Ideally, seeds undergo imbibition with a slow process so that the cells in the seed are not damaged.

Conclusion

The study shows us that seed priming using *Pseudomonas fluorescens* can increase the percentage of seed germination and increase seed vigor compared to control. The recommended seed soaking treatment for seed priming kenaf seeds with low initial viability is soaking for 1 hour.

Acknowlegment

The author extends gratitude to the Jember State Polytechnic for funding this research through the 2024 PNBP Research Grant Junior Lecturer Research (PDP) scheme.

References

[1] Chantiratikul A 2005 Effects of kenaf (Hibiscus cannabinus) as a substitative feed on lactating dairy cattle. Universiti Putra Malaysia; 2005.

[2] Noori Z, Saleh G, Foroughi M, Behmaram R, Kashiani P, Babaei N, et al. 2016 Forage yield and quality of kenaf (Hibiscus cannabinus L.) for consumption as ruminant feed. J Fundam Appl Sci. 8(2):12–30.

[3] Kujoana TC, Weeks WJ, der Westhuizen MM, Mabelebele M, Sebola NA. 2023 Potential significance of kenaf (Hibiscus cannabinus L.) to global food and feed industries. Cogent Food \& Agric. 9(1):2184014.

[4] Krismawati A 2005 Uji adaptasi varietas dan galur kenaf (Hibiscus cannabinus l.) di lahan pasang surut kalimantan tengah Ind Crop Res J. 11(3):107–11.

[5] Roby R, Mentari SD 2018 Pengembangan Tanaman Kenaf (Hibiscus cannabinus L) pada Lahan Pasca Tambang Batubara soilrens 16(1).

[6] Meints PD, Smith CA 2003 Kenaf seed storage duration on germination, emergence, and yield Ind Crops Prod. 17(1):9–14.

[7] RS TH, Marjani M 2020 Peningkatan Mutu Fisiologis Benih Kenaf (Hibiscus cannabinus L) dengan Penerapan Teknologi Seed Priming. Bul Tanam Tembakau Serat dan Miny Ind. 12(2):67–77.

[8] Rhaman MS, Rauf F, Tania SS, Khatun M 2020 Seed priming methods: Application in field crops and future perspectives Asian J Res Crop Sci. 5(2):8–19.

[9] Wijayanti KS, Rahardjo BT, Himawan T 2016 Pengaruh PGPR terhadap penekanan populasi nematoda puru akar (Meloidogyne incognita (Kofoid and White) Chitwood) pada tanaman kenaf (Hibiscus cannabinus L.) Bul Tanam Tembakau, Serat dan Miny Ind.8(1):30–9.

[10] Deshmukh AJ, Jaiman RS, Bambharolia RP, Patil VA 2020 Seed biopriming--a review. Int J Econ Plants 7(1):38–43.

[11] Ananthi M, Selvaraju P, Sundaralingam K 2014 Effect of bio-priming using bio-control agents on seed germination and seedling vigour in chilli (Capsicum annuum L.)'PKM 1.' J Hortic Sci Biotechnol 89(5):564–8.

[12] Moeinzadeh A, Sharif-Zadeh F, Ahmadzadeh M, Tajabadi FH 2010 Biopriming of Sunflower ("Helianthus annuus" L.) Seed with'Pseudomonas fluorescens' for Improvement of Seed

Invigoration and Seedling Growth. Aust J Crop Sci. 4(7):564–70.

[13] Sher A, Sarwar T, Nawaz A, Ijaz M, Sattar A, Ahmad S 2019 Methods of Seed Priming. In: Hasanuzzaman M, Fotopoulos V, editors. Priming and Pretreatment of Seeds and Seedlings: Implication in Plant Stress Tolerance and Enhancing Productivity in Crop Plants [Internet]. Singapore: Springer Singapore;. p. 1–10. Available from: https://doi.org/10.1007/978-981-13-8625-1_1

[14] Wang Y, Qiu Y, Lu Y, Wang X, Zhang H, Wu P 2023 Biocontrol of bacterial fruit blotch by seed priming with Bacillus amyloliquefaciens and Pseudomonas fluorescens. Seed Sci Technol. 51(2):191–203.

[15] Devika OS, Singh S, Sarkar D, Barnwal P, Suman J, Rakshit A 2021 Seed priming: a potential supplement in integrated resource management under fragile intensive ecosystems. Front Sustain Food Syst. 5:654001.

[16] Chin JM, Lim YY, Ting ASY 2022 Biopriming Pseudomonas fluorescens to vegetable seeds with biopolymers to promote coating efficacy, seed germination and disease suppression. J Saudi Soc Agric Sci. 21(8):493–505.

IOP Conf. Series: Earth and Environmental Science **1446** (2025) 012016 doi:10.1088/1755-1315/1446/1/012016

Development of Traditional Indonesian Layer Cake From Pureed Yellow Pumpkin (Cucurbita moschata Durch) And Carrot (Daucus carrota L) as a Highly Nutritious as an Additional Food For Toddlers

Mazarina Devi[1], Chintya Paramita P[1], Nunung Nurjanah[2], Dinar Ajeng Oktaviantry[2], Norrina Din[3]

[1] Faculty of Vocational Studies, Universitas Negeri Malang, Indonesia
[2] Department of Culinary Art and Clothing Education, Universitas Negeri Malang, Indonesia
[3] Faculty of Hotel and Tourism Management, Universiti Teknologi MARA, Permatang Pauh, 13500 Permatang Pauh, Pulau Pinang, Malaysia

mazarina.devi.ft@um.ac.id, chintya.paramita.fv@um.ac.id

Abstract. Yellow pumpkin (Cucurbita moschata Durch) and carrots (Daucus carota L) are readily available and economical in Indonesia, possessing a high nutritional value, particularly in beta-carotene. Sufficient beta-carotene intake is vital for preventing developmental disorders in toddlers. This study employed a Completely Randomized Design (CRD) to assess the chemical properties of layer cake made with pumpkin and carrot puree, subject to three different treatments involving adding 20%, 30%, and 40% puree, each being replicated twice. Data analysis was performed using ANOVA, followed by a further DMRT test. The results indicated that incorporating 40% puree into 100 grams of layer cake yielded the highest nutritional content: protein at 8.89%, fat at 18.31%, water at 48.18%, ash at 1.15%, beta-carotene at 54.41 g, and calcium levels. Conversely, the carbohydrate content decreased by adding 40% puree, resulting in a carbohydrate level of 26.26%. The highest carbohydrate content was observed in the layer cake with 20% puree added. These findings demonstrate that including pumpkin and carrot puree alters the chemical properties of layer cake, rendering it a beneficial complementary food for toddlers due to its enhanced nutritional profile, which may contribute positively to their overall health, growth, and nutritional status.

1. Introduction

Balanced nutrition represents a systematic approach to daily meal planning that ensures the intake of nutrients in quantities aligned with the body's specific requirements [1]. To effectively combat malnutrition in Indonesia, the development of nutritionally standardized supplementary food formulations and advanced processing technologies is underway [2]. Malnourished toddlers experience compromised immune function, hindered development and growth, and a decline in cognitive abilities [3]. An ideal dietary regimen for children provides the calories and nutrients in the appropriate portions required for daily sustenance [4].

In Indonesia, 3.5 percent of children under the age of five and 14.5 percent of adults experience malnutrition, which is classified as a moderate public health issue [2]. Supplemental feeding (PMT),

IOP Conf. Series: Earth and Environmental Science **1446** (2025) 012016 doi:10.1088/1755-1315/1446/1/012016

which focuses on foods not typically considered daily staples, is one effective strategy for combating malnutrition in children under five [5,6].

Pumpkin and carrot plants are widely distributed across Indonesia and appear on most islands. Raising awareness regarding these commodities' nutritional value and potential is essential, as their usage is currently limited [7]. Pumpkin (Cucurbita moschata Durch) is notably rich in provitamin A, with a concentration of 767 g/g, as well as containing carbohydrates, fiber, protein, and various minerals, including calcium, iron, phosphorus, along with B vitamins and vitamin C [7]. The consumption of pumpkin offers several benefits for toddlers, such as supporting eye health, enhancing energy levels, maintaining healthy skin, promoting smooth digestion, and improving sleep quality [8], [9].

Carrots (Daucus carota L) provide 42 calories per 100 grams, 0.9 g of fiber, 1.2 g of protein, and 12,000 SI of vitamin A. The consumption of carrots offers several benefits, particularly for toddlers, such as preventing night blindness, enhancing memory, boosting physical strength, fortifying bones, accelerating wound healing, strengthening hair, and aiding in the elimination of intestinal worms [10]. Pureeing is a food processing technique that transforms raw ingredients into a smooth, soft, and creamy texture [11]. Incorporating pumpkin and carrot purees aims to achieve optimal dough consistency for further ingredient integration during processing [12]. Layer cake is a beloved traditional Indonesian snack characterized by its vibrant colors and multiple layers. Its striking, chewy, and savory appearance is attributed to using coconut milk, and its preparation involves steaming each layer in a stacked manner. The nutritional profiles of pumpkin and carrots are noteworthy; both ingredients are readily accessible across most of Indonesia's islands, cost-effective, and serve as natural alternatives to synthetic dyes in layer cakes [13,14]. Carrots are higher in vitamin A, protein, fat, and energy than pumpkins. Conversely, pumpkins offer more significant amounts of calcium, phosphorus, iron, vitamin B, and vitamin C, enhancing layer cakes' nutritional quality and value. Layer cakes carry substantial cultural importance as a traditional Indonesian delicacy. Through rigorous research, we can develop new variants that preserve traditional elements while incorporating modern innovations using more inventive ingredients. In the culinary industry, innovation is essential to maintaining product relevance. By integrating pumpkin and carrot puree, this research aims to create cake layers with enhanced visual appeal, novel textures, and distinctive flavors, thereby broadening the market appeal of traditional lapis cake, particularly for toddlers.

The study's objective was to analyze various chemical properties, including proximate analysis (protein, carbohydrates, fat, water, ash) and the levels of β-carotene, calcium, and iron.

2. Materials and Methods

2.1. Materials

This research aims to investigate the chemical properties of layer cakes incorporating pumpkin and carrot puree. A completely randomized design (CRD) was employed with three different treatments: 20% puree addition (P1), 30% puree addition (P2), and 40% puree addition (P3), each with two replications. The study analyzed various chemical properties, including proximate levels (protein, carbohydrates, fat, water, ash), β-carotene, calcium, and iron. The acquired data will be analyzed using ANOVA followed by Duncan's Multiple Range Test (DMRT). The equipment for preparing layer cakes includes a thermometer, scales, knife, steamer, blender, stainless bowl, pan, baking dish (16.5 x 7 x 6 cm), cutting board, balloon whisk, and strainer. The primary ingredients for the layer cakes consist of pumpkin and carrot puree, sourced from local markets in Malang, along with supplementary ingredients such as rice flour, tapioca flour, sugar, and coconut milk. These ingredients have been carefully selected based on their local availability and capability to enhance the desired texture and flavor of the cakes.

Table 1. Layer Cakes Formulation Substituting Yellow Pumpkin and Carrot Puree

Ingredients	**A (20%)**		**B (30%)**		**C (40%)**	
	White Dough	**Yellow Dough**	**White Dough**	**Yellow Dough**	**White Dough**	**Yellow Dough**
Rice flour	50 g	40 g	50 g	35 g	50 g	30 g
Tapioca flour	34 g	27 g	34 g	27 g	34 g	27 g
Sugar	30 g	30 g	30 g	30 g	30 g	30 g
Salt	1 g	1 g	1 g	1 g	1 g	1 g
Coconut milk	141 g	90 g	141 g	90 g	141 g	90 g
Pumpkin Puree		34 g		36,5 g		39 g
Carrot Puree		34 g		36,5 g		39 g

2.2. Chemical Analysis Tools

Table 1 outlines the tools utilized for analyzing proximate content, including protein, carbohydrates, fat, water, and ash, as well as β-carotene, calcium, and iron.

Table 2. Chemical Analysis Tools

Parameter	**Tools**
Protein	100 ml kjeldahl flask, distillation apparatus, electric heater/burner, analytical balance, 25 ml burette, distillation flask, measuring cup, 5 ml volume pipette.
Carbohydrate	Carbohydrate content using by difference.
Fat	Filter paper, fat flask, soxhlet apparatus, electric heater, oven, analytical balance, fat-free cotton, tongs.
Water	Closed weighing bottle, desiccator, oven, analytical balance.
Ash	Cups, furnaces, analytical balances, desiccators, ashing furnaces, Bunsen.
Calcium	Spectrophotometer, Erlenmeyer flask, Burette, Mert oven (furnish), electric stove, mettler toledo XP 205 analytical balance, measuring cup, cuvette, electric heater, suction cup, test tube.
β-carotene	Buncher funnel, test tube, thin layer chromatography, 254 nm UV lamp, Pharmaspec UV-1700 Shimadzu UV-VIS spectrophotometer, centrifuge.
Iron	Atomic absorption spectrophotometer (AAS), 0.1 accuracy pH meter, furnace, digital scale, separator funnel, measuring flask, beaker, measuring cup, volumetric pipette, measuring pipette, filter with a pore size of 0.45 m, filter paper, glass bottles and jars with sharpening caps.

3. Result and Discussion

The table below summarizes research findings on layer cakes made with pumpkin and carrot puree.

Table 3. Results of Chemical Properties Analysis of Pumpkin and Carrot Puree Layer Cake

Formula	Protein content	Carbohydrate content	Fat content	Water content	Ash content	β-caroten content	Calcium content	Iron content
20%	7,60	28,07	15,24	45,37	0,91	42,22	34,49	1,45
30%	8,04	26,20	17,40	47,40	1,02	49,53	39,01	1,60
40%	8,89	26,26	18,31	48,18	1,15	54,41	41,60	1,80

3.1. Protein Content

The ANOVA test indicated a statistically significant difference, suggesting that increased treatment levels correspond to a more excellent puree addition. Furthermore, as the puree quantity in the pumpkin and carrot puree layer cake increases, the protein content also rises.

Table 4. Results of ANOVA Analysis of Protein Content of Layer Cakes

	Number of quadran	Middle quadran	F	Sig.
Between groups	1,804	,902	267,188	,000
Within groups	,010	,003		
Total	**1,814**			

The protein content of the layer cake increases with the addition of puree, which aligns with the protein content of pumpkin, quantified at 1.1 grams per 100 grams of ingredients [11]. Carrots exhibit a slightly higher protein content than pumpkin, recorded at 1.20 grams per 100 grams [12]. The combination of pumpkin puree and carrots creates a layer cake with a significantly elevated protein content. The layer cake's protein level is further enhanced by rice flour, which contributes 7 grams of protein per 100 grams of material [15]. Gopalan has noted that the protein content of carrots is 0.9%. Furthermore, including pumpkin plays a vital role in influencing the protein content of layer cakes. Research indicates that the incorporation of pumpkin flour positively impacts the protein content of cookies, whereby an increase in pumpkin inclusion leads to a higher protein content in the final product [16].

For toddlers, the recommended protein intake as a Protein Minimum Target (PMT) is 9 grams per 100 grams of food products [17]. Analysis of Treatment C (containing 40 percent puree) indicates a protein content of 8.89 grams per 100 grams of layer cake. This amount closely approaches the PMT requirements, indicating that the layer cake containing pumpkin and carrot puree meets 40% of the recommended protein intake.

3.2. Carbohydrate Content

The results of the ANOVA analysis indicate that the pumpkin and carrot puree layer cake exhibited the lowest carbohydrate content in the 30% treatment, which was not significantly different from the 40% treatment.

Table 5. Results of ANOVA Analysis of Carbohydrate Content of Layer Cakes

	Number of quadran	df	Middle quadran	F	Sig.
Between groups	4,533	2	2,26 7	16,451	,024
Within groups	,413	3	,138		
Total	**4,947**	**5**			

Adding puree to the layer cake formulation results in a reduction of carbohydrate content, as increased puree allows for a decreased flour requirement.

Specifically, carrots provide 9.30 grams of carbohydrates per 100 grams, while pumpkin offers approximately 10 grams per 100 grams [6,11]. In contrast, tapioca flour contains 88.20 grams of carbohydrates per 100 grams, and rice flour contains 80.1 grams per 100 grams (AKG, 2018). The formulation at 20% puree incorporates the highest carbohydrate levels, comprising 90 grams of rice flour and 61 grams of tapioca flour.

Pumpkin is known for its high concentrations of vitamins A, B, and C, alongside essential minerals and carbohydrates. A study by Hatta [16] indicated that a higher inclusion of pumpkin correlates with decreased carbohydrate levels in pumpkin cookies. It is important to note that high-temperature cooking can cause the degradation of specific carbohydrate molecules. Therefore, as processing temperatures rise, the carbohydrate content tends to diminish [13].

As per PMT guidelines, the recommended carbohydrate intake for toddlers is 71 grams per 100 grams of food products [17]. The analysis of the layer cake variations revealed that formulations with 20%, 30%, and 40% puree contained 28.07, 26.20, and 26.26 grams of carbohydrates per 100 grams, respectively. These values are significantly lower than those associated with Supplementary Feeding, indicating that the layer cake made with pumpkin and carrot puree does not fulfill the recommended protein levels for Supplementary Feeding.

3.3. Fat content

The results of the ANOVA test revealed a significant difference, suggesting that an increase in the amount of puree added correlates with a higher fat content.

Table 6. Results of ANOVA Analysis of Fat Content of Layer Cakes

	Number of quadran	**df**	Middle quadran	**F**	**Sig.**
Between groups	9,900	2	4,950	286,021	,000
Within groups	,052	3	,017		
Total	**9,952**	**5**			

This aligns with the observed increase in fat and protein levels, as a decrease in one nutrient's proximate content typically increases another to maintain nutritional balance [14].

The fat content of the materials contributes to this increase in overall fat levels. For instance, pumpkins contain only 0.3 grams of fat per 100 grams of product [15]. Carrots have a similar fat content of 0.30 grams per 100 grams [17]. In contrast, coconut milk has a higher fat content, providing 10 grams of fat per 100 grams [18]. Adding cooking oil to grease the baking sheet and prevent the dough from sticking enhances the high-fat content of the pumpkin and carrot puree layer cake [19]. Notably, the cooking oil utilized in the recipe is absorbed during processing, further augmenting the fat content [20].

According to dietary recommendations and PMT guidelines, the ideal fat content for toddlers is 15 grams per 100 grams of food products [15]. The analysis indicates that the formulation with 20% pumpkin and carrot puree contains 15.24 grams of fat per 100 grams of layer cake. This measurement aligns with the PMT requirements, signifying that the 20% pumpkin and carrot puree layer cake satisfies the recommended fat content for toddlers.

3.4. Water content

The water content associated with each treatment demonstrates significant variation; an increase in the amount of puree incorporated corresponds to an elevation in water content.

Table 7. Results of ANOVA Analysis of Water Content of Layer Cakes

	Number of quadran	**df**	Middle quadran	**F**	**Sig.**
Between groups	8,393	2	4,196	85,593	,002
Within groups	,147	3	,049		
Total	**8,540**	**5**			

This is attributable to the fact that 100 grams of pumpkin comprise 91 grams of water [21]. Similarly, carrots contain 90.4 grams of water per 100 grams of the vegetable [22]. Incorporating puree into each formulation results in increased water content due to the high water concentration in both pumpkin and carrots [23]. This elevated water content arises from the water added during the puree production process [24], where the ratio of ingredients to water is maintained at 1:1. Additionally, the water content in the dough is believed to influence the dough's rheological properties and the final product's overall quality. In addition to facilitating the gluten network formation, water is essential for the starch gelatinization process [25]. Diluted coconut milk is produced by mixing water in a 1:4 ratio with grated coconut, where coconut milk contains 80 grams of water per 100 grams of the product [18].

3.5. Ash content

ANOVA analysis demonstrated a statistically significant difference, indicating that an increase in puree addition correlates with higher ash content.

Table 8. Results of ANOVA Analysis of Ash Content of Layer Cakes

	Number of quadran	df	Middle quadran	F	Sig.
Between groups	,057	2	,028	66,544	,003
Within groups	,01	3	,000		
Total	**,058**	**5**			

In food processing, the elevated temperatures utilized can lead to an increase in ash content as the water content diminishes [20].

The rise in ash content is attributed to pumpkin, which contains 1.20 grams of ash per 100 grams of the product (AKG, 2018). Carrots have an ash content of 0.6 grams per 100 grams. Combining pumpkin and carrot puree results in a layer cake with notable ash content. Ash represents a component of food other than water and organic matter.

Additionally, ash content encompasses mineral elements. The residue remaining after the combustion of material until it is devoid of carbon elements and non-volatile components retains only the ash from the ignited organic compounds [26].

While there are no specific requirements for the ash content in pumpkin and carrot puree (PMT), the product must contain seven essential minerals, including iron, zinc, phosphorus, selenium, and calcium. These minerals demonstrate insensitivity and resistance to degradation during processing [27].

3.6. β-Carotene Content

The results of the ANOVA analysis indicated a significant difference; notably, an increase in the amount of puree correlated with elevated levels of β-carotene.

Table 9. Results of ANOVA Analysis of β-Carotene Content of Layer Cakes

	Number of quadran	df	Middle quadran	F	Sig.
Between groups	150,705	2	75,353	72,564	,003
Within groups	3,115	3	1,038		
Total	**158,820**	**5**			

This is attributed to the high levels of β-carotene in carrots and pumpkins. Pumpkin contains approximately 180 SI of β-carotene per 100 grams of material [28]. Carrots are significantly richer, with around 12,000 SI, equivalent to 8,285 grams of β-carotene per 100 grams [29]. The elevated β-carotene content in the 40% formula is primarily due to the dominant composition of pumpkin and carrot puree, along with other β-carotene-rich ingredients such as cooking oil.

There is no established requirement for the amount of β-carotene in PMT; the only stipulation is that it includes ten vitamins, with vitamin A being one of them. The processing of layer cakes can reduce β-carotene levels, given that this compound is sensitive to heat at temperatures exceeding 100°C for 120 minutes [30]. Conversely, vitamin A exhibits resilience to temperatures ranging from 145 to 165°C but is prone to degradation at temperatures above 180°C.

Research indicates that incorporating carrots can enhance β-carotene levels in rabbit meat nuggets [31]. Approximately 50% of carrots' composition consists of β-carotene, making them suitable for enriching various products such as cakes, breads, and biscuits [32].

Moreover, pumpkin comprises protein, starch, and β-carotene [33], with its β-carotene levels ranging from 9 to 19.9 mg per 100 grams. It has been observed that the concentration of β-carotene increases with higher percentages of pumpkin puree and arrowroot tuber flour (Putri Mulianananda). In a separate study, the β-carotene content in pumpkin mooncakes increased by 28.42% compared to the control [34].

3.7. Calcium Concentration

The analysis revealed notable differences in calcium content among the various treatments. This conclusion was reached using ANOVA test results.

Table 10. Results of ANOVA Analysis Calcium Concentration of Layer Cakes

	Number of quadran	df	Middle quadran	F	Sig.
Between groups	51,746	2	25,873	81,403	,002
Within groups	,954	3	,318		
Total	**52,699**	**5**			

The quantity of puree incorporated directly influences the calcium content. Pumpkin provides 45 mg of calcium per 100 grams, while carrots offer 33 mg of calcium per 100 grams [22,35]. Additionally, coconut milk is a valuable calcium source, containing 25 mg per 100 grams [18]. Although there are no specific requirements for the calcium content in Premixed Tomato (PMT), PMT needs to include seven minerals, with calcium being one of them. Minerals possess the advantage of being heat-resistant during processing [36]. In children and adolescents, insufficient intake or absorption of calcium and vitamin D can hinder growth in height, making it crucial to ensure an adequate supply of both nutrients during vital developmental phases [37].

3.8. Iron Concentrations

The ANOVA results indicate a statistically significant difference among the treatments, suggesting that increased puree added correlates with a higher iron content.

Table 11. Results of ANOVA Analysis Iron Concentration of Layer Cakes

	Number of quadran	**df**	Middle quadran	**F**	**Sig.**
Between groups	,122	2	,061	35,975	,008
Within groups	,005	3	,002		
Total	**,127**	**5**			

Pumpkins have 1 mg of iron per 100 grams [21], whereas carrots provide 0.66 mg per 100 grams [35]. There are no established requirements for iron in PMT; it comprises only seven minerals, with iron being one of them [17]. Steaming food does not markedly lessen iron content since there is no direct interaction with water, only steam [38]. An increase in iron can occur because the iron from foods released during cooking binds to the non-heme iron group, making it more absorbable [39].

4. Conclusion

The research findings indicate that layer cakes made with pureed pumpkin and carrots can effectively serve as a nutritional supplement for toddlers. The layer cake formulated with purees of pumpkin and carrot is notably rich in protein, carbohydrates, fats, water, β-carotene, calcium, and iron. The cake containing 20% puree demonstrated the highest carbohydrate levels among the variations studied. In contrast, the formulation with 40% purees displayed the most substantial amounts of protein, fat, water, β-carotene, calcium, and iron in the evaluated layer cakes.

5. Acknowledgments

The author wishes to thank the Institute for Research and Community Service (LP2M) at the University of Malang for generously funding this project.

6. References

[1] W. H. Organization, *Vitamin and mineral requirements in human nutrition*. World Health Organization, 2004.
[2] S. Arif, W. Isdijoso, A. R. Fatah, and A. R. Tamyis, *Strategic review of food security and nutrition in Indonesia: 2019-2020 update*. Jakarta: SMERU Research Institute Jakarta, 2020.
[3] Z. A. Bhutta, J. A. Berkley, R. H. J. Bandsma, M. Kerac, I. Trehan, and A. Briend, "Severe childhood malnutrition," *Nat. Rev. Dis. Prim.*, vol. 3, no. 1, pp. 1–18, 2017, doi: 10.1038/nrdp.2017.67.
[4] UNICEF, *Children, food and nutrition*. New York: UNICEF, 2019.
[5] I. Iskandar, "Pengaruh pemberian makanan tambahan modifikasi terhadap status gizi balita," *AcTion Aceh Nutr. J.*, vol. 2, no. 2, pp. 120–125, 2017, doi: 10.30867/action.v2i2.65.
[6] A. S. R. Putri and T. Mahmudiono, "Efektivitas pemberian makanan tambahan (PMT) Pemulihan pada status gizi balita di Wilayah Kerja Puskesmas Simomulyo, Surabaya," *Amerta Nutr.*, vol. 4, no. 1, pp. 58–64, 2020, doi: 10.20473/amnt.v4i1.2020.58-64.
[7] J. Kearney, "Food consumption trends and drivers," *Philos. Trans. R. Soc. B Biol. Sci.*, vol. 365, no. 1554, pp. 2793–2807, 2010, doi: 10.1098/rstb.2010.0149.
[8] New Zealand and Ministry of Health, "Food and nutrition guidelines for healthy children and young people (aged 2-18 years): a background paper."
[9] P. Loelinda and W. S. Windrati, "Substitusi tepung labu kuning (Cucurbita moschata Durch) dan koro pedang (Canavalia ensiformis L.) terhadap terigu pada pembuatan cake," *J. Agroteknologi*, vol. 11, no. 01, pp. 45–54, 2017, doi: 10.19184/j-agt.v11i1.5444.
[10] P. Christian *et al.*, "Vitamin A or β-carotene supplementation reduces but does not eliminate

maternal night blindness in Nepal," *J. Nutr.*, vol. 128, no. 9, pp. 1458–1463, 1998, doi: 10.1093/jn/128.9.1458.

[11] K. Seidel *et al.*, "Quality assessment of baby food made of different pre-processed organic raw materials under industrial processing conditions," *J. Food Sci. Technol.*, vol. 52, pp. 803–812, 2015, doi: 10.1007/s13197-013-1109-5.

[12] H. Raees-ul and K. Prasad, "Nutritional and processing aspects of carrot (Daucus carota)-A review," *South Asian J. Food Technol. Environ.*, vol. 1, no. 1, pp. 1–14, 2015, doi: 10.46370/sajfte.2015.v01i01.01.

[13] M. Hosen *et al.*, "Pumpkin (Cucurbita spp.): a crop to mitigate food and nutritional challenges," *Horticulturae*, vol. 7, no. 10, p. 352, 2021, doi: 10.3390/horticulturae7100352.

[14] N. T. de Castro *et al.*, "Influence of cooking method on the nutritional quality of organic and conventional Brazilian vegetables: A study on sodium, potassium, and carotenoids," *Foods*, vol. 10, no. 8, p. 1782, 2021, doi: 10.3390/foods10081782.

[15] Direktorat Gizi Departemen Kesehatan RI, *Daftar Komposisi Bahan Makanan*. Jakarta: Bhratara Karya Aksara, 2004.

[16] H. Hatta and M. Sandalayuk, "Pengaruh Penambahan Tepung Labu Kuning terhadap Kandungan Karbohidrat dan Protein Cookies," *Gorontalo J. Public Heal.*, vol. 3, no. 1, p. 41, 2020, doi: 10.32662/gjph.v3i1.892.

[17] Kementrian Kesehatan Republik Indonesia, "Perbaikan Gizi Untuk Generasi Agar Mampu Menangkan Persaingan."

[18] R. Anugrah, Sukarno, and F. R. Zakaria, "Minuman santan kelapa (Cocos nucifera L.) rendah lemak dengan penambahan ekstrak daun Stevia rebaudiana sebagai produk diversifikasi pangan berbasis santan kelapa," IPB (Bogor Agricultural University), 2011.

[19] R. Holinesti, "Pengaruh Substitusi Tepung Ubi Jalar Ungu Terhadap Kualitas Roti Tawar," *J. Teknol. Pertan. Andalas*, vol. 20, no. 2, pp. 50–57, 2016, doi: 10.25077/jtpa.20.2.50-57.2016.

[20] D. Sundari, A. Almasyhuri, and A. Lamid, "Pengaruh proses pemasakan terhadap komposisi zat gizi bahan pangan sumber protein," *Media litbangkes*, vol. 25, no. 4, pp. 235–242, 2015, doi: 10.22435/mpk.v25i4.4590.235-242.

[21] D. Ratnasari, Y. Yunianta, and J. M. Maligan, "Pengaruh tepung kacang hijau, tepung labu kuning, margarin terhadap fisikokima dan organoleptik biskuit [in press september 2015]," *J. Pangan dan Agroindustri*, vol. 3, no. 4, 2015.

[22] I. R. Wulansari, M. Devi, and L. Hidayati, "Pengaruh lama blanching terhadap karakteristik fisiko-kimia dan sensorik jus kecambah kedelai dan wortel," *Teknol. dan Kejuru.*, vol. 40, no. 2, pp. 157–168, 2017, doi: 10.17977/um031v40i22017p157.

[23] N. R. Ranonto, N. Nurhaeni, and A. R. Razak, "Retensi karoten dalam berbagai produk olahan labu kuning (Cucurbita moschata Durch)," *Nat. Sci. J. Sci. Technol.*, vol. 4, no. 1, 2015, doi: 10.22487/25411969.2015.v4.i1.4009.

[24] S. Gallego-Castillo and A. A. Ayala-Aponte, "Changes in physical properties of sweet potato due to effects of thermal pre-treatments for puree production," *Dyna*, vol. 85, no. 207, pp. 135–142, 2018, doi: 10.15446/dyna.v85n207.72876.

[25] E. Wilderjans, B. Pareyt, H. Goesaert, K. Brijs, and J. A. Delcour, "The role of gluten in a pound cake system: A model approach based on gluten–starch blends," *Food Chem.*, vol. 110, no. 4, pp. 909–915, 2008, doi: 10.1016/j.foodchem.2008.02.079.

[26] Direktorat Gizi Departemen Kesehatan RI, *Daftar Komposisi Bahan Makanan*. Jakarta: Bhratara Karya Aksara, 2001.

[27] B. Verawati and N. Yanto, "Substitusi tepung terigu dengan tepung biji durian pada biskuit sebagai makanan tambahan balita underweight," *Media Gizi Indones.*, vol. 14, no. 1, pp. 106–114, 2019, doi: 10.20473/mgi.v14i1.106-114.

[28] K. B. D. Permatasari, P. T. Ina, and N. M. Yusa, "Pengaruh penggunaan tepung labu kuning (cucurbita moschata durch) terhadap karakteristik chiffon cake berbahan dasar modified cassava flour (Mocaf)," *J. ITEPA Vol*, vol. 7, no. 2, p. 53, 2018, doi: 10.24843/itepa.2018.v07.i02.p06.

[29] R. Lidiyawati, F. Dwijayanti, and S. F. Pradigdo, "Mentel (permen wortel) sebagai solusi

penambah vitamin A," *J. Ilm. Mhs.*, vol. 3, no. 1, pp. 11–14, 2013.

[30] R. P. Sianturi, S. N. Aritonang, and I. Juliyarsi, "Potensi Tepung Wortel (Daucus carrota L.) dalam Meningkatkan Sifat Antioksidan dan Fisikokimia Sweet Cream Butter," *J. Ilmu dan Teknol. Has. Ternak*, vol. 13, no. 1, pp. 63–71, 2018, doi: 10.21776/ub.jitek.2018.013.01.7.

[31] S. Susanti, A. Katherinatama, D. A. Aryani, D. N. Afifah, T. Nugroho, and F. Arifan, "Effect on carrot substitution on nutrition facts, Beta-Carotene, and hedonic characteristic of rabbit meat nugget," *Prog. Nutr.*, vol. 25, no. 2, 2023, doi: 10.23751/pn.v25i2.12574.

[32] K. D. Sharma, S. Karki, N. S. Thakur, and S. Attri, "Chemical composition, functional properties and processing of carrot-A review," *J. Food Sci. Technol.*, vol. 49, no. 1, pp. 22–32, 2012, doi: 10.1007/s13197-011-0310-7.

[33] F. Yohandrey, K. Kasmiyetti, M. Handayani, D. Dwiyanti, and E. Edmon, "Pengaruh Pemberian Puding Labu Kuning (Cucurbita Moschata Durch) Terhadap Penurunan Tekanan Darah Pada Penderita Hipertensi," *J. Nutr. Coll.*, vol. 12, no. 4, pp. 311–317, 2023, doi: 10.14710/jnc.v12i4.40494.

[34] M. H. Widayati, N. Ratnaningsih, and B. Lastariwati, "The content of proximate, beta carotene, and organoleptic pumpkin mooncake as a snack to improve the immune system in the elderly," *AcTion Aceh Nutr. J.*, vol. 8, no. 4, p. 642, 2023, doi: 10.30867/action.v8i4.1182.

[35] R. Rahmayani, N. Yaumi, and F. Agustini, "Carbed (Carrot Bread) Sebagai Sayuran Instan Untuk Anak Kekurangan Vitamin a," in *Prosiding Industrial Research Workshop and National Seminar*, 2017, pp. 110–116.

[36] S. T. Paramitha, "Optimalisasi pemanfaatan mineral fosfor dalam membentuk kesehatan fisik anak usia dini melalui reedukasi keluarga," *Gladi J. Ilmu Keolahragaan*, vol. 9, no. 1, pp. 24–34, 2018, doi: 10.21009/GJIK.091.02.

[37] A. L. Bueno and M. A. Czepielewski, "The importance for growth of dietary intake of calcium and vitamin D," *J. Pediatr. (Rio. J).*, vol. 84, pp. 386–394, 2008, doi: 10.2223/JPED.1816.

[38] A. D. T. Fabbri and G. A. Crosby, "A review of the impact of preparation and cooking on the nutritional quality of vegetables and legumes," *Int. J. Gastron. Food Sci.*, vol. 3, pp. 2–11, 2016, doi: 10.1016/j.ijgfs.2015.11.001.

[39] N. Abbaspour, R. Hurrell, and R. Kelishadi, "Review on iron and its importance for human health," *J. Res. Med. Sci.*, vol. 19, no. 2, pp. 164–174, Feb. 2014.

IOP Conf. Series: Earth and Environmental Science **1446** (2025) 012017 doi:10.1088/1755-1315/1446/1/012017

Evaluation of Nutritional Composition of Synbiotic Yoghurt from Purple Sweet Potatoes and Fresh Milk

D Fitriyah[1*], M Jannah[1], D P Ayu[1], H Warsito[1] and A E Werdiharini[1]

[1]Departement of Health, Politeknik Negeri Jember, East Java, Indonesia

[*]Email : dinafitriyah@polije.ac.id

Abstract. Purple sweet potatoes contain anthocyanin and oligosaccharides. Anthocyanin has a role as an antioxidant, meanwhile oligosaccharides as a prebiotics. Processing sweet potatoes into synbiotic yoghurt is an alternative functional food to repair intestinal microflora and gastrointestinal function. The objective of this research was to assess the nutritional composition of purple sweet potatoes and fresh milk synbiotic yoghurt. Fresh milk was obtained from the Integrated Agricultural Development Unit of Politeknik Negeri Jember. This study is experimental research with 4 treatments. The proportions of fresh milk and purple sweet potato filtrate at each treatment level were P1(1:4), P2(2:3), P3(3:2), P4(4:1). The synbiotic yoghurt was measured for protein, fat, water, ash content, carbohydrate, energy from fat and total energy. The highest protein, fat, carbohydrate, energy from fat and total energy in synbiotic yoghurt was the P4 treatment, respectively (2.6%, 2.2%, 12.9%, 19.98 kkal/100g, and 80.9 kkal/ 100g. The fat content of all treatments was below 2,9%, so it is low-fat yoghurt. The ash content in all treatments ranges from 0.5-0.67 which is appropriate with SNI 2981:2009 standards.

1. Introduction

Food diversification using local ingredients is one of the efforts to develop functional foods that can increase nutritional value. Functional foods are part of the human diet and are proven to provide health benefits by reducing the risk of chronic diseases. Functional food development can be done by using microorganisms in food and beverages, such as fermented beverages. One of the fermented foods/beverages is yoghurt. In general, yoghurt comes from cow's milk fermented by lactic acid bacteria such as *Lactobacillus* and *Streptococcus thermophilus*. Yoghurt contributes to enhancing the gut environment, increasing immunity, avoiding lactose intolerance, lowering cholesterol, preserving the equilibrium of gut microbiota, and even for the treatment of colorectal cancer [1].

Synbiotic yoghurt is yoghurt with a combination of probiotics and prebiotics that can influence the quantity of microflora in the gastrointestinal tract by enhancing the amount of substrate to promote the growth of probiotic bacteria [2]. Including prebiotics acts as an effective growth environment for lactic acid bacteria such as Lactobacillus [3] so as to increase the growth and viability of these bacteria [4]. Prebiotic sources can be added to yoghurt to increase nutritional value. Prebiotic components are naturally found in ingredients that contain oligosaccharides. Oligosaccharides that cannot be digested are considered beneficial molecules if added to food or beverage products as they promote the development of probiotic bacteria. Some prebiotic sources that can be used to make yoghurt are rice bran [5], organic green banana flour [6], and some tubers [7]. One of the tubers that has the potential to be prebiotic is purple sweet potato.

Purple sweet potato contains oligosaccharides such as raffinose, stachyose, and verbascose. The concentration of raffinose in raw sweet potatoes is 48.04 ppm, while in steamed sweet potatoes, it is 39.50 ppm [8]. Purple sweet potato also contains anthocyanins that act as antioxidants. Oligosaccharides from the raffinose group are functional because they cannot be digested by human digestive enzymes so that they will be fermented by microflora in the colon, which can have a good influence on the growth of probiotics so as to improve gastrointestinal health and can benefit health. Purple sweet potato is used as a food ingredient whose utilization is still relatively limited. The potential of purple sweet potato as a prebiotic and antioxidant source can be used as an alternative in improving the quality, nutritional value, and functional value of yoghurt. Synbiotic yoghurt made from purple sweet potato and fresh cow's milk is expected to be a healthy option to help improve gut microflora. This research focused on examining the nutritional properties of synbiotic yogurt created from purple sweet potato and fresh cow's milk.

2. Materials and Methods

This study used purple sweet potato var. Ayamurasaki, cow fresh milk from State Polytechnic of Jember and commercial yoghurt starter. The study was conducted by varying the formulation of purple sweet potato filtrate and fresh milk ratio. The method of this study is experimental laboratory with development of four formulation. The synbiotic yoghurt from purple sweet potato and fresh milk will be tested for nutrient content (protein, fat, ash, water, carbohydrate, total energy and energy from fat) and compared according to yoghurt SNI 2981:2009 standards.

2.1. Making of Sweet Potato Filtrate

Sweet potatoes were sorted, peeled, washed, cut, and blanched. After that, weighing and crushing using a blender with the addition of water 1: 2 (purple sweet potato: water). After that, filtering is done using a filter cloth and precipitated to obtain the filtrate [9].

2.2. Making of probiotic starter

One liter of fresh milk was pasteurized at 70°C for 15-30 seconds with stirring. When the temperature reached 40°C, one gram of commercial yoghurt starter was inoculated into the pasteurized milk, then incubated for 18 hours at 37°C.

2.3. Formulation of Synbiotic Yoghurt

The formulation of synbiotic yoghurt was a comparison of fresh milk and filtrate of purple sweet potato. The fresh milk/ filtrate of purple sweet potato ratios used were 1:4, 2:3, 3:2, 4:1. Other ingredients for making synbiotic yoghurt was sugar.

2.4. The procedure of making synbiotic yoghurt from the purple sweet potato and fresh milk

The preparation of synbiotic yoghurt refers to previous research [9] [10], which begins with the preparation of sweet potato filtrate. Sweet potato filtrate and milk (following the formulation) were pasteurized at 70 °C for 15-30 seconds. When pasteurized, 8% sucrose was also added, stirred until dissolved, and then cooled. Starter inoculation was carried out after the temperature reached 40°C in a sterile manner, as much as 3%, and stirred until evenly distributed. Incubation was carried out for 18 hours.

2.5. Analysis of protein

The protein content was analyzed using the titrimetric method (Kjeldahl). This method consists of 3 stages, namely deconstruction, distillation, and titration. One gram of the sample was placed into a Kjeldahl tube with a capacity of 300 ml, then incorporated a mixture of selenium and concentrated H_2SO_4 and deconstructed at a temperature of 420 ˚C. The sample of the results of the destruction was then distilled with a distillation unit, added distilled water and NaOH, and installed an Erlenmeyer containing H_3BO_3 as a reservoir in the distillation unit. Distillation continuous until the amount of distillate is three times the volume of the original container. Throughout the distillation process, the color of the container changes from red to green. Titration of the distillate with HCl solution until the end point, which is the color change from green to red. Blank is also done every cycle of destruction. Protein content can be calculated using the formula:

$$\text{Protein Content (\%)} = \frac{(Vp - Vb) \times N \times 1.4007 \times Fk}{W}$$

Annotation:
Vp = Volume of 0.2 N HCL required for sample titration
Vb = Volume of 0.2 N HCL required for sample blank
N = Normality of 0.2 N HCl Solution
Fk = Convertion factor of protein
W = Sample weight (g) or sample volume (mL)

2.6. Analysis of fat

The fat content was evaluated through the hydrolysis method (Weibull) based on SNI 01-2891-1992 [11]. The principle of this method is the extraction of fat using non polar following the hydrolysis of the sample in an acidic environment to release the bound fat. A total of 1 gram of sample was hydrolyzed using HCl solution, distilled water, and several boiling stones and brought to a boil on a hot plate. The residue was filtered in a hot state using gray filter paper, followed by washing the residue with hot distilled water. The residue was heated in an oven at 105 ˚C until dry, and then place into filter paper sleeves (hulls). The hulls were loaded into a Soxhlet apparatus linked to a 300-ml fat flask with dried boiling stones of known weight. Soaking the hulls with hexane in the Soxhlet, then extraction was carried out. The fat residue was dried in an oven at 105˚C, then cooled and weighed to a fixed weight. Fat content was calculated by the formula:

$$\text{Fat Content (\%)} = \frac{(C - A)}{B} \times 100\%$$

Annotation:
A = Weight of empty fat flask (g)
B = Weight of sample (g)
C = Fix weight of fat flask + test sample after heating

2.7. Analysis of ash

The ash content was assessed using the gravimetric technique according to SNI 01-2891-1992 [11]. Weighing the sample (B) into a porcelain cup that has known weight (A), then the charring process is carried out until the smoke disappears. The charring was carried out in a furnace at a temperature of 550 ˚C until complete charring, then cooled in a desiccator. Weighing was carried out until a fixed weight was obtained (C). The ash content was determined by applying the formula:

$$\text{Ash Content} = \frac{(C - A)}{B} \times 100\%$$

Annotation:
A = Weight of empty porcelain crucible (g)
B = Weight of sample (g)
C = Fix weight of porcelain crucible + sample after heating

2.8. Water content analysis
The water content was analysed using the gravimetric method (SNI 01-2891-1992) [11]. Samples were weighed in a container of known weight, thereafter dried in an oven set at 105°C for 3 hours. Cooling was carried out in a desiccator, then weighed and repeated drying until a fixed weight was obtained.Water content was calculated using the formula:

$$\text{Water Content} = \frac{((A + B) - C)}{B} \times 100\%$$

Annotation:
A = Weight of empty container (g)
B = Weight of sample (g)
C = Fix weight of container + sample after heating

2.9. Analysis of carbohydrate, total energy and energy from fat
Carbohydrate and total energy were calculated using the calculation by difference method. The total carbohydrate was determined using the following formula:
Total carbohydrate (%) = 100% - (% protein + % fat + % water + % ash)
While total energy was calculated using the formula:
The Total Energy (kcal/100 g) = (%fat x 9 kcal/g) + (%protein x 4 kcal/g) + (%carbohydrate x 4 kcal/g)
Energy from fat is calculated by the formula:
Energy from fat (kcal/100 g) = % fat x 9 kcal/g.

3. Result and Discussion
Purple sweet potato synbiotic yoghurt and fresh milk in four treatments have been tested for nutritional content, including protein, fat, water content, ash content, carbohydrates, energy from fat, and total energy. The protein contained in milk is not always active; the protein will be active if there is proteolytic activity that converts protein into smaller peptides or active amino acids by fermentation, one of which is yoghurt products. The protein content of synbiotic yoghurt in this study is shown in Table 1. The results showed that the protein content of synbiotic yoghurt ranged from 1.01 to 2.60%. The highest protein content is in treatment 4 with the ratio of purple sweet potato and milk (1:4), which is 2.60%. This shows that the proportion of milk was higher so the protein content was higher; on the other hand, the proportion of sweet potato was higher, the protein content was lower. This is because the higher protein content in milk (3.2%) [12] compared to purple sweet potato protein (2.5%) [13] that is degraded by protease enzymes produced by bacteria into soluble amino acids.

The more protein dissolved in the material, so the protein content was higher. The protein found in yoghurt is the total amount of the protein from the ingredients used as well as the protein from the lactic acid bacteria contained in it. According to the Badan Standar Nasional (SNI yoghurt 2981:2009), the protein content set for yoghurt is at least 2.7%, so the protein content of synbiotic yoghurt in this study did not meet the SNI standards. This can be suspected due to the use of a starter consisting of five types of probiotic bacteria, so that there are bacterial species that can inhibit the growth of other bacterial species. Another study reported that yoghurt made using two bacterial cultures had high protein levels compared with the yoghurt using three or four bacterial cultures [14]. Protein levels can be influenced by the kind of substrate used and the quantity of viable bacterial cells. The increase in the quantity of viable bacterial cells can increase the amount of protease enzymes. When compared with other research on synbiotic yoghurt which used commercial inulin as prebiotic sources with the 3% inoculant contained a protein of 1.90% [15], so the synbiotic yoghurt in treatment 4 in this research still contained a greater protein level (2.60%).

Table 1. Nutrient content of synbiotic yoghurt from purple sweet potato and fresh milk and quality standard of yoghurt (SNI 2981: 2009)

Nutritional Content	P1	P2	P3	P4	SNI 2981: 2009 (Yoghurt)
Protein (%)	1.01	1.67	2.16	2.60	Min 2.7
Fat (%)	0.81	1.71	2.04	2.22	Min 3. (yoghurt) 0.6-2.9 (low fat yoghurt) Max. 0.5 (non fat yoghurt)
Water (%)	85.5	84.3	82.5	81.6	-
Ash (%)	0.49	0.54	0.61	0.68	Max 1.0
Carbohydrate (%)	12.16	11.74	12.71	12.90	-
Energy from fat (kcal/ 100 g)	7.25	15.39	18.32	19.98	-
Total Energy (kcal/ 100 g)	59.9	68.9	77.8	80.9	-

The results of fat content analysis in purple sweet potato synbiotic yoghurt and fresh milk are shown in Table 1. According to the the table, the fat content measured ranges from 0.81 to 2.22%. The P4 treatment has the highest fat content at 2.22%. This indicates that the higher the percentage of milk, the higher the fat content; otherwise, if the percentage of sweet potato is high, the lower the fat content. This result is linear with the protein content. Yoghurt that contains more cow's milk has a denser texture, and the density of yoghurt is related to the higher fat content. The difference in fat content can be affected by the fat level in the raw materials utilized, so that the greater the initial fat content, the greater the fat content of the yoghurt produced. According to BSN (SNI yoghurt 2981: 2009), the fat content of yoghurt for the low-fat category is in the range of 0.6-2.9%, so the fat content of synbiotic yoghurt in all treatments has met the quality requirements of low-fat yoghurt. This is in accordance with other research on synbiotic yoghurt, with the addition of several prebiotic sources, including suweg flour and mocaf flour, that the synbiotic yoghurt has a fat content ranging from 1.219 to 2.083 percent [14]. Low fat content in synbiotic yoghurt products can provide health benefits, one of which can potentially reduce cholesterol levels.

The water content of sweet potato and fresh milk synbiotic yoghurt showed that the water content ranged from 81.6 to 85.5%. The lowest water content was in treatment 4 with the ratio of purple sweet potato and fresh milk (1:4), which was 81.6%. These findings indicate that an increased proportion of fresh milk corresponds to a decrease in water content; on the other hand, the higher the percentage of sweet potato extract, the higher the water content. This is indicated by the higher total solids content in milk compared to sweet potato filtrate. High total solids content in milk as yoghurt raw material can reduce water content and increase viscosity. Low water content will produce a denser yoghurt texture. In another study, it was stated that the addition of prebiotics such as suweg flour, mocaf flour, and commercial inulin caused an increase in total solids, so that the water content decreased [15]. This is in contrast to this study, which used prebiotics in the form of filtrate, which has a lower total solids content than milk.

The ash content analysis results for purple sweet potato yoghurt and fresh milk presented in table 1 indicated that the ash content value varied between 0.49 and 0.68%. The higher the percentage of milk, the more ash content there is. Ash content represents the total minerals present in a food or product. The highest amount of ash was found in treatment 4 with the ratio of sweet potato and milk (1:4), which was 0.68%. This result is according with research on black rice juice synbiotic yoghurt with skim milk variations that the tendency of increasing ash content occurs along with the increasing concentration of skim milk, namely 1.03-1.45% [16]. Synbiotic yoghurt with the addition of prebiotics in the form of inulin, β-glucan, and Hi-maize has an ash content of 0.91–1.05% [17]. Based on SNI yoghurt standard 2981:2009, the ash content set for yoghurt is a maximum of 1.0, so the ash content of sweet potato and cow's milk yoghurt in this research aligns with the SNI yoghurt standards.

The carbohydrate content of purple sweet potato synbiotic yoghurt and fresh milk has been calculated using the by difference method. Carbohydrate is one of the macronutrients needed by humans, which functions to produce the energy needed for the human body. Carbohydrates function as a source of energy, sweetener in food, protein saver, and metabolic regulator. Carbohydrates are also significant in influencing the properties of food ingredients, including flavor, color, and texture. Carbohydrate levels obtained in this study ranged from 11.74 to 12.90% (Table 1). The highest carbohydrate was in the P4 treatment, which was 12.90%. Carbohydrate content depends on the ingredients added to the yoghurt. Carbohydrate sources can come from the fresh milk used and purple sweet potato extract. Fresh cow's milk contains total carbohydrates of about 4.3 g/100 g; after being made into yoghurt, the carbohydrate content becomes 7.75 g/100 g [18]. Carbohydrates contained in fresh cow's milk include lactose (4.14%), glucose (1.41%), and galactose (0.77%). The carbohydrate content of purple sweet potato is around 57.5 g/ 100 g. Although the carbohydrate content of sweet potatoes is higher, what was used in this research was the filtrate, so yoghurt with a high concentration of purple sweet potato filtrate had a lower carbohydrate content compared to yoghurt with a higher concentration of milk.

Purple sweet potato synbiotic yoghurt and fresh milk had their energy calculated using the by difference method. Energy is the result of protein, fat, and carbohydrate metabolism that functions as a power source for the body's metabolic processes, growth, temperature regulation, and physical activity. The energy is obtained from fat and also from components other than fat. Fat serves as a source of energy reserves. When digested, this fat will be broken down into fatty acids, which are then used by the body as fuel to produce energy. Energy from fat in sweet potato synbiotic yoghurt ranges from 7.25 to 19.98 kcal/100 g. The highest energy from fat is in the sweet potato synbiotic yoghurt treatment. The highest energy from fat is in treatment 4 with a higher percentage of fresh milk. The higher the fat content in an ingredient, the higher the energy from fat. According to the findings from the total energy analysis in purple sweet potato synbiotic yoghurt and fresh milk, it ranged from 59.9 to 80.9 kcal per 100 grams of product. The total energy in this synbiotic yoghurt is lower when compared to commercial yoghurt, namely Biokul plain yoghurt, which is 90 kcal/100 g. Therefore, this product can be used as an alternative as a snack for people on a low-calorie diet.

4. Conclusions

The development of functional food in the form of synbiotic yoghurt from purple sweet potato and fresh milk has nutritional content that closely align with the SNI yoghurt standard, especially the protein content, while the ash content in all treatments is in accordance with the SNI yoghurt standard, which is less than 1.0%. The fat content of all treatments is included in the low-fat yoghurt category. Total energy was also lower compared to commercial yoghurt products. Therefore, this synbiotic yoghurt has potential as a food or beverage for health.

5. Acknowledgement

This study was supported by a research grant of PNBP 2024 State Polytechnic of Jember.

References

[1] Yao S. 2023. The Value of Yogurt and Its Function in Health Care. *Theoretical and Natural Science*, **3** (1): 765-771. doi:10.54254/2753-8818/3/20220477

[2] Mawar LA, Aini N, Wijonarko G. 2018. Formulasi minuman sinbiotik dari susu dan ubi jalar menggunakan *Lactobacillus casei*. *JITIPARI*, **3** (1): 74–83.

[3] Riyanto RA, Nafisah DA. 2022. Telaah singkat aplikasi oligosakarida dari umbi-umbian lokal Indonesia sebagai prebiotik. *Journal of Food and Agricultural Product*, **2** (1): 15-22. http://journal.univetbantara.ac.id/index.php/jfap

[4] Fazilah NF, Ariff AB, Khayat ME, Rios-Solis L, Halim M. 2019. Influence of probiotics, prebiotics, synbiotics and bioactive phytochemicals on the formulation of functional yogurt. *Journal of Functional Foods*, (48): 387-399. https://doi.org/10.1016/j.jff.2018.07.039

[5] Demirci T, Aktas K, Sőzeri, Őztürk HI, Akin N. 2017. Rice bran improve probiotic viability in yoghurt and provide added antioxidative benefits. *J. Func. Food*. **36**:396-403

[6] Batista ALD, Silva R, Cappato LP, Ferreira MVS, Nascimento KO, Schmiele M, et.al. 2017. Developing a synbiotic fermented milk using probiotic bacteria and organic green banana flour. *J. Func. Foods.* **38**:242-250

[7] Sari FYK, Septiani, Melati A. 2024. Karakteristik organoleptic dan kimia produk yogurt sinbiotik dari umbi local. *Jurnal Medika Indonesia*, **5** (1): 9-18.

[8] Ricky D, Widanarni, Sukenda, Suprayudi MA, Zairin JM. 2017. Growth Performance and Health Status of Common Carp (*Cyprinus carpio*) Supplemented with Prebiotic from Sweet Potato (*Ipomoea batatas* L.) Extract. *Pakistan Journal of Nutrition*, (16): 155-163

[9] Kristi EE, Santosa B, Wirawan. 2023. Pembuatan dan analisa usaha minuman sinbiotik dari ubi jalar kuning (*Ipomoea batatas* L .) (Kajian konsentrasi starter dan lama fermentasi). *Journal of Industrial Engineering & Technology Innovation*, **1** (1): 43–53

[10] Mawar LA, Aini N, Wijonarko G. 2018. Formulasi minuman sinbiotik dari susu dan ubi jalar menggunakan *Lactobacillus casei. JITIPARI (Jurnal IlmiahTeknologi Dan Industri Pangan UNISRI)*, **3** (1): 74–83. https://doi.org/https://doi.org/10.33061/jitipari.v3i1.1991

[11] Badan Standar Nasional. 2009. SNI 2981:2009. Yoghurt. Badan Standar Nasional. Jakarta

[12] Rahmaniar RP, Aini DWN, Widyawati R, Mardjianto A. 2021. Perbedaan kadar protein, kadar lemak dan nilai pH susu sapi pada daerah dataran tinggi dan dataran rendah di Kabupaten Jombang. *Jurnal Sains Peternakan*, **9** (2): 100-103.

[13] Susanti I, Hartanto ES, Wardayanie NIA. 2012. Studi kandungan oligosakarida berbagai jenis ubi jalar dan aplikasinya sebagai minuman fungsional. *Journal of Agro-Based Industry*, **29** (2): 23-33.

[14] Purwantiningsih TI, Bria MAB, Kia KW. 2022. Kadar protein dan lemak yoghurt yang terbuat dari jenis dan jumlah kultur yang berbeda. *Journal of Tropical Animal Science and Technology*, **4** (1): 66-73. https://doi.org/10.32938/jtast.v4i1.967.

[15] Mukhoiyaroh S, Nurdyansyah F, Ujianti RMD, Affandi AR. 2022. Pengaruh penggunaan berbagai sumber prebiotic terhadap karakteristik kimia yoghurt sinbiotik. *Jurnal Teknologi Pangan*, **16** (1): 124-140.

[16] Wijayanti MI, Purwijantiningsih LME, Pranata FS. 2016. Kualitas yoghurt sinbiotik sari beras hitam (*Oryza sativa* L.) dengan variasi susu skim. *Journal Universitas Atma Jaya Yogyakarta*, 2016: 1-16.

[17] Jaman S, Islam MZ, Sojib MSI, Hasan MS, Khandakar MMH, Bari MS, et al. 2022. Physicochemical characteristics, sensory profile, probiotic, and starter culture viability of synbiotic yogurt. *J Adv Vet Anim Res.* **9** (4): 694-701. doi: 10.5455/javar.2022.i638.

[18] Hadjimbei E, Botsaris G, Chrysostomou S. 2022. Beneficial effects of yoghurts and probiotic fermented milks and their functional food potential. *Foods*, **11** (17): 2691. https://doi.org/10.3390/foods11172691.

IOP Conf. Series: Earth and Environmental Science **1446** (2025) 012018 doi:10.1088/1755-1315/1446/1/012018

Effect of humidity and temperature on the growth of staple stems of robusta coffee plants resulting from rejuvenation of bp 234 and bp 409 clones in the politeknik negeri jember collection garden

D Hartatie[1], H T Handayani[1]*, U Setyoko[1], H A Hudori[2]

[1]Department of Agricultural Production, Politeknik Negeri Jember, Jalan Mastrip Po. BOX 164, Jember, 68101, Indonesia
[2]Department of Agribusiness Management, Politeknik Negeri Jember, Jalan Mastrip Po. BOX 164, Jember, 68101, Indonesia

*Email : hatmiyarni@polije.ac.id

Abstract. The growth of tree stems in coffee plants is influenced by one of the microclimates, namely air humidity and air temperature. The purpose of this study is to determine the influence of air humidity and temperature around BP 234 and BP 409 robusta coffee plants on the growth of rejuvenated stems. The research method uses a descriptive analysis. Observation parameters include the number of leaf pairs, branch diameter, and plant height. Air humidity and temperature checks are carried out every 2 days (morning-noon-evening) at 5 points. Based on the results of research conducted for 2 months (June-July), the air humidity ranged from 50%-85% in June and 42%-83% in July. Meanwhile, the air temperature in June and July ranges from 27°C-34°C. The observation parameters of the number of leaf pairs, stem diameter, and plant height in June were 86-370 pairs of leaves; 7,73-15,75 mm; 83,33-130,00 cm for BP 234 clone and for BP 409 clone is 117-384 pairs of leaves; 8,90-18,80 mm; 63,75-151,00 cm. Meanwhile, in July, 90-484 pairs of leaves were obtained; 8,28-31,85 mm; 94,33-150,33 cm for BP 234, and BP 409 is 140-420 pairs of leaves; 10,80-22,73 mm ; and 93,00-172,50 cm.

1. Introduction

Coffee is one of the plantation crops that has been cultivated for a long time and has significant economic value. The coffee industry in Indonesia has a very important role in the national economic sector, both as a source of foreign exchange income from exports and as a source of community income and to meet domestic coffee needs [1].

In general, robusta coffee is cultivated using clones that have been developed to improve coffee production and quality. The two clones that are often used are BP 234 and BP 409. However, in some cases, robusta coffee can experience stress caused by environmental factors, such as sub-ideal air temperature and humidity.

Rejuvination is the process of restoring the condition of coffee that has been damaged or its quality has decreased. The rejuvination process can be carried out in several ways, with single rods and double rods. This is in accordance with the purpose and condition of the plant. Rejuvinated plants are grown by selected water shoots or wiwilans on the trunk as a place for the growth of primary branches. The growth of water buds and branches in coffee plants is affected by the environment in which they grow, such as temperature and air humidity. The principle of rejuvination is to rejuvenate and improve the quality of old coffee plants by growing production branches on coffee plants that are no longer productive so that they return to being like young coffee plants. According to [2] Rejuvination can form new branches of production on a sustainable basis in sufficient quantities. The coffee plant, after being rejuvated at the age of 2-3 years, can be produced again like the corn phase [3][4][5].

Robusta coffee can grow at a relatively low altitude of up to 800 meters above sea level. In general, robusta coffee can grow well in areas with an average annual temperature between 22°C -26°C. For optimal growth, robusta coffee requires environmental conditions with temperatures between 22°C - 25°C, rainfall between 2,000-2,500 mm/year, and 2-3 months of dry season [6]. However, there are obstacles in coffee cultivation that need to be overcome, namely the water deficit that occurs in the dry season, even in areas that have a lot of rainfall. One of the efforts to overcome this problem is by irrigation using piping.

Based on the above background, it is necessary to conduct research to investigate the effect of temperature and air humidity on the growth of water shoots selected as robusta coffee staples of BP 234 and BP 409 clones as a result of rejuvation. The results of this study are expected to provide useful information to improve the production and quality of robusta coffee.

2. Materials and Methods

2.1. Materials

The materials used in this study are BP 234 clone robusta coffee plants and BP 409 clones. Meanwhile, the tools used are calipers, thermohydrometers, and roll meters.

2.2. Methods

This research consists of several stages, which begin with research preparation, observation, and maintenance.

2.2.1. Preparation of research. The preparations carried out include observation of the growth of selected shoots and observation of the growing environment of BP 234 and BP 409 clone plants as well as the preparation of tools to be used.

2.2.2. Observation. Check the temperature and air humidity, as well as make observations with parameters of the number of leaf pairs, branch diameter and plant height.

2.2.3. Maintenance. Irrigating so that the humidity of the garden can be controlled in the range of 70%-90%.

3. Results and Discussion

3.1. Air Temperature and Air Humidity

Observation of air temperature and humidity is carried out every two days 3 times at 5 points, namely in the morning (at 09.00 WIB), afternoon (at 13.00 WIB), and afternoon (at 15.00 WIB) for 3 months (May, June and July). Air Humidity in May is in the range of 37%-74%, in June in the range of 52%-80% and in July in the range of 51%-82%. The distribution of air humidity for morning, afternoon, and evening in May, June, and July can be seen in figure 1. In May, it can be seen that the air humidity is low both in the morning and during the day, meaning it is drier than June and July. For June, the air humidity experienced a higher value, but it was still included in the air humidity range for robusta coffee plants.

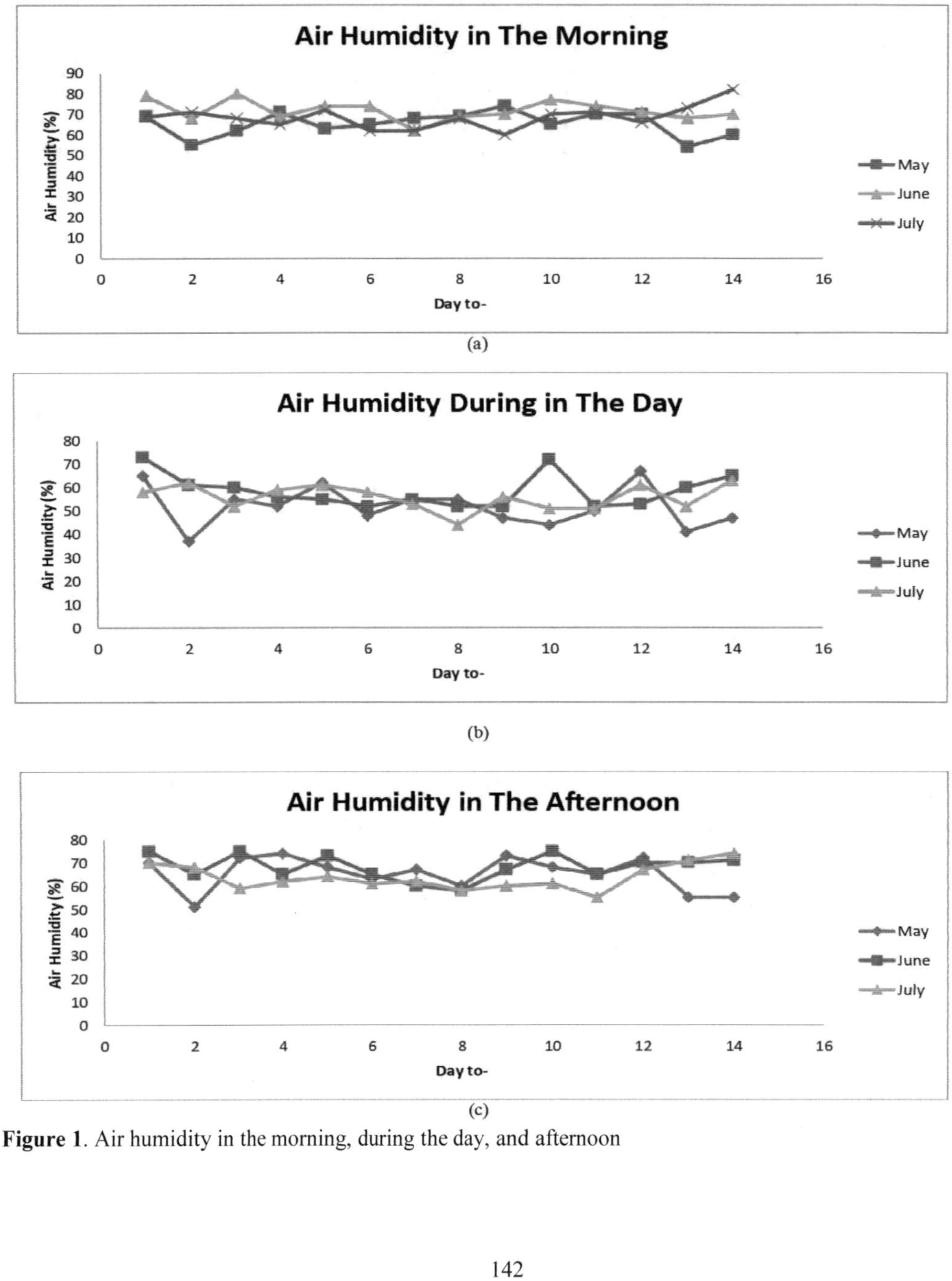

(a) (b) (c)

Figure 1. Air humidity in the morning, during the day, and afternoon

Air humidity is also one of the important factors that affect the development and growth of a plant. In general, plants need low humidity to stay alive. However, humidity conditions that are too high can cause disease attacks on cultivated plants, so they can be very detrimental to plant growth and yield [7]. Pest and disease infestations that are too high can have a significant negative impact on plant growth and expected yields. However, keep in mind that high humidity conditions can also have a positive impact, such as helping the growth of pathogens that are good for plants, so that they can accelerate the growth and development process of plants [8].

Meanwhile, the temperature observation from May to July is between 26°C-35°C. Observations are made in June and July when the weather is extreme so that the air temperature can reach 34°C, because the optimal conditions for the growth of robusta coffee are in the temperature range of 24°C-30°C and the air humidity is in the range of 80-90% [9]. According to [10] that air temperature plays a very important role in the life of a plant, because it can affect the physiological processes that take place in each plant. Temperature can also affect biochemical processes, especially enzymatic processes, which are crucial for the growth and development of a plant. Therefore, suitable or ideal temperature characteristics are needed to produce optimal growth and yields. The distribution of air temperature in the morning, afternoon, and evening is shown in figure 2.

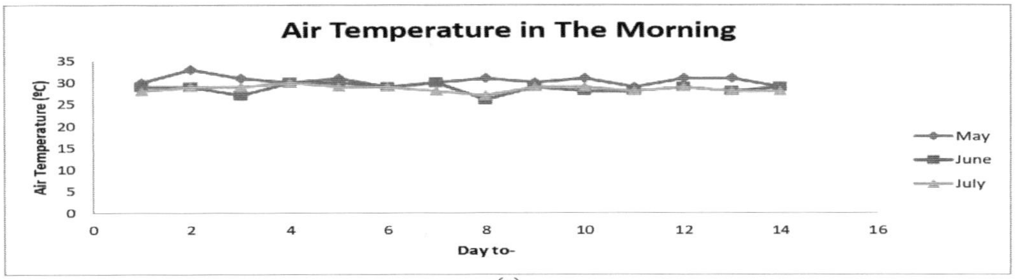

(a)

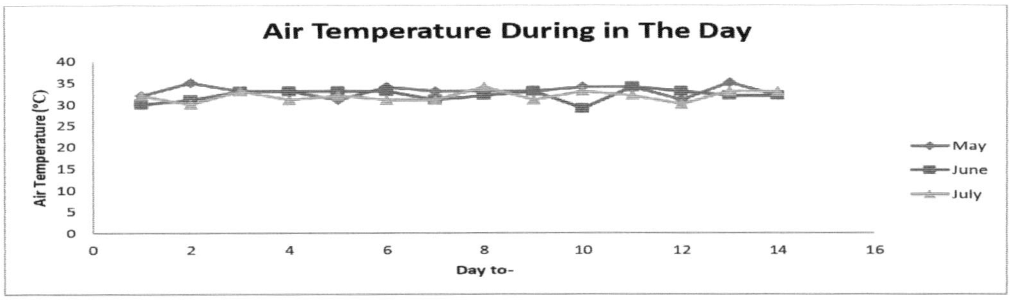

(b)

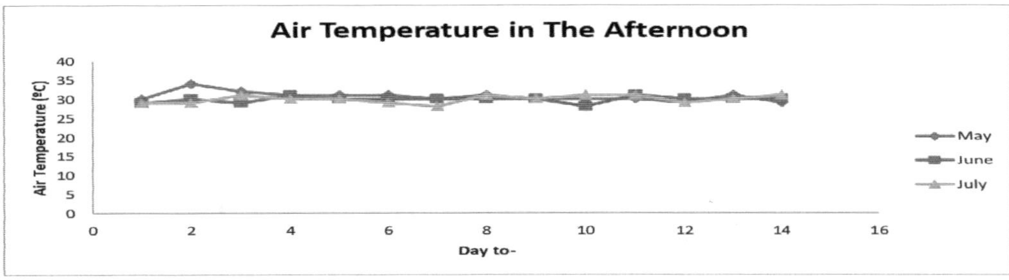

(c)

Figure 2. Air Temperature in The Morning, During The Day, and Afternoon

3.2. Number of Leaf Pairs

The number of leaf pairs in robusta coffee determines the production of coffee beans, affects the quality of coffee beans, affects plant survival, and indicates plant health. Figure 3 shows that the number of leaf pairs in BP 409 clone robusta coffee is more than BP 234. This is because BP 409 is more resistant to extreme temperature and humidity conditions, so the number of leaf pairs produced in June and July is large. In addition, the characteristics of BP 409 are tolerant of difficult environmental conditions, such as drought and high temperatures [11]. Temperatures that are too high can cause the number of robusta coffee leaves to decrease [12]. This is because temperatures that are too high can cause stress on plants, thus inhibiting leaf growth. If the temperature is too low, it can also cause the number of robusta coffee leaves to decrease. Temperatures that are too low can cause leaf growth to slow down. According to [13] the optimal temperature for robusta coffee is 20°C-25°C. In addition to temperature, air humidity also greatly affects the number of leaf pairs. Humidity that is too high can cause the number of leaf pairs to decrease. This is because too high humidity can cause diseases in plants, thus inhibiting leaf growth [14]. Meanwhile, too low humidity can also cause the number of robusta coffee leaves to decrease. This is because too low humidity can cause leaf growth to slow down. According to [15] the optimum air humidity is 60%-80%. The number of pairs of leaves is large, so it can be known that the number of branches is also large so that the percentage of flower growth is also large. The number of BP 409 clone leaf pairs in June is between 117 pairs – 384 pairs and in July is between 92 pairs – 420 pairs.

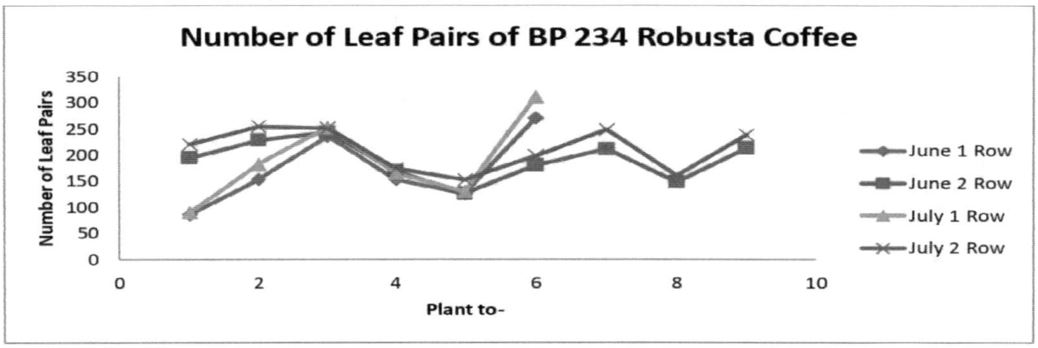

(a)

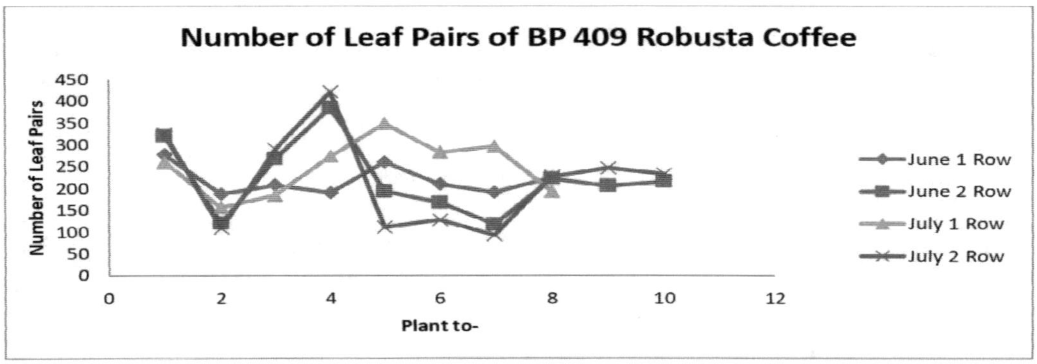

(b)

Figure 3. Number of Leaf Pairs of BP 234 and BP 409 Robusta Coffee During May, June and July

3.3. Branch Diameter

The temperature and humidity of the air have an inversely proportional relationship for the branch diameter parameters. Temperatures that are too high can cause branch diameters [14], while too high humidity can cause the diameter of the branch to increase [15]. Figure 4 shows that BP 409 clone robusta coffee can adapt to high temperatures and high humidity due to its characteristics that are able to withstand extreme conditions. This is evidenced by the larger diameter size compared to the BP 234 clone. This research has been in accordance with the statement [13] that a larger branch diameter can lead to higher coffee production, while a smaller branch diameter can lead to lower coffee production [15].This is because a larger branch diameter can indicate that BP 409 coffee plants have a better root system than BP 234, so they can absorb nutrients and water more effectively. Therefore, higher coffee production can be produced. The branch diameter of BP 409 in June is between 8.90 mm – 18.08 mm and in July is between 10.80 mm – 22.73 mm.

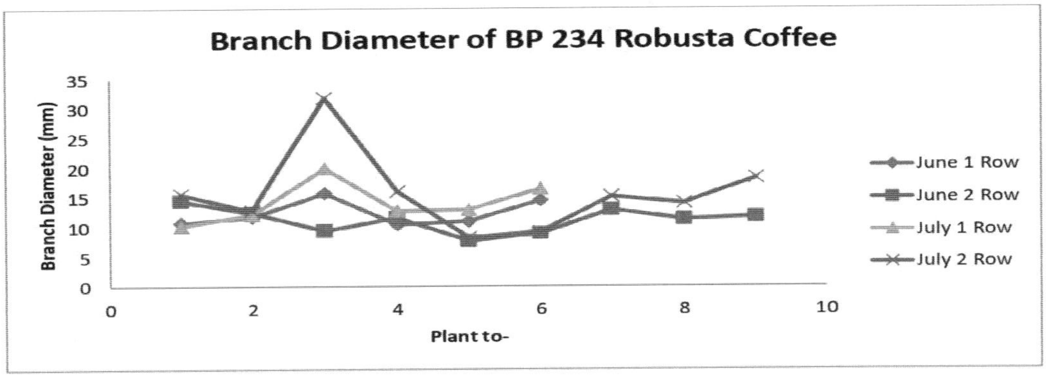

(a)

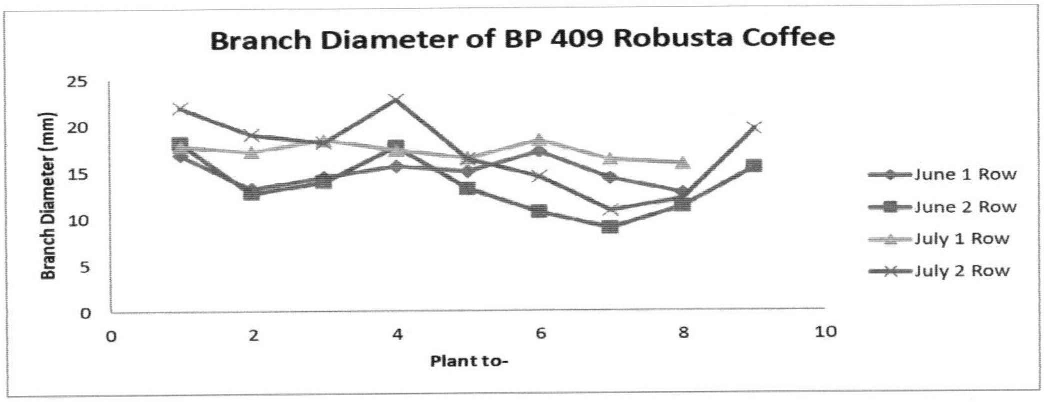

(b)

Figure 4. BP 234 and BP 409 Robusta Coffee Branch Diameter During May, June and July

3.4. Plant Height

The height of robusta coffee plants is influenced by several factors, including temperature and humidity. Temperatures that are too high or too low can cause stress in plants so that plant height decreases. If the temperature is too high (more than 30°C), it can cause the plant to dry out and experience stress so that the plant height decreases[13]. Meanwhile, temperatures that are too low (less than 15°C) can cause plants to become slow and stressed, so that plant height decreases. In addition, humidity that is too high or too low can cause stress on the plant so that the height of the plant decreases [15]. If the humidity is too high (more than 90%), it can cause the plant to dry out and experience stress so that the plant height decreases. Meanwhile, if the humidity is too low (less than 40%), it can cause the plant to slow down and experience stress so that the plant height decreases. Figure 5 shows that the BP 409 clone has a higher plant height than BP 234, namely in June between 63.75 cm – 151.00 cm and in July between 93.00 cm – 172.50 cm.

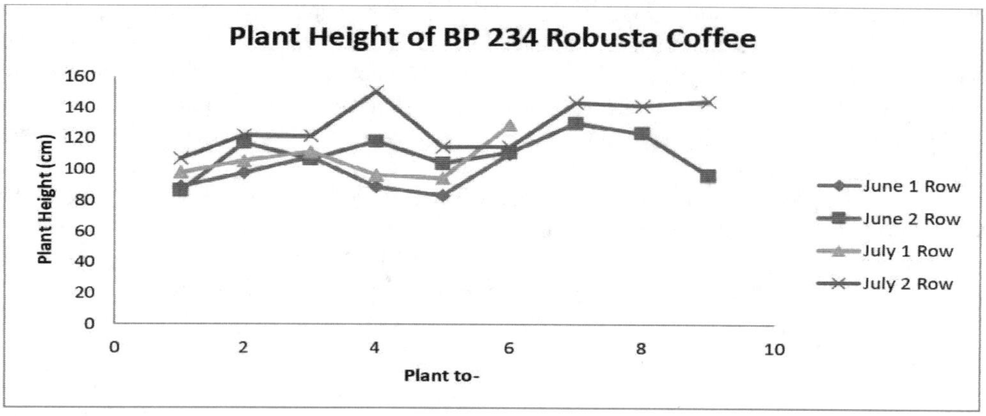

(a)

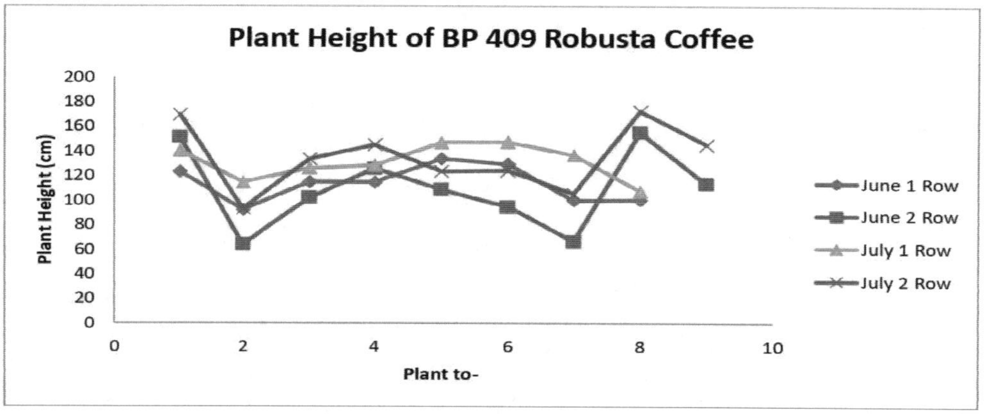

(b)

Figure 5. BP 234 and BP 409 Robusta Coffee Plant Height During May, June and July

4. Conclusion

The results showed that the temperature and air humidity around the robusta coffee plants from the rejuvenation of BP 409 clone supported growth so that it produced a better number of leaf pairs, stem diameter, and plant height (higher yield) than BP 234 robusta coffee. The growth of BP 409 buds in rejuvenation carried out in the previous study had a growth percentage of 100%, so the results obtained in the current study are also higher than BP 234. Therefore, it can be said that the research conducted at the Jember State Polytechnic Collection Garden is said to be successful. The number of BP 409 clone leaf pairs in June is between 117 pairs – 384 pairs and in July is between 92 pairs – 420 pairs. The branch diameter of BP 409 in June is between 8.90 mm – 18.08 mm and in July is between 8.8 mm – 17.23 mm. The height of BP 409 plants in June is between 63.75 cm – 151 cm and in July is between 93 cm – 172.5 cm.

5. Acknowledgement
The author would like to thank P3M Jember State Polytechnic for supporting funding for the applied research scheme through the 2024 PNBP funding source so that this research can be carried out.

References

[1] Rawanda et al 2021 Pengaruh Pengelolaan Kopi Robusta Terhadap Kondisi Sosial Ekonomi Masyarakat di HKM Binawana Register 45B Desa Tri Budisukur , Kecamatan Kebun Tebu , Lampung Barat , Lampung *Jopfe J.* **1** 1–10
[2] Kementerian Pertanian 2014 *Pedoman Teknis Budidaya Kopi Yang Baik (Good Agriculuture Practices/GAP on Coffee)*
[3] Evizal R, Prijambada I D, Widada J, Erry Prasmatiwi F, Fak Pertanian Universitas Lampung D, Lampung B and Fak Pertanian Universitas Gadjah Mada D 2010 Pengaruh Tipe Agroekosistem Terhadap Produktivitas Dan Keberlanjutan Usahatani Kopi *J. Agrotropika* **15** 17–22
[4] PTPN XII 2019 *Robusta Coffee*
[5] Nurhati Atikah, Muhammad Rahmad A W 2024 Pemangkasan Meningkatkan Produktivitas Tanaman Kopi Robusta (Coffea canephora Pierre ex. A. Froehner) di Malang, Jawa Timur *Bul. Agrohorti* **2**
[6] Prastowo, B., Karmawati, E., Indrawanto, C., & Munarso S J 2010 Budidaya dan Pasca Panen Kopi
[7] Pautasso M, Döring T F, Garbelotto M, Pellis L and Jeger M J 2012 Impacts of climate change on plant diseases-opinions and trends *Eur. J. Plant Pathol.* **133** 295–313
[8] Bambang Prastowo 2010 *Budidaya dan Pasca Panen Kopi* (Pusat Penelitian dan Pengembangan Perkebunan)
[9] Arisandi D P and Munandar D E 2015 Respon Karakteristik Fisiologi dan Pertumbuhan Bibit Kopi Robusta (Coffea canephora) Klon BP 358 dan BP 308 Pada Berbagai Tingkat Naungan *Berk. Ilm. Pertan.* **10** 1–5
[10] K. Amutha, & Priya K J 2011 Effect of pH, temperature and metal ions on Amylase activity from Bacillus subtilis KCX 006 *Int. J. Pharma Bio Sci.* **2** 407–13
[11] Wahyuni W 2020 Toleransi Klon Kopi Robusta BP 409 Terhadap Kondisi Lingkungan yang Sulit *J. Ilmu Pertan.* **14** 1–10
[12] Pardiyono D 2015 Pengembangan Klon Kopi Robusta BP 409 *J. Ilmu Pertan.* **10** 1–10
[13] Rachmadi R 2018 Produksi Biji Kopi dari Klon Kopi Robusta BP 409 *J. Ilmu Pertan.* **13** 1–10
[14] Santoso S 2017 Toleransi Klon Kopi Robusta BP 409 Terhadap Penyakit dan Hama *J. Tanam.* **16** 1–8
[15] Purnomo P 2019 Kualitas Biji Kopi dari Klon Kopi Robusta BP 409 *J. Tanam.* **18** 1–8

IOP Conf. Series: Earth and Environmental Science **1446** (2025) 012019 doi:10.1088/1755-1315/1446/1/012019

Growth and production of shallots (*allium cepa* L) at various volumes and watering intervals: case study in nganjuk district

F Krismiratsih[1*], S Fibriani[1], K Ni'mah[1], E R Pristiwaningsih[1], Suyitno[1]

[1]Department of Agribusiness Management, Politeknik Negeri Jember

[*]Email : Fitri.krismiratsih@polije.ac.id

Abstract. Shallots are one of the leading horticultural commodities in Nganjuk Regency. Water is one of the limiting factors in shallot cultivation activities to require sufficient water availability every day to produce optimally. This research aims to provide recommendations for appropriate water supply in shallot cultivation activities. This experiment was carried out from August to September 2024. Red onion planting was carried out in polybags with a diameter of 15 cm with a planting media composition of soil: compost: cocopeat with a volume ratio of 1:1:1. The experimental design used was a split-plot design with 4 replications. The first plot is the irrigation volume based on Field capacity moisture content which includes 4 levels, namely 50% (p1), 100% (p2), 150% (p3), and 200% (p4) and the subplot is the watering interval which consists of 4 The levels are 1 time a day in the morning (p1), 1 time a day in the afternoon (p2), and twice a day (p3). Observation variables in this study included plant height at 3, 4, and 5 WAP, number of leaves at 3.4 and 5 WAP, wet weight of tubers, and number of tubers per cluster.

1. Introduction

Shallots (*Allium Cepa* L.) are one of the important horticultural commodities in Indonesia which are used as spices, have high economic value, and also have good market prospects [1]. In addition, in 2023 shallots will rank first in vegetable production in Indonesia, which is 1.98 million tons or contribute a production percentage of 13.59% of vegetables with the highest distribution area of shallot production, namely East Java Province with a production of 448.000 tons with a percentage of 24% [2]. The five provinces contributing are Nganjuk Probolinggo, Malang, Sampang, and Bojonegoro. Of the five provinces, Nganjuk contributed the most, namely 38.72%, followed by Probolinggo at 11.67% and finally Bojonegoro at 6.63% [3]. The high shallot production in Nganjuk is attributed to the favorable climate and soil conditions for shallot cultivation, with optimal temperatures ranging between 25-32°C and light exposure exceeding 70% during the dry season [4].

Shallot plants have a shallow root system so they are very sensitive to water shortages due to their inability to absorb as much water as support-rooted plants water availability can be a limiting factor in shallot cultivation activities [5], this causes shallots to require sufficient water availability to produce optimally. Therefore, an efficient watering strategy is needed to increase shallot production. During the dry season, shallot plants need additional water to maintain the availability of water in the roots of the plant because if there is a lack of water the plant will respond by inhibiting plant growth. Regulating watering through volume and frequency is one way to regulate the availability of water in shallot plants at each phase of their growth. Based on the research results [5] irrigation treatment of watering every two days gives the best results for the growth of shallot plants, while the research results [6] 100% irrigation volume had significantly higher yields than other treatments.

This study attempts to answer problems related to watering efficiency. By varying the volume and interval of watering, this study will determine the optimal watering pattern to increase the growth and production of shallots, which in turn can make the use of water resources more efficient.

2. Material and Method

2.1. Place and Time of Implementation

Field trials have been carried out from July to September 2024 at the Research Field at Campus 3 Jember State Polytechnic, Nganjuk Regency, located in Ringin Anom Villag, Nganjuk Regency East Java..

2.2. Tools and Materials

The materials used in this research were Tajuk variety shallot seeds, label paper, polybags measuring 20 cm x 25 cm, manure, basic fertilizer (NPK), KNO3 fertilizer, and insecticide. The tools used included digital scales, ovens, calipers, and stationery.

2.3. Research Methods

The experimental design used was a randomized block design with 4 replications. The first treatment was the volume of irrigation based on the value of soil moisture content capacity which included 4 levels, namely 50% (p1), 75% (p2), 100% (p3), and 150% (p4), and the next treatment was the watering interval consisting of 3 levels, namely once a day in the morning (p1), once a day in the afternoon (p2), and twice a day (p3) so that there were 48 experimental units.

2.4. Preparation of growing media

The growing media form of soil, cocopeat, and manure are homogenized with a ratio of 1:1:1. To determine the water requirement based on the moisture content of the soil field capacity, a gravimetric method is used. Then the growing media is inserted into a polybag with a diameter of 25 cm.

2.5. Planting

Planting of shallots is done using a direct planting system, where the prepared planting material is cut off the top 1/3 and planted in polybags one by one with almost uniform sizes.

2.6. Fertilization

The fertilizer used is Phonska 16:16:16 fertilizer applied when the plants are 2 and 4 weeks old, fertilization with Urea and KNO3 at the ages of 3.5 and 6 MST.

2.7. Data Observation and Analysis

lant growth variables were observed, including plant height at 3, 4, and 5 week after planting (WAP), number of leaves at 3, 4, and 5 week after planting (WAP), number of tubers, and 3.4. Wet Weight of Tubers. The data were analyzed using ANOVA and Duncan's Multiple Range Test (DMRT) at a 5% significance level.

3. Result and Discussion

3.1. Plant height (cm)

Plant height observations were made when the plants were 3 week after planting(WAP) and observations were made once a week until the plants were 5 week after planting (WAP). Plant height observations were measured from the surface of the planting medium to the tip of the highest leaf using a ruler. Based on the results of the analysis of variance, it was found that the volume of irrigation did not show significantly different results in the components of plant height observations. Meanwhile, the treatment of watering frequency had a significant effect on plant height at the ages of 4 MST and 5 MST. The highest average plant height was obtained in the treatment of watering frequency 2 times a day in the morning and evening, which were 23.4 cm and 29.0 cm not significantly different from the treatment of watering in the afternoon and significantly different from the treatment of irrigation in the morning (Table 3.1).

Table 1. Average plant height of shallots at 3, 4, and 5 MST due to the influence of volume and frequency of irrigation

Treatment	Observation Time		
	3 WAP	4 WAP	5 WAP
Volumes			
50 % KL	19,2	21,8	27,3
75 % KL	20,0	23,6	28,1
100 % KL	20,5	22,9	28,9
150 % KL	20,2	21,7	27,8
Uji DMRT	tn	tn	tn
Watering Intervals			
1 time a day in the morning	19,4 a	21,1 b	26,3 b
1 time a day in the afternoon	20,1 a	23,1 a	28,9 a
twice a day, morning and evening	20,4 a	23,4 a	29,0 a
Uji DMRT	tn	n	n

Description: The average number followed by the same letter in a column or row indicates no significant difference between treatments based on the DMRT test at the 5% significance level

3.2. Number of leaves (blades)

Plant height observations were carried out when the plants were 3 WAP old and observations were carried out once a week until the plants were 5 WAP old. Observations of the number of leaves were carried out by counting all the leaves on the plant. Based on the results of ANOVA, the effect of the interaction between watering volume and watering interval showed no significant effect on the number of leaves. The independent effect of watering volume showed no significant difference in the number of leaves, while the watering interval showed a significant effect on the number of leaves. The provision of irrigation volumes of 100% KL and 75% Kl had the highest average number of leaves in week 4, namely 24.5 leaves and 23.3 leaves, and in the observation of week 5, the provision of irrigation volumes of 100% KL and 75% Kl had the highest average number of leaves in week 5, namely 27.8 leaves and 26.7 leaves (Table 3.2).

Table 2. Average number of leaves (blades)of shallots at 3, 4, and 5 WAP due to the Various Volumes and Watering Intervals

Treatment	Observation Time		
	3 WAP	4 WAP	5 WAP
Volumes			
50 % KL	18,3 b	21,1 b	24,4 b
75 % KL	18,4 ab	23,3 ab	26,7 ab
100 % KL	19,4 a	24,5 a	27.8 b
150 % KL	17,9 b	21,4 b	24,5 b
Uji DMRT	tn	n	n
Watering Intervals			
1 time a day in the morning	18,1	21,5	24,9
1 time a day in the afternoon	18,8	23,7	27,0
twice a day, morning and evening	18,6	22,5	25,7
Uji DMRT	tn	tn	tn

Description: The average number followed by the same letter in a column or row indicates no significant difference between treatments based on the DMRT test at the 5% significance level

3.3. Number of tubers

The result of number of tubers was carried out at the time of harvest at 60 days after planting (DAP), by counting the number of shallot tubers in each clump. Based on the results of ANOVA, the interaction effect between watering volume and watering interval showed no significant effect on the number of tubers per clump. Independently, each application treatment of volume and frequency showed different results. The watering volume treatment showed the highest number of tubers in the 100% irrigation volume treatment with several tubers of 5.3, while the irrigation frequency treatment showed the highest results in the irrigation volume treatment twice a day in the morning and evening with some tubers of 4.9 but did not show significantly different results with the watering frequency treatment once a day in the afternoon, namely 4.9 tubers.

Table 3. Average number of shallot tubers due to the various volumes and watering intervals

Treatment	Number of tubers
Volumes	
50 % KL	4,3 b
75 % KL	5,1 a
100 % KL	5,3 a
150 % KL	4,4 b
Interfal Penyiraman	
1 time a day in the morning	5,54 b
1 time a day in the afternoon	4,92 a
twice a day, morning and evening	4,94 a

Description: The average number followed by the same letter in a column or row indicates no significant difference between treatments based on the DMRT test at the 5% significance level

3.4. Wet Weight of Tubers,

Observation of wet tuber weight was observed after harvesting, at the age of 60 DAP. Based on the results of the ANOVA analysis, the effect of interaction between volume and frequency of watering showed a significantly different effect on the wet tuber weight. The combination of 75% and 100% volume treatments with watering twice a day showed the best results, namely 46.8 grams and 47.3 grams respectively, and also showed no significant difference with the 100% volume treatment and watering once a day in the afternoon, namely 45.5 grams significantly different from other treatments. The smallest tuber weight was in the combination of 50% volume treatment and watering once a day in the morning (Table 3.4)

Table 4. Average wet weight of shallot tubers based on Various Volumes and Watering Intervals.

Volume	Watering Intervals					
	1 time a day in the morning		1 time a day in the afternoon		twice a day, morning and evening	
50%	27,3	cB	31,0	dA	28,3	cAB
75%	37,3	aC	41,5	bB	46,8	aA
100%	39,8	aB	45,0	aA	47,3	aA
150%	32,5	bA	35,5	cA	35,5	bA

Description: Lowercase letters (vertical) and capital letters (horizontal), numbers followed by different letters show significant differences according to the DMRT test at the 5% level.

3.5. Discussion

The results of the study showed that with a watering volume of 50% of the field capacity, the plant height, number of leaves, number of tubers, and tuber weight were getting lower. Shallot plants require a sufficient water supply for plant growth and development. This is evidenced by the application of water with a volume of 100% of the field capacity showing the best growth and development results compared to the treatments of 75% and 150% of the field capacity. Water is an important limiting factor in plant growth and production which causes changes in plant physiology and morphology [7]. Plants that receive less water volume will have stunted growth, the number of leaves will decrease and their size will become smaller.[8]. Providing sufficient water for shallot plants will provide optimal growth, with the available water affecting the height of the plant, which plays a role in cell division and enlargement if the plant cells experience turgidity or expand due to water pressure.[9]. Moreover according to[10] plant height decreases along with the decreasing availability of water in the soil, because drought stress can stop the cell division process which causes the plant size to become smaller. In addition, plant height is also affected by photosynthesis activity which will be disrupted and will have an impact on plant growth and development.

The frequency of watering shallots twice a day and once a day in the afternoon shows optimal plant height results because increasing the frequency of irrigation will cause high soil moisture and water is available around the root zone. So that the roots can still absorb water in optimal conditions so that they can maintain the availability of water in the plant. The watering interval of once a day in the afternoon and twice a day did not show any significantly different results because the water content in the soil required by the plants was maintained.[11].

The smaller volume of watering shows that the number of leaves is also smaller, while the water volume of 100 KL shows the results of the largest number of leaves and is not significantly different from the water volume treatment of 75% KL. In soil water conditions where water availability is lacking, the focus is more on root growth rather than shoot growth as a response to plants being able to continue to grow in conditions where there is a lack of water.

The results of this study are by [9] the frequency of watering once every 2 days and once every 3 days showed the lowest effect on the number of leaves in shallot plants, where the low frequency of watering can affect the decrease in the number of leaves.

Height growth and number of leaves will greatly affect the formation and filling of tubers. The number of shallot tubers in cultivation activities is greatly influenced by plant varieties and also the density of planting in one area [12]. The availability of water during the fruit formation process will affect the size of the fruit, in optimal irrigation conditions, the size of the fruit or bulb will also be optimal [13]. Another thing is in line with research[6]. The frequency of irrigation twice a day produces the highest total harvest weight. The right volume of watering and frequency according to needs will increase the efficiency of water absorption by the roots so that plants can grow optimally [14]. An important phase in the formation of shallot tubers is at the age of 40-52 DAP, which is a critical period for the formation and enlargement of tubers [15]. Maintaining soil moisture will increase the availability of nutrients, and optimize the absorption process of nutrients so that plants will grow optimally as seen from the height of the plant and also the number of leaves so that the volume of photosynthate produced by the plant is higher so that it can increase the weight of the tubers [16].

With the availability of water will cause better soil moisture, soil that has sufficient water content causes the soil to be healthy so that nutrient transportation in the soil is well established. Water has a function as a medium for nutrient transportation from the roots to be distributed to all parts of the plant [17]. The role of water in supporting the photosynthesis process to produce energy that can be used by shallot plants for the growth process. If shallot plants lack water supply, this can reduce the plant's ability to photosynthesize, which ultimately slows growth, thus affecting the formation of shallot tubers [18]. Water contained in the planting medium in sufficient quantities can make plants grow optimally, water has an important role in the process of photosynthesis, which is the key to providing energy that can be used by shallots in their growth. Water acts as a source of electrons and hydrogen in the process of photosynthesis.

The water contained in the planting medium plays an important role in increasing the height of shallot plants through various mechanisms, including increasing cell turgor, where turgor pressure in the plant cells causes the cell walls to remain tense so that the plants can be supported and stand upright [19]. When plant cells get enough water, the plant cells are able to enlarge and elongate which can directly contribute to the growth of the leaf stems of the shallot plant which can increase the height of the shallot [20].

4. Conclusion

The results of this study indicate that the volume and frequency of watering play an important role in the growth and development of shallot plants, especially in terms of plant height, number of leaves, and bulb formation. Sufficient water volume, especially at 100% field capacity, results in optimal growth, while insufficient watering (50% field capacity) causes a decrease in plant height, number of leaves, and bulb size. Optimal watering frequency also has a positive impact, maintaining soil moisture and facilitating the absorption of nutrients needed for photosynthesis and cell growth. Water shortages inhibit photosynthesis and growth, while sufficient water helps enlarge plant cells and increase turgor pressure, which contributes to vertical growth and plant health.

5. Acknowledgment

The research was founded by PNBP State Polytechnic of Jember Pusat Penelitian dan Pengabdian Kepada Masyarakat (P3M) for fiscal yer 2024 under the "Affirmation Research Program Scheme" Number 746/PL17.4/PG/2024. Thanks you to the director of the State Polytechnic of Jembe for this funding.

References

[1] Suwandi, "Teknologi Bawang Merah Off-Season : Strategi Implementasi Budidaya," Bandung: Balai Penelitian Tanaman Sayur, 2015.

[2] K. P. Direktorat Jenderal Hortikultura and Pengarah, "Angka Tetap Hortikultura Tahun 2023," 2024.

[3] K. Pertanian and S. Hortikultura, "Outlook Komoditas Pertanian Subsektor Hortikultura Bawang Merah," 2023.

[4] M. R. Ghozali and R. Wibowo, "Analisis risiko produksi usahatani bawang merah di desa petak kecamatan bagor kabupaten nganjuk," *J. Ekon. Pertan. dan Agribisnis*, vol. 3, pp. 294–310, 2019.

[5] A. Sumarianti, K. D. Jayanti, and Y. Tanari, "Pengaruh frekuensi penyiraman terhadap pertumbuhan dan hasil bawang merah (Allium cepa L.)," *Agrovigor J. Agroekoteknologi,* vol. 15, no. 1, pp. 39–43, 2022.

[6] R. Fauziah, A. D. Susila, and E. Sulistyono, "Budidaya Bawang Merah (Allium ascalonicum L .) pada Lahan Kering Menggunakan Irigasi Sprinkler pada berbagai Volume dan Frekuensi," *Hort. Indones. 7(1)*, vol. d, no. April, pp. 1–8, 2016.

[7] D. Sopandie, *Fisiologi adaptasi tanaman*. IPB Press, 2013.

[8] F. B. Salisbury and C. W. Ross, *Fisiologi Tumbuhan Jilid 3: Perkembangan Tumbuhan dan Fisiologi Lingkungan*. ITB Press, 1995.

[9] G. . Manurung, Kusumiyati, and J. . Hamdani, "Pengaruh interval penyiraman terhadap pertumbuhan dan adaptasi tiga bawang merah komersial Effect of watering interval to the growth and adaptation of three commercial cultivars of shallot," *J. Kultiv.*, vol. 21, no. April, pp. 24–32, 2022, doi: 10.24198/kultivasi.v21i1.34836.

[10] F. Nugraheni, S. Haryanti, and E. Prihastanti, "Pengaruh perbedaan kedalaman tanam dan volume air terhadap perkecam bahan dan pertumbuhan benih sorgum (Sorghum Bicolor (L.) Moench)," *Bul. Anat. dan Fisiol.*, vol. 3, no. 2, pp. 223–232, 2018.

[11] H. . Adiwijaya, C. I, and R. S. Basuki, "Penentuan kebutuhan air, pengaturan volume dan interval penyiraman untuk mengoptimalkan pertumbuhan dan produksi bawang putih di dataran tinggi," *J. Kultiv.*, vol. 21, no. 517, pp. 33–41, 2022, doi: 10.24198/kultivasi.v21i1.34991.

[12] V. M. Russo, "Plant Density and Nitrogen Fertilizer Rate on Yield and Nutrient Content of Onion Developed from Greenhouse-grown Transplants," vol. 43, no. 6, pp. 1759–1764, 2008.

[13] R. A. Purwanto and Susila, *Seri 1 Hortikultura Tropika Teknologi Hortikultura*. Bogor: IPB Press, 2014.

[14] N. latief, N. Musa, and W. Pembengo, "Pengaruh Frekuensi Pemberian Air dan Dosis Phonska Terhadap Pertumbuhan dan Hasil Tanaman Cabai Rawit (Capsicum frutescens L.)," *JATT*, vol. 8, no. 3, pp. 330–336, 2019.

[15] R. Suryani, S. Gafur, and T. AAbdurraahman, "Respon Tanaman Bawang Merah Terhadap Cendawan Mikoriza Arbuskula (Cma) Pada Cekaman Kekeringan Di Tanah Gambut," *J. Pedon Trop.*, vol. 3, pp. 69–78, 2017.

[16] C. . Marpaung and M. . Maghfoer, "Respon Pertumbuhan dan Hasil Bawang Merah (Allium ascalonicum L .) terhadap Berbagai Kombinasi Sumber Nitrogen dan Plant Growth Promoting Rhizobacteria (PGPR) Growth and Yield Response of Shallots (Allium ascalonicum L .) on Various Combinations of Ni," *J.*

Produksi Tanam., vol. 11, no. 12, pp. 940–951, 2023.

[17] F. Anisyah, R. Sipayung, and C. Hanum, "Pertumbuhan Dan Produksi Bawang Merah dengan Pemberian Berbagai Pupuk Organik," *J. Online Agroekoteknologi*, vol. 2, no. 2337, 2014.

[18] I. Firmansyah and N. Sumarni, "Pengaruh Dosis Pupuk N dan Varietas Terhadap pH Tanah, N-Total Tanah, Serapan N, dan Hasil Umbi Bawang Merah (Allium ascalonicum L.) pada Tanah Entisols-Brebes Jawa Tengah," *J. Hort.*, vol. 23, no. July, 2020, doi: 10.21082/jhort.v23n4.2013.p358-364.

[19] E. Mutryarny, Endriani, and I. Purnama, "Efektivitas Zat Pengatur Tumbuh dari Ekstrak Bawang Merah pada Budidaya Bawang Daun (Allium Porum L.)," *J. Pertan.*, vol. 13, no. April, pp. 33–39, 2022.

[20] Sugianto and K. D. Jayanti, "Pengaruh Komposisi Media Tanam terhadap Pertumbuhan dan Hasil Bawang Merah," *Agrotechnology Res. J.*, vol. 5, no. 1, pp. 38–43, 2021, doi: 10.20961/agrotechresj.v5i1.44619.

Management of waste and water to support rice cultivation in narrow land of residential area

A Anshori[1], T E Suswatiningsih[2], and S W Budiarti[1*]

[1]Research Center for Food Crops, Research Organization for Agriculture and Food, National Research and Innovation Agency, Bogor, Indonesia
[2]Department of Agribusiness, Faculty of Agriculture, Institut Pertanian Stiper, Yogyakarta, Indonesia
[*]Corresponding author: sriwahyuni.budiarti@gmail.com

Abstract. Waste and water are resources, and with proper management will be beneficial for human life and environmental sustainability. Study aimed to examine utilization of canned waste, livestock waste and rice husk, in variations of water management, for supporting rice cultivation, in narrow land of residential area. Study was conducted in Sleman, Indonesia. Canned waste as 'pot'. Livestock waste and rice husk charcoal were functioned as planting media. It's supported by water management. Variation of water management applied were 'pot' without hole (00), holes 5 cm from 'pot' surface (05), hole 10 cm from 'pot' surface (10), and hole at 'pot' base (15), reflect Randomized Completely Block Design with 4 treatments and 5 replications. Results showed that utilization of canned waste, livestock waste and rice husks with proper water management supports development of rice plant. Water management at 00 and 05 gave the best results. Results of treatment 00 were grain of 46.40 gram 'pot'$^{-1}$, straw biomass of 40.54 gram 'pot'$^{-1}$ and root biomass of 25.54 gram 'pot'$^{-1}$, and carbon absorption in grain of 24.90 gram 'pot'$^{-1}$, in straw of 18.18 gram 'pot'$^{-1}$ and in root of 10.95 gram 'pot'$^{-1}$. Waste utilization provides economic and environmental benefits.

1. Introduction

Waste is a discarded material from human activities, natural processes, and not or not yet have economic value, and can have negative value. Waste can be in solid, liquid or gas form. Waste is usually thrown into the environment. Waste can pollute the environment, and to overcome this, waste handling and control are needed [1]. Waste can be organic and inorganic. Inorganic waste is difficult to decompose, for example plastic and canned [2]. Organic waste is easier to decompose, for example agricultural waste, such as animal manure and rice husk [3]. Waste, without proper and appropriate management, has the potential to cause environmental pollution [4]. Proper and appropriate waste management will reduce environmental pollution, further improving the community's economy [5], and environmental health [6]. Waste management is integrative, with the attention to the benefits and impacts on the environment. The strategy is based on efforts to overcome pollution problems, with the hope of improving environmental quality [1], including strategy of waste utilization for rice cultivation in narrow land of residential area.

Rice cultivation in narrow land of residential area can be done by utilizing household and agricultural waste. According to [7] one of the narrow area that is often not utilized is the median road, which has the potential for agriculture, including rice plant. This type of agriculture is beneficial of socially, economically and environmentally, furthermore towards increasing food security. Agriculture can use the pot system [8], for food production in housing. Water availability is needed to support success. Agricultural practices are based on the use of local resources, simple and applicable technology, waste utilization, sustainable and environmental friendly concept and there is support [9]. Rice cultivation in narrow land depends on the condition of the planting media, good of physically, chemically, and biologically. The composition of the planting media and water status affect the growth, yield components and rice yield [10].

Canned is one of the household wastes. Rice husks and livestock waster are waste from agricultural. Livestock waste is processed as organic fertilizer. Rice husk is processed as rice husk charcoal. Canned as 'pot', filled with growing media from a mixture of soil, rice husk charcoal, and organic fertilizer with proper irrigation, has the potential for developing rice cultivation in narrow land of residential area. This research aimed to examine the use of canned waste, livestock waste and rice husk, in variations of water management, in supporting rice cultivation, in narrow land of residential areas.

2. Materials and Method

The study was conducted in Sleman, D.I. Yogyakarta, Indonesia, at the coordinates 7.77529°N and 110.46603°E, 154 meters above sea level, in December 2020 to March 2021, during the rainy season. The rice planting was carried out on 1st December 2020 and harvested on 2nd March 2021. The experiment was placed on the side of a village road, on the narrow land of residential area. Water was available for irrigation.

The research materials came from waste consisting of 1) the canned waste, high of 15 cm and diameter of 15 cm, used as a 'pot', 2) the livestock waste, processed into organic fertilizer, as a planting medium, and 3) the rice husk waste, processed into rice husk charcoal, as a planting medium that also functions as an ameliorant. Other materials are soil as a planting medium and synthetic chemical fertilizers. Soil as a planting medium has the characteristics of silty loam, pH of 6.45, C-organic of 1.41%, N-total of 0.47% and cation exchange capacity of 17.41 cmol kg^{-1}. The characteristics of the organic fertilizer used have C-organic of 27.25%, nitrogen of 1.37%, Phosporus of 3.35%, and Potassium of 1.87%. According to [11] the percentage of elemental content in rice husk charcoal such as C of 47.2%, N of 0.9%, Ca of 1.57%, and several other elements. The synthetic chemical fertilizers used are 46% nitrogen fertilizer and 15-15-15 NPK fertilizer. The planting medium is a mixture of soil, organic fertilizer and rice husk charcoal with ratio of 5:3:2. The fertilizer dosage is equivalent N 46% of 200 kg ha^{-1} and 15-15-15 NPK of 300 kg ha^{-1}. The rice variety used was Inpari 42. Seeds are planted as many as 2 seeds per 'pot'.

The ‘pots’, from waste canned, were arranged to indicate the presence of water management treatment. ‘Pots’ equipped with pipe to measure water level. The water management treatment was in the form of ‘pot’ without hole (00), hole at 5 cm from the ‘pot’ surface (05), hole at 10 cm from the ‘pot’ surface (10), and hole at the bottom of the ‘pot’ (15), repeated 5 times. Thus, the design of experiments was in the form of a randomized completely block design (RCBD) with 4 treatments and 5 replications. Water utilization relied on rainfall and irrigation. If there was no rain, the water requirements were met by irrigation. The plant was not irrigated when rainfall caused puddles in the ‘pot’ to be 2 cm high or level with the surface of the ‘pot’. Pipes were placed in the ‘pot’ to determine the water level, indicating the infiltration that occurred after irrigation.

Agronomic parameters such as plant height and tiller number were observed at 90 days after planting. Harvested grain weight, straw weight, root weight, panicle number, panicle length, grains number per panicle and empty grains percentage were observed at harvest. Weight of carbon in rice grain, root and straw were calculated based on carbon content in the rice grain, root and straw. The data of rice harvest were analyzed using F test followed by DMRT with confidence level of 95%.

3. Results and Discussion

3.1. Management of waste and water

Negative thoughts will say that waste is nothing more than a pollutant that must be removed, thrown away or eliminated, as a material that has no value. On the other hand, positive thoughts will say that waste is a resource, with proper management it will be beneficial for human life and environmental sustainability. Both are choices in life. Waste canned can be thrown away or used as plant 'pot'. Livestock waste can be considered as smelly manure or fertilizer that is beneficial for plant. High volume of rice husk waste will require storage, if burned it will produce side effects in the form of air pollution, or according to [12] and [13] with the pyrolysis process producing rice husk charcoal and according to [14] smoke condensation produces useful liquid smoke.

Utilization of canned as plant 'pot', filled with soil planting media, organic fertilizer and rice husk charcoal, with proper water management, becomes an alternative waste and water management in supporting rice planting in narrow land of residential area. The abundant water during the rainy season is not immediately lost as the surface flow, but it can be used for rice plant in 'pot'. Figure 1 shows the rainfall that occurs when rice is planted in 'pot'. The lack of water is met by irrigation. Rice plants require high rainfall in a month, depending on land conditions [15]. The rainfall that occurred during the growing period was 1,343 mm, which is theoretically sufficient. However, the condition for planting rice in 'pot' are different from the field, so irrigation is still needed to anticipate water shortages.

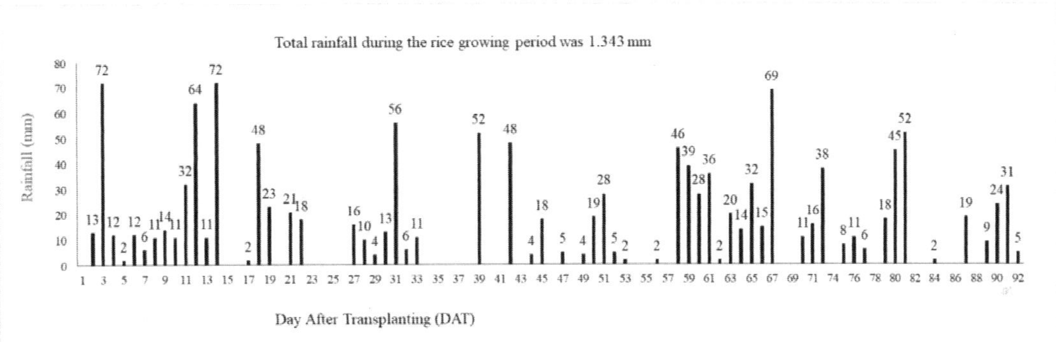

Figure 1. Total rainfall during the rice growing period

Plant growth shows that water sufficiency is very important in determining the growth of rice plant in 'pot'. 'Pot' without hole show the best appearance of rice plant growth, followed by 'pot' with hole 5 cm from the surface, 10 cm from the surface and 15 cm from the surface or holes at the 'pot' bottom resulting in the worst growth (Figure 2). Rice plant in their growth require sufficient water. Irrigation adds to the adequacy of water for rice plant in 'pot' when rainfall was insufficient. Resources from waste in the form of livestock manure, rice husk, and canned, with proper water management support the growth of rice plant.

Figure 2. Rice plant in 'pot'

3.2. *Harvest yield*

The height of rice plants in treatment 00 produced the highest plants, not significantly different from treatment 05. Treatment 15 produced the lowest plants. The number of tillers in treatment 00 was the highest, not different from treatment 05. The number of tillers in treatment 15 was the lowest (Table 1). In treatments 10 and 15, the rice plants lacked water, so the plants were lower and had fewer tillers. Evaporation of water from the canned 'pot' took place quickly because heat energy flowed or conducted faster.

The panicles number, the grains number per panicle, and the percentage of empty grains showed significant differences, except for treatments 00 and 05 (Table 1). The number of panicles and the number of grains per panicle in treatments 00 and 05 were higher than in treatments 10 and 15, with a lower percentage of empty grains. Treatments 00 and 05 produced a lower percentage of empty grains (13.20% and 13.40%) which was significantly different from treatments 10 and 15 (18.00% and 21.00%). The panicle length was not significantly different between treatments. In accordance with [16] and [17] sufficient water availability for plants tends to produce better grain.

Table 1. Tiller number, plant height, panicle, grains number per panicle, and empty grain percentage

Treatment	Tiller number	Plant height (cm)	Panicle number	Panicle length (cm)	Grain per panicle	Empty grain (%)
00	22.20±0.98a	84.48±1.29a	20.00±0.63a	20.58±0.66a	108.80±6.34a	13.20±1.33a
05	20.80±2.93a	83.44±2.34a	19.20±1.72a	20.08±0.99a	108.40±6.71a	13.40±1.62a
10	16.20±0.75b	74.10±0.85b	15.40±0.49b	19.82±1.36a	96.00±2.83b	18.00±0.89b
15	12.20±0.75c	70.78±0.61c	10.00±0.63c	19.36±1.52a	83.80±5.49c	21.00±1.26c

Note: The numbers that followed by same letter show that not significantly different by the DMRT of 5%

Treatment 00 produced grain of 46.40 gram 'pot'$^{-1}$, no different from treatment 05, 45.45% higher than treatment 10 and 127.67% higher than treatment 15. Water management at 00 and 05 produced grain that was significantly different from 10 and 15. Dry straw and dry roots also showed the same pattern. The presence of hole at the bottom of the 'pot' and 10 cm from the surface of the 'pot' caused water to drain out of the 'pot' more quickly, causing the rice plant to lack water, so that growth was inhibited and yields were lower, according to [17]. Water sufficient for rice plant in 'pot' without hole (00) and 'pot' with hole 5 cm from the surface (05).

Table 2. The dry weight of grain, straw and root and the carbon absorption in grain, straw and root

Treatment	Dry grain	Dry straw	Dry root	Carbon in grain	Carbon in straw	Carbon in root
	gram 'pot'$^{-1}$					
00	46.40±1.49a	40.54±3.27a	25.54±2.49a	24.90±0.80a	18.18±1.47a	10.95±1.07a
05	45.74±2.28a	40.76±2.99a	24.90±2.59a	24.55±1.22a	18.28±1.34a	10.68±1.11a
10	31.90±3.14b	32.28±3.30b	21.12±1.54b	17.12±1.69b	14.48±1.48b	9.06±0.66b
15	20.38±2.26c	23.70±2.75c	20.18±1.92b	10.94±1.21c	10.63±1.23c	8.65±0.82b

Note: The numbers that followed by same letter show that not significantly different by the DMRT of 5%

Treatment 00 carbon absorption in grain of 24.90 grams 'pot'$^{-1}$, in straw of 18.18 grams 'pot'$^{-1}$, and in roots of 10.95 grams 'pot'$^{-1}$, was not significantly different from treatment 05. However, it was significantly higher than treatments 10 and 15. Treatments 00 and 05 absorbed more carbon, reduced greenhouse gases, and mitigated climate change. Carbon absorption will reduce the greenhouse gases in atmosphere, further contributing to the realization of environmentally friendly and sustainable agriculture. According to [18], agriculture is one of the sectors that produces greenhouse gases, so it is important to pay attention to this.

3.3. Water dynamic

Infiltration serves to determine the dynamics of water in the 'pot'. Irrigation was done until the water level reaches the surface of the 'pot', then the water level is measured every hour. Infiltration was recorded based on the water level in the 'pot'. The dynamics of the water level in the 'pot', please see in Figure 3.

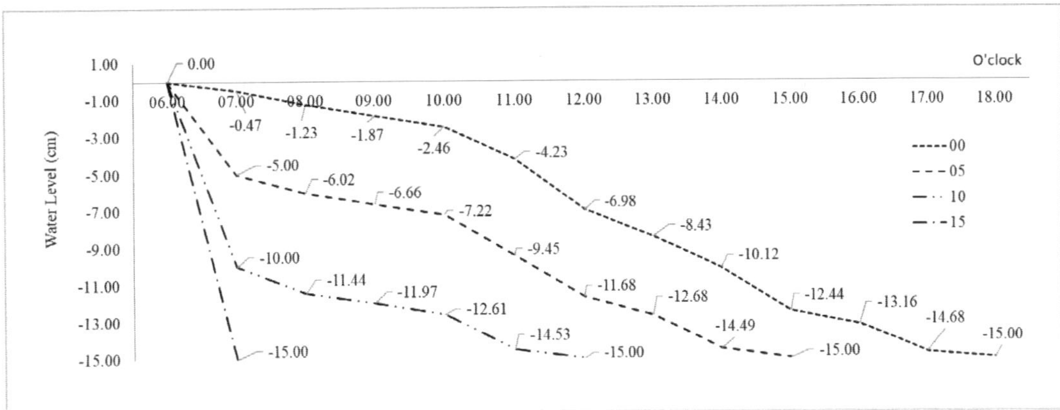

Figure 3. Water levels of the 'pot' experiment for 12 hours

In treatment 15, all the water that filled the drainage pores was completely infiltrated after 1 hour. In treatment 10, the water level reached -10 cm after 1 hour, and no water filled the drainage pores after 6 hours of irrigation. In treatment 05, the water level reached -5 cm after 1 hour of irrigation, and no water filled the drainage pores after 9 hours of irrigation. In treatment 00, the drainage pores were not filled with water after 12 hours of irrigation. The height of the 'pot' hole determines the movement of water, because water in the drainage pores will exit through the hole. The lower the 'pot' hole, the faster the water will be lost. Rapid water loss is not beneficial for rice plant, because it causes rice plant to lack water, also because according to [19] dan [20] this will be followed by movement and loss of nutrients from the soil.

3.4. Environmental economic

Utilization of narrow land in residential areas for rice cultivation in 'pot' has the potential to maintain food security and contribute to family income. The resulting grain supports family food security. The use of canned waste as 'pot' is a form of waste utilization. Livestock waste produces fertilizer and rice husk waste produces rice husk charcoal which is useful as a planting medium. Utilization of waste increases utility and economic value. Rice cultivation with 'pot' in narrow land of residential area increases the aesthetic value and sense of community togetherness. Carbon absorption by rice plant contributes to environmental sustainability. In accordance [21] waste management related to economic and social conditions takes into account current and future conditions. The strategy is comprehensive in terms of economics, social, energy and environment.

Waste utilization provides economic benefits, more efficient use of inputs to reduce raw material purchases. Efforts to reduce and utilize waste can be seen as ecoefficiency, which combines the concepts of economic and environmental efficiency. Ecoefficiency depends on the relationship between output and pollution levels [22]. Ecoefficiency reduces the environmental impact per unit of product. Waste management strategies are based on the approach to utilizing the waste formed. Waste utilization is concentrated on efforts to prevent pollution and environmental damage. Efforts to utilize waste and manage water appropriately for rice cultivation in narrow land of residential area have been based on the principle of ecoefficiency.

4. Conclusion

Utilization of canned waste, livestock waste and rice husks with proper water management support the development of rice plants in narrow land of residential areas. Canned as 'pot', livestock waste as organic fertilizer and charcoal from rice husk waste as ameliorant. Proper water management increases resource efficiency. Water management, for 'pot' without holes and 'pot' with holes 5 cm from the surface, gave the best results. The results of the 'pot' treatment without holes were 46.40 gram 'pot'$^{-1}$ grain, 40.54 gram 'pot'$^{-1}$ of straw biomass and 25.54 gram 'pot'$^{-1}$ of root biomass, and carbon absorption from the air in grain of 24.90 gram 'pot'$^{-1}$, in straw of 18.18 gram 'pot'$^{-1}$, and root of 10.95 gram 'pot'$^{-1}$. A 'pot' without holes inhibits water loss. Guided by ecoefficiency, waste is economically and environmentally beneficial.

References

[1] Rahmadi A, Sari N M, and Indriyani E 2021 *Pemanfaatan Limbah Industri* vol 4
[2] Sunarsih E 2019 Konsep pengelolaan limbah rumah tangga dalam upaya pencegahan pencemaran lingkungan *J. Ilmu Kesehat. Masy.* **5** 68
[3] Simarmata T 2022 Pemanfaatan Limbah Pertanian *Modul Kegiat. Belajar* 21–21
[4] Widjaja G and Gunawan S L 2022 Dampak Sampah Limbah Rumah Tangga Terhadap Kesehatan Lingkungan *J. Heal. Med. Res.* **2** 266–75
[5] Jamilah, Ernita M, Fridarti, and Yevendri 2022 Pemanfaatan Limbah Pertanian Untuk Meningkatkan Ekonomi Masyarakat *Abdimas Galuh* **4** 795
[6] Inderiati S, Syatrawati, Asmawati, Riswan A A Y 2023 Utilization of rice trash in various plant cultivation activities at Pitusunggu Village *J. Apl. Teknol. Rekayasa dan Inov.* **2** 25–33
[7] Septya F, Rosnita, Yulida R, and Andriani Y ulia 2022 Urban Farming Sebagai Upaya Ketahanan Pangan Keluarga Di Kelurahan Labuh Baru Timur Kota Pekanbaru *RESWARA J. Pengabdi. Kpd. Masy.* **3** 105–14
[8] Roidah I S 2020 Konsep Pertanian Kota dalam Mewujudkan Pertanian Berkelanjutan *Progr. Stud. Agribisnis, Fak. Pertanian, Univ. Tulungagung* **6** 1–9
[9] Fauzi A R, Ichniarsyah A N, and Agustin H 2016 Urban Agricuture : Urgency, Role, and Best Practice *J. Agroteknologi* **10** 49–62
[10] Hodiyah I and S D 2011 Rice growth and yield are planted in the pot with planting media composition and different soil *J. Agrivigor* **11** 42–51
[11] Hossain N, Nizamuddin S, Griffin G, Selvakannan P, Mubarak N M and Mahlia T M I 2020 Synthesis and characterization of rice husk biochar via hydrothermal carbonization for wastewater treatment and biofuel production *Nat. Res.* **10:18851**
[12] Bardalai M, Mahanta D K and Da B 2019 Production and characterisation of teak tree saw dust and rice husk biochar *Pollutants from Energy Sources* (Springer Nature Singapore Pte Ltd.) pp 291–306
[13] Cheng X, Tang Y, Wang B, and Jiang J 2018 Improvement of Charcoal Yield and Quality by Two-Step Pyrolysis on Rice Husks *Waste and Biomass Valorization* **9** 123–30
[14] Xin X, Dell K U, Young I A, and Baroutian S 2021 Transforming biomass pyrolysis technologies to produce liquid smoke food flavouring *J. Clean. Prod.* **294** 125368
[15] Oldeman, L R, and Frere, M. 1982 A Study of the agroclimatology of the humid tropics of South-east Asia *WMO Tech. Note No. 179*
[16] Anshori A, Suswatiningsih T E, Al Viandari N, Mujiyo M, Purwaningsih H, Budiarti S W, Pustika A B and Cahyaningrum H 2023 Ameliorant of rice husk charcoal supports rice cultivation in dry land *IOP Conf. Ser. Earth Environ. Sci.* **1168**
[17] Zhang S, Rasool G, Guo X, Sen L and C K 2020 Effects of Different Irrigation Methods on Environmental Factors, Rice Production, and Water Use Effciency *Water* **12**
[18] Anshori A, Sunarminto B H, Haryono E and Mujiyo M 2018 Potential Production of CH4 And N2O in Soil Profiles from Organic and Conventional Rice Fields *SAINS TANAH - J. Soil Sci. Agroclimatol.* **15** 54

[19] van Dijk H W J 1985 The impact of artificial dune infiltration on the nutrient content of ground and surface water *Biol. Conserv.* **34** 149–67

[20] Wang K, Zhang X, Ma J, Ma, Z, Li G, and Zheng J iyong 2020 Combining infiltration holes and level ditches to enhance the soil water and nutrient pools for semi-arid slope shrubland revegetation *Sci. Total Environ.* **729** 138796

[21] Amheka A, Higano Y, Mizunoya T, and Yabar H 2015 An overview of current household waste management in Indonesia: Development of a new integrated strategy *Int. J. Environ. Waste Manag.* **15** 86–98

[22] Chen J X, Zhang Y and Zheng S 2019 Ecoefficiency, environmental regulation opportunity costs, and interregional industrial transfers: Evidence from the Yangtze River Economic Belt in China *J. Clean. Prod.* **233** 611–25

IOP Conf. Series: Earth and Environmental Science **1446** (2025) 012021
doi:10.1088/1755-1315/1446/1/012021

Effect of acetic acid concentration on physicochemical characteristics of collagen from mackerel fish waste (*Euthynnus affinis*)

T Budiati[1*], W Suryaningsih[1], M Ardiyansyah[2], A B D Cahyani[1] and M M D Utami[3]

1 Food Engineering Department, Politeknik Negeri Jember, Mastrip PO Box 164 Jember 68101 East Java, Indonesia
2 Food Technology Department, Politeknik Negeri Jember, Mastrip PO Box 164 Jember 68101, East Java, Indonesia
3 Animal Science Department, Politeknik Negeri Jember, Mastrip PO Box 164 Jember 68101, East Java, Indonesia

*Email: titik_budiati@polije.ac.id

Abstract.

Utilization of mackerel fish waste from the fish processing industry into collagen can increase the value of by-products. The aim this study was to investigate the effect of acetic acid concentration on physicochemical characteristics of collagen from mackerel fish waste (*Euthynnus affinis*). Different concentration of acetic acid (0.1, 0.5, 0.9, 1.5 M) was carried out to extract collagen. The water content was ranging from 1.06 to 1.42%, ash content was ranging from 5.27 to 5.86%, protein content was ranging from 75.22 to 80.87%, white degree was ranging from 73.60 to 81.08%. Amino acids contents were alanine, glutamic acid, glycine, leucine, lysine, methionine, proline, valine, tyrosine, aspartate acid. The best treatment was found on 1.5 M of acetic acid.

1. Introduction

Sustainable production activities by minimizing waste that can produce high-value products are challenges in overcoming environmental problems that are increasing along with the explosion of the human population [1,2]. There are several opportunities that can be reprocessed from the food industry waste include fish processing industry, livestock and agriculture into high-value products such as collagen (protein) which need to be explored further [3, 4]. Collagen is sustainable products because of its characteristics of the physical, structure, safety, availability, biocompatibility, and biodegradability of the material [5]. Collagen can be obtained from fish bones and heads as fish canning waste [6] that decompose organic waste [7]. Collagen is a protein found in animal connective tissue that is stored in the skin and bones [6]. Collagen that undergoes thermal process causing denaturation in its structure with molecular weight between 15 to 250 kDa has been reported to show unique rheological properties. Collagen is an important material in the food industry, especially intended as a microencapsulation material and food additive. In addition, collagen has advantages when used as a packaging material because it has biodegradable properties so it is environmentally friendly [6]. In addition, to obtain peptides that have biological activity where the molecular weight is in the range of 300 and 8000 Da, several processing methods can be carried out, namely by chemical hydrolysis, enzymatic production or/and production through the collagen proteolytic fermentation process. [8]. Of the various products that are part of the peptide group, one of them is collagen, which has been reported to have advantages, especially in terms of its activity as an antioxidant [9].

Collagen has a structure in the form of long protein fibers that function differently from globular proteins such as enzymes. The basic molecules that form collagen are three units of α polypeptide chains that twist together to form a triple helix structure better known as tropocollagen. The amino acid composition contained in collagen is specifically composed of the main amino acids, namely glycine, proline, hydroxyproline and alanine. But the amino acid composition of collagen varies greatly and this depends on the tissue as a source of collagen. Where this can certainly affect the physicochemical characteristics of collagen [10].

Collagen consists of type I (bones and tendons), II (cartilage) and III (reticular fibers, lungs, liver, and spleen) [11]. Collagen is widely extracted from skin and bone tissue [12] using acetic acid which can dissolve collagen [13]. Acetic acid is one of the simplest chemicals that falls into the category of carboxylic acids after formic acid. Acetic acid solution in water is a weak acid group. H+ and CH3COO- ions will be formed when acetic acid undergoes a dissociation process [13]. Soaking the skin in acid will cause skin swelling due to the penetration of water into the collagen fibers. The process of making collagen can be done by soaking the skin in an acid solution where this will cause water to enter the collagen fibers and result in skin swelling. Matinong et al. [13] stated that the entry of water into collagen is caused by the electrostatic force that occurs between polar groups and the presence of H+ ions will form hydrogen bonds in collagen fibers. This skin swelling process will damage the collagen fiber structure due to disruption of non-covalent bonds so that it will dissolve collagen in the acetic acid solution and meet SNI 8706: 2014 [14].

Study of optimization of collagen extraction from fish waste is still very limited. This study aims to investigate the effect of acetic acid concentration on physicochemical characteristics of collagen from mackerel fish waste (*Euthynnus affinis*).

2. Material and methods

2.1. Sample

Mackerel fish waste (*Euthynnus affinis*) were obtained from the local fish canning industry. These were collected in the form of fish bones and heads and washed. These were followed by blanching at 70°C for 3 minutes, drained and naturally dried. Powder of mackerel fish waste was collected by grinding, filtering 50 mesh and kept in glass bottle.

2.2 Collagen extraction

Collagen extraction from fish waste was carried out using the method already carried out by Pamungkas et al. [15]. Powder of mackerel fish waste was processed through 5 stages (deproteination, demineralization, depigmentation, collagen extraction, and precipitation). The deproteination process was carried out by soaking the material in 3.5 N NaOH with a ratio of 1:10 at a temperature of 90°C for 1 hour, followed by the demineralization process by soaking in 1.5 N HCl with a ratio of 1:7 at a temperature of 90°C for 1 hour. This was followed by the depigmentation process to remove the pigment contained, the sample was soaked in 1% KMnO4 solution (w/v) for 1 hour and followed by oxalic acid solution to clean the remaining KMnO4. Collagen extraction was carried out with acetic acid (0.1, 0.5, 0.9, 1.5M) added which was used with a ratio of 1: 10 (w/v) for 24 hours. Precipitation was performed by precipitation in 0.5M NaCl and centrifugation (800 rpm, 30 minutes, 24°C) and dried at 50 °C for 1 hour. Collagen quality testing includes amino acids, proximates (water, ash, protein content) and color test. The chemical characteristic was measured by using AOAC method [16] and color test was measured by using color reader [17].

2.3 Statistical Analysis

Data of different acetic acid concentration on chemical properties (water, ash, protein content) and physical properties (color) will be analyzed by using one way Anova and be followed by Duncan test at a level of 5%.

3. Result and Discussion

Collagen extracted from mackerel fish waste with variations of concentration of acetic acid was used to optimize the extraction with the greatest quality of collagen. Soaking the powder of fish waste at different of acetic acid in 15 minutes with agitation and followed by soaking overnight resulted different chemical properties of collagen as shown in Table 1. The acetic acid solution in this study underwent a partial dissociation process into H^+ and $CH3COO^-$ ions [18]. Acetic acid is widely used as an extractant in collagen extraction because acetic acid dissolve both uncross-linked and cross-linked collagen [10].

Table 1. Characteristics of collagen extracted from mackerel fish waste

Acetic acid Concentration	Chemical properties			Physical properties
	Water content (%)	Ash content (%)	Protein content (%)	White Degree (%)
0.1 M	$1{,}42 \pm 0{,}14^c$	$5{,}27 \pm 0{,}65^a$	$80{,}87 \pm 0{,}96^c$	$73{,}60 \pm 1{,}67^a$
0.5 M	$1{,}38 \pm 0{,}21^{bc}$	$5{,}59 \pm 1{,}50^b$	$80{,}38 \pm 2{,}45^{bc}$	$77{,}20 \pm 1{,}33^{bc}$
0.9 M	$1{,}23 \pm 0{,}09^b$	$5{,}64 \pm 1{,}74^{bc}$	$77{,}00 \pm 0{,}50a^{bc}$	$78{,}18 \pm 0{,}13^c$
1.5 M	$1{,}06 \pm 0{,}02^a$	$5{,}86 \pm 1{,}28^{cd}$	$75{,}22 \pm 1{,}34^{ab}$	$81{,}08 \pm 0{,}60^d$

Note : Different notation means significant different ($P < 0{,}05$)

Table 2. Amino acid Composition of Collagen

Amino acid	Acetic acid Concentration				Comercial Collagen
	0,1 M	0,5 M	0,9 M	1,5 M	
Alanine	+	+	+	+	+
Glutamic acid	+	+	+	+	+
Glycine	+	+	+	+	+
Leucine	+	+	+	+	+
Lysine	+	+	+	+	+
Methionine	+	+	+	+	+
Proline	+	+	+	+	+
Valine	+	+	+	+	+
Thyrocine	+	+	+	+	+
Aspartic acid	+	+	+	+	+

Note : Notation + for the presence of amino acid ; - for the absence of amino acid

There were significant difference for water content, ash content, protein content of all the extracts ($p < 0.05$). There was also significant different for white degree. The brightness level of collagen is influenced by the extraction method used. According to Farooq et al. [19] the acid-soluble collagen method can produce whiter color results compared to using the hydroextraction method, because the acid used in the acid-soluble method has a higher concentration compared to hydroextraction. This statement states that the higher the concentration of acid used, the brighter the resulting color will be. The results of this study are in accordance with the theory. This can be suspected because soaking using a base or acid causes development in the skin so that the pigment can be eliminated [19].

In this study, increasing acetic acid concentration could decrease water content. This might be caused by the increasing of H^+ ions resulting bond of collagen with water to be weak. Thus, water tends to evaporate during the drying process. Noorzai [20] revealed that the water content in collagen tends to be low as H increases, indicating a decrease in the strength of the binding force between water molecules and collagen. This study observed that the increasing acetic acid concentration could increase ash content. The higher acetic acid concentration introduces the higher ash content value due to the binding of $CH3COO^-$ ions with metal ions [21]. Protein content of collagen was found to be decrease with the use of acetic acid as an extracting agent. The higher acetic acid concentration could shrink collagen fibres followed by amino acid polymer chain breaking [21].

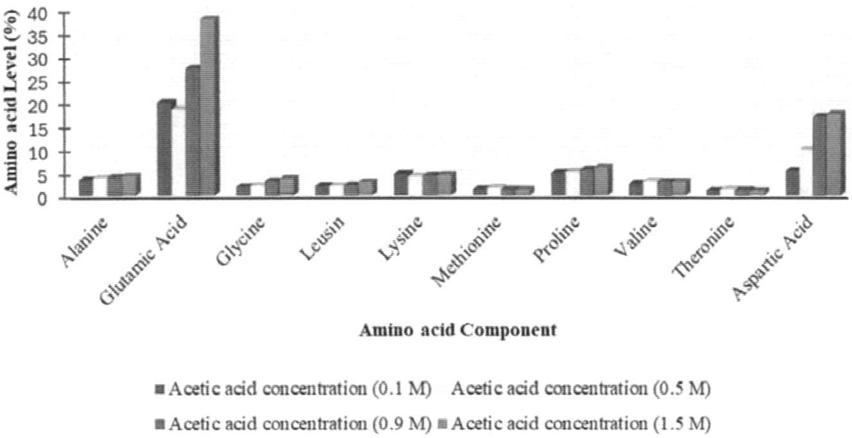

Figure 1. Amino acid component in collagen extracted from mackerel fish waste

Figure 1 shows that there are five highest amino acid contents, namely glutamic acid, aspartic acid, proline, lysine, and alanine in collagen from mackerel fish waste with acetic acid extraction of 0.1 M; 0.5 M; 0.9 M and 1.5 M. The lowest amino acid content is found in threonine and methionine. According to Krukowski et al. [22] that revealed that glycine, proline, hydroxypropylline, and alanine are the main amino acids in collagen. According to Safithri et al. [23] collagen has low levels of tyrosine and histidine and does not contain tryptophan and cysteine. The results of the study are in accordance with the theory because this collagen already contains glycine, proline and alanine as the main components in the formation of the triple helix in the collagen amino acid chain, in addition the sample also does not contain tryptophan and cysteine as amino acids that should not be present in collagen.

According to Safithri et al. [23], the differences in amino acid content of collagen can be caused by the differences in isolation material and methods and amino acid determination methods used. Collagen consists of three polypeptides that polypeptides that are formed by twisting together to form a triple helix consisting of three main amino acids, namely glycine, proline, and alanine. Collagen consists of three polypeptides that are twisted together to form a triple helix composed of three main amino acids, namely glycine, proline, and alanine. Glycine makes up almost a third of the existing collagen structure. Glycine will affect hydrogen bonds and play a role in the formation of alpha chains. Glycine itself plays a role in the formation of the collagen triple helix amino acid structure (Gly-X-Y) which position X is proline and position Y is hydroxyproline that form a triple helix [23].

Amino acids that make up collagen other than glycine are proline and alanine. In this study, the content of proline amino acids was included in the highest amino acids. Proline amino acids or hydroxy proline play a role in influencing the thermal stability value of collagen. In addition, the function of proline amino acids is to help strengthen the thermal stability of the collagen triple helix chain [23]. The amino acid alanine is an amino acid that is similar to the amino acid glycine, namely a non-polar amino acid with an aliphatic R group. Alanine also has the same function as glycine, namely forming three alpha helix chains [24].

The highest amino acid content in this study was glutamic acid. Glutamic acid is a non-essential amino acid. Glutamic acid is found in foods that contain high amounts of protein. Glutamic acid is also widely used as a flavor enhancer to enhance the savory taste (umami) in food [25].

The next highest amino acid content is aspartic acid. Aspartic acid is the same acid as glutamic acid, namely an amino acid that is included in the type of non-essential amino acids. Aspartic acid plays a role in the synthesis of urea, pyrimidine precursors and gluconic precursors [26]. Aspartic acid has an important role and is widely used in the food industry as a creator or enhancer of aroma and flavor characteristics in the food produced. The fourth highest amino acid content is lysine. Lysine is an amino acid component that has many uses and functions. Lysine amino acid plays an important role in overcoming the problem of malnutrition in humans, especially toddlers. Lysine amino acid can be used to increase body weight, especially in children and has been proven by the increase in body weight in children if the child is given additional lysine amino acid in the food they consume [27].

4. Conclusion

The present work highlights the effect of acetic acid concentration on physicochemical characteristics of collagen from mackerel fish waste (*Euthynnus affinis*). The best concentration of acetic acid was found at 1.5 M with collagen characteristics of water content, ash content, protein content and white degree for $1{,}06 \pm 0{,}02\%$, $5{,}85 \pm 1{,}28\%$, $75{,}00 \pm 1{,}34\%$ and $81{,}08 \pm 0{,}60\%$, respectively. Amino acids contents were alanine, glutamic acid, glycine, leucine, lysine, methionine, proline, valine, tyrosine, aspartate acid. Glutamic acid was dominant in collagen extract.

5. Acknowledgment

The financial support was provided by Fundamental research scheme – PNBP of Politeknik Negeri Jember is gratefully acknowledged

References

[1] Xiong X, Iris K M, Tsang D C, Bolan N S, Ok Y S, Igalavithana A D, Kirkham M B, Kim K H and Vikrant K 2019 *Chemical Engineering Journal* **375**
[2] Ong K L, Kaur G, Pensupa N, Uisan K and Lin C S K 2018 *Bioresource technology* **248**
[3] Lee J K, Patel S K S, Sung B H and Kalia V C 2020 *Bioresource technology* **298**
[4] Meena R A A, Banu J R, Kannah R Y, Yogalakshmi K N and Kumar G 2020 *Bioresource Technology* **298**
[5] Nasrollahzadeh M, Sajjadi M, Iravani S and Varma R S 2021 *Carbohydrate Polymers* **251**
[6] Tang C, Zhou K, Zhu Y, Zhang W, Xie Y, Wang Z, Zhou H, Yang T, Zhang Q and Xu B 2022 *Food Hydrocolloids* **131**
[7] Budiati T, Yudiastuti S O N and Wahyono A 2023 *Food Science and Technology Journal* **6** (2)

[8] León-López A, Fuentes-Jiménez L, Hernández-Fuentes A D, Campos-Montiel R G and Aguirre-Álvarez G 2019 *International Journal of Molecular Sciences* **20** (16)
[9] Nguyen T T, Heimann K and Zhang W 2020 *Marine Drugs* **18**(8)
[10] Meyer M 2019 *Biomedical Engineering Online* **18** (1)
[11] Henriksen K and Karsdal M A 2024 *Biochemistry of Collagens, Laminins and Elastin* Academic Press London
[12] Naomi R, Ridzuan P M and Bahari H 2021 *Polymers* **13**(16)
[13] Matinong A M E, Chisti Y, Pickering K L and Haverkamp R G 2022 *Biology* **11**(6)
[14] Badan Standardisasi Nasional 2013 *Badan Standardisasi Nasional* Jakarta
[15] Pamungkas B F, Supriyadi, Murdiati A dan Indrati R 2018 *Jurnal Pengolahan Hasil Perikanan Indonesia* **21** (3)
[16] Association of Official Analytical Chemists 2005 *Official methods of analysis of AOAC International, 2*
[17] Akhter M and Alam M M 2023. *Physical Pharmacy and Instrumental Methods of Analysis* Cham: Springer Nature Switzerland
[18] W J H 2016 Transducer & Microsystem Technologies **35**
[19] Farooq S, Ahmad M I, Zheng S, Ali U, Li Y, Shixiu C and Zhang H 2024 *Collagen and Leather* **6**(1)
[19] Matinong, A M E, Chisti Y, Pickering K L and Haverkamp R G 2022 *Biology* **11**(6)
[20] Noorzai S, Verbeek C J R, Lay M C and Swan J 2020 *Waste and Biomass Valorization* **11**
[21] Chanmangkang S, Maneerote J, Surayot U, Panya A, You S and Wangtueai S 2024 *Journal of Agriculture and Food Research* **15**
[22] Krukowski S and Sztelmach K 2022 *RSC advances* **12**(37)
[23] Safithri M, Tarman K, Suptijah P dan Widowati N 2019 *Jurnal Pengolahan Hasil Perikanan Indonesia* **22** (3)
[24] Mompeán, M, McAvan B S, Félix S S, Trevino M A, Oroz J, López-Sánchez R, Pantoja-Uceda D, Cabrita E J, Doig A J and Laurents D V 2021 *Archives of Biochemistry and Biophysics* **704**
[25] Vasilaki A, Panagiotopoulou E, Koupantsis T, Katsanidis E and Mourtzinos I 2022 *Critical Reviews in Food Science and Nutrition* **62**(32)
[26] Holeček M 2023 *Nutrients* **15**(18)
[27] Aggarwal R and Bains K 2022 *Critical Reviews In Food Science and Nutrition* **62**(9)

IOP Conf. Series: Earth and Environmental Science **1446** (2025) 012022 doi:10.1088/1755-1315/1446/1/012022

Exploring the potential of *Aloe vera* extract for enhancing meatball textural characteristics

Cahyo Indarto[1*], Wahyu Prihanta[2], Millatul Ulya[1], and Supriyanto[1]

1. Agriculture Industrial Technology Department, Trunojoyo University Madura (UTM), Bangkalan, East Java, Indonesia. 69162.
2. Biology Department, University of Muhammadiyah Malang (UMM), Malang, East Java Indonesia. 65144.

*Corresponding author: cahyo.npust@gmail.com

Abstract. Texture is a critical quality parameter for meatballs, influencing consumer preference and marketability. Typically, texture enhancement in meat products is achieved through the use of various synthetic or animal-derived additives, which can raise concerns related to health, cost, and consumer acceptance. Gelling agent is a commonly used to improve texture by promoting protein cross-linking, leading to a firmer and more cohesive product. However, there is growing interest in exploring natural and plant-based alternatives to conventional additives. This study aims to analyze the efficacy of *Aloe vera* extract in improving the texture quality of meatballs. The experiment compares three groups: meatballs prepared with transglutaminase from *Aloe vera*, those without any transglutaminase (control), and those with commercial transglutaminase. Key parameters analyzed include texture (measured with a Texture Analyzer), water content, and water holding capacity (WHC). Results indicate that transglutaminase from *Aloe vera* significantly enhances meatball texture, showing increase in texture strength (Springiness 0.52) compared to the control. Additionally, the *Aloe vera*-treated meatballs demonstrated increase in water content (70.14) and an 23.41% in WHC. These findings suggest that transglutaminase from *Aloe vera* has strong potential as a natural gelling agent for improving meatball quality.

1. Introduction

Meatballs are a type of processed food derived from animal products that are highly favored and widely consumed by the Indonesian population. Based on data from a national socio-economic survey conducted by the Central Statistics Agency in 2017, the daily consumption of beef meatballs in Indonesia grew by an average of 0.58% annually between 2013 and 2017. This growth is attributed to shifting lifestyle trends, with people increasingly opting for convenient and ready-to-eat food products. According to SNI standards, high-quality meatballs should have a distinctive meat aroma, a savory flavor, a typical grayish color, a chewy texture, and must be free from harmful substances or improper additives (1). The appealing flavor and chewy texture of meatballs are key reasons they are enjoyed by people across different demographics.

Gel strength is a parameter related to elasticity, where an increase in gel strength is in line with an increase in the level of elasticity of a food product (2). Firmness is an important indicator in determining the quality of food products (3). Sodium tripolyphosphate (STTP) is a chemical that is often used to increase chewiness in processed meat and fish products. However, continued use of STTP can cause health problems, especially when combined with the very dangerous use of borax. Therefore, many studies have been carried out to look for alternatives that can replace STTP in increasing the gel strength of food products, using transglutaminase as a solution (4) (5) (6).

Efforts to minimize the utilization of chemical additives can enhance the quality of meatball products through protein modification using enzymatic treatments. One notable enzyme is transglutaminase, which facilitates the formation of cross-links between proteins. This enzyme catalyzes the acyl transfer reaction, involving the ε-amino group of lysine residues in one protein and the γ-carboxamide group of glutamine residues, either within the same protein or between different protein molecules (7). Transglutaminase has broad applications spanning various industries, including the food sector, non-food products, and even the healthcare field (8).

Aloe vera leaves are reported to be able to heal and close wounds because they contain active compounds that are able to produce collagen to fill areas damaged by wounds (9) (10). This is an initial indication that *Aloe vera* leaves and iodine contain transglucoaminase because they are able to close wounds by forming cross-links in body tissue protein. Therefore, this study aims to analyze the efficacy of *Aloe vera* extract in improving the texture quality of meatballs.

2. Materials and methods

The research materials included fresh *Aloe vera* leaves obtained from farmers. After being picked, they were stored in a stereofoam box containing crushed ice cubes to keep the leaves fresh. Beef for meatballs is purchased from slaughterhouses. The chemicals used include pure transglutaminase enzyme (Sigma Aldrich), transglutaminase substrate (Merck), ammonium sulfate (Merck), EDTA (Merck) and other chemicals with pro analysis (PA) specifications. The instruments used in this research include a Texture analyzer type TA-XT2i.

2.1. Extraction of transglutaminase

Transglutaminase was extracted from *Aloe vera* leaves using a water solvent containing NaCl salt, buffer H_2PO_4 0.2 Molar, acetic acid 0.2 Molar. Extraction was carried out with a weight ratio of 1:3 material and solvent. Next, blend and filter it with Whatman paper. The filtrate wascentrifuged at 10.000g, temperature was maintined 4℃for 20 minutes. The clear solution was then purified using the precipitation method. Purification of transglutaminase crude extract was carried out by precipitation using ammonium sulphate (NH4)2SO4. Based on the results of previous research, the saturation of the ammonium sulfate used was 60% (11). The addition of ammonium sulfate was carried out gradually and was stirred at low speed. The purification process is carried out at a temperature of 4°C to avoid denaturation of the enzyme protein. Ammonium sulfate is separated from transglutaminase by dialysis (12). The parameter observed was the transglutaminase activity resulting from the extraction to ensure that the transglutaminase extract contained an active enzyme. Next, the transglutaminase extract was stored in a refrigerator at 4 °C until used for next step of this research.

2.2. Transglutaminase activity

The activity of transglutaminase in catalyzing protein cross-linking as a gelling agent was tested in making meatballs. Meatballs are made from beef as basic ingredients (according to treatment) with the addition of 40% tapioca and 5% NaCl, 1% sucrose, then homogenized until smooth with the addition of 10% ice. The dough is formed into rounds of the same size and soaked in water at the optimal temperature for transglutaminase activity, heating at 50 °C and keep the temperature up to 30 minutes, followed by steaming or boiling for 10 minutes in constan temperature 100 °C. The treatments in this study included: control, meatballs with *Aloe vera* transglutaminase and meatballs with commercial transglutaminase of 0.5% and 1.0%.

3. Results and Discussion

3.1 Texture

Meatball texture testing is carried out using springiness analysis which expresses a measure of the elasticity of a product. Results of statistic analysis proved that transglutaminase enzymes, both transglutaminase from *Aloe vera* and commercial transglutaminase, resulted in an increase in springiness values compared to the control (Table 1).

Table 1. Average springiness of meatball due to the addition of *Aloe vera* extract.

Treatment	Springiness
Control	0.35[c]
Transglutaminase *Aloe vera*	0.52[b]
Transglutaminase 0.5%	0.57[b]
Transglutaminase 1.0%	0.68[a]

The average followed by a different superscript letter indicates a difference in the LSD 0.05 test.

Meatballs made with the addition of transglutaminase from *Aloe vera* showed an increase in the springiness value compared to the control treatment, experiment showed that the springiness value of meatballs with *Aloe vera* transglutaminase was not different from meatballs treated with the addition of 0.5% transglutaminase. Transglutaminase has the ability to cross-link protein chains to produce a compact structure (13). Increasing the compactness of this protein structure increases the elasticity of meatballs, which are a processed protein-based product. In protein-rich food products, the development of cross-links leads to alterations in the physical characteristics or texture of the final product. such as elasticity and hardness (14). The use of transglutaminase in a meatball-like product, namely surimi, also increases the gel strength thereby improving the texture quality of surimi (15).

3.2 Water content

Water content is an important factor that can influence the quality of the gel in meatballs. The research result proved that the transglutaminase enzyme increased the average of the meatball's moisture. Table 2 present the average water content of meatballs.

Table 2. Average of meatball's water content due to the addition of *Aloe vera* extract.

Treatment	Water content (%)
Control	68.32[c]
Transglutaminase *Aloe vera*	70.14[b]
Transglutaminase 0.5%	72.07[a]
Transglutaminase 1.0%	72.93[a]

The average followed by a different superscript letter indicates a difference in the LSD 0.05 test.

Based on the results of data analysis, the transglutaminase extracts increased meatball's moisture, where increasing the transglutamiase concentration can produce a higher water content value for the meatballs. This is because the transglutaminase enzyme also helps increase the ability to retain water, as a result of the formation of cross-links that strengthen the protein network (7). The presence of the transglutaminase enzyme induces bonds between the gel matrix, thereby increasing the water holding capacity of the gel (15). Transglutaminase activity increases the complexity of protein cross-links in the meatballs to form a stable net structure, so that it can hold water and result in the water content of the meatballs becoming higher.

3.3 Water holding capacity

This parameter represents the ability of product to avoid water loss due to the effects of cooking. When cooking using high temperatures, some product components can be damaged and reduce value of water holding capacity. Tabel 3 provide the average of meatball's water holding capacity.

Table 3. Average water holding capacity meatball due to the addition of *Aloe vera* extract.

Treatment	Water holding capacity (%)
Control	22.37^{d}
Transglutaminase *Aloe vera*	23.41^{c}
Transglutaminase 0.5%	24.23^{b}
Transglutaminase 1.0%	24.98^{a}

The average followed by a different superscript letter indicates a difference in the LSD 0.05 test.

Data on the average water holding capacity of meatballs has a tendency or trend that is similar to data on water content of meatballs. The water-holding capacity of a product is intrinsically linked to its moisture content, as an enhanced ability to retain water corresponds to an elevated level of bound and free water within the matrix. The treatments in this study showed significant differences in the water content of meatballs, which means that the use of transglutaminase, both *Aloe vera* transglutaminase and commercial transglutaminase, significantly increased the water holding capacity value. Proteins that have complex cross-links have an effect on the high-water holding capacity value, likewise protein damage will reduce the water holding capacity value. Proteins that experience denaturation will have a lower water holding capacity value (16).

Conclusion

Aloe vera has a significant influence on improving the texture quality of meatballs. Transglutaminase in *Aloe vera* is able to increase protein cross-linking in meatballs, thereby increasing elasticity. The use of *Aloe vera* transglutaminase increased springiness (0.52), water content (70.14%) and water holding capacity (23.41%), compared to the control.

Acknowledgment

This publication is derived from research funded by the Directorate of Research and Community Service-Directorate General of Higher Education, Research, and Technology-Ministry of Education of the Republic Indonesia (Contract No:015/UN.46.4.1/PT.01.03/2024), June 13th 2024.

References

[1] National-DSN, D. S. (1995). Meat Meatballs. DSN.

[2] Ardianti Y, Widyastuti S, Rosmilawati, W S, Handito D. 2019. Effect Of Carrageenan On The Physical And Organoleptic Properties Of Fish Ball (*Euthynnus affinis*) Yuli. J Chem Inf Model. 53(9):1689–99.

[3] Herlambang FP, Lastriyanto A, Ahmad AM. 2019. Karakteristik Fisik dan Uji Organoleptik Produk Bakso Tepung Singkong sebagai Substitusi Tepung Tapioka. J Keteknikan Pertan Trop dan Biosist. 7(3):253–8.

[4] Laksono UT, Suprihatin, Nurhayati T, Romli M. 2019. Peningkatan Kualitas Tekstur Surimi Ikan Malong dengan Sodium Tripolifosfat dan Aktivator Transglutaminase. J Pengolah Has Perikan Indones. 22(2):198–208.

[5] Mayasopha AY, Herfianita F, Sutrisno A. Aplikasi Enzim Transglutaminase Pada Produk Pangan: Kajian Pustaka [in Press Juli 2015]. J Pangan dan Agroindustri. 2014;3(3):1145–51.

[6] Ismanto A, Lestyanto DP, Haris MI, Erwanto Y. 2020. Komposisi Kimia, Karakteristik Fisik, dan Organoleptik Sosis Ayam dengan Penambahan Karagenan dan Enzim Transglutaminase. Sains Peternak. 18(1):73.

[7] Nuraisyah, A., Raharja, S., & Udin, F. 2018. Chemical Characteristics of Rice Flour Bread with Added Transglutaminase Enzyme. Journal of Agricultural Industrial Technology, 28(3), 319-330.

[8] Sidauruk, S. W., Nurhayati, T., Suptijah, P., & Laksono, U. T. 2017. Characterization of endogenous transglutaminase enzymes from the liver of crayfish (Congresox talabon)]. Journal of Indonesian Fishery Products Processing, 20(3), 582-591.

[9] Buulolo AJ, D S. 2019. Formulasi Sediaan Gel Sari Lidah Buaya (*Aloe vera* L.) sebagai Obat Luka. J Dunia Farm.1(1):1–6.

[10] Astuti P, Agfiany SR, Abidin KR. 2020. Aktivitas Ekstrak gel *Aloe vera* sebagai Antiinflamasi Untuk Mempercepat Proses Penyembuhan Luka Pada Tikus Sprague dawley Sprague dawley. J Ilm Umum dan Kesehat Aisyiyah.;5(1):50–5.

[11] Indarto C, Prihanta W, Supriyanto. 2024. The beginning study of transglutaminase from plant origin. E3S Web Conf. 499:01030.

[12] Sturaro A, Marchi M De, Masi A, Cassandro M. 2016. Quantification of whey proteins by reversed phase-HPLC and effectiveness of mid-infrared spectroscopy for their rapid prediction in sweet whey. J Dairy Sci. 299(1):68–76.

[13] Zhang, M., P. Wang., M. Zou., R. Yang., M. Tian, dan Z. Gu. 2019. Microbial Transglutaminase Modified Protein Network And Its Importance In Enhancing The Quality Of High-Fiber Tofu With Okara. Food Chemistry. 289: 169-176.

[14] Nugroho, H. C., Amalia, U., & Rianingsih, L. (2019). Karakteristik Fisiko Kimia Bakso Ikan Rucah Dengan Penambahan Transglutaminase Pada Konsentrasi Yang Berbeda. *Jurnal Ilmu Dan Teknologi Perikanan*, *1*(2), 47–55.

[15] Kaewudon, P., S. Benjakul, dan K. Kijroongrojana. 2013. Properties of Surimi Gel As Influenced By Fish Gelatin and Microbial Transglutaminase. Food Bioscience, 1: 39- 47

[16] Olympias, Sujarwanta, R., Setiyono, Supadmo, Suryanto, E., & Rusman. (2016). Quality of Beef Sausages Fortified With Cod and Corn Oil and Processed With Different Cooking Methods. Buletin Peternakan, 40(1), 48–57.

IOP Conf. Series: Earth and Environmental Science **1446** (2025) 012023 doi:10.1088/1755-1315/1446/1/012023

Agronomic performance, genetic variability estimation and mutant selection in M3 generation of sorghum under optimum condition

Anisiyah[1], Wijaya Murti Indriatama[1*], Sihono[1], Taufiq Bachtiar[2], Muhammad Nurcholis[3], Susila Herlambang[3], and Soeranto Human[1]

[1]Research Center for Food Crops, National Research and Innovation Agency, Jl. Raya Jakarta-Bogor KM. 46, Cibinong, Bogor, Jawa Barat 16915, Indonesia
[2]Research Center for Environmental and Clean Technology, National Research and Innovation Agency, Jl. Cisitu, Sangkuriang Bandung, Jawa Barat 40135, Indonesia
[3]Department of Soil Science, Faculty of Agriculture, Universitas Pembangunan Nasional Veteran Yogyakarta, Jl. SWK No. 104, Yogyakarta 55283, Indonesia

*Email : wija003@brin.go.id

Abstract. The population resulting from mutations in the M3 generation generally has a diversity that can be selected to obtain mutants with the desired character. This study aimed to observe the agronomic performances of the M3 generation of sorghum and to estimate its genetic parameters as information for selection. The genetic materials used in this research were 242 sorghum mutant lines of the M3 generation, and six comparison genotypes, namely Numbu, Samurai-1, Kawali, Pahat, B-35, and Watar Hammu Putih (parent). This experiment used an augmented design. The result analysis revealed that the genotype treatments affected the differences in performance of panicle length, stem diameter, and chlorophyll content characters. Significant differences were also seen between the control genotypes on all the observed characters. The appearance of sorghum mutant lines and control genotypes showed substantial differences in panicle length, stem diameter, panicle weight, biomass weight, leaf chlorophyll content, and seed yield characters. The wide genetic diversity and high heritability values were indicated by plant height, panicle length, panicle weight, and chlorophyll content characters. Using panicle weight as a selection character in M3 populations has increased the differential selection of stem diameter, leaf chlorophyll content, biomass, and seed yield characters.

1. Introduction

Global warming-induced climate change results in a range of extreme conditions that are anticipated to intensify extreme precipitation occurrences and raise the danger of flooding [1]. An environment known as a drought check occurs when there is not enough soil water available to enable plants' growth and development to their full potential. It has an impact on the decrease in agricultural yield [2]. Among the cereal crops with a higher resistance to drought is sorghum (*Sorghum bicolor* L. Moech). Compared with other dry cereal crops, sorghum requires less water. Apart from its exceptional resistance to drought, sorghum has other advantages that can be leveraged, such as the ability to use its seeds for food and feed due to their high nutritional content [3].

It takes expertise to create genetic differences in yields and the factors that affect them in order to create high-yielding sorghum varieties. The fundamental tactic utilized to enhance the desired character is to take use of genetic variety found in nature and induction outcomes. Given Indonesia's current situation of restricted germplasm and genetic variability [4], mutagenesis in sorghum is a critical tool for producing new varieties [5]. One useful method for boosting genetic diversity in self-pollinating plants and obtaining a greater range of reliable agronomic traits is mutation induction [6]. Mutagenesis is a breeding strategy used in programs to select individual mutant genotypes with improved traits and for genetic screening [7].

The consequences of mutagenesis varied depending on the genetic background and increased broad-sense heritabilities and variation [8]. In the M2 and M3 populations, mutagenic treatment enhanced the variety of quantitative features to be higher and lower simultaneously. Breeding objectives-driven character selection can begin in the M2 generation and be further pursued for confirmation in the M3 generation [9]. The design of the population development in the first generation of mutation-induced mutations, the strength of selection, and the intended isolation process for the idiotype mutant all affect the efficacy of mutation-selection [10]. In the M3 generation, progeny selection based on the predicted mean and significant genetic variance contributes to improved yields and their constituent parts. The value of the average change between chosen generations to improve plant varieties can be calculated by estimating genetic characteristics such as heritability, coefficient of variation, and genetic progress in the population [11].

Two methods are used to assess a plant's ability to withstand drought insecurity: direct and indirect observations. Observing the decline in crop yields and biomass produced during drought conditions relative to normal conditions can be a direct observation. By observing the morphology and physiology linked to the resistance to drought insecurity, indirect observations were made. The morphological reactions, physiology, and reduced yields can therefore be identified as a foundation for identifying the traits linked to sorghum varieties' drought security tolerance. This study aimed to identify the superior sorghum mutants of M3 generations under optimum conditions and estimate the genetic parameters to decide the characters' selections.

2. Materials and Methods

2.1. Location and Time of Research

The research was carried out from July to December 2020 at the experimental station of the Indonesian Centre for Agricultural Biotechnology and Genetic Resources Research and Development (ICABIOGRAD), Citayam, Bogor, West Jawa. These sites are located in a dry region and present a well-defined dry season, allowing water control to be performed only by irrigation. In this trial case, rainfall came early more than five times in this period in the middle of September as the Lanina effect. The soil of the experimental area is characterized as medium-textured Latosol.

2.2. Research Materials and Tools

The experimental design was augmented design, with treatments as 252 mutant lines M3 generation resulted from gamma rays induced mutation [12], the parent (Watar Hammu Putih), also check variety for drought-tolerant (Numbu, Kawali, B-35 line) and susceptible (Samurai-1). Each row (one row, one genotype) consisted of 20 seeds planted in a 2 m long row with 70 cm between rows and 20 cm between plants within a row. Fertilization was given twice, namely when the plant was 10 days after planting (DAP) with a dose of 150 kg.ha-1 Urea, 200 kg.ha-1 SP36 and KCl 100 kg.ha-1 and the second fertilization with a dose of Urea 150 kg.ha-1 at 30 DAP. Observations were made on the plant height, number of leaves, number of internodes, panicle length, stem diameter, panicle weight, biomass weight, leaf chlorophyll content, and seed yield.

2.3. Data Analysis Procedure

The data was analyzed statistically using variance analysis ($\alpha = 0.05$) [13]. If there was a difference, further tests were carried out with Tukey Test $\alpha = 5\%$.

Table 1. Variance Analysis of Augmented Design [13]

AoV	df	SS	MS	EMS
Block	r-1	SSb	MSb	
Treatment	(g+c)-1	SSt	MSt	
Genotype M3 (g)	g-1	SSg	MSg	$\sigma^2e+\sigma^2g$
Control (c)	c-1	SSc	MSc	$\sigma^2e+r\sigma^2c$
g x c	1	SSgxc	MSmxk	
Error	(c-1)(r-1)	SSe	MSe	σ^2e
Total	(rc+g)-1	SStotal		

Estimation of the broad sense heritability value (h^2bs) was carried out based on the separation of the expected mean square value by dividing the genetic variance of genetic (σ^2g) to the phenotype variance (σ^2p) with the following criteria according to Stansfield [14] : high ($h^2bs > 0.5$), moderate ($0.2 \leq h^2bs \geq 0.5$) and low ($h^2bs < 0.2$). Coefficient of genetic variability (CVG) criteria base on Knight [15] : low (0<10%), medium (10–20%) and high (> 20%). Heritability values and coefficient of genetic diversity (CVG) were calculated by the equation [16]:

$$\sigma^2e = MSe/r;\ \sigma^2g = (MSg - MSe)/r;\ \sigma^2p = \sigma^2e+\sigma^2g;$$

$$h^2bs = (\sigma^2g/\sigma^2p) \times 100\%;\ CVG = ((\sqrt{\sigma^2g}/\bar{x}) \times 100\%);$$

where : MSe= the mean square of error; MSg = mean square of genotypes; r = replication ; s= standard deviation; $\bar{x}$ = mean of all M3 generations.

Based on the outcomes, the best genotype was chosen, and the process of choosing the best individuals in the lines was then carried out. One character (character grain weight panicle^{-1}) with a selection intensity of 10% was used to select good yielding lines. Next, use the following equation to determine the selection differential: S is equal to $((\bar{x}i - \bar{x}o)/ \bar{x}o) \times 100\%$; thus, $\bar{x}i$ stands for average of selected populations, $\bar{x}o$ for average of initial population, and S is for differential selection [17].

3. Results and Discussions

3.1. Climatic Conditions

The climatic data during this study were obtained from the Meteorological and Geophysical Agency in Bogor, West Jawa (Table 2). It showed that from July until December 2020, the total rainfall per month varied from 2,87 to 20,11 mm per day, the average temperature was from 26.09 to 27,24 °c, and the duration of the sun's irradiation was from 2,90 to 7,42 hours every day. The average humidity was between 77,39 to 84.39 %. Naturally this condition was suitable for sorghum growth without watering activity. All genotypes can be optimally grown.

Table 2. Climate data of Bogor District, West Java between July to December 2020

Month	Tmin (°c)	Tmax (°c)	Tavg (°c)	RH_avg (%)	RR (mm/day)	Ss (h)	ff_avg (m/s)
July	22.23	31.93	26.09	81.03	7.45	6.47	1.48
August	22.88	32.88	26.54	77.39	2.87	7.42	1.77
September	22.50	33.16	26.91	78.50	11.89	6.90	1.12
Oktober	22.60	31.92	27.24	84.38	20.11	5.16	2.00
November	22.80	32.14	26.51	83.19	9.56	5.03	1.57
December	21.61	31.62	25.89	83.41	6.42	2.90	1.59

Note : Tmin : Temperature minimum; Tmax : Temperature maximum; Tavg : Average temperature; RH_avg : Average relative humidity; RR : Total rainfall per month ; Ss : ; ff_avg : Average wind speed

3.2. Parent and Control Varieties Performances

The diversity of each observed character was found in control genotypes. The Numbu genotype and Watar Hammu Putih line have the highest plant levels and the best stem juice % brix features. The biomass characters' weight and the most noticeable stem diameter both display the Watar Hammu Putih line. The genotype B-35 possesses the most significant numbers of nodes, leaves at harvest, and chlorophyll content features. The Numbu, Samurai-1, Watar Hammu Putih, and Kawali were Panicle's strongest heavy characteristics. The Pahat variety had the longest panicle feature. The Numbu variety, Samurai-1, and Watar Hammu Putih line had the highest seed weight panicle^{-1} character. Table 3 displays the character variety of control genotypes of sorghum.

Table 3. Performance of the agronomic characters of some sorghum genotypes

Characters	Genotypes					
	Numbu	Samurai-1	Watar Hammu Putih	B-35	Kawali	Pahat
Plant Height (cm)	296.75a	250.25ab	331.25a	113.5c	167.75bc	136.375c
Number of Nodes	14.25ab	14.75a	13.75ab	14ab	11b	12ab
Number of Leaves	8.5abc	9ab	8abc	9.5a	4.75b	4.25c
Panicle Lenght (cm)	26	31.75	29	25.75	29	32.375
Stem Diameter (mm)	18.76ab	19.77ab	21.46a	20.97ab	17.90ab	16.08b
%Brix Stem Juice	13.325a	10.9ab	12.13ab	13.43a	6.4b	8.15ab
Biomass (g)	638.27ab	422.9bc	882.5a	211.93c	295.88c	153.35c
Panicle Weight (g)	93.49a	86.58a	90a	33.57b	87.13a	52.75b
Chlorophyll Content	32.65d	45.58bc	35cd	55.48ab	60.45a	42.1cd
Grain Weight Panicle^{-1} (g)	74.25a	65.43a	66.37a	19.56c	58.69ab	37.38bc

Note: The numbers followed by different letters in the same row are significantly different based on HSD tests at the 0.05 probability level

3.3. Performance of M3 Mutant Sorghum Population Vs Control Genotypes

Variance analysis results revealed that there was a substantial difference between the genotypes of the M3 lines and control in terms of panicle length, stem diameter, and chlorophyll content (Table 4). The M3 sorghum genotypes exhibit observable variations in seed weight per panicle, biomass weight, panicle weight, stem diameter, panicle length character, and chlorophyll content when compared to the control. A higher likelihood of mending particular lines is indicated by the discrepancies between the M3 genotypes and control compared to the parent and national varieties as the control.

Table 4. Mean square value of M3 sorghum population and control genotypes on agronomic characters observed at Citayam, Bogor

Characters	Source of Variance					
	Control Variety		M3 lines		Check Vs lines	
	MSc	F-Hit	MSg	F-Hit	MSgxc	F-Hit
Plant Height	15661.28	2.49	3024.91	0.48	248140.60	39.48
Number of Nodes	5.05	1.89	1.59	0.60	3.81	1.43
Number of Leaves	15.14	2.24	2.45	0.36	24.64	3.65
Panicle Lenght	25.88	3.28*	15.23	1.93*	32.92	4.17*
Stem Diameter	6.78	0.83*	8.29	1.01*	0.74	0.09*
%Brix Stem Juice	22.24	2.69	8.39	1.01	86.74	10.49
Biomass	69354.15	0.79	82251.30	0.94	2495805.92	28.52*
Panicle Weight	894.84	1.08	1098.59	1.32	4134.26	4.97*
Chlorophyll Content	266.02	4.30*	79.87	1.29*	1206.19	19.49*
Grain Weight Panicle^{-1}	861.55	1.36	666.69	1.06	3706.02	5.87*

Note : * : significant different on a level $\alpha = 5\%$; ** : significant different on a level $\alpha = 1\%$

In terms of plant height and grain weight per plant, Table 5 demonstrates that the mean population of the M3 sorghum generation is greater than the mean of the parent and control varieties. It indicates that some lines that were created from possible segregants outperform both the parent and control types. The mean values of M3 sorghum generation were in between the parent and control varieties' mean values for the following characteristics: biomass weight, panicle weight, panicle length, stem diameter, stem juice Brix, and number of nodes and leaves. There are some mutants that perform better than the mean value of the parent and control types, as seen by the population's wide value range. This allows for selection, so that superior mutant lines are obtained that are better than all genotypes tested. Likewise, related to drought resistance, with a wide enough range, there is a chance that the selected mutant lines will be tolerant to drought conditions even though they are currently tested under optimum conditions. In previous studies, drought resistance was related to leaf area and chlorophyll content. The appearance of mutant lines with a smaller average number of leaves and a wider range of total chlorophyll content among the comparators. It indicates the potential for drought-tolerant mutant lines when further selected.

Table 5. The mean of agronomic characters of 3 control varieties and selected sorghum M3 generation

Character	Varieties			Sorghum M3 Generation	Range
	Numbu	Samurai-1	Watar Hammu Putih		
Plant Height (cm)	296.75	250.25	331.25	314.08	135-454
Number of Nodes	14.25	14.75	13.75	13.84	10-17
Number of Leaves	8.5	9	8	8.16	5-12
Panicle Lenght (cm)	26	31.75	29	30.23	17-45
Stem Diameter (mm)	18.76	19.77	21.46	19.46	10.85-28
%Brix Stem Juice	13.33	10.9	12.13	13.3	3.2-18.8
Biomass (g)	638.27	422.9	882.5	832.87	108.5-1625.31
Panicle Weight (g)	93.49	86.58	90	89.54	18.47-225
Chlorophyll Content	32.65	45.58	35	32.42	8.67-62
Grain Weight Panicle^{-1} (g)	74.25	65.43	66.37	69.14	1.94-174.55

Genes linked to economically significant qualities are frequently modified as a result of viable mutations in the quantitative traits that alter the allelic combinations. Numerous characteristics, including panicle area, seed weight, and number of seeds per panicle, contribute to the grain yield, a significant characteristic that determines the crop's economic value [6]. The possibility of obtaining genotypes with high yield productivity will improve with an increase in the variability of seed production in the M3 generation.

3.4. Estimation of Genetic Parameters and Selection Process

Diverse genetic material must be readily available for the chosen tasks in every breeding effort. Genetic variability, as measured by broad-sense heritability, is the main factor influencing the success of selection. Plant height, node count, leaf count, panicle length, panicle weight, and chlorophyll content characters are the traits with the highest heritability among the studied characters (Table 6). Insan et al. [18] also discovered high broad sense heritability values in the RIL sorghum population's characteristics of plant height, leaf number, panicle length, and panicle weight. The plant height, number of nodes, panicle length, panicle weight, chlorophyll content, biomass, and seed weight panicle^{-1} among the quantitative parameters exhibited high genetic diversity; the other characters displayed lower genetic variability. When M3 populations were in optimal condition, sometimes the effects of physical mutagens that enhanced variability for specific quantitative features did not show up.

According to Raina et al. [19], gene clusters that are prone to mutations with similar probability in both directions and have minimal effects on their own govern most quantitative features. Through these micro-mutations in polygenic regulated traits, breeders would be able to improve economically significant aspects without sacrificing the genotype of a parent. Furthermore, a large amount of genetic variety brought about by mutation induction would result in a greater number of markers for morpho-yield characteristics, simplifying linkage analyses to increase sorghum's resistance to climatic change. In M3 mutant sorghum populations, grain yield had the highest coefficient of variability whereas the number of nodes had the lowest. The capacity of mutagens to generate a high degree of variability in mutant populations presents an opportunity to boost productivity [4].

Table 6. Genetic parameters for agronomic characters of Sorghum M3 population in Citayam, Bogor

Characters	σ^2E	σ^2G	σ^2P	h^2_{bs}(%)	CVG (%)
Plant Height	418.96	814.88	1233.84	66.04^H	85.39^W
Number of Nodes	0.18	0.27	0.45	60.00^H	37.37^W
Number of Leaves	0.45	1.08	1.53	70.59^H	12.44^M
Panicle Lenght	0.53	1.83	2.36	77.54^H	44.92^W
Stem Diameter	0.54	0.03	0.58	5.17^S	8.86^N
%Brix Stem Juice	0.55	0.03	0.58	5.17^S	13.66^M
Biomass	5834.86	1317.92	7152.78	18.43^S	45.79^W
Panicle Weight	55.48	66.61	122.09	54.56^H	88.66^W
Chlorophyll Content	4.12	4.50	8.62	52.20^H	57.11^W
Grain Weight Panicle^{-1}	42.10	8.80	50.90	17.29^S	42.46^W

Note: σ^2E = environtment variance; σ^2G = genetic variance; σ^2P = phenotypic variance; h^2bs = Broad sense heritability; CVG: coefficient of genetic variability. h^2bs: S=Low, H=High. CVG: N=Narrow, M=Medium, W=Wide.

Table 7. Differential selection based on the panicle weight character

Characters	Average M3 generation	Average M3 selected by panicle weight	Differential selection
Plant Height	334.31	324.48	-2.94
Number of Nodes	13.90	13.32	-4.20
Number of Leaves	8.35	8	-4.23
Panicle Lenght	30.12	29.34	-2.58
Stem Diameter	19.55	20.7784	6.30
%Brix Stem Juice	12.68	11.712	-7.65
Biomass	792.77	895.0896	12.91
Panicle Weight	92.05	143.108	55.46
Chlorophyll Content	37.14	38.564	3.82
Grain Weight Panicle^{-1}	69.87	111.394	59.42

The goal of this study's pure line development is high yielding variety. A strong heritability and wide genetic diversity of yield's character can lead to significant improvements through direct selection for yield. Due to its wide genetic diversity and high heritability value, the panicle weight can be employed as an efficient selection trait. There were 25 lines selected based on the of panicle weight with selection intensity of 10%. The use of panicle weight as a selection character in M3 populations has increased the differential selection of stem diameter, leaf chlorophyll content, biomass and seed yield characters (Table 7).

4. Conclusion

Due to genetic factors, the M3 sorghum population's performance varies greatly. High heritability values are seen in the plant height, leaf count, internode count, panicle length, chlorophyll content, and panicle weight characteristics. Following selection, M3 populations' mean character values increased, particularly for the panicle weight, biomass, and seed weight panicle^{-1}. In order to increase yield potential, direct selection based on panicle weight proved to be sufficiently successful.

Acknowledgments

Thanks to the IAEA, which has provided funding for this research through the RC-22433 program. Thanks to the National Research and Innovation Agency for financing this research through funding the

Research Organization for Nuclear Energy (RP HITN). Thank you to Mr Tardi Suseno, Mr Parno Suharjo, Ms Marina Yuniawati Maryono, and Mr Kodir for the technical assistance.

References

[1] Tabari H. Climate change impact on flood and extreme precipitation increases with water availability. *Sci Rep*. 2020;1–10. https://doi.org/10.1038/s41598-020-70816-2
[2] Farooq M, Wahid A, Zahra N, Hafeez MB, Siddique KHM. Recent Advances in Plant Drought Tolerance. *J Plant Growth Regul*. 2024; https://doi.org/10.1007/s00344-024-11351-6
[3] Zheng H, Dang Y, Sui N. Sorghum: A Multipurpose Crop. *J Agric Food Chem* [Internet]. 2023 Nov 22;**71**:17570–83. https://doi.org/10.1021/acs.jafc.3c04942
[4] Surashe SM, Kalpande HV., Badiganavar A, More A, Ganapathi TR. Mutation induced phenotypic variation for yield traits in sorghum [*Sorghum bicolor* (L.) Moench]. *Electron J Plant Breed*. 2022;**13**:312–8.
[5] Human S, Sihono, Indriatama WM. Sorghum improvement program by using mutation breeding in Indonesia. *IOP Conf Ser Earth Environ Sci*. 2020;**484**.
[6] Kalpande H V, Surashe SM, Badigannavar A, More A. Induced variability and assessment of mutagenic effectiveness and efficiency in sorghum genotypes [*Sorghum bicolor* (L .) Moench]. Int *J Radiat Biol*. 2022;1–14. https://doi.org/10.1080/09553002.2022.2003466
[7] Jung C, Till B. Mutagenesis and genome editing in crop improvement: perspectives for the global regulatory landscape. *Trends Plant Sci*. 2021;**26**:1258–69. Available from: https://www.sciencedirect.com/science/article/pii/S1360138521002132
[8] Jonas M, Navarro D. Induced mutations alter patterns of quantitative variation, phenotypic integration, and plasticity to elevated-CO2 in *Arabidopsis thaliana*. *J Plant Res*. 2018;**132**:33-47. Available from: http://dx.doi.org/10.1007/s10265-018-1064-3
[9] Ravichandran V, Jayakumar S. Effect of mutagens on quantitative characters in M2 and M3 generation of sesame (*Sesamum indicum* L .). *Int Lett Nat Sci*. 2015;**42**:76–82.
[10] Sakin MA, Yildirim A. Induced mutations for yield and its components in durum wheat (*Triticum durum* Desf .). 2004;**2**:285–90.
[11] Wani MR, Khan S. Estimates of genetic variability in mutated populations and the scope of selection for yield attributes in *Vigna radiata* (L.) Wilczek. *Egypt J Biol*. 2006;**8**:1–6.
[12] Sihono S, Indriatama WM, Anisiyah A, Puspitasari W, Maryono MY, Human S. Increasing genetic variability of local sorghum yield by using gamma-ray induced mutation. *AIP Conf Proc*. 2024;**2967**.
[13] Sharma JR. *Statistical and Biometrical Techniques in Plant Breeding*. New Delhi (IN): New Age International; 2006.
[14] Stansfield WD. *Theory and Problems of Genetics*. 2 Eds. New Delhi (IN): McGraw-Hill; 1969. 281 p.
[15] Knight R. *Practical in Statistic and Quantitative Genetic*. Brisbane, Australia: Australian Vice Chancelors Commite; 1979. 197 p.
[16] Singh, R.K., Chaudhary B. *Biometerical Methods in Quantitative Genetic Analysis*. Vol. 2. New Delhi: Kalyani publishers; 1979.
[17] Falconer DS, Falconer M. *Introduction to Quantitative Genetics*. Textbook Publishers; 2003. 365 p.
[18] Insan RR, Wirnas D. Estimation of genetic parameters and selection of sorghum [*Sorghum bicolor* (L.) Moench] RILS F5 derived from single seed descent. *Int J Agron Agric Res*. 2019;**8**:95–103.
[19] Raina A, Laskar RA, Khursheed S, Amin R. Role of Mutation Breeding in Crop Improvement-Past, Present and Future. *Asian Res J Agric*. 2016;**2**:1–13.

Extraction and identification of phenolic compounds in ethanol extract of *Protium javanicum* leaves at different solvent concentrations

N L A. Yusasrini[1*], I N Kencana Putra[2], D G Mayun Permana[2], and K A Nocianitri[2]

[1]Doctoral Study Program in Agricultural Science, Udayana University Denpasar 80361, Indonesia
[2]Department of Food Science, Udayana University, Denpasar 80361, Indonesia

*Email : ariyusasrini@unud.ac.id

Abstract. *Protium javanicum* is an indigenous Indonesian plant belonging to the Burceraceae family, utilized in traditional medicine. The leaves possess bioactive constituents including phenolics, flavonoids, steroids, tannins, and terpenoids, which have possible antioxidant properties. The isolation of these phenolic compounds can be achieved through extraction using ethanol as a solvent. Phenolic compounds exhibit varying chemical characteristics and polarity. The polarity of the ethanol solvent significantly enhances the solubility of phenolic compounds. This study aimed to determine the ethanol concentration that produces *Protium javanicum* leaf extract with optimal antioxidant activity. This study utilized five distinct levels of solvent concentration: 90%, 80%, 70%, 60%, and 50%. The findings suggested that a 90% ethanol solvent yielded extracts exhibiting the maximum antioxidant activity (IC_{50} = 20.60 ppm), total phenolic concentration of 243.81 mg GAE/g, a yield of 21.90%, and total flavonoids of 33.02 mg QE/100g. The extract contains substances classified as flavonoids, terpenoids, and amino acids. This group of flavonoids exhibits potential as an antioxidant.

1. Introduction

Protium javanicum, belongs to the Burseraceae family which has a straight stem and is classified as a fast-growing species. In the book of Usada Bali, it is mentioned that the stem of *Protium javanicum* can function as an antidote [19], and the leaves are processed into herbal medicine to treat digestive disorders such as diarrhea, abdominal pain, and cough. Based on phytochemical tests from previous research, it is known that the ethanol extract of mature *Protium javanicum* leaves contains total phenols of 22.61%, total flavonoids of 4.33%, and IC_{50} of 21.265 ppm [25]. Other researchers [1], also reported that the EtOAc fraction of methanol extract of *Protium javanicum* leaves contained limonoid compounds 6-desacetylnimbin, quercitrin, and myricitrin, while Suirta *et al.* [24] reported that *Protium javanicum* leaves also contained flavonoids, tannins, steroids, and terpenoids. These phenolic compounds have potential as antioxidants.

Phenolic compounds are distributed in the vacuoles of plant cells so a suitable extraction method is needed to isolate these compounds. Maceration is one of the extraction methods through the process of soaking the material using a solvent that is suitable for the active compounds to be extracted, with a low or no heating process. One aspect influencing extraction efficiency is the type of solvent [26]. The choice of solvent type for extraction must consider selectivity, toxicity, and be adjusted to the polarity of the desired compound [26].

Simamora *et al.*,[22] reported that methanol and ethanol are the right solvents used to extract *Protium javanicum* leaves. Ethanol has lower toxicity than methanol, is polar, easily available, and costs less so it is the right choice to identify phenolic group compounds.

Phenolic substances possess one or more hydroxyl groups (polar component) immediately attached to aromatic rings (non-polar component) and are frequently present in plants as esters or glycosides, rather than as free molecules [9]. Phenolic compounds have diverse chemical characteristics and polarities. In this case, the solubility of phenolic substances is significantly enhanced by the polarity of the ethanol. Determination of the right polarity can be done by adjusting the variation of solvent concentration. Hikmawanti *et al.* [12] reported that the extraction of *katuk* leaves with 50% ethanol produced extracts with higher total phenols and flavonoids compared to 70% and 96% ethanol solvents. Other researchers reported that ideal conditions for the extraction of phenolic substances from olive leaves is 80% ethanol, at 40°C with a solvent/solids ratio of 30:1 [8]. This research aims to achieve the right concentration of ethanol to produce *Protium javanicum* leaf extract with the highest antioxidant activity.

2. Materials and procedures

The main objects utilized in this analysis were mature *Protium javanicum* leaves, while the chemical reagents used were ethanol, Folin Ciocalteu reagent, NaOH, $NaNo_2$, Na_2CO_3, DPPH (2,2-diphenyl-1-picrylhydrazyl), and $AlCl_3$. A rotary vacuum evaporator (IKA RV 10 Basic), an analytical balance (Shimadzu ATY224), an oven, a vortex (Maxi), aluminum foil (Klin Pack), a spectrophotometer (Genesys 10S UV-Vis), a micropipette (Socorex), and glassware were the principal instruments utilized in this research.

Extraction of Bioactive Compounds.

Leaves of *Protium javanicum* were extracted utilizing the maceration technique with ethanol as the solvent. Ten grams of *Protium javanicum* leaf powder were measured and immersed in 100 ml of ethanol at several concentrations (90%, 80%, 70%, 60%, and 50%). The ratio of substance to solvent was 1:10 (b/v). Extraction was conducted for 24 hours at ambient temperature, with stirring occurring every 6 hours, for 5 minutes. The solution was filtered and condensed utilizing a rotating vacuum evaporator at 40°C, 200 mBar, and a speed of 60 rpm.

Observation variables

This study examines the following variables: extract yield [3], total phenolics of the extract by Folin-Ciocalteu method [4], total flavonoid content assessed by spectrophotometric method [4], and antioxidant activity of the extract determined by IC_{50} using the DPPH (2,2-diphenyl-1-picrylhydrazyl) method [18], and identification of phenolic compounds in the extract through LC-MS/MS.

Statistical Analysis.

This study employed a completely randomized experimental design. Every experiment was performed thrice. The data were presented as mean ± standard error. One-way variance analysis was employed to determine the significant difference among the samples at a p-value of less than 0.05. [11].

3. Results and Discussion

The results indicated that varying amounts of ethanol solvent significantly influenced ($P<0.05$) on yield, total phenolics, and antioxidant activity of *Protium javanicum* leaf extract. The average values of these parameters are displayed in Table 1.

Table 1. Average yield values, total phenolics, and antioxidant activity of *Protium javanicum* leaf extracts at various ethanol concentrations.

Treatments	Parameter		
	Yield (%)	Total phenol (mg GAE/g)	Antioxidant activity (IC_{50}) (ppm)
EtOH 50%	25.38 ± 0.405^{b}	206.92 ± 0.861^{a}	30.83 ± 0.085^{e}
EtOH 60%	33.39 ± 0.822^{d}	225.01 ± 0.406^{b}	27.59 ± 0.230^{d}
EtOH 70%	35.97 ± 0.463^{a}	236.19 ± 0.461^{c}	25.58 ± 0.112^{b}
EtOH 80%	27.89 ± 1.049^{c}	225.17 ± 0.361^{b}	27.12 ± 0.119^{c}
EtOH 90%	21.90 ± 0.655^{e}	243.81 ± 0.703^{d}	20.60 ± 0.460^{a}

Values with distinct superscripts in the same column signify statistically significant differences between treatments ($P < 0.05$).

Extract Yield

The information presented in Table 1. indicates that the maximum extract yield was achieved using 70% ethanol, whereas the minimum extract yield was recorded with 90% ethanol. Increasing the concentration of ethanol solvent tends to produce higher extract yields and reaches an optimum value at 70% ethanol concentration. The extract yield decreased from 80% to 90% ethanol concentration. This indicates that the difference in solvent polarity greatly affects the amount of extracted compounds.

Solvent concentration is one of the factors affecting extraction efficiency. Grounded in the concepts of similarity and miscibility, compounds with the same polarity as the solvent are more easily extracted. The extraction yield is influenced by the chemical properties of phytochemicals, the employed extraction process, the solvent utilized, and the presence of interfering compounds [23]. According to the information presented in Table 1, it is evident that 90% ethanol solvent gives the lowest extract yield. Increasing the polarity of the solvent produces higher extract yields and reaches the highest value at 70% ethanol solvent concentration. The sequence of solvent concentrations that produce the highest extract yield is: 70% ethanol > 60% ethanol > 80% ethanol > 50% ethanol > 90% ethanol. Simamora *et al* [22] previously reported that 90% ethanol extract of *Protium javanicum* leaves produced an extract yield of 14.99%. The yield of 90% ethanol extract observed in this study surpassed the results reported in earlier research. This was likely attributable to variations in extraction methodologies and duration employed.

Total Phenol

Table 1. indicates that variations in ethanol concentration significantly influence ($P <0.05$) the total phenol content of *Protium javanicum* leaf extract. The total phenolic content was maximized using a 90% ethanol solvent, showing a significant difference compared to all other treatments. In contrast, the lowest total phenolic content was recorded using a 50% ethanol solvent. The total phenolic content of extracts from 60% and 80% ethanol solvents showed no significant difference. Solvents exhibiting greater solubility possess the capacity to extract a broader range of compounds characterized by varying polarities. Hydroalcoholic solvents, which consist of a combination of water and alcohol, possess the capability to extract compounds that are soluble in either water or ethanol solvents.

Polyphenols typically exhibit greater solubility in less polar organic solvents than in water [10]. This study categorized 90% ethanol concentration as a less polar solvent, 70% ethanol as a moderately polar solvent, and 50% ethanol as a highly polar solvent. The research conducted by [6] discovered that less polar flavonoid compounds, including aglycones of isoflavones, flavones, methylated flavones, and flavonols, exhibit improved solubility in less polar solvents. This means that these compounds are likely to dissolve better in 90% ethanol than in ethanol solvents with higher polarity. Meanwhile, groups of phenolic compounds with highly hydroxylated aglycone forms and phenolic compounds with high molecular weight are more soluble in polar solvents [14].

Antioxidant Activity (IC_{50})

Observations of antioxidant activity indicated that the leaves of *Protium javanicum* extracted with 90% ethanol solvent exhibited the lowest IC_{50} value, whereas the extracts obtained with 50% ethanol resulted in the highest IC_{50} value. A reduced IC_{50} value indicates improved antioxidant activity. The extract with 90% EtOH exhibits the highest total phenolic content, indicating it also possesses the greatest antioxidant activity among the samples analyzed.

The antioxidant activity of *Protium javanicum* leaf extract can be assessed by comparing the DPPH concentration value with the IC_{50}, which is represented as the Antioxidant Activity Index (AAI). The classification of antioxidant activity is as follows: it is considered low when the AAI value is less than 0.5, moderate when the AAI value ranges from 0.5 to 1.0, strong when the AAI value is between 1.0 and 2.0, and very strong when the AAI value exceeds 2.0 [21]. By comparing DPPH concentration and IC_{50} value of the extract, it is observed that the 90% ethanol extract of *Protium javanicum* leaves exhibits an AAI value of 1.94, classifying it as a potent antioxidant.

Flavonoid content of the best treatment extract

The evaluation of flavonoid content in the leaf extract of *Protium javanicum* was performed utilizing the sample that demonstrated the greatest total phenol concentration and the smallest IC_{50} value.. The data presented in Table 1 demonstrates that treatment EtOH 90% satisfies the established criteria. The analysis of the flavonoid concentration in the sample EtOH 90% revealed that the extract comprised flavonoids at 33.02 mg QE/100g.

Identification of Bioactive Compounds of *Protium javanicum* Leaf Extracts

The analysis of selected extracts (90% ethanol extract) for the identification of bioactive compounds was performed using LC MS/MS techniques. Figure 1 presents the chromatogram of the compounds found in the extract, and Table 2 provides a summary of the interpretation of the identified compound types.

According to the information presented in Figure 1 and Table 2, the compounds believed to influence the antioxidant activity of *Protium javanicum* leaf extract encompass flavonoids such as: umbelliferone, coumarin, procyanidin, catechin, rutin, myricitrin, kaempferol-3-0-rutinoside, kaempferol, quercetin, and luteolin.

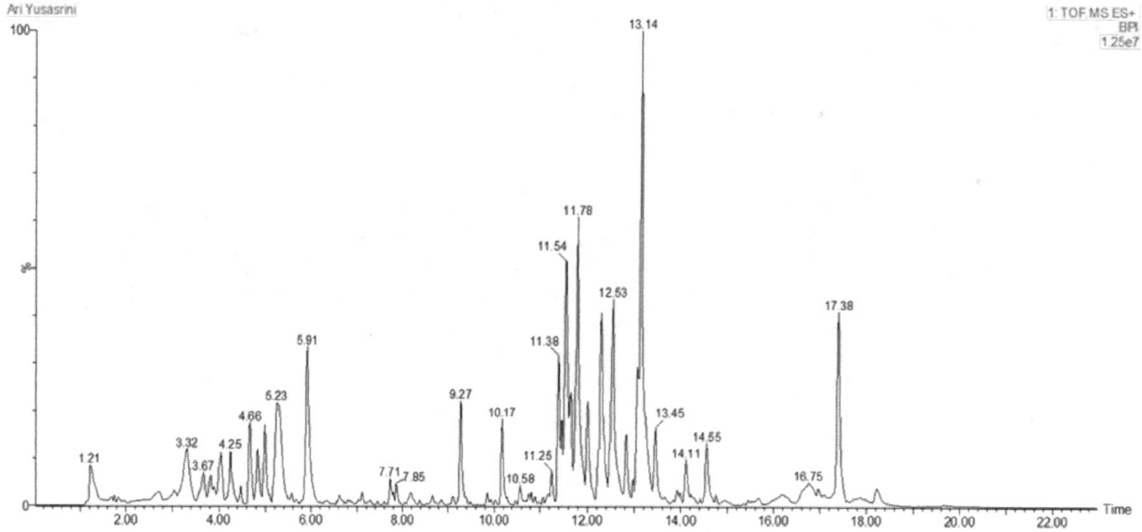

Figure 1. Chromatogram illustrating the various bioactive compounds present in the 90% ethanol extract derived from the leaves of *Protium javanicum*

Table 2. Interpretation of compound types in ethanol extract of *Protium javanicum* leaves

No	RT	Measured mass/parent ion (m/z)	Area (%)	Molecular formula	Predicted compound	References
1	1.26	381.0796	2.19	$C_{17}H_{16}O_{10}$	9-Hydroxy-7-oxo-7H-furo[3,2-g]chromen-4-yl β-D-glucopyranoside	ChemSpider ID 24699994
2	3.32	163.0395	2.66	$C_9H_6O_3$	Umbelliferone	ChemSpider ID 4444774
3	3.86	147.0443	2.66	$C_9H_6O_2$	Coumarin	ChemSpider ID 13848793
4	4.023	579.105	2.62	$C_{30}H_{27}O_{12}$	Procyanidin B2	ChemSpider
5	4.24	291.0864	0.95	$C_{15}H_{14}O_6$	Catechin,	ChemSpider ID 8711
6	4.663	611.1609	2.37	$C_{27}H_{30}O_{16}$	Rutin	ChemSpider ID 4444362
7	4.818	465.1032	2.33	$C_{21}H_{20}O_{12}$	Myricitrin	ChemSpider ID 4444992
8	4.952	595.1633	2.33	$C_{27}H_{30}O_{15}$	Kaempferol-3-O-rutinoside	Chemspider ID 4477257
9	4.973	595.167	2.33	$C_{27}H_{30}O_{15}$	Kaempferol	Chemspider ID 4444395
10	5.232	303.05	6.03	$C_{15}H_{10}O_7$	Quarcetin	Chemspider ID 4444051
11	5.893	287.0556	5.85	$C_{15}H_{10}O_6$	Luteolin	Chemspider ID 4444102

No	RT	Measured mass/parent ion (m/z)	Area (%)	Molecular formula	Predicted compound	References
12	9.27	657.2936	2.83	$C_{35}H_{44}O_{12}$	Segatanin A	PubChem ID 11756133
13	10.17	669.327	2.12	$C_{37}H_{48}O_{11}$	Euphorbiaproliferin	ChemSpider ID 28478696
14	10.576	333.2039	0.40	$C_{20}H_{28}O_4$	Carnosic acid	Chemspider ID 58635
15	11.54	518.3244	13.34	$C_{24}H_{39}N_9O_4$	N-Acetyl-L-phenylalanyl-N-{(1E,2S)-1-(carbamoylhydrazono)-5-[(diaminomethylene)amino]-2-pentanyl}-L-leucinamide	Chemspider ID 8182381
16	11.78	287.0556	8.43	$C_{24}H_{39}N_9O_4$	N-Acetyl-L-Leucyl-N-{(1E,2S)-1-(carbamoylhydrazono)-5-[(diaminomethylene) amino]-2-pentanyl}-L-phenylalaninamide	Chemspider ID8182381
17	12.53	520.3422	5.00	$C_{28}H_{45}N_3O_6$	Boc-3-(Z-amino)-L-alanine (dicyclohexylammonium) salt	Chemspider 2036716
18	13.14	496.3403	15.21	$C_{22}H_{41}N_9O_4$	N5-(Diaminomethylene)-L-ornithyl-L-prolyl-L-lysyl-L-prolinamide	Chemspider ID150667
19	14.11	353.2692	1.57	$C_{21}H_{35}O_4$	Monolinolenin (2-alpha-linolenoylglycerol)	Chemspider ID 4518847
20	14.55	524.3721	1.97	$C_{25}H_{45}N_7O_2$	(2E)-2-{4-[3-(Diethylamino)propoxy] benzyliden}-N'-[(E)-{4-[3-(diethylamino)propoxy]phenyl}methylen] hydrazincarbohydrazonoid	Chemspider ID7876317
21	16.75	684.2035	3.64	$C_{11}H_{13}N_{35}O_3$	Unknown	Unknown
22	17.38	954.6176	7.03	$C_{49}H_{75}N_{15}O_5$	Unknown	Unknown

Phenolic compounds in *Protium javanicum* leaf extract function as antioxidants via various mechanisms. Procházková *et al* [16] reported, the mechanisms by which flavonoids function as antioxidants through various pathways, including: radical scavenging, metal binding, inhibition of enzymes involved in free radical formation, and enhancement of the activity of endogenous antioxidant enzymes.

It has been reported that coumarin and quercetin exhibit antioxidant activity via radical scavenging and metal-chelating mechanisms [13]. Luteolin and catechin are flavonoid compounds from the flavone and flavanol groups, respectively, having antioxidant properties mainly attributed to the 3′,4′-dihydroxy structure (catechol group) in the phenolic B ring as the most active group [15] [2]. Meanwhile, myricetin-3-O-a-rhamnoside (myricitrin), rutin, and kaempferol besides acting as antioxidants also show anti-

inflammatory, antifibrotic [7], antimicrobial, anticarcinogenic, neuroprotective, antithrombotic and antiviral properties [5] [20].
The 90% ethanol extract of *Protium javanicum* leaves also contained terpenoid compounds, namely segetanin A (diterpene), euphorbiaproliferin (monoterpene) and carnosic acid (diterpene). Carnosic acid exhibits potential antioxidant properties [17].

4. Conclusion
The findings indicate that ethanol concentration significantly influences the properties of *Protium javanicum* leaf extract. The extracts derived from a 90% ethanol concentration demonstrate the biggest overall phenolic content and antioxidant activity, as indicated by IC_{50} values. The extract of *Protium javanicum* leaves, comprising 90% ethanol, was found to include various flavonoid compounds such as umbelliferone, coumarin, procyanidin, catechin, rutin, myricitrin, kaempferol-3-0-rutinoside, kaempferol, quercetin, and luteolin. These compounds exhibit potential as antioxidant agents.

5. References

[1]. Adfa, M., Y. Hattori, M. Ninomiya, Y. Funahashi, T. Yoshimura, and M. Koketsu. 2013. Natural Product Research. **27**(3):270–273
[2]. Ahmadi, S.M., R. Farhoosh, A. Sharif, and M. Rezaie. 2020. *Journal of Food Science*. **Vol. 85**, Iss. 2. DOI: 10.1111/1750-3841.14994
[3]. AOAC. 1990. Official Method of Analysis of Association Official Agriculture Chemist. Washington DC.
[4]. Baba, S. A., & Malik, S. A. 2015. *Journal of Taibah University for Science*. **9**(4): 449–454. https://doi.org/10.1016/j.jtusci.2014.11.001
[5]. Choi, S.-S.; Park, H.-R.; Lee, K.-A. 2021. Antioxidants, **10**, 1696. https://doi.org/10.3390/ antiox10111696
[6]. Chaves JO, de Souza MC, da Silva LC, Lachos-Perez D, Torres-Mayanga PC, Machado APdF, Forster-Carneiro T, Vázquez-Espinosa M, González-de-Peredo AV, Barbero GF and Rostagno M.A. 2020. *Front. Chem.* **8**:507887. doi: 10.3389/fchem.2020.507887
[7]. Domitrovic, R., K. Rashed, O. Cvijanovic, S. Vladimir-Knezˇevic, M. Škoda, A.Višnic. 2015. *Chemico-Biological Interactions*, **230** (21-29)
[8]. Elboughdiri, N. 2018. *Engineering, Technology & Applied Science Research*. **8** (2) : 2805-2808
[9]. Galanakis, C.M., V. Goulas, S. Tsakona , G.A. Manganaris and V. Gekas. 2013. *International Journal of Food Properties*, **16**:2, 382-396. doi.org/10.1080/10942912.2010.522750
[10]. Haminiuk, C.W.I., M.S.V. Plata-Oviedo, G. de Mattos, & S.T. Carpes, I.G. Branco. 2014. *J Food Sci Technol*. **51**(10):2862–2866. DOI 10.1007/s13197-012-0759-z
[11]. Harsojuwono, B. A., Arnata, I. W., dan Puspawati, G. A. K. D. 2021. Lintaskata Publishing.
[12]. Hikmawanti, N.P.E., S. Fatmawati and A.W. Asri. 2021. *IOP Conf. Series: Earth and Environmental Science*. **755**. doi:10.1088/1755-1315/755/1/012060
[13]. Islam, M.R., Zaman, A., Jahan, I., Chakravorty, R., Chakraborty, S., 2013. *J. Young Pharm*. **5**, 173–179.
[14]. Kaczorová, D.; Karalija, E.; Dahija, S.; Bešta-Gajevi´c, R.; Pari´c, A.; Cavar Zeljkovi´c, S. 2021. *Molecules* **26**, 1601. https://doi.org/10.3390/ molecules26061601
[15]. Maestri, D. M., Nepote, V., Lamarque, A. L., & Zygadlo, J. A. 2006. *Advance Research*, **37**, 105–135.
[16]. Procházková, D., Boušová, I., & Wilhelmová, N. 2011. *Fitoterapia*, **82**(4), 513–523. https://doi.org/10.1016/j.fitote.2011.01.018
[17]. Mantzourani, C.; Tarantilis, P.A.; Kokotou, M.G. 2023. *Separations*, **10**, 481. https://doi.org/10.3390/separations10090481
[18]. Molyneux, P. 2004. *Songklanakarin J. Sci. Technol*; **26** (2): 211-219.
[19]. Nala, N. 1992. Usada Bali. Penerbit PT. Upada Sastra, Denpasar.
[20]. Sharma, N.; Biswas, S.; Al-Dayan, N.; Alhegaili, A.S.; Sarwat, M. 2021. *Antioxidants*, **10**, 1419. https://doi.org/10.3390/antiox10091419
[21]. Scherer, R. & Godoy, H. T. 2009. *Food Chemistry*, **112**, 654-658. doi:10.1016/j.foodchem.2008.06.026
[22]. Simamora, A.C.Y., N.L.A. Yusasrini dan I. N. K. Putra. 2021 *Itepa*, **10** (4). 681-689.
[23]. Stalikas CD. 2007. *J. Sep. Sci*; **30**:3268-95.
[24]. Suirta, I.W., N. M. Puspawati, I.A.R.A. Asih. 2016. *Cakra Kimia (Indonesian E-Journal of Applied Chemistry)* **4**, (1)

[25]. Yusasrini, N.L.A and Darmayanti, L.P.T. 2024. *Jurnal Teknologi Pertanian* **Vol. 25** No. 1. doi: https://doi.org/10.21776/ub.jtp.2024.025.01.6
[26]. Zhang, Q.W., L.G. Lin, and W.C. Ye. 2018. *Chin. Med.* **13**, 1–26. doi:10.1186/s13020-018-0177-x

Characteristic and prebiotic index of instant multigrain rice

E Kurniawati[*1)], R P Putri[1)], A Wahyono[1)], S O N Yudiastuti[1)], C P Pinandita[2)]

[1)] Food Engineering Technology Study Program, Department of Agricultural Technology, Politeknik Negeri Jember
[2)] Agribusiness Management Study Program, Department of Agribisnis Management, Politeknik Negeri Jember

Email: *elly_kurniawati@polije.ac.id

Abstract. Multigrain rice is a blend of various grains and legumes. This study aimed to develop instant multigrain rice that is rich in protein and dietary fiber, as well as to determine the prebiotic index. Five formulations were tested, all based on 40% brown rice, 20% corn, and 10% sorghum, with varying proportions of protein sources from soybean and edamame: M2 with 40% soybean; M3 with 30% soybean and 10% edamame; M4 with 20% soybean and 20% edamame; M5 with 10% soybean and 30% edamame; and M6 with 40% edamame. M1, consisting of 100% brown rice, served as the control. A non-factorial Completely Randomized Design (CRD), which included five treatments and three replications, was used to assess the treatments. The results indicated that the best formulation, based on both chemical and sensory properties, was M4, which included 30% brown rice, 20% corn, 10% sorghum, 20% soybean, and 20% edamame. This formulation had a water content of 10.52%, protein content of 17.8%, fat content of 10.05%, and dietary fiber content of 10.35%. The sensory properties (aroma, color, taste, and texture) were well-liked by the panelists, and the prebiotic activity was measured at 1.08.

1. Introduction

The increasing recognition of the importance of healthy and safe foods, such as functional foods, in promoting overall health and reducing the risk of specific diseases has driven the demand for novel staple food products. These products must effectively meet dietary requirements and can be prepared with greater convenience and efficiency [1]. Prebiotics are one of the fastest-growing sectors within functional foods. Local food sources, such as corn, brown rice, sorghum, and soybean serve as viable prebiotics sources. These ingredients can be utilized to formulate instant multigrain rice, a blend of different grains and legumes that offers balanced and diverse nutrient profiles compared to regular white or brown rice. The inclusion of multiple grains enhances fiber, protein, vitamins, minerals, and prebiotic content [2].

Whole grains, such as multigrain rice, contain various compounds like dietary fiber, resistant starch, and other indigestible substances that influence the gut environment [3]. Resistant starch (RS) is a form of dietary fiber that the small intestine cannot digest. It has been recognized as a powerful prebiotic that supports a healthy gut microbiome and promote proper function and balance of the digestive system [4]. As a functional ingredient, RS plays a role in reducing postprandial blood glucose levels by slowing the digestion, absorption, and release of glucose into the bloodstream [5]. This delayed glucose response is especially beneficial for improving glycemic control and insulin sensitivity in individuals with diabetes and obesity. Moreover, RS's prebiotic activity contributes to the symbiotic relationship between the host and gut microbiota, offering multiple health advantages, including anti-obesity, anti-diabetic, anti-pathogenic, and immunomodulatory effects. It also selectively stimulates the growth and metabolic activity of specific beneficial bacteria in the colon [6].

A successful advancements in instant multigrain rice formulations from corn, brown rice, sorghum, and a type of soybean, edamame, had been made previously to address the demand for high-fiber and high-protein food products [7]. However, there is a significant gap in research concerning the potential of instant multigrain rice made from these ingredients as a prebiotic. Therefore, in this study we aimed to investigate the characteristics and prebiotic index of instant multigrain rice derived from brown rice, corn, sorghum, soybean, and edamame.

2. Materials and Methods

2.1 Raw Materials

The raw materials such as edamame was acquired from the Mitratani 27 outlet, while brown rice, sorghum, corn, and soybeans were procured from the BMW Organic Store in Jember Regency, Indonesia.

2.2 Equipments

The required equipments such as food dehydrator, analytical balance and other equipments used for proxymate analysis, were procured from the Analytical Laboratory, State Polytechnic of Jember.

2.3 Methods

This research utilized a formulation based on the study of [7], as outlined in Table 1. The raw materials were prepared according to [7], brown rice and sorghum were soaked for 2 hours in sodium citrate, while corn, soybean, and edamame were washed. Then, they were steamed, reduced in size, and freezed. Each material was dried separately to lower water content. Finally, the product was rehydrated by soaking it in hot water in a closed system for 5 minutes.

Table 1. Formulation

Treatment	Brown rice %	Corn %	Sorghum %	Soybean %	Edamame %
M1	100	-	-	-	-
M2	30	20	10	40	0
M3	30	20	10	30	10
M4	30	20	10	20	20
M5	30	20	10	10	30
M6	30	20	10	0	40

Subsequently, the chemical components such as water, fat, protein, and dietary fiber content of the instant multigrain rice were analyzed according to AOAC methods. The determination of the prebiotic activity shown as prebiotic activity score (PAS), was conducted following the method described by [2]. It was calculated using the following equation:

$$PAS = \left[\frac{probioticlog\frac{cfu}{mL}\ on\ the\ prebiotic\ at\ 24h - probioticlog\frac{cfu}{mL}\ on\ the\ prebiotic\ at\ 0h}{probioticlog\frac{cfu}{mL}\ on\ the\ glucose\ at\ 24h - probioticlog\frac{cfu}{mL}\ on\ the\ glucose\ at\ 0h}\right] - \left[\frac{entericlog\frac{cfu}{mL}\ on\ the\ prebiotic\ at\ 24h - entericlog\frac{cfu}{mL}\ on\ the\ prebiotic\ at\ 0h}{entericlog\frac{cfu}{mL}\ on\ the\ glucose\ at\ 24h - entericlog\frac{cfu}{mL}\ on\ the\ glucose\ at\ 0h}\right] \quad (1)$$

To evaluate sensory properties, a randomized method was used. Twenty five untrained panelists, all healthy individuals aged 20–23, comprising both male and female were selected as panelists. The products were labeled with random letters, and the panelists were given written instructions to assess properties such as aroma, taste, color, and texture using a 5-point hedonic scale. The scale ranged from 1 to 5, where 1 represented "Dislike extremely," indicating strong negative preference, and 5 represented "Like extremely," indicating the highest level of preference. The intermediate values, 2 and 4, corresponded to "Dislike moderately" and "Like moderately," respectively, allowing for more nuanced ratings. A score of 3, or "Like slightly," was used to indicate a mild preference. The study followed a Completely Randomized Design with one factor, involving six treatment levels, each replicated three times. Statistical analysis was performed using IBM SPSS 20.

3. Results And Discussions

3.1 Nutritional analysis of instant multigrain rice

The nutritional analysis of instant multigrain rice, as presented in Table 2, revealed significant variations across different formulations in terms of water, protein, fat, and dietary fiber content. Water content is a key quality factor, it influencing its shelf life and the conditions under which microorganisms that could lead to spoilage thrive [8]. Water content was significantly higher in formulations M2 to M6 compared to M1, with M2 exhibiting the highest average water content at approximately 11.72% and M1 having the lowest at around 9.38%. This finding aligns with [9], which indicates that instant multigrain rice with low water content maintains quality and supports long-term storage. Instant multigrain rice experienced a significant reduction in water because of dehydration process [7].

Table 2. Nutritional content

Treatment	Water content (%wb)	Protein content (%wb)	Fat content (%wb)	Total dietary fiber (%wb)
M1	9.38±0.20^{a}	3.15±0.17^{a}	1.59±0.24^{a}	3.48±0.05^{a}
M2	11.72±0.02^{c}	16.11±0.16^{b}	9.70±0.06^{b}	9.47±0.14^{b}
M3	10.70±0.10^{b}	16.49±0.12^{b}	9.88±0.12^{b}	9.42±0.07^{b}
M4	10.52±0.18^{b}	17.80±0.42^{b}	10.05±0.03^{c}	10.35±0.23^{b}
M5	10.25±0.08^{b}	18.01±0.14^{c}	10.26±0.21^{c}	11.22±0.16^{d}
M6	9.65±0.04^{a}	18.22±0.24^{c}	10.32±0.21^{c}	11.58±0.16^{d}

Note: (% wb = % wet base)

The different concentrations of soybean and edamame in the formulations of instant multigrain rice led to variations in water content. During steaming, gelatinization occurs as starch granules are infiltrated by water molecules. The distinct cotyledon structures of soybean and edamame contribute to increased water content. This increase is due to the gelatinization process, during which the hydrogen bonds that maintain the structural integrity of the starch granule weaken. The free hydroxyl groups in amylopectin draw in water molecules, resulting in granule enlargement [10]. Because edamame has a lower cotyledon content than soybean, formulations with a higher proportion of edamame and a lower proportion of soybean, such as M6, showed the lowest water content.

In terms of protein content, there was a significant difference across the treatments. Treatment M1, consisting only of brown rice, had the lowest average protein content at 3.15%, while M6, with 40% edamame, exhibited the highest protein content at 18.22%. Formulations M2 to M6, which included high concentrations of soybeans and edamame, showed a notable increase in protein content, attributed to the high nutritional value of these legumes, both known for their protein-rich composition [13].

The fat content varied significantly among the formulations, with M1 (brown rice) having the lowest fat content at 1.59%. Formulations M2 to M6, particularly M6, exhibited significantly higher fat content, ranging from 9.70% to 10.32%. This increase is mainly due to the inclusion of soybean and edamame, dry edamame contains 22% fat [12] and soybean contains 19.9% fat [11], which includes beneficial polyunsaturated fatty acids such as linoleic acid and α-linolenic acid [13]. These fatty acids contribute to the overall higher fat content in the instant multigrain rice formulations.

Finally, dietary fiber content showed notable differences across treatments. Dietary fiber is the component of plant material in the diet that resists enzymatic digestion. It is classified into two main types: soluble and insoluble fibers. Insoluble fibers, such as cellulose, hemicellulose, and lignin, do not dissolve in water, while soluble fibers, including pectins, gums, and mucilages, turn gummy when they come into contact with water [14]. Formulation M2 had the lowest average dietary fiber content at 9.47%, while M6 had the highest at 11.58%. The incorporation of legumes, particularly soybean and edamame, contributed to the increased dietary fiber content, aligning with previous research [15] that highlights the fiber-enhancing effects of legumes in food formulations.

3.2 Sensory characteristics

The instant multigrain rice, made by incorporating varying ratios of soybean to edamame, underwent sensory evaluation. This instant multigrain rice was assessed for sensory quality based on attributes such as aroma, color, taste, and texture. The results are presented in Table 3.

Table 3. Sensory analysis

Treatment	Aroma	Color	Taste	Texture
M1	4.53 ± 0.34^{b}	3.86 ± 0.63^{ab}	3.57 ± 0.65^{a}	4.46 ± 0.81^{c}
M2	3.36 ± 0.47^{a}	4.06 ± 0.64^{b}	3.33 ± 0.68^{a}	3.50 ± 0.81^{a}
M3	3.67 ± 0.53^{b}	3.88 ± 0.61^{ab}	3.56 ± 0.44^{a}	3.52 ± 0.80^{a}
M4	4.20 ± 0.14^{b}	3.90 ± 0.87^{ab}	4.25 ± 0.65^{b}	3.77 ± 0.51^{ab}
M5	3.55 ± 0.32^{a}	3.36 ± 0.45^{a}	4.05 ± 0.53^{b}	3.92 ± 0.04^{b}
M6	3.43 ± 0.27^{a}	3.66 ± 0.24^{a}	4.15 ± 0.12^{b}	3.84 ± 0.38^{b}

The sensory evaluation of instant multigrain rice revealed varying preferences across different attributes, including aroma, color, taste, and texture. Aroma preference analysis indicated scores ranging from 3.36 to 4.53, suggesting a generally pleasant aroma, as shown in Table 3. Significant differences were observed between formulations M1 and M4 compared to M2, M3, M5, and M6. Formulations M1 and M4, with average scores of 4.20 to 4.53, were "liked" by the panelists, while the others received lower scores, indicating that they were "slightly liked." The inclusion of 40% soybean in certain formulations enhanced the beany aroma, which was preferred by the panelists.

For color, the hedonic test showed that treatment M5 received the lowest score of 3.36, indicating a dislike, while treatment M2 achieved the highest preference with an average score of 4.06. This suggests that a balanced mixture of grains and legumes produced a more attractive color brightness. The average score for M2 indicated that panelists liked the color, while the scores for M1, M4, M5, and M6 ranged from 3.36 to 3.88, suggesting that panelists "somewhat liked" the color of these formulations. The results suggest that color brightness plays a key role in overall acceptability.

In terms of taste, treatment M4 received the highest score of 4.25, indicating that it was the most preferred by the panelists. Treatment M5, containing 20% soybean and 30% edamame, also scored highly with an average of 4.05, likely due to the lighter beany flavor from the edamame. On the other hand, M2 recorded the lowest taste score at 3.33. Treatments M1 and M6 received average scores of 3.57 and 3.75, respectively, indicating that the taste of these formulations was generally favorable.

Lastly, the texture evaluation, shown in Table 2, revealed that treatment M1 had the highest preference score of 4.46, while treatments M4, M5, and M6, with a dominant edamame protein source, also received high preference scores ranging from 3.77 to 3.92. This indicates that panelists preferred the texture of these formulations.

According to the sensory evaluation findings, M4 emerged as the best treatment among the formulations tested. It received the highest score for taste, indicating that panelists found it the most favorable in terms of flavor. Additionally, M4 achieved strong preference scores in aroma, where it was rated as "liked" by the panelists, and performed well in texture, reflecting a positive reception. While M2 received the highest score for color, M4's overall performance across multiple sensory attributes—especially its superior taste—makes it the most preferred formulation. These results suggest that M4 offers a balanced combination of desirable sensory characteristics, making it the best formulation for instant multigrain rice.

3.3 Prebiotic activity

The prebiotic activity reflects the correlation between changes in beneficial microbes and unwanted microbes in relation to their initial concentration. It provides a quantitative score that indicates the prebiotic effect, which shown as Prebiotic Activity Score (PAS). In this study, the prebiotic activity of instant multigrain rice combinations with varying protein sources from soybean and edamame was evaluated based on their content of fructooligosaccharides (FOS) and galactooligosaccharides (GOS). The different concentrations of soybean and edamame (10%, 20%, 30%, and 40%) resulted in varying prebiotic activity scores, as presented in Table 4.

Tabel 4. Prebiotic activity score for instant multigrain rice

Treatment	Instant multigrain rice					Prebiotic activity score
	Brown rice %	Corn %	Sorghum %	Soybean %	Edamame %	
M1	100	-	-	-	-	0.46 ± 0.12
M2	30	20	10	40	0	0.68 ± 0.31
M3	30	20	10	30	10	0.73 ± 0.35
M4	30	20	10	20	20	1.08 ± 0.27
M5	30	20	10	10	30	1.13 ± 0.26
M6	30	20	10	0	40	1.20 ± 0.12

The prebiotic activity score of instant multigrain rice without any prebiotics was measured at 0.46 ± 0.12. In contrast, the prebiotic activity score for instant multigrain rice with the protein source combinations from soybean and edamame ranged from 0.68 ± 0.31 to 1.20 ± 0.12. This indicates that the control sample (brown rice) exhibited prebiotic activity even without the addition of sorghum, corn, soybean, or edamame, with a score of 0.46. However, the inclusion of soybean and edamame as protein sources enhanced the prebiotic activity score to between 0.68 and 1.20. This score indicates the ability of resistant starch in legumes to promote the replication of specific microorganisms in the presence of competing microbes.

Naturally occurring resistant starch (RS) is commonly found in cereal grains, legumes, and seeds. Resistant starches in legumes are categorized to Class 1, which consists of starch that is physically inaccessible. The resistance arises from the dense molecular structure, which prevents digestive enzymes from accessing the starch [16]. According to [13], the starch content in edamame ranges from 11.99% to 16.58%, while [17] reported that soybeans contain approximately 11% starch. With up to 40% of their starch being amylose, legumes have a substantially greater amylose concentration than other high-starch foods like corn, wheat, and rice, which enhances their resistance to digestion. The resistance is caused by the dense molecular structure, which hinders digestive enzymes from accessing the resistant starch [18]. Research has demonstrated that starches with higher amylose content exhibit greater resistance to enzymatic digestion due to the compact linear structure of amylose and the hydrogen bonds linking the glucose chains within amylose [19]. As shown by [20], one of the requirements for utilizing starch as a prebiotic is its ability to withstand gastric acidity, mammalian enzyme hydrolysis, and digestive absorption in the gastrointestinal tract. This may explain the high prebiotic activity score observed in multigrain formulations with an increased percentage of edamame, as formulation F6, containing 40% edamame, exhibited the highest prebiotic activity score.

4. Conclusion

It can be concluded that instant multigrain rice, which incorporated 30% brown rice, 20% corn, 10% sorghum, 20% soybean, and 20% edamame (formulation M4), was the best treatment among the formulations studied in terms of both chemical and sensory qualities. It exhibited 10.52% water, 17.8% protein, 10.05% fat, and 10.35% dietary fiber. Additionally, the sensory properties (aroma, color, taste, and texture) of M4 were well received by the panelists, and its prebiotic index was measured at 1.08. Given its rich protein and dietary fiber content, along with its strong prebiotic activity, this instant multigrain rice formulation shows great potential as a healthy food option. It can be particularly beneficial for individuals with diabetes and obesity, as its prebiotic effects may help improve overall metabolic health.

Acknowledgement

The authors expressed their appreciation for the financial grants provided by P3M Politeknik Negeri Jember by means of 2024 PNBP funding, which enabled the completion of this research.

References

[1] Hartanti L D and Kurniawati E 2023 *J. Food Eng.* **2**
[2] Arya S S and Shakya N K 2021 *LWT* **135** 109991
[3] Slavin J, Jacobs D and Marquart L 1997 *Nutr. Cancer* **27** 14–21
[4] DeMartino P and Cockburn D W 2020 *Curr. Opin. Biotechnol.* **61** 66–71
[5] Sajilata M G, Singhal R S and Kulkarni P R 2006 *Compr. Rev. Food Sci. Food Saf.* **5** 1–17
[6] Megur A, Daliri E B-M, Baltriukienė D and Burokas A 2022 *Int. J. Mol. Sci.* **23** 6097
[7] Kurniawati E, Putri R P and Wahyono A 2024 *IOP Conference Series: Earth and Environmental Science* **1338** 012049
[8] Sasmitaloka K Sanggrami, Widowati S and Sukasih E 2020 *J. Penelit. Pascapanen Pertan.* **17** 1–14
[9] Mohd Ramli N A, Fazalul Rahiman M H, Kamarudin L M, Mohamed L, Zakaria A, Ahmad A and Rahim R A 2021 *Sensors* **21** 1–19
[10] Bartolomiussihosa J A and Amrinola W 2023 *IOP Conf. Ser. Earth Environ. Sci.* **1169** 1–8
[11] USDA 2019 Soybeans, mature seeds, raw
[12] USDA 2019 Edamame, frozen, prepared *US Dep. Agric.*
[13] Yu D, Lin T, Sutton K, Lord N, Carneiro R, Jin Q, Zhang B, Kuhar T, Rideout S, Ross J, Duncan S, Yin Y, Wang H and Huang H 2021 *Front. Sustain. Food Syst.* **5** 1–14
[14] Dhingra D, Michael M, Rajput H and Patil R T 2012 *J. Food Sci. Technol.* **49** 255–66
[15] Kurniawati E 2015 *Tepung Edamame (Glycine max (L) Merrill) Sebagai Sumber Serat Pangan dan Oligosakarida: Karakterisasi Sifat Kimia dan Fisikokimia Serta Efek Fisiologisnya.* (Gadjah Mada University)
[16] Huebner J, Wehling R L and Hutkins R W 2007 *Int. Dairy J.* **17** 770–5
[17] Stevenson D G, Doorenbos R K, Jane J and Inglett G E 2006 *Starch - Stärke* **58** 509–19
[18] Haralampu S G 2000 *Carbohydr. Polym.* **41** 285–92
[19] Sajilata M G, Singhal R S and Kulkarni P R 2006 *Compr. Rev. Food Sci. Food Saf.* **5** 1–17
[20] Zaman S A and Sarbini S R 2015 *Crit. Rev. Biotechnol.* 1–7

Assessment of insect population abundance and its effect on the production value of fresh fruit bunches of oil palm (*Elaeis guineensis* Jacq.)

R Taufika*, D N Erawati, U Fisdiana, S Humaida, I Harlianingtyas

Department of Agricultural Production, Politeknik Negeri Jember, Jl. Mastrip 164, Jember, East Java, Indonesia

*ramadhantaufika@polije.ac.id

Abstract. Over the last three years, Crude Palm Oil (CPO) production in Indonesia has fluctuated. One of the factors causing fluctuations in CPO production is resistance and resurgence, which causes an explosion in the population of secondary pests and a decrease in the population of pollinator insects and natural enemies. This study aimed to analyze the insect population based on the abundance level and insects' role in oil palm plantations. It also analyzed the estimated production of oil palm fresh fruit bundles based on fruit census based on age. The research was conducted in the 7-year-old oil palm collection garden of Politeknik Negeri Jember. Insect population data were collected for 90 consecutive days in all oil palm trees. Production estimation was carried out by census of fruits aged 0-1, 1-2, 2-3, 3-4, 4-5, and 5-6 months. Parameters observed were an abundance of each insect species and estimated and realized production. Data analysis used the Shannon-Wienner diversity index, production estimation equation, and production distribution. The results showed that insects caught in oil palm plantations were ten families, with a total of 614 individuals, with the family of Formicidae having the highest species abundance. Estimated oil palm production 2024 for 12 consecutive months is 266; 252.5; 234.1; 215.7; 169.8; 156.1; 133.1; 128,5; 105,6; 192,8; 252,5; 270,8 kg. This shows that insect populations affect the value of fresh fruit bunch production.

Keywords: estimation, palm oil, production, insects, taxation

1. Introduction

Palm oil (*Elaeis guineensis* Jacq.) is one of plantation crop with total crude palm oil (CPO) production in 2021-2023 of 40,121,480 tons, 37,580,892 tons, and 42,235,425 tons. Based on these data, there are fluctuations in national CPO production. Fluctuations in CPO production occur because the implementation of Good Agricultural Practice (GAP) has not been fully optimized [1]. Based on the cultivation aspect, the implementation of GAP that has not been optimally implemented in oil palm cultivation is the continuous use of chemical pesticides to reduce pest populations. The continuous use of synthetic pesticides causes resurgence and resistance in target pests [2]. By 2024, chemical insecticides will be used for pest control in oil palm, which has many negative impacts, namely increasing the main pest population due to resistance and resurgence, secondary pest outbreaks, decreasing non-target insect populations, disturbing human health and ecosystem balance [3]. Insect populations are critical for increasing production of Fresh Fruit Bunches (FFB), CPO, and the continuity of the oil palm ecosystem [4]. According to [5] the growth and development of oil palm plants is strongly influenced by the diversity of insect populations of pollinators, predators, decomposers, and parasitoids.

In addition to technical cultivation factors, climate anomalies affect insect populations in oil palm plantations [6]. Increased temperature increases the main pests' feeding activity, *Oryctes rhinoceros* and Adoretus sp. [7]. The population abundance of the pollinating insect *Elaeidobius kamerunicus* decreased due to the increase in temperature, thus reducing FFB production [8]. According to [5] insects are poikilothermic creatures, so the biochemical and physiological processes of insects are highly dependent on environmental temperature [9]. [7] stated that the speed of insect life development increases directly proportional to the increase in temperature until the optimum development speed. Increasing temperature past the optimum development point can cause death in some insect species.

Strategies that can be carried out to manage the insect population of oil palm plantations to increase FFB production and better ensure product continuity are through insect population detection [9] . Through this strategy, the level of abundance and the role of insects that make up the ecosystem will be analyzed as an effort to manage insects in oil palm plantations that are environmentally sound. Insect populations are known to be effective in increasing FFB production [10] . Therefore, it is necessary to analyze and study the abundance and diversity of insect populations so as to obtain green economy-based FFB product continuity that is in line with sustainable and environmentally sound agriculture. So far, information on insect diversity in oil palm plants is still limited, so insect control and conservation activities in oil palm plants are still less effective.

One way to measure the success of oil palm cultivation is to use oil palm FFB production forecasts. Production forecasting is done for one year, one semester, and three semesters. Production forecasting is done once every six months through flower counting by observing and recording flower bunches' development and radiance. This study aims to analyze the insect population based on the level of insect abundance in the oil palm collection garden, analyze the estimated production of oil palm FFB based on fruit census based on age, and analyze the realization of oil palm FFB production in the Politeknik Negeri Jember Collection Garden.

2. Material and Method

2.1 Place and Time of Implementation

The research was conducted in the oil palm plant collection garden in Juli-September 2024 with an altitude of ±89 meters above sea level.

2.2 Tools and Materials

The materials used were all types of insects living in the Polije oil palm plant collection garden, oil palm fruit, distilled water, 70% alcohol, detergent, and chloroform. The tools used were a flask, insect identification book, drying box, hand counter, insect net, digital camera, label paper, killing bottle, brush, specimen box, lup, meter, microscope, ruler, pitfall trap, plastic clip, thermohygrometer.

2.3 Activity Implementation

Data collection of insect populations was carried out for ten times with an interval of ten days in all plots of oil palm vines. Insect inventory and temperature and humidity measurements were conducted at 06.00-10.00 am and 14.00-17.00 pm. Estimating oil palm FFB production was done by census of fruits aged 0-1, 1-2, 2-3, 3-4, 4-5, and 5-6 months based on physical characteristics in all vines. The parameters observed in the study were population abundance, estimation, and production realization. Data analysis used the population abundance equation, FFB production estimation equation, and FFB production distribution.

2.4 Data analysis

The insects that were inventoried were calculated based on the insect population abundance equation

$$= \frac{\textbf{number of individuals of a species}}{\textbf{total number of individuals}} \; x\, 100\%$$

Estimation of FFB production using the equation $= \frac{\textbf{total estimation}}{\textbf{monthly production distribution}}$

Distribution of FFB production using the equation $= \frac{\textbf{number of FFB each months}}{\textbf{total number of FFB}} \; x\, 100\%$

3. Result and Discussion

3.1 Insect Abundance

The observation results showed that 478 individuals of insects were successfully counted from all research plots in the Oil Palm collection garden of Politeknik Negeri Jember. Details of the number of individuals and population abundance of each insect order counted in the Oil Palm collection garden of Politeknik Negeri Jember are presented in Figure 1.

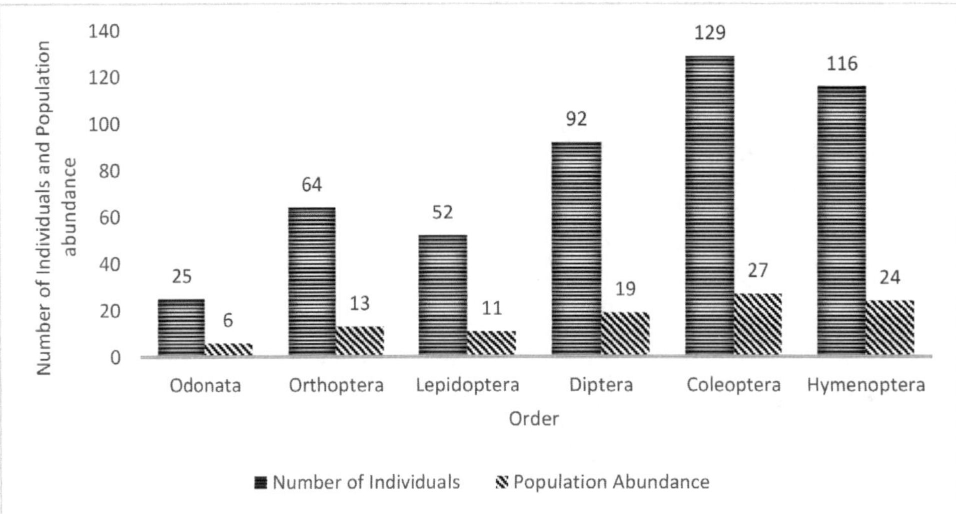

Figure 1. Details of the number of individual and abundance of insect populations

Based on the data, it is known that insect orders with the lowest to the highest order of abundance are Odonata, Coleoptera, orthoptera, Diptera, Hymenoptera, and Lepidoptera. The abundance of insect species is determined by two factors, namely biotic potential and environmental barriers [11]. Biotic potential is the ability of insects to reproduce under optimal conditions. Environmental barriers are various biotic and abiotic elements in the ecosystem that tend to reduce insect fertility and survival. According to [12], insect abundance is influenced by environmental factors such as temperature and humidity. The results showed that the air temperature and humidity in the garden ranged from 31-34^{0}C with relative humidity of 74-82%. According to [5] , insects can live optimally at an ambient temperature of 26.5-31^{0}C with a humidity of 64.25-65.25%. These data show that the temperature and humidity at the time of research activities in the collection garden are not in accordance with the limiting factors of insects. The abiotic factor of temperature influences the insect body's metabolism because temperature is a biocatalyst for enzymes in metabolism [9]. The insect life cycle is positively correlated with the increase in temperature until it reaches the optimal life cycle. Environmental temperatures that exceed insect limiting factors cause death in some insect species [13].

Based on the data, the highest abundance of insect species in oil palm plants is in the Coleoptera order. This result is in line with the research of [14] which states that the Coleoptera order insects have the highest abundance in oil palm plants with different soil types. The movement and population of Coleoptera order insects are influenced by abiotic factors, namely temperature and humidity. [11] stated that air temperature positively correlates with all insect populations, especially the order Coleoptera in oil palm plants. The insect population of the order Coleoptera with the species *Elaeidobius kamerunicus* dominates the collection results. This occurs because *E. kamerunicus* is attracted by the volatile compound estragole secreted by female flowers at the time of receptivity [8] . *E. kamerunicus* can know the presence and direction of receptive female flowers because of the presence of chemoreceptors on the antenna [9] .

The number of *E. kamerunicus* insects obtained during collection on 45 7-year-old oil palm trees was 129 individuals with a population abundance of 28%. This result is different from the research **[5]** which states that population of *E. kamerunicus* in 140 7-year-old oil palm trees is 803 individuals. The difference in the population of *E. kamerunicus* can occur due to abiotic factors, namely temperature and humidity, which are not influenced by the living conditions of *E. kamerunicus*. During the observation in the collection garden, there was a very significant increase in temperature due to the influence of the long dry season. According to [9] apart from temperature, the population density of *E. kamerunicus* is also influenced by the age of oil palm plants. Oil palm plants ages four, seven, and nine years have *E. kamerunicus* populations that increase with plant age. This is due to the larger size of the male flower cluster at older oil palm ages. The older the age of the oil palm, the larger the size of the male flower cluster, and the number of spikelets will also increase. Male flower spikelets contain pollen as a food source for *E. kamerunicus*.

The insect population of the Hymenoptera order consisted of 116 individuals with a population abundance of 24%. The insects that dominate are from the Formicidae family. These insects were found during the research behind the fronds of oil palms that had rotted and at the base of the plant especially around the oil palm flowers. Insects of the order Formicidae are found in ecosystems with climatic factors that are suitable for the life of these species and are able to provide food sources for these insects such as flower nectar or volatile compounds secreted from oil palm flowers. The temperature range between 25-33°C is optimal for the activity of Hymenoptera order insects. So it can be stated that the oil palm collection garden provides an environment that is suitable for the living conditions of the insect order Hymenoptera. This is also evidenced by the higher abundance of Hymenoptera order insects compared to insects from other orders. Some research from [6] states that the population of Formicidae members is also strongly influenced by the age of oil palm plants [15]. The more productive the age of the oil palm plant, the higher the population of the insect order Hymenoptera because it provides a food source. This statement aligns with the condition of plants in oil palm plantations aged seven years entering the Producing Plant phase with a relatively high population of the insect order Hymenoptera found compared to other insect populations.

3.2 Estimated Production and Distribution of Oil Palm FFB Production

See on the results from census of oil palm FFB for the last three years, the estimated production of oil palm FFB this year can be projected. Details of the estimated production of oil palm FFB and the distribution of oil palm fresh fruit bunch production for one year in the collection garden of Politeknik Negeri Jember are presented in Table 2.

Table 2. Estimated production and percentage distribution of oil palm fresh fruit bunch production

Parameters	Month												Total
	1	2	3	4	5	6	7	8	9	10	11	12	
a	45	45	45	45	45	45	45	45	45	45	45	45	45
b	0,58	0,55	0,51	0,47	0,37	0,34	0,29	0,28	0,23	0,42	0,55	0,59	5,18
c	10,2	10,2	10,2	10,2	10,2	10,2	10,2	10,2	10,2	10,2	10,2	10,2	122,4
d	266,2	252,5	234,1	215,7	169,8	156,1	133,1	128,5	105,6	192,8	252,5	270,8	2377,6
e	11,20 %	10,62 %	9,85 %	9,07 %	7,14 %	6,56 %	5,60 %	5,41 %	4,44 %	8,11 %	10,62 %	11,39 %	100%

Note
a = Number of Staples
b = Bunches Per Principal
c = Average Tier Weight
d = Estimated Production
e = Percentage Distribution of Production

The data above shows that there are months with the lowest and highest production for one year. The lowest production occurred in September and the highest production occurred in December. This is in line with the results of Rizkiani's research 2023 which states that the month with the highest FFB production lasts for 2-3 months a year and is near the end of the year, namely November and December. According to [7] the monthly production distribution is determined based on the trend of last year's Fresh Fruit Bunch production and the circumstances or environment around the company. One of them is related to the provision of labor for harvesting. Fresh Fruit Bunch production in January also had almost the same Fresh Fruit Bunch production as in December. This is similar to the results reported by [5] that Fresh Fruit Bunch production at the beginning and end of the year is always higher than in other months. This could be due to the influence of daily temperatures in these months that are suitable for the living conditions of all insect species. Environmental temperature is closely related to the distribution of monthly rainfall. The higher the rainfall, the lower the environmental temperature which can affect insect population density which can affect the production of FFB from oil palm. The average rainfall in Sumbersari District from January to December is 422; 197; 552; 273; 135.5; 213; 61;8 41.5 550; 556.5; 442 mm [16]. Based on these data, it can be stated that there was an increase in the production of FFB in October, November, December and January. The increase in oil palm fresh fruit bunch production is directly proportional to the distribution of monthly rainfall in one year.

The estimated production of oil palm FFB for one year is 105.6 kg to 270.8 kg. This production result differs from the oil palm production results reported by [6] which is 25.60 tons/ha produced from 140 plants. The low FFB production can be caused by the fact that GAP oil palm cultivation has not been optimally implemented such as fertilization, pest control, and weeding [7] .

Apart from cultivation factors, temperature and humidity factors affect the production of fresh fruit bunches. During the observation, the temperature and humidity in the collection garden were 31-34^{0}C with a relative humidity of 74-82%. The ideal temperature and humidity for plantation insects to live and carry out their biological functions is at a temperature of 26.5-31^{0}C with a humidity of 64.25-65.25%. Environmental temperature and humidity greatly affect the diversity of insects in oil palm plantations. Insects significantly influence the sustainability of oil palm agroecosystems, with each role as an ecological function, including pollinating insects, predatory insects, parasitoids, and natural enemies. These insects can carry out their ecological functions when abiotic and biotic factors support insects in living and reproducing according to their biotic potential [4]. When abiotic and biotic factors can regulate the insect population, it is certain that the agroecosystem of plantation crops, especially oil palm, will produce optimal Fresh Fruit batch production while still paying attention to the concept of sustainable agriculture.

4. Conclusion
The abundance of the insect population in the collection farm from August to September was low. This had an impact on the low production of Oil Palm FFB in that month. The low abundance of insects can be caused by biotic and abiotic factors, not by insect growth requirements, so the biotic potential is not optimal.

5. Acknowledgments
The author would like to thank the Politeknik Negeri Jember through the PNBP 2024 Fund so that this research can be carried out.

References
[1. Egonyu JP, Sisye SE, Baguma J, Otema M, Ddamulira G. Insect flower-visitors of African oil palm Elaeis guineensis at different sites and distances from natural vegetation in Uganda. Int J Trop Insect Sci [Internet]. 2021 Dec [cited 2024 Mar 20];41(4):2477–87. Available from: https://link.springer.com/10.1007/s42690-021-00426-6

2. Yahya S, Ariyanti M, Asbur Y. Perpektif Baru: Manajemen Vegetasi Bawah Tegakan Pada Budidaya Kelapa Sawit Berkelanjutan. J Agron Indones Indones J Agron. 2022;50(3):343–56.

3. Egonyu JP, Baguma J, Martínez LC, Priwiratama H, Subramanian S, Tanga CM, et al. Global advances on insect pest management research in oil palm. Sustainability [Internet]. 2022 [cited 2024 Mar 20];14(23):16288. Available from: https://www.mdpi.com/2071-1050/14/23/16288

4. Lumentut N, Alouw J, Santosa B. The Role Several Types of Pollinator Insects in Pollination to Improve Fruit Setting in Oil Palm (Elaeis Guineensis) in the Province of Central Sulawesi and North Sulawesi. In: IOP Conference Series: Earth and Environmental Science [Internet]. IOP Publishing; 2022 [cited 2024 Mar 20]. p. 012083. Available from: https://iopscience.iop.org/article/10.1088/1755-1315/974/1/012083/meta

5. Rizali A, Rahardjo BT, Karindah S, Wahyuningtyas FR, Sahari B, Clough Y. Communities of oil palm flower-visiting insects: investigating the covariation of Elaeidobius kamerunicus and other dominant species. PeerJ [Internet]. 2019 [cited 2024 Mar 20];7:e7464. Available from: https://peerj.com/articles/7464/

6. Azhar A, Tawakkal MI, Sari A, Rizali A, Tarigan SD, Nazarreta R, et al. Tree diversity enhance species richness of beneficial insect in experimental biodiversity enrichment in oil palm plantation. Int J Oil Palm [Internet]. 2022 [cited 2024 Mar 20];5(2):39–49. Available from: http://ijop.id/index.php/ijop/article/view/82

7. Yeo JG, Julien KN, Edoukou EF, Ahui JLD. Changes in surface soil properties and macroinvertebrate communities with the conversion of secondary forests to oil palm (Elaeis guineensis) plantations. Crop Pasture Sci [Internet]. 2020 [cited 2024 Mar 20];71(9):837–49. Available from: https://www.publish.csiro.au/cp/cp19370

8. Zulkefli MHH, Jamian S, Adam NA, Jalinas J, Mohamad SA, Masri MMM. Beyond four decades of Elaeidobius kamerunicus Faust (Coleoptera: Curculionidae) in the Malaysian oil palm industry: A review. J Trop Ecol [Internet]. 2020 [cited 2024 Mar 20];36(6):282–92. Available from: https://www.cambridge.org/core/journals/journal-of-tropical-ecology/article/beyond-four-decades-of-elaeidobius-kamerunicus-faust-coleoptera-curculionidae-in-the-malaysian-oil-palm-industry-a-review/6FFC1A36C6235C2AAE97CE3224CFABC8

9. Yousefi M, Rafie ASM, Abd Aziz S, Azrad S. Introduction of current pollination techniques and factors affecting pollination effectiveness by Elaeidobius kamerunicus in oil palm plantations on regional and global scale: A review. South Afr J Bot [Internet]. 2020 [cited 2024 Mar 20];132:171–9. Available from: https://www.sciencedirect.com/science/article/pii/S0254629919304697

10. Arsi A, Pujiastuti Y, Kusuma SSH, Gunawan B. Eksplorasi, isolasi dan identifikasi Jamur entomopatogen yang menginfeksi serangga hama. J Prot Tanam Trop [Internet]. 2020 [cited 2023 Oct 28];1(2):70–6. Available from: https://jurnal.unej.ac.id/index.php/JPTT/article/view/18554

11. Ishara J, Ayagirwe R, Karume K, Mushagalusa GN, Bugeme D, Niassy S, et al. Inventory reveals wide biodiversity of edible insects in the Eastern Democratic Republic of Congo. Sci Rep [Internet]. 2022 [cited 2024 Mar 20];12(1):1576. Available from: https://www.nature.com/articles/s41598-022-05607-y

12. Manjeri G, Muhamad R, Tan SG. Oryctes rhinoceros beetles, an oil palm pest in Malaysia. Annu Res Rev Biol. 2014;3429–39.

13. Hosang MLA, Maskromo I, Alouw JC. Comparative effectiveness of three types of pheromones against Rhynchophorus vulneratus (Panzer) and Oryctes rhinoceros (L.) in Indonesia. In: IOP conference series: earth and environmental science [Internet]. IOP Publishing; 2022 [cited 2023 Oct 28]. p. 012090. Available from: https://iopscience.iop.org/article/10.1088/1755-1315/974/1/012090/meta

14. Bakara RDJ, Buchori D, Kusumah RYM, Sahari B, Apriyanto A. Genetic diversity of Elaeidobius kamerunicus Faust.(Coleoptera: Curculionidae) across geographic areas in Indonesia: implications for oil palm sustainability. In: IOP Conference Series: Earth and Environmental Science [Internet]. IOP Publishing; 2019 [cited 2024 Mar 20]. p. 012003. Available from: https://iopscience.iop.org/article/10.1088/1755-1315/325/1/012003/meta

15. Harahap FRS, Afrianti S, Situmorang VH. Keanekaragaman serangga malam (nocturnal) di kebun kelapa sawit PT. Cinta Raja. Perbal J Pertan Berkelanjutan [Internet]. 2020 [cited 2024 Mar 20];8(3):122–33. Available from: http://journal.uncp.ac.id/index.php/perbal/article/view/1542

16. Badan Pusat Statistik. Jember Dalam Angka. Jember: Badan Pusat Statistik Kabupaten Jember; 2022.

Yield test of oriental melon in Tasikmalaya, West Java

Anung Wahyudi[1]*, Siska Yunita Sari[1], Onny Chrisna Pandu Pradana[1]

[1)]Seed Technology Study Program, Food Crop Cultivation Department, Politeknik Negeri Lampung. Jl. Soekarno Hatta, No. 10, Rajabasa, Bandar Lampung, Lampung

*Corresponding Author: anung@polinela.ac.id

Abstract. Melon plants (*Cucumis melo* L) are horticultural commodities cultivated in Indonesia. This study aims to describe five oriental melon genotypes in Tasikmalaya (500 meters above sea level) as a multilocation test. The research was carried out from December 2023 to February 2024. Multilocation testing is one of the stages of plant breeding before the release of new superior varieties. The genetic material used in this research is Ginsen Makuwauri; F1 MM 0201-02; Ougan Makuwauri; F1 MM 0102-16; and Golden Langkawi (comparison variety). This research design used a Randomized Block Design (RBD) with five treatments and three replications to obtain 15 experimental units. Observation of research data uses the standards of the International Union for the Protection of Plants (UPOV), the International Plant Genetic Resources Institute (IPGRI), and the Decree of the Indonesian Minister of Agriculture. Data were analyzed using the F-test and continued with a LSD test at the 5% level. The results showed that there was a diversity of quantitative and qualitative characteristics from all the genotypes tested. Based on the diversity of quantitative and qualitative characteristics of the five genotypes tested, three superior genotypes were selected, namely F1 MM 0201-02; F1 MM0102-16; and Ougan Makuwauri. Genotype F1 MM 0201-02 has superior quantitative characters in the parameters of internode length (7.33 cm), fruit weight per plant (151.28 g), fruit length (9.21 cm), fruit width (5.30 cm), and the fruit flesh is thicker (1.17 cm) than the comparison variety.

1. Introduction

Melon plants are suitable for cultivation in tropical and subtropical areas so the production time is relatively short. The distribution of melon plants in Indonesia has reached several areas such as Ngawi, Madiun, Ponorogo, Sukabumi, and other areas [1]. In cultivating melon plants, careful planning and careful monitoring of the seeds used, location selection, soil conditions, and proper plant management are needed [2]. Melon plants can grow well at an altitude of 300 - 1,000 meters above sea level. In the lowlands and middle plains, melon plants will grow quickly at an altitude of <400 - 700 meters above sea level. If the altitude is more than 900 meters above sea level, the melon plants will not produce optimally [3]. The suitable temperature for melon cultivation ranges from 25 °C - 30 °C, rainfall ranges from 83.00 - 208.33 per month, and air humidity ranges from 75 - 86 per month [4].

Hybrid melons have the advantages of fruit uniformity, fast growth rate and are able to obtain the desired combination in one plant [5]. Before releasing a variety, testing is needed by identifying the appearance of prospective hybrid varieties compared to commercial varieties as a comparison in determining the varieties to be released [6]. Seed availability is one of the obstacles or problems in melon production efforts. This is because the fulfillment of seeds in Indonesia still relies on imports from abroad and is still limited in number [7].

Therefore, a plant breeding program is needed to meet the needs of the community. The stages of plant breeding are a series of plant testing activities with agroecology in the field to determine the advantages and interactions of varieties being tested in several places [8]. There are three stages, namely preliminary yield testing, advanced yield testing, and multi-location testing. Multi-location testing is one of the stages of plant breeding carried out before the release of a variety. This test is carried out in several places to see the adaptation test and stability test on plants that will be released as varieties. Multi-location testing as a requirement for releasing new varieties aims to see the stability of melon varieties when planted in new locations before being cultivated by farmers [9]. The requirements and guidelines for releasing candidate varieties are to comply with the regulations of the Ministry of Agriculture with clear pedigrees, complete descriptions, superior, and sufficient availability of seed types [10].

This study used the Ginsen Makuwauri variety obtained through selfing pollination from generation one to generation five. The resulting characters are golden yellow skin with stripes, rough fruit surface texture, thin fruit skin, fragrant fruit aroma, white fruit flesh, high water content, low sugar content, and resistance to pests and diseases [11].

Selfing was carried out on the Ougan Makuwauri variety obtained from generation one to generation six. The resulting characters include golden yellow skin, no stripes, thin fruit skin, fragrant fruit aroma, white flesh, high sugar content, smooth fruit surface texture, and tolerance to high temperatures [12].

In the F1 MM 0102-16 genotype, the results of the cross between the pure line parents ♀ (MM 01-01-06-01-18-19) x ♂ (MM 02-01-09-09-08-08) which have the characteristics of golden yellow fruit skin, there are thin stripes on the skin, with white fruit flesh. Hybridization of the F1 MM 0201-02 genotype was obtained from the pure line parents ♀ (MM 02-01-09-09-08-08) x ♂ (MM 01-01-06-01-18-19) which have the characteristics of golden yellow fruit skin, thin fruit skin, white fruit flesh [13]. While the comparison seeds used the Golden Langkawi variety from the KNOW YOU SEED company. To complete the description of melon plants in variety registration, variety protection of the four melon genotypes, it is necessary to conduct a multilocation test. Multilocation testing is one of the stages of breeding before the release of varieties. Five melon genotypes were tested in multiple locations to determine the qualitative and quantitative characteristics of each genotype studied.

Release of varieties is an effort to provide quality seeds in sufficient and sustainable quantities. In superior varieties, the results of breeding must be released and registered for circulation before being traded. The testing method is used to compare the characteristics of new varieties with varieties that have been released. The tests used are adaptation tests and stability tests. Adaptation tests are carried out to obtain data on the superiority of varieties and interactions with the environment carried out in several locations, while stability tests are used to see genotypes in various environments so that their phenotypes do not change [10].

2. Material and Method

This research was conducted from December 2023 to February 2024. This research took place at the Greenhouse of the Manbaul Ulum Islamic Boarding School, Tanjungmekar, Jamanis District, Tasikmalaya Regency, West Java. The research location was carried out on a medium plain with an altitude of 500 meters above sea level, a tropical climate type C with an air temperature between 21.4-30 °C. The average rainy days are 11 days/month, with a wind speed of 18 km/hour which generally blows from the south and southeast, but from April to July it blows from the northwest at a speed of 4-8 km/hour. The humidity at this location is an average of 84.74%.

The tools used are potray for melon seedling media, bucket for fertilizer container, sprayer for spraying pesticide, scissors for trimming tendrils and leaves, watering can for watering melon seedlings, planter bag with a volume of 50 liters for melon planting container, caliper for measuring stem diameter, roll meter for measuring (plant height, leaf width, leaf length), RHS Color Chart for color observation (flowers, leaves, stems, and fruit), stationery for writing observation data (books, pencils, markers, rulers, and erasers). The materials used are Ougan Makuwauri Melon seeds, Ginsen Makuwauri Melon, F1 MM 0102-16 Melon, F1 MM 0201-02 Melon, Golden Langkawi Melon, soil, compost, dolomite lime, water, cocopeat, NPK fertilizer, urea fertilizer, insecticides (Curacron active ingredient: Profenofos 500 EC, Alika active ingredient: Lambda cyhalothrin, Abacel active ingredient: Abamectin, Regent active ingredient: Fipronil, Decis active ingredient: Deltametrin), fungicides (Ridomil gold active ingredient: mefenoxam 4% mancozeb 64%, Antracol 70 WP active ingredient: Propinep 70%, Previcure active ingredient: Propamokarb hydrochloride), bactericides (Bactocyn 150 AL active ingredient: Oxytetracycline), vine ropes, plastic labels, fruit covers, thread, wire, and tissue.

The experimental design used in this study was a non-factorial Randomized Block Design (RBD) with melon genotypes as treatments, consisting of:

1. B: Ginsen Makuwauri (Inbrid)
2. B1: F1 MM 0201-02 (Hybrid)
3. B2: Ougan Makuwauri (Inbrid)
4. B3: F1 MM 0102-16 (Hybrid)
5. B4: Golden Langkawi (Hybrid)

This study used two melon lines and two melon varieties with one comparison variety, which was repeated three times. Oservation data were analyzed using the F test, then continued with the Least Significant Difference (LSD) test with a level of 5% if the analysis results showed a significant difference in the smallest value of the treatment. LSD 5% is used to determine the confidence interval of 95%.

2.1. A Quantitative observation.

A quantitative data, namely germination, cotyledon length, cotyledon width, tendril length, Number of leaves, leaf length, leaf width, petiole length, internode length, stem diameter, flowering age, harvest age, fruit weight, fruit length, fruit width, fruit flesh thickness, sugar content, the number of seeds per fruit, seed length, seed width, and seed weight per fruit.

2.2. A Qualitative observation.

A qualitative data, namely stem color, leaf color, leaf shape, leaf lobe shape, flower corolla color, primary skin color of ripe fruit, secondary skin color pattern, fruit flesh color, fruit shape, and seed shape.

3. Result and Discussion

An important factor in melon cultivation is temperature. The optimal temperature for melon growth is between 27-35 °C. Temperatures above 35 °C can affect the success of melon pollination [14]. Light intensity plays a role in the process of photosynthesis and the development of melon fruit. Light intensity affects the structure and function of photosynthetic organs which affect plant growth and production [15]. The use of a greenhouse as a place to cultivate melon plants protects plants which aims to avoid high rainfall intensity and can protect plants from pests and diseases [16]. The length of sunlight can affect plant growth and productivity. The good air temperature for melon cultivation is around 20-30 °C. The seeds used in this study were selected from previous work, which will be continued in the multi-location test stage in several locations [11]. Melon seed sowing was carried out for 11 Days After Sowing (HSS) which will then be transplanted into planter bags. The results of the percentage of oriental melon germination presented in Table 1.

Quantitative observation variables are measurable characters and can be expressed using units. The results of the recapitulation of the analysis of variance on quantitative characters are presented in Table 2. The coefficient of variation values ranged from 2.8-19.2%. The analysis of variance showed that there was an effect of each treatment (genotype) that was significantly different or not significantly different. The observation variables that were very significantly different were the length of the tendril, cotyledon length, cotyledon width, leaf stalk length, internode length, number of leaves, leaf length, leaf width, fruit weight, fruit length, fruit width, fruit flesh thickness, seed weight, seed length, seed width, and number of seeds. The variables that were significantly different were stem diameter, while the variables that were not significantly different were flowering age, harvest age, and total dissolved solids. Table 2 shows the results of the recapitulation of the analysis of variance on melon observation parameters.

Table 1. Germination percentage of five melon genotypes.

No	Genotype	Germination		
		No. of seeds	No. of germinated	(%)
1.	Ougan Makuwauri	50	48	96
2.	Ginsen Makuwauri	50	46	92
3.	F_1 MM 0102-16	50	43	86
4.	F_1 MM 0201-02	50	44	88
5.	Golden Langkawi	50	30	60

The length of the cotyledon from the analysis results ranged from 2.04-3.01 cm, while the width of the cotyledon ranged from 1.37-1.83 cm (Table 3). The genotype that had the minimum cotyledon length and cotyledon width was found in Ginsen Makuwauri which was significantly different from other genotypes, while the genotype that had the maximum cotyledon length and cotyledon width was found in Golden Langkawi. The length of the melon cotyledon in the Ougan Makuwauri genotype was equivalent to F1 MM 0201-02 and F1 MM 0102-16. Cotyledons are a place to store food reserves before the first leaf is formed, so the size of the cotyledon affects plant growth in the early stages [17].

Table 2. Results of the recapitulation of variance analysis on melon observation parameters.

No	Observation parameters	F-test	CV (%)
1	Cotyledon length	35.12 **	4.14
2	Cotyledon width	7.07 **	8.51
3	Tendril length	68.2 **	8.28
4	Stem diameter	5.24 *	7.82
5	Petiole length	13.60 **	9.49
6	Internode length	15.58 **	13.63
7	No. of leaves	10.53 **	13.71
8	Leaf length	21.09 **	8.33
9	Leaf width	21.34 **	9.11
10	Flowering age	3.42 ns	3.51
11	Harvest age	0.89 ns	2.80
12	Fruit weight	138.12 **	9.18
13	Fruit length	14.45 **	13.73
14	Fruit width	17.19 **	10.63
15	Fruit flesh thickness	22.26 **	10.46
16	Sugar content	2.98 ns	9.97
17	Seed weight	70.39 **	19.20
18	Seed length	50.35 **	6.70
19	Seed width	13.16 **	11.58
20	No. of Seeds	92.03 **	14.84

Note : (CV): coefficient of variance, (*): significantly different at the 5% level of significance, (**): very significantly different at the 1% level, (ns): not significantly different.

Table 3. Variance analysis of cotyledon length and cotyledon width observations.

No	Genotype	Cotyledon length (cm)	Cotyledon width (cm)
1	Ougan Makuwauri	2.43 b	1.73 b
2	Ginsen Makuwauri	2.04 a	1.37 a
3	F_1 MM 0201-02	2.39 b	1.48 a
4	F_1 MM 0102-16	2.41 b	1.42 a
5	Golden Langkawi	3.01 c	1.83 b
	LSD 5%	0.19	0.25
	CV %	4.14	8.51

Note : Numbers followed by the same letter in the same coloum indicate no significant difference according to least significant different (LSD) analysis at the 5% level, CV: coefficient of variance

Table 4. Variance analysis of tendril, stem, petile, and internode observations.

No	Genotype	Tendril length (cm)	Stem diameter (mm)	Petiole length (cm)	Internode length (cm)
1	Ougan Makuwauri	85.56 b	7.13 b	10.65 b	6.67 bc
2	Ginsen Makuwauri	31.66 a	5.57 a	6.28 a	2.75 a
3	F_1 MM 0201-02	115.72 c	7.37 b	10.71 b	7.33 c
4	F_1 MM 0102-16	107.56 c	6.76 b	11.00 b	6.28 bc
5	Golden Langkawi	77.28 b	6.86 b	9.83 b	5.61 b
	LSD 5%	13.05	0.94	1.73	1.44
	CV %	8.28	7.42	9.49	13.1

Note : Numbers followed by the same letter in the same coloum indicate no significant difference according to least significant different (LSD) analysis at the 5% level, CV: coefficient of variance

The length of the tendrils from the analysis results ranged from 31.66-115.72 cm. The Ginsen Makuwauri genotype was significantly different from the Ougan Makuwauri, Golden Langkawi, F1 MM 0201-02 and F1 MM 0102-16 genotypes, while the F1 MM 0201-02 genotype was equivalent to F1 MM 0102-16 (Table 4). Three factors that influence the length of the tendrils, namely genetic factors, environmental conditions, and processing methods [18]. Growth and development play a very important role, namely the availability of nutrients needed by plants. The nutrients used are N, P, K where the N element can stimulate the growth of plant height on plant tendrils. Melon cultivation cannot be separated from various obstacles during growth [19]. The presence of plant pests can cause plant growth and development to be less than optimal, this is because it disrupts several functions of plant organs [20].

The stem diameter from the analysis results ranged from 5.57-7.37 mm. The genotype that has the minimum stem diameter is found in Ginsen Makuwauri, while the maximum stem diameter is found in F1 MM 0201-02 which is not significantly different from F1 MM 0201-02, Ougan Makuwauri and Golden Langkawi (Table 4). There is secondary meristem tissue activity that causes the stem diameter to grow larger in dicotyledonous plants [12]. This occurs when the cambium tissue divides inward to form xylem (wood) and divides outward to form phloem (bark). Nutrients absorbed by plants can cause the accumulation of carbohydrates and proteins. This can affect the thickness of the xylem and phloem due to the nature of the plant which actively divides, thus affecting the size of the stem diameter [21].

The length of the leaf stalk from the analysis results ranged from 6.28-11 cm. The genotype with the maximum leaf stalk was found in Ginsen Makuwauri, which was significantly different from the other genotypes (F1 MM 0201-02, Ougan Makuwauri, F1 MM 0102-16, and Golden Langkawi), while the maximum leaf stalk was found in F1 MM 0102-16, which was significantly different from Ginsen

Makuwauri (Table 4). The leaf shape consists of various variations consisting of flat blades with one leaf stalk (petiole) in general, Cucurbitaceae leaf organs play a role in the photosynthetic process. The length of the internode was observed using a ruler [22].

The length of the internode from the analysis results ranged from 2.75-7.33 cm. The genotype that has the minimum internode is found in Ginsen Makuwauri which is significantly different from the other genotypes (F1 MM 0201-02, Ougan Makuwauri, F1 MM 0102-16, and Golden Langkawi), while the maximum internode is found in F1 MM 0201-02 which is not significantly different or equivalent to Ougan Makuwauri, and F1 MM 0102-16. The length of the internode in Golden Langkawi is significantly different from the other genotypes (Table 4). The length of the tendril is related to the length of the internode of each stem, female and male flowers will appear on each stem internode [23]. The number of flowers affects the number of internodes on the plant stem, so that it can increase the number of fruits per plant. The number of nodes is related to the length of the tendril, if the length of the tendril is longer, the number of nodes will increase [24].

Table 5. Variance analysis of leaf, flower, harvesting age.

No	Genotype	No. of leaf	Leaf length (cm)	Leaf width (cm)	Flowering age (DAT)	Harvest age (DAT)
1	Ougan Makuwauri	12.33 a	10.65 bc	13.97 bc	19.67 a	55.33 a
2	Ginsen Makuwauri	10.11 a	6.59 a	7.41 a	21.67 a	57.00 a
3	F_1 MM 0201-02	19.33 c	12.33 d	14.15 bc	20.33 a	56.00 a
4	F_1 MM 0102-16	17.11 bc	11.78 cd	15.26 c	20.00 a	57.00 a
5	Golden Langkawi	13.44 ab	9.81 b	12.81 b	20.67 a	58.00 a
	LSD 5%	3.73	1.6	2.18	ns	ns
	CV %	13.71	8.33	9.11	3.51	2.8

Note : Numbers followed by the same letter in the same coloum indicate no significant difference according to least significant different (LSD) analysis at the 5% level, CV: coefficient of variance

The number of leaves from the analysis results ranged from 10 to 19. Genotypes that have a small number of leaves in Ginsen Makuwauri are not significantly different from Ougan Makuwauri and Golden Langkawi, while the largest number of leaves in F1 MM 0201-02 is not significantly different or equivalent to F1 MM 0102-16 (Table 5). Plant growth is influenced by environmental conditions which can provide factors that can influence plant development [25]. If one factor does not give and receive each other, the factor can inhibit the growth of the plant. One factor has a stronger influence than another factor, the influence of that factor is covered up and if each factor has very different properties and working characteristics, it will produce a relationship that has no real effect on plant growth [26]. The presence of carbohydrates obtained from the results of the photosynthesis process can be used for fruit formation and influence the increase in the number of leaves by plant genetics [12].

The leaf length from the analysis results ranged from 6.59-12.33 cm. The genotype with the minimum leaves was found in Ginsen Makuwauri which was significantly different from the other genotypes (F1 MM 0201-02, Ougan Makuwauri, F1 MM 0102-16, Golden Langkawi), while the maximum leaf length was found in F1 MM 0201-02 getting a value of 12.33 cm which was not significantly different from F1 MM 0102-16. The leaf length of Ougan Makuwauri was not significantly different or equivalent to Golden Langkawi (Table 5). The leaves are a part of the plant that contributes to plant growth and development [13]. Increasing leaf area can make light capture more efficient for the normal photosynthesis process in conditions where plants have low light intensity [27]. The growth and development will increase if plants are able to absorb nutrients that have been provided by the soil through the fertilization process [28]. Leaf width measurements were carried out starting from the widest leaf point.

The leaf width from the analysis results ranged from 7.41-15.26 cm. Genotypes that have minimum leaves in Ginsen Makuwauri are significantly different from other genotypes (F1 MM 0201-02, Ougan Makuwauri, F1 MM 0102-16, and Golden Langkawi), while leaves that have maximum in F1 MM 0102-16 are not significantly different or equivalent to Ougan Makuwauri and F1 MM 0201-02. Leaf width in Golden Langkawi is significantly different from Ginsen Makuwauri and F1 MM 0102-16 (Table 5). Plants can grow well and are able to produce high production, so sufficient nutrients are needed [29]. The nutrients needed by plants are nitrogen (N) which have a role in accelerating plants. The N element is very much needed by plants when they have entered the vegetative phase for root, stem, and leaf growth. The availability of nitrogen and phosphorus nutrients in the growing medium can have an impact on the leaf formation process. Leaf width affects fruit weight, this is because the photosynthesis process is very helpful in the growth and development of plants [30].

Table 6. Variance analysis of fruit and sugar content.

No	Genotype	Fruit weight (g)	Fruit length (cm)	Fruit width (cm)	Flesh thickness (mm)	Sugar content (brix)
1	Ougan Makuwauri	113.67 c	7.78 b	4.92 c	1.14 b	9.62 a
2	Ginsen Makuwauri	18.33 a	4.18 a	2.62 a	0.49 a	11.27 a
3	F_1 MM 0201-02	151.28 d	9.21 b	5.30 c	1.17 b	9.43 a
4	F_1 MM 0102-16	126.22 c	8.69 b	5.00 c	1.04 b	9.31 a
5	Golden Langkawi	44.78 b	5.59 a	3.88 b	1.06 b	11.31 a
	LSD 5%	15.7	1.83	0.87	0.19	ns
	CV %	9.18	13.73	10.63	10.46	9.97

Note : Numbers followed by the same letter in the same coloum indicate no significant difference according to least significant different (LSD) analysis at the 5% level, CV: coefficient of variance

The fruit weight from the analysis results ranged from 18.33-151.28 g (Table 6). The genotype that has the minimum fruit weight is found in Ginsen Makuwauri which is significantly different from other genotypes (F1 MM 0201-02, Ougan Makuwauri, F1 MM 0102-16, and Golden Langkawi), while the heavier fruit weight is found in F1 MM 0201-02 which is significantly different from other genotypes. The fruit weight in Ougan Makuwauri is not significantly different or equal to F1 MM 0102-16. This is due to the influence of external and internal factors on plants that can influence plant growth and development. Plant growth is influenced by external factors, namely environmental factors around the plant and internal factors in plants related to the plant's ability to adapt to changes in the environment around it [28]. It shows that the growth of melon fruit begins before pollination. The size of the ovary before pollination can determine the growth of the next fruit, if the ovary has a small size then the growth of the next fruit will be slow or experience miscarriage [31]. The weight of the melon fruit is influenced by the results of photosynthesis, if assimilation is able to meet the needs of the plant then it will encourage the formation of flowers and fruit. If the results of photosynthesis are not sufficient then it will affect the growth of the melon plant.

The length of the fruit from the analysis results ranged from 4.18-9.21 cm (Table 6). The genotype that has the minimum fruit length is Ginsen Makuwauri which is not significantly different from Golden Langkawi, while the maximum fruit length is in F1 MM 0201-02 which is not significantly different from Ougan Makuwauri and F1 MM 0102-16. It shows that the size of the fruit affects the length and width of the fruit, the larger the fruit size, the longer and wider the fruit obtained. Melon fruit that has an elongated shape will have a higher length than round fruit [33].

The width of the fruit from the analysis results ranged from 2.62-5.30 cm (Table 6). Genotypes that have smaller fruit widths are found in Ginsen Makuwauri, while larger fruit widths are found in F1 MM 0201-02, which are not significantly different from Ougan Makuwauri, F1 MM 0102-16, and Golden Langkawi. Fruit shape affects fruit diameter, the greater the weight and size of the fruit, the greater the fruit diameter [34].

The flesh thickness from the analysis results ranged from 0.49-1.17 cm. Genotypes that have thinner fruit flesh are found in Ginsen Makuwauri, significantly different from other genotypes (F1 MM 0201-02, Ougan Makuwauri, F1 MM 0102-16, and Golden Langkawi), while thicker fruit flesh is found in F1 MM 0201-02, not significantly different from Ougan Makuwauri, F1 MM 0102-16 and Golden Langkawi (Table 6). Fruit quality can be determined by the thickness of the fruit flesh, this is because of the high public interest in melons with thick fruit flesh [35]. The character of the thickness of the fruit flesh is related to the weight of the fruit, with both being related to the shelf life of the fruit. The shelf life is relatively long if it has thick fruit flesh and large fruit weight [13].

Table 7. Variance analysis of seeds.

No	Genotype	Seed weight (g)	Seed length (mm)	Seed width (mm)	No. of seeds
1	Ougan Makuwauri	1.61 b	5.14 b	2.48 b	235 b
2	Ginsen Makuwauri	0.02 a	2.68 a	1.56 a	20 a
3	F_1 MM 0201-02	2.05 c	5.28 b	2.43 b	296 c
4	F_1 MM 0102-16	2.14 c	5.22 b	2.38 b	311 c
5	Golden Langkawi	0.03 a	6.35 c	3.21 c	22 a
	LSD 5%	0.42	0.62	0.53	49.37
	CV %	19.2	6.7	11.58	14.48

Note : Numbers followed by the same letter in the same coloum indicate no significant difference according to least significant different (LSD) analysis at the 5% level, CV: coefficient of variance

The seed weight from the analysis results ranged from 0.02-2.14 g (Table 7). Genotypes that have lower seed weights are found in the Ginsen Makuwauri and Golden Langkawi varieties, which are significantly different from other genotypes (F1 MM 0201-02, Ougan Makuwauri, and F1 MM 0102-16), while higher seed weights are found in F1 MM 0102-16 which are not significantly different from F1 MM 0201-02 and Ougan Makuwauri has a lower weight. Seed yield is the net weight of seeds after extraction and sorting according to the physical and characteristics of each genotype. The seed length from the analysis results ranged from 2.68 mm to 6.35 mm, while the seed width ranged from 1.56 mm to 3.21 mm (Table 7). The genotype that had the minimum seed length and seed width was found in Ginsen Makuwauri which was significantly different from the other genotypes (F1 MM 0201-02, Ougan Makuwauri, F1 MM 0102-16, and Golden Langkawi), while the maximum seed length and seed width were found in Golden Langkawi which was significantly different from the other genotypes. The F1 MM 0201-02 genotype was not significantly different or equivalent to Ougan Makuwari and F1 MM 0102-16.

The number of seeds from the analysis results ranged from 20-311 seeds. The genotype with the lowest number of seeds in the Ginsen Makuwauri variety was not significantly different from Golden Langkawi, while the genotype with the highest number of seeds in F1 MM 0102-16 was not significantly different from F1 MM 0201-02, but significantly different from Ougan Makuwauri (Table 7). The formation of the number of seeds is influenced by several factors, namely environmental temperature and air humidity. If low temperatures can interfere with the growth of pollen tubes, while high temperatures can cause changes such as pollen not sticking to the stigma [34].

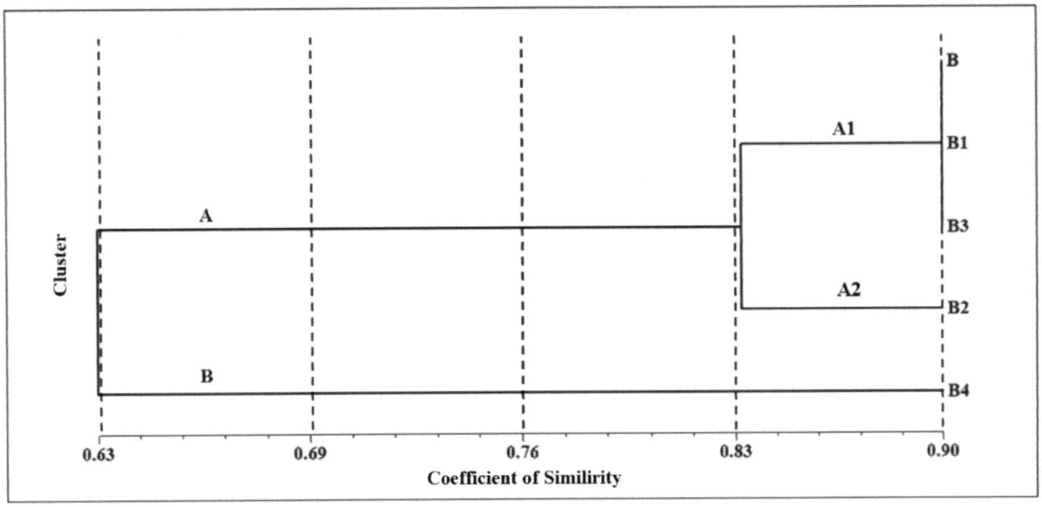

Figure 1. Dendogram analysis. B= Ginsen Makuwauri; B1= F_1 MM 0201-02; B2= Ougan Makuwauri; B3= F_1 MM 0102-16; B4= Golden Langkawi

Based on the qualitative characteristics of the four genotypes tested with one comparison, they were analyzed using dendogram analysis, so that a similarity coefficient of 0.63-0.90 or 63-90% was obtained (Figure 1). If the kinship similarity index is 0.60, it indicates that the number is almost the same as 1.00, then the kinship value is uniform, but if the number is almost the same as zero, it indicates a distant kinship value. Genotype kinship can be divided into two large clusters, namely cluster A and cluster B. Cluster A consists of four melon genotypes (Ginsen Makuwauri, Ougan Makuwauri, F1 MM 0102-16, and F1 MM 0201-02), while cluster B consists of one comparison genotype, namely Golden Langkawi.

The results of the dendogram analysis (Figure 1) showed eleven observation variables that were included in the 100% similarity level, indicating that the variables were uniform. The higher the percentage similarity value, the closer the kinship or uniformity [36]. The more qualitative character similarities between the compared individuals, the closer the kinship, while the fewer character similarities, the further the kinship.

Cluster A is divided into two sub-clusters, namely sub-clusters A1 and A2. Sub-cluster A1 consists of three genotypes, namely Ginsen Makuwauri, F1 MM 0201-02, and F1 MM 0102-16. Sub-cluster A2 consists of one genotype, namely Ougan Makuwauri. Sub-cluster A1 has a similarity level of between 0.83 - 0.90 or between 83% - 90% in the same six characters, namely stem color, leaf shape, leaf lobe shape, fruit shape, seed color and seed shape. Subcluster A2 has the same eleven characters, namely stem color, leaf color, leaf shape, leaf lobe shape, flower crown color, mature fruit color, fruit skin color pattern, fruit flesh color, fruit shape, seed color and seed shape. Cluster B consists of one comparison genotype, namely Golden Langkawi, which has a similarity to cluster A of 0.63 or 63% in the same characters, namely stem color, leaf color, leaf shape, leaf lobe shape, flower crown color, old fruit color, fruit skin color pattern, fruit flesh color, fruit shape, seed color and seed shape (Figure 1).

Conclusion

It can be concluded that, of the five genotypes tested, the Ougan Makuwauri variety has a fruit length equivalent to F1 MM 0201-02, F1 MM0102-16, and Golden Langkawi. The thickness of the fruit flesh in genotypes F1 MM 0102-16, F1 MM 0201-02, and Ougan Makuwauri is equivalent to Golden Langkawi. Genotypes F1 MM 0201-02, F1 MM 0102-16, Ginsen Makuwauri, and Ougan Makuwauri have a sweetness level equivalent to the Golden Langkawi variety. There is diversity in qualitative characters such as stem color, leaf color, leaf shape, leaf lobe shape, crown color, ripe fruit color, secondary skin color pattern, fruit shape, seed color, and seed shape. The level of similarity for all genotypes is 0.63-0.90 or 63-90%. The F1 MM 0201-02 genotype is a hybrid line that has the potential to be released as a new superior hybrid variety with advantages in fruit weight, fruit length, fruit width, and fruit flesh thickness.

Acknowledgments
This study was supported by INOVOKASI Scheme 1st batch (2024) from The Minister of Education, Culture, Research and Technology, Directorate General of Vocational Study, contract number 313/BAP/D.D4/PPK.01.APTV/VIII/2024

References

[1] Ginting AP, Asil B and Rosita S 2017 *Agroekoteknologi Journal FB USU.* **5** (4) 784-798.
[2] Durroh B and Dawud MY 2022 *Journal of Sociology of Agriculture and Agribusiness*. **4** (2) 01-13.
[3] Yani L 2022 *Golden Melon Seed Production (Cucumis melo L.) at PT East West Seed Indonesia,* Politeknik Negeri Lampung, Bandar Lampung.
[4] Syahputri R, Prasetyo BD, Taufik, Andriyadi A, Nugroho HW and Trisnawati S 2023 *Design of A Mobile-Based To-Do List Application for Melon Fruit Cultivation with Genetic Algorithms*, IBI Darmajaya, Bandar Lampung.
[5] Sa'diyah H and Suhartono 2022 *Journal of science and technology* **15** (2) 247-252.
[6] Sari IP 2018 *Appearance of 9 Candidate Hybrid Melon Varieties (Cucumis melo L),* Brawijaya University, Malang.
[7] Daryono BS and Nofriarno N 2018 *Biosfera.* **35** (1) 44-48.
[8] Wahyudi A and Syukur M 2021 *Biodiversitas*. 22 (10) 4246-4253.
[9] Taus I, Hamakonda UI and Puspita VA 2022 *Agriovet.* **15** (1) 112-122.
[10] Syukur M, Sujiprihati S and Yunianti R 2018 *Plant Breeding Technic*, Penebar Swadaya. Jakarta.
[11] Wahyudi A, Putri R, Maila S and Zaini AH 2023 *E3S Web of Conferences* **444**, 04038.
[12] Anggraini SP 2023 *Comparison of Quantitative Characters of Fifth Generation Oriental Makuwauri Melon*, Politeknik Negeri Lampung, Bandar Lampung.
[13] Sari IP 2018 *Appearance of 9 Candidate Hybrid Melon Varieties (Cucumis melo L),* Universitas Brawijaya. Malang.
[14] Minarni EW and Ulinnuha Z 2023 *Agricultural Science Journal* **25** (1) 145-151.
[15] Shafiq I, Husaain S, Raza MA, Iqbal N, Asghar MA, Raza A, Fan YF, Mumtaz M, Shoaib M, Ansar M, Manaf A, Wen-yu Y and Feng Y 2021 *Journal of Integrative Agriculture* **20** (1) 4-23.
[16] Savitri OM, Puspitorini P, Serdani AD and Pitaloka D 2023 *Scientific Journal of Agricultural Sciences* **13** (2) 59-65.
[17] Silalahi M 2018 *Plant Morfology*, Universitas Kristen Indonesia, Jakarta.
[18] Bilalang AC and Maharia D 2021 *Student Scientific Journal of the Faculty of Agriculture* **1** (3): 119-124.
[19] Khuluq M, Phabiola TA and Wijaya IN 2020 *Tropical Agrotechnology Journal* **9** (1) 76-86.
[20] Hawiyah AN, Afifah L, Abadi S, Prabowo DP, Irfan B and Widiawan AB 2022 *Agrotech Journal* **12** (2) 79-86.
[21] Serdani AD, Puspitorini P, Wibowo AS and Ariani IF 2020 *Buana Sains* **20** (2) 171-176.
[22] Zufahmi, Dewi E and Zuraida 2019 *Agroristech Journal* **2** (1) 7-14.
[23] Sukarsa, Bhagawati D and Priyono RE 2017 *Proceedings of the National Seminar on Biology and Science and Technology Education* **2** 274-283.
[24] Yono S and Putri SD 2023 *Agroplasma Journal* **10** (1) 300-310.
[25] Subli M, Peran SB and Rudy GS 2019 *Sylva Scienteae Journal* **2** (5) 922-929.
[26] Siregar M and Sulardi ES 2020 Journal of Animal Science and Agronomy Panca Budi **5** (1) 46-52.
[27] Mahardika S and Adiredjo AL 2020 Journal of Plant Production **8** (11) 1074-1081.

[28] Fitriani DN, Musa N and Pembengo W 2022 *Journal of Tropical Agriculture Land* **1** (2) 5-9.
[29] Izlin BR, Nurrachman and Isnaini M 2022 *Agrocomplex Student Scientific Journal* **1** (1) 57-65.
[30] Carsidi D, Saparso, Kharisun and Febrayanto CR 2021 *Agro Journal* **8** (1) 68-83.
[31] Christy J 2020 *Agrium* **22** (3) 151-156.
[32] Rahayu A, Serhalawan RJPJ and Munandar E 2011 **Agriculture Journal 2** (2) 139-144.
[33] Handayani DR and Ashari S 2019 *Journal of Plant Production* **7** (11) 2010-2017.
[34] Badriyah and Amzeri A 2022 *Journal of Science and Technology* **15** (2) 233-241.
[35] Saputra HE, Salamah U, Herman W and Mustafa M 2021 *Indonesian Journal of Agricultural Science* **23** (1) 61-65.
[36] Hidzroh F and Daryono BS 2021 *Journal of Biospecies* **14** (2) 11-19.

Analysis of superior agricultural sector commodities to increase economic growth in Sumenep District

I Maflahah[1]*, S Joko Utomo[2], D Farida Asfan[1], Fathor AS[3]

[1] Department of Agroindustrial Technology, Faculty of Agriculture, Universitas Trunojoyo Madura, Bangkalan 69162, Indonesia
[2] Departement of Development Economic, Faculty of Economic and Business, Universitas Trunojoyo Madura, Bangkalan 69162, Indonesia
[3] Departement of Management, Faculty of Economics and Bussines, Universitas Trunojoyo Madura, Bangkalan 69162, Indonesia

*E-mail: iffanmaflahah@gmail.com

Abstract. Economic growth can be increased according to the natural potential of a region. Sumenep district has relatively high agricultural potential. Economic improvement must be made wisely, especially in determining superior agricultural commodities. This study aims to analyze the leading commodities in food crop agricultural commodities and the classification of economic growth in Sumenep District. The method approach used is Location Quotient and Shift Share analysis. The data is on the food crop production (paddy, corn, soybeans, and sweet potatoes) per subdistrict in Sumenep District in 2020 - 2023. The analysis shows that the distribution of superior paddy commodities in Sumenep District is in Batuan, Lenteng, Manding, Gapura, Batang-Batang, Raas, and Sapeken Districts. Corn commodities are in Pragaan, Bluto, Saronggi, Giligenting, Talango, Kalianget, Lenteng, Ganding, Pasongsongan, Ambunten, Rubaru, Dasuk, Batuputih, Gayam, Raas, Kangean, and Masalembu Subdistricts. The superior commodity of sweet potato was spread in Pragaan, Saronggi, Giligenting, Kalianget, Batuan, Lenteng, Rubaru, Dasuk, Batuputih, Gapura, Batang-Batang, Dungkek, Nonggunong, Gayam, Raas, and Masalembu Subdistricts.

1. Introduction

Sumenep District is one of the areas on Madura Island that is rich in natural potential. Some of the natural potential found in the area is in the form of agricultural commodities, livestock commodities, fishery commodities, and the potential of many beaches along the Sumenep Regency. Economic growth is a process of economic activity that increases the number of goods and services made by the community, increasing the community's prosperity. One indicator that can be used to measure the rate of regional economic growth is the rate of increase in GRDP, without considering shifts in economic structure or population growth. Based on the economy's structure, Sumenep Regency is dominated by agriculture, forestry, and fisheries. This is indicated by the contribution of these fields to GRDP at current ppaddys in Sumenep District in 2022, amounting to 36.12 percent, and in 2023, amounting to 38.23 percent [1], [2]. Consistent economic growth in 2022 and 2023 indicates the potential for positive economic recovery and development in the future.

The agricultural sector's potential as a producer of food crops produced by the Sumenep District is paddy, corn, soybeans, and sweet potatoes. The agricultural land in paddy fields reaches 25,686 hectares and 117,341 hectares [3]. Paddy production in 2020-2023 reached 182,581, 157,819.47, and 242,556.61 tons respectively. Meanwhile, corn production was 361,737.66 tons, 378,091.99 tons, and 447,143.16. Soybean production was 347.15 tons, 83.48 tons, and 179.25 tons. Sweet potato production was 41,889.51 tons, 138,742.11 and 145,963.02 tons [2]–[4]. The potential of these food crops can be used as a source of livelihood for the community to increase the amount of income. The competitiveness of a region can use its potential. However, the potential of a region is also owned by other regions, so decision-makers must take definite steps to determine the superior potential. The competitiveness of a region can be used as an indicator of economic growth. Regional and national economic competitiveness will increase when the economic welfare of the people of a region also increases [5]. The government needs to increase competitiveness based on the results of evaluations of potential sectors.

. The development of a region must pay attention to the development of potential sectors [6]. A sector can be categorized as an economic base when it can meet the internal needs of the region and then export the sector outside the producing region.

Conversely, non-base sectors can only meet a region's goods and services needs. The base sector's contribution to a region's economy makes this base sector often referred to as the leading sector. The base sector influences the non-base sector to develop into the base sector in the future. The economic base sector can provide a multiplier effect for non-base sectors in a region's economy. The regional economic base can be used to assess the region's export activity, thus serving to project the growth and its influence on other sectors [7].

The determination of base or non-base sectors can be done using the Location Quotient (LQ) approach. The location quotient is a method used to determine the level of specialization of economic sectors in a region using the base sector [8]. The LQ method is used to identify superior and non-superior sectors in a region as a reference for regional development strategies [9][10]. LQ calculation is based on data on the proportion of a commodity in an area with the same proportion of industries at the national level. If the LQ value is more significant than one, it indicates that the commodity in the region is higher than the national average, indicating a comparative advantage in that commodity. LQ measures the degree of specialization of economic activities through a comparative approach. Some researchers use the LQ approach to determine the leading commodities in the region, such as food crop agricultural commodities, to evaluate economic development [11]–[15]. LQ can also determine leading horticultural and agricultural commodities [16]–[25]. Analysis of the leading sectors of construction and transportation due to the advantages of tourism that sustainably contribute to economic growth [9], [26]–[29].

The base sector cannot necessarily be used as a leading commodity. To increase competitiveness, the commodity used is a superior commodity. A commodity that is said to be superior can be seen from the consistency of its availability over time. Economic growth can be measured using the shift-share method. The shift-share analysis is used to divide the economic growth of a region into three components and measure the contribution of each component. Shift-share is an analysis that compares the economic growth rate of a location/region with that of a broader scope of locations/regions. The shift-share of a district can be determined by comparing the economic growth rate of a district with the provincial economic growth rate. The components in conducting a Shift-share analysis include (1) provincial growth in an observed area that can be seen from changes in regional production caused by changes in the production of reference areas with the assumption that there are no differences in economic characteristics between sectors and between regions; (2) proportional growth measured by the growth of regional economic activity in a certain period; and (3) regional share growth measured by the rate of economic growth of a region compared to other regions. Shift share analysis analyzes economic growth based on base commodities [30]–[32].

Sumenep District has the potential for superior agricultural commodities [32], so it is necessary to analyze each food crop agricultural commodity in each sub-district. This study aims to analyze the leading commodities in food crop agricultural commodities and the classification of economic growth in Sumenep District.

IOP Conf. Series: Earth and Environmental Science **1446** (2025) 012028 doi:10.1088/1755-1315/1446/1/012028

2. Methodology

This research is a quantitative research with a quantitative descriptive analysis approach. The data is on the food crop production (paddy, corn, soybean, and sweet potato) per sub-district in Sumenep District. The data used are from the Central Bureau of Statistics of Sumenep Regency in 2020 - 2023. Sumenep regency has 27 subdistricts, namely (1) Pragaan, (2) Bluto, (3) Saronggi, (4) Giligenting, (5) Talango, (6) Kalianget, (7) Sumenep, (8) Batuan, (9) Lenteng, (10) Ganding, (11) Guluk-Guluk, (12) Pasongsongan, (13) Ambunten, (14) Rubaru, (15) Dasuk, (16) Manding, (17) Batuputih, (18) Gapura, (19) Batang-Batang, (20) Dungkek, (21) Nonggunong, (22) Gayam, (23) Raas, (24) Sapeken, (25) Arjasa, (26) Kangean, and (27) Masalembu. The analysis methods used are location quotient and shift-share analysis.

1. *Location quotient analysis*

The LQ calculation method can be used to see the leading sector owned by each sub-district in Sumenep Regency. The LQ method analyzes commodities that show base or non-base commodities. LQ > 1 indicates a relative concentration in a region compared to the whole region. The commodity i in an area is a base sector or in the area has a comparative advantage. LQ = 1 is a non-base sector, meaning commodity *i* in a region has no comparative advantage. The commodity production produced is only enough to fulfill its own needs in the region. LQ < 1. is a non-base sector, meaning commodity *i* in a region has no comparative advantage; the production of commodity *i* in the region cannot meet its own needs and must obtain supplies from outside the region [33]. The LQ formulation is as follows:

$$LQ = \frac{p_i/p_t}{P_i/P_t} \quad (1)$$

Where, LQ is Location Quotient.

pi : Production (production quantity) of commodity *i* at the district level.
pt : Production (production quantity) of all commodities t at the district level.
Pi : Production (production quantity) of commodity i at Sumenep District.
Pt : Production (production quantity) of commodity t at Sumenep District.

2. *Shift share analysis*

The shift-share method is used to compare the differences in the economic growth rates of various regional sectors with the national region sector [34]. The following formula of shift-share is as follows:

$$D_{ij} = N_{ij} + M_{ij} + C_{ij} \quad (2)$$
$$D_{ij} = E^*{}_{ij} - E_{ij} \quad (3)$$
$$N_{ij} = E_{ij}\, x\, r_n \quad (4)$$
$$M_{ij} = E_{ij}(r_{in} - r_n) \quad (5)$$
$$C_{ij} = E_{ij}(r_{ij} - r_{in}) \quad (6)$$
$$r_{ij} = \frac{(E^*{}_{ij} - E_{ij})}{E_{ij}} \quad (7)$$
$$r_{in} = \frac{(E^*{}_{in} - E_{in})}{E_{in}} \quad (8)$$
$$r_{in} = \frac{(E^*{}_{n} - E_{n})}{E_{n}} \quad (9)$$

Description:
i : Food crop commodity sector studied
j : District variable Sumenep
Nij : Change in sector *i* in the district
Mij : National growth of sector *i* in the district
Cij : Industry mix of sector *i* in the district
Dij : Competitive advantage of sector i in the district
rij : Growth rate of sector *i* at sub-district level
rin : Growth rate of sector *i* at district level
rn : Growth rate of food crop commodity sector at district level

The LQ and shift share calculation results will analyze the commodities studied, including one of the 4 sector criteria of shift-share analysis. The sector division can be seen in **Table 1.**

Table 1. Shift share analysis sector division

	LQ<1	LQ>1
Shift share >0	Quadrant II Slightly superior sector	Quadrant I Superior sector
Shift share <0	Quadrant IV Backward sector	Quadrant III Slightly backwards

3. Results and Discussion

3.1. Paddy Commodity

Sumenep District produces food crops in the form of paddy. Sumenep District has a paddy plantation area of 39,942 ha [2]. Based on data from BPS, Sumenep District states that almost all sub-districts have the potential to produce paddy, except for Talango, Gili Genting, and Masalembu. An analysis of the leading commodities in each sub-district uses the LQ and shift shares approach. A commodity is considered superior in an area if it has a commodity in a primary sector (LQ> 1) and the shift-share value is more than one. Data on paddy crop production in each sub-district in 2020-2022, LQ value analysis, shift share, and sector criteria analysis of shift-share analysis can be seen in **Table 2.**

Table 2 shows that paddy is a primary sector in Sumenep City, Batuan, Lenteng, Guluk-Guluk, Manding, Gapura, Batang-Batang, Raas and Sapeken Subdistricts. However, not all areas with an essential sector commodity can be used as a superior commodity. One of the factors determining the leading commodity is the consistency of the amount of production each period. Based on the shift-share analysis results, the areas with superior paddy commodities are Batuan, Lenteng, Manding, Gapura, Batang-Batang, Raas, and Sapeken Subdistricts.

Analysis of LQ and shift share of paddy commodities in Sumenep Regency shows differences in the level of excellence between sub-districts in 2020-2022. Gapura, Sapeken, and Arjasa Subdistricts are included in Quadrant I as a leading sector with an average LQ of more than 1 and a high shift share value. For example, Arjasa District has an average LQ of 2.09 and the highest shift share contribution of 14,390.93, reflecting its significant role in the regional economy. In contrast, Pragaan and Bluto Subdistricts are in Quadrant IV, indicating that the region is a lagging sector with low LQ and shift share values.

Regarding production, Arjasa Subdistrict recorded the highest yield of 64,641.61 tons in 2022, while Giligenting and Talango Subdistricts showed minimal or insignificant production. Several sub-districts show a positive growth trend, such as Lenteng Subdistrict, whose LQ value increased from 0.85 in 2020 to 1.28 in 2022. However, some sub-districts have experienced a decline in competitiveness, such as Dungkek Subdistrict, whose LQ value dropped from 0.83 to 0.52 in the same period.

Table 2. LQ and shift-share analysis of paddy commodities in Sumenep District

Sub district number	Paddy production (Ton)			LQ			Average of LQ	Shift share	Quadrant
	2020	2021	2022	2020	2021	2022			
1	2,052.25	1,677.28	1,997.29	0.30	0.36	0.35	0.33	-54.96	IV
2	142.12	157.53	450.38	0.03	0.05	0.11	0.06	308.26	II
3	6,755.20	5,431.60	8,673.86	0.97	0.42	0.79	0.73	1918.66	II
4	0.00	0	54.49	0.00	0.00	0.02	0.01	0	-
5	0.00	0	0	0.00	0.00	0.00	0.00	0	-
6	347.84	264.10	525.84	0.55	0.55	0.65	0.58	178	II
7	7,177.85	7,461.08	8,659.17	2.33	3.16	2.24	2.58	1481.32	I
8	7,641.62	9,751.48	9,739.55	2.20	3.04	2.08	2.44	2097.93	I
9	7,663.52	6,388.35	17,271.02	0.85	0.88	1.28	1.00	9607.5	I
10	5,196.63	4,317.14	6,544.10	0.76	0.77	0.91	0.81	1347.47	II
11	13,930.24	9,381.38	13,567.24	1.08	1.59	1.34	1.34	-363	III
12	4,330.76	2,405.13	9,849.84	0.40	0.36	0.81	0.52	5519.08	II
13	3,850.79	4,309.50	5,427.04	0.56	0.98	0.72	0.75	1576.25	II
14	3,515.79	3,496.93	5,061.65	0.62	0.60	0.50	0.57	1545.86	II
15	4,171.88	3,411.66	4,540.38	0.66	0.71	0.61	0.66	368.5	II
16	9,884.02	9,902.09	11,686.04	1.40	1.92	1.53	1.62	1802.02	I
17	1,988.93	1,929.76	4,649.55	0.29	0.25	0.36	0.30	2660.62	II
18	15,822.96	16,587.14	20,747.08	1.98	1.82	1.89	1.90	4924.12	I
19	8,245.45	8,045.32	8,699.66	1.21	1.38	1.12	1.23	454.21	I
20	4,587.66	4,043.66	4,766.85	0.83	0.62	0.52	0.66	179.19	II
21	1,841.89	1,692.26	2,813.31	0.60	0.25	0.31	0.39	971.42	II
22	3,238.01	2,887.82	4,257.31	0.50	0.43	0.44	0.46	1019.3	II
23	1,212.26	1,318.24	2,906.12	0.36	0.59	0.66	0.54	1693.86	II
24	8,473.97	8,079.16	12,124.85	1.24	1.56	1.24	1.35	3650.88	I
25	50,250.68	36,452.10	64,641.61	2.13	2.12	2.02	2.09	14390.93	I
26	10,259.11	8,428.76	12,586.69	0.88	0.89	0.74	0.84	2327.58	II
27	0.00	0	315.69	0.00	0.00	0.07	0.02	0	-
Total	182,581	157,819.47	242,556.61						

Description: I (Superior Sector); II (slightly superior Sector); III (slightly backward); IV (Backward sector)

The analysis provides an overview of the priorities for developing the paddy sector in Sumenep Regency. Subdistricts in Quadrant I need to continue to be supported as superior production centers with improved agricultural infrastructure and technology. Meanwhile, Subdistricts within Quadrant IV need policy interventions such as farmer training, increased access to capital, and strengthening marketing networks to boost their productivity and competitiveness. The right approach will help increase the contribution of the paddy farming sector to the economy of Sumenep Regency.

3.2. Corn Commodity

Corn commodity in the Sumenep District is produced in all areas [2]. Corn production in Sumenep District in 2020, 2021, and 2022 is 361,738 tons, 378,092 tons, and 447,143 tons, respectively. The analysis of superior commodities uses LQ and shift-share analysis in each region. Corn production data for each sub-district in 2020-2022, LQ value analysis, shift share, and sector criteria analysis of shift-share analysis can be seen in **Table 3.**

Table 3. LQ and shift-share analysis of corn commodities in Sumenep District

Sub district number	Corn production (ton)			LQ			Average of LQ	Shift share	Quadrant
	2020	2021	2022	2020	2021	2022			
1	17,960.35	17,754.63	15,479.41	1.32	1.44	1.26	1.34	-2480.94	III
2	13,608.90	13,572.35	13,471.19	1.60	1.57	1.52	1.56	-137.71	III
3	13,653.63	12,892.70	17,663.72	0.99	0.38	0.75	0.71	4010.09	II
4	4,647.08	4,716.74	8,342.69	1.41	1.24	1.33	1.33	3695.61	I
5	12,824.78	15,688.51	13,391.21	1.07	1.23	1.11	1.13	566.43	I
6	1,423.10	1,102.28	1,787.07	1.13	0.87	1.04	1.01	363.97	I
7	1,543.77	2,509.38	3,789.28	0.25	0.40	0.46	0.37	2245.51	II
8	2,081.35	3,968.96	4,874.69	0.30	0.47	0.49	0.42	2793.34	II
9	19,476.17	21,706.86	25,164.87	1.08	1.14	0.88	1.03	5688.7	I
10	15,034.89	18,235.60	16,789.94	1.11	1.23	1.10	1.15	1755.05	I
11	25,991.12	15,305.64	17,855.59	1.02	0.98	0.83	0.94	-8135.53	IV
12	22,431.46	23,653.43	30,088.36	1.06	1.33	1.16	1.18	7656.9	I
13	17,175.91	12,956.28	18,509.00	1.27	1.12	1.15	1.18	1333.09	I
14	13,755.79	16,599.70	21,937.92	1.23	1.08	1.02	1.11	8182.13	I
15	12,310.90	13,806.50	17,201.78	0.98	1.08	1.08	1.05	4890.88	I
16	9,312.15	9,213.56	13,462.81	0.66	0.68	0.83	0.72	4150.66	II
17	19,977.74	25,529.28	29,639.09	1.45	1.26	1.08	1.26	9661.35	I
18	9,096.19	10,508.84	12,974.87	0.58	0.44	0.56	0.52	3878.68	II
19	12,258.76	14,253.33	14,821.85	0.91	0.93	0.90	0.91	2563.09	II
20	11,357.93	14,431.37	16,564.88	1.04	0.84	0.84	0.91	5206.95	II
21	7,876.75	8,059.98	9,473.62	1.30	0.46	0.49	0.75	1596.87	II
22	17,423.43	19,763.52	23,768.89	1.36	1.13	1.15	1.21	6345.46	I
23	8,650.46	8,184.77	9,570.97	1.29	1.40	1.02	1.24	920.51	I
24	13,082.09	14,012.69	17,490.43	0.97	1.03	0.84	0.95	4408.34	II
25	25,371.71	24,723.87	30,538.26	0.54	0.54	0.45	0.51	5166.55	II
26	27,100.48	27,152.69	32,790.82	1.18	1.09	0.91	1.06	5690.34	I
27	6,310.77	7,788.53	9,699.95	1.59	1.23	0.96	1.26	3389.18	I
Total	361,737.66	378,091.99	447,143.16						

Description: I (Superior Sector); II (slightly superior Sector); III (slightly backward); IV (Backward sector)

Table *3* shows that the corn commodity is a base sector in almost all sub-districts, except for the Subdistrict of KotaSumenep, Batuan, Guluk-Guluk, Manding, Gapura, Batang-Batang, Dungkek, Nonggunong, Sapeken, and Kecamtan Arjasa is a non-base sector. The results of LQ analysis and shift share of corn commodities in Sumenep Regency in 2020-2022. Sub-districts included in Quadrant I (superior sectors) have an average LQ above 1, such as Lenteng, Pasongsongan, and Gayam Subdistricts. Pasongsongan Subdistrict recorded the highest shift-share value of 7,656.9, while Batuputih Subdistrict showed a significant contribution with a shift-share of 9,661.35 and production reached 29,639.09 tons in 2022. Guluk – Guluk Subdistrict is in Quadrant IV with an average LQ of 0.94 and a negative shift share value of -8,135.53, reflecting a decline in competitiveness.

Sub-districts in Quadrant II (relatively superior sectors) show development potential, such as Saronggi Subdistrict, with a positive shift share of 4,010.09 even though the average LQ is only 0.71. Some sub-districts in Quadrant III (slightly underdeveloped sectors), such as Pragaan and Bluto Subdistricts, have higher average LQ but negative shift share values, signaling the need for improved production efficiency. Pragaan Subdistrict shows an average LQ of 1.34, but its shift share value decreased by -2,480.94.

The analysis shows the importance of different interventions in each sub-district. Sub-districts in Quadrant I need to be maintained as a leading sector through technological improvements and expanded market access. Sub-districts in Quadrant IV require special attention to increase productivity, such as farmer training and capital support. Meanwhile, sub-districts in Quadrants II and III can be optimized through better supply chain management to maximize their potential contribution to the maize sector in Sumenep Regency.

3.3. Soybean Commodities

The food crop commodity in the Sumenep District is soybean. Based on data from BPS Sumenep District, soybean production in 2022 reached 179.25 tons. Soybean production is only produced by the Pragaan, Batuan, and Guluk-Guluk Subdistricts. Soybean production data for each sub-district in 2020-2022, LQ value analysis, shift share, and sector criteria analysis can be seen in **Table 4.**

Table 4 LQ and shift-share analysis of soybean commodities in Sumenep District

Sub district number	Corn production (ton)			LQ			Average of LQ	Shift share	Quadrant
	2020	2021	2022	2020	2021	2022			
1	18.87	0	6.74	1.45	0.00	0.57	0.67	-12.13	IV
2	32.94	0	0	4.04	0.00	0.00	1.35	-32.94	III
3	0.00	0	0	0.00	0.00	0.00	0.00	0	-
4	0.00	0	0	0.00	0.00	0.00	0.00	0	-
5	2.66	0	0	0.23	0.00	0.00	0.08	-2.66	IV
6	0.00	0	0	0.00	0.00	0.00	0.00	0	-
7	0.00	0	0	0.00	0.00	0.00	0.00	0	-
8	13.59	0	10.87	2.06	0.00	1.14	1.07	-2.72	III
9	71.21	49.97	0	4.13	2.72	0.00	2.28	-71.21	III
10	51.51	4.17	35	3.96	0.29	2.38	2.21	-16.51	III
11	92.28	29.34	75.82	3.78	1.96	3.67	3.14	-16.46	III
12	30.89	0	0	1.52	0.00	0.00	0.51	-30.89	IV
13	7.09	0	17.02	0.54	0.00	1.10	0.55	9.93	II
14	0.00	0	0	0.00	0.00	0.00	0.00	0	-
15	0.00	0	33.8	0.00	0.00	2.22	0.74	0	-
16	2.62	0	0	0.19	0.00	0.00	0.06	-2.62	IV
17	6.41	0	0	0.48	0.00	0.00	0.16	-6.41	IV
18	2.51	0	0	0.17	0.00	0.00	0.06	-2.51	IV
19	14.57	0	0	1.12	0.00	0.00	0.37	-14.57	IV
20	0.00	0	0	0.00	0.00	0.00	0.00	0	-
21	0.00	0	0	0.00	0.00	0.00	0.00	0	-
22	0.00	0	0	0.00	0.00	0.00	0.00	0	-
23	0.00	0	0	0.00	0.00	0.00	0.00	0	-
24	0.00	0	0	0.00	0.00	0.00	0.00	0	-
25	0.00	0	0	0.00	0.00	0.00	0.00	0	-
26	0.00	0	0	0.00	0.00	0.00	0.00	0	-
27	0.00	0	0	0.00	0.00	0.00	0.00	0	-
Total	347.15	83.48	179.25						

Description: I (Superior Sector); II (slightly superior Sector); III (slightly backward); IV (Backward sector)

Table 4 shows soybean productivity in this region, which shows significant fluctuations, with a total production of 347.15 tons in 2020, which decreased to 179.25 tons in 2022. Most of the sub-districts are included in Quadrant III (slightly underdeveloped) and Quadrant IV (underdeveloped), reflecting the low contribution of the soybean sector to the local economy. For example, guluk – Guluk Subdistrict, despite having a high LQ average of 3.14, recorded a negative shift share of -16.46, signaling a decline in competitiveness.

Several sub-districts in Quadrant III show potential for improvement, such as Ambunten Subdistrict, which has a positive shift share of 9.93 even though the average LQ is only 0.55. However, most other sub-districts, such as Bluto and Batang – Batang Subdistrict, are in Quadrant IV, showing an average LQ below 1 with a negative shift share of -12.13 and -14.57, respectively. This condition indicates the need for special attention to improve the efficiency of soybean production and competitiveness in these regions.

Overall, this analysis shows that the soybean sector in Sumenep Regency is still in the early stages of development and needs various strategic interventions. Efforts such as crop diversification, provision of agricultural technology, and better market support are needed to drive the growth of this sector. With the right moves, the sub-districts in Quadrants III and IV have the potential to move towards a more superior sector in the future.

3.4. Sweet Potatoes Commodities
A commodity that is widely found in the Sumenep District is sweet potato. The local community widely uses sweet potatoes. Sweet potato is not only consumed as a staple food but is also produced in other processed products. Planting sweet potatoes is relatively easy, so there are sweet potato plants in every region. Data on sweet potato commodity production in each sub-district in 2020-2022, LQ value analysis, shift share, and sector criteria analysis of shift-share analysis can be seen in **Table 5.**

Table 5 shows the analysis of LQ and shift share of sweet potato commodities in Sumenep Regency for 2020-2022. Sweet potato production significantly increased from 41,889.51 tons in 2020 to 145,963.02 tons in 2022. Some sub-districts, such as Saronggi, Giligenting, and Talango, have high LQ averages (in Quadrant I) and show a positive shift share, signaling that they are flagship sectors with high growth potential. Saronggi Subdistrict recorded an average LQ of 4.95 and a shift-share of 9,760.14, indicating its strategic position as the main sweet potato production area.

However, some sub-districts are facing challenges. Sumenep and Ganding Subdistricts are included in Quadrant IV, with a low LQ average (below 1) and a negative shift share, indicating that the sector's contribution to the local economy is still weak. In contrast, Nonggunong Subdistrict recorded a significant surge in production with the highest average LQ (5.96) and largest shift-share (18,874.47), making it one of the major players in the sector in 2022. Several other sub-districts, such as Gapura and Gayam Subdistrict, are areas with slightly superior sectors (Quadrant II), albeit with significant variations in production.

Overall, this analysis indicates that the sweet potato sector in Sumenep Regency has an excellent opportunity to grow, especially with a focus on sub-districts in Quadrants I and II. Strategic measures such as land optimization, application of agricultural technology, and market development need to be strengthened to harness the growth potential in sub-districts that have shown significant production increases. Sub-districts in Quadrants III and IV also need more attention through productivity and competitiveness improvement programs to keep pace with more advanced sub-districts.

Table 5 LQ and shift-share analysis of sweet potato commodities in Sumenep District

Sub district number	Sweet potato production (ton)			LQ			Average of LQ	Shift share	Quadrant
	2020	2021	2022	2020	2021	2022			
1	1,981.19	585.35	2,461.48	1.26	0.41	1.73	1.13	480.29	I
2	-	315.65	480.99	-	0.31	0.47	0.26	0	-
3	1,967.16	36,407.59	11,727.30	1.23	9.31	4.31	4.95	9760.14	I
4	699.46	1,442.63	1,763.21	1.83	3.28	2.43	2.51	1063.75	I
5	6,617.98	5,019.50	6,229.56	4.77	3.39	4.45	4.20	-388.42	III
6	276.49	683.94	480.22	1.89	4.67	2.41	2.99	203.73	I
7	1,187.52	133.60	860.95	1.68	0.19	0.91	0.92	-326.57	IV
8	1,432.89	-	1,478.13	1.80	-	1.29	1.03	45.24	I
9	1,922.57	2,839.49	4,170.50	0.92	1.28	1.25	1.15	2247.93	I
10	1,688.62	1,402.16	1,432.32	1.08	0.82	0.81	0.90	-256.3	IV
11	1,250.23	587.46	3,419.30	0.42	0.33	1.37	0.71	2169.07	II
12	7,611.69	2,871.48	2,172.60	3.10	1.39	0.72	1.74	-5439.09	III
13	953.88	1,453.54	2,195.45	0.61	1.09	1.18	0.96	1241.57	II
14	857.47	4,853.90	7,747.87	0.66	2.72	3.12	2.17	6890.4	I
15	3,923.07	3,468.66	3,998.81	2.69	2.35	2.17	2.40	75.74	I
16	3,509.46	2,904.38	1,210.16	2.16	1.85	0.64	1.55	-2299.3	III
17	420.76	5,304.53	10,368.63	0.26	2.27	3.25	1.93	9947.87	I
18	702.79	11,794.62	4,094.50	0.38	4.25	1.52	2.05	3391.71	I
19	1,440.39	2,638.19	3,305.32	0.92	1.48	1.73	1.38	1864.93	I
20	1,705.06	9,460.80	10,547.21	1.35	4.74	4.63	3.58	8842.15	I
21	89.74	18,964.21	18,964.21	0.13	9.25	8.50	5.96	18874.47	I
22	134.61	5,818.27	5,504.18	0.09	2.86	2.30	1.75	5369.57	I
23	1,018.10	-	2,739.88	1.31	-	2.52	1.28	1721.78	I
24	378.85	-	4,015.81	0.24	-	1.67	0.64	3636.96	II
25	-	12,389.18	15,283.01	-	2.36	1.94	1.43	0	-
26	-	4,937.71	12,961.49	-	1.71	3.11	1.61	0	-
27	119.53	2,465.27	6,349.93	0.26	3.37	5.43	3.02	6230.4	I
Total	41,889.51	138,742.11	145,963.02						

Description: I (Superior Sector); II (slightly superior Sector); III (slightly backward); IV (Backward sector)

The analysis data presented in **Table 2, Table 3, Table 4**, and **Table 5** on the LQ value and shift share of various agricultural commodities in Sumenep Regency show a variation in the level of excellence and contribution between sub-districts. Paddy, corn, sweet potato, soybeans, and other commodities have different development patterns depending on production factors, infrastructure, and regional competitiveness. Subdistricts such as Arjasa, Gapura, Lenteng, and Saronggi in Quadrant I show great potential as superior production centers with significant economic contributions. On the other hand, sub-districts such as Guluk-Guluk, Pragaan, and Bluto in Quadrant IV need special attention to increase productivity and competitiveness through farmer capacity-building programs, capital support, and technology development.

Considering each region's characteristics, the agricultural sector development strategy needs to be focused on optimizing the potential in Quadrants I and II while improving conditions in Quadrants III and IV. Leading areas need continuous support in infrastructure improvements, modern technology, and broader market access to maintain their contribution to the local economy. On the other hand, the underdeveloped regions need policy interventions such as crop diversification, increased production efficiency, and strengthened marketing networks. An integrated approach will help drive equitable growth in the agricultural sector, improve farmers' welfare, and strengthen the sector's contribution to the economic development of Sumenep Regency.

4. Conclusion

Based on the results of the analysis using LQ and shift-share analysis, it shows that the distribution of superior paddy commodities in Sumenep District is in Batuan, Lenteng, Manding, Gapura, Batang-Batang, Raas, and Sapeken Subdistricts. Corn commodities are in Pragaan, Bluto, Saronggi, Giligenting, Talango, Kalianget, Lenteng, Ganding, Pasongsongan, Ambunten, Rubaru, Dasuk, Batuputih, Gayam, Raas, Kangean, and Masalembu Subdistricts. The leading commodity of sweet potato is spread in Pragaan, Saronggi, Giligenting, Kalianget, Batuan, Lenteng, Rubaru, Dasuk, Batuputih, Gapura, Batang-Batang, Dungkek, Nonggunong, Gayam, Raas, and Masalembu Subdistricts.

Acknowledgments
The author would like to thank the Directorate of Research, Technology and Community Service (DRTPM), Ministry of Education, Culture, Research and Technology for providing financial support through the Fundamental Research Schemes 2024 contract number 023/UN46.4.1/PT.01.03 /2024.

References

[1] BPS, "Kabupaten Sumenep dalam Angka," Kabupaten Sumenep, 2024.
[2] BPS, "Kabupaten Sumenep dalam Angka," Kabupaten Sumenep, 2023.
[3] BPS, "Kabupaten Sumenep dalam Angka," Kabupaten Sumenep, 2022.
[4] BPS, "Kabuapten Sumenep dalam Angka," Kabupaten Sumenep, 2021.
[5] J. D. Buendía Azorín, R. Martínez Alpañez, and M. del M. Sánchez de la Vega, "A new proposal to model regional input–output structures using location quotients. An application to Korean and Spanish regions," Pap. Reg. Sci., vol. 101, no. 5, pp. 1219–1237, 2022.
[6] A. A. Nur, "Analisis Sektor Unggulan Sebagai Daya Saing Dalam Pembangunan Wilayah Kabupaten," TIN Terap. Inform. Nusant., vol. 4, no. 3, pp. 211–217, 2023.
[7] Gunawan, A. E. Cahyono, and A. Santoso, "Local superior commodities, regional specializations and regional economic contributions," J. Distrib. Sci., vol. 16, no. 9, pp. 35–41, 2018.
[8] G. D. Djira, F. Schaarschmidt, and B. Fayissa, "Inferences for selected location quotients with applications to health outcomes," Geogr. Anal., vol. 42, no. 3, pp. 288 – 300, 2010.
[9] D. F. Meyer and F. Niyimbanira, "Formulation and application of a multi-variable location quotient index in the Mpumalanga Province, South Africa," Local Econ., vol. 36, no. 4, pp. 273 – 286, 2021.
[10] I. Maflahah, D. F. Asfan, S. Joko Utomo, F. AS, and R. Arief Firmansyah, "Agglomeration Agricultural Zone Analysis Based on the Hybrid Hierarchical Clustering Approach in Madura, Indonesia," Int. J. Ind. Eng. Prod. Res., vol. 35, no. 4, 2024.
[11] R. Harini, R. D. Ariani, B. Susilo, and E. H. Pangaribowo, "Regional Potential Analysis Based on Agricultural Commodities of Food Crops and Their Contribution to the Economy in the Karangsambung-Karangbolong Geopark Area, Indonesia: A Location Quotient Approach," E3S Web Conf., vol. 325, 2021.
[12] H. S. P. Rahayu, Saidah, Risna, Muchtar, and Syafruddin, "Determination of Leading Commodities and Its Development Area in Sigi District Central Sulawesi," E3S Web Conf., vol. 316, pp. 4–11, 2021.
[13] A. M. Sausan et al., "Location Quotient Analysis of the Agricultural Sector in Yogyakarta, Indonesia," Proc. 2nd Int. Conf. Smart Innov. Agric. (ICoSIA 2021), vol. 19, no. ICoSIA 2021, pp. 5–9, 2022.

[14] R. Sinabutar, S. N. Lubis, and H. Harahap, "Analysis of Leading Agricultural Commodities and Their Role in Regional Development in North Tapanuli Regency, Indonesia," Asian J. Adv. Res. Reports, vol. 17, no. 12, pp. 116–126, 2023.

[15] W. Hanim, H. A. Prasca, W. N. Pertiwi, H. G. Yudawisastra, and S. Sugiartiningsih, "Determination Analysis of Leading Commodity in the Melonguane National Border Strategic Area (NBSA)," J. Econ., vol. 19, no. 1, pp. 55–67, 2023.

[16] D. Yanti, E. Stiyanto, N. R. Yanti, C. Batubara, and F. Ariyadi, "Identification of Leading Commodity Areas in the Agricultural Sector Based on Historical Data and Land Suitability," J. Tek. Pertan. Lampung (Journal Agric. Eng., vol. 12, no. 3, p. 698, 2023.

[17] L. N. Amalia and U. Galuh, "Analisis Komoditas Sayuran Unggulan di Kabupaten Blora Lidya," Mimb. Agribisnis J. Pemikir. Masy. Ilm. Berwawasan Agribisnis, vol. 10, no. 1, pp. 893–898, 2024.

[18] A. WIJAYA, Z. ILMI, and D. C. DARMA, "Economic Performance: Leading Sector, Economic Structure and Competitiveness of Export Commodities," J. Bus. Econ. Environ. Stud., vol. 10, no. 3, pp. 23–33, 2020.

[19] P. Diansari, S. R. Yunus, M. Arsyad, N. Lanuhu, M. H. Jamil, and Mahyuddin, "Analysis of agricultural based commodities in Bolaang Mongondow Regency, North Sulawesi, Indonesia," in IOP Conference Series: Earth and Environmental Science, 2021, vol. 681, no. 1.

[20] A. Latif, A. Karim, S. Sugianto, Romano, M. F. Ikhwali, and M. Rusdi, "Top Agricultural Commodities for Agropolitan Development in Nagan Raya District, Aceh, Indonesia," Green Energy Technol., pp. 505 – 527, 2023.

[21] D. Novita, T. Rinanda, and M. Ilham Riyadh, "Determination of superior agriculture commodities in North Sumatra Province," in E3S Web of Conferences, 2022, vol. 339.

[22] N. F. Gaffar, Mahyuddin, and A. N. Tenriawaru, "Commercial Advantages of Soybeans Commodities South Sulawesi," in IOP Conference Series: Earth and Environmental Science, 2021, vol. 921, no. 1.

[23] Gunawan, A. E. Cahyono, and A. Santoso, "Local superior commodities, regional specializations and regional economic contributions," J. Distrib. Sci., vol. 16, no. 9, pp. 35 – 41, 2018.

[24] Jusni and A. Aswan, "Identifying sustainable agricultural commodities in Wajo regency," in IOP Conference Series: Earth and Environmental Science, 2020, vol. 473, no. 1.

[25] E. Humaidi, I. Putu Ajie Windu Kertayoga, and Analianasari, "Preparation of a Map of Leading Food Commodities in the Lampung Province Using the Location Quotient (LQ) Method," in IOP Conference Series: Earth and Environmental Science, 2021, vol. 1012, no. 1.

[26] Y. E. Wahyuningsih, L. P. Ansari, Y. Yasrizal, S. R. Sani, T. Zulham, and J. Saputra, "Analyzing the Regional Leading Sectors with Location Quotient and Its Effect on Economic Growth in Aceh Jaya, Indonesia," Front. Bus. Econ., vol. 1, no. 1, pp. 35–42, 2022.

[27] M. Pominova, T. Gabe, and A. Crawley, "The Stability of Location Quotients," Rev. Reg. Stud., vol. 52, no. 3, pp. 296 – 320, 2022.

[28] Z. Tian, P. D. Gottlieb, and S. J. Goetz, "Measuring industry co-location across county borders," Spat. Econ. Anal., vol. 15, no. 1, pp. 92 – 113, 2020.

[29] T. Nie, N. Li, and F. Yan, "Distortion and Improvement Method of Location Quotient in Water Consumption Evaluation," Math. Probl. Eng., vol. 2022, 2022.

[30] Mursidi and S. Arief, "Analisis Pertumbuhan Ekonomi dan Penetapan Sektor Ungggulan Dengan Menggunakan Location Quotient, Shift Share dan Klassen Typology di Madura," Int. J. Endocrinol., vol. 16, no. 4, pp. 327–332, 2021.

[31] A. Swantari, H. Ratnaningtyas, and F. Festivalia, "Agrotourism Development Using Location Quotient Analysis in Cibuntu-Kuningan Village," Tour. Res. J., vol. 5, no. 2, p. 165, 2021.

[32] Y. S. Maulana, A. H. Munawar, D. Hadiani, Ratningsih, and T. Wibisono, "Location Quotient Analysis (LQ) in Determining the Excellent Commodity," in Proceedings of the 1st International Conference on Science, Health, Economics, Education and Technology, 2020, vol. 27, no. ICoSHEET 2019, pp. 65–68.
[33] S. Chiang, "Location quotient and trade," Ann. Reg. Sci., vol. 43, no. 2, pp. 399–414, 2009.
[34] M. N. Iglesias, "Measuring size distortions of location quotients," Int. Econ., vol. 167, pp. 189 – 205, 2021.

Nutritional content of gluten-free cookies as an alternative snack for undernourished children

A D Elisanti[1*], Rindiani[1], G Purnasari[1], D I Amareta[1], M R Permadi[1]

[1]Clinical Nutrition, Health Departement, Politeknik Negeri Jember, East Java, Indonesia

[*]Email: alinea@polije.ac.id

Abstract. Undernourishment is one of the nutritional problems that often occurs in developing countries. In Indonesia, the undernourished is defined as underweight (17.1%), stunting (21.6%), and wasting (7.7%). The rising number of children's food allergies is one of the factors. One symptom of the body's immunological system is allergies. Food sources can act as triggers, with gluten being the most prevalent. The purpose of this study was to determine the amount of gluten, protein, vitamin C, sodium, and calcium in cookies. Through laboratory analytical research, an analysis of the nutritional content of cookies will be carried out based on SNI 01-2891;1992, Iodimetry, AAS, while the best organoleptic test results are analyzed descriptively. The results showed that 100 g of cookies contain 7.48 g of protein, 988.15 mg/kg of calcium, 90.03 mg/100g of vitamin C, 67.81 mg of sodium, and no quantitative gluten was detected. Nutritional content of gluten-free cookies are source of protein, high in calcium, high in vitamin C, low in sodium, and gluten-free, so it is safe for consumption by children, in 100 grams of cookies can fulfill the daily needs of vitamin C and calcium for children.

1. Introduction

Undernourishment is one of the nutritional problems that often occur in developing countries. About 663 million people are suffering from malnutrition in the world [1] . Globally 148.1 million children under 5 are affected by stunting, and 45 million are wasting. In 2022, 52% of children under 5 were affected by stunting, 70% were affected by wasting, and 2.1 % were suffering from severe wasting lived in Asia [2]. In Indonesia, undernourished is described as underweight (17.1%), stunting (21.6%), and wasting (7.7%), In East Java prevalence of underweight (10,6%), stunting (7,68%), and wasting (8,93%) [3].

Multifactorial causes of malnutrition. Such as social, environmental, and biological factors, including poverty, inadequate nutrition and care practices, poor health services, and infectious diseases [4]. In high-poverty countries, malnutrition remains a major cause of health problems, premature death, and disease in children [5], estimated to affect about one-third of children worldwide [6]. Acute malnutrition is considered an emergency because of the immediate threat to life. Children with severe acute malnutrition and moderate acute malnutrition have a 5-20 times higher risk of death compared to normal nutritional status [7]. Malnutrition in low- and middle-income communities contributes to 45% of deaths in children under the age of five [8].

Based on its short-term and long-term impacts, reducing the prevalence of malnutrition is recognized as a global priority in the UN Sustainable Development Goals and the WHO Global Nutrition Targets for 2025 [9].

Reducing the prevalence of malnutrition can be achieved through various interventions. Those that have been implemented include supplementation, formula distribution, education for mothers, nutrition counseling, and dietary regulation. Malnutrition interventions in Indonesia include specific interventions and sensitive interventions. Specific nutrition interventions aim to address various factors causing malnutrition, including food supplementation and fortification, with an emphasis on iron, zinc, folic acid, or micronutrients as well as activities to support exclusive breastfeeding and dietary diversification [10]. Meanwhile, sensitive nutrition interventions aim to address the underlying causes of malnutrition, such as increasing food security through agricultural support, nutrition education, access to clean water, sanitation, hygiene, and improving access to health services [11]. In this case, an innovation has been made in the form of local food-based snack products as a solution to nutritional problems.

Cookies can be made from non-gluten ingredients. One of the triggers for the undernourished is allergies, food allergies have an impact on the digestion system of under-five children. Gluten allergies and lactose intolerance often occur in under five children. There is an interaction between gluten and lactose diets with the immune system in the intestines, with the most common impacts being diarrhea, weight loss due to mal absorption, and anemia which have an impact on the undernourished. The purpose of this study was to identify the nutritional content of non-gluten cookies as an alternative snack for undernourished chlidren.

2. Materials and Methods

This section will give information the materials and methods used in this research.

2.1. *Research materials*

The materials in making cookies include a large bowl for mixing the dough (mixing bowl), whisk, spatula, spoon, measuring spoon, measuring cup, mixer, baking paper, baking pan, oven and timer. The ingredients used in making ½ recipe cookies, namely sago flour (200-300 g), soy flour (100-125 g), egg yolks (70-78 g) margarine (125-130 g), powdered sugar (125-150 g) and vanilla powder (3-5 g).

2.2. *Research method*

This laboratory study used a one-factor Complete Randomized Design (CRD) combined the raw materials of sago flour: soy flour as follows: 9: 2 (Formula-1), 7: 3 (Formula-2), 5: 5 (Formula-3), 3: 7 (Formula-4). Each treatment was repeated twice, and organoleptic and acceptability tests were carried out, where the best cookies would be tested for nutritional and mineral content. The best treatment is determined descriptively by the stages of 1). Ranking the results of organoleptic tests including taste; 2). texture, color, and aroma using a hedonic form; 3). Then calculating the average of each variable (with a score of 1-5); 4). Calculating the variable weight, the variable with the highest average value is given a value of 1; another variable is obtained from the average of the variable divided by the average of the ranked variable 1; 5). Calculating the normal score by dividing the weight of each variable by the total score of the variables; 6). Calculating the effectiveness value (Ne) with the formula Ne = (treatment value - worst value) / (best value - worst value); 7). Calculating the result value (Nh) by multiplying the effectiveness value (Ne) by the normal weight of each variable, the result value (Nh) of all variables for each treatment is added up, and the highest Nh value is the highest value - the lowest value [12].

The nutritional content examined is protein, calcium, vitamin C, sodium, and gluten levels. Sodium testing uses the SNI 01-2891; 1992 testing method, calcium uses an Atomic Absorption Spectrometer (AAS), vitamin C testing uses Iodimetry and gluten uses the AOAC 2012.01 method (quantitative gluten).

3. Result and discussion

This section will explain the results of determining the best treatment based on organoleptic tests using the hedonic form as in Table 1 will be explained and then the results of the nutritional content test of the best treatment of cookies for undernourished toddler snacks will be described.

3.1. Determine the best formula

Table 1. List of scores to determine the best formula

Variable	Variable score	Normal score	F1		F2		F3		F4	
			Ne	Nh	Ne	Nh	Ne	Nh	Ne	Nh
Taste	1	0.36	0.23	0.08	0	0	0.46	0.16	1	0.36
Texture	0.79	0.28	0	0	0.25	0.07	0.55	0.15	1	0.28
Color	0.56	0.20	0.72	0.14	1	0.20	0.31	0.06	0	0
Aroma	0.46	0.16	0	0	0.25	0.04	1	0.16	0	0
Total	2.81	1.00		0.23		0.31		0.54		0.64

Source : organoleptic data

Table 1 explains the organoleptic test data score based on a hedonic form. There are 4 variables to determine the best formula of cookies including taste, texture, color, and aroma. The formula of calculating the score of each variable based on the category have been explained in the research method (1.2). It is known that formula F4 emerged as the best treatment, with organoleptic quality scores for taste, texture, aroma and color having the highest Nh value of 0.64.

3.2. Gluten free cookies

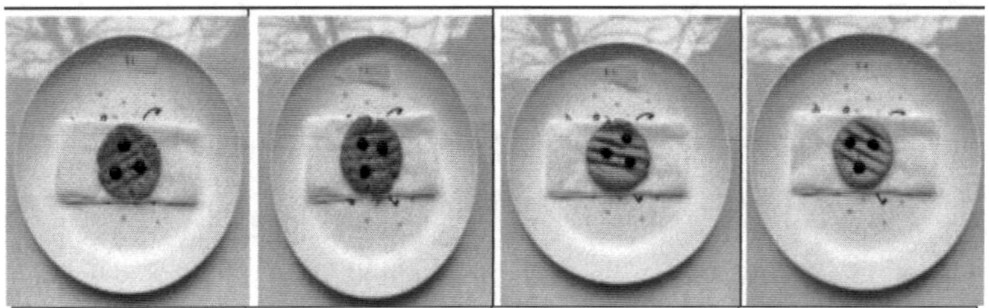

F1 (9 :2)	F2 (7:3)	F3 (5:5)	F4 (3:7)

Figure 1. Gluten-free cookies based on formula

Figure 1 showed the appearance of 4 gluten-free cookies with different formula. Every 100 g of gluten-free cookies contain 7.48 g of protein, 988.15 mg/kg of calcium, 90.03 mg/100g of vitamin C, and 67.81 mg of sodium. This means that as a snack food, gluten-free cookies are source of protein, high in calcium, high in vitamin C, and low in sodium. This cookies is safe for consumption by toddlers, 100 grams of cookies can meet the daily needs of vitamin C and calcium for toddlers.

As a source of protein cookies, it is recommended to be consume. Protein is a very vital substance for the body, as a source of amino acids containing the elements C, H, O, and N, which are not found in fat or carbohydrates. In the body, protein acts as the main source of energy besides carbohydrates and fat. In addition, protein functions as a building and regulatory substance [13]. The protein content analysis test on gluten-free cookies showed that it is contain 7.48 grams of protein in 100 grams of cookies. Based on National Agency of Drug and Food Control Regulation No. 1 of 2022 concerning the supervision of claims on Processed Food Labels and Advertisements (PFLA), the protein content in cookies, which is 28.76%, is included in the category of food with a protein source. This is known by calculate the protein content in 100 grams of cookies divided by the Nutrition Label Reference (NLR) protein for children aged 1-3 years. Food products can be claim as source of protein if they contain 20% of NLR per 100 grams in solid form or 10% of NLR per 100 ml in liquid form. Meanwhile, food products can be categorized as high-protein products if the contain 35% of NLR per 100 grams in solid form or 17.5% of NLR per 100 ml in liquid form. So the gluten-free cookies of this research becomes a snack source of protein.

Several things can affect the protein content of a product. Based on previous research conducted, the protein content in a food product, especially in cookies [14]. The protein content is influenced by the water content in the cookies, the lower of water content will give affect to the higher protein content in the cookies. This study are in line with research conducted by Pratama et al [15], that the measured protein content in the product can be influenced by the dehydration conditions of the product, namely how much water content is lost in the material. The measured protein value will increase along with the increasing amount of water lost. Factors that affect protein levels are volatile nitrogen compounds, so they evaporate during the processing process. Heating can cause protein in the product to degrade, which not only reduces nutritional value, but also reduces protein activity, so the enzymes and hormones contained will be lost [16]. In addition, protein damage in food ingredients will increase along with increasing temperature and processing time [17].

The cookies have a high calcium content, in 100 grams of cookies contain 988.15 mg/100 grams of calcium. According to Food and Drug Administration Regulation No. 1 of 2022 concerning the supervision of claims on PFLA [18], the source of vitamins and minerals product if it is not less than 15% of the NLR per 100 g (in solid form) and is be high/rich if it is not less than 2 times the amount for the source. Food in solid form can be claimed as a source of calcium if it is not less than 97.5 mg/100 grams in the 1-3 year age group and 165 mg/100 grams in the general group, while it can be claimed as high/rich in calcium if it is not less than 195 mg/100 grams in the 1-3 year age group and 330 mg/100 grams in the general group. The results of the calcium content analysis of cookies were 988.15 mg/100 grams, so they can be categorized as high/rich in calcium.

Cookies can be categorized as high in calcium because the ingredients used in the cookie-making process contain quite high levels of calcium. Calcium is essential, especially for children who are malnourished. Calcium is one of the micronutrients needed for bone growth in children. Lack of calcium in the body can affect linear growth in children. Therefore, sufficient calcium intake plays an important role in children's growth, especially when entering peak bone mass [19].

The dominant ingredients contains calcium are sago flour, soy flour, and egg yolks. In 100 g, sago flour contains 0.3 gram of protein, 209 kcal of energy, 0.2 gram of fat, 51.6 grams of carbohydrates, 27 mg of calcium, 13 mg of phosphorus, and 0.6 mg of iron. According to Indonesian Food Composition Table (IFCT) 2017 [20], the nutritional content of egg yolks is 355 kcal of energy, 16.3 grams of protein, 31.9 grams of fat, 0.7 gram of carbohydrates, 147 mg of calcium, and 111 mg of sodium. According to the US Department of Agriculture, the nutritional content of soybean flour includes 372 kcal of energy, 49.8 grams of protein, 8.9 grams of fat, 3.6 grams of carbohydrates, 285 mg of calcium, and 9 mg of sodium [21].

The main ingredients to replace wheat flour in making cookies are sago flour and soy flour. A hundred gram of sago flour contains 355 kcal of energy, 85.6 g of carbohydrates, 0.6 g of protein, 1.1 g of fat, 91 g of calcium, 167 g of phosphorus, 2.2 mg of iron, 0.3 g of fiber and 0.01 mg of vitamin B1 [22]. While 100 g of soy flour contains 347 kcal of energy, 35.9 g of protein, 29.9 g of carbohydrates, 20.6 g of fat, 195 mg of calcium, 554 mg of phosphorus, and 8 mg of iron. In addition, soy flour also contains vitamin A, vitamin B1, and vitamin C [23]. The nutritional content of both flours greatly affects the results of the analysis of protein, calcium, and vitamin C levels in cookies. Cookies contain 7.48 g of protein because soy flour contains quite high protein in 100 g. The calcium content of the cookie product, which is 988.15 mg/kg, is obtained from both ingredients that contain quite high calcium.

The vitamin C content in the product is 90.03 mg/100g obtained from soy flour which contains vitamin C in 100 g. In addition, the gluten-free content of the cookies product is obtained from sago flour as the basic ingredient for making cookies. The low protein content of sago flour makes sago flour free of gluten. Thus, sago flour is used as an alternative to wheat flour and can be consumed by consumers who are allergic to gluten [24].

The vitamin C content in cookies is quite high, which is 90.03 mg/100 grams. There are ingredients that contain vitamin C which are dominant in making cookies, namely soy flour. The nutritional content in soy flour includes protein, carbohydrates, fat, calcium, phosphorus, and iron. Not only that, soy flour also contains vitamin A, vitamin B1, and vitamin C [25]. According to research, it shows that germination of soybean seeds can increase protein levels, protein digestibility, and vitamin C contained in soy flour. The average protein content increased by 0.15%, protein digestibility increased by 1.26%, and vitamin C increased by 7.67 mg [26].

Based on the laboratory tests, the sodium content in cookies is 67.81 grams which is classified as low sodium. Several things that can cause low sodium content in products are the main ingredient containing sodium used such as margarine and there are no other additional ingredients. In addition, several factors affect the sodium content in a product such as the water content in a food. The food product experiences an increase in water content, and the sodium content in the food product will decrease. The high levels of sodium in food products can affect the renin-angiotensin system in the kidneys if the product is consumed too often. High sodium levels will affect the sodium and potassium balance system in the body, which can cause hypertension [27][28].

Cookies are not identified as gluten because this cookies used gluten-free ingredients (soy and sago flour). Soy flour is an alternative to wheat flour that has a high protein content and is gluten-free. Soy flour is a product made from whole soybeans or soybeans that have skin. Soy flour can be used as a thickener and baked product. Soy flour has a high water content and is gluten-free, so if it is substituted with wheat or other flour, modification is needed [29]. In addition, another ingredients used in making this cookies is sago flour. Similar to soy flour, sago flour (Metroxylon sp.) is also an alternative to wheat flour. For certain groups who avoid foods containing gluten, sago flour can be an alternative. The content of sago flour is carbohydrates, fiber, calcium, iron, and protein (less) so it is gluten-free.

4. Conclusion

The best formula of gluten-free cookies has obtained on F4 (3:7) with best treatment based on organoleptic quality scores for taste, texture, aroma and color having the highest Nh value of 0.64. The nutritional content of 100 g cookies have obtained on 7.48 g of protein, 988.15 mg/kg of calcium, 90.03 mg/100g of vitamin C, and 67.81 mg of sodium. This means gluten-free cookies are snack food which is source of protein, high in calcium, high in vitamin C, and low in sodium. It is become a safe snack and meet the daily needs of vitamin C and calcium for toddlers.

5. Acknowledgments

Authors acknowledge to the Ministry Of National Education, Culture, Research and Technology of The Republic Of Indonesia through the PNBP Competitive Research Grant Politeknik Negeri Jember 2024.

References

[1] Roser HRPR. and M. "Hunger and Undernourishment," *OurWorldinData.org. [Online Resour* [Internet]. 2023. Available from: https://ourworldindata.org/hunger-andundernourishment

[2] UNICEF, WHO, and World Bank, "Level and trend in child malnutrition," *World Heal.Organ* [Internet].p. 4, 2023. Available from:https://www.who.int/publications/i/item/9789240073791

[3] East Java Provincial Health Servise, "East Java Province Health Profile in 2023," https://dinkes.jatimprov.go.id. Surabaya: 2024. p.1-17

[4] Akombi BJ., Agho KE., Hall J J., Wali N., Renzaho AMN., and Merom D."Stunting, wasting and underweight in Sub-Saharan Africa: A systematic review," *Int. J. Environ. Res.Public Health*, vol. 14, no. 8, pp. 1–18, 2017, doi: 10.3390/ijerph14080863.

[5] Rice AL., L. Sacco, A. Hyder, and R. E. Black, "Malnutrition as an underlying cause of childhood deaths associated with infectious diseases in developing countries," *Bull. World Health Organ.*, vol. 78, no. 10, pp. 1207–1221, 2000.

[6] Black RE. *et al.*, "Maternal and child undernutrition and overweight in low-income and middle - income countries," *Lancet*, vol. 382, no. 9890, pp. 427–451, 2013, doi: 10.1016/S0140-6736(13)60937-X.

[7] H. Professionals, C. Harkensee, K. Olness, and B. E. Esmaili, *Child Refugee and Migrant Health*. 2021. doi: 10.1007/978-3-030-74906-4.

[8] WHO. Malnutrition [Internet]. Reports World Health Statistics. 2024. Available from:https://www.who.int/news-room/fact-sheets/detail/malnutrition

[9] United Nations. Transforming our world: the 2030 Agenda for Sustainable Development [Internet].2015. Available from: https://sdgs.un.org/2030agenda

[10] Arnold T. Nutrition-Specific and Nutrition-Sensitive Interventions. In: *Chapter 5.4*, Downloaded., SUN Movement, 2015.

[11] Ruel MT. and Alderman H. Nutrition-sensitive interventions and programmes: How can they help to accelerate progress in improving maternal and child nutrition?.LancetI [Internet],vol. 382, no.

9891, pp. 536–551, 2013, doi: 10.1016/S0140-6736(13)60843-0.

[12] Ernest Paul DeGarmo JRC. William G. Sullivan, Engineering Economy (7th ed.). Macmillan: Michigan University, 1984. [Online]. Available: https://books.google.co.id/books/about/Engineering_Economy.html?id=yoNRAAAAMAAJ&source=sh/x/srp/wr/m1/2&redir_esc=y

[13] Natsir NA. Analysis of Total Protein Content of Red Snapper and Duck Grouper. Biosel Biol. Sci. Educ., vol. 7, no. 1, p. 49, 2018, doi: 10.33477/bs.v7i1.392.

[14] Hairunnisa H, Suherman S, Supriadi S, Analylis of Macro Nutrients from Cocoa Combination Flour (Theobroma Cacao L) and Cassava (Manihot Utilissima) as a Basic Ingredient for Biscuits. *J.* Akad. Kim., vol. 6, no. 4, p. 200, 2018, doi: 10.22487/j24775185. 2017.v6.i4.9449.

[15] Intan Pratama R., I. Rostini, and D. E. Liviawaty. Rusky Intan Pratama, Characteristics of Biscuit with Jangilus (Istiophorus sp.) Fish Bone Flour Su. J. Akuatika, vol. 5, no. 1, pp. 30–39, 2014.

[16] Iqbal A, Rochima E, Rostini I. The Addition of Sharkminow's Eggs on Preference of Stick Product,. *J.* Perikan. Kelaut., vol. VII, no. 2, pp. 150–155, 2016.

[17] Sundari D, Almasyhuri, Lamid A. Effect of the Cooking Process on Protein. *Media litbangkes*, vol. 25, no. 4, pp. 235–242, 2015.

[18] National Agency of Drug and Food Control. Food and Drug Supervisory Agency Regulation Number 1 of 2022 concerning Supervision of Claims on Processed Food Labels and Advertisements. Peratur. BPOM, vol. 11, pp. 1–16, 2022.

[19] Bueno AL, M. Czepielewski A. A importância do consumo dietético de cálcio e vitamina D no crescimento. J. Pediatr. (Rio. J)., vol. 84, no. 5, pp. 386–394, 2008, doi: 10.2223/JPED.1816.

[20] Ministry of Health of the Republic of Indonesia, Indonesian Food Composition Table. Jakarta: Indonesian Ministry of Health. Directorate General of Public Health, 2018.

[21] US. Departement of Agriculture. Soy flour, low fat. [Internet]. USDA, USDA Agricultural Research Service. Available from: https://fdc.nal.usda.gov/fdcapp.html#/food-details/172444/nutrients

[22] Makmur SA. Adding Sago Flour and Wheat Flour to Making Sweet Bread. *Gorontalo Agric. Technol. J.*, vol. 1, no. 1, p. 1, 2018, doi: 10.32662/gatj.v1i1.161.

[23] Taufik Y. Effect of Concentration of Fruit Pulp and Soybean Flour (Glycine max) on Characteristics of Fit Bar Black Mulberry (Morus nigra L.). *Pas. Food Technol. J.*, vol. 5, no.1, p. 10, 2018, doi: 10.23969/pftj.v5i1.805.

[24] Rahmawati A, Putranto K, Tristianne A. Effect of Sago Starch Substitution (Metroxylon sagu R) on Flour on the Characteristics of Dry Cakes. *AGRITEKH* (Jurnal Agribisnis dan Teknol. Pangan), vol. 4, no. 1, pp. 1–10, 2023, doi: 10.32627/agritekh.v4i1.777.

[25] Wala J, Ransaleleh T., I. Wahyuni, and M. Rotinsulu. Water Content, pH and Total Meat Microbes Chicken with added White Turmeric (Curcuma mangga Val.). *Zootec*, vol. 36, no. 2, p. 405, 2016, doi: 10.35792/zot.36.2.2016.12567.

[26] Antarlina SS. Utilization of soybean sprout flour as raw material in weaning food processing [Internet]. AGRIS - International System for Agricultural Science and Technology. Malang Indonesia. 2000. Available from: https://agris.fao.org/search/en/providers/122456/records/6472425553aa8c8963039f46

[27] Ariyanto R and Dwi Kartika AG. The Effect of Different Methods for Making Healthy Salt Low Sodium in NaCl, Water and Sodium Levels. Juv. Ilm. Kelaut. dan Perikan., vol. 3, no. 1, pp. 32–36, 2022, doi: 10.21107/juvenil.v3i1.15344.

[28] Emi Inayah Sari Siregar. Systematic Review of the Relationship between Diet and Events Hypertension in the Elderly. J. Ilm. *PANNMED* (Pharmacist, Anal. Nurse, Nutr.Midwivery, Environ. Dent., vol. 17, no. 1, pp. 202–209, 2022, doi: 10.36911/pannmed.v17i1.1296.

[29] Oktaviani. Effect of Substituting Wheat Flour with Soy Flour on Properties Physycochemistry and Organoleptics of Donuts. *Skripsi*. Universitas Semarang, 2024.

IOP Conf. Series: Earth and Environmental Science **1446** (2025) 012030 doi:10.1088/1755-1315/1446/1/012030

The effect of adding cassava leaves and fermentation duration on the mocaf quality

M A Hamid[1,2*], W T Ho[2], C H Yeap[2], M Devi[3], In`am A[4] and M M Effendi[4]

[1] Food Security Research Laboratory, Faculty of Food Science and Nutrition (FSMP), Universiti Malaysia Sabah (UMS), UMS Street, 88400 Kota Kinabalu, Sabah, Malaysia
[2] FSMP, UMS, UMS Street, 88400 Kota Kinabalu, Sabah, Malaysia
[3] Department of Industrial Technology, Faculty of Engineering, Universitas Negeri Malang, Jalan Semarang 5, 65145 Malang, Indonesia
[4] Department of Mathematics Education, Universitas Muhammadiyah Malang, East Java, 65145 Malang, Indonesia

*Email: chot@ums.edu.my

Abstract. Protein content is an important indicator of mocaf for its better substitution on wheat flour in product development. The fermentation is found able to improve the protein content of cassava flour but the rise is unsatisfied and inconsistent due to a multitude of influencing factors. This research was aimed to produce the mocaf with higher protein by adding the cassava leaves as alternative protein source. The different concentration of cassava leaves (3% and 5%) were added into the mocaf production in the *L. plantarum* fermentation of 24 hrs and 48 hrs. The yield of final product, proximate and physicochemical analysis were carried out to all the formulation and the storage analysis was only done on the best formulation. The addition of cassava leaves into the mocaf had resulted in higher ash and crude fiber contents, similar moisture and crude fat contents and lower carbohydrate content as compared to the native mocaf. Overall, it was reported that there was significantly ($p<0.05$) increase of crude protein content in the mocaf with 5% cassava leaves and the highest crude protein content was recorded for 48 hours fermentation period. Thus, F6 was the best formulation and it was safe consumed over the 8 months storage period.

1. Introduction

Mocaf stands for modified cassava flour that produced by microbial fermentation on the cassava roots. According to International Atomic Energy Agency [1], cassava (*Manihot esculenta* Crantz) which also known as Brazilian arrow root, manioc, tapioca or yucca is a tropical woody shrub native to South America. In Malaysia, it is recognized as Tapioca or in Bahasa Malaysia *"Ubi Kayu"*. Cassava, being a highly perishable root crop, requires immediate processing upon uprooting to reduce the harvesting loss and increase the shelf life. Cassava roots are rich in carbohydrate in terms of starch. The carbohydrate content of fresh cassava roots is 32-38%, of which about 80% are starch, comprising approximately 83% amylopectin and 17% amylose. Due to this composition, cassava roots are primarily processed into flour, and mocaf is one such flour variant.

The invention of mocaf is mainly to substitute wheat flour to reduce the reliance on imported wheat. Thus, the protein content of mocaf becomes a critical indicator to ensure the competitive nutritional value in food products after the substitution. With this, various protein-rich sources, such as leaves and legumes are recommended into the food products made of mocaf to increase the protein value [2]. In this case, the cassava leaves are one of the options, while decreasing the waste of cassava leaves in the cassava plantation. The cassava leaves are highly nutritious, containing relatively high crude protein, minerals and vitamins, compared to other green leafy vegetables. The protein content of cassava leaves on dry weight basis ranges 19.9% to 25.9% [3], depending on the varieties. Thus, incorporating cassava leaves in the cassava root products, for example mocaf, should enhance the overall nutritional value. This research was aimed to determine the best mocaf formulation from cassava roots and leaves, with fermentation time, to have the mocaf with higher protein value compared to native mocaf. Then, the storage quality of best formulation was studied.

2. Materials and methods

2.1. Materials

This research was conducted in the laboratories of FSMP, UMS. The cassava roots and cassava leaves in the *Putih* varieties were bought at local market of Kota Kinabalu, Sabah. The *L. plantarum* was obtained from the Microbiology Laboratory, FSMP, UMS.

2.2. Sample preparation

The preparation of starter culture followed [4] with minor modifications. The culture from stock culture was transferred onto the MRS agar, followed by 37°C of incubation for 48 hrs to obtain the single colonies of bacteria. Then, it was kept at 4°C for further use. The culture was sub-culturing at two-week intervals to maintain the freshness. To prepare the starter culture, 18 mL of sterile MRS broth was used to culture the single colony bacteria with 37°C of incubation for 20 hrs.

The flour samples were prepared by following the methods of [4] with some modifications. The cassava roots were peeled and the cassava leaves were plucked from the stalk. The leaves were cut into small fragments. They were put in the plastic basin and washed cleanly. Then, they were soaked in 1 L water for 15 mins. After that, the water was drained and the chopper was used to slice the cassava roots into 1-2 mm thickness. The fermentation process was only applied on mocaf, by mixing well with 2 L water and the starter culture. The basin was covered with aluminium foil and placed at room temperature for fermentation. After that, the water was drained and the cabinet dryer with 45°C was used to dry the samples for 24 hrs. Then, they were ground followed by sieving (sieve shaker) to produce the flour with 250 µm particle size. Formulations were made based on combination percentage of cassava roots and cassava leaves in range of 95 - 100% and 0 - 5% respectively with fermentation time of 24 - 48 hrs. As in sequence, the control sample was consist of 100% cassava roots without fermentation, F1 (100% root; 24 hrs), F2 (97% root + 3% leaves; 24 hrs), F3 (95% root + 5% leaves; 24 hrs), F4 (100% root; 48 hrs), F5 (97% root + 3% leaves; 48 hrs) and F6 (95% root + 5% leaves; 48 hrs). After completed the fermentation process, all samples were packed in the zip-lock polyethylene pouch with one-sided aluminium foil and kept in the room temperature at the place without the direct sunlight.

2.3. Yield, proximate and physicochemical analysis of all the flour formulations

The yield of final product was calculated as the equation 1. In proximate analysis, it was conducted according to AOAC method [5]. In physicochemical analysis, the colour was determined by colorimeter (HunterLab ColorFlex). The total starch content analyzed by the method of [2]. The hydrogen cyanide (HCN) content analysis followed the method of [6] with minor modifications of using the ammonium chloride. The pH was measured by pH meter (Eutech Instruments pH 2700).

$$Yield\ of\ final\ product\ (\%) = \frac{\text{Final product weight}}{\text{Initial weight of ingredients}} x\ 100 \quad \text{-----------------} \quad (1)$$

2.4. Storage analysis on the best formulation

Storage analysis was conducted once every two weeks by using the accelerated method. The colour was analysed by colorimeter (HunterLab ColorFlex). The water activity was determined by water activity meter (Hygrolab). The microbiological test in terms of total plate count (TPC) via plate count agar (PCA) and yeast and mould count (YMC) via potato dextrose agar (PDA) was carried out.

2.5. Statistical analysis

The analysis was done in triplicate and the data was presented as mean ± standard deviation. The results were statistically analysed by SPSS Version 28 and the sample means were compared at a 95% confidence level ($p<0.05$) by using one-way ANOVA and Tukey's HSD post-hoc test.

3. Results and discussion

3.1. Yield of final product

All the mocaf demonstrated a significant reduction ($p<0.05$) of yield in the range of 32.58% - 36.51% compared to control (40.74%). The control and native mocaf indicated the fermentation process resulting in the reduction of yield by 10.38% to 20.02%. This was because the processing steps during the production of mocaf will naturally contribute to the loss of final weight of mocaf. For example, the removal of cassava peels contributes to the loss of about 15%, and the subsequent process of soaking, fermentation, pressing, drying, grinding and sieving will continuously reduce the yield of mocaf as the liquid waste is being drained. Furthermore, the addition of leaves to the mocaf reported no significant differences ($p>0.05$) of yield with their native mocaf (F1 and F4).

3.2. Proximate analysis

All the mocaf had the similar moisture content ($p>0.05$) ranging from 7.89% to 8.89%. The control and the native mocaf demonstrated that there was no significant difference ($p > 0.05$) of moisture content after the fermentation process. Furthermore, the addition of leaves into mocaf also did not have a significant impact ($p>0.05$) on the moisture content when compared to mocaf without leaves. All samples were fulfill the requirements of the SNI 7622:2011 for mocaf, in which the permitted maximum moisture content is 13% [7]. The ash content was significantly reduced from 0.82% (control) to 0.49% (F1) and further reduced to 0.43% (F4) after 48 hrs of fermentation. The loss of ash content could result from the leaching of soluble minerals into the liquid medium during the production process [8]. Moreover, all the mocaf with added leaves indicated evident rise ($p < 0.05$) in ash content by the range of 76.74% to 134.88% compared to samples without leaves, as the cassava leaves are rich in minerals compared to cassava roots. The results of proximate analysis including moisture, ash, crude protein, crude fat, crude fiber and carbohydrate content were shown in Table 1.

In this research, all the mocaf met the SNI requirement with no exceed maximum ash content of 1.5% [7]. The fermentation had significantly improved ($p < 0.05$) the crude protein content of control to F4 by 40.54% after 48 hrs of fermentation. Similar results were obtained by [9] as the protein content of mocaf was increased by 110.88% after 48 hrs of fermentation. In addition, the study by [10] demonstrated an increase of 34.91% in the protein content of cassava leaves itself after 48 hrs of *L. plantarum* fermentation. Moreover, [11] also reported that *L. plantarum* had the antifungal properties in which the participated antipatogen compounds were a kind of proteins, and *L. plantarum* also helped the release of peptidoglycan compounds (lipoprotein and glycoprotein) that found in the cell walls of cassava leaves.

The crude fat content had significantly decreased by 28.38%, from 0.74% (control) to 0.53% (F4), due to the lipase from *L. plantarum* breaking down the fat components in cassava flour. [12] also recorded the decrease of crude fat content in mocaf, compared to unfermented cassava flour.

Table 1. Proximate composition including moisture, ash, crude protein, crude fat, crude fiber and carbohydrate content of control and mocaf samples.

Sample	Composition (%)					
	Moisture	Ash	Crude protein	Crude fat	Crude fiber	Carbohydrate
Control	9.45 ± 0.20^{a}	0.82 ±0.13^{ab}	1.11 ± 0.11^{e}	0.74 ± 0.09^{a}	2.03 ± 0.02^{c}	85.84 ± 0.23^{bc}
F1	8.89 ± 0.23^{ab}	0.49 ±0.04^{c}	1.38 ± 0.06^{de}	0.59 ± 0.06^{ab}	1.74 ± 0.03^{e}	86.92 ± 0.25^{ab}
F2	8.30 ± 0.53^{b}	0.77 ± 0.06^{ab}	1.76 ± 0.08^{cd}	0.73 ± 0.05^{a}	2.12 ± 0.03^{b}	86.30 ± 0.61^{ab}
F3	8.54 ± 0.50^{ab}	1.01 ± 0.05^{a}	2.38 ± 0.29^{b}	0.75 ± 0.10^{a}	2.37 ± 0.02^{a}	84.98 ± 0.33^{c}
F4	8.74 ± 0.58^{ab}	0.43 ± 0.08^{c}	1.56 ± 0.07^{cd}	0.53 ± 0.08^{b}	1.66 ± 0.03^{f}	87.08 ± 0.49^{a}
F5	8.30 ± 0.28^{b}	0.76 ± 0.06^{b}	1.83 ± 0.04^{c}	0.6 ± 0.05^{ab}	1.94 ± 0.03^{d}	86.56 ± 0.32^{ab}
F6	7.89 ± 0.24^{b}	0.88 ± 0.15^{ab}	3.20 ± 0.15^{a}	0.66 ± 0.04^{ab}	2.34 ± 0.03^{a}	85.02 ± 0.53^{c}

*Mean score with different superscripts on the same column indicates there is significant difference ($p < 0.05$)

Furthermore, the addition of leaves into mocaf did not demonstrate the significant difference of crude fat content compared to native mocaf. Generally, the low fat content is preferred as it could decrease the risk of rancidity and preserve the quality of flour. All samples had significant differences ($p<0.05$) in crude fiber content. The crude fiber content in F1 and F4 recorded a decrease by 14.29% to 18.23% from the control. The study by [13] demonstrated that the amylolytic activity of *L. plantarum* was 20 U/mL and cellulolytic activity was 12 U/mL after 18 hrs of fermentation, and the crude fiber content of cassava flour was decreased by 45.14%. The addition of cassava leaves significantly increased ($p<0.05$) the crude fiber content of F1 and F4 as the leaves contained high fiber. Basically, the native mocaf fulfilled SNI requirements of maximum 2.0% crude fiber content, while mocaf with leaves did not meet the requirements. The carbohydrate content significantly increased from 85.84% (control) to 87.08% (F4) after 48 hrs of fermentation. This may be due to the loss of ash, resulting in relatively higher carbohydrates remained in the mocaf. The study of [14] also reported an increase of carbohydrates content from 84.91% to 90.86% after the fermentation. Furthermore, the F3 and F6 had significant decreased ($p<0.05$) of carbohydrate compared to their respective native mocaf, as the cassava leaves contained lower carbohydrate than the cassava roots. Overall, the carbohydrate content of mocaf F1 to F6 was 84.98% to 87.08%. The similar carbohydrate content was recorded in study of [8] in which mocaf had 81.00% to 87.08% carbohydrate, depending on the varieties.

3.3. Physicochemical analysis

The results of physicochemical analysis included colour property, total starch, hydrogen cyanide and pH were shown in Table 2.

3.3.1. Colour. Both native mocaf did not possess significant changes ($p > 0.05$) in whiteness with control, reporting the whiteness of 93.95 to 95.27. Similar results was obtained from [15], with the whiteness of 93.65 to 94.69. The addition of leaves substantially reduced ($p<0.05$) the whiteness of mocaf. The lowest whiteness was 88.81 (F3) and 88.91 (F6). All the mocaf samples meet the SNI minimum 87 degree of whiteness [7]. In terms of a* value, F1 and F4 had significant increase ($p<0.05$) of redness from 0.22 to 0.70 after the fermentation. This may due to the increase of sugars during fermentation, which may participate in non-enzymatic browning reaction as the protein compounds also presented in mocaf [14]. Furthermore, the mocaf with leaves had decreased a* value due to the presence of chlorophyll pigments, masking the changes of a* value, which recorded a* value in the range of -0.23 to -0.07. F1 and F4 demonstrated a significant decrease in b* value ($p < 0.05$) compared to control. This may be due to the degradation of carotenoids during fermentation, thus reducing the yellowness of mocaf [14]. The addition of leaves did not significantly affect the b* value of mocaf.

Table 2. Physicochemical properties including colour property, total starch, hydrogen cyanide and pH of control and mocaf samples.

Sample	Physicochemical analysis					
	Colour			Total starch content (%)	Hydrogen cyanide content (mg/kg)	pH
	L*	a*	b*			
Control	93.95 ± 0.08[a]	0.22 ± 0.05[c]	9.60 ± 0.60[a]	73.91 ± 2.08[a]	3.60 ± 0.16[a]	5.74 ± 0.13[a]
F1	94.87 ± 0.19[a]	0.59 ± 0.08[a]	6.99 ± 0.46[b]	57.14 ± 2.64[b]	2.70 ± 0.22[bc]	3.91 ± 0.07[b]
F2	90.16 ± 0.56[bc]	-0.13 ± 0.01[d]	7.94 ± 0.40[ab]	44.28 ± 2.30[c]	2.74 ± 0.16[bc]	3.67 ± 0.06[c]
F3	88.81 ± 1.02[c]	-0.23 ± 0.04[d]	8.15 ± 1.17[ab]	35.76 ± 1.47[de]	2.88 ± 0.16[b]	3.61 ± 0.07[cd]
F4	95.27 ± 0.20[a]	0.70 ± 0.05[a]	6.74 ± 0.03[b]	39.97 ± 2.96[cd]	2.38 ± 0.22[c]	3.66 ± 0.02[c]
F5	91.66 ± 0.89[b]	-0.03 ± 0.01[d]	6.84 ± 0.76[b]	33.58 ± 2.38[ef]	2.59 ± 0.11[bc]	3.62 ± 0.06[cd]
F6	88.91 ± 1.50[c]	-0.07 ± 0.04[d]	6.98 ± 0.70[b]	28.80 ± 1.04[f]	2.74 ± 0.06[bc]	3.43 ± 0.06[d]

[*]Mean score with different superscripts on the same column indicates there is significant difference ($p < 0.05$)

3.3.2. Total starch content. The total starch content decreased significantly from 73.91% (control) to 57.14% (F1) and further decreased to 39.97% (F4) after 48 hrs of fermentation. [8] stated that the fermentation may break down the starch into less complex sugar so the *L. plantarum* can use the sugar as energy to carry out the metabolism activity. Furthermore, the addition of leaves into mocaf contributed decrease of total starch content, with the lowest total starch content was 28.80% in F6.

3.3.3. Hydrogen cyanide content (HCN). The HCN content decreased significantly from 3.60 mg/kg (control) to 2.70 mg/kg (F1) and further decreased to 2.38 mg/kg (F4) after 48 hrs of fermentation. This was due to the activity of linamarase produced by *L. plantarum* that hydrolysed and released the cyanide compounds from the cassava flour. In addition, [8] also demonstrated that the processing steps such as washing, drying and fermentation contributed to the great reduction of cyanide content. The addition of leaves into mocaf possessed slightly higher but not significantly different ($p>0.05$) HCN content. According to SNI 7622:2011 (mocaf) and Codex Stan 176-1989 (edible cassava flour), the HCN content shall not exceed 10 mg/kg [7][16]. In this current research, all the samples fulfilled the requirement of standard.

3.3.4. pH. The pH had dropped significantly from 5.74 (control) to 3.91 (F1) and further dropped to 3.66 (F4) after 48 hrs of fermentation. This may due to the production and accumulation of lactic acid as a result of fermenting carbohydrate compounds by *L. plantarum*. The study from [9] also indicated that the pH decreased with time from 5.6 to 3.9 after the fermentation. The mocaf with leaves recorded significant differences ($p<0.05$) of pH from their respective native mocaf, due to the lower pH of cassava leaves (6.5-7) compared to cassava roots (7-8) [17].

3.4. Storage analysis

The storage analysis was carried out on the best formulation, which was F6 that recorded the highest protein value. The results of storage analysis including colour, water activity (a_w) and microbiological test were shown in Table 3. The longer the storage period indicated the colour significantly changes over the month of storage. The L* value was decreased from 89.36 to 88.60, due to the formation of darker compounds as a result of the browning effect during storage. The a* value was increased from -0.13 to 0.19, due to the degradation of chlorophyll pigments while the b* value was decreased from 8.24 to 7.66, because of the degradation of carotenoids as the temperature and oxygen were the key factors that contributed to the depletion of carotenoids [14].

Table 3. Colour property, water activity (a_w) and microbial status of best mocaf sample over 8 months storage period.

Storage period (month)	Storage analysis					
	Colour			Water activity (a_w)	Microbiological test	
	L*	a*	b*		Total plate count (CFU/g)	Yeast and mould count (CFU/g)
0	89.36 ± 0.17^a	-0.13 ± 0.04^b	8.24 ± 0.05^a	0.31 ± 0.01^e	TFTC	TFTC
2	89.21 ± 0.13^a	-0.11 ± 0.04^b	8.22 ± 0.04^{ab}	0.36 ± 0.01^d	TFTC	TFTC
4	89.17 ± 0.28^{ab}	-0.04 ± 0.06^b	8.13 ± 0.05^{ab}	0.39 ± 0.01^c	3.0×10^2	TFTC
6	88.97 ± 0.19^{ab}	0.11 ± 0.03^a	8.01 ± 0.10^b	0.43 ± 0.01^b	7.2×10^2	TFTC
8	88.60 ± 0.27^b	0.19 ± 0.05^a	7.66 ± 0.12^c	0.49 ± 0.01^a	4.5×10^3	TFTC

[*]Mean score with different superscripts on the same column indicates there is significant difference *($p < 0.05$)*

The water activity (a_w) of F6 was notably increased ($p < 0.05$) throughout the storage period, from 0.31 to 0.49. The study from [18] reported similar results on the increase of water activity in mocaf that packaged in low density polyethylene plastic with the increase of storage period. The water activity (a_w) of below 0.6 is preferable, as this indicates that there will be no microbial proliferation since microbes generally will only grow at the water activity (a_w) of at least 0.6. The TPC was recorded from too few to count (TFTC) to 4.5×10^3 CFU/g , while YMC was recorded TFTC over the eight months of storage, which indicated that the low water activity is not favourable for the growth of yeast and mould. [18] obtained the similar results with no significant difference in the amount of colonies during the storage. According to SNI 7622:2011 for mocaf, the maximum of TPC is 10^6 CFU/g while the maximum of YMC is 10^4 CFU/g. Therefore, the mocaf in the current research was safe to consume.

4. Conclusion

In conclusion, the best formulation was 95% cassava roots with the addition of 5% cassava leaves fermented for 48 hrs with significantly increase the protein content. It also had higher ash and crude fiber content compared to native mocaf. In physicochemical properties, the outcomes of lower starch content and lower pH were obtained. Besides that, the whiteness and HCN content of F6 also met the requirement. The sample has recorded the increase of water activity, changes in colour and increase of microbial load during storage, but it was within the range of food safety. For future research, it is suggested to cooperate with local farmers to get the fresh and good quality cassava roots and leaves, and also analyse the composition of amino acids and carbohydrate compounds to determine the bioavailability of these nutrients for digestion.

References

[1] International Atomic Energy Agency (IAEA) 2018 *Cassava Production Guidelines for Food Security and Adaptation to Climate Change in Asia and Africa* (Vienna, Austria: IAEA)
[2] Nilusha R A T, Jayasinghe J M J K, Perera O D A N, Perera P I P and Jayasinghe C V L 2021 *International Journal Food Science.* **2021** 1-13
[3] Leguizamon A J, Rompato K M, Hoyos R E and Audisio M C 2021 *Journal of Food Composition and Analysis.* **101** 103986
[4] Seveline, Heldyana R and Kurniawati S 2020 *Ind. J. Technol. Mgmt. Agroindustry.* **9** 163-72
[5] AOAC 2000 *Official Methods of Analysis* (Gaithersburg, MD: Association of Official Analytical Communities)
[6] Moriasi G A, Olela B O, Waiganjo B W, Wakori E W T and Onyancha J M 2017 *J. Food Nutr. Res.* **5** 817-23
[7] *Standard Nasional Indonesia (SNI) 7622:2011 - Mocaf flour*
[8] Lambebo T and Deme T 2022 Evaluation of Nutritional Potential and Effect of Processing on Improving

Nutrient Content of Cassava (*Mannihot esculenta crantz*) Root and Leaves.
[9] Gunawan S, Widjaja T, Zullaikah S, Ernawati L, Istianah N, Aparamarta H W and Prasetyoko D 2015 *International Food Research Journal.* **22** 1280-87
[10] Terefe Z K, Omwamba M and Nduko J M 2022 *Journal of Food Safety.* **42** 1-12
[11] Eko Trapsilo P A, Martinah Hariati A and Dwi Sulistiati T 2020 *JIPK.* **12** 324
[12] Sulistyo, J., Lee, J. S., Mamat, H. & Abdul Wahab, N. 2016. *AIP Conference Proceedings.* 1744.
[13] Frendiansyah A and Kurniadi M 2016 *Nusantara Bioscience.* **8** 207-14
[14] Odey G N and Lee W Y 2019 *International Journal of Food Science and Technology.* **5** 813-22
[15] Chisenga S M, Workneh T S, Bultosa G and Alimi B A 2019 *Journal of Food Science and Nutrition.* **56** 2799-813
[16] Codex Alimentarius Commission 1995 *CODEX STAN 176-1989 Edible Cassava Flour* Rev 1–1995 (USA: Codex Alimentarius Commission)
[17] Kobawila S C, Louembe D, Keleke S and Hounhouigan D J 2005 *African Journal of Biotechnology.* **4** 689-96
[18] Lestari D, Liandro Y, Ang E and Kresnowati P 2019 *Journal of Engineering and Technological Sciences.* **51** 64

Development of Food Seasoning using Sub-standard Fish Cracker, Pumpkin (*Cucurbita moschata*) and Seaweed (*Kappaphycus alvarezii*)

J M Hoy[1], **M A Hamid**[1,2*], **M S Ayub**[3], **R Suratman**[4], **M M Effendi**[5] **and A In`am**[5]

[1] Faculty of Food Science and Nutrition, Universiti Malaysia Sabah, Jalan UMS,Kota Kinabalu 88400, Sabah, Malaysia
[2] Food Security Research Laboratory, Faculty of Food Science and Nutrition, Universiti Malaysia Sabah, Jalan UMS 88400 Kota Kinabalu, Sabah, Malaysia
[3] Faculty of Management, Economy & Accountancy, Universiti Malaysia Sabah, Jalan UMS 88400 Kota Kinabalu, Sabah, Malaysia
[4] Rosmie Snack Industry Sdn Bhd, TB19643, 1.5 Storey Light Industrial Building, Taman Unipark, 91000 Tawau, Sabah, Malaysia
[5] Department of Mathematics Education, Universitas Muhammadiyah Malang, East Java 65145 Indonesia

*Email : chot@ums.edu.my

Abstract. The study focuses on sub-standard fish crackers (amplang) that do not meet quality standards but are sold at reduced prices which affecting business profitability. It aims to create a seasoning powder using sub-standard fish crackers, pumpkin (*C. moschata*), and seaweed (*K. alvarezii*). Six formulations with varying proportions of sub-standard fish cracker (55% and 65%), pumpkin powder (20%) and seaweed powder in proportions of 1.5%, 3%, and 4.5% were developed and evaluated for physicochemical properties and sensory acceptance. 65% addition of sub-standard fish cracker to the seasoning powder produced a brighter colour. Formulation (F5) with 65% sub-standard fish cracker and 3% seaweed powder obtained the highest acceptance. F5 exhibited improved nutritional quality with higher ash, protein and dietary fibre content compared to control. The seaweed powder also helped to improve the seasoning powder's mineral profile, which is low in sodium and high in potassium. An eight-week storage study on F5 showed stable properties throughout, including water activity, colour, peroxide value, free fatty acid and microbiological status. Overall, the study offers a strategic approach to enhance the utilization of sub-standard fish crackers, improving acceptability and nutritional value for consumers.

1. Introduction

Seasoning powders, known for enhancing the taste and flavour of various foods, play a vital role in culinary experiences. They transform ordinary dishes like porridge and instant noodles, which may lack flavour into enjoyable meals. This study used fish crackers, known as amplang, as the primary ingredient for food seasoning powder. Amplang was originally popular in East and North Kalimantan, but has also gained popularity in parts of Sabah, Malaysia. Food seasoning powders found in the market primarily concentrate on improving on its taste and deficient in certain nutrients, particularly fibre. Pumpkin was

Published under licence by IOP Publishing Ltd

incorporated in this research to enhance the seasoning powder's quality. Pumpkin is a widely consumed vegetable in many countries, and is typically rich in dietary fibre, vitamin C, minerals, and carotenoids. It has several health advantages, such as anti-inflammatory, antioxidant, anticancer, antiangiogenic, antilipogenic and glucose-reducing properties [20]. There were strategies have been explored to reduce salt content that includes the use of naturally salty-tasting ingredients like seaweed to develop seasoning powder that align with nutritional guidelines. It is well known that seaweeds contain high levels of vitamins, minerals, dietary fibre and trace elements like iodine. [21]. They are also notable for their umami flavour, which not only enhances taste but also helps reduce salt intake, an important factor for maintaining a healthy diet [21]. This study addresses these issues by exploring new markets for sub-standard fish crackers, aiming to enhance their value through transformation into seasoning powder. The inclusion of pumpkin aims to boost its commercial appeal, while seaweed powder offers a healthier alternative with lower sodium content.

2. Materials and Methods

Sub-standard fish crackers were supplied by Rosmie Snack Industry Sdn. Bhd., Tawau, Sabah. Pumpkins (*Cucurbita moschata*) and fresh *Kappaphycus alvarezii* were bought from the local market in Kota Kinabalu, Sabah, Malaysia. Salt, garlic powder, onion powder and black pepper powder were locally purchased. Pumpkin powder was prepared using [1] while seaweed powder by [2] method. The basic formulation of the seasoning powder consisted of 6% salt, 2% garlic powder, 2% onion powder and 2% black pepper powder. Two independent variables were the sub-standard fish cracker (SFC) and seaweed powder (SP); which were tested at two and three levels respectively (SFC: 55% and 65%; SP: 1.5%,3.5% and 4.5%) with 20% pumpkin powder.

Colour of seasoning powder were measured with Hunter Lab colorimeter. Hedonic test was performed using 9-point scale. Proximate analysis and total dietary fibre determined by [3]. Mineral quantification of sodium and potassium was performed according [4]. Accelerated shelf-life test of seasoning powder was conducted described by [5] for 8 weeks. Freshly produced seasoning powder were packed in aluminium laminated zipper bag and kept in incubator (45°C). Water Activity was measured using water activity meter. Peroxide value and free fatty acid were measured according [6] and [7] respectively. Microbiological test was conducted to determine yeast and mould count. One way ANOVA, post hoc Tukey's Honestly Significant Difference (HSD) test and independent t-test with a 95% confidence interval was used to analyze data.

3. Results and Discussion

3.1. Colour

The values of seasoning powder samples (F1 to F6) had significantly different ($p < 0.05$) from the control as shown in table 1. Frying process of the fish cracker causes lighter in colour due to high temperature that denatures and oxidizes fish protein [8]. This explains why fish crackers are light in colour, as higher levels of sub-standard fish crackers (F4, F5, F6) result in greater L* values than those with lower levels (F1, F2, F3). The presence of beta-carotene in pumpkin increased the a* and b* values [9].

Table 1. Colour profile (L*, a* and b* values) of the seasoning powder.

Formulation	Colour		
	L^*	a^*	b^*
Control (F0)	74.06 ± 0.11^a	3.49 ± 0.053^d	21.05 ± 0.34^d
F1	55.65 ± 0.17^d	10.85 ± 0.59^{bc}	57.69 ± 0.05^a
F2	55.50 ± 0.47^d	11.64 ± 0.07^{ab}	56.37 ± 0.03^b
F3	55.47 ± 0.15^d	10.83 ± 0.14^{bc}	56.54 ± 0.03^b
F4	57.66 ± 0.17^{bc}	11.79 ± 0.13^a	55.48 ± 0.12^c
F5	57.88 ± 0.31^c	10.79 ± 0.13^{bc}	55.83 ± 0.49^{bc}
F6	57.72 ± 0.24^b	10.68 ± 0.51^c	55.81 ± 0.38^{bc}

*Mean score with different superscripts on the same column indicates there is significant difference ($p<0.05$)

3.2. Sensory evaluation

The sensory data of the seasoning powder was illustrated in table 2. Significant changes (p>0.05) were observed for all attributes, except for colour and aroma. The unnoticeable colour difference among the formulations was due to the pumpkin powder added in a fixed percentage (20%). All the six formulations achieved a mean liking score of seven on a nine-point scale, indicating that its colour quality is acceptable to consumers. Similarity in the aroma of all samples where the aroma of the formulated seasoning powders was mainly contributed by the spices. It is generally anticipated that seaweed will give the finished products a distinctly fishy scent which is linked to the presence of amines and pyridines that contribute to marine, fatty, and seafood scents [23]. However, the strong and distinct aroma of the spices overpowered the aroma of seaweed as seaweed was used in a relatively low proportion (1.5 - 4.5%) and thus its aroma was believed to be masked by the other ingredients used.

Table 2. Sensory evaluation attributes of food seasoning powder samples

Formulations	Attributes				
	Colour	Aroma	Taste	Saltiness	Overall acceptance
F1	7.58 ± 1.30^a	6.70 ± 1.42^a	6.60 ± 1.34^b	6.32 ± 1.44^b	6.80 ± 1.16^b
F2	7.56 ± 1.39^a	6.88 ± 1.29^a	6.64 ± 1.70^b	6.62 ± 1.21^b	6.78 ± 1.35^b
F3	7.52 ± 1.33^a	6.72 ± 1.54^a	5.64 ± 1.61^c	5.40 ± 1.67^c	6.14 ± 1.44^b
F4	7.53 ± 1.36^a	6.64 ± 1.61^a	6.58 ± 1.64^b	6.34 ± 1.59^b	6.84 ± 1.39^b
F5	7.59 ± 1.32^a	6.90 ± 1.06^a	7.52 ± 0.91^a	7.48 ± 0.81^a	7.66 ± 1.04^a
F6	7.56 ± 1.31^a	6.66 ± 1.69^a	5.68 ± 1.25^c	5.46 ± 1.56^c	6.10 ± 1.45^b

*Mean score with different superscripts on the same column indicates there is significant difference (p<0.05)

F5 was statistically different (p<0.05) from formulation 1, 2, 3, 4, 5 and 6. Both formulations 3 and 6 recorded the lowest scores in terms of taste, owing to the highest incorporation percentage of seaweed powder (4.5%). Seaweeds have an uncommon taste that may be less accepted by consumers, as dimethylsulfide released from their decomposition interacts with tongue's taste receptors, resulting in a fishy taste [10]. F5 received the highest mean score from the aspect of saltiness whereas the lowest mean score was seen in formulation 3. The ratio of salt and seaweed play an important role in giving the saltiness of the formulations. According to the research done by [11], soup with ratio 1:1 of salt to seaweed powder was the most acceptable and resembled to soup with ordinary salt. Based on all attributes, sample F5 exhibited significant differences compared to the other samples and was selected as the best formulation.

3.3. Proximate composition

The best formulation of seasoning powder, F5 (65% sub-standard fish cracker + 3% seaweed powder) was used to determine proximate composition and compared with control sample, F0. The proximate composition of F5 and F0 were shown in table 3. Significant changes (p < 0.05) were observed in table 3 for all nutritional components between F5 and control. The increased protein content in F5 compared to F0 is likely due to the sub-standard fish cracker, which is made from fish. F5 had a significantly higher crude fat content (p<0.05) than the control, most likely as a result due to the deep-frying process which elevates fat content in the product [12]. The dietary fibre content in F5 was significantly higher compared to control, aligning with study conducted by [13] where adding seaweed and pumpkin powder increases dietary fibre in seasoning powders.

Table 3. Proximate composition, dietary fibre and energy content of F5 and control sample (F0)

Composition	F5	F0
Moisture content (%)	5.63 ± 0.04^{a}	3.89 ± 0.08^{b}
Ash content (%)	3.36 ± 0.06^{a}	1.43 ± 0.08^{b}
Crude protein (%)	12.29 ± 0.17^{a}	1.03 ± 0.07^{b}
Crude fat (%)	22.20 ± 0.12^{a}	1.20 ± 0.16^{b}
Crude fibre (%)	3.06 ± 0.22^{a}	0.41 ± 0.06^{b}
Total carbohydrate (%)	53.46 ± 0.52^{b}	92.04 ± 0.42^{a}
Total dietary fibre (%)	10.89 ± 0.42^{a}	1.52 ± 0.30^{b}
Energy content (kcal)	462.80 ± 2.13^{a}	383.08 ± 1.01^{b}

*Mean score with different superscripts on the same row indicates there is significant difference (p<0.05)

3.4. Mineral analysis

There was significant difference as illustrated in table 4 on sodium and potassium content in F5 and control. The sodium content in both formulations were mainly contributed by salt added. The salt added into control (6%) was 2 times greater than that of in F5 (3%). The high mineral composition of seaweeds may present a promising alternative to sodium in food. Therefore, inclusion of seaweed powder in F5 had contributed to a higher potassium content. The World Health Organization (WHO) suggested that a Na/K ratio of less than 1.0 is ideal for maintaining cardiovascular health.

Table 4. Sodium and potassium composition of F5 and control (F0) seasoning powder

Element	F5	F0
Sodium, Na (mg/100g)	140.02 ± 0.61^{b}	157.70 ± 0.63^{a}
Potassium, K (mg/100g)	153.21 ± 0.49^{a}	113.35 ± 0.59^{b}
Ratio Na/K	0.91	1.39

*Mean score with different superscripts on the same row indicates there is significant difference (p < 0.05)

3.5. Storage study

3.5.1. Water activity (a_w).

A significant difference ($p < 0.05$) of in every two weeks of storage was depicted in table 5. The result can be attributed to the elevated storage temperature (45°C). Elevated temperature often relates to the increase of water activity (a_w). The hydrogen bonds that hold water and hydrophilic molecules together are broken by high temperature, increases the amount of free water molecules and water activity [14].

Table 5. Water activity (a_w) of F5 throughout 8 weeks storage period

Storage week	Water activity (a_w)
0	0.43 ± 0.01^{e}
2	0.45 ± 0.01^{d}
4	0.47 ± 0.01^{c}
6	0.48 ± 0.01^{b}
8	0.50 ± 0.01^{a}

*Mean score with different superscripts on the same column indicates there is significant difference (p < 0.05).

3.5.2. Colour.

There was a significant change ($p < 0.05$) of brightness, redness, and yellowness of F5 as shown in table 6. The reduced brightness, redness, and yellowness of the seasoning powder may be due to carotenoids degradation likely caused by the oxidation of these pigments during storage, influenced by light, oxygen, water activity, pH and temperature [16]. Additionally, non-enzymatic browning reactions may occur alongside with oxidation and isomerization of beta-carotene [15].

IOP Conf. Series: Earth and Environmental Science **1446** (2025) 012031 doi:10.1088/1755-1315/1446/1/012031

Table 6. Colour of F5 throughout 8 weeks storage period

Storage week	Colour		
	L*	a*	b*
0	56.66 ± 0.48^{a}	11.16 ± 0.70^{a}	53.95 ± 0.07^{a}
2	54.32 ± 0.24^{b}	10.77 ± 0.07^{a}	51.30 ± 0.05^{b}
4	52.90 ± 0.22^{c}	10.61 ± 0.05^{a}	50.13 ± 0.10^{c}
6	52.60 ± 0.67^{c}	10.50 ± 0.04^{a}	48.65 ± 0.03^{d}
8	52.04 ± 0.10^{c}	10.48 ± 0.08^{a}	48.44 ± 0.20^{d}

*Mean score with different superscripts on the same column indicates there is significant difference ($p < 0.05$)

3.5.3. Peroxide value (PV).

A significant difference ($p < 0.05$) in peroxide value (PV) is observed, as shown in Table 7. The rise in PV is caused by the production of hydroperoxides from the oxidation of carbon atoms close to double bonds in the triacylglycerol structure [17]. A high peroxide value (PV) can lead to off-flavours and decreased nutritional value in food products, however, the seasoning sample remained in good condition after 8 weeks of storage, with a PV not exceeding 10 meq/kg, which is within the acceptable range [18].

Table 7. Peroxide value of F5 throughout 8 weeks storage period

Storage week	Peroxide value (meq/kg)
0	1.02 ± 0.07^{c}
2	1.30 ± 0.03^{b}
4	1.35 ± 0.05^{b}
6	1.39 ± 0.07^{ab}
8	1.50 ± 0.02^{a}

*Mean score with different superscripts on the same column indicates there is significant difference ($p < 0.05$)

3.5.4. Free fatty acid (FFA).

The free fatty acid result of F5 was shown in table 8. There was a significant difference ($p < 0.05$) in FFA contents throughout the study. The low increase of FFA can be attributed to the shielding effect by aluminium laminated packaging that hindered light from passing through and causing the product to become rancid since the seasoning powder was packed in an aluminium laminated zip lock bag and exposure to light was very minimal in the enclosed environment of incubator [19].

Table 8. Free fatty acid of F5 throughout 8 weeks storage period

Storage week	Free Fatty Acid (%)
0	0.48 ± 0.03^{d}
2	1.06 ± 0.04^{c}
4	1.23 ± 0.04^{b}
6	1.33 ± 0.04^{b}
8	1.57 ± 0.04^{a}

*Mean score with different superscripts on the same column indicates there is significant difference ($p < 0.05$)

3.5.5. Microbiological study.

Table 9 shows the microbial count from the microbiological analysis conducted on 8 weeks. According to the obtained results, the seasoning powder in this research remained microbiologically safe for consumption for up to 8 weeks of storage as the microbial counts stayed within the acceptable maximum range (10^5 CFU/g) [20].

Table 9. Colony count of F5 throughout 8 weeks storage period

Storage week	Total yeast and mould (CFU/g)
0	1.12×10^{1}
2	1.78×10^{2}
4	3.68×10^{2}
6	3.91×10^{2}
8	4.01×10^{2}

*Mean score with different superscripts on the same column indicates there is significant difference ($p<0.05$)

4. Conclusion

Seasoning powder with sub-standard fish cracker, pumpkin powder and seaweed powder were formulated and their sensory characteristics, colour were investigated and nutritional composition of best formulation was compared with control seasoning powder. Shelf-life stability was conducted for 8 weeks. Seasoning powders developed with 65% sub-standard fish cracker with 3% seaweed powder were found to be organoleptically acceptable. Addition of seaweed to seasoning powder can provide health benefits by enhancing its dietary fibre and mineral content. Furthermore, the seasoning powder sample remained stable during storage due to its low water activity and within safe microbial limits making it safe for consumption. Food seasoning is a good alternative for improving sub-standard fish crackers and reducing their under utilization. Seaweed powder could be used as salt-reduction ingredient to develop lower sodium content food seasoning with enhanced nutritional properties and achieve a sustainable food supply chain.

References

[1] Hussain A 2023. *Pure and Applied Biology* **12** 48-60
[2] Munsu E, Mohd Zaini H, Matanjun P, Ab Wahab N, Sulaiman NS and Pindi W 2021. *Applied Sciences* **11** 11347
[3] AOAC 2000 Association of Official Analytical Chemists 17th Edition, AOAC, Washington DC.
[4] Kılıç Altun S, Dinç H, Paksoy N, Temamoğulları F K, and Savrunlu M 2017 *International Journal of Analytical Chemistry* 6391454
[5] Ng H K, Hamid M A, Mamat H, Akanda J H and Ahmad F 2017 *1st International Conference on Social, Applied Science and Technology in Home Economics (ICONHOMECS 2017)*
[6] Pidatala P K, Bellmer D and McGlynn W 2021 *International Journal of Food Science* 5528315
[7] Hashem A M, Sakhare S D and Kudre T G 2023. *Biocatalysis and Agricultural Biotechnology* 51 102761
[8] Ibadullah W Z W, Idris A A, Mustapha N A, Shukri R, Saari N and Abedin N H Z 2019 *Current Research in Nutrition and Food Science Journal* **7** 369-381
[9] Ghendov-Mosanu A, Netreba N, Balan G, Cojocari D, Boestean O, Bulgaru V, Gurev A, Popescu L, Deseatnicova O, Resitca V, Socaciu G, Pintea A, Sanikidze T and Sturza R 2023 *Foods* **12** 3907
[10] Neta M T S and Narain N 2018 *Examines Marine Biology & Oceanography* **2** 195–201
[11] Notowidjojo L, Ascobat P, Bardosono S and Anggadiredja J T 2020 *World Nutrition Journal* **5** 47
[12] Lumanlan J C 2019. Western Sydney University, Australia Master of Thesis
[13] Huang M and Yang H 2019. *LWT* **110** 262–268
[14] Syamaladevi R M, Tadapaneni R K, Xu J, Rojas R V, Tang J, Carter B, Sablani S and Marks B 2016. *Food Research International* **81** 163–170
[15] Zhou C L, Liu W, Zhao J, Yuan C, Song Y, Chen D and Li Q H 2013. *Innovative Food Science & Emerging Technologies* **21** 24–34
[16] Sahni P and Shere D M 2018. *Asian Journal of Dairy and Food Research* **37** 202-211
[17] Emelike N J T, Ujong A, and Achinewhu S C 2020. *Agriculture and Food Sciences Research* **7** 89-96
[18] *Codex Alimentarius International Food Standard* 2011 World Health Organization (WHO)
[19] Setyawan N, Dewandari K T and Prabawati S 2023. *IOP Conference Series: Earth and Environmental Science* **1172** 012056
[20] FDA 2022 BAM Chapter 18: Yeasts, Molds and Mycotoxins [online] https://www.fda.gov/food/laboratory-methods-food/bam-chapter-18-yeasts-molds-and-mycotoxins
[21] Kaur S. Panghal A, Garg M K, Mann S, Khatkar S K, Sharma P and Chhikar, N 2019. *Nutrition & Food Science* 50(**2**) 384-401
[22] Milinovic J, Mata P, Diniz M S and Noronha J P 2021. *International Journal of Gastronomy and Food Science* **23** 100301
[23] Vilar E G, O'Sullivan M G, Kerry J P and Kilcawley K N 2020. *Algal Research-Biomass Biofuels and Bioproducts* **45**: 101740.

IOP Conf. Series: Earth and Environmental Science **1446** (2025) 012032 doi:10.1088/1755-1315/1446/1/012032

Electric agriculture hand tractor conversion kits: a technological design, performance analysis, and innovative conversion techniques

A Budiprasojo[1], D A Tyagita[1], A Rofii[2]

[1]Automotive Engineering Study Program, Engineering Department, Politeknik Negeri Jember
[2]Mechatronics Engineering Technology Study Program, Engineering Department, Politeknik Negeri Jember

*E-mail: azamataufiq@polije.ac.id

Abstract. The shift to electric tractors is vital for sustainable agriculture, yet developing efficient conversion kits for two-wheeled tractors presents technical challenges in battery technology, motor performance, and system integration. This study aims to synthesize recent advancements in electric tractor conversion kits, focusing on technological designs, performance metrics, and innovative conversion methods using recent literature, deployed based on required formulas. Key parameters evaluated include battery energy density, motor torque, and efficiency of power control systems. High-capacity lithium iron phosphate (LiFePO4) 72V 100Ah batteries offer energy densities up to 200 Wh/kg and rapid charging rates of 1C. Electric motors max power 5kW (6,7 HP). Sustainable for agricultural applications deliver up to max 150 Nm or 19 Nm of torque at low speeds, with power densities of 3 kW/kg. Advanced power control systems employ adaptive algorithms, achieving 80% efficiency in dynamic power distribution. Electric tractor conversion kits demonstrate substantial improvements in efficiency and performance. The integration of high-performance batteries, robust motors, and sophisticated power control systems offers a viable path toward enhanced sustainability in agriculture. The adaptive power control system optimized performance across various operational conditions.

1. Introduction

Changes has occurred in agricultural with the development of agricultural machinery into various forms. The widespread use of agricultural machinery has led to increased reliance on internal combustion engine vehicles, resulting in increasing of air pollution which becomes environmental damage. Research Statistics indicate that global reserves of fuel oil expected to last for another hundred years, respectively. As global environmental awareness rises, many sector are actively promoting the electric vehicle industry, achieving significant progress in electric vehicles technology development.

One of the agricultural machineries is a Tractor. Tractor are essential vehicles in agriculture, primarily used to tow various unpowered farming equipment. Unlike other common vehicles, tractors don't need to excel in high-speed performance or strength for traction. Their main role is to pull agricultural implements and support farming activities with high torsion. Although tractors in Indonesia are typically big motor-driven tractor, there is a growing trend among small-scale farmers towards focuses on small-scale tractor like hand tractor.

Driven by increasing environmental awareness and the demand for greater efficiency, today many tools are transitioning from conventional engines to electric motors, including electric tractors, which are now reflecting efforts to make significant changes in the agricultural industry in line with the desires of the government, particularly the Ministry of Agriculture.

The electric tractor was advancements primarily developing relying on batteries and primarily designed supporting rotary tillage featured a gearbox and speed control, allowing it to move forward and backward without the need for steering. Developed of the electric tractors also need current batteries technology to power a permanent magnet brushless motor. Advancements in control and battery technologies have enhanced electric tractor performance, leading to increased research into electric tractors and their electromechanical designs. Additionally, numerous studies focus on tractor performance and stability, with an emphasis on improving tillage effectiveness.

In conclusion, this research aims to develop an electric converter kit that can be easily utilized by farmers, facilitating their transition to electric tractors while retaining their existing body and transmission systems. This approach not only offers a cost-effective solution but also aligns with the growing emphasis on sustainability in agriculture. Additionally, this study serves as the initial step in the conversion process, including preliminary calculations and initial performance testing results. By enabling farmers to adopt electric tractors without the need for significant investment in new machinery, we hope to contribute to a more environmentally friendly agricultural sector. Further investigation into this technology is essential to ensure its practicality and effectiveness in real-world applications.

2. Method

The design of the electric tractor conversion kit is categorized into three main components: the power and control systems, which deliver the necessary energy and integrate various systems for unmanned operation; the mechanism design, which enhances mechanical performance and assesses the tractor's driving conditions; and the field testing and overall implementation of the tractor.

For the small electric tractor, its dimensions, weight, and motor capacity are reduced compared to larger, heavy-duty tractors. This tractor is primarily intended for tasks such as rotary tillage, so resistance calculations are based on those specific operations. Since this electric tractor will mainly function in relatively flat surfaces, factors like slope resistance and air drag are not considered.

Additionally, the tractor needs to accommodate a wide range of force variations, particularly when handling heavy loads, necessitating higher torque output. As a result, a reducer is implemented to slow down and amplify torque for varying operational conditions. The drive motor is adjustable and allows for both forward and reverse rotations to meet the requirements for reversing. During field transport, the system can switch to a high gear to enhance speed. Moreover, to prevent wheel slip on uneven terrain. Based on this analysis, the power system configuration of the electric tractor in this research includes a motor, reducer, differential, and controller, as illustrated in Figure 1.

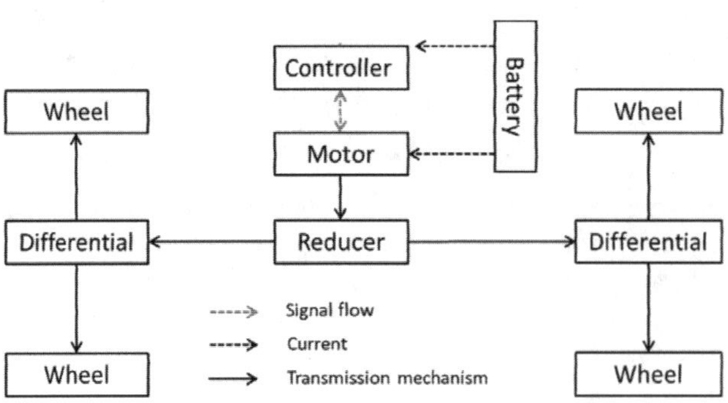

Figure 1. Power system configuration.

3. Results and discussions

3.1. Power and control system integration

Because of varying operating conditions and load distributions, the performance requirements for the tractor differ. The power system of the electric tractor needs to be tailored to accommodate these diverse load distributions during different operational scenarios. Calculations were made based on the specific demands of the tractor in various situations, ensuring that an appropriate drive motor and reducer were selected.

3.1.1. Force analysis and power system design. The hand tractor's resistance is directly linked to the specifications of its power system, allowing for the calculation of total resistance during both transportation and rotary tillage tasks. The electric tractor utilizes two induction motors for its rotary tillage operations: one motor is responsible for propelling the tractor, while the other is dedicated to operating the rotary plow and similar implements. When the tractor is engaged in transport, the motor powering the rotary plow remains inactive, resulting in a lower overall driving force for the tractor.

In transportation operations, the necessary driving force can be achieved through a combination of factors that enhance efficiency and effectiveness. This driving force typically stems from a well-coordinated system of logistics, which includes the optimal scheduling of tractors, effective route planning, and the integration of advanced technology for real-time tracking and management. Furthermore, the skilled workforce plays a crucial role, as trained drivers and operational staff contribute to minimizing delays and maximizing productivity. Additionally, maintaining tractors in top condition and utilizing fuel-efficient practices can significantly impact the overall performance of transportation operations. By addressing these components collectively, organizations can ensure that they harness the required driving force to meet their operational goals.

$$Initial\ Required\ Force = Normal\ Force = friction\ coeficient \times tractor\ weight \quad (1)$$

$$F_{Normal} = f_{resistance} \times W \quad (2)$$

The weight of a hand tractor varies depending on the brand and model, but generally, a hand tractor weighs between 100 and 200 kg. To calculate its weight in Newtons, use the formula:

$$Weight\ = mass \times gravity$$

$$W\ = m \times g \quad (3)$$

With the acceleration due to gravity being approximately 9.81 m/s^2, the weight of the hand tractor would range as Approx. 981N for 100 kg and approx. 1962N for 200 kg. The coefficient of rolling resistance for tractors ($f_{resistance}$) typically ranges from 0.1 to 0.2. However, this value can vary based on several factors, including: Tire type (e.g., agricultural vs. standard tires), Surface conditions (e.g., asphalt, dirt, or gravel) or Load carried by the tractor. Therefore, the F_{Normal} can be calculated as

$$F_{Normal} = f_{resistance} \times W = 0.2 \times 1962N = 392.4\text{N} \quad (4)$$

Before determining the driving force $\text{F}_{\text{driving}}$ needed for rotary tillage operations, it is essential to establish the rotary tillage speed ratio $R_{tillage}$ and the soil cutting pitch (S_{pitch}). Once these two parameters are computed, it becomes possible to derive the specific resistance $KR_{tillage}$ and the soil resistance F_{soil}, allowing for the calculation of the required driving force $\text{F}_{\text{driving}}$ for the operation. The rotary tillage speed ratio $R_{tillage}$ can be calculated using formula.

$$R_{tillage} = \frac{2\pi \times r_{turning} \times n_{\text{rotary}}}{60{,}000 \times \text{V}_{\text{tractor}}} = \frac{2\pi \times 200 \times 230}{60{,}000 \times 0.50} = 9.62 \quad (5)$$

With a turning radius of $r_{turning} = 200$ mm for the scimitar, a rotary knife shaft speed of $n_{\text{rotary}} = 230\text{r/min}$, and a tractor speed of $V_{\text{tractor}} = 0.50\ \text{m/s}$.

$$S_{pitch} = \frac{\pi \times r_{turning}}{5 \times N_{Sci} \times \lambda} = \frac{\pi \times 200}{5 \times 2 \times 9.6} = 6.5\ \text{cm} \tag{6}$$

$N_{Sci} = 2$ represents the number of scimitars located in the same vertical plane.

To determine the soil resistance, it is essential first to calculate the standard specific resistance of rotary tillage ($KR_{tillage}$), which corresponds to the soil cutting pitch (S_{pitch}). Afterward, the appropriate correction coefficient, accounting for the working conditions, should be applied to obtain the specific rotary tillage resistance ($KR_{tillage}$). The specific resistance ($KR_{tillage}$) for the rotary tillage is:

$$KR_{tillage} = \text{K}_1 \times \text{K}_2 \times \text{K}_3 \times \text{K}_4 \times \text{K}_5 = 16 \times 0.70 \times 0.80 \times 0.70 \times 0.60 = 3.76\ \text{N/cm}^2 \tag{7}$$

Where $\text{K}_1 = 16\ \text{N/cm}^2$ is the standard specific resistance of rotary tillage, , $\text{K}_2 = 0.70$ is the correction coefficient for tillage depth, $\text{K}_3 = 0.8$ is the correction coefficient for soil moisture content, $\text{K}_4 = 0.70$ is the correction coefficient for stubble vegetation, and $\text{K}_5 = 0.60$ is the correction coefficient for the operational mode. With the specific resistance of the rotary tillage known, the soil resistance F_{soil} can be calculated as follows:

$$\text{F}_{\text{soil}} = 100 \times KR_{tillage} \times \text{W}_d \times \text{D} = 100 \times 3.76 \times 1.1 \times 13 = 5381.37\ \text{N} \tag{8}$$

$Wd = 1.1$ m is the width of the rotary tillage, and $\text{D} = 13$ cm is the depth of the rotary tillage. Following the calculations of Formulas (5) through (8), we determine the rotary tillage speed ratio, the soil cutting pitch, the rotary tillage specific resistance, and the soil resistance. Subsequently, the operational time for the rotary tillage can be computed using Formula (9). The necessary driving force $\text{F}_{\text{driving}}$ is given by:

$$\text{F}_{\text{driving}} = \text{k}_{11} \times \text{F}_{soil} + f_{resistance} \times (\text{W} + \text{k}_{22}\ \text{F}_{\text{soil}}) \tag{9}$$
$$\text{F}_{\text{driving}} = 0.7 \times 5381.37 + 0.2 \times (1962 + 0.8 \times 5381.37) = 5020.38\ N \tag{10}$$

$\text{k}_{11} = 0.7$ is the horizontal coefficient of component, and $\text{k}_{22} = 0.8$ is the vertical coefficient of component.

3.1.2. Planning Of Power Operation for BLDC motor Selection. The specifications of the electric vehicle power system play a crucial role in determining the overall performance of the vehicle. The motor within the vehicle power system are determined based on the power demands associated with rotary tillage operations, which require higher energy output.

Power when Transportation operations:

$$P_{\text{tranportation}} = \left(\frac{1}{3600 \times \eta_{transmission}}\right) \times f_{resistance} \times F_{Normal} \times V_{\max} \tag{11}$$
$$= \left(\frac{1}{3600 \times 0.81}\right) \times 0.2 \times 1962 \times 18 = 2.4\ \text{kW} \tag{12}$$

$\eta_{transmission} = 0.81$ is the efficiency of the power transmission of the whole tractors because of any loses of friction or heat due operation, and $\text{V}_{\max} = 18\ \text{km/h}$ is the highest tractor speed during transportation.

Power when tillage operations:

$$P_{operation} = \left(\frac{1}{3600 \times \eta_{transmission}}\right) \times [k_{11}\ F_{soil} + f_{resistance} \times (F_{Normal} + k_{22}\ F_{soil})] \times V_{tillage} \quad (13)$$

$$P_{operation} = \left(\frac{1}{3600 \times 0.81}\right) \times [0.7 \times 5381.37 + 0.2 \times (1962 + 0.8 \times 5381.37)] \times 2 = 3.4\ kW \quad (14)$$

where the highest tractor speed during rotary tillage operation is $V_{tillage} = 2$ km/h. Generally, the tractor used in the field requires 3.4 kW = 4 kW, so two motors with a rated power of 4 kW were finally selected as the driving motors in this study. Table 1 presents the specifications of the selected motor with proper sustainable specification according to calculation.

Table 1. Selected BLDC motor specifications.

100B3/5-72V5.0-3200 Rated voltage DC 48-72V	
Rated power	3 kW @ 48 VDC
Peak power	5 kW @ 72 VDC
Max. motor speed	3200 rpm
Maximum torque	19 Nm

3.2. *In-depth analysis of BLDC motor performance for electric tractor conversion*

The conversion of traditional hand tractors into electric tractors using BLDC (Brushless Direct Current) motors is an emerging trend in agricultural machinery modernization. The key to a successful conversion lies in selecting a motor that optimally balances power, torque, and efficiency for practical agricultural applications. Understanding the performance characteristics of BLDC motors is essential, as it directly impacts the operational efficiency and overall reliability of the electric tractor. This section will analyze the performance relationships between speed, torque, power, current, and efficiency, drawing on relevant literature and fundamental motor principles. Table 2 presents the motor's performance data, including efficiency across various speeds, output power, and input current. These data, derived from experimental testing or simulations, showcase the motor's performance under operational scenarios.

Table 2. BLDC motor performance and efficiency data.

Motor revolution (RPM)	Torque (Nm)	Power (Watt)	Efficiency (%)	Current (A)
0,00	100,00	0,00	80,00	80,00
5,26	97,37	53,66	79,47	77,89
10,53	94,74	104,41	78,95	75,79
15,79	92,11	152,26	78,42	73,68
21,05	89,47	197,22	77,89	71,58
26,32	86,84	239,27	77,37	69,47
31,58	84,21	278,43	76,84	67,37
36,84	81,58	314,68	76,32	65,26
42,11	78,95	348,03	75,79	63,16
47,37	76,32	378,49	75,26	61,05
52,63	73,68	406,04	74,74	58,95
57,89	71,05	430,69	74,21	56,84
63,16	68,42	452,44	73,68	54,74

68,42	65,79	471,30	73,16	52,63
73,68	63,16	487,25	72,63	50,53
78,95	60,53	500,30	72,11	48,42
84,21	57,89	510,45	71,58	46,32
89,47	55,26	517,70	71,05	44,21
94,74	52,63	522,05	70,53	42,11
100,00	50,00	523,50	70,00	40,00

3.2.1. Torque-speed relationship

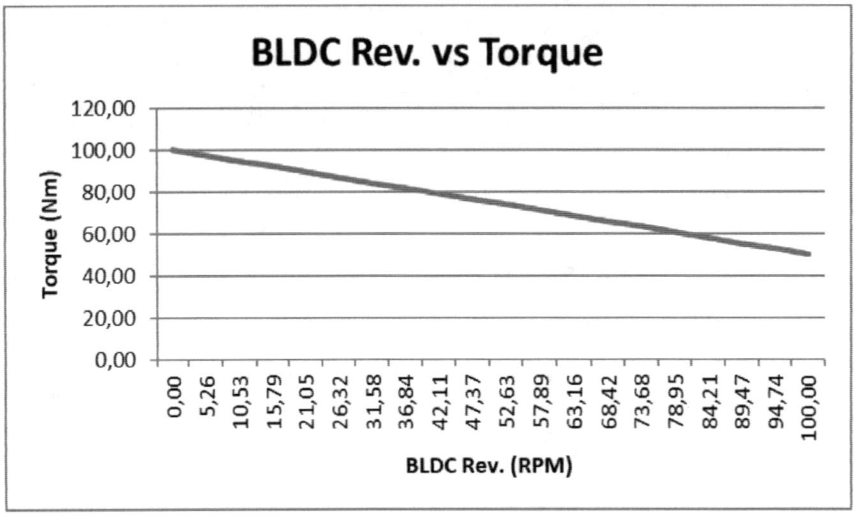

Figure 2. BLDC torque curve.

Figure 2 depicts the relationship between the motor's torque and speed, showing a linear decrease in torque as speed increases up to a certain point, reflecting the typical characteristics of a BLDC motor. One of the most critical characteristics of BLDC motors is the inverse relationship between torque and speed.

As speed increases, the motor's torque tends to decrease, which is a well-documented phenomenon in motor theory (Slemon, 1996). This occurs because of the back Electromotive Force (EMF) generated in the motor coils, which opposes the input current, reducing the effective torque output at higher speeds. According to research by Krishnan (2001), the back EMF increases proportionally with speed, thereby limiting the amount of current available to generate torque. In practical applications like electric hand tractors, which operate at low to moderate speeds, selecting a motor that can maintain sufficient torque at these speeds is vital. The relationship can be expressed as:

$$T = T_{max} - (K \times N) \tag{15}$$

Where Tmax is the maximum torque at low speeds, K is a constant dependent on the motor design, and N is the rotational speed. Understanding this relationship allows for optimizing the motor performance to meet the mechanical demands of tractor operations. In general, the torque of a BLDC motor decreases as speed increases. This is because in BLDC motors, maximum torque is usually generated at low speeds and decreases as motor speed increases.

At low speeds, higher current flows through the motor windings, producing greater torque. However, as speed increases, the effect of back EMF (electromotive force) increases, which reduces the current that can flow into the motor, resulting in a decrease in torque. This phenomenon is important in electric tractor applications, as low speeds are more commonly used in agricultural operations that require high torque to pull heavy loads.
The selected BLDC motor must be designed to provide sufficient torque at low to medium speed ranges, as torque at higher speed ranges tends to be less relevant in this application.

3.2.2. Power-speed relationship

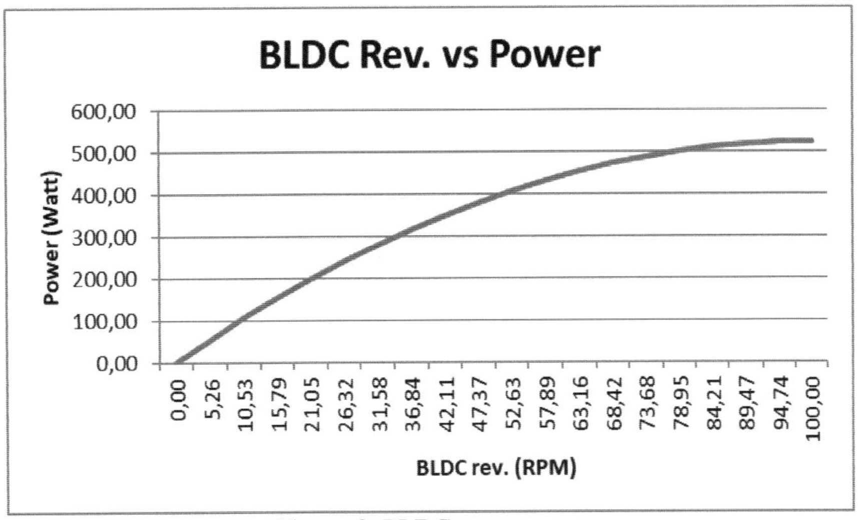

Figure 3. BLDC power curve.

Figure 3 illustrates the relationship between motor output power and operational speed, showing an increase in power until reaching a maximum point before decreasing at high speeds. Power output is another key performance metric, and it increases with speed up to a certain point before declining due to efficiency losses (Kolar & Friedli, 2010). The power-speed relationship is governed by the equation:

$$P = \frac{T \times N}{9550} \tag{15}$$

In agricultural machinery, maintaining sufficient power at low speeds is essential for ensuring that the tractor can perform heavy-duty tasks such as plowing or towing. However, excessive speeds may result in power surges that go beyond the motor's designed capacity, leading to efficiency degradation. Research by Gerling (2015) suggests that maintaining an optimal balance between speed and power is necessary for energy-efficient operation. Electric tractors benefit from power optimization, particularly when operating under variable load conditions. An increase in speed tends to raise the motor's power output, but only up to a certain point, where the motor's efficiency starts to decline.

The relationship between speed and power aligns with a basic mechanical law, which states that power is the product of torque and angular velocity. In hand tractor applications, high power at low to medium speeds is necessary to drive the wheels and agricultural implements. However, at very high speeds, the power generated may become less relevant, as hand tractors typically operate at low speeds with heavy loads.

In electric tractor applications, it is important to select a motor with appropriate power output for the speeds used in agricultural operations. The motor must remain efficient even when operating at low speeds with relatively high-power output.

3.2.3. Efficiency-speed relationship

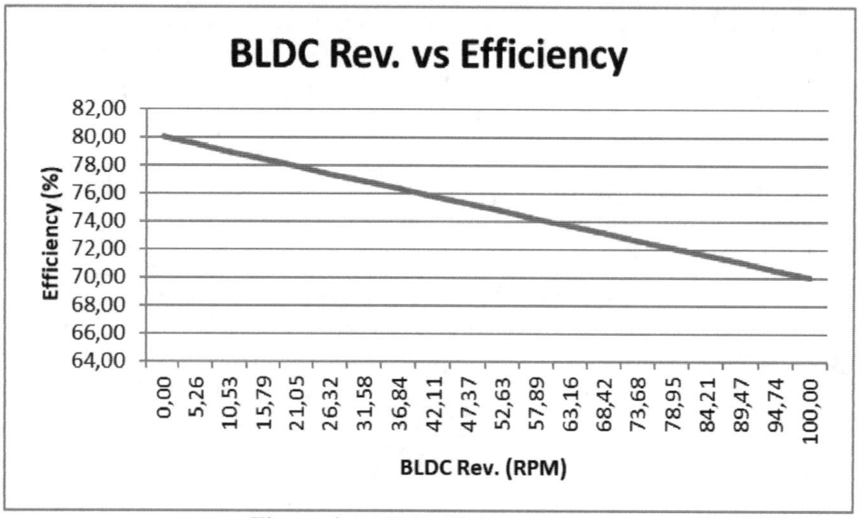

Figure 4. BLDC efficiency curve.

Figure 4 shows the motor's efficiency relative to operational speed, typically indicating that peak efficiency is achieved at moderate speeds. Motor efficiency is inversely related to speed, with efficiency typically decreasing at higher speeds due to increased losses in the motor's winding and magnetic core (Lipo, 2017). Efficiency is an essential factor in electric vehicles, including electric tractors, as it directly affects battery life and overall energy consumption. According to research by Hansen and Wang (2018), the decrease in efficiency can be attributed to factors such as core losses, copper losses, and switching losses in the motor controller. In practical terms, a motor that operates efficiently at low to medium speeds is preferred for tractors, where sustained operation over several hours is required. This relationship is governed by the equation:

$$\eta = \frac{P_{out}}{P_{in}} \times 100\% \quad (16)$$

Where P_{out} is the power output, and P_{in} is the power input. The ability to maintain high efficiency at operational speeds ensures that the motor consumes less energy, thereby extending the operational time per battery charge. Motor efficiency tends to decrease as speed increases, due to the rise in power losses (such as winding losses and friction losses).

BLDC motors are typically more efficient at low to medium speeds, but at higher speeds, internal losses such as iron loss and copper loss start to dominate. Efficiency can also be influenced by the quality of the motor controller and cooling system. However, the significant efficiency drop at high speeds may not be as critical in tractor applications, as tractors generally operate at low speeds.
In practical applications, prioritizing higher efficiency at lower speeds should be a key design consideration. This will minimize power losses, extend battery life, and reduce the need for additional cooling systems.

3.2.4. Current-torque relationship

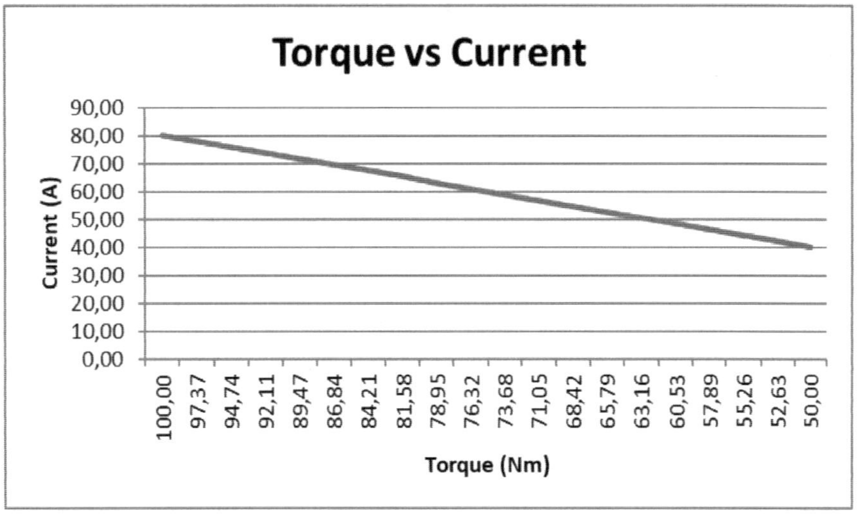

Figure 5. Relationship between torque (Nm) and current (A) in BLDC motor.

Figure 5 depicts the relationship between electrical current and the torque generated, showing a linear increase in torque with current up to a certain limit. The relationship between current and torque is linear, with torque increasing as current increases, as described by Lee (2004). In BLDC motors, the current passing through the motor windings generates a magnetic field, which in turn produces torque. The relationship can be described as:

$$T = Kt \times I \tag{17}$$

Where T is the torque, Kt is the motor constant, and I is the current. However, higher current also increases resistive losses, leading to more heat generation within the motor. Efficient thermal management systems are essential for preventing overheating, which can otherwise degrade motor performance and lifespan. In electric tractors, ensuring that the motor can handle high torque demand without excessive current draw is critical for sustained operations, especially when navigating uneven or challenging terrain. The torque generated by a BLDC motor is directly proportional to the current flowing through its windings.

In a BLDC motor, higher current produces a stronger magnetic field within the motor, which in turn generates greater torque. However, higher current also leads to an increase in I^2R losses (resistive losses) and heat buildup within the motor. Therefore, the maximum torque that can be produced is typically limited by the maximum current the motor controller can supply.

The motor control system must be capable of handling sufficient current to produce the required torque in electric tractor applications. Additional cooling systems or the use of low-resistance components may be necessary to prevent overheating and maintain optimal performance.

3.2.5. Power-efficiency correlation

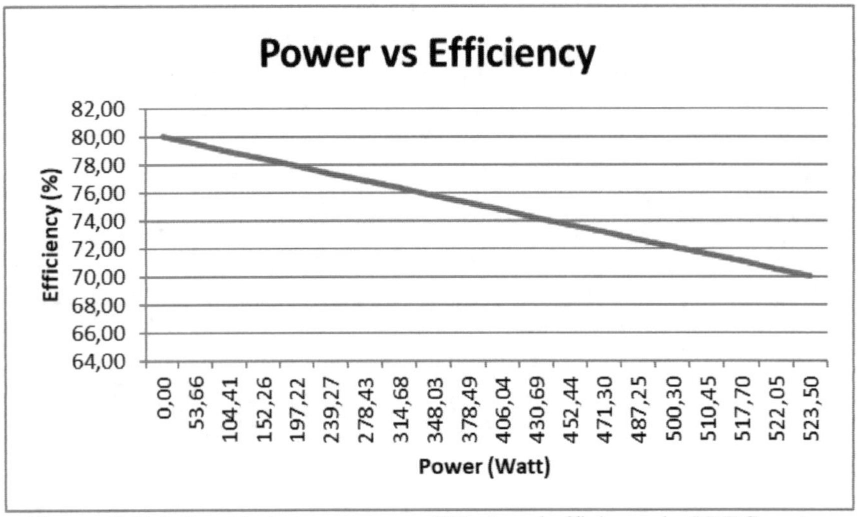

Figure 6. Relationship between power (Watt) and efficiency in BLDC motor.

Figure 6 demonstrates the correlation between the motor's output power and its efficiency across various operational scenarios, highlighting the trade-off between power and efficiency. The correlation between power output and efficiency follows a typical performance curve, where efficiency peaks at a specific power level before declining. Higher power outputs often lead to efficiency losses due to the increased electrical and thermal stresses on the motor (Bose, 2017). For electric tractors, operating at maximum efficiency is crucial, as this reduces the load on the battery and prolongs operational time. According to Jahns and Soong (2012), matching the power output to the load demands is essential to optimize efficiency. The following relationship captures the importance of power efficiency balance:

$$\eta = f(P) \tag{18}$$

This equation highlights that efficiency is a function of the power delivered, and selecting a motor with a well-matched power curve for the desired application is key to ensuring long-term operational efficiency. Motor efficiency tends to decrease at higher power levels, as increased power output is often accompanied by higher power losses.

As the motor's power output increases, losses due to resistance, friction, and hysteresis also rise. Additionally, motor controllers may struggle to maintain high current levels without significant switching losses. In hand tractor applications, prolonged operation at high power levels could lead to overheating and a substantial drop in efficiency. Effective thermal management systems and the selection of efficient motor components are crucial to ensure reliable operation at higher power ranges, preventing overheating and maintaining consistent performance.

3.2.6. Application to electric tractor design. In the context of electric tractor design, the above relationships suggest several considerations for selecting an appropriate BLDC motor. The motor should be optimized for delivering high torque at low to medium speeds, as these are the operating conditions commonly encountered in agricultural settings. Furthermore, the motor's efficiency should be maximized within this operational range to minimize energy consumption and extend battery life. According to Fernandez and Dawson (2019), careful balancing of these performance metrics is essential to ensuring reliable and sustainable operation in electric tractors. Power management and thermal design also play critical roles in preventing performance degradation during prolonged use under varying load conditions.

Based on the relationships described above, the selected BLDC motor must be capable of generating high torque at low to medium speeds, as agricultural applications frequently operate under these conditions. Energy efficiency is critical for maximizing battery usage, especially in electric tractor applications. A motor that remains efficient at low and medium speeds will reduce power consumption and extend the tractor's operational time per battery charge. In line with power and torque calculations, an appropriately sized battery must be selected to ensure continuous operation for an adequate period.

4. Conclusion

The analysis presented in this paper provides a comprehensive understanding of the performance characteristics of BLDC motors in the context of electric tractor conversion. By examining the relationships between speed, torque, power, current, and efficiency, it is possible to identify the key factors that influence motor selection and design. Electric tractors benefit from motors that can maintain high torque at low speeds while operating efficiently under variable load conditions. The insights from this analysis can guide the development of more energy-efficient and reliable electric tractor systems, contributing to the broader goal of modernizing agricultural machinery through sustainable technologies.

The analysed phenomena highlight the importance of selecting a BLDC motor that aligns with the operational needs of electric tractors. Speed, torque, power, and efficiency are key factors that must be considered comprehensively. By understanding the relationships between these variables, researchers can design more efficient and reliable electric tractor systems, minimizing energy losses and enhancing operational time.

5. Acknowledgement

The authors would like to express their sincere gratitude to the Ministry of Education, Culture, Research, and Technology of the Republic of Indonesia, particularly the Directorate of Vocational Higher Education, for funding this research under the Vocational Product Scheme. This support has been instrumental in enabling the development of the "Electric Agriculture Hand Tractor Conversion Kits" project.

The authors also extend appreciation to the Politeknik Negeri Jember, specifically the Automotive Engineering Study Program and Mechatronics Engineering Technology Study Program, for providing facilities and technical support throughout the research process.

Lastly, the authors thank go to all team members and collaborators who contributed their expertise and dedication to the successful completion of this study.

6. References

[1] M. Soerijadji, "Indonesia's Electric Vehicle Outlook Supercharging Tomorrow's Mobility," Jakarta, Jul. 2023.
[2] I. - International Energy Agency, "Electricity 2024 - Analysis and forecast to 2026," 2024. [Online]. Available: www.iea.org
[3] Energy Efficiency & Renewable Energy, "Electric Vehicle Basics," Washington DC, Aug. 2023.
[4] F. Alanazi, "Electric Vehicles: Benefits, Challenges, and Potential Solutions for Widespread Adaptation," *Applied Sciences (Switzerland)*, vol. 13, no. 10, May 2023, doi: 10.3390/app13106016.
[5] F. Alamsjah *et al.*, "Potential factors affecting adoption of electric vehicle by indonesia market," *ICIC Express Letters, Part B: Applications*, vol. 12, no. 11, pp. 1081–1090, Nov. 2021, doi: 10.24507/icicelb.12.11.1081.

[6] T. Chen *et al.*, "Applications of Lithium-Ion Batteries in Grid-Scale Energy Storage Systems," Jun. 01, 2020, *Tianjin University*. doi: 10.1007/s12209-020-00236-w.
[7] K. W. Wong and W. K. Chow, "Principle for the Working of the Lithium-Ion Battery," *Journal of Modern Physics*, vol. 11, no. 11, pp. 1743–1750, 2020, doi: 10.4236/jmp.2020.1111107.
[8] M. Rahimi, "Lithium-Ion Batteries: Latest Advances and Prospects," 2021, doi: 10.3390/batteries.
[9] A. T. Zain, D. D. Suranto, C. N. Karimah, F. A. Azhar, and D. A. Tyagita, "Analysis of Lithium-Ion Battery Consumption for Three-Wheeled Electric Vehicle with Variations in Weight and Speed †," *Engineering Proceedings*, vol. 63, no. 1, 2024, doi: 10.3390/engproc2024063013.
[10] B. Xu, L. Kong, G. Wen, and M. G. Pecht, "Protection Devices in Commercial 18650 Lithium-Ion Batteries," *IEEE Access*, vol. 9, pp. 66687–66695, 2021, doi: 10.1109/ACCESS.2021.3075972.
[11] G. Pratap Singh, Y. Lehri, L. Bhatia, and Y. Sehgal, "Monitoring of lithium-ion cells using a microcontroller," *Clean Energy*, vol. 6, no. 1, pp. 853–860, Feb. 2022, doi: 10.1093/ce/zkab058.
[12] P. Harahap, F. I. Pasaribu, and M. Adam, "Prototype Measuring Device for Electric Load in Households Using the PZEM-004T Sensor," *Budapest International Research in Exact Science*, vol. 2, no. 3, pp. 347–361, Jul. 2020, doi: 10.33258/birex.v2i3.1074.
[13] T. T. V. Hong, B. Delinchant, J. Ferrari, and Q. D. Nguyen, "Autonomous Electrical System onitoring and Control Strategies to Avoid Oversized Storage Capacity," in *IOP Conference Series: Earth and Environmental Science*, Institute of Physics Publishing, Jul. 2020. doi: 10.1088/1755-1315/505/1/012045.
[14] P. M. Ramos, N. B. Bras, and A. C. Serra, "A New Calibration Method for Current and Voltage Sensors Used in Power Quality Measurements," Sorrenti: Institute of Electrical and Electronics Engineers (IEEE), Mar. 2007, pp. 2283–2288. doi: 10.1109/imtc.2006.328579.
[15] J. Lu, Y. Hu, G. Chen, Z. Wang, and J. Liu, "Mutual calibration of multiple current sensors with accuracy uncertainties in IPMSM drives for electric vehicles," *IEEE Transactions on Industrial Electronics*, vol. 67, no. 1, pp. 69–79, Jan. 2020, doi: 10.1109/TIE.2019.2896320.

Combination of thermal and non-thermal pasteurization of siam semboro orange juice

M J Wibowo[1], M F Kurnianto[2], F C Kusumasari[2], A Bakri[2], R E Rachmanita[1], D Triardianto3, B Hariono[4]*

[1] Renewable Energy Engineering Study Program, Engineering Department, Politeknik Negeri Jember
[2] Food Industry Technology Study Program, Agricultural Technology Department, Politeknik Negeri Jember
[3] Agricultural Engineering Study Program, Agricultural Technology Department, Politeknik Negeri Jember
[4] Food Engineering Technology Study Program, Agricultural Technology Department, Politeknik Negeri Jember

*E-mail: budi_hariono@polije.ac.id

Abstract. Siam Semboro orange juice is a processed product that is in great demand. Heating is a conventional method that is commonly used to reduce microbial contamination and ensure product safety, but the possibility of damage to taste, nutrition and physicochemical properties cannot be avoided, so there is a need for non-thermal processing alternatives. This research combines processing technology with minimal heat treatment with pasteurization temperatures of 40oC, 50oC, 60oC and pulsed electric field application times of 30, 60 and 90 seconds with a high voltage HPEF source in the form of a flyback TV type BSC25-T1010A with a voltage of 30 kV installed in 2 units in series and the pasteurization heat source is a 600watt electric heating element. Research parameters include total bacteria (TPC), TDS, TSS, pH and Vitamin C. The combination of A3B2 treatment (90 seconds; 60°C) can reduce the total microbes of 2.7 log cycles. The treatment of physical and chemical property parameters on the combination of heat and HPEF generally had a significant influence

1. Introduction

Processed orange products such as orange juice have good market prospects because of the public's need for fresh drinks, especially recently as the weather has been very hot. Apart from having a typical orange appearance, color and taste, this product also contains vitamin C and minerals which are good for health [1]. Orange juice contains vitamin C such as ascorbic acid of 20 - 60 mg per 100 mL, vitamin A of 250 - 420 IU, thiamine of 70 - 120 µg, niacin of 200 - 220 µg, riboflavin of 30 µg, folic acid of 1, 2 µg, and inositol of 135 mg in every 100 mL of orange juice [2]. Vitamin C can also help increase the body's immunity, especially during a pandemic [3]. Orange juice also contains high levels of flavonoids such as hesperidin, tangeritin, naringenin and nobiletin. Research results state that oranges are an excellent herbal ingredient for fighting viruses [4].

Orange juice is a diluted solution of the essence of the fruit flesh, so it has the same taste as the original fruit. Based on the level of turbidity, there are two types of fruit juice, namely clear fruit juice and cloudy fruit juice. Orange juice is included in the cloudy fruit juice category because it contains high levels of pectin. According to [5], orange juice made with the addition of 150 grams of sugar and 1 gram of citric acid received the best consumer preference response with a total dissolved solids content of 25.47%Brix; vitamin C; 28.99 mg/ 100 gr; Total sugar 7.07% and total acid 0.16% w/w.

The problem with the production of orange juice is a short shelf life, therefore pasteurization is needed so that the product has a longer shelf life. Orange juice can generally be damaged due to the growth of pathogenic microbes, such as *Escherichia coli*, *Salmonella sp, Pseudomonas sp*. The uniqueness of this product is the bitter taste which becomes more pronounced when the orange fruit is squeezed hard or heat-treated, due to the presence of *limonene*. The heating process (pasteurization temperature 70-90°C) causes the oxidation process and produces a bitter tasting substance [6].

The thermal process causes the nutritional content, taste, color, and texture of orange juice to be decreased, and its vitamin C is easily damaged [7]. Therefore, it is necessary to diversify orange juice products that remain fresh, have high nutritional content, and have a better shelf life, so the High Pulsed Electric Field (HPEF) non-thermal pasteurization process for orange juice production needs to be carried out.

HPEF is a non-thermal food preservation method that is widely used in the food industry with the aim of improving product quality [8]. The application of HPEF to food is through the provision of high voltage electrical pulses for a short time where the sample or material being tested is placed between two electrodes [9]. The principle of HPEF is the electroporation mechanism, namely the process of increasing cell membrane permeability due to HPEF treatment, causing the components contained in the cells to be released more easily into the media [10]. The aim of this research is to find the best combination of non-thermal and thermal treatment, namely HPEF exposure time and pasteurization temperature on the physical, chemical, and microbiological quality of orange juice.

2. Method

2.1. *Equipment*

The equipment used is batch system HPEF equipment (Figure 1), Hanna Instruments HI2003-02 Total Dissolved Solids (TDS) test equipment; Total Suspended Solid (TSS) test equipment type TB200; VIS-SWNIR spectrometer; NIR spectrometer; HPLC; KERN brand analytical balance type ABJ 220-4NM, Pyrex measuring flask, filter paper, funnel, Erlenmeyer Pyrex, Burette Pyrex, pH meter type PH600L.

Figure 1. Batch system HPEF technology

2.2. Materials

The materials used are Siam Semboro orange fruit, iodine, Plate Count Agar (PCA) media, Buffer Peptone Water (BPW) media, Natrium Agar media, Eosin Methylene Blue Agar (EMBA) media,

Beird Park Agar (BPA) media, tellurite, egg yolk (egg yolk), physiological NaCl 0.85%, crystal violet, gram iodine. Lugol, safranin, as well as 70% alcohol, and 90% alcohol.

2.3. Research stages
The stages of the research conducted can be seen in Figure 2. The research steps taken were as follows:

a. Make orange juice by squeezing orange fruit using manual press equipment.
b. Enter the orange juice into the HPEF batch system equipment with a combination of HPEF pasteurization times of 30, 60, 90 seconds and pasteurization temperatures of 50°C, 60°C, 70°C.
c. Observing the physical properties of orange juice including TDS, TSS, pH; chemical properties of orange juice including sucrose and vitamin C levels; as well as microbiological properties with TPC.

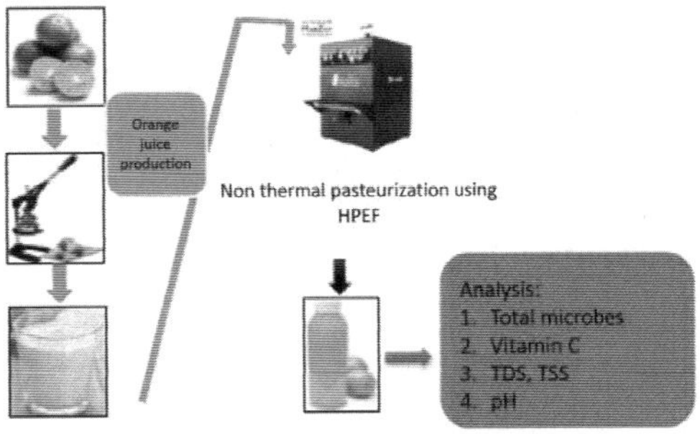

Figure 2. Research procedures

2.4. Research methods
This research uses two factors, namely the temperature of orange juice with heating temperature (A) which consists of four levels, namely 50°C (A1), 60°C (A2), and 70°C (A3) and HPEF treatment time (B), namely 30 second (B1), 60 second (B2), and 90 seconds (B3), at a constant voltage of 30 kV/ cm. The resulting sample was compared with a sample thermally pasteurized at 80°C for 3.5 seconds.

The research parameters that will be tested as a basis for analysis are the effectiveness of killing microbes, Total dissolved solids (TDS), Total Suspended Solid (TSS), acidity (pH), sucrose, and vitamin C levels. The results of these parameter tests will be seen for their conformity with SNI Orange Juice 01-6019-1999, and then the data will be processed statistically using ANOVA to determine whether there is an effect of heating treatment and HPEF.

3. Result and discussion

3.1. Total microbes
The total microbes in the initial fresh orange juice (control) was 2.7 log cycles. After heat treatment and HPEF of 30 (kV/cm), the highest number of surviving microbes was seen in the combination (A1B1) of 30second HPEF treatment and 50°C heating, which was achieved in A3B2 treatment; A2B3 and A3B3 are 0.7 log cycles. This means the combination of treatments A3B2; A2B3 and A3B3 are able to reduce total microbes by 2.0 log cycles or an effectiveness of 99%. This is in accordance with research [11] which states that the lower the heating temperature and exposure time to HPEF, the lower the microbial inactivation. On the other hand, the higher the heating temperature and exposure time to HPEF, the higher the microbial inactivation. [12]; [13] stated that the additional indication of the rate of death by the heating process could be due to the relationship between temperature and the phase transition of the phospholipid membrane from a gel to a crystalline liquid which is because the membrane loses its elastic components with increasing temperature so that it becomes more fluid and therefore makes it easier to damage with the application of HPEF.

Based on the two-way ANOVA analysis, it can be seen that time treatment has a significant influence on the TPC value. Apart from that, DMRT analysis shows that the treatment time values of

90s^{a}, 60s^{b}, and 30s^{c} have significantly different values. Based on Figure 3, the longer the HPEF exposure time given to the samples, the lower the mean total of microbes present in the sample. PEF can inactivate microbes because the electroporation process that occurs results in damage to the microbial cell membrane [14]. Meanwhile, temperature treatment also has a significant influence on the TPC value. Based on DMRT analysis, it shows that the temperatures of all treatments, namely 50^{c}, 60^{b}, and 70^{a}, have significantly different values as shown in Figure 3. The higher the temperature used, the lower the number of microbes contained in the sample. High temperatures cause denaturation of proteins which are an important part of the cell membrane. Therefore, heating at high temperatures causes damage to microbial cells [15].

TPC

Duncan[a,b]

Time	N	Subset 1	Subset 2	Subset 3
90s	9	.3529		
60s	9		.9992	
30s	9			1.2853
Sig.		1.000	1.000	1.000

Means for groups in homogeneous subsets are displayed.
Based on observed means.
The error term is Mean Square(Error) = .040.
a. Uses Harmonic Mean Sample Size = 9.000.
b. Alpha = .05.

TPC

Duncan[a,b]

Temp	N	Subset 1	Subset 2	Subset 3
70C	9	.4444		
60C	9		.7742	
50C	9			1.4188
Sig.		1.000	1.000	1.000

Means for groups in homogeneous subsets are displayed.
Based on observed means.
The error term is Mean Square(Error) = .040.
a. Uses Harmonic Mean Sample Size = 9.000.
b. Alpha = .05.

Figure 3. DMRT analysis at the temperature and exposure time of HPEF for all treatments is significantly different

Based on one-way ANOVA analysis, it can be seen that the TPC values between samples have significant differences between several samples. The notation for each sample can be seen in Table 1. Based on Table 1, it is known that the combination of HPEF and temperature exposure treatment can optimally reduce the total microbes present in the sample. HPEF treatment for 60s and 90s at 70^{0}C gave the lowest total microbes from TPC analysis. These results do not give a significant difference when compared to pasteurization treatment. Therefore, HPEF combined with thermal treatment can be an option to reduce the total microbes in the sample.

Table 1. Treatment data for the combination of temperature and HPEF exposure time on TPC parameters

Sample	TPC
60s – 70^{0}C	0 ± 0^{a}
90s – 70^{0}C	0 ± 0^{a}
90s – 60^{0}C	0.34 ± 0.57^{b}
90s – 50^{0}C	1 ± 0^{c}
30s –70^{0}C	1.05 ± 0.10^{a}
60s – 60^{0}C	1.09 ± 0.88^{a}
Pasteurization	1.09 ± 0.89^{a}
60s – 50^{0}C	1.23 ± 0.06^{a}
30s – 60^{0}C	1.57 ± 0.29^{d}
30s – 50^{0}C	1.62 ± 0.02^{d}
Control	2.63 ± 0.11^{e}

3.2. Total dissolved solids (TDS)

The initial total dissolved solids of orange juice was 10% brix, giving a combination of heat treatment and HPEF to the material caused a decrease ranging from 9.1 – 9.6% brix [11]. This value is not much different from the quality requirements for orange juice (SNI 01-6019-1999) where the total amount of minimal dissolved solids is 10 – 11% brix.

Based on the two-way ANOVA analysis, it can be seen that time treatment has a significant influence on the TDS value. Apart from that, DMRT analysis shows that the 90s[a] and 60s[a] time treatment values have significantly different values from the 30s[b] time treatment values. However, temperature treatment did not have a significant effect on the TDS value. The DMRT results are shown in Figure 4. Neither HPEF nor temperature treatment had a significant effect on the TDS value. The treatment given does not break down the organic or inorganic components contained in the sample that contribute to the cloudiness of the sample [16].

TDS

Duncan[a,b]

Time	N	Subset 1	Subset 2
90s	9	158.3333	
60s	9	164.0556	
30s	9		179.5556
Sig.		.083	1.000

Means for groups in homogeneous subsets are displayed.
Based on observed means.
The error term is Mean Square(Error) = 43.880.

a. Uses Harmonic Mean Sample Size = 9.000.

b. Alpha = .05.

TDS

Duncan[a,b]

Temp	N	Subset 1
70C	9	165.8889
50C	9	167.8333
60C	9	168.2222
Sig.		.489

Means for groups in homogeneous subsets are displayed.
Based on observed means.
The error term is Mean Square(Error) = 43.880.

a. Uses Harmonic Mean Sample Size = 9.000.

b. Alpha = .05.

Figure 4. DMRT TDS results

The results of the one-way ANOVA analysis show that the TDS values between samples have significant differences between several samples. The notation for each sample can be seen in Table 2 and Figure 5. Based on Table 2 and Figure 5, it can be seen that HPEF treatment combined with temperature does not give a significant difference to the total dissolved solid of pasteurized products. However, PEF treatment with an exposure duration of 30s, 60s, and 90s with temperatures of 50^{0}C had a significant effect on the TDS value compared with pasteurization treatment.

Table 2. Treatment data for the combination of temperature and HPEF exposure time on TDS parameters

Sample	TDS
90s – 70^{0}C	153,00 ± 4^{a}
30s – 70^{0}C	161,00 ± 4^{ab}
60s – 70^{0}C	161,00 ± 4^{ab}
90s –60^{0}C	162,34 ± 1,15^{ab}
60s –60^{0}C	164,67 ± 2,88^{ab}
30s –60^{0}C	165,16 ± 4,53^{ab}
Pasteurization	168,34 ± 23,09^{ab}
30s – 50^{0}C	177,34 ± 15,27^{bc}
Control	178,34 ± 23,09^{bc}
60s – 50^{0}C	179,00 ± 8,66^{bc}
90s – 50^{0}C	182,34 ± 14,43^{c}

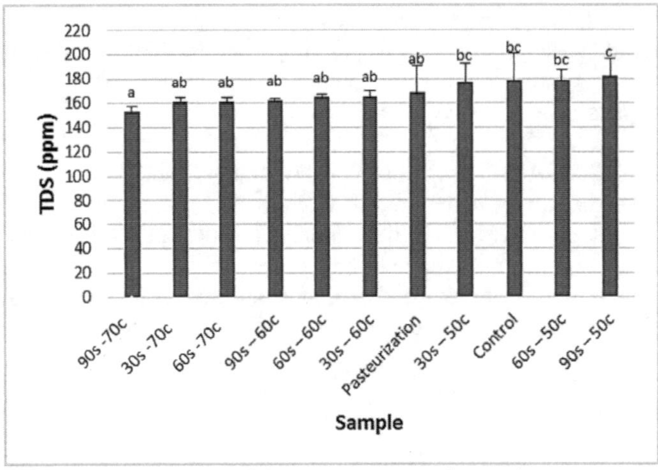

Figure 5. TDS parameter sample results

3.3. Total Suspended Solid (TSS)

Total suspended solids are solids suspended in orange juice in the form of organic materials that can be filtered with millipore paper with 0.45 μm pores. Based on the two-way ANOVA analysis, it can be seen that time treatment has a significant influence on the TSS value. Apart from that, DMRT analysis shows that the 30s[a] and 60s[a] time treatment values have significantly different values from the 90s[b] time treatment values. Meanwhile, temperature treatment also has a significant influence on the TSS value. Based on DMRT analysis, it shows that all temperatures have significantly different values. The DMRT results are shown in Figure 6.

TSS

Duncan[a,b]

Time	N	Subset 1	Subset 2
60s	9	10665.5556	
30s	9	11648.3333	
90s	9		17319.4444
Sig.		.293	1.000

Means for groups in homogeneous subsets are displayed.
Based on observed means.
The error term is Mean Square(Error) = 3702231.481.
a. Uses Harmonic Mean Sample Size = 9.000.
b. Alpha = .05.

TSS

Duncan[a,b]

Temp	N	Subset 1	Subset 2
50C	9	12131.6667	
60C	9	13033.8889	13033.8889
70C	9		14467.7778
Sig.		.333	.131

Means for groups in homogeneous subsets are displayed.
Based on observed means.
The error term is Mean Square(Error) = 3702231.481.
a. Uses Harmonic Mean Sample Size = 9.000.
b. Alpha = .05.

Figure 6. DMRT results of TSS treatment

The results of the one-way ANOVA analysis show that the TSS values between samples have significant differences between several samples. The notation for each sample can be seen in Table 3. Based on Table 3, it is known that PEF treatment combined with thermal produces TSS values that are significantly different from the pasteurization treatment. PEF treatment combined with temperatures of 50^{0}C and 60^{0}C produced smaller TSS values than pasteurization. The smaller the TSS value, the less cloudy the solution is and the more desirable for consumers [16].

Table 3. Treatment data for the combination of temperature and HPEF exposure time on TDS parameters

Sample	TSS
Control	8845 ± 1265.53[a]
30s – 50⁰C	9038 ± 702.78[a]
60s –60⁰C	10150 ± 638.35[a]
90s – 60⁰C	10391.67 ± 669.11[a]
30s – 60⁰C	11455 ± 768.48[ab]
60s – 50⁰C	11793.34 ± 732.21[abc]
90s – 50⁰C	14113.34 ± 866.33[bcd]
Pasteurization	14983.34 ± 1265.53[cd]
30s – 70⁰C	15901.67 ± 3578.33[de]
60s – 70⁰C	17158.34 ± 2027.98[de]
90s – 70⁰C	18898.34 ± 3629.94[e]

3.4 Acidity (pH)

The control Siam Semboro orange juice had a pH of 4.32 ± 0.16 after being treated with a combination of heat treatment and HPEF, the average pH experienced changes ranging from 4.40 ± 0.29. Based on the two-way ANOVA analysis, it appears that time treatment has a significant influence on the pH value. Apart from that, DMRT analysis shows that all 30s[a] treatment values have significantly different values from the 60s[b] and 90s[b] treatment values. Meanwhile, temperature treatment did not have a significant effect on the pH value. This is in line with the research of [17] which showed that PEF treatment did not have a significant effect on the pH value of milk. This phenomenon can be caused by the duration of PEF treatment that is too short thus there is no change in the composition of the resulting beverage. Based on Figure 7, in the application of HPEF for 30s, the significant effect may be caused by the fact that there are still many active bacteria in the sample compared to the 60s and 90s treatments, causing the environment to become more acidic [18].

pH

Duncan[a,b]

Time	N	Subset 1	Subset 2
30s	9	3.9194	
60s	9		4.3294
90s	9		4.4128
Sig.		1.000	.455

Means for groups in homogeneous subsets are displayed.
Based on observed means.
The error term is Mean Square(Error) = .054.

a. Uses Harmonic Mean Sample Size = 9.000.

b. Alpha = .05.

pH

Duncan[a,b]

Temp	N	Subset 1
70C	9	4.2039
60C	9	4.2094
50C	9	4.2483
Sig.		.705

Means for groups in homogeneous subsets are displayed.
Based on observed means.
The error term is Mean Square(Error) = .054.

a. Uses Harmonic Mean Sample Size = 9.000.

b. Alpha = .05.

Figure 7. DMRT results of pH value

The results of the one-way ANOVA analysis show that the pH values between samples have significant differences between several samples. Table 4 shows that at 50⁰C, the pH of orange juice was lower when compared to 60⁰C and 70⁰C. This can be caused by microbial activity that produces organic acids that can affect the pH of orange juice because there are still many microbes present in the orange juice.

Table 4. Treatment data for the combination of temperature and HPEF exposure time on pH parameters

Sample	pH
90s – 50^{0}C	3.88 ± 0.31
60s – 50^{0}C	3.90 ± 0.29
30s – 50^{0}C	3.96 ± 0.40
60s – 60^{0}C	4.30 ± 0.11
90s – 60^{0}C	4.31 ± 0.09
Control	4.32 ± 0.16
Pasteurization	4.33 ± 0.16
30s – 60^{0}C	4.36 ± 0.05
90s – 70^{0}C	4.40 ± 0.29
30s – 70^{0}C	4.41 ± 0.83
60s – 70^{0}C	4.41 ± 0.11

3.5. Vitamin C

[19] regulates the vitamin C content in orange juice, namely trace. According to [20] the vitamin C content in orange juice ranges from 20 - 60 mg/100 ml. The vitamin C levels of fresh orange juice in this study ranged from 12.95 mg/100ml, while the combined heat and HPEF treatment were between 13.13 – 14.23 mg/100 ml, indicating an increase in vitamin C levels. In general, vitamin C levels were lower than SNI.

4. Conclusion

The results showed that the combination of HPEF and heat treatment in the A3B2 treatment (90 seconds; 60°C) could reduce total microbes by 2.7 log cycles, while testing TDS, TSS, pH and Vitamin C on the combination of heat and HPEF generally had a significant effect.

Acknowledgments

The author would like to thank PNBP Politeknik Negeri Jember for providing funding for research with contract number SP DIPA 023.18.2.677607/2024 as well as APTV research with contract number No 68/SPK/D.D4/PPK/0.1 APTV/III/2024 which has been funded this research activity.

References

[1] Kujawińska M, Kawulok I, Szczyrba A, Grot M, Bielaszka A, Nieć-Leśniak J and Grajek M 2022 Vitamin C content in orange juices obtained by different methods *Journal of Education, Health and Sport* **12** 253–67

[2] Ndife J, Awogbenja D and Zakari U 2013 Comparative evaluation of the nutritional and sensory quality of different brands of orange-juice in Nigerian market *African Journal of Food Science* **7** 479–84

[3] Shahbaz U, Fatima N, Basharat S, Bibi A, Yu X, Hussain M I and Nasrullah M 2022 Role of vitamin C in preventing of COVID-19 infection, progression and severity *AIMS Microbiol* **8** 108–24

[4] Bellavite P and Donzelli A 2020 Hesperidin and SARS-CoV-2: New light on the healthy function of citrus fruits *Antioxidants* **9** 1–18.

[5] Ebrahimi A, Csonka L N and Alam M A 2018 Analyzing Thermal Stability of Cell Membrane of Salmonella Using Time-Multiplexed Impedance Sensing Biophys J 114 609–18

[6] Siddiqui A, Kulkarni D, Kulkarni K and Mulla M 2013 Studies on Debitteringof Sweet Orange Juice *World Journal of Dairy & Food Sciences* **8** 185–9

[7] Aghajanzadeh S, Ziaiifar A M and Verkerk R 2023 Effect of thermal and non-thermal treatments on the color of citrus juice: A review *Food Reviews International* **39** 3555–77

[8] M. Ali, L. Liao, X. A. Zeng, M. F. Manzoor, and M. Mazahir, "Impact of sustainable emerging pulsed electric field processing on textural properties of food products and their mechanisms: An updated review," J Agric Food Res, vol. 15, Mar. 2024, doi: 10.1016/j.jafr.2024.101076.

[9] K. Miloudi, A. Hamimed, Y. Bouhadda, Y. Benmimoun, K. Belhouala, and B. Benarba, "Impact of pulsed electric field treatment for extracting essential oil from Mentha Spicata L," Int J Electrochem Sci, vol. 17, 2022, doi: 10.20964/2022.08.41.

[10] E. Demir, S. Tappi, K. Dymek, P. Rocculi, and F. Gómez Galindo, "Reversible electroporation caused by pulsed electric field – Opportunities and challenges for the food sector," Trends Food Sci Technol, vol. 139, p. 104120, Sep. 2023, doi: 10.1016/j.tifs.2023.104120.

[11] L. C. Hawa, N. Komar, and D. Wirayanti, "Kombinasi Pasteurisasi Termal Dan Non Termal Pada Sari Buah Jeruk," J. Keteknikan Pertan. Trop. dan Biosist., vol. 4, no. 3, pp. 242–249, 2016.

[12] Stanley TL, Grinspoon SK. "Effects of growth hormone-releasing hormone on visceral fat, metabolic, and cardiovascular indices in human studies". Growth Horm IGF Res 25: 59–65, 2015. doi:10.1016/j. ghir.2014.12.005

[13] Budi Hariono, A Brilliantina, EKN Sari, MF Kurnianto, F Erawantini, S Kautsar 2022. "Pulsed Electric Field Application on Pasteurization of Orange Milk from Low Grade Orange: Study on Nutritional, Physical, Chemical Properties, and Total Microorganism." IOP Conference Series: Earth and Environmental Science 980(1): 012041

[14] Lytras F, Psakis G, Gatt R, Cebrián G, Raso J and Valdramidis V 2024 Exploring the efficacy of pulsed electric fields (PEF) in microbial inactivation during food processing: A deep dive into the microbial cellular and molecular mechanisms *Innovative Food Science and Emerging Technologies* **95**

[15] Espinosa M F, Sancho A N, Mendoza L M, Mota C R and Verbyla M E 2020 Systematic review and meta-analysis of time-temperature pathogen inactivation *Int J Hyg Environ Health* **230** 113595

[16] Wibowo M, Bakri A, Hariono B, Wijaya R and Brilliantina A 2022 Application of pulsed electric field in pasteurization of orange juice of siam cultivar: study on nutritional, physical, chemical properties, and total microorganism *IOP Conf Ser Earth Environ Sci* **980** 012007

[17] Priyanto A D, Estiasih T, Putranto A W, Widyasari W and Tola P S 2023 Pulsed electric field and pre-heating treatment effect on free fatty acid (FFA), pH, vitamin C, and organoleptic properties of milk *Advances in Food Science, Sustainable Agriculture and Agroindustrial Engineering* **6** 163–9

[18] Kayalvizhi V, Pushpa A J S, Sangeetha G and Antony U 2016 Effect of pulsed electric field (PEF) treatment on sugarcane juice *J Food Sci Technol* **53** 1371–9

[19] BSN 2014 SNI 3719:2014 Minuman Sari Buah Badan Standarisasi Nasional

[20] N. Martí, P. Mena, J. A. Cánovas, V. Micol, and D. Saura, "Vitamin C and the role of citrus juices as functional food," *Nat. Prod. Commun.*, vol. 4, no. 5, pp. 677–700, 2009, doi: 10.1177/1934578x0900400506.

Application of Combined Coffee Skin Husk Biochar and Rice Husk Charcoal as Planting Media to Improve on the Growth Seedling Coffee

E Rosdiana[1*], S A Nugroho[1], R Wardana[1], A Salim[1] and F Y Ali[1],

[1]Department of Agricultural Production, Politeknik Negeri Jember, Indonesia

*Email : eva_rosdiana@polije.ac.id

Abstract. Coffee is a plantation commodity in great demand and has a fairly good economic value. One way to get good coffee is to see good cultivation methods, especially in its nursery. Good nurseries can utilize Biochar coffee skin and rice husks. This study aims to obtain quality coffee seeds by creating an alternative biochar planting media formulation derived from coffee skin waste and rice husk charcoal. The main ingredients for making biochar are agricultural and plantation wastes such as rice husk charcoal and coffee fruit skin. This study used a non-factorial completely randomized design (CRD). The treatment factor used was a mixture of rice husk charcoal media with coffee husk biochar comprising 5 treatment levels. Each treatment was repeated 5 times, so there were 25 experimental units. The experimental analysis used was the Analysis of variance. The further test used the LSD test. The results showed that the treatment of a mixture of coffee husk and rice husk biochar affected the stem diameter plant and the number of leaf plants.

1. Introduction

The largest coffee production produced by Large Plantations in 2022 will come from East Java Province amounting to 3.39 thousand tons (85.15%) of the total PB production in Indonesia [1]. If coffee production is higher, the waste production will also be higher [2]. In 2022 coffee production will decrease to 774.96 thousand tons (1.43%). Farmers often throw away most of the coffee skin waste without making full use of it. Coffee skin waste that is not properly handled or discarded can lead to pollution in the nearby area. Meanwhile, coffee skin waste has quite high benefits and is very good for plants because it contains Nitrogen, Phosphorus and Potassium which can help plant growth.

The proportion of coffee skins produced during processing is quite large, namely 40-45% [3]. Coffee skin waste is used as biochar material. Application of biochar to the soil as much as 20% can increase water holding capacity from 32% to 52% [4]. Providing coffee fruit skin biochar on planting land can also increase soil fertility [5].

The biochar used besides coffee husk waste is rice husk waste. Utilization of husk waste is still very low and if processed naturally it is slow so it causes environmental pollution and has an impact on human health. The key advantage of rice husk lies in its consistent availability, as rice plants are cultivated in a sustainable and ongoing manner [6]. The Central Bureau of Statistics (BPS) reports that in 2021, the production of dry unhulled rice reached 54.42 million tons. When dry unhulled rice undergoes milling (resulting in 67%), it generates by-products such as hulls (20%), bran (11%), and rice crumbs (2%). The significant volume of waste, particularly husks, is believed to contribute to environmental issues, including alterations in soil structure and other soil-related problems. Rice hulls consist of organic materials that contain lignin (26-31%), cellulose (25-35%), and hemicellulose (18-21%). The substantial presence of lignin, cellulose, and silica renders hulls resistant to decomposition by soil microorganisms.[7]. Rice husks can be used as a planting medium that can improve soil aeration and drainage capabilities [8]. This research aims to determine the formulation of alternative planting media for coffee husk waste and rice husk charcoal for optimal growth of coffee seedlings.

2. Materials and Methods

2.1 Experimental Site

The study took place from June to September 2024 The study was conducted at the Teaching Factory nursery coffee of the Department of Agriculture Productions, Politeknik Negeri Jember, Jember-East Java. Tests on the treatment of biochar and compost on soil media in polybags were carried out in the greenhouse of Laboratory Lapang, which is located at an altitude of 300 m above sea level.

2.2 Tools and Materials

The tools and materials needed for planting robusta coffee (BP 42 Clon) plant seeds, polybag measuring 25 × 30 cm, 70% paranet, coffee fruit skin waste, rice husk charcoal, NPK, Propikonazol fungicide. The tools used consist of hoes, machetes, wire rams, fire starters, saws, soil sieves, nails, ovens, meters, cameras, digital scales and stationery.

2.3 Research Procedures

Making Biochar: Using a retort kiln with the manufacturing steps: Prepare tools (oil barrel and sieve) and materials (coffee husk waste and rice husk charcoal); Put coffee husk waste and rice husk charcoal into oil barrels until the barrel is ¾ full; The top of the oil barrel was filled with corn cobs and newspapers; Burn the corn cobs at the top of the oil barrel and then close the lid until there is no oxygen; Burning (pyrolysis process) is carried out for 16 hours; Sift the resulting charcoal with a 10 mesh sieve. Preparation of Planting Media in Polybags: Planting media (biochar) is put in polybags according to the treatment. Planting: Carry out planting by placing the seeds in Polybag Maintenance: Replanting, weeding and pest control and fertilization. So the result of the procedure is planting media mixture. the mixture is Biochar from coffee fruit skin waste 25g/polybag+25g/polybag rice husk charcoal; Biochar from coffee fruit skin waste 50 g/polybag+75 g/polybag rice husk charcoal; Biochar from coffee fruit skin waste 75 g/polybag+50 g/polybag rice husk charcoal; Biochar from coffee fruit skin waste 100 g/polybag+50 g/polybag rice husk charcoal.

2.4 Desain Experimental

This research used a Randomized Block Design (RCBD) with one factor, namely the dose of coffee fruit peel waste biochar and rice husk charcoal using 5 levels, namely:
P1 = Without Biochar coffee fruit skin waste and rice husk charcoal,
P2 = Biochar from coffee fruit skin waste 25 g/polybag+25 g/polybag rice husk charcoal,
P3 = Biochar from coffee fruit skin waste 50 g/polybag+75 g/polybag rice husk charcoal,
P4 = Biochar from coffee fruit skin waste 75 g/polybag+50 g/polybag rice husk charcoal,
P5 = Biochar from coffee fruit skin waste 100 g/polybag+50 g/polybag rice husk charcoal.
Each treatment was repeated 5 times to obtain 25 experimental units. Repeated 5 times, 25 experimental units were obtained. If there are differences between treatments, a further test (LSDT) at a level of 5% is carried out.

2.5 Statistical analysis

The research data collected were processed statistically by using an analysis of variance. The real effect on compost and biochar treatment was followed by the Least Significant Difference Test (LSDT) at the 5% level.

2.6 Research Variable
The plant variables observed in this study were height per plant, number of leaves per plant, stem diameter per plant, total wet weight per plant, and dry weight of total per plant.

3. Result and Discussion
Based on the analysis of various fingerprints Applications of combined coffee skin husk biochar and rice husk chorcoal as planting media to improve on the growth seeedling coffee can see Table 1 as follows:

Table 1. Anova Test Results of Robusta Coffee Seed Growth for 4 Observations

Variable	Observations To-	notation
Plant Height	1	ns
	2	ns
	3	ns
	4	ns
Stem diameter per plant	1	ns
	2	**
	3	**
	4	ns
The number of leaves per plant	1	-
	2	**
	3	**
	4	**
Total wet weight of biomass per plant	4	ns
The total dry weight of biomass per plant	4	ns

- ns= not significant ($P \geq 0.05$), **= very significant ($P<0.01$), *= significant ($P<0.05$)
- Numbers followed by the same lowercase letters in the same column are not significantly different at 5% LSDT.

Based on the results of the recapitulation shown in table 1. that the variable of plant height observation did not show a significant difference from the first observation to the fourth, while the stem diamater showed a significant effect on the 2nd and 3rd observation. The observation of the number of leaves also showed a significant difference in the second to fourth gardening, and for the fresh weight of seedlings and the dry weight of seedlings did not show any real difference. The results were different and then tested using the LSD follow-up test.

3.1 Plant height
Based on the results of the analysis of the anova test for the observation variable of plant seedling height, there are not significant difference from the first to the fourth planting, but the highest seedling was found in the control treatment (P0) with a seedling height of 12.31 cm (figure 1). This is because robusta coffee seeds have not shown a significant increase because they take time to carry out physiological processes. This is in accordance with the results of research from [9] which states that the optimal condition of the planting medium both in terms of physical, biological, and chemical can affect plant growth. In addition, According to the findings of the study referenced in [10], the use of biochar on coffee fruit peel waste did not significantly impact the growth rate of coffee seedlings.

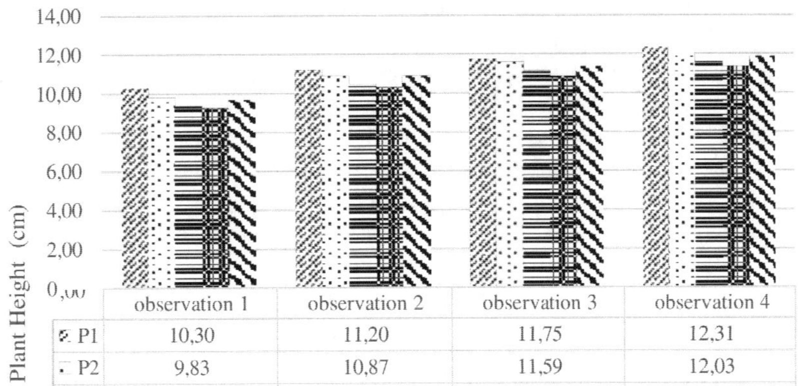

	observation 1	observation 2	observation 3	observation 4
P1	10,30	11,20	11,75	12,31
P2	9,83	10,87	11,59	12,03
P3	9,37	10,43	11,13	11,54
P4	9,27	10,27	10,85	11,29
P5	9,67	10,87	11,33	11,84

Figure 1. Result of Robusta Coffee seedling Plant Height

Based on Figure 1, it can be seen that the effect of the Biochar used was not visible until the 4th observation. According to [5], it also states that biochar has no effect on the height of coffee seedlings. However, as seen in Figure 1, plant height growth continues to increase from observation 1 to observation 4. And the variety of seeds used is quite good, namely Robusta coffee clone BP42.

3.2 Stem diameter per plant

On the diameter of the plant there was a significant influence on the second and third observations. This is shown in table 2. below.

Table 2. Results of further tests of Applications of combined coffee skin husk biochar and rice husk chorcoal as planting medium to improve on the growth of seeedling coffee on stem diameter.

Treatment	Average	
	Observations to-2	Observations to -3
P1 = Without Biochar coffee fruit skin waste and rice husk charcoal	1.86 a	1.86 a
P2 = Biochar from coffee fruit skin waste 25 g/polybag+25 g/polybag rice husk charcoal	1.97 b	1.97 b
P3 = Biochar from coffee fruit skin waste 50 g/polybag+75 g/polybag rice husk charcoal	1.95 b	1.95 b
P4 = Biochar from coffee fruit skin waste 75 g/polybag+50 g/polybag rice husk charcoal	1.81 a	1.81 a
P5 = Biochar from coffee fruit skin waste 100 g/polybag+50 g/polybag rice husk charcoal	1.85 a	1.85 a

According to ANOVA results in Table 1, Steam diameter becomes evident that is significant. This means that the null hypothesis (H0) is rejected and showing a tangible impact resulting from the inclusion of combined coffee skin husk biochar and rice husk chorcoal as planting medium to improve on the growth of seedling coffee on stem diameter.

In Table 2. Shown that the LSD test was conducted to assess the impact of in corporating coffee exocarp and rice husk charcoal as a growth medium for growth of seeedling coffee. The results of the LSD test demonstrate notable differences in test values between various treatments. Notably, treatment P2 emerged as the most optimal treatment to increase the diameter steam.

In the P2 treatment, namely Biochar from coffee fruit skin waste 25 g/polybag+25 g/polybag rice husk charcoal, the highest stem diameter was 1.97 mm compared to other treatments, and the lowest in the P3 treatment Biochar from coffee fruit skin waste 75 g/polybag+50 g/polybag rice husk charcoal. Likewise, in the 3rd week, the P2 treatment = Biochar from coffee fruit skin waste 25 g/polybag+25 g/polybag rice husk charcoal showed the highest stem diage, 1.97 mm compared to other treatments.

This is because the increase in crop yield due to the nutrient content of coffee husk compost can be a stimulator in plant growth which is coupled with husk charcoal which functions as a temporary store of nutrients in the soil so that it is not easily washed by water and will be very easy to release when needed or taken by plant roots [10]. This is also in line with the statement [11] which states that the application of coffee husk biochar shows a more significant improvement to improve the chemical properties of the soil.

3.3 The number of leaves per plant

The results of the follow-up test showed that the variable of observation of the number of leaves showed a significant difference in the second week to the 4th week, which can be seen in table 3.

Table 3. Results of further tests of Applications of combined coffee skin husk biochar and rice husk chorcoal as planting medium to improve on the growth of seeedling coffee on the number of leaves.

Treatment	Average		
	Observations ke- 2	Observations ke-3	Observation ke-4
P1 = Without Biochar coffee fruit skin waste and rice husk charcoal	6 a	6.07 a	6.17 a
P2 = Biochar from coffee fruit skin waste 25 g/polybag+25 g/polybag rice husk charcoal	6.87 d	7.03 b	7.90 c
P3 = Biochar from coffee fruit skin waste 50 g/polybag+75 g/polybag rice husk charcoal	6.4 c	6.40 a	6.80 b
P4 = Biochar from coffee fruit skin waste 75 g/polybag+50 g/polybag rice husk charcoal	6.2 b	6.20 a	6.40 ab
P5 = Biochar from coffee fruit skin waste 100 g/polybag+50 g/polybag rice husk charcoal	6 a	6.10 a	6.17 a

Based on Table 3. shown that the LSD test was conducted to assess the impact of in corporating coffee exocarp and rice husk charcoal as a growth medium for growth of seeedling coffee. The results of the LSD test demonstrate notable differences in test values between various treatments. Notably, treatment P2 emerged as the most optimal treatment to increase the number of leaves.

The highest number of leaves was found in the treatment of P2 = Biochar from coffee fruit skin waste 25 g/polybag+25 g/polybag rice husk charcoal starting from the second, third to fourth week, which was 7.90 pieces. This is because the influence of applying 25 g of coffee husk biochar and 25 g of rice husk is the best dose that can suffice the growth of plant leaves. This is in line with the diameter of the stem which has also increased so that the number of leaves also increases. Planting media derived from coffee husk biochr and rice husk is a good medium and is able to help the growth of robusta coffee seedlings so that when applied plant growth shows real results in the number of leaves [12]. Biochar also supports the growth and development of soil microbes [13]. In addition, biochar as charcoal that improves soil function to avoid short-term and long-term adverse impacts on the environment and human and animal health [14].

3.4 Total wet weight of per plant

The weight of wet coffee seeds did not very much in the fingerprint analysis results. Figure 2 shows this. However, in robusta coffee seedlings, although there is no significant difference in analysis, the P4 treatment is the treatment that has the highest seedling wet weight, which is 1.26 g compared to other treatments. This is because in coffee seedlings in the vegetative period, the roots are still able to absorb nutrients through the soil.

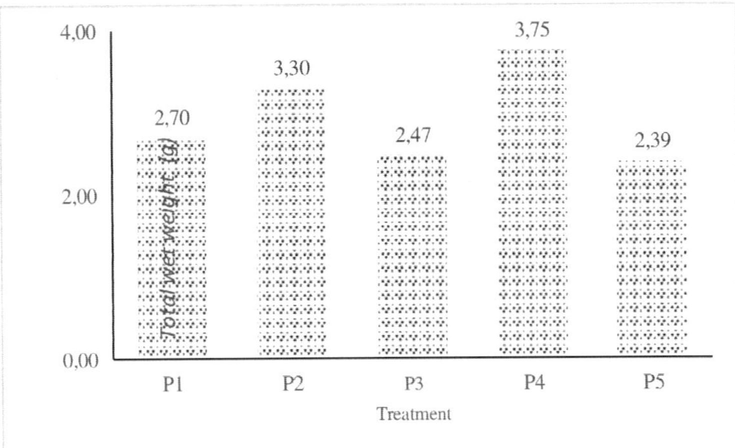

Figure 2. Total Wet Weight Robusta Coffee Seedling

Based on Figure 2, it can be seen that the effect of the Biochar used was not visible on the total of wet weight the plant. However, it can be seen that visually the P4 treatment has the largest average of 3.75 g, compared to the P1 control of 2.70 g. In line with [5], it also states that biochar has no effect on the total wet weight.

3.5 The total dry weight per plant

The dry weight of coffee seedlings showed an insignificant difference based on the analysis of the fingerprints. However, the P2 treatment had the highest root weight of 1.26 g compared to other treatments, shown in figure 3 below. This is in line with the wet weight of the seedlings which also shows an unreal difference.

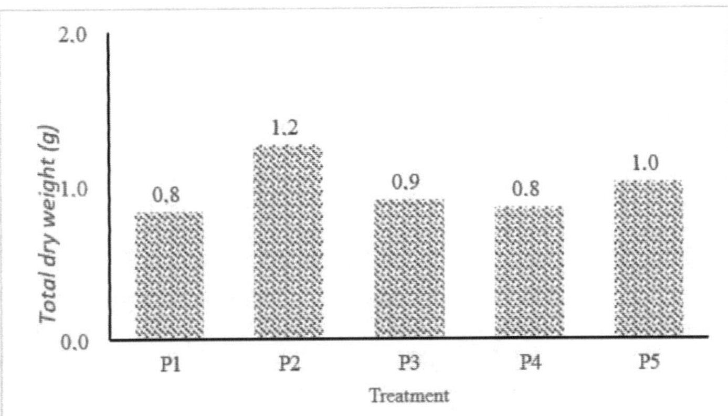

Figure 3. Total Dry Weight Robusta Coffee Seedling

In the process of growing coffee seedlings, increasing the growth of seedlings requires time and sufficient nutrients. Meanwhile, the nutrients contained in coffee husk and husk charcoal waste are not enough for significant seed growth. This aligns with the findings in statement [16], which indicate that plants achieve optimal growth when biochar is used alongside cow manure fertilizer.

Nutrients absorbed by plants that are utilized for various metabolic processes are to maintain the physiological functions of plants (including leaves, stems, and roots). Physiological symptoms can be observed through plant parameters, one of which is dry weight. Dry weight serves as an indicator of plant growth and development since it represents the total organic compounds that plants have effectively produced. The dry weight of a plant indicates its nutritional health and can signal the success of a plant's growth and development.

4. Conclusion

The conclusion of this study is that the application of coffee husk waste biochar and husk charcoal given as a planting medium for robusta coffee seedlings can increase the growth of stem diameter and number of leaves in the second to fourth week with the best dose of 25 g of coffee husk and 25 g of rice husk charcoal, which is 1.97 mm and 7.90 pieces.

Acknowledgment

The authors express their gratitude for the financial backing of this research provided by grants from PNBP, Politeknik Negeri Jember. The author also extends thanks to P3M and Agriculture Production, Politeknik Negeri Jember, for their support and help in finalizing this study..

References

[1] BPS, *STATISTIK KOPI INDONESIA Indonesian Coffee Statistics*. Jakarta: Badan Pusat Statistik/BPS-Statistics Indonesia, 2023.

[2] E. Rosdiana, "KARAKTER FISIOLOGIS BENIH KAKAO (Theobroma cacao L.) PADA BEBERAPA KONDISI SUHU DAN MEDIA SIMPAN YANG BERBEDA," *Agrium*, vol. 17, no. September, 2020.

[3] S. A. Nugroho, Suharjono, and R. N. Kusumaningtyas, "Pemanfaatan Limbah Kulit Kopi sebagai Tepung Roti untuk Pemberdayaan Ibu Rumah Tangga di Desa Kemuning Lor Kabupaten Jember," *Sentrinov Vii*, vol. 7, no. 3, pp. 39–49, 2021.

[4] N. Kiggundu and J. Sittamukyoto, "Pryloysis of Coffee Husks for Biochar Production," *J. Environ. Prot. (Irvine,. Calif).*, vol. 10, no. 12, pp. 1553–1564, 2019, doi: 10.4236/jep.2019.1012092.

[5] F. Yelli, H. Hanisah, R. Evizal, and S. Sugiatno, "Pengaruh Formulasi Biochar Dan Limbah Kulit Kopi Terhadap Pertumbuhan Bibit Kopi," *J. Agrotropika*, vol. 19, no. 2, p. 102, 2020, doi: 10.23960/ja.v19i2.4544.

[6] S. Suryaningsih, O. Nurhilal, Y. Yuliah, and E. Salsabila, "Fabrication and characterization of rice husk charcoal bio briquettes," *AIP Conf. Proc.*, vol. 1927, 2018, doi: 10.1063/1.5021237.

[7] D. Budi Setyawan, "Effects of Pyrolysis on Characterization of Rice Hull Charcoal as Growing Media for Plant," *Eur. J. Biol. Biotechnol.*, vol. 4, no. 1, pp. 33–36, 2023, doi: 10.24018/ejbio.2023.4.1.446.

[8] F. Supriyanto dan Fiona, "Pemanfaatan Arang Sekam untuk Memperbaiki Pertumbuhan Semai Jabon (Anthocephalus cadamba (Roxb Miq) pada Media Subsoil," *Silvikultur Trop.*, vol. 2, no. 2086, pp. 24–28, 2010.

[9] R. R. Sari, A. Marliah, and A. I. Hereri, "Pengaruh komposisi media tanam dan dosis NPK terhadap pertumbuhan bibit kopi robusta (Coffea chanephora L.)," *J. Agrium*, vol. 16, no. 1, pp. 28–37, 2019.

[10] M. A. Sefano and G. Gusmini, "Efek Abu Hasil Erupsi Gunung Marapi dan Biochar Kulit Kopi Terhadap Perubahan Sifat Kimia Andisol," *J. TOP Agric. (TOP JOURNAL)*, vol. 2, no. 2, pp. 102–106, 2024.

[11] G. M. Elebiyo and R. T. Bachmann, "Effect of Biochar Type and *Bradyrhizobium japonicum* Seed Inoculation on Soybean Growth, Nodulation and Yield in a Tropical Ferric Acrisol," *Agric. Sci.*, vol. 15, no. 06, pp. 635–675, 2024, doi: 10.4236/as.2024.156036.

[12] H. Zhang, *Biochar Effects on Soil Microbial Biomass and Activity*. Canada: University of Guelph, 2014.

[13] E. Verbruggen, M. G. A. van der Heijden, M. C. Rillig, and E. T. Kiers, "Mycorrhizal fungal establishment in agricultural soils: Factors determining inoculation success," *New Phytol.*, vol. 197, no. 4, pp. 1104–1109, 2013, doi: 10.1111/j.1469-8137.2012.04348.x.

[14] K. Nantre, O. Oksilia, and T. Syamsuddin, "PENGARUH PEMBERIAN BIOCHAR SEKAM PADI DAN PUPUK KOTORAN SAPI TERHADAP PERTUMBUHAN DAN PRODUKSI TANAMAN KACANG BUNCIS TEGAK (Phaseolus vulgaris L)," *AGRONITAS*, vol. 5, no. 2, pp. 363–371, 2023.

Improvement Nutrient Uptake of Robusta Coffee Seedlings (*Coffea canephora*) through the Application of Plant Growth Promoting Rhizobacteria and Mycorrhiza

F Y Ali[1]*, R N Kusumaningtyas[1], D G Pratita[1], S I Kusuma[1], N B Arifiana[1], E Rosdiana[1]

[1]Department of Agricultural Production, Politeknik Negeri Jember
PO BOX 164 Jember, Indonesia

*fandyka.yufriza@polije.ac.id

Abstract. Improving the quality of coffee plant growth is one of the keys to a coffee plantation's success in Indonesia, as it strives to produce high-quality coffee. Good nutrient uptake in plants is a determining factor in producing plants that have good growth. The research was carried out with the aim of increasing nutrient uptake in coffee seedlings through the application of Plant Growth Promoting Rhizobacteria (PGPR) and Vesicular Arbuscular Mycorrhiza (VAM). The research was carried out from April to September 2024 at the Wire House and Bioscience Laboratory, Politeknik Negeri Jember. The study employed a Randomized Complete Block Design (RCBD) using a combination of Plant Growth Promoting Rhizobacteria and Vesicular Arbuscular Mycorrhiza at several doses as a treatment. The results showed that both PGPR and VAM enhanced nutrient absorption efficiency. The combination of PGPR (Plant Growth-Promoting Rhizobacteria) and Vesicular Arbuscular Mycorrhiza (VAM) significantly increases the wet and dry weight of plants, the effectiveness of nutrient assimilation, encompassing vital components such as nitrogen, phosphorus, and potassium, represents a significant aspect of consideration. Combinations of PGPR 10 g.plant^{-1} with mycorrhiza 10 g.plant^{-1} (P4) and PGPR 15 g.plant^{-1} with mycorrhize 10 g.plant^{-1} (P5), showed optimal synergy that supported the growth of coffee plants and their resistance to environmental stress.
Keyword: Mycorrhiza, Nutrient uptake, PGPR, Robusta coffee seedlings

1. Introduction

Increasing the productivity of coffee plants is very important considering the increasing demand for coffee globally. One way that can be used to increase the productivity and quality of coffee plants is through optimizing nutrient uptake, especially during the nursery phase. Because by getting quality coffee seeds will produce coffee plants with good growth and will support good plant productivity as well. One of the stages that determines it is fertilization or the availability of nutrients. Nutrients are very much needed from the nursery stage to the plants producing [1].

In the process of coffee plant cultivation, sufficient nutrient supply is very important for optimal growth and creating quality seedlings. Improvement of soil biological properties is one of the supporting factors in the growth of quality coffee plant seedlings. Efforts that can be made to help improve soil biological properties are inoculation of Vesicular Arbuscular Mycorrhiza (VAM), one of the soil microbes widely found in the root area (rhizosphere). Soil microbes have enormous potential in efforts to increase plant production and productivity with problems of low nutrient uptake [2]. In this context, the use of a consortium of microorganisms, such as PGPR and VAM can be used as an alternative that is worthy of use. The combination of these two microorganisms can improve the ability of plants to uptake essential nutrients, improve soil structure, and increase resistance to environmental stress.

PGPR has a significant contribution to enhancing the accessibility of essential nutrients, especially nitrogen, through the process of nitrogen fixation and phosphorus solubilization. In addition, PGPR can also produce growth regulators that stimulate root growth and increase plant biomass [3]. On the other hand, VAM can form a symbiosis with plant roots, expanding root tissue and increasing the capacity of water and nutrient absorption, especially phosphorus.

Research shows that combining PGPR and VAM can provide a greater synergistic effect in improving nutrient uptake and plant growth. Research [4] indicated that the use of a consortium of these two microorganisms increased nutrient availability, improved soil health, and ultimately increased crop yields. Thus, research on increasing nutrient uptake in coffee seedlings through the PGPR and VAM consortium is very relevant. Through this approach, it is hoped that it will not only increase the productivity of coffee plants, but also support in sustainable agricultural practices.

2. Material and Methods

2.1 Time and Material

The research was carried out from April to September 2024 at the Wire House and Bioscience Laboratory, Politeknik Negeri Jember. The tools and materials employed in this research comprise: Kepelan phase coffee seedlings from Raung-Ijen, Bondowoso Regency. Polybag 20×30 cm, paranet with intensity 70%, Plant Growth Promoting Rhizobacteria (PGPR) with a composition of microorganisms including: *lignocellulosic*, *penicillium sp* 10^8 - 10^{10} cfu.gr^{-1}, *bacillus sp* and *trichoderma* 10^6 - 10^8 cfu.gr^{-1}. Mycorrhiza (*Glomus* sp.), NPK Mutiara fertilizer (16:16:16), Dithane M.45, hoe, saw, oven, research label, stakes, watering can, meter, camera, digital scales, and stationery.

2.2 Methods

The research employed a Randomized Complete Block Design (RCBD) with single-factor namely a consortium of PGPR and Mycorrhiza. Using 6 combinations of treatments, including: P0 = Without PGPR and Mycorrhiza (control); P1 (PGPR 10 gr.plant^{-1} + VAM 5 gr.plant^{-1}); P2 (PGPR 15 gr.plant^{-1} + VAM 5 gr.plant^{-1}); P3 (PGPR 20 gr.plant^{-1} + VAM 5 gr.plant^{-1}); P4 (PGPR 10 gr.plant^{-1} + VAM 10 gr.plant^{-1}); P5 (PGPR 15 gr.plant^{-1} + VAM 10 gr.plant^{-1}). Each experimental treatment was conducted in triplicate, culminating in a total of 18 distinct experimental units. The parameters measured encompassed shoot wet weight, root wet weight, total wet weight, shoot dry weight, root dry weight, total dry weight, and Nutrient Absorption (total nitrogen, phosphorus, and potassium). The resulting data were subsequently subjected to statistical analysis utilizing the F test or ANOVA (Analysis of Variance). Treatments that exhibited statistically significant differences were further examined using the least significant difference (LSD) at a significance level of 5%.

3. Result and Discussion

Measurement of wet weight and dry weight are essential in plant growth studies, including research on the effectiveness of a consortium of microorganisms in improving the growth of coffee seedlings. These two parameters provide important information regarding the physical and physiological conditions of plants. Table 1 shows the effect of the use of a combination of PGPR and Mycorrhiza on the wet and dry weight of coffee seedlings. Based on Table 1, the use of PGPR and mycorrhiza in a consortium has a significant effect on the wet weight and dry weight of coffee seedlings, in the shoots, roots and total of plants.

Table 1. Effect of combined application of PGPR and mycorrhiza on wet and dry weight on shoots, roots and total seedlings of coffee plants

Treatment	Wet Weight (gr.plant^{-1})			Dry Weight (gr.plant^{-1})		
	Shoots	Roots	Total	Shoots	Roots	Total
P0 (control)	7,64 a	2,29 a	9,93 a	2,47 a	1,64 a	4,11 a
P1 (PGPR 10 gr.plant^{-1} + VAM 5 gr.plant^{-1})	7,68 a	2,30 a	9,98 a	3,09 b	1,68 a	4,77 b
P2 (PGPR 15 gr.plant^{-1} + VAM 5 gr.plant^{-1})	8,62 a	2,45 a	11,07 ab	3,66 c	1,86 b	5,51 c
P3 (PGPR 20 gr.plant^{-1} + VAM 5 gr.plant^{-1})	9,30 b	2,38 a	11,68 b	4,15 d	1,74 ab	5,89 c
P4 (PGPR 10 gr.plant^{-1} + VAM 10 gr.plant^{-1})	10,02 b	3,91 a	13,93 c	5,20 e	2,95 c	8,15 d
P5 (PGPR 15 gr.plant^{-1} + VAM 10 gr.plant^{-1})	10,42 b	4,24 b	14,67 c	5,31 e	3,13 d	8,44 d

Note: Numerical values that are accompanied by identical letters within a singular column exhibit no statistically significant difference according to the Least Significant Difference (LSD) test at the 0.05% significance level.

The ratio between wet and dry weight can provide insight into the moisture level of the plant. This ratio can also indicate how the plant adapts to environmental conditions, such as water and nutrient availability. Healthy plants generally show a balanced ratio, while stressed plants will show significant differences [5].

The combination of PGPR and VAM treatments in general can significantly increase wet weight. It was seen that the consortium treatment between PGPR and VAM in higher doses (P4 and P5) gave the highest total wet weight, 13.93 gr.plant^{-1} and 14.67 gr.plant^{-1}. Meanwhile, the treatment without PGPR and VAM (Control) had the lowest total wet weight with a value of 9.93 gr.plant^{-1}. The wet weight of the plant reflects the total mass of the plant which includes all components, including the water contained in the plant cells. So that it can describe the water absorption capacity, which is related to the process of photosynthesis and plant growth [6]. At the dry weight parameter, all treatments showed a significant increase compared to the control. Treatments P4 and P5 again showed the highest values with a total dry weight in each treatment of 8.15 gr.plant^{-1} and 8.44 gr.plant^{-1}, indicating that increasing the dose of PGPR and VAM effectively increased dry plant growth. Dry weight reflects the biomass produced by coffee seedlings and serves to project plant productivity. The dry weight of the plant reflects the amount of organic matter produced, which is important for further growth and development [7], and can describe the efficiency of nutrient use.

The simultaneous use of PGPR and VAM, especially at higher doses, can increase the wet and dry weight of the plants. Treatment with PGPR provides a considerable growth in the wet and dry weight of the plants. According to Goswami *et al* [8], PGPR functions by increasing the availability of nutrients, producing growth hormones such as auxins, and helping plants cope with abiotic stress. This is reflected in the increase in wet weight from P1 to P5 treatment containing PGPR. Mycorrhiza, which is a symbiosis between fungi and plant roots, also has an important role in improving nutrient absorption. Mycorrhiza can increase the surface area of the roots, thereby improving the ability of plants to absorb water and nutrients, especially phosphorus. The observed increase in dry weight in treatment with mycorrhiza supports these findings.

Combination of PGPR and mycorrhiza can produce a synergistic effect that improves overall plant growth. This combination not only promotes root growth but also increases nutrient availability, which are essential for plant growth. The data in the table show that the treatment combining PGPR and mycorrhiza gave better results compared to single treatments. With the increasing wet and dry weight of plants, the application of a combination of PGPR and mycorrhiza can contribute to an increase in agricultural yields. Research by Mahmood *et al* [9], showed that plants treated with PGPR and mycorrhiza had higher yields compared to controls. This shows the important potential of using biotechnology in agriculture to increase productivity. The integration of PGPR and mycorrhiza demonstrated a notably positive impact on both the wet and dry weight of coffee plants. Both function synergistically in increasing nutrient availability and supporting root growth, which ultimately has a positive impact on the yield of the coffee seedlings produced. Therefore, management strategies involving the use of these two microorganisms are highly recommended to increase the productivity of coffee plants and agricultural sustainability in coffee plantations.

Table 2. showed the effect of various combined treatments of PGPR and mycorrhiza on nutrient uptake (N, P, K) in plants. These data provide an overview of how effective the treatment is in increasing nutrient availability for plants. The results presented in the table suggest that the various treatments involving the application of PGPR (Plant Growth-Promoting Rhizobacteria) and Vesicular Arbuscular Mycorrhiza (VAM) have a significant impact on the absorption of essential nutrients such as nitrogen (N), phosphorus (P), and potassium (K) by the plants. These treatments appear to enhance the plant's nutrient uptake, which is crucial for their growth and overall development. In general, increasing doses of PGPR and their combination with mycorrhiza contributed to increasing nutrient uptake, which had a positive impact on plant growth and productivity.

Table 2. Nutrient absorption (N Total, P and K) in coffee seedlings due to the application of PGPR and mycorrhiza as a consortium

Treatment	Nutrient Absorption (mg.plant^{-1})		
	N Total	P	K
P0 (control)	101,83 a	63,91 a	44,03 a
P1 (PGPR 10 gr.plant^{-1} + VAM 5 gr.plant^{-1})	132,13 b	65,03 a	54,05 b
P2 (PGPR 15 gr.plant^{-1} + VAM 5 gr.plant^{-1})	154,48 c	77,25 b	63,05 c
P3 (PGPR 20 gr.plant^{-1} + VAM 5 gr.plant^{-1})	167,64 d	85,37 c	80,76 d
P4 (PGPR 10 gr.plant^{-1} + VAM 10 gr.plant^{-1})	223,90 e	113,30 d	84,87 d
P5 (PGPR 15 gr.plant^{-1} + VAM 10 gr.plant^{-1})	224,29 e	114,12 d	90,10 e

Note: Numerical values that are accompanied by identical letters within a singular column exhibit no statistically significant difference according to the Least Significant Difference (LSD) test at the 0.05% significance level.

The table shows the effects of various treatments on nutrient uptake (N, P, K) in plants. The control treatment (P0) had the lowest nutrient uptake, with a total N of 101.83 mg.plant^{-1}, P 63.91 mg.plant^{-1}, and K 44.03 mg.plant^{-1}. Gradual increases in PGPR and VAM doses (P1 to P5) showed significant results, with treatments P4 and P5 providing the highest uptake for all nutrients: N (223.90; 224.29) mg.plant^{-1}, P (113.30; 114.12) mg.plant^{-1}, and K (84.87; 90.10) mg.plant^{-1}. This shows that the right combination of PGPR and VAM Can improve the effectiveness of nutrient absorption by plants. Treatment with higher doses showed a greater increase, indicating synergy between PGPR and mycorrhiza in increasing the growth of coffee plants[10]. This combination of treatments not only contributes to increased nutrient uptake but also has the potential to increase productivity in coffee plants produced from seedlings with continuous PGPR and VAM treatments.

Coffee seedlings with application of PGPR overall provide better nutrient absorption compared to coffee seedlings without PGPR. This happens because PGPR can increase the availability of nitrogen and phosphorus by increasing the activity of soil microbes that support the mineralization and solubilization process [11]. In addition, PGPR can stimulate root growth, which leads to increased nutrient absorption capacity [12]. Mycorrhiza also plays an important role in improving the efficiency of nutrient absorption by plants, especially phosphorus, and helps plants cope with environmental stress. Mycorrhiza improves the ability of roots to absorb phosphorus, which is an essential but often limited element in the soil [13]. Mycorrhiza also contributes to plant resilience to stressful conditions, such as drought and salinity, through improved root health and better interaction with the soil [14]

The provision of a consortium of PGPR and VMA has proven effective in increasing nutrient uptake by coffee seedlings, with synergies that optimize the accessibility of nutrients like phosphorus and nitrogen. Recent studies have shown that this combination not only expands root tissue and improves nutrient uptake efficiency, but also supports overall plant growth [15][16]. In addition, the addition of mycorrhizal also acts as Mycorrhizal Helper Bacteria (MHB). MHB is a group of bacteria that play a role in supporting and enhancing mycorrhizal effectiveness. They can increase invasion of roots by mycorrhizal fungi. increase their growth and activity in accessing nutrients as well as increase the effectiveness of mycorrhiza and its impact on nutrient absorption. MHB not only increases mycorrhizal colonization but also promotes plant growth through stronger symbiotic interactions[17]. PGPR (Plant Growth Promoting Rhizobacteria) and Mycorrhiza interact in order to promote the growth and development of coffee seedlings through complimentary processes. PGPR increases nutrient availability by fixing atmospheric nitrogen and solubilizing phosphorus, develops growth hormones such as auxins to promote root elongation, and aids plants in dealing with abiotic challenges such as drought and salinity. Mycorrhiza, by its symbiotic interaction with roots, expands the root system with its hyphal networks, enhancing the plant's ability to absorb water and nutrients, notably phosphorus, while also improving soil structure and plant resistance to environmental stress. They constitute an effective consortium that considerably enhances nutrient uptake, as observed in treatments with higher doses (P4 and P5), resulting in significant increases in wet and dry weights of coffee seedlings.

4. Conclusion

From the findings of the conducted research, it can be concluded that the combination of PGPR (Plant Growth-Promoting Rhizobacteria) and Vesicular Arbuscular Mycorrhiza (VAM) significantly enhances both the wet and dry weight of plants, along with the efficiency of nutrient uptake, including nitrogen, phosphorus, and potassium. The combination of PGPR 10 g.plant^{-1} with mycorrhiza 10 g.plant^{-1} (P4) and PGPR 15 g.plant^{-1} with mycorrhiza 10 g.plant^{-1} (P5), showed an ideal synergy that promoted the growth of coffee plants and improved their resilience to environmental stress. This approach not only increases the availability of nutrients, but can also be used as an effort to increase productivity in coffee plants. Therefore, the integration of PGPR and mycorrhiza is an effective strategy to increase the yield of environmentally friendly and sustainable coffee plantations.

Despite the significant potential of PGPR and Mycorrhiza in enhancing coffee seedling growth and nutrient uptake, several challenges need to be addressed, such as the variability in their effectiveness due to environmental factors, soil properties, and coffee plant varieties, as well as compatibility with local soil microbes. Additionally, the production and distribution of high-quality bio-inoculants remain limited, hindering their accessibility to farmers. Future research should focus on identifying PGPR and Mycorrhiza strains suited to specific soil and climatic conditions, as well as conducting long-term field trials to assess their real-world effectiveness. Developing cost-effective production methods for bio-inoculants and improving farmer education through training programs will also support broader adoption and sustainable agricultural practices.

Acknowledgment
The authors wish to express their gratitude for the financial support of this work through grants from PNBP, Politeknik Negeri Jember. The authors also extend their thanks to P3M and the Agricultural Production department of Politeknik Negeri Jember for their support and assistance in completing this research.

References

[1] F. Y. Ali, E. Rosdiana, and A. Kusumaningtyas, Rizky Nirmala; Budianto, "Pengaruh Inokulasi Cendawan Mikoriza Arbuskular (CMA) terhadap Pertumbuhan dan Perkembangan Bibit Kopi Robusta (Coffea canephora) Effect of Arbuscular Mycorrhizal Fungus (AMF) Inoculation on the Growth and Development of Robusta Coffee," *Agropross Natl. Conf. Proc. Agric.*, pp. 5–7, 2023.

[2] E. Verbruggen, M. G. A. Van Der Heijden, M. C. Rillig, and E. T. Kiers, "Minireview Mycorrhizal fungal establishment in agricultural soils : factors determining inoculation success," pp. 1104–1109, 2013.

[3] B. R. Glick, "Plant Growth-Promoting Bacteria : Mechanisms and Applications," vol. 2012, 2012, doi: 10.6064/2012/963401.

[4] L. N. A. Barajas, Y. E. N. Noya, and M. L. L. Guido, "Impact of a bacterial consortium on the soil bacterial community structure and maize (Zea mays L .) cultivation," *Sci. Rep.*, pp. 1–13, 2021, doi: 10.1038/s41598-021-92517-0.

[5] D. Ghosh, A. Gupta, and S. Mohapatra, "Dynamics of endogenous hormone regulation in plants by phytohormone secreting rhizobacteria under water-stress," 2018.

[6] C. Violle, M. Y. Bourget, M. Scherer-lorenzen, and F. Fort, "Allocation , morphology , physiology , architecture : the multiple facets of plant above- and below-ground responses to resource stress," 2018, doi: 10.1111/nph.15225.

[7] I. Hussain, A. Khan, and H. Akbar, "Research Article Maize Growth in Response to Beneficial Microbes, Humic Acid and Farmyard Manure Application," 2021.

[8] D. Goswami, J. N. Thakker, and P. C. Dhandhukia, "Portraying mechanics of plant growth promoting rhizobacteria (PGPR): A review Portraying mechanics of plant growth promoting rhizobacteria (PGPR): A review," *Cogent Food Agric.*, vol. 19, no. 1, 2016, doi: 10.1080/23311932.2015.1127500.

[9] S. Mahmood, M. Ahmad, Z. Ahmad, A. Javaid, and M. Ashraf, "The role of mycorrhizae and plant growth promoting rhizobacteria (PGPR) in improving crop productivity under stressful environments," vol. 32, pp. 429–448, 2014.

[10] I. S. M. Mosaad, D. E. M. Gaafar, M. A. T. Al-anoos, A. K. Seadh, and S. Fertility, "Egyptian Journal of Soil Science," vol. 64, no. 2, pp. 443–457, 2024, doi: 10.21608/EJSS.2024.256247.1702.

[11] N. Khan, S. Ali, M. A. Shahid, A. Mustafa, R. Z. Sayyed, and A. Cur, "Rhizobacteria for Improving Plant Growth and Tolerance to Abiotic Stresses : A Review," 2021.

[12] V. Kumar, M. Kumar, S. Sharma, and R. P. Editors, *Probiotics and Plant Health.*

[13] Y. Khan, S. Shah, and H. Tian, "The Roles of Arbuscular Mycorrhizal Fungi in Influencing Plant Nutrients , Photosynthesis , and Metabolites of Cereal Crops — A Review," 2022.

[14] F. Y. Ali and D. Giatri Cahyaningrum, "Efektifitas Penggunaan Pupuk Hayati terhadap Pertumbuhan dan Produksi Cabai Rawit (Capsicum frutescens L.) pada Tingkat Naungan yang Berbeda," *RADIKULA J. Ilmu Pertan.*, vol. 1, no. 2, pp. 61–69, 2022, doi: 10.33379/radikula.v1i2.1872.

[15] A. A. Belimov *et al.*, "Arbuscular Mycorrhizal Fungus Makes Pea Mutant SGECd t Comparable with Indian Mustard in Cadmium Tolerance and Accumulation," pp. 1–20.

[16] D. Hariyono, F. Y. Ali, and A. Nugroho, "Increasing the Growth and Development of Chili-Pepper Under Three Different Shading Condition in Response to Biofertilizers Application," *AGRIVITA*, vol. 43, no. 1, pp. 198–208, 2021.

[17] S. K. Gupta and A. P. Chakraborty, "Mycorrhiza Helper Bacteria : Future Prospects," vol. 7, no. March, pp. 387–391, 2020.

IOP Conf. Series: Earth and Environmental Science **1446** (2025) 012036 doi:10.1088/1755-1315/1446/1/012036

Formulation And Characterization Of Polychar Plus Granule Organic Fertilizer For Agronomic Testing On Tobacco Plants (*Nicotiana tabacum* L.)

DG Cahyaningrum[1*], Sugiyarto[1], EDI Wilujeng[1], NBE Sulistyono[1], LD Soelaksini[1], NNP Adnyani[1]

Department of Agricultural Production, Politeknik Negeri Jember, Jl. Mastrip PO BOX 164, Jember, East Java, Indonesia

[*]Email : descha.giatri@polije.ac.id

Abstract. Polychar plus granule was manufactured in granulated form from biochar that enriched with beneficial fungi act as decomposers for operational convenience. The preceding research substantiates the potential of utilizing organic materials, specifically coconut shells, for the production of biochar in Jember to augment plant growth. The study encompassed the formulation of a composite comprising coconut shell biochar and sheep manure compost, integrated with a decomposer containing *Trichoderma harzianum* DT 38, *T. pseudokoningii* DT 39, *Aspergillus* sp, and white rot fungi. The resultant product, denominated. The investigation centered on tobacco plants, the principal commodity of Jember Regency, and was conducted from May to September 2024 using a randomized block design featuring two variables: Polychar Plus Granule Dosage (G) (0.00; 12.5; 25.0; 37.5 gr/kg soil) and anorganic fertilizer dosage (P) (0;25;50;100)% from standard operating procedures (S0P). Data analysis was executed utilizing ANOVA and DMRT at a 5% significant level. The findings revealed that the application of anorganic fertilizer doses significantly influenced leaf width but yielded insignificant effects on leaf length and count. Conversely, the application of polychar plus granules showed an insignificant effect, thus requiring dose recombination in further research.

1. Introduction

One of the primary plant commodities in Jember known for its high market value is tobacco. However, the cultivation of tobacco faces various challenges, including uncertain weather conditions and soil saturation with inorganic fertilizers. As reported by [1], efforts to enhance quality through fertilization have proven ineffective. While inorganic fertilization increases tobacco plant productivity, it also leads to soil fertility degradation. To address this, the application of polychar plus organic fertilizer, a granule fertilizer combined with sheep manure (srinthil) compost, biochar, and soil microbes, has been introduced to minimize the residual effects of inorganic fertilizers and support sustainable agriculture. As stated in [2], sheep manure contains significant nutrient content, including 0.70% N, 0.40% P2O5, 0.25% K2O, C/N 20-25, and organic matter 31%. Another viable approach involves utilizing agricultural waste processed into biochar and enriched with non-pathogenic microbes or fungi.

Biochar, a porous form of charcoal derived from the partial carbonization of lignin-cellulose materials via pyrolysis [3], [4], [5], can be sourced from materials such as coconut shells. Aside from its role as a soil amendment, biochar contributes significantly to addressing global warming, climate change, and environmental challenges by facilitating carbon sequestration, stabilization, reduction of greenhouse gas emissions, and soil pollutant remediation [6]. Notably, its application to mustard greens has demonstrated a positive influence on various growth parameters, including plant height, leaf count, total wet weight, wet weight of roots, total dry weight, and dry weight of roots. Moreover, research has indicated that the combined use of biochar and biofertilizer can enhance production yield, resulting in a higher average weight of harvested fresh fruit [7]. Furthermore, the application of biochar mixed with compost, either pre or post-composting, has exhibited more favorable effects compared to the separate application of biochar, compost, or manure [8]. These findings underscore the necessity of investigating biochar formulations to enhance practical applicability, with the dual objective of bolstering long-term land fertility and augmenting current-year production. This endeavor holds the potential for downstream product development emanating from the Politeknik Negeri Jember Organic House Tefa Startup.

2. Material and Methods

2.1. Material and tools

The tools used include pyrolysis reactors, drum disk mill, strainers, mixers, granulators, scales, hoes, buckets, analytical scales, meters, water hoses, water reservoirs, padlocks, stationery, and cameras. The materials used are dried coconut shells, srinthil domba, decomposers, starch, plastics, tarpaulin, polybags, tobacco seeds, and anorganic fertilizers (SP36, Urea, ZA, and KNO3).

2.2. Research method

The study utilized the RAK experimental design, which involved 2 factors or treatments: Polychar Plus Granule Dosage (G) (0.00; 12.5; 25.0; 37.5 gr/kg soil) and anorganic fertilizer dose (P) (0 ; 25; 50; 100) % from standard operating procedures (S0P). (SP36 20 g/plant ; Urea 5 g/plant, ZA1 10 g/plant, ZA2 10 g/plant ; KNO3 5 g/plant). There are 16 treatment combinations, with each treatment repeated 3 times, and each experimental unit containing 5 plants. The preparation involved 3 steps; making biochar, using srinthil domba compost (sheep manure compost), mixing biochar and sheep manure compost, and characterizing polichar plus granule fertilizer. The parameters measured included the number of leaves, as well as leaf length and width (cm).

a. Biochar Preparation
Dried coconut shells have been put into (pyrolysis) burned in closed conditions (low oxygen content) at a temperature of 300-350 °C for approximately 5 hours. Pyrolysis has been declared complete if the smoke production is low/stops and then put into the drum disk mill to be ground. The crushed biochar is sieved with a 1 mm sieve to obtain uniform-sized materials.

b. Sheep Composting Method
Srinthil Domba (sheep manure) has been supplemented with this decomposer containing bacteria *Trichoderma harzianum* DT 38, T. *pseudokoningii* DT 39, *Aspergillus* sp and white rot fungi that can accelerate the composting process as well as a biological fertilizer. In order to be given evenly, this decomposer is dissolved in 100-200 liters of water. Srinthil domba have been spread on the ground floor covered with thick plastic with a size of 1.5 m x 2.5 m with a thickness of 10-15 cm then evenly watered with decomposer solution and covered with tarpaulin. The composting process takes 2-3 weeks and its success is marked by the disappearance of the smell of manure, and on the surface of the srinthil grains, white rot fungi can be found growing.

c. Polychar Plus Granule Method and Formulation
Polychar Plus formulation has been done through several stages, namely: mixing between biochar (Step a) and enriched sheep srintil compost (Step b) with a planned ratio based 1 : 2. The mixing results are sprayed with water mixed with 20 g/l starch and sieved until the mixing results clump into granules.

3. Results and Discussion

Observations on the parameters of leaf number, leaf length (cm), and leaf width (cm) were observed at the age of 30 hst, namely in plants that entered the vegetative phase. The results showed that the applying of anorganic fertilizer gave very significant results (**) on the leaf width parameter, and the application of polychar plus had an insignificant effect (ns) on all parameters (Table 1).

Table1. Summary of results (F-value) on all observation variables

No	Parameter	Polichar Plus Granule (G)		Anorganic Fertilizer (P)		Interaction P X G	
1	Number of leaves	2,19	ns	1,09	ns	0,88	ns
2	Width of leaf (cm)	140,20	ns	0,53	**	0,80	ns
3	Length of leaf (cm)	0,53	ns	0,03	ns	0,90	ns

Note: ns = Non Significant, **= Very Signifinant using DMRT at a 5%

Productivity refers to the evaluation of how effectively resources are handled and used to attain the best outcomes. The productivity of tobacco plants can be assessed by examining the quantity of leaves produced, as well as their length, width, and overall quality. Internal and external factors influence the productivity of tobacco plants. Internal factors are factors that affect growth and come from inside the plant body, such as genetics and hormones. In contrast, external factors are factors that come from outside the plant body, namely the environment, which includes light, humidity, temperature, water availability, media and nutrients [9].

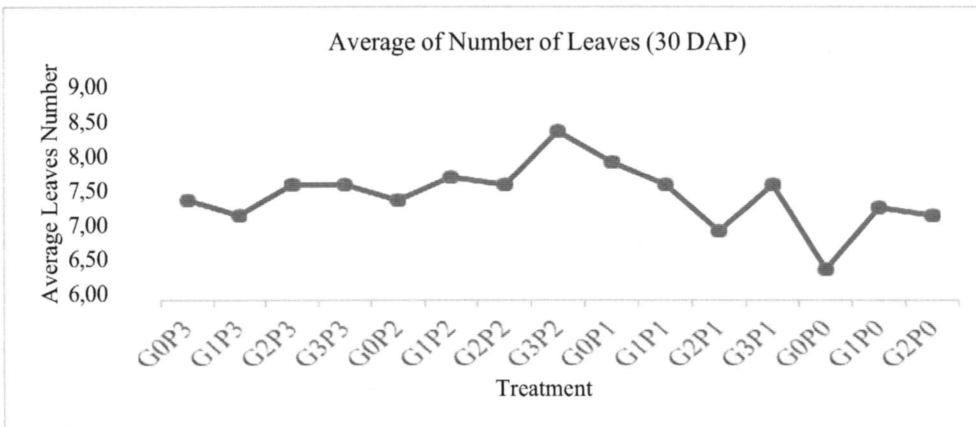

Figure 1. Effect of Polychar Plus Granul (G) Treatment and Anorganic Fertilizer (P) on Parameters of Leaves Number Average

Figure 1 shows the results of applying polychar plus granule and anorganic fertilizer on the leaf count parameters observed at 30 hst, which have not had a significant influence (ns). It is suspected that the observations made are still in the early vegetative stage, so the effects of the treatment on tobacco plants are still not clearly visible. The combination of G3P2 application, namely Polychar Plus Granul (G) Treatment of 37.5 g and anorganic fertilizer as much as 50% of the SOP, gave the highest leaf count yield. The application of Polychar Plus Granul (G) Treatment had a positive effect on the leaf count parameters. This was proven in the treatment of G1P0, G2P0, and G3P0 had a higher number of leaves compared to the control (G0P0). Biochar has a large carbon content of about 85-89%; according to [10], biochar is beneficial in maintaining the moisture of the medium and can help plants retain nutrients in the soil. Sheep manure compost enriched with soil microbes can help plants absorb phosphorus elements so that plants can receive enough nutrients from the soil. Other factors can cause unreal results between treatments and interactions. This can occur due to several factors, including the genetics of the plant itself. According to [11], genetic factors greatly affect the number of leaves formed. In addition to [12] continuing that the number of leaves is also affected by the age of the plant so in addition to the number of leaves, long and wide leaves can also be affected by the genes of the plant.

Table 2. Effect Anorganic Fertilizer Dosage Treatment on Parameters of Leaf Width (cm).

Treatment	Average of Leaf Width (cm)	Notation
P0	9,74	c
P1	20,20	b
P2	21,26	a
P3	9,83	c

Note: P = Anorganic fertilizer dose (0 % (P0) ; 25 % (P1); 50 % (P2); 100 % (P3))
Different letters in the mean column indicate significant differences at the 5% level ($P < 0.05$).

The ANOVA test results indicated that the application of anorganic fertilizer had a highly significant effect (**) on the leaf width (cm) measurement. Furthermore, DMRT tests were carried out, which resulted in anorganic fertilizer treatment at P2; namely, the administration of anorganic doses of 50% of the SOP had the best value on the leaf width (cm) parameter. Table 2 shows that the 100% SOP treatment (P3) provides a smaller leaf width compared to P2. The results of laboratory tests on soil samples showed low N and Organic content, namely 0.09% and 0.95% (Table 3). Element N is necessary for the formation and growth of vegetative parts of plants and also plays a role in the process of photosynthesis, which enhances the growth of plants such as roots, stems, and leaves [13]

The higher the dose of anorganic fertilizer, when given together with manure, will decrease the absorption of plant K. Inorganic fertilizers that contain excessive macronutrients will disturb the balance in the soil, namely inhibit the uptake of micronutrients by plant roots so that it will interfere with plant growth because the metabolic process in plant tissues is disturbed [14].

Table 3. Results of Soil Analysis at the Physics, Chemistry and Biology Laboratory of the Sukosari Research Center. (Author data, 2024)

No	Type	Analysis Result						
		KA (%)	pH (H2O)	Corg (%)	N (%)	P2O5 (ppm)	K20 Cmol(+)K	KTK Cmol(+)K
1	Soil TS	10.24	6.75	0.95	0.09	78.60	0.28	31.31

Apart from improving soil properties, the use of biochar in tropical soils can increase the availability of soil nutrients in the long term, increase soil productivity through improving the physical, chemical and biological properties of soil, improve soil structure, and increase soil water storage capacity. This research was carried out by examining a mixture of coconut shell biochar and sheep manure compost made with a decomposer containing the bacteria *Trichoderma harzianum* DT 38, T. *Pseudokoningi* DT 39, *Aspergillus* sp and white rot fungi. [15] The nutrient element N plays a crucial role in the creation of proteins and chlorophyll. The element P serves as an energy source that aids plants during the vegetative growth stage. Calcium (Ca) is essential for promoting the development of root hairs and reinforcement of the stem. Potassium (K) acts as a catalyst in protein formation, cell division, and carbohydrate synthesis, manages the activity of various minerals, enhances the growth of meristematic tissue, controls stomatal movement, supports the vertical growth of the stem to prevent collapse, and activates enzymes in both direct and indirect ways.

Figure 2 shows the shortest tobacco leaf length, namely in the control (G0P0) Polychar Plus Granule Fertilizer treatment at P2, which is 25 g, gives a better effect compared to P3. Several factors can influence Annova test results that are not significant. The application of organic fertilizers has many advantages but also disadvantages. The application of sheep manure in a relatively short time, namely 2 weeks, still cannot be decomposed properly, so plants will not be maximized in absorbing nutrients [16].

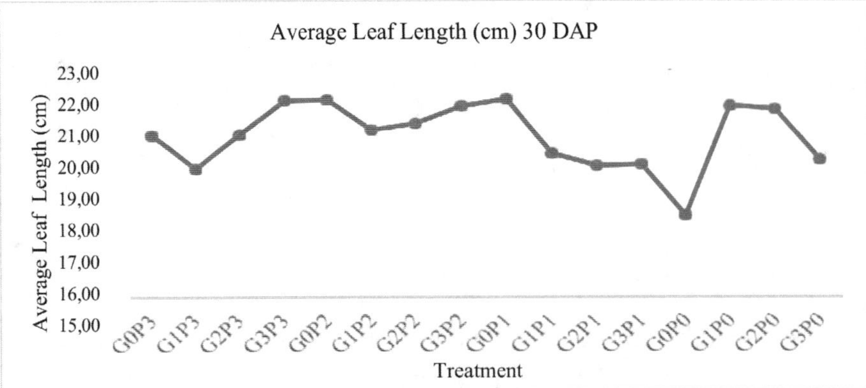

Figure 2. Effect of Polychar Plus Granule Fertilizer Dosage Treatment and Anorganic Fertilizer on Parameters of Leaf Length (cm)

The surplus of carbon in decomposition process stimulates the growth of soil microorganisms and triggers microbial scavenging of soil nutrients, resulting in nutrient immobilization within microbial cells [17]. Those make plants will not be maximized in absorbing nutrients. This will have an impact on plant production. Potassium and phosphorus play an important role in crop yield [17]. There was no difference between treatments due to potassium absorption, which was directly affected by the element N. The lower the N content in the soil, the smaller the absorption of K, so it would have an impact on plant production [18]. Many studies have shown that manure such as sheep and chicken manure is also good when combined with *Trichoderma* [19] [20], and manure applied with *Trichoderma* can provide additional soil bacteria that can increase plant growth [21].

4. Conclusion

Application of polichar plus granules formulation has a positive effect on the growth of tobacco plants. The anorganic fertilizer dose of 50% of the SOP gave very significant results on the leaf width parameter. Further research requires recombination of polychar plus granules organic fertilizer doses to provide significant results on plants.

5. Acknowledgment

Thank you to P3M Politeknik Negeri Jember for funding this research from PNBP scheme No:7934/PL17/KP/2024

References

[1] A. Abdullah and Soedarmanto, *Budidaya Tembakau*. Jakarta: CV Yasaguna, 1986.

[2] Badan Penelitian dan Pengembangan Pertanian, *Pupuk Organik dan Pupuk Hayati*. Bogor: Balai Besar Penelitian dan Pengembangan Sumberdaya Lahan Pertanian, 2006.

[3] D. Widiastuti and M. Maghdalena, "Analysis Benefit Cost Ratio of Biochar in Agriculture Land To Increase Income Household in Merauke Regency," *Jurnal Sosial Ekonomi Kelautan dan Perikanan*, vol. 13, no. 2, pp. 135–143, 2016.

[4] Irvan Andriko Sinaga, Josi Ali Arifandi, and Marga Mandala, "The Effect of Planting Media of Several Formulations of Solid Sandy Soil Biochar on the Quality of Besuki Na-oogst Tobacco Seeds (*Nicotiana tabacum*)," *Agritrop*, vol. 15, no. 2, pp. 277–292, 2017.

[5] Indah Nurul Safitri, Tricandra Setiawati, and Cahyoadi Bowo, "Biochar and Compost for Improving Soil Physical Properties and Water Use Efficiency," *Techno: Jurnal Penelitian*, vol. 7, no. 1, pp. 116–127, 2018.

[6] Sugiyarto, A. Salim, and R. Firgiyanto, "The effect of the use of various kinds of biochar and soil nutrients on pakcoy (brassica rapa l.)," *IOP Conf Ser Earth Environ Sci*, vol. 672, no. 1, 2021.

[7] Y. Alianti, S. Zubaidah, and D. Saraswati, "Response of Tomato Plants (Lycopersicum esculentum Mill.) to the Application of Biochar and Biofertilizer on Peat Soil," *Jurnal AGRI PEAT*, vol. 17, no. 2, pp. 115–125, 2016.

[8] Rombel, Aleksandra, Patrycja Krasucka, and Patryk Oleszczuk, "Sustainable Biochar-Based Soil Fertilizers and Amendments as a New Trend in Biochar Research," *Science of the Total Environment*, vol. 816, pp.

1–17, 2022.

[9] J. Maghfiroh, "Pengaruh Intensitas Cahaya Terhadap Pertumbuhan Tanaman," *Pendidikan Biologi dan Biologi*, vol. 1, no. 1, pp. 51–58, 2017.

[10] A. Kurniawan, B. Haryono, M. Baskara, and S.Y. Tyasmoro, "The Effect of Biochar Use in Planting Media on the Growth of Sugarcane Seedlings (Saccharum Officinarum L.)," Plant Production, vol. 4, no. 2, pp. 153–160, 2016.

[11] B. Lakitan, *Dasar-Dasar Fisiologi Pertumbuhan dan Perkembangan Tanaman*. Jakarta: Raja Grafindo Persada, 1996.

[12] Y.E. Fauzi, W. Yustiana, S. Iman, and R. Hartono, *Kelapa Sawit, Cultivation Utilization of Results and Waste Business Analysis and Marketing*. Jakarta: Penebar Swadaya, 2008.

[13] S. Bahri, B.R. Juanda, and H. Maulida, "Pengaruh Jenis Biochar dan Pupuk Za Terhadap Pertumbuhan dan Produksi Tomat (Lycopersicum esculentum Mill.)," *Agrosamudra*, vol. 5, no. 2, pp. 46–60, 2018.

[14] Elizabeth Kaya, "The Effect of Organic Fertilizer and ANORGANIC Fertilizer on Soil PH and K Availability and K Absorption, Growth, and Yield of Lowland Rice (Oryza sativa L)," *Buana Sains, Jurusan Budidaya Pertanian Fakultas Pertanian Universitas Pattimura*, vol. 14, no. 2, pp. 113–122, 2014.

[15] S. Hadisuwito, *Membuat Pupuk Kompos Cair*. Jakarta: PT. Agromedia Pustaka Jakarta, 2007.

[16] M. Iqbal Abdillah, Titin Setyorini, and Pauliz Budi Hastuti, "Pengaruh Waktu Dekomposisi dan Dosis Pupuk Kandang Kotoran Kambing Terhadap Pertumbuhan dan Hasil Tanaman Terong (Solanum melongena)," *Agroista: Jurnal Agroteknologi*, vol. 7, no. 1, pp. 1–8, 2023.

[17] M. Gannett, A. D. Tommas, J. P. SparksJenny Kao-Kniffin, "a Microbial nitrogen immobilization as a tool to manage weeds in agroecosystems. *J. Agriculture, Ecosystems and Environment* 366 (2024).

[18] L. Liferdi, "Efek Pemberian Fosfor terhadap Pertumbuhan dan Status Hara pada Bibit Manggis," *Jurnal Hortikultura*, vol. 20, no. 1, pp. 18–26, 2010.

[19] Irma Harlianingtyas, Sugiyarto, Cherry Triwiarto, and Supriyadi, "Making Liquid Smoke, Biochar, and Activated Charcoal with Detachable Pyrolysis Equipment at the Jember State Polytechnic Nursery Teaching Factory Pilot Project," *Agrimas: Jurnal Pengabdian Masyarakat Bidang Pertanian*, vol. 2, no. 2, pp. 97–104, 2023.

[20] Cahyaningrum D G, Hayu K, Siwi M, Harlianingtyas I and Dinata G F 2024 Growth response of cocoa seedlings (*Theobroma cacao* L .) ICCRI 06 clone with application of manure and **17** 55–63

[21] Rohman H F, Sulistyono N B E and ... 2022 Pengembangan Produk Pupuk Organik yang Diperkaya Mikoriza (Miko-Pukan) Di Gapoktan Makmur Desa Kemuning Lor, Kecamatan Arjasa, Kabupaten Jember *NaCosVi Polije ...* 265–70

[22] Siswadi E, Sulistuono N, Firgiyanto R, Dinata G and Suharjono 2023 Exploration of bacterial diversity from the soil of citrus plantations applied with organic fertilizer and salicylic acid Exploration of bacterial diversity from the soil of citrus plantations applied with organic fertilizer and salicylic acid *IOP Conf. Ser. Earth Environ. Sci. 1168*

IOP Conf. Series: Earth and Environmental Science **1446** (2025) 012037 doi:10.1088/1755-1315/1446/1/012037

Production of tapai contained probiotic-candidate yeast as emerging food made from cassava for enhanced microflora gut health

N Nurhayati[1,2], A Subagio[2], I Wahyuni[3], U Purwandari[4]

[1] Research Group of Pangan ASUH (Aman Sehat Utuh Halal), Faculty of Agricultural Technology, Universitas Jember, Indonesia
[2] Department of Agricultural Products Technology, Faculty of Agricultural Technology, Universitas Jember, Indonesia
[3] Alumni of Magister Biotechnology, Universitas Jember, Indonesia
[4] Department of Agroindustrial Technology, Faculty of Agriculture, Universitas Trunojoyo Madura, Indonesia
E-mail: nurhayati.ftp@unej.ac.id

Abstract. Emerging food is developed from local commodities to support food security and consumer health from natural ingredients. The aim of these research was to produce the probiotic tapai from cassava using native strains isolated from Indonesian tapai. The research stages carried out were formulating tapai probiotics using local strains, then carrying out inoculation, fermentation and quality analysis. There were six treatments in this study and four treatments were designed for fermentation using local probiotics from preliminary research. Sensory analysis was conducted to determine the preferred tapai as the best product. Based on the research showed the best tapai probiotic formula is the F6 formula which is prepared with the addition of NKL tapai yeast and 1% *Kodamaea ohmeri*. The survival of probiotic yeast on tapai was 52.68-82.71%. It can be concluded that tapai products can be functional food products that have the potential to increase probiotic bacteria.

1. Introduction

It is a wise effort to develop emergency food based on local food commodities. Emergency disaster conditions, whether natural disasters such as floods, landslides, volcanic eruptions, droughts, fires or wars, require emergency food that meets the needs of victims or affected people. Emergency food is not only sufficient for energy but also able to provide good effects on health. Like probiotic food that can provide healthy effects for the body.

One of the fermented foods based on starch as an energy source is tapai which is made from cassava fermented by yeast. Tapai can generally be made from starchy foods such as cassava, sweet potatoes, and glutinous rice, and some are made from plantain var. kepok. Cassava tapai is made from cassava. Rice tapai is made from glutinous rice. The process of making tapai requires yeast as a fermentation starter that will convert starch into simple sugars, then converted again into alcohol and organic acids[1].

The fermentation process in tapai can increase the content of Vitamin B1 (thiamine) up to three times. Thiamine is needed by the digestive system, nervous system and muscle cells in the body. Several studies have reported that fermented food products can be useful for improving the digestive system by increasing the number of good bacteria and reducing the number of unfavorable bacteria in the digestive tract. The presence of yeast and mold in fermented tapai can change complex nutritional compounds such as carbohydrates into simple compounds such as glucose. Changes in these compounds will make tapai easier to digest and absorb by the intestines.

Tapai production generally uses yellow cassava compared to white cassava. The method of making cassava tapai is to choose cassava that is in good condition, not too young or too old. Cassava is peeled, sorted, then cut and washed with clean water before being steamed until cooked for about 30 minutes. The cooked cassava is left to cool and then sprinkled with 1% w/w tapai yeast. Then put in a closed porous container such as woven bamboo ("besek" Javaness language) and fermented for 2-5 days before being served/consumed[2].

2. Methodology

2.1. Materials
The yellow variety of cassava was purchased from local Jember farmers. NKL tapai yeast was purchased from the Jember traditional market. The indigenous tapai probiotic isolate was taken from the ASUH Food collection.

2.2. Design Research
The production process of probiotic tapai uses a completely randomized design (CRD) research design with one factor[3], namely the type of starter to produce probiotic tapai with three replications of treatment. There are eight probiotic tapai fermentation treatments, namely: F1is NKL 0.5%; F2 is NKL 1%; F3 is NKL + *Pichia kudriavzevii* 0.5%; F4 is NKL + *Pichia kudriavzevii* 1%; F5 is NKL + *Kodamaea ohmeri* 0.5%; F6 is NKL + *Kodamaea ohmeri* 1%; F7 is NKL + *Pichia kudriavzevii* + *Kodamaea ohmeri* 0.5%; F8 is NKL + *Pichia kudriavzevii* + *Kodamaea ohmeri* 1%. The overall research stages are presented in Figure 1.

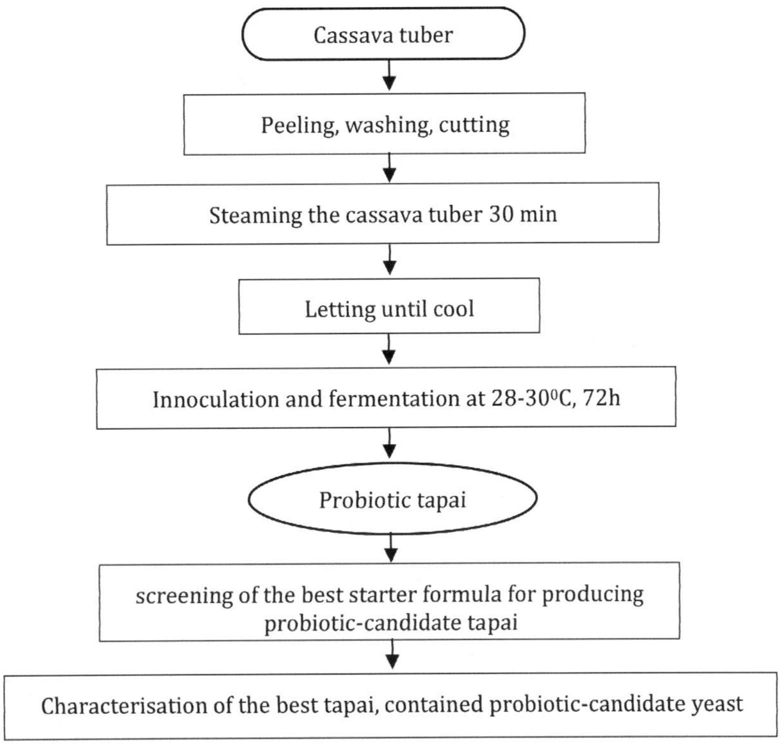

Figure 1. The research stage of tapai contained probiotic-candidate yeasts [4]

2.3. Sensory quality of probiotic tapai

Sensory analysis of the quality of probiotic tapai was carried out using a preference test. The most preferred tapai means the best product produced by a certain starter formula. This evaluation was conducted by 30 trained panelists[5].

2.4. Survival analysis of probiotic yeast

Survival analysis of probiotic yeast was carried out *in vitro* method on volunteer faecal[6]. The inoculated yeast population was calculated using the fertilization method on malt extract agar media. Furthermore, after the tapai fermentation process, the yeast population was recalculated on the same media. The difference in the initial population and the final population will determine the number of yeast that increased or decreased after 24h fermentation process.

3. Results and discussion

3.1. The preferred tapai

Tapai contained probiotic-candidate yeasts was produced from yellow cassava var Malang 2 (Figure 2; Table 1). The sensory attributes assessed in tapai include pale to yellow color, bland to alcoholic aroma, hard to soft texture, watery to dry texture, sour to non-sour taste, sour to sweet taste and a total of very dislike to very like tapai contained probiotic-candidate yeasts.

Figure 2. Fresh yellow cassava var. Malang 2 (a), steamed cassava (b), tapai contained probiotic-candidate yeast (c)

Table 1. Nutrition value of cassava tapai contained probiotic-candidate yeast per 100g

Nutrition	Cassava tapai
Energy (kcal)	173.0
Carbohydrate (g)	42.5
Fat (g)	0.1
Protein (g)	0.5
Calsium (mg)	30.0
Water (g)	56.1
Ash (g)	2.8
Phosphor (mg)	30.0
Vitamin B1 (mg)	0.07

Figure 3. showed the sensory attributes of probiotic tapai with variations in probiotic isolates and starter concentrations, which shows the color of the tapai is quite yellow with a panelist preference level ranging from 4.78-5.09. The color of the resulting tapai is influenced by the ingredients used, such as the variety and age of cassava, the yeast or starter used and the length of the tapai fermentation. Tapai

yeast has a composition of garlic, galangal, glutinous rice flour, pepper and sugar cane juice which causes the formation of a distinctive color in cassava tapai. In this study, the difference in probiotic yeast isolates and starter concentrations did not significantly affect the color of the cassava tapai produced ($p>0.05$).

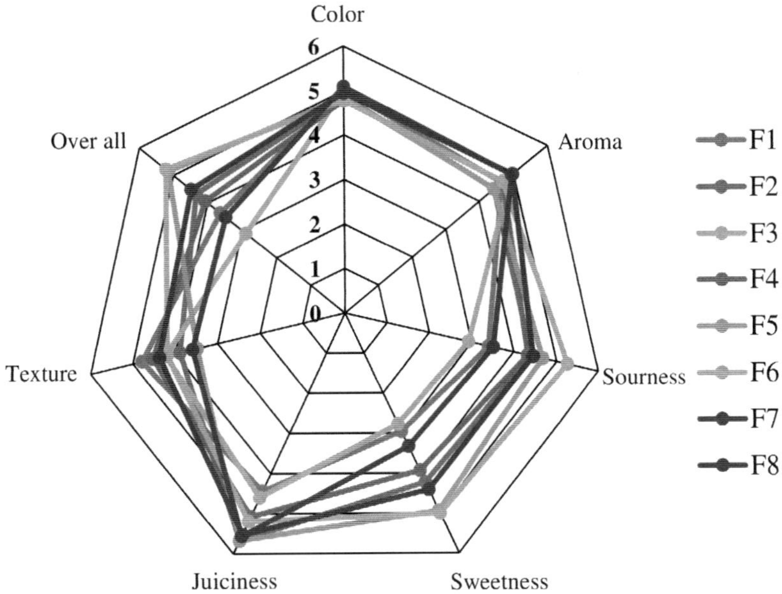

Figure 3. Sensory attributes of tapai contained probiotic-candidate yeast with starter: NKL 0.5% (F1), NKL 1% (F2), NKL+ *P. kudriavzevii 0.5%* (F3), NKL+ *P. kudriavzevii* 1% (F4), NKL+ *K. ohmeri* 0.5% (F5), NKL+ *K. ohmeri* 1% (F6), NKL+ *P. kudriavzevii* + *K. ohmeri* 0.5% (F7), NKL+ *P. kudriavzevii* + *K. ohmeri* 1% (F8)

The panelists liked the color of the cassava tapai produced, with a range of values that were close to each other. The results of the sensory test of the texture of tapai sample F1 with control treatment (NKL 0.5%) or without the addition of starter had the lowest value of 4.44 so that the texture was harder and less preferred by panelists. Sample F1 also showed a rougher texture than the other samples. Panelists tend to prefer a softer tapai texture with the addition of 1% starter.

The higher the percentage of starter used, the more microbes will break down glucose, resulting in a softer tapai texture, and so panelists prefer tapai with the addition of yeast. In addition to improving the taste of tapai products, *K. ohmeri* also has a function for body health. *K. ohmeri* isolated from fish intestines has high phytase activity which is used as a supplement faecal[7; 8]. *Kodamaea* sp can be used as a source of probiotics and consumed as a fermented drink by the Malaysian people[9]. The statement above can be concluded that tapai with the addition of *K. ohmeri* starter also plays a role in improving the resulting taste and has the potential as a probiotic candidate for processing probiotic tapai.

3.2. The survival of probiotic tapai yeast

Probiotic survival of yeast was analyzed in vitro method to determine the modulation of microflora from volunteer feces. Probiotic tapai was inoculated with fecal microflora for 24 hours, then the initial and final microbial populations were calculated. The initial microbial population was compared with the final microbial population (after 24 hours of incubation) which included total microbial growth, enteric bacteria and probiotics (yeast and BAL). Probiotics will play a role in reducing the growth of enteric bacteria and can stimulate the growth of probiotics. The statement above explains that the treatment of probiotic resistance evaluation and probiotic index values in vitro in volunteer faecal need to be further analyzed.

Yeast is a microorganism that plays an important role in the fermentation process of cassava tapai and can stimulate the growth of other probiotics. This statement correlates with the study of probiotic tapai which showed the highest microbial population was dominated by yeast compared to LAB. Table 2 showed the changes of the microflora population in yeast ranged from 13.06-23.25%, while the LAB population decreased with a range of values of 15.02-27.62%. The role of yeast in tapai also increases the growth of yeast in feces. The highest increase in yeast population was sample F7 (NKL + *P. kudriavzevii* + *K. ohmeri* 0.5%) of 23.25%. Sample F3 (NKL + *P. kudriavzevii* 0.5%) had a yeast population close to sample F7 with a value of 20.51%, while F1 and F5 had values close to each other, namely 13.06% and 14.21% respectively. *P. kudriavzevii* is a yeast commonly found as an indigenous microbe, especially in fermented foods and beverages containing alcohol such as kefir, kimchi, cheese, beer[10; 11]. Another research reported that 1200 isolates of *P. kudriavzevii* isolated from food could potentially be probiotic candidates[12].

Table 2. Population of microflora on faecal before and after 24h fermenting tapai contained probiotic-candidate yeast using in vitro model

Tapai contained probiotic-candidate	Population of microflora (Log_{10} cfu/g)							
	Total microbe		Lactic acid bacteria		Yeast		Enteric bacteria	
	Before	After	Before	After	Before	After	Before	After
F1	8.98	9.85	5.77	4.41	6.63	7.49	5.78	5.82
F4	8.80	9.91	5.41	3.91	6.41	7.83	5.76	5.63
F6	8.11	8.93	4.71	4.00	6.76	7.72	5.48	5.53
F7	7.80	8.78	4.56	3.56	6.41	7.90	5.32	5.18

Noted: (+) is increasing population; (-) is decreasing population
F1 = probiotic tapai with starter NKL 0.5%
F4 = probiotic tapai with starter NKL+ *P. kudriavzevii* 1%
F6 = probiotic tapai with starter NKL+ *K. ohmeri* 1%
F7 = probiotic tapai with starter NKL+ *P. kudriavzevii* + *K. ohmeri* 0 .5%

Table 3. The changes of microflora population on faecal after 24h fermenting tapai contained probiotic-candidate yeast using in vitro model

Tapai contained probiotic-candidate	Changes of microflora (%)			
	Total microbe	LAB probiotic	Yeast	Enteric bacteria
F1	(+) 9.65±0.61	(-) 23.66±0.97	(+) 13.06±0.61	(+)0.57±0.05
F4	(+) 10.16±0.58	(-) 15.02±0.50	(+) 14.21±0.68	(+) 0.85±0.02
F6	(+) 12.54±0.78	(-) 27.62±0.50	(+) 20.51±0.94	(-) 2.28±0.09
F7	(+) 12.58±0.69	(-) 21.93±0.71	(+) 23.25±1.05	(-) 2.53±0.10

Noted: (+) is increasing population; (-) is decreasing population
F1 = probiotic tapai with starter NKL 0.5%
F4 = probiotic tapai with starter NKL+ *P. kudriavzevii* 1%
F6 = probiotic tapai with starter NKL+ *K. ohmeri* 1%
F7 = probiotic tapai with starter NKL+ *P. kudriavzevii* + *K. ohmeri* 0 .5%

The enteric bacterial population in samples F4 and F7 decreased by 2.28% and 2.53% respectively, but there was an increase in enteric bacteria in samples F1 (0.57%) and F6 (0.85%). The presence of probiotic yeast can modulate the microflora in feces, namely by increasing total microbes and yeast, and decreasing enteric bacteria. According to another research[13], probiotic yeast plays a role in inhibiting the growth of pathogenic bacteria. Yeast acts against enteric bacteria through many different mechanisms. For example, when in the digestive tract, yeast will prevent bacterial adhesion and translocation in intestinal epithelial cells. Yeast will also produce or release metabolites to neutralize toxins in bacteria[14]. The highest decrease in the enteric bacterial population was the F7 tapai (NKL + *P. kudriavzevii* + *K. ohmeri* 0.5%), but the F6 tapai (NKL + *P. kudriavzevii* 1%) also experienced a decrease that was not much different. *P. kudriavzevii* plays a very important role in inhibiting enteric bacteria and its properties as a probiotic agent are better than tapai with the addition of *K. ohmeri*. This statement is supported by another research[14; 15] that *P. kudriavzevii* is the best potential candidate for probiotic yeast with non-pathogenic characteristics, high alcohol tolerance, temperature resistance of 37° C and bile salt resistance.

4. Conclussion

The most preferred tapai contained probiotic-candidate yeast using sensory analysis is tapai fermented by starter formula 6 (F6) made from NKL yeast and the addition of 1% *K. ohmeri* that increase in yeast population from in vitro probiotic resistance testing by 13.06-23.25%, decrease in LAB population by 15.02-27.62% and decrease in enteric bacterial population by 0.54-2.53%. The addition of both yeasts has been proven to have many benefits in the field of fermented food and biotechnology, especially its role as probiotic-candidated yeast.

5. Acknowledgement

Thank you to Universitas Jember for the facilities provided for this research through Accelerated Research Grant for Professors 2024 with strengthening of Applied Research DRPM 2024

References

[1] Anisa, F. (2017). Mutu kimia dan organoleptik tape hasil fermentasi umbi talas kimpul (*Xanthosoma sagittifolium*) dengan berbagai konsentrasi ragi. *Jurnal Aplikasi Teknologi Pangan*, 6(1), 43–47.

[2] Islami, R. (2019). Making yeast tape and tape [Pembuatan ragi tape dan tape, Indonesian]. *Journal of Chemical Information dan Modeling*, 53(9), 56–62

[3] Ali, F. (2020). Detection and treatment of outliers in experimental design: real data for completely randomized design. *Journal of Al-Rafidain University College or Sciences (Print ISSN: 1681-6870, Online ISSN: 2790-2293)*, (1), 420-431.

[4] Wahyuni, I., Purwandari, U., Subagio, A. and Nurhayati, N. (2023). Isolation and identification of gastric acid-tolerant yeast from tapai. *Food Research* 7 (Suppl. 1) : 276 - 282

[5] Yusmarini, Y., Johan, V. S., Fitriani, S., Rahmayuni, R., Artanti, V. F., & Pato, U. (2019). Characteristics of probiotic tapai made by the addition of *Lactobacillus plantarum* 1. *Intern* J *of AgriTech*. Vol. 15(1): 195-206

[6] Dalu, K. C. A., Nurhayati, N., & Jayus, J. (2019). In vitro modulation of fecal microflora growth using fermented "pisang mas" banana and red guava juices. *Current Research in Nutrition and Food Science Journal*, 7(2), 449-456.

[7] Hirimuthugoda, N. Y., Chi, Z., Li, X., Wang, L. dan Wu, L. (2006). Diversity of phytase-producing marine *yeasts*. *Ciencias Marinas*, 32(4), 673-682.

[8] Kanti, A., Sumerta, I. N., Idris, I., Napitupulu, T. P., dan Ikhwani, A. Z. N. (2020). Isolation and identification of ascomycetous *yeast* producing phytase. *IOP Conference Series: Earth and Environmental Science*, 572(1).

[9] Azhar, M. A., Sakinah, M., dan Munaim, A. (2019). Isolation and molecular identification of potential probiotic *yeast* strains found in Malaysian kefir drinks samples. *International Journal of Food Engineering*, 8(4), 128–131.

[10] Greppi, A., Saubade, F., Botta, C., Humblot, C., Guyot, J. P., & Cocolin, L. (2017). Potential probiotic *Pichia kudriavzevii* strains and their ability to enhance folate content of traditional cereal-based African fermented food. *Food Microbiology*, *62*, 169-177.

[11] Helmy, E. A., Soliman, S. A., Abdel-Ghany, T. M., & Ganash, M. (2019). Evaluation of potentially probiotic attributes of certain dairy yeast isolated from buffalo sweetened Karish cheese. *Heliyon*, *5*(5).

[12] Gil-Rodríguez, A. M., Carrascosa, A. V., dan Requena, T. (2015). *Yeast*s in foods dan beverages: In vitro characterisation of probiotic traits. *LWT - Food Science dan Technology*, 64(2), 1156–1162.

[13] Binetti, A., Carrasco, M., Reinheimer, J., dan Suárez, V. (2013). *Yeast*s from autochthonal 396 cheese starters: technological and functional properties. *Journal Apply Microbiology*. 115 (397), 434–444.

[14] Tiago, F. D. C. P., Martins, F. D. S., Souza, E. L. S., Pimenta, P. F. P., Araújo, H. R. C., and Castro, I. D. M., and Nicoli, J. R. (2012). Adhesion to the *yeast* cell surface as a mechanism for trapping pathogenic bacteria by *Saccharomyces* probiotics. *Journal of Medical Microbiology,* 61(9), 1194–1207

[15] Wulan, R., Astuti, R. I., Rukayadi, Y., dan Meryandini, A. (2021). Evaluation of indigenous *Pichia kudriavzevii* from cocoa fermentation for a probiotic candidate. *Biodiversitas*, 22(3), 1317–1325.

Pellet formulation of peppermint essential oil *Mentha piperita* L as repellent againts *Araecerus fasciculatus* (De Geer) (Coleoptera: Anthribidae)

Zulfadly[1], Sylvia Sjam[1*] and Melina[1]

[1] Department of Plant Pest and Disease, Faculty of Agriculture, Hasanuddin University, Makassar, Indonesia 90245

[*] Email: sylviasjam@yahoo.com

Abstract. *Araecerus fasciculatus* is a pest that can cause damage to various commodities in storage. This pest can significantly impact crop quality and quantity when infestations are high. *Mentha piperita* L is a plant that contains various bioactive compounds known for their repellent and toxic properties, making it a potential environmentally friendly solution for controlling pests in warehouses. The research aimed to testing the repellent influence of *M. piperita* essential oil on *A. fasciculatus* using a pellet media. The research was conducted in the laboratory using a completely randomized design with pellet media and four treatment concentrations: 1%, 3%, 5%, and 7%, as well as a control, each with four replications. The total number of insects in each treatment consisted of 32 adults. An olfactometer was used for the test, and the repellent effect was observed daily for seven days after the essential oil application. The results demonstrated that *M. piperita* essential oil, when packaged in pellet media, repelled *A. fasciculatus*. The highest repellent effect, reaching 86.43%, was observed at a 7% concentration, while the lowest repellent effect, at 1% concentration, was 32.5%.

1. Introduction

Crop yields can be significantly impacted by commodity damage during storage, leading to economic losses and affecting the community's well-being. One of the primary contributors to postharvest commodity damage is the presence of biological factors, specifically pests that harm stored crops. [1]

Araecerus fasciculatus is common postharvest pest found in storage that attacks areca nut, brown coffee, nutmeg, and dried cassava [2]. It is classified as a polyphagous pest [3].

Chemical fumigation is a common method for pest control. Phosphine (PH_3) and methyl bromide (CH_3Br) are two common fumigants, but the use of methyl bromide is limited to quarantine purposes due to its classification as an ozone-depleting compound. Inappropriate use of phosphine can result in insect resistance in stored grains. Therefore, it is important to explore natural plant-based products as eco-friendly alternatives for managing postharvest pests in storage facilities.

Plants serve as natural producers of various bioactive organic chemicals that primarily function protectively, particularly against insects pests. Such bioactive natural compounds release fragrances that are usually volatile in nature called as plant volatile organic compounds (PVOC) [4]. Many plant species are rich sources of phytochemicals for use as insecticides [5]. Plant metabolites have a portion in plant defense instrument making them one of the foremost curiously viewoints of plant security [6]. In addition, they also act as attractants or repellents.

Mentha piperita L contains bioactive secondary metabolites such as terpenes, as well as other substances that are toxic, repellent, and have antifeedant qualities against various insects and pests. In addition to repelling *Tribolium castaneum* adult beetles, *M. piperita* also has a fumigant effect that causes the death of the beetles [7]. The menthol component found in *M. piperita* has the ability to repel *Anopheles tessellatus* and *Culex quinquefasciatus* [8].

This research aims to testing the repellent effect of *M. piperita* L oil against *A. fasciculatus* using pellet media.

2. Material and methods

2.1 Insect multiplication

A. fasciculatus were reared in plastic containers measuring 20 cm x 17 cm x 17 cm. 200 grams of dried cassava were put into each plastic container with 30 imago [9]. After seven days of infestation, the imago was removed from the plastic container. Dried cassava was then stored for two months indoors at 26-30°C with 60-80% humidity.

2.2 Essential oils

The essential oil of *M. piperita* used was obtained from one of the herbal product stores in Yogyakarta.

2.3 Preparation of pellet media

The process of making pellets involves using 50 ml water, 10 gram Carboxymethyl Celulose (CMC), and 50 gram sawdust [10]. The resulting mixture was formed into a round shape weighing 1 gram.

2.4 Repellency test

Repellent effect of *M. piperita* essential oil against *A. fasciculatus* imago was evaluated using the two-choice method with concentrations of 1%, 3%, 5%, and 7%. The essential oils were diluted with acetone solvent to their respective concentrations, evenly distributed on the pellet media, and air-dried for 3-4 minutes to evaporate the solvent. The olfactometer used was modified from [11] (Figure. 1). This tool comprises two plastic tubes (length = 10 cm; diameter = 2 cm). In the center, there is a container (length = 10 cm; width = 10 cm). Two containers (length = 8 cm; width = 5 cm) are connected to both ends of the tube.

To conduct the test, we placed pellet media with different concentrations of essential oils in one container, while the other container contained pellet media with solvent as a control. We released eight test insects into the olfactometer through a hole in the center connected to a plastic tube. A 10 cm section of the tube connected to the plastic container containing the treated pellet media was designated as the treatment area, while the other section was the untreated area. We observed the number of insects in each treatment and control area every 24 hours for seven days after release [12] and then calculated using the percentage of repellency to assess the tendency of insect choice to approach or move away from the treatment area.

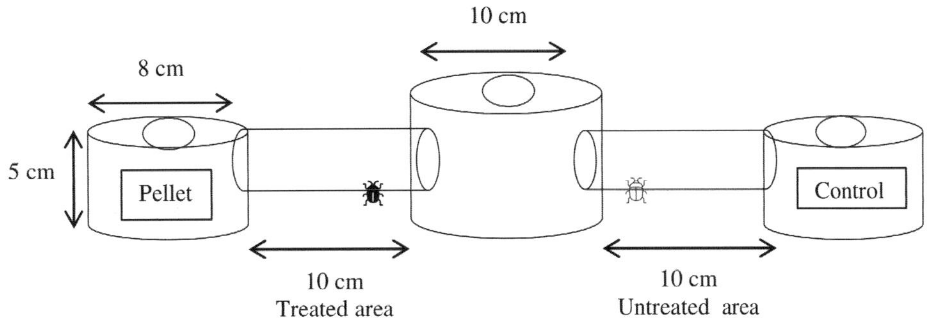

Figure 1. Olfactometer

2.5 Test design

The study utilized a completely randomized design with four concentration treatments (1%, 3%, 5%, and 7%) and one control, each with four replications. There were a total of 32 adult insects in each treatment. The observation parameters included the average number of arrival by *A. fasciculatus*, fluctuation, percentage of arrival, and percentage of repellency. Repellency was counted using the formula [13]:

$$\text{Percentage of Repellency}=\frac{(\text{Ta-Tb})}{\text{Ta}}\times 100 \quad (1)$$

where T_a = insects in control, T_b = insects in treatment

The repellent effect of a plant extract can be assessed using the following categories [14]:

Category 1 (Very low) : 0,1-20%
Category 2 (Low) : 20,1-40%
Category 3 (Medium) : 40,1-60%
Category 4 (High) : 60,1-80%
Category 5 (Very high) : 80,1-100%

The percentage of arrival is calculated using the following formula [15]:

$$\text{Percentage of Arrival} =\frac{\text{Ca}}{\text{Tb}}\times 100 \quad (2)$$

where C_a= insects in treatmenr, T_b = total insects used

Observational information were analyzed using ANOVA, and if treatment affected the analysis advance tests were conducted utilizing the DMRT at 5%.

3. Results and discussion

3.1 Average arrival of Araecerus fasciculatus

The test results using *M. piperita* essential oil showed a repellent effect on *A. fasciculatus* in all treatments except the control. The analysis of variance results indicated a significant effect on the arrival of *A. fasciculatus*, as shown in (Table 1).

Table 1. Average number of *A. fasciculatus* arrivals at four concentrations of *M. piperita* during a seven-day observation period

Treatment	Days						
	1	2	3	4	5	6	7
Control	1,44^{a}	2^{a}	1,68^{a}	1,88^{a}	1,12^{a}	0,75^{a}	0,06^{b}
1%	0,25^{b}	0,75^{b}	0,75^{ab}	0,75^{b}	0,5^{b}	0,75^{a}	0,75^{a}
3%	0,25^{b}	0,75^{b}	1^{ab}	1^{ab}	0^{b}	0^{a}	0^{b}
5%	0,25^{b}	0,25^{b}	0^{b}	0^{b}	0,25^{b}	0,25^{a}	0^{b}
7%	0^{b}	0^{b}	0^{b}	0^{b}	0,25^{b}	0^{a}	0,25^{b}
statistics (df = 4,15)	F=5.6;P <0.05	F=5,58; P<0,05	F=3,30; P<0,05	F=5,02; P<0,05	F=4,31; P<0.05	F=2,37; P>0.05	F=3,90 P<0.05

*Number taken after the same letter are within same column show comes about that are not essentially diverse at P≤ 0.05 agreeing to DMRT

Based on Table 1, the control treatment shows the highest average arrival, while the 7% concentration treatment has the lowest average arrival. This is because *M. piperita* contains menthol (59%) and menthone (19%), which are volatile and repellent. The strong aromatic components of *M. piperita* can affect the insect nervous system, acting as a neurotoxin that causes insects to avoid the scent. The watched side effects, such as hyperactivity and tremors are similar to those caused by insecticides like organophosphates [16]. The vapor pressure of essential oil molecules can affect their availability to the trachea during respiration as they vaporize and become accessible [17]. Essential oil molecules with high vapor pressure are more volatile and generally more harmful than molecules with low vapor pressure.

3.2 Fluctuation of Araecerus fasciculatus arrival number

The test results displayed fluctuations in the arrival of *A. fasciculatus* across various treatments during the 7-day observation period, as shown in (Figure 2).

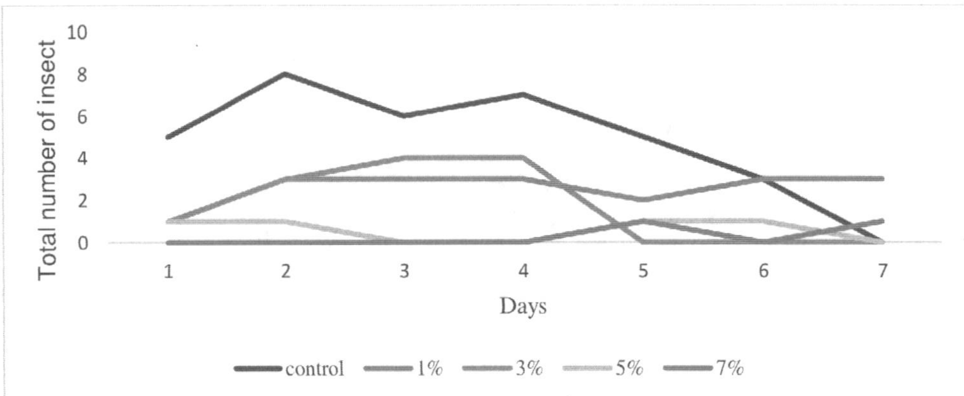

Figure 2. Fluctuations in the number of *A. fasciculatus* arrivals were observed in each treatment over a period of 7 days.

Based on Figure 2, the highest fluctuation occurs in the control and then low in the concentration of 3%, 1%, 5% and 7%. The highest number of *A. fasciculatus* arrivals was eight in the control group, while the 3%, 5%, and 7% treatments had zero arrivals. In the control group, fluctuations increased from day one to day two and decreased on the day four, whereas in the 1%, 5%, and 7% treatments, fluctuations began to increase on the fourth and fifth days. This is due to the repellent effect that increases and decreases as the observation time increase. The oils have volatile physical at higher temperatures. The temperature during observation ranged from 26-29°C. They are naturally liquid at room temperature and easily vaporize at higher temperatures without decompos [18].

3.3 Percentage of Araecerus fasciculatus arrival
The test results indicate the percentage of *A. fasciculatus* arrival in the control is higher and shows a significant difference compared to all concentrations of *Mentha piperita* essential oil, as shown in (Figure 3).

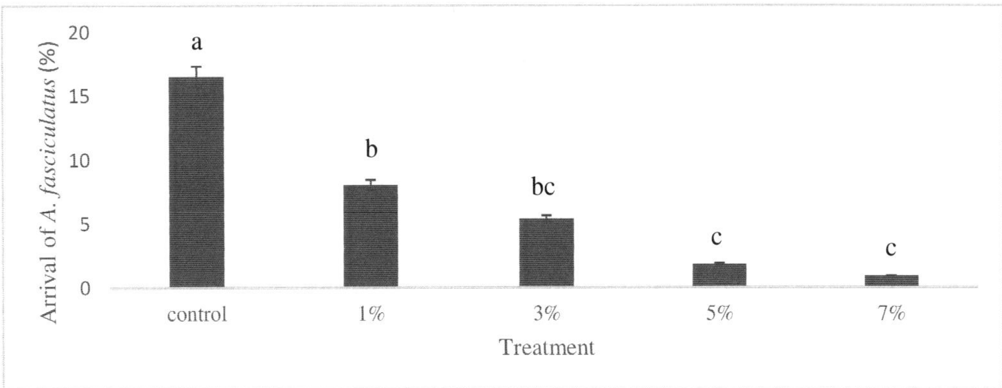

Figure 3. Percentage of *A. fasciculatus* arrivals in each treatment over a seven-day observation period.

According to Figure 3, it is evident that the highest percentage of *A. fasciculatus* arrival is observed in the control group at 16.5%, and the lowest at 7% concentration, at 0.89%. This indicates that *M. piperita* essential oil has a strong repellent effect, leading to *A. fasciculatus* avoiding pellets with concentrations of 1%, 3%, 5%, and 7%, and preferring the control. The percentage of arrival decreases as the concentration of *M. piperita* essential oil increases. It is suggested that the repellent effect can be strengthened and extended by increasing its concentration..

3.4 Percentage of Mentha piperita repellency against Araecerus fasciculatus
The results of the percentage of repellency of *M. piperita* against *A. fasciculatus* showed varying outcomes in each treatment, as detailed in Table 2.

Table 2. Percentage and classification of repellency level of *M. piperita* essential oil against *A. fasciculatus*.

Treatment	Repellency (%)	Class	Category
1%	$32{,}50^{b}$	II	Low
3%	$65{,}09^{ab}$	IV	High
5%	$78{,}49^{a}$	IV	High
7%	$86{,}43^{a}$	V	Very High
statistics (df= 3, 24)	F=3,02; P<0.05		

*Number taken after the same letter are within same column show comes about that are not essentially diverse at P≤ 0.05 agreeing to DMRT

According to Table 2, the highest repellency percentage is at a 7% concentration, at 86.43%, which falls into the very high category. The lowest repellency percentage is at a 1% concentration, at 32.50%, falling into the low category. This is because higher concentrations of essential oils contain more active biochemicals. *M. piperita* has main compound components, namely mentol [19] which exhibits strong insecticidal and repellent properties. [20].

4. Conclusion

Mentha piperita essential oil with pellet formulation provides a repellent effect against *Araecerus fasciculatus* with a very high category at a concentration of 7% by 86.43% and a low category at a concentration of 1% by 32.5%. It has no negative effects on the environment due to its volatility nature so oil and constituent is enviromentally non persistent with outdoor half lives of 24 hour on surfaces.

5. Acknowledgments

The authors express their appreciation to all parties included in this research. They also acknowledge the Ministry of Agriculture and the Indonesian Quarantine Agency for providing educational assistance through the Civil Servant Study Task Scholarship..

References

[1] F.X. Wagiman. Hama Pascapanen Dan Pengelolaannya. D.I.Yogyakarta: Gadja Mada University Press; 2019. 6–16 p.

[2] Dharmaputra OS, Sunjaya, Retnowaty I, Nurfadila N. Keanekaragaman serangga hama pala (Myristica fragrans) dan tingkat kerusakannya di penyimpanan. Entomologi. 2018 Jul;15:57–64.

[3] Koval AG, Makarov KV, Korotyaev BA. On a finding of Araecerus fasciculatus in natural habitats of different regions of Southern Russia. Entomol Rev. 2018 Jan;99((1)):129–32.

[4] Isman MB. Botanical insecticides, deterrents, and repellents in modern agriculture and an increasingly regulated world. Entomol. 2006 Jul 11;51:45–66.

[5] Green PW, Davis AP, Cosse AA, Vega FE. Can coffee chemical compounds and insecticidal plants be harnessed for control of major coffee pests? Agric Food Chem. 2015 Oct 12;63:9427–34.

[6] Bennet RN, Wallsgrove RM. Secondary metabolites in plant defence mechanisms. New Phytol. 1994 Apr 4;127:617–33.

[7] Lashgari A, Mashayekhi S, Javadzadeh M, Marzban R. Effect of Mentha piperita and Cuminum cyminum Essential Oil on Tribolium castaneum and Sitophilus oryzae. Archivesof Phytopathology and Plant Protection. 2013 May 24;47(3):324–9.

[8] Samarasekera R, Weerasinghe IS, Hemalal KP. Insecticidal Activity of Menthol Derivatives Against Mosquitoes. Pest Manag Sci. 2008 Dec 20;64:290–5.

[9] Eduku A, Maalekuu BK, Kaledzi PD, Tandoh PK. Development of bait for the management of Araecerus fasciculatus. Asian Research Journal of Agriculture. 2018 Apr 7;8(3):1–11.

[10] Sjam S, Melina, Thamrin S. Pengujian Ekstrak Tumbuhan Vitex trifolia L., Acorus colomus L., dan Andropogon nardus L. terhadap Hama Pasca Panen Araecerus fasciculatus De Geer (Coleoptera: Anthribidae) pada Biji Kakao. J Entomol Indones. 2010 Jan;7:1–8.

[11] Boeke SJ, Baumgart IR, van Loon JJA, van Huis A, Dicke M, Kossou DK. Toxicity of African plants traditionally used for the protection of stored cowpea against Callosobruchus maculatus. . J Stored Prod Res. 2004 Apr 8;40:423–38.

[12] Caballero-Gallardo K, Olivero-Verbel J, Stashenko E. Repellency and toxicity of essential oils from Cymbopogon martinii, and Lippia origanoides against Tribolium castaneum. Journal of Stored Products Research. 2012 May 9;50:62–5.

[13] Govindarajan M, Krishnappa K, Anandan A. Ovicidal and repellent activities of botanical extracts against Anopheles stephensi (Diptera: Culicidae). Asian Pac J Trop Biomed. 2011 Feb 28;1(1):43–8.

[14] Inyang UE, Emosairue SO. Laboratory Assessment of The Repellent and Anti- Feedant Properties of Aqueous Extracts of 13 Plants Against Sordidus Germar (Coleoptera: Curculionidae). Tropical and Subtropical Agroecosystems. 2005 Mar 24;5:33–44.

[15] Susanto A, Nasahi C, Rumaisha YK, Murdita W, Puji Lestari TM. Penambahan Essens Buah terhadap Keefektifan Metil Eugenol dalam Menarik Bactrocera spp. Drew & Hancock. Jurnal Agrikultura. 2019;30(2):53–62.

[16] Kostyukovsky M, Rafaeli A, Gileadi C, Demchenko N, Shaaya E. Activation of octopaminergic receptors by essential oil constituents isolated from aromatic plants. Pest Manag Sci. 2002 May 20;58:1101–6.

[17] Philips AK, Appel AG, Sims SR. Toxicity and repellency of essential oil to the German cockroach (Dictyoptera; Blattellidae). J Econ Entomol. 2010 Apr;103(2):448–59.
[18] Koul O, Walia S, Dhaliwal GS. Essential oils as green pesticides:. Biopestic Int. 2008;4(1):63–84.
[19] Politeo O, Jukic M, Milos M. Chemical composition and antioxidant activity of essential oils. Croatica Chemica Acta. 2006 Mar 16;79(4):545–52.
[20] Aggarwal KK, Tripathi AK, Ahmad A, Prajapati V, Verma N, Kumar S. Toxicity of L-Menthol against storage insects. Insect Science Applic. 2001 Jul 2;21(3):229–35.

IOP Conf. Series: Earth and Environmental Science **1446** (2025) 012039 doi:10.1088/1755-1315/1446/1/012039

Strategies for Increasing Rice Productivity in Lowland Rainfed Fields Environment-Friendly Systems

V E Aristya[1], W A Nugroho[1], Samijan[1], S Minarsih[1] and Y Hindarwati[1*]

[1]Research Center for Food Crops, Agricultural and Food Research Organization, National Research and Innovation Agency (BRIN), Jl. Raya Jakarta-Bogor KM 46, Cibinong, Indonesia 16911

[*]Email: yuli059@brin.go.id

Abstract. Indonesia's efforts to increase rice production are driven by the goal of becoming the world's food granary by 2045. Rainfed lowlands currently occupy 46% of all rice cultivation and hold the potential for expanding national rice production but encounter biophysical constraints and low rice productivity. Land management neglecting environmental sustainability threatens sustainable rice production. This study examines alternative technologies that can be applied to lowland rainfed paddy fields to increase productivity and environmental sustainability. This systematic literature review uses 40 references, 81% of which are primary journals related to rainfed agroecosystems, rice cultivation, and environmentally friendly systems. Research showed that efforts to advance rice productivity in rainfed fields were pursued through environmentally friendly technologies and optimization of land management, including using superior varieties, planting technology, timing of planting, water resource management, balanced fertilization, pest/disease control, and appropriate harvesting practices. Drought-adaptive varieties were the solution to increasing plant populations and shortening harvest time. Balanced organic and inorganic fertilization was also essential in managing rainfed lowland rice fields. Furthermore, the farmers' active role, infrastructure, social, economic, and environmental support have encouraged environmentally friendly technological innovations to attain food self-sufficiency.

1. Introduction

Indonesia aims to become the world's food granary by 2045. Continuous efforts to achieve sustainable rice self-sufficiency are being intensified to realize the vision of food sovereignty. Indonesia's rice productivity is 5,226 tonnes/ha, and a total production of 54.42 million tonnes from a harvested area of 10.41 million hectares. Rice production is derived from 7.4 million hectares of land, of which 46% comprises rainfed lowland fields [1].

Considering the substantial proportion of available land, rainfed lowland areas hold the potential for expanding rice production. However, the challenge lies in the significantly lower rice productivity in rainfed lowlands than in irrigated rice fields [2]. Rice cultivation in rainfed lowlands yields productivity ranging from 3.0 to 3.5 tonnes/ha, substantially lower than in irrigated fields (6-7 tonnes/ha). Furthermore, rice cultivation in rainfed lowlands can only be during the rainy season (September-December).

The availability of land resources plays a pivotal role in agricultural management. As a production factor, land is a determinant of crop productivity success. Optimal land management should prioritize environmental sustainability and the application of eco-friendly technologies [3].

Environmental resources also play a crucial role in achieving national food security. Proper environmental management ensures sustainable agricultural systems in rainfed lowland fields. Enhancing rice productivity in rainfed field areas requires interdisciplinary efforts in designing and implementing agricultural management practices and efficient use of resources.

Imbalances in combining organic and inorganic fertilizer applications can lead to hazardous residues in both land and agricultural products. The use of environmentally unfriendly technologies contributes to land degradation and environmental harm. Additionally, unwise chemical use in agriculture pollutes the environment and adversely affects public health [4].

Improving soil quality has a positive impact on increasing soil organism populations. Innovations in adaptive eco-friendly technologies in rainfed lowlands can boost rice productivity. Developing environmentally friendly agricultural practices by combining organic and inorganic fertilizers in rainfed lowlands can reduce environmental degradation and support the sustainability of soil biological resources [5, 6].

The synergy between climate, soil, and crops also determines food production in rainfed lowland agroecosystems. Efforts to optimize rainfed lowland productivity, which supports food self-sufficiency, are influenced by environmental factors and non-physical components such as social, cultural, and economic aspects [7].

The productivity of rainfed lowland rice is determined by environmental improvements (land management, fertilization, and cultivation techniques) [8]. Rice productivity is influenced by seed quality, seed treatment, fertilization, the application of technological innovations, capital, infrastructure, education, and farmers' knowledge. Other contributing factors to food security include agricultural efficiency and crop productivity optimization.

This paper examines alternative technologies in rainfed lowland fields to enhance rice productivity and environmental sustainability. The research elucidates efforts to optimize the use of rainfed lowland areas through eco-friendly technology and analyzes strategies for increasing rice production to support the national food supply.

2. Research Method

This systematic literature review employed a bibliographic approach, using keywords such as rainfed agroecosystem, rice cultivation, and eco-friendly systems. The research involved a systematic literature search from online publishing sources, namely Scopus, ScienceDirect, and Google Scholar, to find articles published over the last decade, from 2019 to 2024. Full-text articles accessible and available in English or Indonesian were included as references.

The study was conducted from September 2023, and the data was updated to 2024. The electronic search resulted in 300 relevant scholarly publications. The authors examined the relevant articles, and 40 manuscripts were included in the bibliography. Data synthesis from the publications consisted primarily of primary journals (81%) among all references, with data validity identified and utilized in writing the paper as part of the research instrument.

3. Results and Discussion

The limiting factors of rainfed land

The low average productivity of rice is attributed to various biophysical constraints such as poor soil fertility, low rainfall, and the limitation of tolerant varieties suitable for rainfed agriculture. The biophysical constraints of rainfed land make it suboptimal. Suboptimal land is a field with low natural productivity in biological, chemical, and physical soil, climate factors, high temperatures, and low rainfall [9].

Poor soil fertility and limited water resources are challenging issues that are not quickly resolved. Poor soil fertility coincides with a lack of natural resources and the social support of local farmers in managing and enhancing soil fertility. Innovative water resource utilization often only aids in stabilizing rice production but has not yet significantly increased rice productivity.

The main problems with suboptimal land are its poor biological, physical, and chemical soil properties (high soil density and low water-holding capacity) for plants. In addition to its low fertility, rainfed land is highly acidic, low in organic matter, and prone to erosion. The extent of suboptimal land is estimated to reach 55.62 million hectares [10].

Based on land suitability standards for agricultural commodities, land with a pH content of <5, cation exchange capacity <5, base saturation <35, organic carbon <0.8%, and low to shallow nutrient content of N, P, and K can make it difficult for rice crops to grow. Even the estimated production on marginal land is only about 60% of its maximum potential [11].

Increasing the sustainability of rainfed rice farming not only benefits farmers but also plays a vital role in preserving the delicate ecosystems in which these farms are situated, ultimately contributing to the long-term health of the environment. Although most farmers consider rainfed rice field management relatively more straightforward, farmers' income in this area is much lower than that of irrigated rice field farmers. Rainfed rice field farmers' income was only about one-third of irrigated rice field farmers. This income gap is attributed to the difference in rice productivity between the two agroecosystems [12].

The pronounced income gap between farmers in rainfed rice fields and those in irrigated areas highlights the critical necessity for focused interventions and support programs to enhance rainfed rice productivity, improve market access, and empower farmers to bridge this income gap effectively.

The perception that rice cultivation in rainfed fields is more accessible than in irrigated fields is due to limited options for developing farming patterns. Climate uncertainty and the limited cultivation timeframe impact farmers' priorities, focusing more on crops that can grow and reach the harvest stage. Cultivation plans, including determining the proper planting time, often dominate farmer concentration.

Efforts to achieve maximum rice production require environmental management to address land constraints. Increased rice productivity is significantly influenced by technological innovation, including adopting superior varieties, balanced fertilization, using agricultural machinery, water and soil management, and farmer and stakeholder partnerships. As technological innovations continue to play a pivotal role in enhancing rainfed rice productivity, farmers, researchers, and agricultural stakeholders must collaborate closely, sharing insights and knowledge to fine-tune and adapt these innovations for local agroecosystems.

This synergistic approach promotes innovation while guaranteeing that the solutions are contextually relevant, driving sustainable practices and increased rice yields. In pursuing improved rainfed rice farming practices, it is crucial to recognize that each region's unique agroecosystem demands localized adaptation, making it essential for farmers to exchange their valuable on-ground experiences and insights with fellow farmers, researchers, and policymakers.

Technologies for increasing rice productivity require the involvement of farmers for social, cultural, and economic acceptance. The complexity of agricultural issues requires adaptive testing with farmers in local agroecosystems. Rice cultivation activities using the same technology package, during the same season, and on the same land can still result in productivity variations. Economic factors (rising prices and limited inorganic fertilizer availability) also contribute to differences in production yields [13].

The Implementation of Environmentally Friendly Technology

Adopting environmentally sustainable management technologies is essential for enhancing the productivity of food crops, particularly rice. Environmentally friendly agriculture practices incorporating agricultural waste can address environmental pollution issues, increase nutrient availability, and reduce soil nutrient loss [14]. These practices aimed to increase production and minimize negative environmental impacts, promoting the sustainability of rainfed lowland agriculture.

Implementing eco-friendly technology is urgent in rainfed lowland agriculture, given the marginal conditions of the existing land. One alternative to minimize environmental pollution, especially heavy metal contamination in paddy fields, is the application of organic farming practices [15]. Organic farming techniques not only help protect the environment but also promote the health of the soil, contributing to the long-term sustainability of rainfed rice cultivation.

Another recommended introduction of technology in rainfed lowland fields is controlled aerobic organic-based rice cultivation. This specific rainfed land rice cultivation technology is based on water management and optimizing organic matter use. Aerobic technology is crucial for crops in water-deficient environmental conditions, aiming to achieve high productivity and improve the crop index.

The 33.5% increase in rice productivity by adopting the controlled aerobic rice system. Research conducted in rainfed lowland areas at 740 meters above sea level increased rice productivity by 40.8%, reaching 6.59 tonnes/ha [16]. The controlled aerobic rice system is technically feasible and increases the population of N-fixing bacteria, enhancing growth, yield components, and rice production.

Implementing eco-friendly technology effectively reduced blast disease with an intensity 40.91% higher than chemical pesticide application. Adopting these eco-friendly systems not only enhances rice production by 8.31-12% but also leads to a substantial 10.52-22.85% increase in farm profits, illustrating the economic viability of these sustainable approaches [17].

To further enhance rainfed lowland rice cultivation, a holistic management strategy should encompass environmentally friendly components that promote agricultural productivity and sustainability. Adaptive techniques that can potentially increase rainfed lowland rice productivity include using superior varieties, planting schedules based on a planting calendar, water resource management with retention basin technology, integrated pest/disease control, and land improvement [6].

Rainfed lowland rice cultivation development involves adopting planting patterns, constructing reservoirs, utilizing additional irrigation, soil amelioration, and accelerated land preparation [18]. Adapting short-duration varieties to suit environmental conditions, efficient planting schedules, and water use are practical steps in increasing the rice crop index.

The management strategy for rainfed lowland fields involves adopting environmentally friendly components to enhance agricultural productivity. Eco-friendly technologies encompass superior varieties, planting technology, planting schedule determination, water management, balanced fertilization, integrated pest/disease control, and appropriate harvesting practices (Figure 1). By embracing these eco-friendly approaches, rainfed lowland agriculture can thrive while minimizing its environmental footprint.

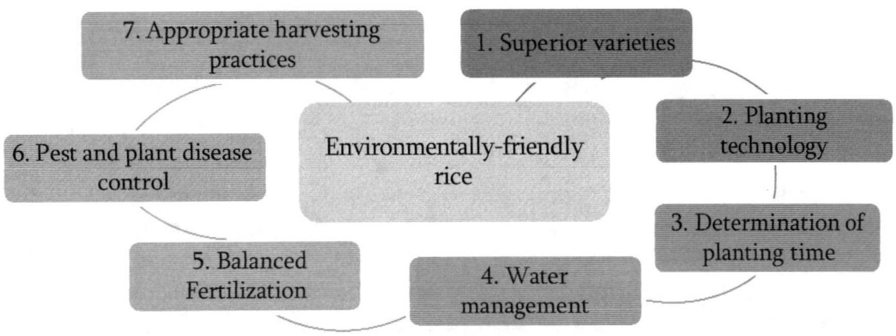

Figure 1. Enhancing rice productivity in rainfed paddy fields with eco-friendly technology

1. The Use of Superior Varieties

The selection of rice varieties for rainfed lowland agroecosystems should prioritize adaptive and tolerant varieties to natural constraints such as limited rainfall and low soil fertility. Early-maturing superior varieties have the potential for high yields and can be considered for cultivation in rainfed lowlands. The selection of early-maturing varieties is an effort to shorten the harvesting period, increase productivity, and achieve double cropping in rainfed lowland fields [19].

Superior varieties can be combined with other technologies to optimize their performance. The contribution of superior varieties integrated with irrigation and fertilization technology can increase rice productivity by up to 75%. The performance of the genotype is based on genetic traits and is influenced by the nitrogen fertilizer rate [20].

The main principle of selecting superior varieties is to address water availability issues during cultivation in rainfed lowlands. Genotype selection is also prioritized to suit rainfed conditions and tolerate uncertainties in rainfall. The main issue in rainfed lowland agroecosystems is the lack of water supply, making them susceptible to drought. Drought stress can be identified both morphologically and physiologically in plants. Furthermore, root characteristics determine various responses to drought stress, including root reach mechanisms and increased root weight [21].

Planting drought-tolerant varieties with high yield potential is an alternative solution for cultivating areas prone to drought stress. Drought-tolerant assembly and high-yielding varieties enhance rice productivity in rainfed agroecosystems. Some superior varieties in rainfed ecosystems include Gamagora 7, Inpari 41, Inpari 40, Inpari 39, Inpari 38, Cisaat, Silugonggo, and Situbagendit.

In addition to drought-tolerant varieties, low methane (CH_4) emission varieties also support increased productivity in environmentally friendly rainfed lowland fields. Inpari 17 and Inpari 18 are high-yield and low-methane-emission rice cultivars with an emission index of 0.01 kg CH_4/kg rice yield [6].

Low-methane-emission varieties can be combined with organic materials. Applying environmentally friendly technology innovations through selecting organic materials is expected to reduce CH_4 emissions without reducing rice yields [9].

2. Planting Technology

Agricultural challenges, especially in rainfed lowland rice cultivation, are closely related to climate variability, particularly rainfall patterns. Climate affects water availability, water sources, and planting patterns. The irrigation conditions in rainfed lowland fields are highly dependent on rainfall and significantly impact planting areas and rice crop indices [2].

The critical period for a plant's water supply determines the planting time. Choosing the proper planting time is a vital factor in the success of rainfed lowland rice cultivation [7]. Supportive technology to prevent rice crop failures due to waterlogging and drought includes proper water management.

Water supplementation is needed during water deficit periods in the rice growth stage. Water management models are effective as backup water sources, such as rainwater storage during high rainfall and constructing water reservoirs like retention basins and long storage systems. Alternative irrigation sources can increase land capacity and support rice during the flowering or grain-filling phase and up to the harvest stage. Rice can utilize stored irrigation sources throughout its growth cycle until harvest.

Planting in low-fertility rainfed fields can reduce rice productivity. Adjusting plant populations and managing planting patterns and spacing can help overcome water availability constraints. However, the jajar legowo 2:1 system in rainfed lowlands does not always yield significantly higher results than conventional systems [22]. Optimizing planting systems combined with intermittent irrigation can enhance rice growth and yields in rainfed lowland fields.

Adopting water-saving technology in rainfed lowland fields plays a significant and necessary role, as it aligns with water scarcity conditions and improves fertilizer efficiency. Applying fertilizer under minimal water conditions can prevent nutrient loss/leaching processes and enhance the plant roots' ability to absorb applied nutrients.

3. Determination of planting time

The uncertainty of rainfall variability necessitates technology for anticipating drought stress during low rainfall and excess water during high rainfall, achieved through timing and planting pattern adjustments. The primary factor in determining planting patterns is the availability or supply of water to establish optimal planting patterns and intensities for crop growth [23].

A locally relevant planting technology is gogo rancah and walik jerami systems. The hill planting system during the rainy season is a mitigation strategy to prepare for the monsoon, thus saving time and labor. On the other hand, the walik jerami system is an effort to anticipate the short rainy season.

Gogo rancah and walik jerami systems are effective in intensively utilizing the limited rainy season by shortening the rice planting cycle at the beginning of the season through dry planting system management (gogo rancah) and rapid soil preparation and seedling management outside the field (walik jerami).

The hill planting system at the start of the rainy season (gogo rancah) is an adjustment to the physical conditions of rainfed lowland fields, which tend to have more massive soil structures than irrigated fields [24]. Gogo rancah cultivation reduces labor costs for planting and crop maintenance while shortening the planting period. To address the risk of a short rainy period during the planting season, the walik jerami planting system is applied to expedite the planting preparation process. Gogo rancah and walik jerami enhance rice productivity contributions.

The planting process at the beginning of the rainy season by farmers is intended as a response to water availability issues. The timing for planting environmentally friendly gogo rancah is predicted based on anticipating the onset of rain and observing natural signs. Typically, natural signs include the growth of certain plant species after the dry season.

The sprouting of the kapok and mahogany trees serves as a signal to begin hill rice cultivation. The walik jerami system also considers the average remaining wet months after the harvest of gogo rancah. This planting pattern aligns with the rainfed lowland field conditions, with an average of five consecutive wet months, allowing the rice to grow for two seasons through gogo rancah and walik jerami systems [25].

4. Water Management

The prerequisite for successful rice cultivation in hot climate regions is ample sunshine and high humidity. An average monthly rainfall of 200 mm or more, distributed over four months, and a total annual rainfall of 1500-2000 mm are suitable conditions for rice development [23].

However, even in regions with favorable climatic conditions, water availability remains the primary limiting factor in rainfed lowland rice agroecosystems. Optimizing the use of limited water in rainfed lowland fields and identifying potential water sources are crucial considerations. Identifying water source potential can be achieved through exploration, exploitation, water distribution design, and irrigation technique selection [11].

Identifying water requirements in rainfed lowland fields is essential for efficient utilization. Optimizing land irrigation during the flowering and grain-filling phases significantly enhances rice productivity. Additionally, sufficient water during the dry season increases rice production, often surpassing the rainy season. Rice production during the dry season with adequate water can reach 6-6.5 tonnes/ha [19].

Water management through pumps and reservoirs can provide plant water and enrich soil nutrients. Reservoirs are designed to optimize the use of limited water resources efficiently. Soil management coupled with organic fertilizers also affects soil physical properties, such as soil moisture.

One approach to optimize water use in rainfed lowland fields is the application of intermittent irrigation systems, where alternating wet and dry conditions are employed. Intermittent irrigation in rainfed lowland fields, with intervals of up to five days, does not lead to a significant decrease in yield. Irrigation every three days yields higher rice production than daily irrigation [26].

Integrating intermittent irrigation using the 2:1-row planting system can also enhance rice growth and yield. From an environmental perspective, intermittent irrigation, when integrated with technological components such as organic matter application, balanced fertilization, plant-based pest and disease control, row planting spacing, and the use of low methane emission varieties, can reduce greenhouse gas emissions by 18-26% [27].

5. Balanced Fertilization

Fertilization is a fundamental factor in rainfed lowland rice cultivation. Soil issues related to nutrient deficiencies are addressed by implementing balanced fertilisation and cultivating superior varieties. Disparities in rice yields exist between farmer's production in rainfed lowland fields and the potential yield of rice varieties. One of the causes is improper fertilizer application in terms of rate and timing. Many farmers apply excessively at the early growth stage with nitrogen fertilizer [28]. Ineffective and inefficient fertilization practices can make crops susceptible to pests and diseases.

To maximize the effectiveness of fertilization, it is crucial to adjust fertilizer application rates to align with the crop's specific nutrient needs at each growth stage. One of the consequences of improper timing and rates of chemical fertilization is wastage due to nutrient leaching caused by rainfall and the plant's inability to absorb nutrients. Application of chemical nitrogen fertilizer should ideally be carried out in three stages: during early vegetative growth, grain-filling phase, and primordial development.

Rainfed lowland fields generally have lower organic matter levels than irrigated fields. High organic matter decomposition rates due to environmental factors, without sufficient organic matter replenishment, lead to declining organic matter content in rainfed lowland soils [29].

The low soil fertility can be addressed through regular applications of organic fertilizers during each planting season. The long-term application of organic matter enhances soil fertility. Organic matter is crucial in land productivity, particularly improving physical, chemical, and biological fertility [5].

The application of organic fertilizers in rainfed lowland fields has the benefit of improving soil fertility. On a physical level, organic fertilizers raise water-holding capacity when groundwater is excessive and help maintain soil moisture during dry periods. Chemically, organic fertilizers provide macro and micronutrients and generate humus colloid, enhancing cation exchange capacity and making nutrients more readily available to plants. On a biological aspect, organic fertilizers serve as food for earthworms and microbes, contributing to soil fertility [30].

Soil infertility in rainfed fields is not solely due to low nutrient content but also soil acidity, which makes nutrients less accessible to crops. Environmentally friendly fertilization technology, such as biofertilizers, natural spices, and associated plants, is needed. Beneficial microorganisms enhance growth, biomass, and crop yield [4].

Biofertilizer formulas must exhibit environmental friendliness (non-polluting, non-toxic, and biodegradable), allow nutrient addition, and be adaptable to soil pH. These natural-based fertilizers are derived from inexpensive, readily available materials, have a long shelf life, and maintain metabolic quantities under unfavorable conditions. Biofertilizers are typically effective when applied to soils with high organic matter content [31].

The adoption of innovative biofertilizers boosts the productivity of rainfed lowland fields. Biofertilizers such as Agrimeth and Biofert are microorganism-based fertilizers that are non-pathogenic. These nutrients enhance soil health and fertility through enzymatic rush by microbes. Microbes play a role in nitrogen fixation, solubilizing insoluble phosphates, and producing phytohormones (growth promoters). Microbes stimulate rice growth, flowering, and seed ripening and increase seed vigor, viability, productivity, and the efficiency of inorganic fertilizers.

Environmentally friendly farming through biochar is beneficial for crops and the environment. Combining biochar with rice straw compost effectively provides nutrients to Ultisol soil, improves pH, and increases N, P, K, Ca, Mg, and S levels by up to 75%. Treatment with biochar, husk, compost, and straw increases the number of productive rice tillers, producing up to 8.23 tonnes/ha [14].

In rainfed lowland agroecosystems, biochar, combined with husks and manure, enhances rice yields by 11.5%. Furthermore, biochar can mitigate adverse climate anomalies such as drought and waterlogging. Biochar can also ameliorate soil degradation, ensuring sustained crop production [32].

6. Integrated Pest and Disease Management

Integrating pest and disease management innovations can be achieved through environmentally friendly technologies. Pest and disease control technologies using chemical pesticides that leave residues in the soil can be substituted with organic fertilizers. Precise and continuous application accelerates the degradation of toxic residue elements remaining in the soil.

Farmers still practice pest control in crops using synthetic materials without considering their environmental impact. Increased rates and the use of various pesticides not following recommended guidelines leave behind residues that harm biodiversity. As various pests and diseases emerge, cost-effective, environmentally friendly technologies become an alternative for environmentally friendly management. Planting refugia in rainfed lowland rice significantly reduces the population of stem borer pests [33].

7. Appropriate Harvesting Practices

Proper timing of harvest and modern agricultural machinery expedite the harvesting process. Agricultural equipment becomes essential with the increasing difficulty of farm labor. Farm mechanization significantly aids farmers in improving rice cultivation processes, from land preparation and seeding to planting, crop maintenance, harvesting, and post-harvest processing [34].

Agricultural mechanization has also been proven to reduce costs and increase the income of rainfed lowland rice farmers. Another equally important technology that needs development in rainfed lowland fields is post-harvest handling, which results in various processed products [35].

The Participation of Farmers in Environmentally Friendly Rice Farming

Adopting environmentally friendly cultivation technologies in rainfed lowland fields must begin with farmer awareness and participation. Addressing marginal environmental conditions requires land quality improvement and enhancement (through organic fertilizers and supplementary water sources) as part of sustainable cultivation practices.

The adoption of environmental farming innovations needs to be supported by positive perceptions and preferences from farmers. Accurate technology information allows farmers to make informed decisions to support sustainable agriculture. The success of technology adoption requires breakthroughs that make it easy for farmers to adopt through technology transfer and dissemination processes [36].

The success of increasing rainfed lowland rice productivity through environmentally friendly technology is also dependent on infrastructure support. One of the prominent supports is the village irrigation network and the availability of organic and inorganic fertilizers. Using superior varieties, quality seeds, fertilizer application, and appropriate pesticide use tailored to the texture of rainfed lowland fields also supports implementing sustainable, environmentally friendly technology [37].

Generally, rice agronomy recommendations should focus on combining innovative technology choices rather than single practices. Additionally, farmers are encouraged to choose technologies to be applied to their fields based on knowledge and preferences [38]. Adaptive testing with optimal results encourages farmers to innovate in rainfed lowland farming.

The agricultural extension also contributes to sustainable food security by disseminating knowledge to farmers for appropriate decision-making. The agricultural extension workers' role is necessary in the agricultural development process. The success of introducing technological innovation to society can be seen from several communication effects. The communication effects received were in the form of cognitive (increasing knowledge or insight), affective effects (changes in attitudes and views of farmers), and conative (behavioral changes or actions) that occur in communicants [39].

Strengthening farmer-based organizations should be prioritized to enhance farmers' access to agricultural extension. they serve as crucial channels for delivering extension services and disseminating information [40].

4. Conclusion

This study used a systematic literature review on rainfed agroecosystems, rice cultivation, and environmentally friendly systems. Fifty-two articles (81% from primary journals) were used as research instruments. The literature study results show that environmentally friendly technology in agriculture is an efficient strategy for increasing rice productivity in rainfed lowland fields. Land management that considers environmental sustainability by introducing environmentally friendly innovations can coexist with farmer application awareness. The components of environmentally friendly technology for increasing rice productivity in rainfed lowland fields include using superior varieties, planting technology, timing of planting, water resource management, balanced fertilization, pest/disease control, and appropriate harvesting practices. Technologies tested with farmers in their fields can lower the likelihood of crop failure and optimize rainfed lowland fields sustainably.

5. Acknowledgments

Gratitude to the reviewers who have provided valuable suggestions for the refinement of the paper, enabling its publication.

References

[1] Badan Pusat Statistik 2022 *Rice harvest area, production, and productivity by province 2020-2022* https://www.bps.go.id/indicator/53/1498/1/luas-panen-produksi-dan-produktivitas-padi-menurut-provinsi.html [25 Januari 2021]

[2] Arifin, Biba M A and Syafiuddin 2021 Determinants of Production and Income Risks of Rainfed Lowland Farming: A Case Study in Maros Regency, Indonesia *Caraka Tani: Journal of Sustainable Agriculture* 36(2): 319-328

[3] Nurlaily R., Samijan, Aristya VE and Lestari F 2019 Effect of biofertilizer application on the population of nitrogen-fixing bacteria and yields of chili pepper at Temanggung, Central Java *Proceeding of the International Workshop and Seminar Innovation of Environmental-Friendly Agricultural Technology Supporting Sustainable Food Self-Sufficiency* pp 346–354

[4] Kumar S, Sindhu S S and Kumar R 2022 Biofertilizers: an Eco-Friendly Technology for Nutrient Recycling and Environmental Sustainability *Current Research in Microbial Sciences* 100094

[5] Kasno A, Setyorini D and Suastika I W 2020 Pengelolaan Hara Terpadu pada Lahan Sawah Tadah Hujan sebagai Upaya Peningkatan Produksi Beras Nasional *Jurnal Sumberdaya Lahan* 14(1):15- 24

[6] Wihardjaka A, Pramono A and Sutriadi M T 2020 Peningkatan Produktivitas Padi Sawah Tadah Hujan melalui Penerapan Teknologi Adaptif Dampak Perubahan Iklim *Jurnal Sumberdaya Lahan* 14(1):25-36

[7] Erythrina, E, Anshori A, Bora C Y, Dewi D O, Lestari M S, Mustaha M A, Ramija K E, Rauf A W, Mikasari W, Surdianto Y, Suriadi A, Purnamayani R, Darwis V and Syahbuddin H 2021 Assessing Opportunities to Increase Yield and Profit in Rainfed Lowland Rice Systems in Indonesia. *Agronomy* 11: 777

[8] Kour D, Rana K L, Yadav A N, Kumar M N, Kumar V and Saxena A K 2020 Microbial Biofertilizers: Bioresources and Eco-Friendly Technologies for Agricultural and Environmental Sustainability *Biocatalysis and Agricultural Biotechnology* 23 101487

[9] Yunianti I, Haryono E, Hanudin E and Sutriadi M 2022 The Feasibility of Rice Farming With Panca Kelola Ramah Lingkungan Technology at Rainfed Rice Fields in Pati Regency *AGRIC* 34(2):143-154

[10] Mahrup I G M, Kusnartha, Padusung, Soemenaboedy N and Fahrudin 2020 Inovasi dalam Pemberdayaan Petani Lahan Tegalan Guna Kesetaraan Ekonomi *Jurnal Pepadu* 1(2):235-244

[11] Cahyana D, Barus B, Darmawan, Mulyanto B and Sulaeman Y 2019 Ragam Konteks Skala dalam Perspektif Kajian Sumberdaya Lahan *Jurnal Sumberdaya Lahan* 13(2):63-71

[12] Rahmadiah R., Tanjung F and Hariance R 2019 Analisis Perbandingan Usahatani Padi Sawah Irigasi Dengan Padi Sawah Tadah Hujan di Kecamatan Koto Tangah Kota Padang *Journal of Socio Economic on Tropical Agriculture* 1(3):9-23
[13] Aristya V E, Trisyono Y A, Mulyo J H and Taryono 2021c Participatory Varietal Selection for Promising Rice Lines *Sustainability* 13(6856):1-18
[14] Kenawy M H M, Osman M A and Seddik W M A 2024 Impact of Rice Straw and Biochar Compost Treated with Olive Mill Wastewater on some Sandy Soil Properties, Nutritional Status and Yield Productivity *Journal of Soil Sciences and Agricultural Engineering* 15(7): 165-173
[15] Kharel M, Dahal B M and Raut N 2022 Good agriculture practices for safe food and sustainable agriculture in Nepal: A review *Journal of Agriculture and Food Research* 10: 100447
[16] Sutrisna N, Ruswandi A, Surdianto Y and Liferdi 2019 Kajian Rakitan Teknologi Patbo Super pada Lahan Sawah Tadah Hujan di Provinsi Jawa Barat *CR Journal* 5(1):11-22
[17] Kurniawati S, Astuti Y, Hidayat Y S and Susanti E Y 2021 Implementation of Eco-friendly Technologies to Control Bacterial Leaf Blight of Rice Disease (*Xanthomonas oryzae pv. oryzae*). *IOP Conf. Ser.: Earth Environ. Sci.* 715(1):012041
[18] Aristya V E and Samijan 2021 Rice breeding breakthrough for rainfed fields. Rainfed Rice Field Management, Technology, and Institutional (UNS Press: Surakarta) pp 245-265
[19] Tando E and Asaad M 2020 Keragaan Varietas Padi Musim Tanam Melalui Inovasi Teknologi pada Lahan Sawah Tadah Hujan di Kabupaten Konawe Selatan Provinsi Sulawesi Tenggara *Jurnal Pengkajian dan Pengembangan Teknologi Pertanian* 23(1):93-106
[20] Aristya V E and Samijan 2022. The yield gap of maize under intensive cropping system in Central Java *Planta Tropika: J. Agrosains (J. of Agro Science)* 10(1):1-12
[21] Aristya V E, Taryono, Trisyono Y A, Mulyo JH 2021b The morphological fingerprint of new rice genotypes *IOP Conf. Ser.: Earth Environ. Sci.* 662:012022
[22] Adnyana I N S 2020 Efektifitas Sistem Tanam Jajar Legowo 2:1 Dengan Sistem Tegel Terhadap Produktivitas Padi Sawah di Subak Babakan Cangi, Desa Batuan Kaler, Kecamatan Sukawati, Kabupaten Gianyar *dwijenAGRO* 10:2
[23] Sari Y, Nasution I S and Syahrul S 2021 Pengaruh Perubahan Iklim Terhadap Jadwal Tanam Dan Produktivitas Padi Sawah Di Daerah Irigasi (DI.) Krueng Aceh Kabupaten Aceh Besar *Jurnal Ilmiah Mahasiswa* 6: 3
[24] Wihardjaka A, Pramono A and Sutriadi M S 2020 Peningkatan Produktivitas Padi Sawah Tadah Hujan Melalui Penerapan Teknologi Adaptif Dampak Perubahan Iklim *Jurnal Sumberdaya Lahan* 14:1
[25] Fatmawati P 2019 Pengetahuan Lokal Petani dalam Tradisi Bercocok Tanam Padi oleh Masyarakat Tapango di Polewali Mandar *Walasuji: Jurnal Sejarah dan Budaya* 10(1):85-95
[26] Zhang L, Li L, Tang Q, Xu H, Zheng H, Wang F and Tang J 2024 Intermittent irrigation as a solution for reduced emissions and increased yields in ratoon rice systems *Plant Soil* 501, 225–236
[27] Wihardjaka A 2022 Technologies to Reduce Emission of Methane And Nitrous Oxide in Rice Fields: A Brief Review *AGRIC* 34(2):197-210
[28] Agus F, Andrade J F, Rattalino Edreira J I, Deng N, Purwantomo D K G, Agustiani N, Aristya V E, Batubara S F, Herniwati, Hosang E Y, Krisnadi L Y, Makka A, Samijan, Cenacchi N, Wiebe K and Grassini P 2019 Yield Gaps in Intensive Rice-Maize Cropping Sequences in The Humid Tropics of Indonesia *Field Crops Research* 237:12–22
[29] Samijan, Minarsih S, Jauhari S, Basuki S, Susila A, Nurwahyuni E, Hindarwati Y, Supriyo A and Aristya V E 2023 Revitalizing sub-optimal drylands: Exploring the role of biofertilizers *Open Agriculture* 8(1):20220214
[30] Supriyo A, Hindarwati Y and Nurlaily R 2020 Pengelolaan Tanaman Terpadu Padi Ramah Lingkungan terhadap Hasil Padi dan Emisi Gas Rumah Kaca di Lahan Sawah Irigasi *Jurnal Ilmu Lingkungan* 18(1):15-22
[31] Hindarwati Y and Supriyo A 2022 Inovasi Teknologi Pupuk Hayati Untuk Meningkatkan Hasil Padi Pada Lahan Sawah di Jawa Tengah in *Inovasi Teknologi Mendukung Pengembangan Kawasan Pertanian Berbasis Wilayah Jawa Tengah* (Loka Aksara: Banten) pp 155-215

[32] Hasnain M, Munir N, Abideen Z N, Zulfiqar F Z, Koyro H W F, El-Naggar A H W, AI Caçador, Duarte B, Rinklebe J and Yong J W H 2023 Biochar-Plant Interaction and Detoxification Strategies Under Abiotic Stresses for Achieving Agricultural Resilience: A Critical Review *Ecotoxicology and Environmental Safety* 249:114408

[33] Rahmawasiah and Suriyanto R 2020 Efektivitas Refugia terhadap Populasi Penggerek Batang Padi Putih (*Schirpophaga innotata*) pada Sawah Tadah Hujan di Kecamatan Malangke Barat Kabupaten Luwu Utara *Jurnal Pertanian Berkelanjutan* 8(2):87-92

[34] Mulyono J 2020 Prospek Pengembangan Alat dan Mesin Pertanian Mendukung Peningkatan Indeks Pertanaman Padi in *Bukti nyata peningkatan indeks pertanaman: fondasi lumbung pangan masa depan* (IAARD Press: Jakarta) pp 37-48

[35] Prasetio T, Sari Y and Oktariana Y 2022 Penerapan Mekanisasi Pertanian terhadap Peningkatan Pendapatan Petani Padi Sawah Tadah Hujan di Kecamatan Semidang Aji Kabupaten Ogan Komering Ulu *Journal of Agriculture Social and Economic* 1(1):20-32

[36] Aristya V E, Taryono, Trisyono Y A and Mulyo J H 2021a Stakeholder Preferences on Major Characteristics of Promising Rice Lines *IOP Conf. Ser. Earth Environ. Sci.* 686:012056

[37] Hashim N, Ali M M, Mahadi M R, Abdullah A F, Wayayok A, Kassim M S M and Jamaluddin A 2024 Smart Farming for Sustainable Rice Production: An Insight into Application, Challenge, and Future Prospect *Rice Science* 31(1): 47-61

[38] Eldon J, Bairdb G, Sidibeh S, Dobasin D, Rapapor P, Cheng W and Shennan C 2020 On-Farm Trials Identify Adaptive Management Options For Rainfed Agriculture in West Africa *Agricultural Systems* 182:102819

[39] Raidimi E N and Kabiti H M 2019 A review of the role of agricultural extension and training in achieving sustainable food security: A case of South Africa *South African Journal of Agricultural Extension* 47(3):120-130

[40] Anang B T, Bäckman S and Sipiläinen T 2020 Adoption and income effects of agricultural extension in northern Ghana *Scientific African* 7:e00219

EFFECT OF HUMIC ACID AND BLOTONG APLICATION ON VEGETATIVE GROWTH OF CHILI (*Capsicum annum* L.) PLANT

N. B. E. Sulistyono [1*], H. F. Rohman [1], M. Z. Sukri [1], M. Bintoro [1], F Rohman [1]
[1] Departemen of Agricultural Production, Politeknik Negeri Jember Mastrip Street PO BOX 164 Jember, Indonesia
besjember@polije.ac.id

Abstract. Chili has high economic value. The increasing demand for chilies in Indonesia leads to an intensification of land use for cultivation, which can lead to a decrease in soil fertility and quality. Adding organic materials such as humic acid and blotong is expected to help improve fertility and land quality. Humic acid showcases a multifaceted molecular arrangement and demonstrates the potential to support biological and physiological mechanisms in organisms inhabiting the soil. Blotong, with its high levels of P2O5, CaO, and K2O, can potentially enhance the nutrients P, K, and Ca. This research was carried out at Teaching Factory Rumah Organik Politeknik Negeri Jember from June to November 2024. This research utilized a non-factorial Randomized Complete Block Design approach, combining humic acid doses of 40g, 60g, and 80g per plant with blotong of 100g, 125g, and 150g per plant. Additionally, a control group received no treatment, resulting in 10 treatments. Each treatment was replicated six times, resulting in 60 experimental units. Parameters observed included plant height and stem diameter. We processed the resulting data using the SAS 9.4 application. The analysis of variance revealed a statistically significant result in the F test at the alpha level of 5%, leading to a more detailed exploration using Duncan's Multiple Range Test (DMRT). The research results indicated that the application of humic acid and blotong resulted in an increase in the height of chili plants during the early vegetative phase, as well as in the stem diameter parameters of the Humic Acid 80 g/plant + Blotong 100 g/plant treatment, which yielded the best results.

Keywords: *chili, humic acid, blotong, organic*

1. Introduction

Chili (Capsicum annum L.) is one of the horticultural plants with high economic value. The strategic position of chili commodities in Indonesia leads to fluctuations in chili prices, which in turn influence the national inflation rate. The requirement for substantial chili consumption on a per capita basis is estimated to be approximately 3 kilograms per individual annually; therefore, considering Indonesia's population of 250 million, 750,000 tons are needed per year, and that amount is predicted to not be met by domestic production, so the government partly imports [1]. The production of large chilies in Indonesia in 2022 amounted to 1.4 million tons, an increase of 8.47% or 115.25 thousand tons from 2021 [2]. The organic house, one of the pioneering teaching factories (Tefa) located in the Jember State Polytechnic environment, is a place for learning and business of Tefa organic products that lead to increased productivity so that it can increase economic benefits.

Chili, which has high economic value, will be very profitable and become an interesting learning experience if they can increase the production of organic vegetables in Tefa Rumah Organik. However, some soil conditions used for vegetable cultivation experience a decrease in quality, so production is not optimal. Improving soil quality by providing organic soil conditioners can be a solution to help and support optimal plant growth.

Less fertile soil can be characterized by low organic matter content, poor nutrients, acidic soil, and soil physical properties that do not support plant growth, such as texture that is too sandy or clayey [3]. Improvement of less fertile soil can be done by adding organic and inorganic fertilizers, but excessive use of inorganic fertilizers and pesticides can reduce the carbon value in the soil [4]. The correct application of organic fertilizer will affect soil improvement and minimize the application of synthetic fertilizers..

This study used organic fertilizer humic acid and blotong. Humic acid represents a prominent category of organic compounds and is known as a soil conditioner because it has a complex molecular structure and has the faculty to initiate and interact with biological and physiological mechanisms inherent in the soil's living organisms [5]. Humic acid can be utilized in fertilizer efficiency because humic acid is able to bind negatively charged soil particles with positively charged nutrients [6]. Increase the availability of phosphate, nitrogen, cation exchange capacity (CEC), pH, and several micronutrients in the soil that are easily absorbed by roots [7]. Using humic acid at a rate of 20 kg/ha greatly impacts plant height, leaf count, cob diameter, cob weight, the length of four cobs, and the sweetness level. Application of humic acid 20 kg/ha provides a cob weight of 185 kg per 120 m2 or 15.42 tons/Ha can increase the weight of corn cobs, dry weight of corn kernels, and seed weight per hectare [8].

Blotong contains quite high P2O5, CaO & K2O, which are respectively 8.03%; 14.64%; & 0.82% so that the use of blotong can increase the nutrients P, K, and Ca [9]. Blotong compost is crucial in improving the physical properties of the soil, resulting in increased water retention capacity. Every ton of blotong with a moisture content of 70% provides nutrients equivalent to 28 kg of ZA, 22 kg of TSP, and 1 kg of KCl. These nutrients include 5.88 kg of nitrogen, 9.9 kg of phosphorus, and 0.6 kg of potassium. [10]. It was mentioned that utilizing blotong can enhance soil fertility, which is reflected in the improvements of both the physical and chemical characteristics of the soil [11]. Therefore, this treatment is expected to provide a positive contribution to chili production at Tefa Rumah Organik. To support the development of Tefa Rumah Organik,

2. Method

This research was conducted at Teaching Factory Rumah Organik Politeknik Negeri Jember at an altitude of 90 meters above sea level. The materials used are Imola variety chili seeds, humic acid, and blotong. The equipment used includes polybags for planting, buckets for mixing the ingredients, scales for weighing humic acid and blotong, hoes for preparing the soil, cans, watering hoses for irrigation, and rulers for measuring plant height. This study used a non-factorial Randomized Complete Block Design (RCBD) with a combination of humic acid doses of 40g, 60g, and 80g per plant and blotong doses of 100g, 125g, and 150g per plant, and one control without treatment, obtaining ten treatments, namely P0 (control), P1 (humic acid 40g/plant + Blotong 100g/plant), P2 (humic acid 40g/plant + Blotong 125g/plant), P3 (humic acid 40g/plant + Blotong 150g/plant), P4 (humic acid 60g/plant + Blotong 100g/plant). P5 (humic acid 60 g/plant + Blotong 125 g/plant), P6 (humic acid 60 g/plant + Blotong 150 g/plant), P7 (humic acid 80 g/plant + Blotong 100 g/plant), P8 (humic acid 80 g/plant + Blotong 120 g/plant), P9 (humic acid 80 g/plant + Blotong 150 g/plant) with six replications in each treatment so that there were a total of 60 experimental units. Parameters observed included plant height and stem diameter. The resulting data were processed using the SAS 9.4 application. The variance analysis findings indicated a significant effect on the F test at the α level of 5%. These results were subsequently examined using Duncan's Multiple Range Test (DMRT).

3. Results and Discussion

The results of the F test analysis at significant levels of 1% and 5% levels showed that humic acid and blotong had no significant effect on the variables of plant height and diameter of stem.

Tabel 1. The Effect of Humic Acid and Blotong on the Height of Chili Plants

Treatment	Average Plant Height (cm)		
	2 WAP	4 WAP	6 WAP
Control (Without Treatment)	15,56	36,78	65,50
Humic Acid 40g/plant + Blotong 100g/plant	14,67	35,50	66,78
Humic Acid 40g/plant + Blotong 125g/plant	16,11	39,83	70,89
Humic Acid 40g/plant + Blotong 150g/plant	16,89	37,17	61,33
Humic Acid 60g/plant + Blotong 100g/plant	14,56	39,06	69,94
Humic Acid 60g/plant + Blotong 125g/plant	15,89	40,00	67,83
Humic Acid 60g/plant + Blotong 150g/plant	15,78	36,83	66,39
Humic Acid 80 g/plant + Blotong 100 g/plant	15,89	39,22	74,39
Humic Acid 80 g/plant + Blotong 120 g/plant	14,94	40,00	70,94
Humic Acid 80 g/plant + Blotong 150 g/plant	15,00	40,56	69,22

Note: WAP = Week After Planting

Table 2 shows that the use of humic acid and blotong influenced the height growth of chili plants at 2 weeks after planting (WAP), 4 WAP, and 6 WAP. Multiple treatments demonstrated a notable rise in plant height compared to the control (P0). At 2 WAP, treatment P3 (16.89 cm) showed the highest growth, followed by P2 (16.11 cm). The increase in plant height at this early stage indicated that treatments P3 and P2 provided better conditions for chili plants to start their growth compared to other treatments. At 4 WAP, treatments P9 (40.56 cm) and P5 and P8 (each 40.00 cm) produced the highest plant growth. This suggests that using blotong and humic acid had a more pronounced impact on plant growth after weeks of consistent application. During this period, the observed growth rate can be linked to the enhanced availability of macro and micronutrients promoted by blotong and humic acid, particularly through enhancing the soil's physical and chemical characteristics. At 6 WAP, the P7 treatment (74.39 cm) produced the most optimal plant height growth compared to other treatments. The P7 treatment also consistently showed good results in each measurement period (2 WAP, 4 WAP, and 6 WAP). This may be due to the ability of humic acid and blotong to increase the cation exchange capacity of the soil and increase nutrient absorption, which ultimately increases the growth of chili plants.

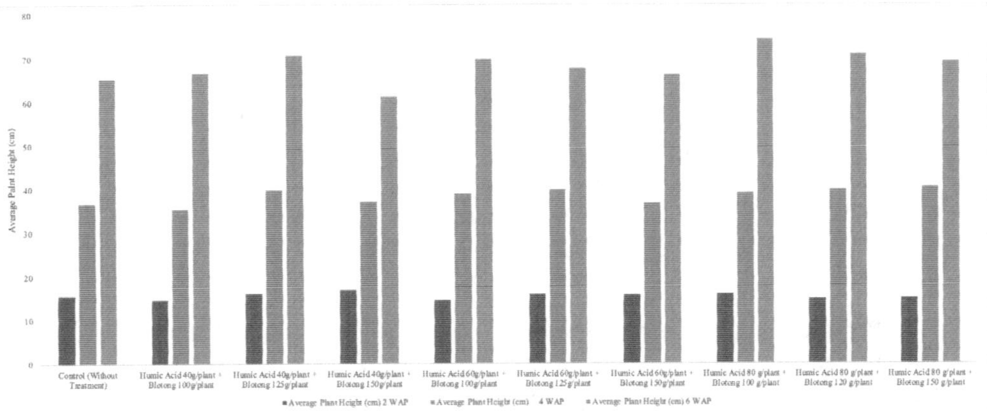

Figure 1. Diagram of the effect of humic acid and blotong treatment on the height of chili plants

Humic acid is recognized as a material that can enhance nutrient accessibility and improve soil structure, resulting in healthier plant growth. Humic acid is crucial for enhancing plants' absorption of nutrients as it enhances soil structure and boosts cation exchange capacity, accelerating the rate of plant vegetative growth. Additionally, humic acid enhances microbial activity in the soil, fostering a healthier rhizosphere. This increased microbial population not only aids in decomposing organic matter and releasing essential nutrients but also forms symbiotic relationships with plant roots, further promoting nutrient absorption. Humic acid can also help mitigate soil compaction, allowing roots to penetrate more easily and access water and nutrients. Consequently, plants treated with humic acid often exhibit stronger root systems, greater resilience to environmental stressors, and improved overall health, ultimately leading to higher yields and better-quality produce. Humic acid also improves soil structure and water retention capacity, supporting better plant growth [12]. In addition, blotong, as organic waste from the sugar industry, contains organic matter rich in essential nutrients such as phosphorus and potassium that support optimal growth of chili plants [13]. Blotong, as organic matter, is also rich in microelements that are important for plant growth, such as calcium and magnesium, which significantly increase chili growth [14].

Tabel 2. The Effect of Humic Acid and Blotong on the Stem Diameter of Chili Plants

Treatment	Average Stem Diameter (cm)		
	2 WAP	6 WAP	8 WAP
Control (Without Treatment)	3,22	36,78	65,50
Humic Acid 40g/plant + Blotong 100g/plant	3,19	35,50	66,78
Humic Acid 40g/plant + Blotong 125g/plant	3,58	39,83	70,89
Humic Acid 40g/plant + Blotong 150g/plant	3,32	37,17	61,33
Humic Acid 60g/plant + Blotong 100g/plant	3,27	39,06	69,94
Humic Acid 60g/plant + Blotong 125g/plant	3,63	40,00	67,83
Humic Acid 60g/plant + Blotong 150g/plant	3,09	36,83	66,39
Humic Acid 80 g/plant + Blotong 100 g/plant	3,32	39,22	74,39
Humic Acid 80 g/plant + Blotong 120 g/plant	3,23	40,00	70,94
Humic Acid 80 g/plant + Blotong 150 g/plant	3,49	40,56	69,22

Note: WAP = Week After Planting

Table 3 illustrates the impact of humic acid and blotong on the height growth of chili plants. At 2 WAP, the average value of the stem diameter of chili plants showed quite small variations among the various treatments. Treatment P0 (control) had an average diameter of 3.22 mm, while the highest treatment was in P5, with a diameter of 3.63 mm, while the lowest treatment occurred in P6, with a diameter of 3.09 mm. Based on this information, it is noticeable that the incorporation of humic acid and blotong showed some effect, although it still needs to be significant at the initial growth stages (2 WAP). At 4 WAP, the variation in stem diameter became more obvious. P0 had an average of 36.78 mm, while P9 showed the largest increase in stem diameter with 40.56 mm. This shows that adding humic acid and blotong began to show a more significant effect on stem diameter during this growth phase. Humic acid is known to increase nutrient absorption and soil moisture, which can encourage vegetative plant growth [15]. The P7 treatment resulted in the highest average stem diameter at 6 WAP, measuring 74.39 mm, showing a significant increase compared to the control treatment (P0) which only reached 65.50 mm. This is in line with previous research 16] It was reported that blending humic acid with organic fertilizers like blotong can enhance the soil's physical characteristics and boost nutrient availability, resulting in faster plant growth.

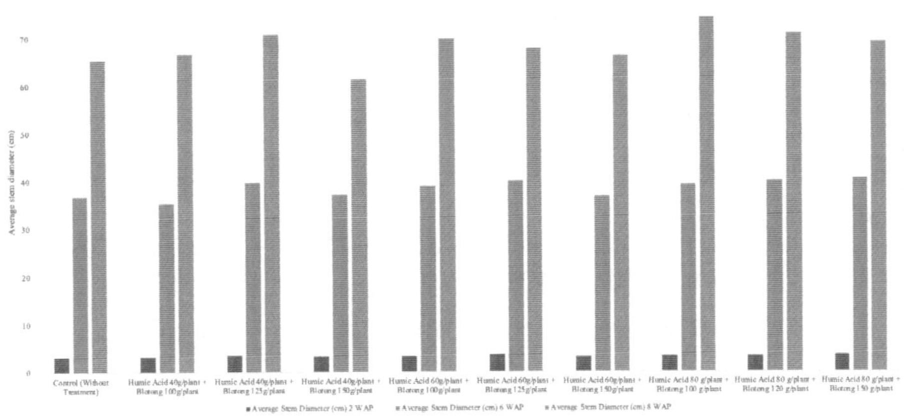

Figure 2. Diagram of the effect of humic acid and blotong treatment on the diameter of chili plant stems

The use of humic acid as an organic material has the potential to improve soil quality by improving soil structure and increasing cation exchange capacity (CEC). This can affect the availability of nutrients for plants, increasing plant growth. Blotong, as a by-product of the sugar processing process, also has benefits as an organic material rich in nutrients such as phosphorus and potassium. Adding organic substances like humic acid and blotong can enhance plant growth by promoting better soil structure and nutrient availability. Furthermore, over time, organic materials can elevate the availability of key macronutrients, including nitrogen, phosphorus, and potassium, which are crucial for plant vegetative development. This rise in nutrient accessibility is likely to result in notable differences in growth when comparing treatments with humic acid and blotong to the control group.

The intensive and continuous use of chemical fertilizers has been shown to impact soil quality and the environment negatively. The accumulation of chemical fertilizer residues in the soil can lead to soil structure degradation, reduced soil capacity to absorb and retain water, and diminished diversity of soil microorganisms that play a crucial role in nutrient cycling. Furthermore, excessive use of chemical fertilizers can trigger nutrient imbalances, such as high nitrogen levels that inhibit phosphorus and potassium uptake, ultimately reducing plant productivity. Economically, reliance on chemical fertilizers poses challenges for farmers due to price fluctuations, which are often unaffordable, especially for small-scale farmers.

Alternatively, using organic materials such as humic acid and blotong offers a more economical and sustainable solution. Humic acid improves soil structure and increases cation exchange capacity (CEC), while blotong, rich in macronutrients like phosphorus and potassium, has the potential to enhance nutrient availability in the soil. Substituting chemical fertilizers with organic materials supports sustainable plant productivity and minimizes negative environmental impacts.

Conclusion

Applying humic acid and blotong showed a significant increase in the height of chili plants at 2, 4, and 6 weeks after planting, especially in certain treatments that provided better conditions for growth. While there was no significant difference in the initial measurements (2 WAP), at 4 WAP and 6 WAP, there was a clear increase in stem diameter, with humic acid 80 g/plant + Blotong 100 g/plant showing the best results. The application of humic acid and blotong enhances the accessibility of essential macro and micronutrients for plant growth while also improving the soil's physical characteristics, leading to healthier vegetative growth. Overall, the use of humic acid and blotong as organic materials can increase the growth of chili plants by improving soil structure and increasing the availability of nutrients.

Acknowledgment
On this occasion the author would like to thank P3M Politeknik Negeri Jember for the 2024 PNBP research program grant.

References

[1] Siahaan, D,S, Tarigan K, Sebayang T. 2016. Efisiensi Penggunaan Faktor Produksi Usahatani Cabai Besar. Agribisnis Faperta Universitas Sumatera Utara.
[2] Badan Pusat Statistik. 2023. Statistik Hortikultura 2022. Jakarta: Badan Pusat Statistik.
[3] Delsiyanti, Danang, W. & Rajamuddin, U.A. (2016). Sifat fisik tanah pada beberapa penggunaan lahan di desa olobuoju kabupaten Sigi. E-J. Agrotekbis, 4(3), 227-234.
[4] Tao, H.H., Snaddon, J.L., Slade, E.M., Caliman, J.P., Widodo, R.H., Suhardi & Willis, K.J. (2017). Long-term crop residue application maintains oil palm yield and temporal stability of production. Agronomy for Sustainable Development, 37(4).
[5] Badan Litbang Pertanian. 2013. Asam Humat, Senyawa Organik Penghemat. Pemakaian Pupuk Anorganik. Jakarta.
[6] Saptana, 2011. Efisiensi Produksi dan Perilaku Petani terhadap Risiko Produktivitas Cabai Besar di Jawa Tengah, IPB.
[7] Tan, K. H. 2003. Humic Matter in Soil and the Environment. New York: CRC Press.
[8] Nuswantooro N. A., Sulistyono, N.B.E. 2023. Optimasi Jarak Tanaman dan Pemberian Asam Humat untuk Meningkatkan Produksi Benih Jagung (Zea mays L.). Agropross, National Conference Proceedings of Agriculture Vol 5.
[9] Kasmadi, Nugroho, B., Sutandi, A., & Anwar, S. (2020). Optimalisasi Pemanfaatan Limbah Blotong Untuk Peningkatan Serapan Hara Tanaman pada Formulasi Produksi Pupuk Majemuk Granul. Jurnal Ilmu Lingkungan. 18(1): 1–7. Daryono BS, Ibrohim AR, Maryanto SD. 2015. Biogenesis. 3(1):39-46.
[10] Leovici, H, 2012. Pemanfaatan blotong pada budidaya tebu (saccharum officinarum L.) di lahan kering. Jurnal penelitian pertanian 4 (1) : 13 – 18.
[11] Sukri, M. Z., Firgiyanto R., Sugiyatro, Rohman H. F. 2021. The increasing fertility of sandy soil and chili production through the application of organic fertilizer, zeolite and cane blotong. IOP conference series : Eartn and Environment Science Sci. 672 012101. IOP Publishing.
[12] Kusuma, N. I., & Sulistyono, N. B. E. (2023). Penggunaan Asam Humat dan Pupuk SP-36 terhadap Produksi Kacang Hijau (Vigna radiata L.). Agropross : National Conference Proceedings of Agriculture, 248–251.
[13] Sulistyono, N. B. E. & Sri rahayu (2018). Pengaruh blotong sebagai media tanam terhadap pertumbuhan bibit tebu (Saccharum officinarum L.) tiga varietas sistem bud chips. Journal of Applied Agricultural Science, Agriprima, Vol. 2, No. 2, Hal. 87-97
[14] Chairani. 2005. Pengaruh pemberian pupuk organik blotong dan pupuk sulfomag plus terhadap sifat kimia tanah, pertumbuhan dan produksi tanaman jagung (Zea mays L.) pada Tanah Typic Paleudult. Jurnal Penelitian Bidang Ilmu Pertanian. 3(3): 73-78.
[15] Sukri, M. Z., Firgiyanto, R., Sari, V. K., & Basuki, B. (2020). Kombinasi Pupuk Kandang Sapi, Asam Humat Dan Mikoriza Terhadap Infeksi Akar Bermikoriza Tanaman Cabai Dan Ketersediaan Unsur Hara Tanah Udipsamments. *Jurnal Penelitian Pertanian Terapan*, *19*(2), 141-145.
[16] Sukri, M.Z., Sari, V. K. & Firgiyanto, R. (2018). Improving soil fertilizer through application of organic fertilizer humid acid and mikoriza in supporting growth and production of chilli plants in sand land .IOP conference series : Eartn and Environment Science Sci. 672 012101. IOP Publishing.

Aplication of Humic Acid and Fulvic Acid on the Growth of Chrisant (*Chrysanthemum cinerariaefolium*) Plants

HF Rohman[1]*, MZ Sukri[1], M Azizah[1], DE Putra[2], H Prasetyo[1], AR Fadilah[1]

[1] Departemen of Agricultural Production, Politeknik Negeri Jember Mastrip Street PO BOX 164 Jember East Java, Indonesia
[2]Departemen of Agribussiness Management, Politeknik Negeri Jember Mastrip Street PO BOX 164 Jember East Java, Indonesia

*Email : haniffaturrohman@polije.ac.id

Abstract. One of the popular houseplants in Indonesia, Chrysanthemum flowers offer great prospects for use as vase flowers, room decorations, and bouquets. However, Chrysanthemum cultivation still faces challenges in increasing fertilization efficiency through proper soil fertility management and improving the growth and quality of cutting Chrysanthemum flowers. Adding humic acid and fulvic acid, which can improve plant metabolism and soil fertility, is one method to improve fertilization efficiency. This research aimed to assess the effect of humic acid and fulvic acid on the growth of cutting chrysanthemum plants. The research was carried out at Teaching Factory Nursery and Cut Flowers, Politeknik Negeri Jember, from May to December 2024, with an altitude of ± 650 masl and a temperature of 18-25°C. We employed a Completely Randomized Design (CRD) research design, which included nine treatment combinations: control, which did not include humic and fulvic acids, humic acid at 500, 1000, 1500, and 2000 ppm, and fulvic acid at 500, 1,000, 1500, and 2000 ppm, which we repeated three times. Each experiment unit consisted of 100 plants with 30 observation samples. The study observed two key parameters: the height of the plants and the number of leaves. The analysis showed that the deployment of fulvic acid and humic acid at certain concentrations had a considerable effect on the growth of chrysanthemums. In particular, fulvic acid at a concentration of 500 ppm increased plant height during the early stages of growth, while humic acid at a concentration of 1500 ppm increased leaf number over a longer period of growth.

1. Introduction

Chrysanthemum is an ornamental plant highly favored due to its flowers, which come in various shapes and colors. Chrysanthemums as cut flowers are classified into two types: standard chrysanthemums and spray chrysanthemums. The characteristic of standard chrysanthemums is that each stem only has one flower with a uniform diameter of around 8 to 12 cm and a variety of colors. Spray chrysanthemums exhibit multiple flowers on each stem, each containing more than eight flower heads. The overall appearance of the spray chrysanthemums forms a rounded flower arrangement. The flowers come in various colors and have a disk or layered shape [1]. Chrysanthemum production in Indonesia from 2019 to 2022 decreased by 15.23%, from 465,359,952 stems to 394,502,028 stems [2].

Teaching Factory (Tefa) Kebun Inovasi dan Nursery Bunga Potong Politeknik Negeri Jember (Polije) is the main producer of chrysanthemums in Jember Regency. This potential is supported by an agroclimate condition suitable for chrysanthemum growth, as it is located at an altitude of 650 meters above sea level with a temperature range of 18-25°C. However, to support the sustainable production of high-quality cut flowers, we need to implement effective and efficient technology to enhance the growth and flowering of cut chrysanthemums. Enhancing fertilization effectiveness through proper soil fertility management is one way to improve the growth and quality of cut chrysanthemums at the Tefa Innovation Garden and Cut Flower Nursery of Polije. The application of materials that can increase fertilization effectiveness, such as soil amendments and plant biostimulants, is crucial. Soil amendments include artificial and natural materials, whether made of minerals or organic matter and can be solid or liquid. They can improve the soil's physical, chemical, and biological properties [3]. Fundamental constituents for soil modification and phytostimulant applications encompass humic acid and fulvic acid [4].

This research aims to investigate the application of humic acid and fulvic acid, which constitute integral components of humic substances (soil organic matter) that play a pivotal role in the promotion of plant growth, facilitated by both direct pathways (plant metabolism) and indirect pathways (soil fertility) [5]. The use of humic and fulvic acids has been proven to enhance nutrient uptake and promote vegetative growth in plants. Researchers have specifically shown that applying humic acid to the Jinba variety of chrysanthemums increases chlorophyll content, boosts photosynthesis, improves flower size, and increases the fresh weight of harvested chrysanthemums [6]. Adding 1500 mg/l of fulvic acid or 1000 mg/l of humic acid to Flyer cultivar chrysanthemums effectively raises the plants' height, number of branches, root/shoot ratio, fresh and dry weight, leaf area, flower diameter, flower number, flower number per plant, flower chlorophyll and carotenoid content, and leaf carbohydrate, nitrogen, phosphorus, and potassium content [7].

2. Method

This research was conducted at the Teaching factory Kebun Inovasi dan Nursery Bunga Potong Politeknik Negeri Jember in Kemuning Lor Village, Arjasa District, Jember Regency, located at an altitude of approximately 650 meters above sea level with a temperature range of 18-25°C, as well as at the Bioscience Laboratory of Polije. Chrysanthemum fiji varieties seedlings were planted on beds using hoes, and urea, TSP, and KCl base fertilizers were given at predetermined doses, while NPK and KNO3 were given in the vegetative and generative phases. Application of humic and fulvic acids was carried out using an electric sprayer. While watering and other maintenance, such as weeding and pest control, were also carried out routinely.

This study was carried out in Complete Random Design (CRD), which was prepared in a non-factorial design with treatment consisting of controls, humic acid 500, 1000, 1500, and 2000 ppm, and fulvic acid 500, 1000, 1500, and 2000 ppm. We replicated each level of treatment three times, resulting in a total of 27 experimental units. Each unit consisted of 100 plants, with 30 observational samples taken from them. The growth parameters comprised plant height and number of leaves, which were assessed periodically utilizing a ruler and documented by writing instruments.

The data obtained in this study was analyzed by One-Way Analysis of Variance (One-Way ANOVA) based on the F test at a significance level of 5%. Furthermore, the mean values of the various levels of treatments were compared using Duncan's Multiple Range Test (DMRT) at a significance level of 5% .

3. Results and Discussion

The findings derived from the observational study conducted at 8 weeks after planting (WAP) concerning the growth metrics of chrysanthemum, specifically focusing on plant height and leaf count, were subjected to F-test data analysis at both the 5% and 1% significance levels, resulting in the data delineated in Table 1.

Table 1. Effect of application humid acid and fulvic acid on the plant height

Treatment	Observation Variables			
	2 WAP	4 WAP	6 WAP	8 WAP
Controls	15,50 de	28,33 d	51,97	62,97
Humic Acid 500 ppm	17,20 b	32,00 c	56,50	67,53
Humic Acid 1000 ppm	16,47 c	30,10 cd	51,30	61,37
Humic Acid 1500 ppm	18,57 a	35,17 b	57,00	65,70
Humic Acid 2000 ppm	15,83 d	30,10 cd	51,07	58,90
Fulvic Acid 500 ppm	18,73 a	37,60 a	55,40	66,63
Fulvic Acid 1000 ppm	16,50 c	34,03 b	55,50	65,97
Fulvic Acid 1500 ppm	14,63 f	28,77 d	51,40	62,37
Fulvic Acid 2000 ppm	15,46 e	30,23 cd	50,50	60,87

Note: SH = Shoot Height, NS = Number of Shoots, NL = Number of Leaves

There were big differences in the average height of the plants across the different treatments. This suggests that raising the concentration of humic acid and fulvic acid to a certain level would have an impact on plant growth. Table 1 presents the findings of the DMRT analysis. The treatment groups with identical letters indicated no statistically significant differences, whereas those with dissimilar letters demonstrated significant differences. At 2 weeks after planting, the treatment of humic acid with a concentration of 1500 ppm and fulvic acid with a concentration of 500 ppm did not differ significantly in plant height. These results show that the 1500 ppm humic acid treatment (18.57 cm) or the 500 ppm fulvic acid treatment (18.73 cm) yielded the highest plant height. Four weeks after planting, the treatment of fulvic acid with a concentration of 500 ppm resulted in the highest plant height (37.6 cm).

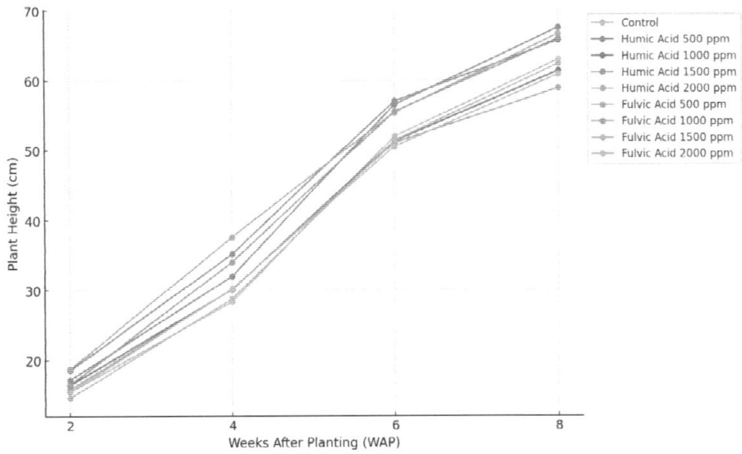

Figure 1. Graph of Effect of application of humid acid and fulvic acid on the plant height

Overall, the treatment of fulvic acid with a concentration of 500 ppm is quite effective in increasing the height of Chrysanthemum plants up to 4 weeks after planting. Supported by Sootahar [8] it is posited that fulvic acid has the potential to augment root development and modulate root functionality to facilitate the exudation of organic acids, culminating in an elevated capacity for nutrient absorption and, consequently, enhanced growth of Chrysanthemum species. This is supported by research [10] which states that the growth of mustard greens can be increased by providing organic fertilizers. In addition, based on the results of research [11], the combination of cattle manure, humic acid, and mycorrhizal inoculants to the root systems of Capsicum annuum resulted in a significant enhancement of the mycorrhizal colonization of the chili plants, concomitantly augmenting the bioavailability of essential nutrients in the soil, specifically an increase of 191.84% in nitrogen levels and 392.93% in phosphorus concentrations.

Table 2. Effect of application humid acid and fulvic acid on the number of leave.

Treatment	Observation Variables			
	2 WAP	4 WAP	6 WAP	8 WAP
Controls	12,67	16,00 cd	23,33 cd	32,33 d
Humic Acid 500 ppm	13,33	15,00 d	22,67 d	32,67 cd
Humic Acid 1000 ppm	13,67	17,00 bc	24,67 bc	34,00 bc
Humic Acid 1500 ppm	14,00	18,33 a	26,00 ab	35,67 a
Humic Acid 2000 ppm	14,33	17,33 ab	24,33 c	33,00 cd
Fulvic Acid 500 ppm	14,00	17,67 ab	26,67 a	35,00 b
Fulvic Acid 1000 ppm	12,33	16,00 cd	23,33 cd	33,00 cd
Fulvic Acid 1500 ppm	12,00	15,00 d	20,67 e	30,67 e
Fulvic Acid 2000 ppm	12,67	16,00 cd	23,33 cd	32,67 cd

The difference in the average number of leaves between the treatments for humic acid and fulvic acid concentration showed that as the concentrations of humic acid and fulvic acid rose to a certain level, the number of leaves would also rise. Table 2 shows the result of the DMRT test, in which the treatment groups with the same letters showed no significant differences. In contrast, the treatment groups that had different letters showed significant differences. At 4 weeks after planting, the treatment of humid acid with a concentration of 1500 ppm resulted in the highest number of leaves (18.33 pieces). At 6 weeks after planting, the treatment of fulvic acid with a concentration of 500 ppm resulted in the highest number of leaves (26.67 pieces). Eight weeks after planting, the treatment of humid acid with a concentration of 1500 ppm resulted in the highest number of leaves (35.67 pieces).

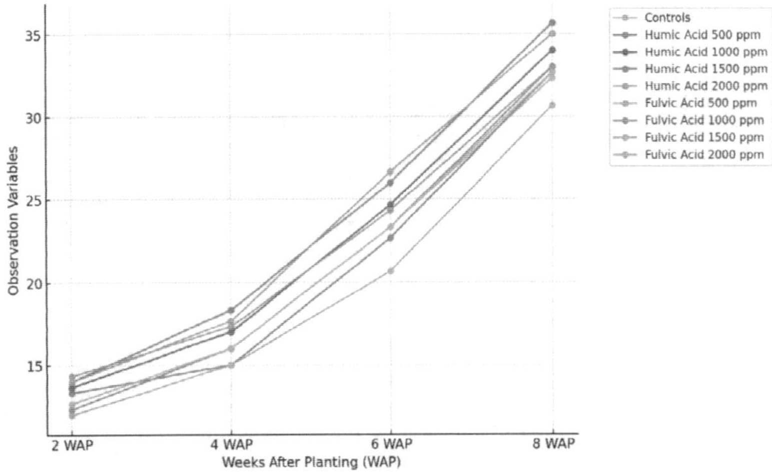

Figure 2. Graph of Effect of application of humid acid and fulvic acid on the number of leave

Overall, the treatment of humic acid with a concentration of 1500 ppm was quite effective in increasing the number of leaves of the Chrysanthemum plant. It also supported by Martins [9] that the function of humic substances is improving nutrient uptake, chlorophyll content, photosynthesis, and flowering of the Chrysanthemum plant. Humic acid stimulates chlorophyll biosynthesis, leading to enhanced photosynthetic efficiency, which provides the energy and assimilates needed for leaf production. Additionally, humic substances act as natural growth stimulators by influencing plant hormones like auxins and cytokinins, which are critical for leaf initiation and expansion. They also improve soil microbial activity, promoting nutrient cycling and stress tolerance, while enhancing enzymatic activity to support cellular processes. These combined effects result in a significant increase in the number of leaves, as well as improved flowering attributes like flower size and number.

4. Conclusion

The provision of fulvic acid and humic acid at certain concentrations significantly affected chrysanthemum growth. Fulvic acid 500 ppm effectively increased plant height in the early phase, while humic acid 1500 ppm was more influential in increasing the number of leaves in a longer growth period.

5. Acknowledgment

On this occasion the author would like to thank P3M Politeknik Negeri Jember for the 2024 PNBP research program grant.

References

[1] S.B. Indrajati, L.D. Saputro, and A.R. Yuniar, *Panduan Teknis Budidaya Krisan Potong*. Bogor: PertanianPress, 2023.
[2] Badan Pusat Statistik, "Statistik Indonesia 2023," Jakarta, 2023.
[3] A. Dariah, S. Sutono, N. L. Nurida, W. Hartatik, and E. Pratiwi, "Pembenah Tanah untuk Meningkatkan Produktivitas Lahan Pertanian The Use of Soil Conditioners to Increase Agricultural Land Productivity," vol. 2, no. 9, pp. 67–84, 2015.
[4] P.P. Hegde, M. Shivaprasad, M. Ganapathi, and S. Y. Chandrashekar, "Influence of Biostimulants on Growth and Physiology of Chrysanthemum (Dendranthema grandiflora TZVELEV.) var. Kolar Local under Protected Cultivation," *Int J Pure Appl Biosci*, vol. 6, no. 2, pp. 1259–1264, 2018, doi: https://doi.org/10.18782/2320-7051.6288.
[5] C.N. Sible, J.R. Seebauer, and F.E. Below, "Plant biostimulants: A categorical review, their implications for row crop production, and relation to soil health indicators," *Agronomy*, vol. 11, no. 7, pp. 1–20, 2021.
[6] H. Khobragade, S. Badge, and D. Gudhekar, "Foliar application of humic acid and salicylic acid on growth, flowering and yield of chrysanthemum cv. Padari revadi," *Int J Sci Res*, vol. 11, no. 12, pp. 3–5, 2022.
[7] M.M.A El-Baset and M.M. Kasem, "Improving Growth Characteristics and Vase Life of Dendranthema Grandiflorum 'Flyer' Using Humic and Fulvic Acids As Biostimulants Substances," *Scientific Journal of Flowers and Ornamental Plants*, vol. 9, no. 2, pp. 87–102, 2022.
[8] Mahendar Kumar Sootahar *et al.*, "The Effect of Fulvic Acids Derived from Different Materials on Changing Properties of Albic Black Soil in the Northeast Plain of China," *Molecules*, vol. 24, no. 8, pp. 1–12, 2019.
[9] Evili Marai Martins, Juan Quijia Pillajo, and Michelle L. Jones, "Humic and Fulvic Acids Promote Growth and Flowering in Petunias at Low and Optimal Fertility," *HortScience*, vol. 59, no. 2, pp. 235–244, 2024.
[10] M. Zayin Sukri,T.R. Kusparwanti, R. Firgiyanto, Hanif F, Gallyndra F.D, and Ella A.S. "Aplikasi Pupuk Urea dan Pupuk Organik Cair terhadap Peningkatan Pertumbuhan dan Hasil Tanaman Sawi Pakcoy (*Brassica rapa* L.)". *Poceedings Agropross*. 2022. Doi: 10.25047/agropross.2022.293
[11] M. Zayin Sukri, R. Firgiyanto, V. Kartikasari, and Basuki. "Combination of Cattle Manure, Humic Acid and Mycorrhiza on Mycorrhizal Root Infection in Chili Pepper and Availability of Udipsamments Soil Nutrients". *Jurnal Penelitian Pertanian Terapan*, Vol. 19 (2): 141-145, 2019. Doi: http://dx.doi.org/10.25181/jppt.v!9i2.1513.

IOP Conf. Series: Earth and Environmental Science **1446** (2025) 012042 doi:10.1088/1755-1315/1446/1/012042

Physical characteristics of various green feed crackers as a small-ruminant supplement

TM Syahniar[1*], M Andriani[1], Nurkholis[1], N Ningsih[1], and D Lestari[2]

[1]Department of Animal Science, Politeknik Negeri Jember, Mastrip Street PO BOX 164, Jember Regency 68101, East Java, Indonesia
[2]Department of Agribusiness Management, Politeknik Negeri Jember, Mastrip Street PO BOX 164, Jember Regency 68101, East Java, Indonesia

*Corresponding author's email: mahiseta@polije.ac.id

Abstract. Purposes of this research were to evaluate dimensions and physical characteristics of various green feed crackers as a small-ruminant supplement. It was made using Odot grass (T1), Gliricidia leaves (T2), Leucaena leaves (T3), mixture of Gliricidia and Leucaena (T4), and 2 other formulations of legume mixture with other feed ingredients (T5 and T6). They were designed using a pattern of Completely Randomized Design (CRD) into 6 treatment levels with 3 replications. Each level treatment was mixed with molasses, homogenized, molded using a quadrangle-shaped mold containing 9 pieces with pressure, then baked in the oven of 60°C overnight. Green feed crackers were analysed the physical characteristic which using parameters observed on dimension, specific density, bulk density, tapped bulk density, and hardness. Data were assessed using ANOVA and followed using DMRT to identify significant differences between treatment levels. Each piece of green feed crackers was ranged 20-27 g/piece with dimensions average of (61*61*11) mm. Results denoted that the treatment affected ($P<0.0001$) dimensions, bulk density and tapped bulk density of green feed crackers however presented similar number on the specific density and hardness. It was implied that the variation and properties of feedstuffs used influenced physical characteristics of the green feed crackers.

1. Introduction

Small ruminants such as goats and sheep are potential animal protein sources for humans. Small ruminants in Indonesia were mostly raised by smallholder farmers than by industry [1]. The advantages of small ruminant livestock farming do not need a lot of capital and large fields, has socio-cultural positions, develops relatively faster, is easy to maintain in small cages or shepherded on empty land or natural grasslands, can give birth every year with twins, the produce faeces to fertilize the farming land, and wide market opportunities because of the demand for small ruminants for meat and animal qurban always increasing every year [2]. On the other hand, feed for goats and sheep is widely available around the field. However, the quality of feed is often insufficient nutrient adequacy, especially on a smallholder farm scale.

The feed available in small livestock farming environments is commonly based on agricultural waste, such as rice straw, corn cobs, or other coarse feed. In addition to the low nutrient content, feed ingredients from these wastes also have a low level of palatability then causes reduced feed consumption. The one of strategy to achieve the target improvement of animal productivity was using feed supplements from valuable crops and agroindustry by-products. The creation of enhanced feed through supplements offered was a potential auxiliary to small-scale livestock farming [3]. Feed supplements are an important part of effective nutritional management, especially when the available feed conditions are not sufficient for their nutritional needs.

One of the supplements that can be used to improve small ruminant productivity is feed supplement in the form of crackers which function as feed toppings. Crackers are a form of innovation that packages supplementary feed in small quantities or only as feed toppings but is expected to increase palatability and basal feed consumption. Small ruminants selected feed with greater time than cattle, commonly they choose feeds or parts of plant (young stems, leaves, and buds) of good quality. The selection process was helped by their narrow muzzle and by the high mobility of their tongue and lips [4]. Therefore, it is necessary to make feed supplements with good physical quality, that can be accepted by small ruminants and can increase animal consumption and productivity. This research was conducted to observe the physical quality of green feed cracker as a feed supplement for small ruminants.

2. Methods

2.1. Experimental design and supplement making

This experiment used two types legume as the main raw material for making various green feed crackers supplements. It was Gliricidia and Leucaena leaves. Both leaves were obtained from the area around Jember Regency. Each or mixed of both leaves were formulated to meet the nutrient quality for small ruminant requirement as a feed supplement. They were constructed using a pattern of Completely Randomized Design (CRD) into 6 treatment levels with 3 replications. The level treatment of various green feed crackers supplement was made using Odot grass (T1), Gliricidia leaves (T2), Leucaena leaves (T3), mixture of Gliricidia and Leucaena leaves (T4), and 2 other formulations of legume mixture with other feedstuffs (T5 and T6).

The green feed of each level treatment was sun-dried and milled first before mixed in with molasses of 350 g/kg. Then, it was well-homogenized manually. The admixture was molded using a quadrangle-shaped mold containing 9 pieces with pressure, then baked in the oven of 60°C overnight. Specific formulations for T5 and T6, as well as the types and quantities of other feedstuffs mixed in, cannot be mentioned in this paper because they will be used in the intellectual property rights application.

2.2. Parameters and data analysis

Parameters determined were the dimension, specific density, bulk density, tapped bulk density, and hardness. Dimensional measurements of the green feed crackers supplement were carried out by determining the weight, length and width of both sides and the thickness formed. Specific density, degrade and tapped bulk density were measured using method that cited to Farda, et al. [5] and Nurrofingah, et al. [6]. The hardness was measured using Hardness Tester number 174886 by Kiya Seisakusho, Ltd. made in Tokyo, Japan. The weighing measurement was carried out using a digital scale with one number after the decimal point by KERN & Sohn GmbH Ser. Number W082043, D-72336 Balingen, Germany.

Data were assessed using analysis of variance (ANOVA) in PASW Statistics 18. Then, data were further analysed with Duncan's Multiple Range Test (DMRT) to identify significant differences between the treatment groups.

3. Results and Discussion

The result of supplement dimensions that was made from various green feed crackers as supplement for small ruminant were provided in **Table 1.** It demonstrated highly significant ($P<0.0001$) in all dimensions ranging from weight, length, width to thickness of green feed crackers. The weight of crackers T2, T3 and T4 showed the similar results of the lowest value and the weight of crackers T5 and T6 showed the highest number. Crackers T1 have a greater weight compared to T2-T4 but was lighter compared to T5-T6. On the contrary, the dimensions of length, width and thickness of crackers showed the similar tendency that crackers T1 have the largest dimensions compared to other levels. This was due to the characteristic of Odot grass which was bulkier than legumes. This was also comparable to the nutrient value contained in Odot grass. The nutrient value of legumes was higher than grass and rice straw as in results of Antari et al. [7] and Syahniar et al. [8].

Table 1. Dimensions of green feed crackers as supplement for small-ruminant

Treatments	Weight (g)	Length (mm)	Width (mm)	Thickness (mm)
T1	23.37^{bc}	81.7^{c}	81.6^{c}	21.4^{b}
T2	21.82^{ab}	61.3^{a}	61.3^{a}	11.6^{a}
T3	20.03^{a}	61.5^{ab}	61.5^{a}	11.7^{a}
T4	21.34^{ab}	61.4^{a}	61.4^{a}	11.6^{a}
T5	25.38^{cd}	61.7^{b}	61.5^{a}	12.0^{a}
T6	27.07^{cd}	61.7^{b}	62.0^{b}	11.6^{a}
SEM (Standart Error Mean)	0.433	0.004	0.007	0.014
P-value	0.0001	0.0001	0.0001	0.0001

$^{a-d}$ Variations in superscripts within columns indicated a statistically significant effect between treatment levels at the specific P-value<0.0001

Physical characteristics which accommodated specific density, degrade and tapped bulk density, and the hardness of green feed crackers supplement for small-ruminant were shown in **Table 2.** The specific density of various green feed crackers was statistically similar (P>0.05) ranged from 965.10 to 1317.45 (kg/m^3). The value of bulk density and tapped bulk density tend to increase start from T1, T2, T3, T4, and T5, but decreases at T6. Each treatment presented a very strong influence (P<0.0001) on bulk density and tapped bulk density. The level treatment of T2, T3 and T4 were similar but treatments T1, T5 and T6 showed very significant differences. It showed that the materials of T2, T3, T4 have the similar characteristics but different from materials of T1, T5 and T6. Bulk density and tapped bulk density were also influenced by the nutrient content and particle size distribution [9] [10]. The results of Farda et al. presented palm kernel cake with high fat content and uniform small particle size distribution tends to have low bulk density [5]. The value of tapped bulk density would be greater than bulk density due to the vibration that caused compaction so that the volume per m^3 of material was getting smaller. Bulk density and tapped bulk density were closely related and played a crucial role in determining the capacity of silo and mixing of materials. It is important to know the value of bulk density and tapped bulk density of materials in filling and packaging into a still container but vibrates which caused the weight of the material per unit volume to increase [11].

Table 2. Physical characteristics of green feed crackers supplement for small-ruminant

Treatments	Specific density (kg/m^3)	Bulk density (kg/m^3)	Tapped bulk density (kg/m^3)	Hardness (kgf)
T1	1174.17	119^{a}	228^{a}	nd
T2	961.37	342^{b}	385^{b}	8.58
T3	1028.06	347^{b}	405^{b}	6.17
T4	965.10	372^{b}	408^{b}	8.01
T5	1317.45	563^{d}	609^{d}	7.38
T6	1161.64	475^{c}	524^{c}	7.00
SEM (Standart Error Mean)	33.293	0.006	0.006	0.342
P-value	0.057	0.0001	0.0001	0.087

$^{a-d}$ Variations in superscripts within columns indicated a statistically significant effect between treatment levels at the specific P-value<0.05

The hardness of T1 was not detected due to a fibrous and bulky characteristic of the main feedstuff material. The fiber content in Odot grass of T1 also affected the hardness value. The fiber bonds that exist in the crude fiber of the material would bind the compounds in it, so that feedstuffs with high crude fiber content indicated bouncy and not hard texture [12]. The level treatment of T2-T6 did not result in a significant change (P>0.05). The feed ingredients of each level treatment did not affect the hardness value. However, the hardness level of crackers tent to increase starting from T3, T5, T6, T4 then T2. The highest number indicated a stronger adhesion process of the material so that the level of cracking was increasingly arduous. The adhesion of materials was influenced by molding process, starch or sugar content as a natural adhesive. In alignment with the perspective of Juniyanto et al., one of the things that affected the hardness value was the starch content. It showed the similar functions as a natural adhesive with the gelatinization process in the pellets during the pelleting process [13]. Meanwhile, the sugar in the form of sucrose contained in molasses could bind other ingredients through a crystallization process when heated.

4. Conclusions

Results was implied that feedstuffs used influenced the dimension and physical characteristics of the green feed crackers as supplement for small ruminants. This was also comparable to the nutrient value contained in the feedstuff and influenced by processing mechanism applied.

Acknowledgment

Authors acknowledge Pusat Penelitian dan Pengabdian kepada Masyarakat (P3M) Politeknik Negeri Jember for conceding Research Grant of PNBP 2024 with contract number of 789/PL17.4/PG/2024.

5. References

[1] Sujarwanta R O, Afidah U, Suryanto E, Rusman, Triyannanto E and Hoffman L C 2024 Review: Goat and Sheep Meat Production in Indonesia *Sustainability (Switzerland)* **16**
[2] Kardiana A, Suherlan E, Hayurani H and Khotijah L 2017 Exploration Data of Animal Protein Fulfillment from Small Ruminant *International Journal of Electronics Communication and Computer Engineering* **8** 2249–071
[3] Boudalia S, Smeti S, Dawit M, Senbeta E K, Gueroui Y, Dotas V, Bousbia A and Symeon G K 2024 Alternative Approaches to Feeding Small Ruminants and Their Potential Benefits *Animals* **14** 1–21
[4] Tedeschi L O, Molle G, Menendez H M, Cannas A and Fonseca M A 2019 The assessment of supplementation requirements of grazing ruminants using nutrition models *Transl Anim Sci* **3** 811–23
[5] Farda F T, Syahniar T M, Wijaya A K and Ermawati R 2020 Sifat fisik bungkil inti sawit hasil ayakan *Jurnal Peternakan Sriwijaya* **9** 21–6
[6] Nurrofingah U, Sumiati, Retnani Y and Dek M S P 2019 Physical characteristics of duck pellet and antioxidant activity of sweet potato leaves and cassava leaves *AIP Conference Proceedings* vol 2120 (American Institute of Physics Inc.)
[7] Antari R, Syahniar T M, Mayberry D E, Pamungkas D, Anderson S T, Poppi D P and others 2014 Evaluation of village-based diets for increasing the weight and condition of Ongole (Bos indicus) and Bali (Bos javanicus) cows in Indonesia *Anim Prod Sci* **54** 1368–73
[8] Syahniar T M, Antari R, Pamungkas D, Mayberry D E, Poppi D P and others 2012 The level of tree legumes required to meet the maintenance energy requirements of Ongole (Bos indicus) cows fed rice straw in Indonesia *Anim Prod Sci* **52** 641–6
[9] McDonald P, Edwards R, Greenhalgh J, Morgan C, Sinclair L and Wilkinson R 2010 *Animal Nutrition* (England: Pearson)
[10] Pranata A 2015 Pengaruh pemberian bungkil inti kelapa sawit yang difermentasi menggunakan isolat selulolitik dari belalang kembara pada pakan terhadap penampilan produksi puyuh jantan *Buletin Peternakan* **39** 49–56

[11] Jaelani A, Dharmawati S and Wacahyono 2016 Pengaruh tumpukan dan lama masa simpan pakan pelet terhadap kualitas fisik *ZIRA'AH* **41** 261–8
[12] Utama C S, Sulistiyanto B and Rahmawati R D 2020 Kualitas fisik organoleptik, hardness, dan kadar air pada berbagai pakan ternak bentuk pellet *Jurnal Litbang Provinsi Jawa Tengah* **18** 43–53
[13] Juniyanto M I R, Susilawati I and Supratman H 2015 Ketahanan dan kepadatan pelet hijau rumput raja (Pennisetum purpuphoides) dengan penambahan berbagai dosis bahan pakan sumber karbohidrat *Jurnal Universitas Padjajaran* **4** 1–13

Potential Loss of Body Weight in Bull Transportation and the Effect of Adding *Acacia mangium* Bark Extract During Recovery Period

TUP Sujarnoko[1*], D Budiono[1], NA Sholeha[1], R Siskandar[1], TM Syahniar[2], M Andriani[2], and AU Prastujati[3]

[1]College of Vocational Studies, Bogor Agricultural University (IPB University), Jalan Kumbang No. 14, Bogor 16151, Indonesia
[2]Department of Animal Science, Politeknik Negeri Jember, Mastrip Street PO BOX 164, Jember Regency 68101, East Java, Indonesia
[3]Department of Livestock Product Processing Technology, Politeknik Negeri Banyuwangi, 68461, East Java, Indonesia

*Email: tekadurip21@apps.ipb.ac.id

Abstract. The transportation process of the bull can induce stress, affecting decreasing body weight and causing a loss for farmers. This study focuses on body loss affected by transportation from Magetan to Bogor and recovery after transportation. This research used 10 cross-breed bulls from Magetan East Java and transported them to Bogor West Java for 16 hours. The bull weighed on Magetan before being transported and Bogor after a 10-hour recovery. The first experiment compares transported bulls and non-transported Bulls. The second experiment focused on recovery after transportation with treatment on the ransom (T0: Control, T1: Control + 0.4% Acacia extract) The experimental design was a randomized block design with two treatments and a tree block based on body weight. The parameters tested included body weight loss, blood metabolism, economic loss, Average daily gain, and opportunity recovery using acacia extract. Results showed that transportation decreased body weight until 9.47±0.76% or Rp. 473,507.00/tail. The transportation significantly increased blood urea nitrogen ($p<0.05$) and did not affect cholesterol, creatinine, or Uric acid $p>0.05$. Added acacia extract of 4% significantly decreased body losses and creatinine after 10 days of recovery time ($P<0.05$). The added extract did not affect cholesterol, uric acid, and blood urea nitrogen ($p>0.05$). In conclusion, transportation caused the loss, and adding 0.4% *Acacium mangium* bark with a pressure hot water extraction method improved the body recovery of the Bull after transportation.

Introduction

Transportation of bulls during Eid Al-Qurban from livestock-producing areas to Jabodetabek is extensive. The transportation process causes stress, fatigue, loss of body weight and decreased meat quality [1]. Therefore, it is necessary to carry out several treatments so that the livestock can recover quickly [2]. Livestock stress due to transportation is reflected in blood metabolism [2]. Accelerating recovery can be done by adding feed ingredients with high antioxidant levels [3]. One of them is phenol found in acacia bark [4].

Acacia mangium is a plant that is widely found in Indonesia. The stems of acacia are used as paper. Acacia bark becomes waste which contains secondary metabolite compounds in phenols and tannins [5]. Phenol is a compound that can act as an antioxidant and antibiotic [6,7]. Tannin is a type of phenol that can bind protein. Tannin can speed up recovery by increasing By-pass protein [8]. Becomes a natural antibiotic [6] and an antioxidant [7]. The bark was extracted with pressure hot water extraction and adding to the feed. The addition of acacia bark extract may reduce stress and make recovery time more quicker than bull transported without acacia extract addition

The research processes this time tested the effect of transportation stress from Magetan to Bogor as far as 630 km, for 16 hours, on body weight loss, blood metabolites, and losses due to body weight loss. Next, the livestock were kept for 10 days with the addition of acacia bark extract to speed up recovery after transportation.

1. Method

1.1. Bull transportation
A total of 10 cows from Magetan were weighed and transported standing up with a density of approximately 13.26 Kg/M^2. The cow's journey covers a distance of 630 km, for 16 hours. The cattle depart at 2.00 in the afternoon and arrive in Bogor at 06.00 in the morning. The livestock are then rested, given water, forage, and concentrate, at 2 pm the livestock are weighed again to measure body weight loss due to shipping.

1.2. Acacia bark extraction
The acacia bark extraction process was carried out using the hot water extraction presser method [9], with a modified pressure of 0.7 bar, and a temperature of 1200C for 30 minutes. The acacia bark extract was then tested for phenol and tannin levels using a UV FIS spectrophotometer at 740 nm.

1.3. Sampling and measurement of blood metabolites
Blood was taken from transported and non-transported cattle from the jugular vein. Blood from the bull after a 10-day recovery period was taken from the jugular vein. Bull is defined into two treatments, with and without acacia extract. The blood is then stored in an ice box which is then immediately centrifuged and separated between supernatants and serum to be tested for blood metabolism using a photometer.

2. Results and Discussion
The transportation process in bull causes stress [2]. The impact of this stress is a decrease in body weight, decreased meat quality, and changes in blood metabolism [1,2]. These changes can be seen in table 1 and table 2. Transportation prosses need a lot off energy, due to some metabolic seconder on the blood changing. Blood metabolite should become good parameter to sporting this research.

Table 1. The effect of cattle transportation on body weight and economic losses

Parameter	With Transportion	Without Transportation	Difference
Mean of bull weight (Kg)	330.00±25.12	297.30±23.90	31.30±3.96
Price (Rp/head)	18,150,000	17,676,493	473,507

Transportation of male cattle from Magetan to Bogor caused a loss in body weight of around 31.30 ± 3.96 kg or around 9.47 ± 0.76%, this loss in body weight caused a loss of around Rp. 1,798,500.00/head.

The blood metabolism of transported and non-transported bull was collected. Transportation increase blood urea nitrogen ($p<0.05$) and did not affect cholesterol, creatinine, or Uric acid $p>0.05$ Tabel 2. The increase in blood urea nitrogen (BUN) is caused by the conversion of muscle mass which contains protein into energy where the byproduct is urea [10].

Table 2. Effect transportation prosses on bull blood metabolite

Parameter	Without Transportation	With Transportion	P-Value
Total Blood protein	4.90±0.61	4.55±0.41	0.40
Albumin	2.90± 0.26	2.80±0.57	0.79
BUN	10.00± 2.64	15.50± 1.29*	0.01
Uric Acid	0.60±0.00	0.50±0.08	0.09
Creatinine	1.50±0.26	1.68± 0.29	0.45
Cholesterol	103.33±1.86	105.25±2.12	0.91

*The symbol mean that transportation has significant effect on blood metabolite with p value<0.05

The use of phenol compounds as antioxidants can accelerate recovery from stress due to illness, travel and other causes [11]. The administration of acacia bark extract significantly accelerated the recovery process by reducing the difference in initial weight and after 10 days of maintenance with a p value <0.05. The percentage difference in body weight recovery after 10 days of rearing shows that the use of acacia bark extract is able to approach the initial body weight compared to treatment without acacia bark extract with a p<0.05 level. Table 3. Giving acacia bark extract reduces transportation losses per bull compared to not amounting to IDR 465,000.00/day.

Table 3. Effect extract acacia on the body weight during recovery time off bull

Parameter	Non-extract	Extract	P-Value
Difference in weight during recovery (Kg)	14.333±8.50	5.00±4.09*	0.03
Recovery difference percentage/ BW (%)	4.23±2.28	1.56±1.25*	0.02

*The symbol mean that extract acacia have significant effect on the body weight with p value<0.05

The accelerated recovery process can occur due to acacia bark extract containing phenol. Phenol acts as an antibiotic that protects livestock from pathogens [12]. Phenol can also act as an antioxidant which can protect cells from free radicals caused by stress [13]. Acacia bark contains tannin [4]. Tannin can increase recovery speed because may protect the protein from damage in the rumen [8]. Tannin is capable of being a protein by-pass agent [4].

Table 4. Effect of adding acacia extract on bull blood metabolite during recovery time.

Parameter	Non-extract	Extract	P-Value
Total Blood Protein	4.133±0.12	4.367±4.09	0.22
Albumin	2.501±2.28	2.433±0.35	0.81
BUN	11.687±0.58	8.67±2.552	0.11
Uric Acid	0.633±0.057	0.600±0.10	0.64
Creatinine	1.77±0.115*	1.400±0.10	0.01
Cholesterol	82.23±12.05	106.00±16.52	0.11

*The symbol mean that acacia bark extract addition have significant impact on blood metabolite with p value<0.05

Added acacia extract of 4% significantly decreased body losses and creatinine after 10 days of recovery time (P<0.05). The added extract did not affect cholesterol, uric acid, and blood urea nitrogen (p>0.05). In conclusion, transportation caused the loss, and adding 0.4% *Acacium mangium* bark with a pressure hot water extraction method improved the body recovery of the Bull after transportation table 4. Creatinine is a substance that arises due to muscle breakdown, the higher it shows that the muscle breakdown is greater. The addition of acacia bark extract is able to maintain muscle remodeling. One mechanism for reducing creatinine turnover can be reduced due to protein bypass due to binding of tannin protein [4].

3. Conclusions

The transportation process from Magetan to Bogor, a distance of 630 km, for 16 hours causes stress which can reduce body weight and reduce farmer profits. The addition of 0.4% dry matter to feed, acacia wood extract, can speed up livestock recovery and reduce losses due to transportation.

Acknowledgment

Authors acknowledge Direktorat Akademik Perguruan Tinggi Vokasi Kementerian Pendidikan, kebudayaan, Riset dan Teknologi for conceding Research Grant of BOPTN 2024 with PTP scheme.

4. References

[1] Bhatt, N., Singh, N. P., Mishra, A. K., Kandpal, D., & Jamwal, S. (2021). A Detailed Review of Stress in Livestock and its Management Techniques. International Journal of Livestock Research, 11(1), 30-41.

[2] Retnani, Y., Jayanegara, A., Risyahadi, S. T., Firdaus, M., Wicaksono, R. I., Barkah, N. N., Taryati, Baihaqi, M., & Sujarnoko, T. U. P. (2022). Feeding by Logistic Feed (Mash, Silage, Pellet and Wafer) Compared Transportation with Conventional Feed on Physiological, Blood Biochemical and Body Weight Gain Recovery in Tropical Sheep Pasca Transportation. American Journal of Animal and Veterinary Sciences, 17(2). https://doi.org/10.3844/ajavsp.2022.172.179

[3] Li, R., Wang, L., Chen, B., Zhang, Y., & Qi, P. (2024). Effects of Transportation on Blood Indices, Oxidative Stress, Rumen Fermentation Parameters and Rumen Microbiota in Goats. Animals, 14(11). https://doi.org/10.3390/ani14111616

[4] Sujarnoko, T. U. P., Ridwan, R., Nahrowi, & Jayanegara, A. (2020). Extraction of Tannin from Acacia (Acacia mangium) Bark and its use as a Feed Additive for Protecting in vitro Ruminal Degradation of Tofu Dregs. Advances in Animal andVeterinarySciences,8(7). https://doi.org/10.17582/journal.aavs/2020/8.7.761.765

[5] Osman, Z., Pizzi, A., Elbadawi, M. E., Mehats, J., Mohammed, W., & Charrier, B. (2024). Effect of Technological Factors on the Extraction of Polymeric Condensed Tannins from Acacia Species. Polymers, 16(11). https://doi.org/10.3390/polym16111550

[6] Farha, A. K., Yang, Q. Q., Kim, G., Li, H. bin, Zhu, F., Liu, H. Y., Gan, R. Y., & Corke, H. (2020). Tannins as an alternative to antibiotics. In Food Bioscience (Vol. 38). Elsevier Ltd. https://doi.org/10.1016/j.fbio.2020.100751

[7] Dey, A., Dutta, N., & Sharma, K. (2015). Antioxidant status, metabolic profile and immune response of lambs supplemented with tannin rich Ficus infectoria leaf meal. Japanese Journal of Veterinary Research, 63(1), 15–24. https://doi.org/10.14943/jjvr.63.1.15

[8] Jayanegara, A., Sujarnoko, T. U. P., Ridla, M., Kondo, M., & Kreuzer, M. (2019). Silage quality as influenced by concentration and type of tannins present in the material ensiled: A meta-analysis. Journal of Animal Physiology and Animal Nutrition, 103(2), 456–465. https://doi.org/10.1111/jpn.13050

[9] Sujarnoko, T. U. P., Jayanegara, A., Ridwan, R., & Nahrowi. (2020). Tannin characteristic from Hevea brasiliensis and Durio zibethinus with pressure and hot water extraction. IOP Conference Series: Earth and Environmental Science, 462(1). https://doi.org/10.1088/1755-1315/462/1/012010

[10] Santinello, M., Lora, I., Villot, C., Cozzi, G., Penasa, M., Chevaux, E., Martin, B., Guerra, A., Righi, F., & de Marchi, M. (2024). Metabolic profile of Charolais young bulls transported over long-distance. Preventive Veterinary Medicine, 231. https://doi.org/10.1016/j.prevetmed.2024.106296

[11] Motta, S., Guaita, M., Cassino, C., & Bosso, A. (2020). Relationship between polyphenolic content, antioxidant properties and oxygen consumption rate of different tannins in a model wine solution. Food Chemistry, 313. https://doi.org/10.1016/j.foodchem.2019.126045

The Effect of Using Solid Decanter Bioconversion Products on the Performance of Laying Hens in the Pullet Period

Asriani Hasanuddin*, Rusdi Rusdi, Minarny Gobel and Fatmawati Fatmawati

Faculty of Animal Husbandry and Fishery, Tadulako University, Palu, Indonesia

*) asrianiuntad60@yahoo.com

Abstract. A study aimed at looking at the effect of using palm mud (solid decanter) as a bioconversion product on the performance of laying hens during the pullet period. Two hundred pullet laying hens were reared for 18 weeks, given 5 levels of bio converted palm sludge, namely; R0 (control treatment without using bio converted palm sludge); R1 (use of 6% bio converted palm sludge; R2 (use of 12% bio converted palm sludge; R3 (use of 18% bio converted palm sludge) and R4 (use of 24% bio converted palm sludge). Rations were mixed with iso protein and iso calories at 18% of protein. Five treatments and four replications were used in the research design, resulting in 20 experimental units with ten laying hens each. The research findings indicated that while the treatments had no significant effect on feed conversion ratio and ration intake (P>0.05), they had a substantial impact on egg production (HDP), egg mass, and income over feed cost (IOFC) (P<0.05). One bird in R1 group died during experimental period. In coclusion, the use of bio converted palm sludge waste (solid decanter) could be included in the ration up to 24% and can support the production performance of laying hens.

Keywords: palm sludge, bioconversion, laying hens, pullet period

1.Introduction.

Increasing efficiency in poultry production, especially laying hens, is one of the main focuses on research in the field of animal husbandry. Various efforts have been made to increase efficiency by using industrial waste as alternative feed. Palm mud or solid decanter, is a waste from the palm oil industry which has the potential to be used as animal feed, because it contains a number of important nutrients such as protein, fat and fiber. They have potential to support the growth and productivity of laying hens. Bioconversion is a biotechnological process that involves the use of microorganisms to increase the nutritional value of raw materials, including plantation waste such as palm oil sludge. *Neurospora sp* is one type of mold that has been widely used in this bioconversion process. Through bioconversion, palm oil sludge can be converted into feed ingredients with higher nutritional value and more easily digested by the poultry.

The pullet period is a critical phase in the life cycle of laying hens, where best growth and development is particularly important because it will find high egg productivity in the future. Therefore, the goal of the present study is to explore the effect of using bio converted palm mud-based rations on the production performance, feed conversion and health of laying hens.

2. Materials and Methods

This research used 200 Loghman Brown laying hens in the pullet period. They were kept in the 20 individual wire-battery cages, each equipped with a drinking nipple and eating trunk. The basal ration mainly consisted of corn, soybeans, fish meal, rice bran, cooking oil, minerals and amino acids and bio converted palm sludge (LSB) were included accordingly. The rations were prepared and formulated based on NRC [1], with 18% of protein and 2727 - 2735 Kcal/kg of metabolizable energy.

Five treatments and four replicates, and each experimental unit consisted of ten birds using a Completely Randomized Design (CRD) according to [2]. The experimental rations were 5 levels of bio converted palm sludge, namely; R0 (control treatment without using bio converted palm sludge); R1 (use of 6% bio converted palm sludge; R2 (use of 12% bio converted palm sludge; R3 (use of 18% bio converted palm sludge) and R4 (use of 24% bio converted palm sludge). Data were analyzed using an ANOVA [2] and Duncan test for means comparison.

3.Results and Discussion

The mean values of intake, hen day production (HDP), egg mass, feed conversion ratio (FCR), income over feed cost (IOFC) and mortality of laying hens given rations treated using *Neurospora sp* bio converted palm sludge during the study are presented in Table 1.

Table 1. Performance of laying hens given bio converted solid decanter

Parameter	R0	R1	R2	R3	R4	SEM	P-Value
Intake, g/head/day	103,76	104,07	105,94	105,02	110,89	0,93	0,067
HDP, %	70,86^{a}	77,92^{ab}	68,01^{a}	75,60^{ab}	82,173^{b}	1,62	0,014
Egg Mass, g/head/day	39,81^{a}	42,76^{ab}	38,54^{a}	42,84^{ab}	46,36^{b}	0,94	0,044
FCR	2,63	2,43	2,76	2,47	2,39	0,06	0,165
IOFC, rupiah	13335^{a}	14763^{ab}	12768^{a}	14303^{ab}	15577^{b}	323,18	0,014
Mortality, (%)	0	0,025	0	0	0	-	-

HDP: hen day production, FCR= feed convertion ratio, IOFC = income over feed cost. Different superscripts within the line are significantly different

3.1. Ration Intake. Analysis of variance data indicated that ration consumption was unaffected by treatment. Table 1 shows that the average consumption ranged from 103.76 to 110.89 g/head/day. The absence of differences in feed consumption shows that the level of preference of the chickens for the feed given is the same, but it is also suspected because the average consumption of the treated feed is not much different compared to the control feed (R0). The results of this study are in line with research conducted by [3] which used palm oil waste fermented with the *Pleurotus ostreatus* fungus in quail rations which did not show a significant difference in ration consumption up to a usage level of 24%, because the nutritional content in each treatment was not much different compared to control feed. Additionally, the energy content in the feed for all treatments were similar to the control treatment, so that energy is adequate to meet energy requirement, and therefore the birds reduced or stopped to consume more rations.

Furthermore, Amrullah [4] said that energy content and environmental temperature have a significant role in regulating livestock ration intake. Meanwhile, Susanto [5] states that the level of livestock ration consumption will increase if the ration given has a low energy content. In contrast, intake will decrease when energy content is high. In addition to the level of preference for rations, other factors that influence ration consumption are color, smell, taste [6] as well as age and physiological condition of livestock [7].

3.2. Hen Day Production. A group of birds' daily egg production at a certain age is determined by calculating their Hen Day Production (HDP). The results of HDP variance analysis showed that the use of solid decanter bioconversion products in the rations, resulting a significant difference between

R4 and R0, while treatment R1 (6%); R2 (12%) and R3 (18%) there was no difference (P>0.05). This is thought to be because the solid decanter bioconversion product has increased the nutritional content and reduced crude fiber content so that the feed is more easily digested by digestive enzymes. During the egg-laying phase, laying hens need around 18% protein and amino acids, where this protein has a few amino acids which are used in forming body tissue, replacing damaged tissue, meeting production needs and the excess is converted into energy [8].

The protein in feed having bio converted solid decanter consumed by livestock will be broken down into amino acids to be absorbed by the body and then arranged into tissue and egg protein. This occurs because of the use of *Neurospora sp* mold as an inoculant in solid decanter bioconversion which can work to break down organic materials, and the results of the bioconversion process produce several other organic compounds such as amino acids and saccharides in dissolved form which are easily absorbed and produce organic acids, hormones, vitamins and antibiotics. All these compounds are needed for egg production. The mechanism for absorbing feed nutrients, especially protein, is preceded by the breakdown of protein into amino acids by enzymes. If amino acids are available in sufficient quantities in the feed, it will be possible to improve egg production [9]. It is believed that the feed provided in solid decanters as a result of this bioconversion contains certain amino acids that are crucial to the laying hens' growth and effective feed use.

According to [10] egg mass is influenced by egg weight, livestock population and daily egg production. Egg mass in the treatment with bioconversion solid decanter up to 24% in the diet was better than the diet without bioconversion solid decanter (control treatment, R0). The results of this study show that the use of solid decanter bioconversion up to 24% has a good influence on egg mass production even with a 100% reduction in rice bran. This result is almost the same as the egg mass production of purebred chickens obtained by [11], namely 46.26 g/head/day. There is no significant difference in the egg mass production of purebred chickens also due to the consumption of the same ration.

3.3. Feed Conversion Ratio (FCR). One of the determining factors in the quality of feed to meet the nutritional needs of livestock is the feed conversion rate (Feed Conversion Ratio) [12]. FCR describes how much ration is needed to produce one unit of livestock product, for example one kilogram of meat or eggs. The average FCR values in laying hens in the treatments of 0%, 6%, 12%, 18% and 24% bioconversion product palm sludge were obtained respectively: 2.63 (R0); 2.43 (R1); 2.76 (R2); 2.47 (R3) and 2.39 for R4 treatment or with a range of 2.39 – 2.76. Table 1 shows that the higher the use of bio converted palm sludge in the ration, the lower the FCR value, although statistically it does not show a significant difference (P>0.05). This is believed to result from a solid decanter that has experienced a bioconversion process, utilized in ration preparation, enhancing the nutritional value and improving the digestibility of the ration. Additionally, this solid decanter bioconversion product can act as a provider of single cell protein which has high nutritional value, especially the essential amino acid content which is needed in egg biosynthesis. Increasing the efficiency of ration use with the bioconversion process by *Neurospora sp* which can degrade crude fiber and increase the protein content in feed ingredients [13]. Apart from that, it is also suspected that because of using solid bioconversion decanters in rations, it does not increase feed consumption, but increases egg weight and egg production, thereby reducing feed conversion and increasing production efficiency.

3.4. Income Over Feed Cost. Income Over Feed Cost (IOFC) represents the difference between overall income and the expenses incurred for feed during maintenance. Utilizing bioconversion solid decanters can enhance overall revenue and lower feed expenses, consequently boosting profits. The IOFC calculation for R0, R1, R2, R3 and R4 is Rp. 13335; Rp. 14763; Rp. 12768: IDR 14303 and Rp. 15577, respectively. Treatment R4 has the highest value while R2 has the lowest value. Berliana and Azis [14] reported that IOFC is the difference between total income and the cost of feed used during maintenance. The increase in the IOFC value can reduce the total feed costs. Bioconversion solid decanters are waste that has high economic value and still has good enough nutrition for laying hens. Solikin [15] pointed out that the elevated and reduced IOFC values resulted from a greater or lesser gap between sales and feed expenses incurred during the data collection period. The analysis of variance results showed that the use of bio converted solid decanters in the diets had a significant effect (P<0.05) on IOFC (Table 1). Based on Table 1, the highest IOFC was obtained in the treatment using a bioconversion solid decanter (24% or R4 treatment), namely Rp. 15,577 and the lowest one was Rp. 12,768 per day at the rate 12% bioconversion solid decanter (L2). Interestingly, the highest values of IOFC were achieved, when the inclusion rate of solid decanter was at the rate of 24%, due to the high eggs production. The IOFC value is also influenced by the price of the feed. However, income over feed cost only shows the revenue value over feed costs, excluding others costs of production [16].

3.5. Mortality. Mortality is a particularly important factor for the success of livestock business development. Mortality in livestock businesses is kept as low as possible. In the present study, the mortality was 0.025% on R1 group, and the other treatments group were zero mortality. Moreover, mortality in this study was not caused using bio converted solid decanter flour but rather due to stress factors, the chickens died suddenly. This indicated that the use of bio converted solid decanter flour in laying hen rations produces not a negative effect on the chicken health status.

4. Conclusion

From the results and discussion presented above, it can be concluded that incorporating palm sludge bioconversion products with *Neurospora sp* at a level of 24% in the diet positively influences the performance of laying hens, particularly in terms of egg production, egg mass, and breeder income, without any significant effect on the feed efficiency.

References

[1] NRC. 1994. Nutrient Requirements of Poultry. Ninth Revised Edition.

[2] Steel RGD and Torrie JH. 1993. Prinsip dan Prosedur Statistika. : Suatu Pendekatan Biometrik. Edisi Kedua. Dialihbahasakan oleh Bambang Sumantri. Jakarta: PT. Gramedia Pustaka Utama.

[3] Gusril L, Nuraini dan Mirzah. 2020. Performa Puyuh yang Diberi Limbah Sawit Fermentasi dengan *Pleurotus ostreatus* dalam Ransum. Jurnal Peternakan Indonesia, 22 (2): 228-235.

[4] Amrullah, I. K. 2003. NutrisiAyam Broiler. Lembaga Satu Gunung Budi: Bogor.

[5] Susanto, R. S. 2004. Pengaruh perbedaan tingkat protein dalam ransum terhadap produktivitas itik Indian Runner. Skripsi. Fakultas Pertanian. Universitas Sebelas Maret, Surakarta. page 32-33.

[6] McDonald, P., Edwards, R. A., Greenhalgh, J. F. D., Morgan, C. A., Sinclair La, and Wilkinson, R. G. 2010. Animal Nutrition. Ed ke-7. New York (US): Prentice Hall.

[7] Akbarillah, T., Kususiyah, dan Hidayat. 2011. Pengaruh suplementasi tepung daun Indigofera pada tepung gaplek sebagai sumber energy pengganti jagung kuning dalam ransum puyuh (*Coturnix coturnix japonica*) terhadap produksi dan warna kuning telur. Jurnal Sains Peternakan Indonesia. 6(1): 33-40

[8] Aziz, F., G. A. M. K. Dewi, dan M. Wiraparta. 2020. Kualitas telur ayam isa brown umur 100-104 minggu yang diberi ransum komersial dengan tambahan tepung kulit kerang. Universitas Udayana. Denpasar.

[9] Fristanti, F., M. F. Wadjdi, dan I. Dinasari. 2016.Pengaruh tingkat pemberian kupang (musculata senhausia) terhadap konsumsi, produksi telur, konversi pakan dan income over feed cost (IOFC) pada Itik Mojosari, Jurnal Peternakan 1 (1): 1-5.

[10] Latif, S., E. Suprijatna dan D. Sunarti. 2017. Performans produksi puyuh yang diberi ransum tepung limbah udang fermentasi. Jurnal Ilmu-ilmu Peternakan. 27 (3): 44-53

[11] Nuraini., Y. S. Nur, and A. Djulardi. 2020. Response of laying quail to a diet enriched with

Cocoa pods fermented by Pleurotus ostreatus. Int. J. Poult. Sci., 10 (1): 96-101. DOI: https:// dx.doi.org/10.36380/jwpr.2020.13 URL : https://jwpr.science-line.com/ attachments/article/52/JWPR%20 10(1)%2096-101,%202020.pdf

[12] Deko, M. K., H. Irfan., Djunaidi dan M. H. Natsir. 2018. Efek penggunaan tepung umbi dan kulit bawang putih (Allium sativum) sebagai feed additive terhadap penampilan produksi ayam petelur. Jurnal Ilmu-Ilmu Peternakan 28 (3): 192 – 202. DOI: 10.21776/ ub.jiip.2018.028.03.02

[13] Bidura, I G.N.G., Warmadewi, D.A., Candrawati, D.P.M.A., Aryani, I G.A.I., Putri Utami, I. A., Partama, I.B.G. and Astuti, D.A.. 2009. The effect of ragi tape fermentation products in diets on nutrients digestibility and growth performance of Bali drake. Proceeding. The 1st International Seminar on Animal Industry 2009. Sustainable Animal Production for Food Security and Safety. 23- 24 November 2009. Faculty of Animal Science, Bogor Agricultural University. Pp:180-187

[14] Berliana, S dan Aziz, A. 2016. Pengaruh waktu pemberian pakan selama periode pertumbuhan ayam broiler terhadap efisiensi pengunaan protein. Jurnal Ilmu Peternakan. 19 (2): 71-76.

[15] Solikin, T. 2016. Bobot Akhir, Bobot Karkas dan Income Over Feed and Chick Cost Ayam Sentul Barokah Abadi Farm Ciamis. Fakultas Peternakan, Universitas Padjajaran, Bandung. (Skripsi)

[16] Zulfan dan Zulfikar. 2020. Evaluasi Performa dan Income Over Feed & Chick Cost (IOFCC) Tiga Strain Ayam Broiler yang Beredar di Aceh. Jurnal Agripet, 20 (2), 136- 142.

Eggs production yield and physical quality eggs of laying quail by inclusion of *Clitoria ternatea* flour as a phytobiotic in the diets

Hafsah[1*], Dwi Sulistiawati[1], Muhammad Tahir[1], Andi Pertiwi Damayanti[1], Andi Nurfadillah[2], Muhammed Rasyiq Al Shahab[3]

[1] Animal Science Department, Faculty of Animal Husbandry and Fishery Tadulako University, Central Sulawesi, Indonesia
[2] Graduate Student of Animal Science Department, Faculty of Animal Husbandry and Fishery Tadulako University, Central Sulawesi, Indonesia
[3] Post Graduate Student of Tadulako University, Central Sulawesi, Indonesia

[1*] Corresponding author email: hafsahuntad62@gmail.com

Abstract. Feed ingredients need to be managed well to maximize quail production potential. In addition to having the right amount and quality of nutrients to maximize egg production, that feed can serve as a source of phytobiotics to help maintain health. The butterfly pea (*Clitoria Ternatea*), which has flowers that has vitamins, minerals, and bioactive compounds, is one feed that can be used as a source of phytobiotics. The study was conducted to determine the quantity and quality of eggs produced by laying quail (*Coturnix coturnix* japonica) by adding butterfly flower (BF) flour to diets as a source of phytobiotics. At total of 180 laying quails were treated in this study, which 6 treatments with 3 replicates. Experimental diets were offered ad libitum and water were available at all time. The study was resulted highly significant ($P<0.01$) affected quail day production(QDP) and index of yolk color. Whereas feed intake and feed conversion ratio were found to be significantly ($P<0.05$) affected. Egg weight, egg index, egg yolk, yolk index, egg albumen, shell weight, and shell thickness, however, were not significantly ($P>0.05$) impacted. Butterfly flower (*Clitoria Ternatea*) flour can be used as a natural phytobiotics and can increase the number of eggs and the yolk's color score without affecting the physical quality of the eggs.

Keywords: *clitoria ternatea*, eggs quality, eggs yield, phytobiotics, quail

1. Introduction

The quail (*Coturnix-coturnix* Japonica) is a poultry species that has the potential to be improved and produced in greater quantities because to its short life cycle and effectiveness in generating animal protein, such as meat and eggs. There are a number of benefits to quail farming, including the fact that sexual maturity, or the ability to lay eggs, is reached at 6 weeks of age, with a reasonably high egg production of 250–300 eggs annually [1], low cost, easy maintenance, and a small area needed to create an animal house.

Animal productivity, particularly the quantity and quality of eggs produced, is mostly influenced by feed. Optimizing egg production is possible if the standard nutrient needs are met. The quality of the ingredients used to formulate the meal determines how well the quail receive the nutrients they need. When high-quality feed meets the necessary nutrient requirements, it can yield maximum productivity. Nutrient requirements for quail are relatively higher compared to other types of poultry. Nutritional requirement of quail in the starter phase (1-3 mg ages) are protein (P) 24% and EM (2900 kcal/kg), the grower phase (3-6 mg ages) is protein (20%) and EM (2700 kcal/kg) and layer phase (> 6mg- culling) needs protein (21-22%) and EM (2900 kcal/kg)[2]. The nutrient elements must be fulfilled from the feed consumed. However, in their application the farmers used synthetic antibiotics in feed with the function as a growth promotor. The compound of that antibiotics could be cause residues in eggs and meat. The adverse effects of using antibiotics as Antibiotic Growth Promoter (AGP) in diets were the cause of the food safety issues that society faces [3]. By using phytobiotics as Natural Growth Promoters (NGP), which have been found to be an efficient substitute for antibiotics, the usage of AGP in quail feed can be predicted. As NGPs, phytobiotics have evolved as feed additives that improve immunity, egg production, and keep poultry's digestive tracts healthy [4].

Herbal plants known as phytobiotics contain chemicals that are good for living things. When added to animal feed, phytobiotics can affected the productivity of the animals. Giving animals herbal plants may not have the same impact as using antibiotics or other synthetic medications. In the wild, animals spontaneously seek out and consume herbal herbs to cure specific ailments [5].

One kind of herbal plant with significant bioactive components and potential for usage is the butterfly pea flower (*Clitoria ternatea*), also known as the Telang plant in the local name. The purplish blue flower of this plant are a morphological feature that suggests the presence of plant pigments, such as anthocyanin. The majority of plants' natural hues are anthocyanins, a type of flavonoid molecules [6]. There are 128.25 ppm of antioxidants, 249.69 mg of anthocyanins per 100g [7], 11.82% of ash, and 5.51% of crude fiber [8]. Being a source of antioxidants in plants is another advantage of anthocyanins.

Antioxidants are substances that, by interacting with reactive free radicals to create unstable, unreactive free radicals, prevent oxidation. All compounds that can postpone or stop damage brought on by target molecules oxidizing are known as antioxidants. The biological definition of antioxidants includes all substances that can lessen the harmful effects of oxidants, such as metal-binding proteins and enzymes, while the chemical definition includes substances that donate electrons. Free radical-induced oxidative stress has also been linked in numerous studies to a number of serious illnesses, including cancer, heart disease, and degenerative diseases. In these conditions, antioxidants are also beneficial [9].

There are two types of antioxidants: natural and synthetic. *BHT* (*Buthylated Hydroxytoluene*, *BHA* (*Buthylated Hydroxyanisole*, and *TBHQ (Tert-butylhydroquinone)* were examples of synthetic antioxidants that are effective at preventing oxidation. Because synthetic antioxidants can generate poisons in the body over time and cause cancer, safer natural antioxidants are required. Natural antioxidants are found in a variety of herbal plants that include phytochemicals such vitamin C, flavonoids, isoflavins, flavones, and anthocyanins [10]. The antioxidant activity of dried *Clitoria ternatea* flower extract at pH 4 (392.91 ppm), pH 5 (371.00 ppm), and pH 6 (344.37 ppm) was also found. which include cyanidin, isoquercetin, cyanine, and kaempferol, which are anthocyanin-active chemicals.

The body constantly produces free radicals as a result of inflammation, starvation, regular cellular metabolic activities, and reactions to external factors such ambient pollution, ultraviolet (UV) light, and cigarette smoke. Free radicals are a naturally occurring byproduct of the body's metabolic activities when they form. A molecule or atom with an unpaired electron is called a free radical. These unpaired electrons create very reactive free radicals that, in order to neutralize themselves, will subsequently seize or steal electrons from other substances like proteins, lipids, carbohydrates, and DNA. Free radicals have the ability to infiltrate the body, damage healthy cells, and destroy their structure and function [11].

The study aims to determine effects of inclusion flower butterfly (*Clitoria ternatea*) flour as a phytobiotic in laying quail diets on quantity and quality of eggs.

2. Materials and Methods

2.1. Study Area

The experiment was conducted in South Sibalaya Village, Sigi Regency, Central Sulawesi Province, Indonesia, Experimental Farm of Animal Science Department, Faculty of Animal Husbandry and Fishery Tadulako University.

2.2. Birds and Cages

This study utilized 180 laying quail in total, during the observation period, quails were housed in 18 pens, each pen were assigned 10 quails. Pens size dimensions were: 95 cm (length), 50 cm (width), and 40 cm (height). Then, quails were kept up to 14 weeks. Quails were unlimited access to diets and water were available at all time. Data collection started from the quail were 8 weeks to 14 weeks ages.

2.3. Experimental Diets

Flowers of the butterfly pea (*Clitoria ternatea*), also known locally as the Telang plant, are harvested from farmers in the Palu region. Fresh butterfly flowers are turned into flour after being dried 24 hours at 40 - 50°C in the oven.
The treatments were arrangement with Completely Randomized Design (CRD) in 6 treatments and 3 replications, each replicate pen filled with 10 quails. Butterfly pea (*Clitoria ternatea*) flowers experiment is given in the form of flour as call a butterfly flower (BF). Feed treatments consist of.:

P_1: 100% commercial diets;
P_2: 100% local diets + tetracycline 0.005%/kg diets + egg stimulant 0.01%/kg diets;
P_3: 99.7% local diets + BF 0.3%/kg diets;
P_4: 99.4% local diets + BF 0.6%/kg diets;
P_5: 99.1% local diets + BF 0.9%/kg diets;
P_6: 98.8% local diets + BF 1.2%/kg diets.

The feed formulation is prepared based on the standard nutrient requirements of quail for layer phase [12], that is 21% protein and ME 2900 kcal/kg. The nutrient composition of the treatment diets as shown in Table 1. Quail were kept for 8 weeks with 1 week for adaptation to the cage environment, 1 week for adaptation to the treatment feed, and 6 weeks for observation of the effect of treatment. Diets and drinking water were provided ad libitum. Feed consumption was calculated every week and egg production was calculated every day (hen day production) during the study. Observations of the physical quality of the eggs were carried out 3 times, with the 9, 10, 11 weeks of quail age. For each replication, 10 eggs samples were taken from all treatments.

2.4. Variables and Data Analysis

Variables observed: eggs production yield (feed intake, quail day egg production, feed conversion ratio) and physical egg quality (egg weigh, egg index, yolk index, yolk composition, yolk color score, albumen composition, egg shell composition and egg shell thickness). Quail egg production yield were collected every day for each pen, the calculated quail day production (QDP) during the observation.

All data were analyzed by using one way analysis of IBM SPSS Statistics 22 in Completely Randomize Design. Significant effects detected by analysis of variance, where further tested by Duncan's Multiple Range Test (DMRT) [13].

Table 1. Ingredients of local feed and nutrient content of the diets of laying quails

Ingredients	Quantity (%)
Corn	61.40
Soybean cake	14.16
Fish meal	12.15
Rice bran	6.50
Palm oil	0.59
Mineral and vitamin mix	3.00
Dicalcium phosphate (DCP)	2.00
Lysine	0.10
Methionine	0.10
Calculated Nutrients:	
Metabolizable Energy, (kcal/kg)	2964.78
Crude Protein (CP)	21.00
Crude Fat (CF)	3.76
Crude Fiber (C.Fiber)	4.47
Calcium	0.99
Phosphorus	0.85
Lysine	1.31
Methionine	0.54

Note: * Nutrient content of laying quail commercial diet: CP (21.69%), CF (4.08%), C.Fiber (3.25%), Ash (3.99%) (Proximate analysis at Feed Nutrition Laboratory, 2024)

3. Results and Discussion

3.1. Egg Production Yield

Data on egg production yield as shown in Table 2.

Table 2. Egg production yield during observation of each treatment

	Treatments					
Eggs Production Yield	P_1	P_2	P_3	P_4	P_5	P_6
QDP, (%)**	97.62^a	76.43^b	89.05^a	74.76^b	77.97^b	84.53^a
Feed Intake, (g/d)*	24.23^a	18.97^b	20.93^a	18.53^b	20.17^a	19.67^b
Feed Conversion Ratio (g/ g)*	2.22^a	1.66^b	1.84^a	1.51^b	1.77^a	1.61^b

Note: ** Means treatment showed high significant effects ; QDP (Quail Day Production)
* Means treatment showed significant effects

Results of the treatments on feed intake and feed conversion ratio showed significant impacts ($P<0.05$) and quail day egg production (QDP) showed high significant effects ($P<0.01$) as result of the treatments. The DMRT test (α:5%) on QDP revealed that P1; P3; and P6 treatments shown significant different ($P<0.05$) with P_2, P_4, and P_5. P_1 (97.62%), P_3 (89.05%), and P_6 (84.53%) all had comparable average QDP. However, in treatments of P_2 (76.43%), P_4 (74.76%), and P_5 (77.07%) had lower rates. The condition indicates that the QDP of quails fed a local diet including 0.03% and 1.2% butterfly flower (BF) flour might be equal to the QDP of commercial diets (P_1). Due to poor feed intake, which impacted nutritional intake, QDP decreased. Yield production were affected by strain, age, feed and water intake, minerals and protein content of the feed [14]. The low feed and energy intake during the production phase of quails can be decreased yield production [15].

DMRT test results showed that feed intake was substantially different (P<0.05) greater in treatments P_1 (24.23 g/day), P_3 (20.93 g/day), and P_5 (20.17 g/day) than in treatments P_2 (18.97 g/day), P_4 (18.97 g/day), and P_6 (19.67 g/day). Similarly, the feed conversion ratio showed the following average results: P_1 (2.22), P_3 (1.84), and P_5 (1.77), with P_2 (1.66), P_4 (1.51), and P_5 (1.61). The larger weight of eggs produced and the reduced feed intake resulted in a lower feed conversion ratio. Feed conversion ratio is strongly related to the feed's capacity to produce eggs. Temperature, feed nutrient content, egg weight, and egg yield all had an impact on the feed conversion ratio [14].

3.2. Physical Quality of Egg

Data on physical quality of egg as shown in Table 3.

Table 3. Physical quality egg of laying quail at 9-11weeks of age for each treatment

	Treatments					
Physical quality eggs	P_1	P_2	P_3	P_4	P_5	P_6
Egg weight, (g)	11.48	10.39	10.31	10.20	9.64	9.71
Egg index	1.27	1.26	1.27	1.28	1.27	1.27
Yolk composition, (%)	28.31	25.60	26.45	28.79	29.74	27.31
Yolk index	0.32	0.31	0.34	0.34	0.35	0.34
Yolk color score *	7.07^a	8.50^a	9.03^b	9.40^b	9.43^b	9.53^b
Albumen composition, (%)	61.77	63.94	63.19	61.31	59.74	62.06
Egg shell composition, (%)	9.93	10.47	10.36	9.90	9.64	10.63
Egg shell thickness, (mm)	0.227	0.207	0.213	0.207	0.183	0.197

Note: * Means treatment showed significant effects

The physical quality of the egg (egg weight, egg index, yolk index, yolk composition, albumen composition, egg shell composition, and egg shell thickness) was unaffected (P>0.05) by the addition of butterfly flower (*Clitoria ternatea*) flour to laying quail diets, though the yolk's color score was affected (P<0.05). The yolk color score was increased with the amount of butterfly flower (*Clitoria ternatea*) flour increased in their diet. On the yolk color score a hight effects. Average of treatments for P_1 (7.07), P_2 (8.50), and P_4 (9.40) was significantly (P<0.05) different from the yolk color score of P_3 (9.03), P_4 (9.40), P_5 (9.43), and P_6 (9.53), according to the DMRT test (α:5%).

The average yolk color score was 7.07 -9.53 (Table 3), the higher yolk color score was founded in the hight level of the treatment (1.2%). The increase in yolk color value is caused by the luteolin or flavone content of *Clitoria ternatea* flowers. Luteolin is a type of flavonoid, with a yellow crystal appearance. It is the main yellow coloring compound which has been used as a dye source [10]. It also contains anthocyanin which is a compound that plays a role in giving red, purple and blue colors of flower [16], carotenoid [17]. Another, influenced by zeaxanthin, lutein, alpha-carotene, beta-carotene, and carotenoids [18]. However, higher the level used of butterfly flower (*Clitoria ternatea*) flour in laying quails diet decreased egg weigt. Average egg weight is in the range of 9.71-11.48. There were standar that ranges 6 -16 [19], protein intake [20], hormones [21], and age [19]. Furthermore, protein or amino acid is a nutrient that play an important role in controlling the egg weight [14]. Even though trend of treatment was that the higher the level of use of butterfly flower flour, decreased egg weight.

Yolk composition in the range 25.60-29.74% (Table 3), it was similar results of 28-29% of the egg weight [22]. There, another study was lower than yolk composition was 30.79-32.67% [23]. Egg size related to the size of the yolk compared with albumen, despite the fact that albumen was still important to determine on egg [24]. Yolk composition was influenced by genotype and age as well as the cholesterol [25]. The decrease of yolk cholesterol levels could be decrease also yolk weight [26].

Albumen composition was in range 59.74-62.06% (Table 3), higher than 54.91-56.74% [23]. Albumen weight is generally influenced by egg weight, age, genetics [25] and hormone [21]. The percentage of normal quail egg albumen ranged between 60-63% of the egg weight [22], 52-62% of the egg weight [26].

egg shell composition in the range 9.64-10.63% and egg shell thickness 0.183-0.227 mm (Table 3). It was lower than 11.13-12.79% and higher in shell thickness 0.155-0.168 mm [23]. The content of calcium and phosphorus in the diets can affect the weight and thickness of the egg shell [22]. Egg shell quality and egg shell shape could be influenced by age and the contents of mineral in the diets, such as calcium, magnesium and phosphorus as in organic compound [24].

4. Conclusion

Quail egg day productivity (QDP) could be matched with the use of commercial feed (P_1) if laying quail diets contained 0.3%–1.2% flour from butterfly flowers (*Clitoria ternatea*). Nevertheless, the addition of feed additives (egg stimulants and antibiotics) to P_2 led to decrease in the quantity and quality of eggs produced.

Acknowledgement

The authors would like to express our sincere thanks to Rector of Tadulako University for providing financial supports, enabling to do this research. We would also like to thank the Faculty of Animal Husbandry and Fisheries for the research facilities used in this study. Our gratitude also goes to the students involved in this study for their supports in taking care of the quails and cleaning the pens.

References

[1] Pradikdo, B. A., Sudjarwo, E. and Muharlien. 2016. Effect of quail types with different commercial feeds on the percentage of carcass and internal organs of quail (*Coturnix-coturnix* japonica). *Jurnal Ternak Tropika*, **17(2)**: 23-33

[2] SNI (Standard National Indonesia). 2006. National Standard Nutrients of Non Ruminant Animals. Medion

[3] Saeid, J., and A. Al-Nasry. 2010. Effect of dietary coriander seeds supplementation on growth performance carcass traits and some blood parameters of broiler chickens. *Int. J. Poult. Sci.* **9**:867-870

[4] Panda, A. S.R. Rao, and M. Raju. 2009. Phytobiotics a natural growth promoter. *Poultry International*. pp:10-11

Biyatmoko, D., U.Santoso, Juhairiah. 2021. The Use of herbal jamu phytobiotics as an antibiotic replacement growth promoter to improve growth performance and meat quality of Alabio Duck. Proc. Seminar Nasional Lingkungan Lahan Basah, Agricultural Faculty of University Lambung Mangkurat. South Kalimantan, Indonesia. **6 (2)**:1-6

[5] Hariadi, H., M. Sunyoto, B. Nurhadi, and A. Kurniawan. 2018. Comparison of phytochemical characteristics pigmen extract (anthocyanin) sweet purple potatoes powder (L.) and *Clitoria ternatea* flowers as natural dry powder. *J. of Pharmacognosy and Phytochemistry*. **7(4):**3420-3429

[6] Iznilillah. W. , A. Jumiono, M. Z. Fanani. 2023. Proximate comparison of antioxidants and anthocyanins in various processed food products with the addition of natural coloring from Clitoria ternatea. *J. Ilmiah Pangan Halal*. **5(2)**: 163–174

[7] Neda, G.D., M.S. Rabeta and M.T. Ong. 2013. Chemical composition and anti proliferative properties of *Clitoria ternatea* flowers. *J. Inter. Food Research*. **20(3)**: 1229-1234

[8] Barhe, T.A. and Tchouya, G.R. 2014. Comparative study of the anti-oxidant activity of the total polyphenols extracted from *Hibiscus habdariffa* L. Glycinemax L. , Yellow tea and Red wine through Reaction with DPPH Free Radical. *Arabian Journal of Chemistry*

[9] Sumartini, Y. Ikrawan, F. M. Muntaha. 2020. Analysis of *Clitoria ternatea* flowers with variation of pH by methods of Liquid Chromatograph-Tandem Mass Spectrometry (LC-TMS). *Pasundan Food Technology Journal*. **7(2)** : 70-77

[10] Liochev, S.I.,2013. Reactive Oxygen Species and the Free Radical Theory of Aging, Free Radical Biology and Medicine. E-Book

[11] Mattjik, A. A., and I. M. Sumertajaya. 2006. Perancangan Percobaan. Edisi ke-2. Institut Pertanian Bogor (IPB)-Press. Bogor

[12] Leeson, S. and J. D. Summers. 2005. Commercial Poultry Nutrition. 3rd ed. Nottingham University Press, Nottingham, England

[13] Widjastuti, T. and R. Kartasidjana. 2006. The effect of restricted feeding and its implication on the performance of *Coturnix-coturnix* japonica at the first production phase. *J. I. Trop. Anim. Agric*. 3:162-166

[14] Parwata, I. 2016. Flavonoid. *Journal FMIPA*.Denpasar: Universitaas Udayana

[15] Hermana, W., T. Toharmat, Sumiati, and W. Manalu. 2014. Performances and egg quality of quail offered feed containing sterol from Katuk (*Sauropus androgynus*) and Mulberry (*Morus alba*) leaf meal. *Int. J. Poult. Sci*. **13**:168-172

[16] Hammershoj, M., U. Kidmose, and S. Steenfeldt. 2010. Deposition of carotenoids in egg yolk by short-term supplement of coloured carrot (*Daucus carota*) varieties as forage material for egg-laying hens. *J. Sci. Food. Agric*. **90**:1163-1171

[17] Tserveni-Goussi, A. and P. Fortomaris. 2011. Production and quality of quail, pheasant, goose and turkey eggs for uses other than human consumption. Woodhead Publishing, United Kingdom. p. 509-537.

[18] Tuleun, C., and A. Adenkola. 2013. Performance and erythrocyte osmotic membrane stability of laying Japanese quails (*Coturnix-coturnix* japanica) fed varying dietary protein levels in a hot-humid tropics. *Agric. Biol. J. N. Am*. **4** :6-13

[19] Latifa, R. 2007. The increasing of afkir duck's egg quality with pregnant mare's serum gonadotropin (pmsg) hormones. *J. Prot*. **14** :21 – 30

[20] Li-Chan, E. C. Y., and H. O. Kim. 2007. Structure and Chemical Compositions of Eggs. In: Y. Mine. Egg Bioscience and Biotechnology. John Wiley and Sons, Inc, Canada. pp.1-95

[21] Darmawan, A., K. G. Wiryawan, and Sumiati. 2013. Egg production and quality of Magelang duck fed diets containing different ratio of omega 3, omega 6 and organic Zn. *Med. Pet*. **36**: 197-202

[22] Zita, L., E. Tůmová, and L. Štolc. 2009. Effects of genotype, age and their interaction on egg quality in brown-egg laying hens. *Acta. Vet. Brno*. **78** :85-91

[23] Subekti, S., W. G. Piliang, W. Manalu and T. B. Murdiati. 2006. The use of powder and extract leaves of katuk (*Sauropus androgynus* L. Merr) as feed substitution to reduce the cholesterol of japanese quail product. *J. Am. Vet*. **11** :254-259

[24] Nys, Y., and N. Guyot. 2011. Egg Formation and Chemistry. In: Y. Nys, M. Bain and F. V. Immerseel (Eds.). Improving the Safety and Quality of Eggs and Egg Products. Woodhead Publishing, United Kingdom. pp. 83-132

[25] Kebreab, E., J. France, R. P. Kwakkel, S. Leeson, H. D. Kuhi, and J. Dijkstra. 2009. Development and evaluation of a dynamic model of calcium and phosphorus flows in layers. *J. Poult. Sci*. **88:**680-689

Supplementation varying level Nano Se-Yeast on blood haematological profile of laying quails

D Pantaya[1], N Ningsih[1], A P Yulinarsari[1], M Hanif[1], NAS Ningrum[1]

[1]Department of Animal Science, Politeknik Negeri Jember, Jember, 68121

Corresponding email: dadik_pantaya@polije.ac.id

Abstract: The research was conducted to evaluate the haematological profiles of laying quails before and after receiving a Newcastle disease (ND) vaccine. The varying doses of Nano Se-yeast (Selenium-yeast) probiotics were tested for their effects on hematological profile such us, hemoglobin (Hb), erythrocytes, hematocrits, MCV, MCH, MCHC, leucocyte (WBC), neutrophile, lymphocytes, monocytes, and trombosit of laying quails. Two hundred laying quails were randomly allocated into four treatments and each treatment consisted of five replicates, with ten birds in each replicate. The first treatment was a control (T0), basal diet without supplementation Nano Se-Yeast, while the second (T2), third (T3), and fourth (T4) treatments were administered basal feed supplemented with Se-yeast containing 1x 10^8 CFU/g at levels of 0.5, 1.0, and 1.5 g/per kg of feed. The quails received vaccinations with ND vaccine strain La Sota seven days following treatment with yeast supplementation. Blood samples were taken forty-eight hours before vaccine and three days after the vaccine. The results show MCV and MCHC profiles were found to be significantly affected by yeast treatment, but hemoglobine, erythrocytes, hematocrites, MCH, RDW-CV, leucocytes, eosinophils, basophils, neutrophils, lymphocytes, and thrombosit were not significantly affected, particularly before vaccine treatment. This finding suggests that Se-yeast supplementation is beneficial under little-stress conditions.

1. Introduction

The consumption of animal protein by the Indonesian population increases every year, recorded in 2021 at 16.61 grams/capita/day and in 2022 it increased to 17.74 grams/capita/day [1]. Sources of animal protein come from meat and egg products. One source of animal protein for the Indonesian people is quail. Quail farming has an important socio-economic role in rural communities. In general, this type of local livestock has low production performance and is susceptible to disease, thus affecting its productivity. One alternative to overcome production performance is with mineral supplementation, including Selenium (Se) [2].

A trace mineral called Selenium supports biological processes like reproduction, production, and antioxidant defense [3, 4]. Selenium is a very vital element for normal body function, because it has an important role in improving livestock health. A deficiency of selenium in poultry feed is associated with selenium role in antioxidant defense through the enzyme glutathione peroxidase (GPx), which is a crucial component of cellular antioxidant defense [5]. Disruption of this role will affect productivity, reproduction and immune response [6].

Organic selenium is more easily absorbed than inorganic[7]. Selenium-Yeast (Se-Yeast) biomass contains protein and some amino acids can bind to selenium to form organic compound bonds [6]. One of the indicators used in the immune system in livestock is through the hematologic profile. Hematology is also used as a response used for medical purposes and can be used to diagnose stress conditions caused by vaccination treatment. This study is used to determine the hematologic profile in laying quails with the supplementation of Nano Se-yeast to the basal diet.

2. Materials and Methods

The research was due on Politeknik Negeri Jember farm. The colony cage was filled with 200 quails (body weight (BW) = 166± 10 g) that were supplied from a local quails pullet farm. The diets were formulated to meet the nutrient requirements for layer quails (Table 1). The feed and water were provided ad libitum. The 200 quail in all were divided among four treatments at random. Each treatment has five replicates, each of which has ten quails. During the two-week observation period, the first treatment was given a basal feed as a control (T0), the second (T1), third (T2), and fourth (T3) treatments were given basal feed supplemented with Se-yeast that contains 1x 10^8 CFU/g at doses of 0.5, 1.0, and 1.5 g/per kg of diets, respectively. The product of Se yeast contained of 440 ppm Selenium. Seven days following Se-yeast supplementation, Quails received oral drops of the ND (New Castle disease) vaccination strain La-Sota.

2.1. Blood sampling and analysis

The blood were taken in 3 d before and 3 d after vaccine administration. The blood samples were drawn from the wing vein and placed in heparinized containers (BD Vacutainer) using a 1 mL syringe. Tube contained blood were centrifugated (3500 x g 10 min and) stored -2°C until analysed. The hematology analyzer was used to analyze the WBC and RBC concentrations (cells/μl). The leukocytes and lymphocytes were separated using an optical microscope [8].

2.2. Statitic analysis

All the statistical analysis were performed using INITAB procedure (Minitab inc USA) with the following models:

$$Y_{ij}=\mu+\alpha_i+\varepsilon_{ij}$$

Where Y= dependent variables, μ: overall mean, αi: the effect of treatment, εij: residual, The significant differences among the treatment groups were analysed by Tukey range test at 0.05 significant level.

Table 1 . Feed Basal Formulation

Parameter	%
Ingredient (% of DM)	
Corn meal	50
Mix concentrate feed Quails layer	35
Rice Bran	15
Total	100
Calculated composition	
Metabolism energy (Kcal/kg of DM)	3123
Crude Protein (%)	19,5
Fibre (%)	6
Fat (%)	3
Calcium (%)	3,2
osphor (%)	0,8
Lysine (%)	1
Methionine (%)	1

Note: Minerals (%), P (0.25), Ca (2.0), Mg (0.45), Na (0.35); trace elements (mg/kg): Cu (15) vitamins (IU/kg): vitamin A (6,000), vitamin D3(1,250), and vitamin E (10 mg/kg)

3. Results and discussions

3.1. Feed consumption

Body weight and feed intake were weekly measured from week 7 to week 10. In the current study, the quails body weight was 166.94 ±1.24 g at 18 weeks old. According to the statistical analysis, feed intake and body weight did not differ significantly with the supplementation Nano Se-Yeast (Table 2). The major yeast cell wall contained oligosaccharide, glycoprotein, peptides, minerals, and vitamins [9] [10]. Besides that, selenium was expected to improve health status and immunity against disease[11]. Nano Se-yeast supplementation at varying levels has no effect on feed consumption. According to another study, quail feed consumption was unaffected by the inclusion of probiotics [12]. The formulation of yeast in the feed, which makes it challenging to compare the effects of organic minerals, may be the cause of variations in animal responses.

Table 2 . Feed intake and body weight of layer Quails with supplementation of Nano Se-Yeast

Parameter	Se Yeast supplementation (g/kg feed)				
	T0	T1	T2	T3	P Value
Feed intake (g)/ bird	23.32±0.57	24.15±0.48	23.99±0.59	22.45±0.75	0.256
Weight (g)	168.12±3.03	165.02±3.96	166.26±1.43	168.36±0.67	0.777

Note: T0: Control (basal diet without supplementation Nano Se-Yeast), T2, T3, and T4: basal feed supplemented with Nano Se-yeast containing 1x 10^8 CFU/g at levels of 0.5, 1.0, and 1.5 g/per kg of feed.

3.2. Blood parameters

The purpose of this study is to evaluate the haematological profiles under high-stress conditions (following vaccination treatment) with those under normal circumstances. Table 3 displays the hematological blood profiles before and three days following vaccination. According to statistical analysis, the addition of Nano Se-Yeast had a significant effect on MCV and MCHC (P<0.05), and did not significantly different on hemoglobine, MCH, MCHC, leucocytes (WBC), neutrophils, lymphocytes, monocytes, trombosit, or lymphocytes in the pre-vaccination.

Table 3. Blood hematological profile of laying quails with supplementation of Nano Se-Yeast

	Se yeast Supplementation				
Parameter	T0	T1	T2	T3	*P value*
4 day before La sota vaccine					
Haemoglobin (Hb) (g/dl)	22.12 ± 1.01	20.04 ± 2.38	20.980 ± 2.08	20.64 ±2.78	0.505
Erytrocytes (RBC) (x10^6)	3.746 ± 0.26	3.262 ± 0.32	3.508 ± 0.35	3.496 ± 0.45	0.242
Haematocrit (%)	60.70 ± 2.43	54.32 ± 6.40	55.92 ± 6.32	52.72 ± 6.09	0.167
MCV (fl)	162.44 ± 8.21^{ab}	166.82 ± 14.86^{a}	159.06 ± 2.52^{ab}	150.98 ± 2.64^{b}	0.050
MCH	59.22 ± 2.39	61.46 ± 3.68	59.800 ± 1.38	58.980 ± 1.08	0.371
MCHC (g/dl)	36.48 ± 0.71^{b}	36.920 ± 1.44^{b}	37.560 ± 0.93^{ab}	39.060 ± 1.01^{a}	0.008
Leucocyte (WBC) (x10^3/μl)	95.96 ± 3.62	83.73 ± 10.46	85.94 ± 10.39	86864 ± 8993	0.176
Neutrophil (%)	0.8400 ± 0.11	0.900 ± 0.25	1.040 ± 0.36	0.8200 ± 0.08	0.460
Lymphocyte (%)	93.600 ± 0.53	93.520 ± 0.68	92.980 ± 0.46	93.320 ± 0.19	0.248
Monocyte (%)	5.760 ± 0.42	5.580 ± 0.54	6.040 ± 0.25	5.8800 ± 0.19	0305
Trombosit (PLT)	8000 ± 2000	8000 ± 1225	7400 ± 1517	7600 ± 1817	0.920
3 d post La sota vaccine					
Haemoglobin (Hb (g/dl)	21.44 ± 0.81	20.040 ± 1.49	19.66 ± 3.39	19.380 ± 1.26	0.544
Erytrocytes (RBC) (x10^6)	3.674 ± 0.33	3.568 ± 0.27	3.380 ± 0.45	3.394 ± 0.30	0.532
Haematocrit (%)	61.88 ± 7.87	55.08 ± 5.30	54.08 ± 10.14	52.38 ± 4.52	0.222
MCV (fl)	168.18 ± 4.72	154.54 ± 12.13	159.30 ± 11.02	154.04 ± 4.25	0.075
MCH	58.300 ± 1.33	56.24 ± 2.36	58 ± 2.34	57.140 ± 1.14	0.335
MCHC (g/dl)	34.680 ± 0.32	36.480 ± 1.68	36.50 ± 2.35	37.100 ± 1.21	0.127
Leucocyte (WBC) (x10^3/μl)	84632 ± 8676	79568 ± 5274	82596 ± 7490	78152 ± 4747	0.444
Neutrophil (%)	93.400 ± 0.54	92.360 ± 1.11	92.70 ± 0.95	92.740 ± 0.48	0.285
Lymphocyte (%)	0.900 ± 0.42	1.800 ± 0.89	1.500 ± 0.60	1.480 ± 0.36	0.166
Monocyte (%)	5.700 ± 0.33	5.840 ± 0.27	5.720 ± 0.47	5.7800 ± 0.16	0.906
Trombosit (PLT)	7.800 ± 1483	8.200 ± 1789	7600 ± 1140	7600 ± 548	0.875

means in a row with no common superscript differ significantly (p<0.05).
MCV: mean corpuscular volume, MCH: mean corpuscular hemoglobin, MCHC: mean corpuscular hemoglobin concentration, RBC: red blood cell, , MCV: Mean corpuscular volume, RDW-CV: red blood celln width coefficient variation
T0: Control (basal diet without supplementation Nano Se-Yeast), T2, T3, and T4: basal feed supplemented with Nano Se-yeast containing 1x 10^8 CFU/g at levels of 0.5, 1.0, and 1.5 g/per kg of feed.

The effects of Selenium minerals organic (Se-yeast) included in laying quail feed on hematological blood profile are showed in Table 3. It was shown that diets supplemented with Se yeast had significant effect on MCV and MCHC on before vaccine tratment. The lowest (P< 0.008) values for MCHC (39.06 fl) as compared to control diets. On the other hand, Se -Yeast had lower values of MCV (P< 0.05) value MCV (!50,98 g/dl) compared to control. MCV (*Mean Corpuscular Volume)* was calculates average size of the red blood cells. [13] whereas MCHC *Mean Corpuscular Hemoglobin Concentration* (MCHC) was concentration Hb red blood cell. In this experiment, the addition of Se Yeast caused an increase of Hb concentration that accompanied a decrease in Hb value. This is thought to be due to the increase in selenium minerals in plasma and their role as a balancer of membrane cells and regulation of their function [5]. The Se content in feed has a positive correlation with animal physiology and health which can influence oxygen binding which will cause levels of oxidative stress in animals [11].

Total leukocyte count and peripheral blood leukocyte of the layer quail with supplementation organically complexed minerals were in the normal range [14] the decrease of mineral levels in forms organically complexed did not influence these parameters (Table 3). This results in agreement with [14], who stated that leukocyte count and peripheral blood leukocyte type were unaffected by addition of selenium on feed. Supplementation Nano Se Yeast after vaccine there was no effect on the blood profile, this may be due to the ability to regulate balance in metabolism.

4. Conclusion

According to the current study, Nano Se-yeast supplementation is beneficial in low-stress situations, but it is less effective in preventing experimental stress following vaccination.

5. Acknowledgments

This study supported by DAPTV (Direktorat Akademik Pendidikan Tinggi Vokasi) grant numbers 0815/D4/AL.04/2024 and SK 489/PL.17.4/PM/2024 Politeknik Negeri Jember.

References

[1] BPS, *Integrated Agricultural Survey*, ed. Indonesia. 2023.
[2] Kletikova, L.V., et al., *Effect of Feed Additive on the Mineral Composition of Quail Blood.* Journal of Experimental Biology and Agricultural Sciences, 2022. **10**(1): p. 117-123.
[3] Kowalczyk, A., et al., *Effect Of Feed Supplementation With Organic Selenium And Vitamin E On Freezability Of Japanese Quail (Coturnix Coturnix) Semen.* 2011. **LXII**: p. 285-295.
[4] Nemati, Z., et al., *Assessment of Dietary Selenium and Vitamin E on Laying Performance and Quality Parameters of Fresh and Stored Eggs in Japanese Quails.* Foods, 2020. **9**(9): p. 1324.
[5] El-Allawy, H., et al., *Effect of Dietary Organic Selenium Supplement on Growth and Reproductive Performance of Japanase Quail Breeders and Their Progeny and its Relation to Antioxidation and Thyroid Activity.* International Journal of Poultry Science, 2015. **14**: p. 317-324.
[6] Reda, F.M., et al., *Biological Selenium Nanoparticles in Quail Nutrition: Biosynthesis and its Impact on Performance, Carcass, Blood Chemistry, and Cecal Microbiota.* Biol Trace Elem Res, 2024. **202**(9): p. 4191-4202.
[7] Brown, K.M., et al., *Effects of organic and inorganic selenium supplementation on selenoenzyme activity in blood lymphocytes, granulocytes, platelets and erythrocytes.* Clin Sci (Lond), 2000. **98**(5): p. 593-9.
[8] Reda, F.M., et al., *Does the use of different oil sources in quail diets impact their productive and reproductive performance, egg quality, and blood constituents?* Poultry Science, 2020. **99**(7): p. 3511-3518.
[9] Jouany, J.P., *Optimizing rumen functions in the close-up transition period and early lactation to drive dry matter intake and energy balance in cows.* Animal Reproduction Science, 2006. **96**(3): p. 250-264.
[10] Pantaya, D., et al., *Fermentation of soybean meal-hydrolysates as the medium that treated by papain enzyme with Saccharomyces cerevisiae for biomass production.* IOP Conference Series: Earth and Environmental Science, 2022. **980**(1): p. 012025.
[11] Khan, M.T., et al., *Effects of selenium supplementation on the growth performance, slaughter characteristics, and blood biochemistry of naked neck chicken.* Poult Sci, 2023. **102**(3): p. 102420.
[12] Chinrasri, O., et al., *Productivity and selenium concentrations in egg and tissue of laying quails fed selenium from hydroponically produced selenium-enriched kale sprout (Brassica oleracea var. alboglabra L.).* Biol Trace Elem Res, 2013. **155**(3): p. 381-6.
[13] Aksu, D., T. Aksu, and B. Özsoy, *The Effects of Lower Supplementation Levels of Organically Complexed Minerals (Zinc, Copper and Manganese) Versus Inorganic Forms on Hematological and Biochemical Parameters in Broilers.* Kafkas Universitesi Veteriner Fakultesi Dergisi, 2010. **16**: p. 553-559.

[14] Dönmez, N., et al., *Effects of Zinc Supplementation to Ration on Some Hematological Parameters in Broiler Chicks.* Biological trace element research, 2002. **87**: p. 125-31.

Semen quality of Indonesian native rooster with single and multi-strain probiotic dietary supplements

Niati Ningsih[1], Alditya Putri Yulinarsari[1], Aryanti Candra Dewi[1], Nur Muhamad[1], Gayuh Syaikhullah[1],

[1]Animal Husbandry Department, Politeknik Negeri Jember, Indonesia

e-mail: niatiningsih@polije.ac.id

Abstract. The aims was to decide the semen quality of Indonesian native roosters with single and multi-strain probiotic dietary supplements from the *Lactobacillaceae* and *Bacillaceae* families. Twenty Indonesian native roosters were randomly distributed to four experimental treatments with similar energy and protein levels. The treatments are T0: basal diet (control), T1: basal diet with supplementation 1% single probiotic *Bacillus subtilis*, T2: basal diet with supplementation 1% single probiotic *Lactobacillus bulgaricus,* and T3: basal diet with supplementation multi-strain probiotics 1% *Bacillus subtilis* and *Lactobacillus bulgaricus.* Each dietary treatment was replicated five times. In this study, an individual cage equipped with complete semen collection and analysis tools was utilized. The native rooster was given a one-week rest period before the treatment. Semen collection was done using the massage method. The evaluated of parameters included volume, color, consistency, pH, and odor. The data analysis applied one way ANOVA and followed by Duncan if the data significantly different. The results revealed a significant difference (P<0,05) in the semen volume of native Indonesian roosters with supplementation multi-strain probiotics 1% *Bacillus subtilis* and *Lactobacillus bulgaricus*. The research concluded that the multi-strain probiotics from the family *Lactobacillaceae* and *Bacillaceae* can improve the quality of sperm in Indonesian native roosters by increasing the volume total of semen.

1. Introduction

Indonesian native chicken is developed to meet the community's animal protein needs such as meat and eggs. Native chicken is expected to support the independence of the national food supply in line with Government Regulation No 18 of 2012 on Food Security, it emphasizes the nation's capacity to produce a variety of foods based on local resources.The native chicken has the potential to develop because local chickens are very adaptive to Indonesian environmental conditions, are more resistant to disease, and the meat more favorable compared to broiler chicken [1]. The availability of prime livestock of native rooster was still deficient [2]. This is thought to be because the feed nutrient cannot yet support the development of the reproductive system in native roosters. Insufficient nutrient content in feed results in impaired decreased semen production by disrupting the function of the testes and accessory glands. One effort to improve feed nutrients is the supplementation of feed additives. Feed additives are growth promoters that can improve animal production [3]. Feed additives were supplements to improve feed nutrition and have the effect of increasing feed digestibility and metabolism, as well as increasing animal production, the quality and food safety of animal products. [4].

Feed additives in the form of probiotics have been shown to have a positive effect on the productivity and reproductive performance of chickens [5]. Probiotics used for feed additives and proven safe for consumption are *Bacillus subtilis* and *Lactobacillus bulgaricus*. Probiotics have a role in animal digestion, being able to stick to the mucosa and epithelium to eliminate microbes, improve the immune system, and produce various enzymes such as protease, amylase and lipase [6][7], then with the function mentioned supplementation probiotics such as *B. subtilis* and *Lactobacillus bulgaricus*. can improve sperm quality in the form of an increased volume of ejaculate, concentration, entire quantity, and reducing morphological cells defects of spermatozoa [8]. Native rooster sperm motility can be decreased by administering *Lactobacillus bulgaricus* so that it can increase productivity [9].

The aim of this research was evaluate quality of native roosters semen by administering feed additives with single and multiple probiotics. Semen quality is one of the parameters for the evaluation of the reproductive traits of native chickens.

2. Materials and Methods

2.1 Housing

Indonesian Native Roosters one-year-old with amount twenty birds were purchased from a traditional market in Jember Regency. All chickens are divided randomly among four dietary treatments, with five replications for each treatment. The chicken was reared in a battery cage, completely with feed and water place that can be accessed throughout the trial. Before treatment, all of the chicken was vaccinated.

2.2. Diet

The basal diet of the research was shown in Table 1 was adjusted to the nutritional needs of the feed for Indonesian native roosters chicken.

Table 1. The basal diet of treatment

Ingredients	Proportion (%)
Yellow corn	62.00
Soybean meal	21.00
Fish meal	5.00
Rice brand	9.00
Palm oil	0.50
Vitamin premix*	1.00
L-lysine HCl (98%)	0.30
DL-methionine (99%)	0.10
$CaCO_3$	0.60
NaCl	0.50
Total	**100.00**
Calculated Nutrient Values	
Crude protein %	19.52
Metabolizable energy (ME Kcal/Kg diet)	2937.7
Calcium %	0.78
Available P%	0.36
Lysine %	1.31
Methionine %	0.48
Threonine%	0.73
Methionine + Cystine %	0.98

Note: Feed formulation using trial error method

2.3 Probiotic Preparation

All of the probiotics (*Bacillus subtilis* and *Lactobacillus bulgaricus*) were rejuvenation with MRS Broth and then incubated in an incubator for 24 hours. After rejuvination, all of the probiotics were a dilution with coconut water (1:5). The solution was shaken in an incubator shaker for 24 hours, and then the population of bacteria was counted with Total Plate Count (TPC). The TPC shows the population of

Bacillus subtilis and *Lactobacillus bulgaricus* with coconut water was 8.9 x 10^9 dan 1.6 x 10^9 respectively.

2.4 Experimental design and bird management
Experimental diets were T0: control or basal diet, T1: basal diet with supplementation 1% single probiotic *Bacillus subtilis*, T2: basal diet with supplementation 1% single probiotic *Lactobacillus bulgaricus*, and T3: basal diet with supplementation with multi-strain probiotics 1% *Bacillus subtilis* and *Lactobacillus bulgaricus*. Each dietary treatment was replicated five times. The native rooster was rested for one week before treatment. Semen collection was done using the massage method. The parameters observed were volume, pH, color, consistency, and odor.

2.5 Evaluation of Semen quality
Semen collection was carried out using massage method, which is one method in semen collection where the male is stimulated by gently massaging the area around the chicken back. The massage was carried out for several minutes until the male was stimulated and ready to release semen, which was indicated by the lifting of the feathers on the tail and the release of fluid. Semen was collected using a 1.50 ml tube to determine the volume of semen produced. The parameters used in assessing the color and consistency of semen were showed on Table 2.

Table 2. Semen quality assessment scores

Scores	Assessment Description	
	Color	Consistency
1	Clear	Liquid
2	Milky white	Slightly thick
3	Cloudy white	Thick

Macroscopic evaluation is done by observing 1) The volume of semen is observed through the scale listed on the wall of the container; 2) The color of semen is observed directly because the semen container is made of glass or transparent plastic; 3) The odor of semen is directly in the container by passing the semen container under the nostrils; 4) The consistency of semen is checked by tilting the tube containing sperm slowly. Good semen with a good degree of viscosity will move very slowly following the tilt of the container where almost all or slightly thicker than milk; and 5) pH of semen is known with dipping litmus paper into the semen that has been collected until it shows the acidity number.

2.6 Statistical Analysis
The data of the research applied one-way ANOVA with Completely Randomized Design (CRD) and followed by Duncan test if the data significantly different (P value <0.05)

3. Results and Discussion
The research data were showed on Table 3. The results revealed the supplementation of single and multiple probiotic families provided significant results on semen volume (P value < 0.05), however no significant difference in the macroscopic quality such as color, pH, consistency and odor of native roosters semen (P value > 0.05).

Table 3. Semen Quality of Indonesia Native Rooster with supplementation of probiotic

Variabel	T0	T1	T2	T3	P-Value
Volume (ml)	0.39 + 0.53[a]	0.36 + 0,49 [a]	0.35 + 0.63 [a]	0.45 + 0,53 [b]	0.02
Color	2.4 + 0,55	2.2 + 0.45	2.2 + 0.47	2.4 + 0.55	0.85
Consistency	2.6 + 0.55	2.6 + 0.55	2.8 + 0.45	2.8 + 0.44	0.85
pH Semen	7.0 + 0.37	7.4 + 0.11	7.1 + 0.18	7.0 + 0.18	0.09
Odor	distinctive semen	distinctive semen	distinctive semen	distinctive semen	-

Supplementation of single and multiple probiotic as feed additives on the semen volume of local roosters gave significant results (P value <0.05). The addition of multiple probiotics gave the highest results compared to other treatments, which was 0.47 ml, compared to other treatments P0, P1, and P2, 0.39, 0.36, and 0.35 ml, respectively. The semen volume of the research results was the same as the research conducted by [10] which observed a volume of semen from Gaok chickens was 0.4 ml. Supplementation of multiple probiotics from the family *Lactobacillaceae* and *Bacillaceae* has been proven to improve native chicken semen quality, especially for total volume parameter. The addition of probiotics to feed can improve the digestion process and absorption of feed nutrients that can promote male reproductive organs to produce semen. This study's results were align with [6] which explained that probiotic supplementation in rooster feed can improve the quality, volume, concentration, and motility of sperm, and can reduce the percentage of abnormalities and death of spermatozoa. The semen average volume of Indonesian native roaster was 0,25 - 0,5 ml.

Supplementation of single and multiple probiotic as feed additives did not provide significant results on the color, consistency, and pH of native roosters semen. The semen color from all treatments was milky white, the pH was neutral 7, and the consistency was solid. This study's results showed color score ranged from 2.25-3.00. Semen with a cloudy white color has a higher concentration of spermatozoa. The average semen consistency score in this study ranged from 2.25-3.00. Semen consistency can be used as a simple parameter in assessing semen quality. Semen with high consistency, namely semen with a thick form, indicates a high concentration of spermatozoa [11].

The average semen pH of native male chickens produced from the study ranged from 7.83-8.50. The results of the study align with [12] reported that the pH value of chicken average in neutral pH (7). The pH value of semen indicates the ability of sperm to survive or not, the higher the pH value indicates the greater ability to persist after ejaculation, a decrease in pH leads to lower survival potential. Apart from that, acidic environments contribute to elevated sperm mortality. pH value of rooster Semen reflects sperm activity, normal pH will improve sperm motility, viability, and concentration and is a factor that affects spermatozoa vitality. pH semen had normal range from 6.3 - 6.9. The survival of spermatozoa is determined by the degree of acidity, if the pH was too high or low it can also caused the death of spermatozoa. [13]. [14] claim that semen pH value plays a crucial role in viability.

The odor of native roosters' semen from the research with the addition of feed additives probiotics in the feed gave the same results in all treatments, namely the distinctive odor of chicken semen was slightly fishy. The odor of sperm in this study was the same as the study conducted by [15] who gave phytobiotics to chicken feed and produced a distinctive odor of semen. The distinctive odor of sperm in the collected semen indicates that the quality of the semen is good and normal. The normal odor of semen was generally fishy and tipycal from the animal's characteristic, but if the smell tends to be rotten, it can be confirmed that there was inflammation or infection in the reproductive tract. [16].

4. Conclusion

The research concluded that the multi-strain probiotics from the family *Lactobacillaceae* and *Bacillaceae* can improve the quality of sperm in Indonesian native roosters by increasing the volume total of semen

5. Acknowledgments
The research team gratefully thanks P3M Politeknik Negeri Jember for the fully funded grant project by PNBP 2024 with contract number 781/PL17.4/PG/2024.

References
[1] Hambu E K, Arifiantini R I, Purwantara B and Darwati S 2016 Raw Semen Characteristics of Three Different Indonesian Local Roosters *Anim. Prod.* **18** 165
[2] Kilatsih R and Daryono B S 2020 The inheritance of phenotype character of second generation (F2) on chicken (gallus gallus domesticus, linnaeus 1758) derived from crossing of pelung and broiler *AIP Conf. Proc.* **2260**
[3] Jadhav K, Sharma K S, Katoch S, Sharma V and Mane B G 2015 Probiotics in Broiler Poultry Feeds: A Review *J. Anim. Nutr. Physiol.* **1** 4–16
[4] Pirgozliev V, Rose S P and Ivanova S 2019 Feed additives in poultry nutrition *Bulg. J. Agric. Sci.* **25** 8–11
[5] Mahfudin W and Prabewi N 2023 Penggunaan Spirulina (Arthrospira platensis) sebagai Feed Additive terhadap Produktivitas dan Kinerja Reproduksi Induk Ayam Ras Petelur Fase Layer *J. Pengemb. Penyul. Pertan.* **20** 95–107
[6] Ningsih N, Respati A N, Astuti D, Triswanto T, Purnamayanti L, Yano A A, Putra R P, Jayanegara A, Ratriyanto A and Irawan A 2023 Efficacy of Bacillus subtilis to replace in-feed antibiotics of broiler chickens under necrotic enteritis-challenged experiments: a systematic review and meta-analysis *Poult. Sci.* **102** 102923
[7] Javanshir N, Hosseini G N G, Sadeghi M, Esmaeili R, Satarikia F, Ahmadian G and Allahyari N 2021 Evaluation of the Function of Probiotics, Emphasizing the Role of their Binding to the Intestinal Epithelium in the Stability and their Effects on the Immune System *Biol. Proced. Online* **23** 1–17
[8] Mazanko M S, Gorlov I F, Prazdnova E V., Makarenko M S, Usatov A V., Bren A B, Chistyakov V A, Tutelyan A V., Komarova Z B, Mosolova N I, Pilipenko D N, Krotova O E, Struk A N, Lin A and Chikindas M L 2018 Bacillus Probiotic Supplementations Improve Laying Performance, Egg Quality, Hatching of Laying Hens, and Sperm Quality of Roosters *Probiotics Antimicrob. Proteins* **10** 367–73
[9] Haines M D, Parker H M, McDaniel C D and Kiess A S 2013 Impact of 6 different intestinal bacteria on broiler breeder sperm motility in vitro *Poult. Sci.* **92** 2174–81
[10] Komarudin K, Sartika T, Kostaman T, Sopiyana S and Zainal H 2020 Analisis Bobot Badan dan Karakteristik Semen Ayam Gaok Terseleksi Generasi ke-6 *J. Agripet* **20** 143–9
[11] Haryuni N, Lidyawati A, Khopsoh B and Hasanah N 2020 Pengaruh Level Energi Dalam Pakan Terhadap Kualitas Spermatozoa Ayam Kampung Secara Mikroskopis *J. Ilmu Peternak. Terap.* **4** 7–13
[12] Mustafa M, Dasrul D, Yaman M A, Wahyuni S and Sabri M 2017 Pengaruh Pemberian Kombinasi Pakan Fermentasi dengan Multi Enzim dan Vitamin E dalam Ransum terhadap Peningkatan Kualitas Semen Ayam Arab *J. Agripet* **17** 43–52
[13] Soi M N J 2016 Uji Viabilitas Spermatozoa Sapi Bali Jantan dengan Menggunakan Larutan Natrium Clorida (NaCl) yang Berbeda Level *J. Anim. Sci.* **1** 28–9
[14] dos Santos M N, Ramachandran R, Kiess A S, Wamsley K G S and McDaniel C D 2018 The Impact of Dietary Yeast Fermentation Product Derived from Saccharomyces cerevisiae on Semen Quality and Semen Microbiota of Aged White Leghorn roosters *J. Appl. Poult. Res.* **27** 488–98
[15] Anggriani Y, Pagala M A, Auza F A and Sandiah N 2024 Optimizing Super Kampung Chicken Semen Quality : Impact of Holding Frequency on Macroscopic and Microscopic Attributes **6** 80–7
[16] Ramadhanty D, Nugraha A, Purnomo N and Fausiah A 2021 Kualitas Makroskopis Semen Ayam Kampung yang Diberi Ekstrak Kulit Buah Naga *AGROVITAL J. Ilmu Pertan.* **6** 43

Physiological conditions of bendi horse in Surakarta City

M A Syarifuddin[1], W Pawestri[2]*, A K Wati[2], E T Rahayu[3], Y Yanti[3] and K Kharismawati[1]

[1]Student of Animal Science Program Study, Faculty of Animal Science, Universitas Sebelas Maret, Surakarta, 57126, Indonesia
[2]Animal Production Laboratory, Animal Science Program Study, Faculty of Animal Science, Universitas Sebelas Maret, Surakarta, 57126, Indonesia
[3]Socio Economics of Livestock Laboratory, Animal Science Program Study, Faculty of Animal Science, Universitas Sebelas Maret, Surakarta, 57126, Indonesia

Corresponding author's e-mail: wari.pawestri@staff.uns.ac.id

Abstract. This study aimed to determine the physiological conditions (frequency of heart rate, respiration, and body temperature) of bendi horses in Surakarta City. The research was conducted from May to November 2023. The method used in the study was a survey with purposive sampling technique. Data collection used interview, observation, and literature study methods. The samples used were 24 bendi horses in Surakarta City. Data were processed using quantitative descriptive analysis with Microsoft Excel software and t-Test was conducted on physiological data of horses against activities. The results showed that the most widely horse breed is Javanese horses, the average horse kept is 5-10 years old, and female sex. The physiological frequency of bendi horses according to age and sex shows normal results. The t-Test results showed that bendi horses that were doing activities had higher heart rate and respiration frequency than bendi horses that were resting in the stable. The conclusion from this study is that the frequency of heart rate, respiration, and body temperature of bendi horses in Surakarta City are in normal condition.

1. Introduction

Horses are domestic livestock in Indonesia, playing various roles such as transportation for humans and goods, as well as providing recreation facilities [1]. A form of using horses as a recreation can also be seen in several cities in Indonesia, one of which is bendi. Bendi is a two-wheeled tourist carriage pulled by a horse [2]. In Surakarta, a city located in Central Java and often referred to as "The Spirit of Java", bendi rides are popular among local residents who enjoy exploring the city centers for leisure. Surakarta has numerous attractions that draw visitors seeking entertainment and relaxation, contributing to a vibrant atmosphere. The enthusiasm of the community attracted bendi horse farmers to the Surakarta City area, which can be found at Jayawijaya Park, Manahan Stadium, Surakarta Palace Square, and Car Free Day.

Horses as pullers bendi must be in good body condition so they can work optimally. Horse performance can be observed from its physiological state. Physiological parameters consist of heart rate, respiration, and body temperature [3]. The heartbeat of horses frequency ranges from 25-45 times/minute [4]. Research on the general performance and physiological status of Sumba horses stated that the respiration frequency ranges from 16-44 times/minute [5]. The frequency of the body temperature ranges between 37-38.5°C [6]. Horses with abnormal physiological conditions should not be used for work, as forcing them can lead to issues such as unstable walking, collapse, and increased disease risk. Farmers must always pay attention to the physiological condition of the horse.

It is very important for farmers to know the physiological condition of horses before using their horses. Horses with normal physiological conditions can be optimally used as bendi pullers. Research regarding the physiological condition and health of bendi horses in Surakarta City has never been carried out. There is a need to investigate the heart rate frequency, respiration, and body temperature of bendi horses in Surakarta City.

2. Materials and method

2.1 Research tools

The tools for data collection included a stethoscope, thermometer, stopwatch, alcohol, cotton, and questionnaire.

2.2 Research materials

This research used 24 samples of bendi horses and 11 farmers from three sub-districts in Surakarta City, which include Banjarsari Sub-district with 16 horses, Jebres Sub-district with 2 horses, and Laweyan Sub-district with 6 horses.

2.3 . Method

The research was conducted from May to November 2023 in Surakarta City, Central Java Province. This research used a survey method, performed in two stages including pre-survey and survey. Subsequently, pre-survey activities were carried out by examining the population of horses data in Surakarta City from the Department of Agriculture, Food Security and Fisheries and looking for addresses of horse farmers through the chairman of the Surakarta bendi association. Survey activities were carried out by interviewing farmers and examination of horses physiological conditions. The sampling method used in the research was purposive sampling. The sample criteria used in this research were horse farmers domiciled in Surakarta City, the farmers willingness to be interviewed, healthy horses, and horses that could be examined. This research received approval from the ethical clearance team of the Faculty of Veterinary Medicine, Gadjah Mada University with Number: 106/EC-FKH/Eks./2023.

2.3.1. Interviews with farmers

Interviews were conducted with horse farmers in Surakarta City using a questionnaire. The questionnaire contains questions regarding the horses identity including horse breed, age, and sex.

2.3.2. Heartbeat frequency

Heartbeat frequency was measured using the auscultation method using a stethoscope. The measurements were taken on the left side of the lower thorax for 1 minute and were repeated three times. The measurement results were recorded in the questionnaire and averaged.

2.3.3. Respiration frequency

Respiration frequency was measured by observing inspiration and expiration on the horse's abdomen for 1 minute, and the observations were repeated three times. The measurement results were recorded and the average was calculated.

2.3.4. Body temperature

The body temperature of the horses was checked using a digital thermometer. The thermometer to be used is washed with 70% alcohol. The thermometer is inserted into horse's rectum until the thermometer shows a stable reading. The temperature was measured three times. The results of temperature checks are recorded and averaged.

2.4. Data analysis

Horses' characteristics and physiological conditions based on age and sex were processed using descriptive analysis. Physiological conditions based on the horses activities were processed using t-test analysis.

3. Results and discussions

The city of Surakarta is one of the cities in Central Java that has many local tourist attractions. Local tourism in the city of Surakarta includes the Surakarta Palace, the Great Mosque, the Surakarta Palace Square, Jayawijaya Park, and the Solo Manahan Stadium. Bendi is used by the community as a recreational facility to get around the city of Surakarta. Horses used to pull carts must be supervised and their performance maintained. Horse performance can be influenced by the characteristics and physiological conditions.

3.1. Horses characteristics

Characteristics are factors that can influence a horse's performance. Horses used to pull bendi must have appropriate characteristics. Horses pulling bendi are very dependent on their strength, thereby there is a need to use horses that are strong and have good power. In this research, characteristics of horses included breed, age, and sex, and the data obtained are presented in Table 1.

Table 1. Characteristics of horses

Parameter	Number	Percentage
Breed		
a. *Thoroughbred*	6	25
b. Sandel	5	21
c. Javanese	13	54
Age		
a. <5	8	33,3
b. 5-15	16	66,7
Sex		
a. Mare	20	83
b. Stallion	4	17

3.1.1. Breed

Breed of bendi horses in Surakarta City are Thoroughbred, Sandel, and Javanese (Figure 1). They are local horses that stronger than imported horses. Research conducted by [7] stated that imported horses have a higher heart rate and respiration frequency than local horses. The higher frequency of heartbeat and respiration of imported horses indicates that imported horses require more effort to carry out homeostatic processes. This is because the environmental conditions in Indonesia are different from the environmental conditions where imported horses originate. Thoroughbred horses are horses that come from crossing local horses and Thoroughbreds. This horse has physical characteristics and strength like local horses with a taller and bigger body. The Sandel horse has a taller body posture than the Javanese horse and is smaller than the Thoroughbred horse.

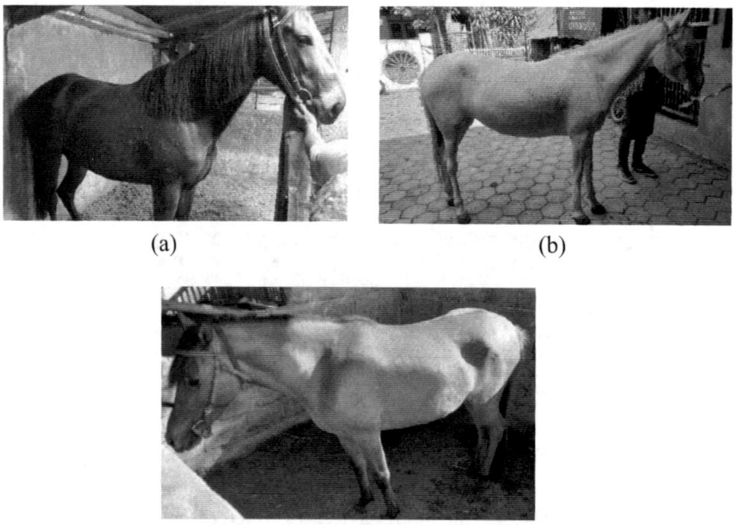

(a) (b) (c)

Figure 1. Horses breeds in Surakarta City Thoroughbred (a), Sandel (b), and Javanese (c)

3.1.2. Age

According to research carried out in the city of Surakarta, bendi horse farmers keep horses of various ages. The results showed that average farmers kept horses aged 5-15 years (66.7%), which is a productive age after puberty. On average, horses can live until 25-30 years old and mature at 5 years old [8]. Bendi horse farmers in the city of Surakarta usually started training horses as bendi pullers between 1 and 2 years old, although at 2 years old, horses can be used for work purposes [9]. Farmers tend to sell horses when they are old or more than 20 years old because of a reduction in their performance and getting sick easily. Horses between the ages of 16 or even 20 are considered old and those aged 30 are considered very old [8].

3.1.3. Sex
Almost all of the horses in Surakarta City are mare (83%) and only a small percentage keep stallion (17%). The mare is preferable because of the reproducing ability of a foal, which can be sold and serve as a source of additional income [10]. According to observations, the stallions are more difficult to control when experiencing heat, and are difficult to control when they see a mare. Stallion have a high level of courage and strong competitiveness [11].

3.2. Physiological conditions
Heart rate frequency, respiration, and body temperature are parameters of physiological conditions that can observe horse's performance. The physiological condition of a horse can be influenced by several factors, including age, sex, and horse activity. These factors must be considered to maintain the horse's physiological condition stable and normal.

3.2.1. Age
Bendi horses in Surakarta City have various ages, ranging from <5 to over 10 years old. Table 2 shows that horses aged <5 years had an average heart rate of 46.83 ± 9.06 times/minute, while horses aged 5-15 years had an average heart rate of 40.17 ± 7.67 times/minute. The results of the physiological examination of the horse according to age showed that the horse's heart rate was normal. Normal heart rate in horses is in the range of 28-48 beats/minute [12]. Young horses are generally more enthusiastic and panic easily because of environmental conditions. Heartbeat frequency of foal is faster than that of young or old horses [5]. As a horse ages, organ performance also decreases, which can cause reduced stimulation and decreased reactions to the environment [10].

The average respiration frequency of bendi horses in the <5 years old group were 41.92 ± 10.13 times/minute, while in the 5-15 years age group it was 35.63 ± 10.74 times/minute. These results indicate that the respiration frequency of bendi horses in Surakarta City is normal. Normal respiratory frequency of horses is between 14-48 times/minute [13]. Differences in measurement results occur due to differences in the ages of young and adult horses. Aging and decreasing in physiological function occur as horses age. As horses get older, their physiological conditions will be different [14]. As a horse ages, lung performance decreases so that respiration frequency also decreases [15].

Table 2. Physiological data of horses according to age

Physiological Parameters	Age	
	<5 years (n=8)	5-15 years (n=16)
Heart rate (times/minute)	46,83±9,06	40,17±7,67
Respiration (times/minute)	41,92±10,13	35,63±10,74
Body temperature (°C)	37,24±0,37	37,48±0,18

The results of body temperature examination in horses aged <5 years and 5-15 years showed the same results (Table 2). When the horses are at rest and in comfortable stable conditions, their body temperature tends to be constant. The factors that influence a horse's body temperature are environmental conditions and the activities carried out by the horse [16].

3.2.2. Sex
Horse farmers use mare and stallion to pull carts. Mare bendi showed an average heart rate of 40.33 ± 7.88 times/minute and stallion bendi 42.80 ± 8.84 times/minute (Table 3). These results showed that the heart rate was normal, ranging from 28-48 beats/minute [12]. Heart rate frequency can be influenced by the horses sex, age, and activity [17]. Stallions have higher fitness than the mares, thereby they control their heart rate better [18]. Mares are rated as more tense, more aggressive and excitable, panic easily, and show more anxiety compared to stallions [20].

Table 3. Physiological data of horses according to sex

Physiological Parameters	Sex	
	Stallion (n=4)	Mare (n=20)
Heart rate (times/minute)	40,33±7,88	42,80±8,84
Respiration (times/minute)	36,67±10,43	37,96±11,08
Body temperature (°C)	-	37,36±0.31

The average respiration frequency of stallion bendi was 36.67 ± 10.43 times/minute and mare 37.96 ± 11.08 times/minute. The results of this examination are in accordance with research by [13] which states that mare and stallion have relatively similar physiological conditions. The sex of the horse does not have much influence on the horse's respiratory frequency. Respiration frequency is more influenced by the horse's physical activity, stress, and environmental temperature [20].

3.2.3. Horse activities
Horse activity increase heart rate significantly (P<0.05). The energy released by a horse during its activities causes organ performance to increase. The average heart rate of bendi horses that are active is higher than that of bendi horses resting in a stable (Table 4). The heart rate of the horses can increase due to several factors, including activity, fear, and hot weather [5]. Horses that do activities need more oxygen supply, and high activity causes the heart rate to increase, resulting in vasodilation and faster blood flow [21]. Active horses have an average heart rate above the normal range, which can increase to approximately 110 beats/minute when active [20]. The heart rate will return to normal when bendi horses rest, and the heart rate will return to normal after 10-15 minutes of rest [13].

Table 4. Horse physiological data on activity (n=13)

Physiological Parameters	Horse Condition		P-*Value*
	Rest	Activities	
Heart rate (times/minute)	42,67±6,67[a]	62,46±7,28[b]	0,00001454
Respiration (times/minute)	38±10,28[a]	91,49±34,34[b]	0,0005631
Body temperature (°C)	37,31±0,33[a]	37,89±0,98[a]	0,252

Active horses require more oxygen to be distributed throughout the body, and the average respiration frequency of bendi horses active was significantly (P<0.05) higher than that of bendi horses resting in the stall. The respiration of horses during activity can increase to approximately 120 times/minute [12]. Activity causes heat in the horse's body to increase due to the burning of energy. For the body temperature to remain normal, excessive body heat is expelled by increasing respiration, thereby maintaining normal body temperature [13]. Active horses have an average respiratory frequency above the normal range. Respiration frequency will return to normal when the horses rest, which will return to normal after 10-15 minutes of rest [13].

Body temperature between horses at rest and activity tended to be constant and did not differ significantly (P>0.05). This constant temperature is due to the horse having a good thermoregulation mechanism. The thermoregulation mechanism used by horses is in the form of increased respiration and evaporation to release heat. The cardiovascular system has an important role in transferring heat from sites of heat production (such as muscles), to sites of heat dissipation (skin) [22]. Factors that influence a horse's body temperature are environmental conditions, training, hot weather, illness and disease. A horse is declared to be suffering from a mild fever if the body temperature is 38.9°C when the horse is resting or not active [5].

4. Conclusion
The conclusion of this study is that the heart rate, respiration frequency, and body temperature of bendi horses in Surakarta City are normal. Horses doing activity have an average heart rate and respiration frequency that is higher than horses when resting in the stall.

5. Acknowledgments
Authors thanks to LPPM UNS for funding this publication trought contract number 194.2/UN27.22/PT.01.03/2024.

6. References
[1] Harisatria, D. Surtina, J. Hendri, dan Jaswandi. 2017. Respon estrus kuda lokal dengan induksi hormon PGF2α di Kota Payakumbuh. Jurnal Peternakan 14(2):65-69. doi:10.24014/jupet.v14i2.3674.
[2] Yefriza, M., I. K. Ardhana, dan A. A. A. R. Wahyuni. 2020. Bendi transportasi budaya di Kota Solok Sumatera Barat antara tantangan dan peluang tahun 1970-2000. Humanis: Journal of Arts and Humanities. 24(1):46-52. doi:10.24843/JH.2020.v24.i01.p06.
[3] Utomo, I. H. dan D. Budiono. 2022. Gambaran parameter fisiologi sebelum dan setelah penanganan pada kuda dengan kasus kolik ringan. Jurnal Arshi Veterinary Letters. 6(4):73-74. doi:10.29244/avl.6.4.73-74.
[4] Camargo, F. and R. Dwyer. 2010. Evaluating the health of your horse. Cooperative extension service. University of Kentucky College of Agriculture, Lexington, KY, 40546.
[5] Gaina, C. D. dan N. D. F. K. Foeh. 2018. Studi performa umum tubuh dan status fisiologis kuda Sumba. Jurnal Kajian Veteriner. 6(2):38-44. doi:10.35508/jkv.v6i2.01.
[6] Stewart, J. 2019. Vital life signs in horses. Hoofbeats magazine. 40(5):1-5.

[7] Aushaf, N. M., I. G. A. A. Putra, dan E. Puspani. 2019. Status fisiologi dan tata laksana pemberian pakan kelompok kuda lokal dan impor di royal sporthorse Bali. Journal of Tropical Animal Science. 7(3):1018-1024.
[8] Wisniewska, M., I. Janczarek, and D. Piwczynski. 2018. The aging phenomenon of horses with reference to humane-horse relations. Journal of Equine Veterinary Science. 73(2019):37-42. doi:10.1016/j.jevs.2018.11.005.
[9] Turangan, S. H. 2017. Penampilan ternak kuda bendi di Kecamatan Tompaso Kabupaten Minahasa. Zootek. 37(1):186-198.
[10] Lestari, A., H. I. Susanti, S. A. Ananda, dan Rusny. 2023. Manajemen pemeliharaan kuda di Desa Gantarang, Kabupaten Jeneponto. ANOA: Journal of Animal Husbandry. (2)2:88-94. doi:10.24252/anoa.v2i2.39954.
[11] Pogach, T. 2023. Male vs female horses: 5 Key differences. Amazing Animals. Accessed 29 January 2024.)
[12] Liburt, N. R., K. Malinowski, and C. Williams. 2017. Measuring temperature, pulse, & respiration (TPR): what's normal for my horse. Rutgers, The State University of New Jersey 88 Lipman Drive, New Brunswick, NJ 08901-8525. p. 1-2.
[13] Saefullah, Y. Y., S. B. K. Prajoga, dan A. A. Yulianti. 2016. Keadaan faali kuda Sumba sebelum dan sesudah pertandingan (perlombaan pacuan kuda di lapangan Rihi Eti, Kabupaten Sumba Timur, Provinsi Nusa Tenggara Timur). Student e-journal. 5(4):1-13.
[14] McKeever, K. H. 2003. Aging and how it affects the physiological response to exercise in the horse. Clinical Techniques in Equine Practice. 2(3):258-265. doi:10.1053/S1534-7516(03)00068-.4.
[15] Pacheco, A. P. M., R. Paradis, A. M. Hoffman, P. Hermida, A. Sanchez, J. A. Nadeau, M. Tufts, and M. R. Mazan. 2014. Age effects on blood gas, spirometry, airway reactivity, and bronchoalveolar lavage fluid cytology in clinically healthy horses. Journal of veterinary internal medicine. 28(2):603-608. doi:10.1111/jvim.12318.
[16] Langobelen, H. S., F. U. Datta, dan M. U. E. Sanam. 2021. Parameter kuda Sandelwood di Kecamatan Katikutana, Kabupaten Sumba Tengah. Jurnal Veteriner Nusantara. 4(1):1-8. doi:10.35508/jvn/vol4iss1pp7.
[17] Ohmura, H. and J. H. Jones. 2017. Changes in heart rate and heart rate variability as a function of age in Thoroughbred horses. Journal of Equine Science. 28(3):99-103. doi:10.1294/jes.28.99.
[18] Stefánsdóttir, G. J., S. Ragnarsson, V. Gunnarsson, and A. Jansson. 2014. Physiological response to a breed evaluation field *test* in Icelandic horses. Animal. 8(3):431-439. doi:10.1017/S1751731117000556.
[19] Fenner, K., G. Caspar, M. Hyde, C. Henshall, N. Dhand, F. P. Rapsey, K. Dashper, A. McLean, and P. McGreevy. 2019. It's all about the sex, or is it? Humans, horses and temperament. Plos ONE 14(5):1-18. doi:10.1371/journal.pone.0216699.
[20] Freeman, D. W. 2017. Physical conditioning of horses. Oklahoma Cooperative Extension Service. Oklahoma State University, Stillwater. p. 1-6.
[21] Rahman, M. A., S. B. Komar, dan A. A. Yulianti. 2016. Kajian status faali kuda polo sebelum dan sesudah dilatih di nusantara polo club. Students e-journal 5(4):1-8.
[22] Wood, C. and A. Griffin. 2019. Equine thermoregulation: How do horses stay cool during work and exercise?. Exercise and Performance, Horses. University of Kentucky, Lexington.

IOP Conf. Series: Earth and Environmental Science **1446** (2025) 012049 doi:10.1088/1755-1315/1446/1/012049

Development of a Hybrid Energy-Powered IoT-Based Monitoring and Control System for Smart Fish Farming

Fitri Wibowo[1*], Ika M. Deviarni[2], Nabila[1] and Ahmad Choirudin[1]

[1] Electrical Engineering Department, Politeknik Negeri Pontianak, Indonesia
[2] Marine and Fisheries Department, Politeknik Negeri Pontianak, Indonesia

*fitri.wibowo@polnep.ac.id

Abstract. Fish farming plays an important role in aquaculture as water quality is a crucial factor in fish health and growth. This article presents the design and development of an Internet of Things (IoT)-based Smart Fish Farming (SHIFA) system that automates the monitoring and control of key water parameters such as pH, temperature and turbidity. The system is powered by hybrid energy sources and integrates both PLN (the national electricity grid) and solar energy to ensure continuous operation even during power outages. The SHIFA system consists of sensors connected to an ESP32 microcontroller that collects water data in real time. This data is processed and displayed through the Home Assistant platform, enabling remote monitoring and automated actions, including water circulation and aeration, to maintain optimal water quality. Additionally, the system offers a user-friendly web interface for remote access and control. The SHIFA system demonstrated sensor accuracy of 98% for pH, 97% for temperature and 95% for turbidity measurements, ensuring precise monitoring of water quality parameters. This level of accuracy is expected to increase the efficiency and sustainability of fish farms by reducing manual intervention and optimizing resource utilization through intelligent automation.

1. Introduction

Indonesia, as one of the largest archipelagic nations in the world, is endowed with vast marine resources, with the fisheries sector being a fundamental pillar of its economy. West Kalimantan, with its expansive and diverse water bodies, serves as a strategic area for the cultivation of various fish species[1]. However, fish farming in this region faces major challenges, particularly in maintaining sustainable aquatic environmental management[2, 3]. Increasing market demand has increased the economic value of aquaculture, but this intensification also poses threats to environmental sustainability and resource management.

To address these issues, the Indonesian government has adopted various regulations to ensure the sustainable management of marine and fisheries resources[4]. These efforts aim to regulate the exploitation of fish stocks and ensure that fishing and aquaculture practices are conducted in a responsible and sustainable manner, while protecting aquatic ecosystems from further degradation. Given these challenges, innovations in fish farming are critical to improve both the efficiency and sustainability of aquaculture systems. The introduction of hardware technology with an Internet of Things (IoT) technology offers an innovative solution that focuses on the design and implementation of systems that can automatically monitor key environmental parameters such as pH levels, turbidity and water temperature. This approach enables precise, real-time monitoring to ensure optimal conditions for fish health and growth[5].

In addition, the operational continuity of such monitoring systems is highly dependent on a stable energy supply. In areas with inconsistent energy infrastructure, it is critical to have energy management strategies in place that ensure system sustainability. Several similar studies have not yet integrated hybrid energy sources into their systems and focus exclusively on single-source energy solutions, which often prove inadequate in regions with unreliable power supplies. This limitation is particularly evident in West Kalimantan, where the power grid is not consistently stable and power outages are frequent. Therefore, the integration of hybrid energy systems that combine electricity from PLN (the national power grid) and solar energy represents an important solution for maintaining the operation of monitoring and control systems during power outages. The hybrid approach ensures continuous operation, compensates for the unreliability of the grid and is essential in regions such as West Kalimantan to ensure system functionality without interruption. Based on this overview, the proposed design and implementation of an IoT-based SHIFA monitoring and control system using ESP32 represents a meaningful step toward enhancing the sustainability of fish farming in West Kalimantan, in alignment with the government's environmental conservation policies.

2. Literature Review

2.1 IoT in aquaculture

The integration of the Internet of Things (IoT) into aquaculture is revolutionizing the industry through improved monitoring and management practices. IoT systems enable real-time monitoring of critical water quality parameters such as pH, turbidity, temperature and dissolved oxygen, which are critical to the health of aquatic life[6, 7]. For example, a framework that leverages Arduino and various sensors enables efficient data collection and storage in the cloud, facilitating timely interventions[8]. In addition, innovative architectures combining IoT with machine learning and edge computing are proposed to improve autonomous control of smart marine farms and address productivity and sustainability challenges. Furthermore, the development of intelligent aquaponics systems shows the potential to combine fish farming with crop cultivation, optimizing resource utilization and reducing environmental impact[9]. Overall, IoT technologies are critical to advancing aquaculture practices, ensuring food security and promoting economic growth in this sector.

2.2 Water quality monitoring technologies

The development of water quality monitoring technologies in fish farming has made significant progress through the integration of Internet of Things (IoT) systems, which have significantly contributed to both increased productivity and sustainable aquaculture practices. These technologies use interconnected networks of sensors that monitor key water parameters such as temperature, pH, dissolved oxygen, turbidity and salinity in real time. By providing continuous and immediate feedback, these systems enable aquafarmers to take immediate corrective action when monitored parameters exceed preset thresholds, thereby maintaining optimal conditions for fish health and growth[10, 11].

An example of such systems is AquaBot, which not only automates the data collection process but also provides tailored recommendations for specific fish species based on real-time water quality data, thereby optimizing operational efficiency in fish farming[12]. Furthermore, advances in sensor technology have led to improvements in the accuracy of low-cost sensors through methods such as simple linear regression, ensuring that even economically viable solutions can provide reliable data for aquaculture monitoring[13]. This is particularly beneficial for small-scale farmers who may not have access to more expensive, high-precision equipment but still need reliable data to maintain water quality standards.

Furthermore, the integration of cloud platforms for data storage and analysis plays a crucial role in modern aquaculture management. These platforms not only enable the aggregation of large amounts of sensor data, but also facilitate the visualization of trends and anomalies over time, enabling stakeholders to make informed decisions about farm operations and environmental management[11, 14]. By analysing long-term data trends, farmers can anticipate potential problems, optimize feeding schedules and reduce the risk of disease outbreaks, thereby promoting more sustainable agricultural practices. Thus, these technological advances in water quality monitoring not only improve fish health and increase farm productivity, but also contribute significantly to the broader goal of sustainable aquaculture by reducing environmental impacts and promoting more efficient use of resources.

2.2.1 pH and temperature sensors
Various pH and temperature sensors are used in aquaculture systems, each with different properties. Optical pH sensors, such as those using lipophilic carboxyfluorescein derivatives, provide high accuracy with low cross-sensitivity to ionic strength, making them suitable for marine environments[15]. Wireless passive sensors with temperature compensation can achieve measurement accuracy of less than 0.1 pH with a fast response time of less than 1 second, which is critical for dynamic water conditions[16]. Furthermore, ISFET pH sensors exhibit variable response times (0.3 to 2.0 seconds) depending on the flow conditions and sensor geometry, indicating their adaptability in different configurations[17]. Furthermore, differential pH FET sensors enable precise measurements in biological systems, while magnetic nanofluid-based sensors provide rapid colorimetric response to temperature and pH changes with a linear response range[18, 19]. These diverse technologies highlight the importance of selecting appropriate sensors based on the specific requirements of aquaculture.

2.2.2 Turbidity and dissolved solids sensors
Common types of turbidity sensors used in water quality monitoring include nephelometric sensors, fibre optic-based sensors, and submersible infrared probes. Nephelometric sensors, as described by Mersch, utilize methods to minimize interference from air bubbles and scattered light, making them suitable for various environments, although they may require frequent calibration[20]. By Bilro et al. highlighted optical fibre-based sensors offer robustness and cost-effectiveness, making them user-friendly for various applications, but may have limitations under extreme conditions[21]. Submersible infrared probes provide precise measurements by controlling stray light interference, which is beneficial in murky waters but can be complex to maintain. Furthermore, intelligent turbidity converters designed for distributed measurement systems such as those described by Tai et al. have been presented to provide continuous monitoring with high accuracy, although they may be more expensive and require technical expertise to set up[22]. Each type of sensor has its strengths and weaknesses, so the choice depends on specific environmental conditions and monitoring needs.

2.3 Research opportunities
While IoT-based aquaculture systems have made significant progress, gaps remain, particularly in integrating hybrid energy systems and open-source platforms for real-time monitoring and control. Current systems often depend on a single energy source, risking disruptions in areas with unstable grid power or inadequate solar energy. Hybrid energy systems combining solar and grid power offer enhanced resilience but remain underexplored. Additionally, most systems rely on

costly commercial cloud platforms, posing challenges for small or remote farms due to subscription fees and connectivity issues. Open-source platforms like Home Assistant present a viable alternative by enabling local data processing and reducing dependence on cloud services, yet research on their integration with hybrid energy solutions in aquaculture is limited. Addressing these gaps could improve the resilience, cost-efficiency, and sustainability of smart aquaculture systems.

3. Methods

The development of the Hybrid Energy-Powered IoT-Based Monitoring and Control System for Smart Fish Farming (SHIFA) took place in several phases, including hardware design, software integration and system testing. This section describes the methodology used, highlighting hybrid energy system integration, IoT platform configuration, and system evaluation.

3.1 System Design

The system design for SHIFA was divided into two main parts: hardware and software.

3.1.1 Hardware Design

Hardware design included the development of the SHIFA controller, which is housed in a custom-built enclosure containing several key components for monitoring and control. The hardware consists of several key components. The ESP32 microcontroller acts as the central processing unit for the system, managing data acquisition from pH, temperature (DS18B20), and turbidity (TDS SEN0244) sensors, while also controlling the system's actuators. The sensors provide continuous real-time monitoring of the aquaculture environment, transmitting data to the ESP32 for analysis and decision-making. Actuators in the system include two water pumps (for draining and refilling), a water filter, an aerator (for oxygen enrichment), and a servo motor for automatic feeding. The pumps, filter, and aerator are AC-powered devices controlled via relays connected to the SHIFA controller, enabling automatic or manual operation through a web interface. The feed servo motor is directly managed by the ESP32 using pulse width modulation (PWM) for precise control of feeding angles. A hybrid power system ensures uninterrupted operation by combining solar energy, supplied through a solar charge controller (SCC) and battery, with grid power. The system dynamically switches between these energy sources based on availability and consumption, ensuring continuous functionality even during power outages.

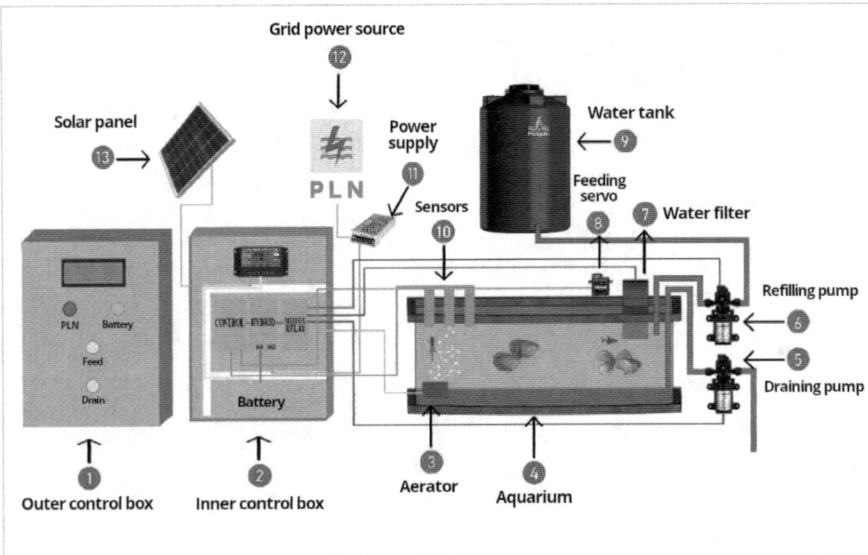

Figure 1. Block diagram of the Smart Fish Farming (SHIFA) system.

Figure 1 presents the block diagram of the SHIFA system, illustrating the integration of the various components. The SHIFA controller integrates these components and is responsible for automated control of water conditions, using a relay module to switch between different actuators. The power system has a backup battery to ensure continuous operation in the event of a power failure. Additionally, Figure 2 shows the detailed wiring connections between the ESP32 microcontroller, sensors, actuators, and power components, providing a comprehensive overview of the system's electrical structure.

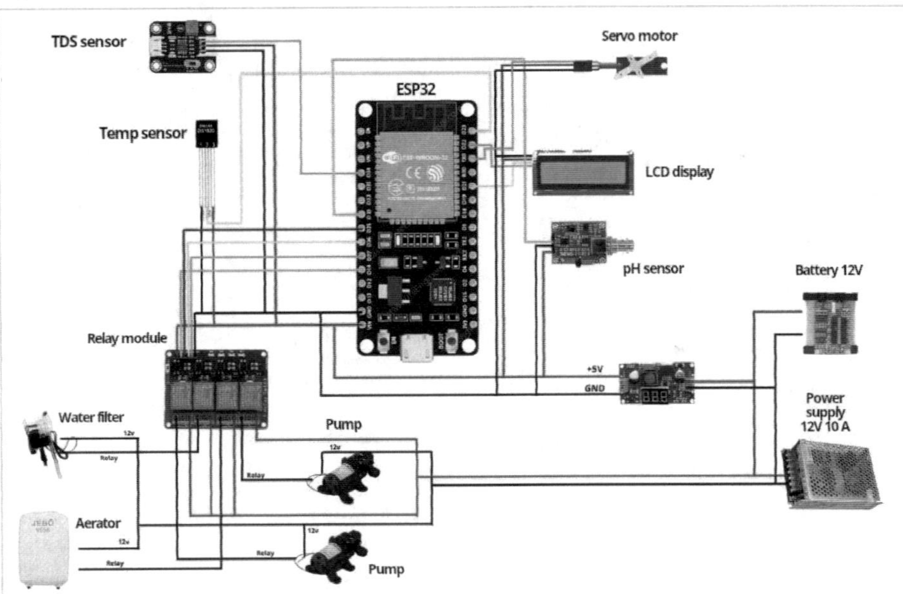

Figure 2. Wiring connections of the Smart Fish Farming (SHIFA) system.

3.1.2 Software Design

The software design leveraged the Home Assistant IoT platform in supervisor mode installed on a device powered by an Amlogic S905X System on Chip (SoC) with Armbian Server. Home Assistant was chosen for its open-source nature, flexibility, and ability to function as a local edge processing platform, eliminating the need for external cloud services. The software enables real-time monitoring and automation of fish farming operations, such as water flow management and aerator activation, based on customizable rules. The integration of the ESP32 microcontroller with Home Assistant was achieved by flashing it with the ESPHome firmware. ESPHome enables seamless integration with Home Assistant, allowing the ESP32 to collect sensor data and send it to Home Assistant for processing and visualization. ESPHome was configured to manage data collection and processing from connected pH, temperature, and turbidity sensors, sending the data to Home Assistant, where automation rules trigger control actions such as water drainage, refilling, and ventilation based on predefined thresholds.

Additionally, ESPHome provided a customizable firmware configuration interface, enabling precise adjustments to data reporting and control logic. Home Assistant, in turn, was configured to collect and display sensor data from ESP32 devices via ESPHome in real time, automate system responses based on specified pH, temperature, and turbidity levels, and offer a web-based interface for remote monitoring and control of the aquaculture system. This local edge processing architecture, supported by Home Assistant and ESPHome, minimizes reliance on paid cloud services, providing greater control over data protection and operational costs. The system supports analysis of historical data for decision making, making it an efficient tool for managing fish farms.

3.2 Hybrid energy integration

To ensure uninterrupted operations, a hybrid energy system was implemented that combines solar energy and grid power. Solar panels provide renewable energy during the day, while grid power serves as backup during times of insufficient solar power. A solar charge controller manages energy supply and distribution, optimizing the use of solar energy and ensuring that the system switches seamlessly between energy sources. The system design includes a battery backup that stores excess energy from the solar panels, ensuring the system remains operational even during power outages. This hybrid setup significantly reduces reliance on grid power and provides a more sustainable solution for aquaculture operations, particularly in rural or remote areas where grid stability can be a challenge.

3.3 System Testing and Evaluation

The system was deployed in a controlled aquaculture environment to evaluate its performance through real-time monitoring and automated control. Two primary testing methods were used: black box testing and sensor accuracy assessment. To evaluate the functionality and accuracy of the SHIFA system, black-box testing and sensor accuracy assessments were conducted. Black-box testing validated the control mechanisms by simulating changes in pH, temperature, and turbidity. When these parameters deviated from ideal thresholds, the system successfully triggered the routine water draining process, sequentially activating the draining and refilling pumps. This method ensured the system performed the expected actions based on sensor inputs without analysing the underlying code.

Additionally, sensor accuracy was assessed by comparing the pH, temperature, and turbidity readings with laboratory-grade instruments. Multiple experiments under varying environmental conditions demonstrated that the sensor outputs remained consistent with reference measurements, confirming their reliability in maintaining optimal water quality. The overall performance of the system was evaluated by analysing the success rate of automated control measures during black box testing and the accuracy of the sensors in maintaining the aquaculture environment within acceptable thresholds.

4. Result and Discussion

4.1 Hardware and software implementations

Figure 3 shows the installation of key components in the control panel box and the entire system at development scale. It includes the ESP32, which acts as the main microcontroller to control the system and is mounted on a specially designed circuit board to allow for easier wiring connections. The panel also includes the Solar Charge Controller (SCC), which controls the charging of the battery via the solar modules, as well as the power supply, which supplies power from the grid (PLN). A relay module is used to control devices such as drain pump, refill pump, water filter and aerator. The Step Down LM2956 module reduces the voltage from 12V to 5V to meet the voltage requirements of the ESP32 and other components.

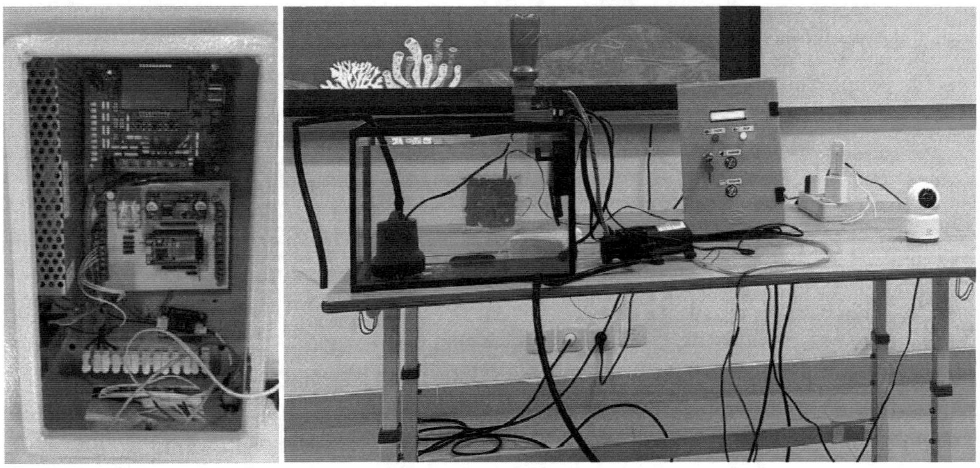

Figure 3. SHIFA hardware and system development-scale setup.

In the software configuration, the ESPHome firmware was successfully flashed to the ESP32 microcontroller to manage the main hardware components of the SHIFA system, including the relay-controlled pumps, aerator, water filter and feed servo motor. This configuration allows for seamless interaction between the hardware and Home Assistant, allowing the system to automate water changes, manage feeding schedules, and control water conditions in real time. The ESPHome setup also includes services to perform routine actions such as automatic water draining and refilling, which are critical to maintaining a stable aquaculture environment.

Additionally, the firmware configures multiple sensors – including pH, temperature and TDS sensors – connected to the ESP32, allowing the system to continuously monitor water quality and adjust the hardware accordingly. Wi-Fi integration provides remote monitoring and control capabilities with fallback options to ensure system stability in the event of network issues. Figure 5 shows the web interface of SHIFA, allowing users to monitor and control the SHIFA system through a web interface.

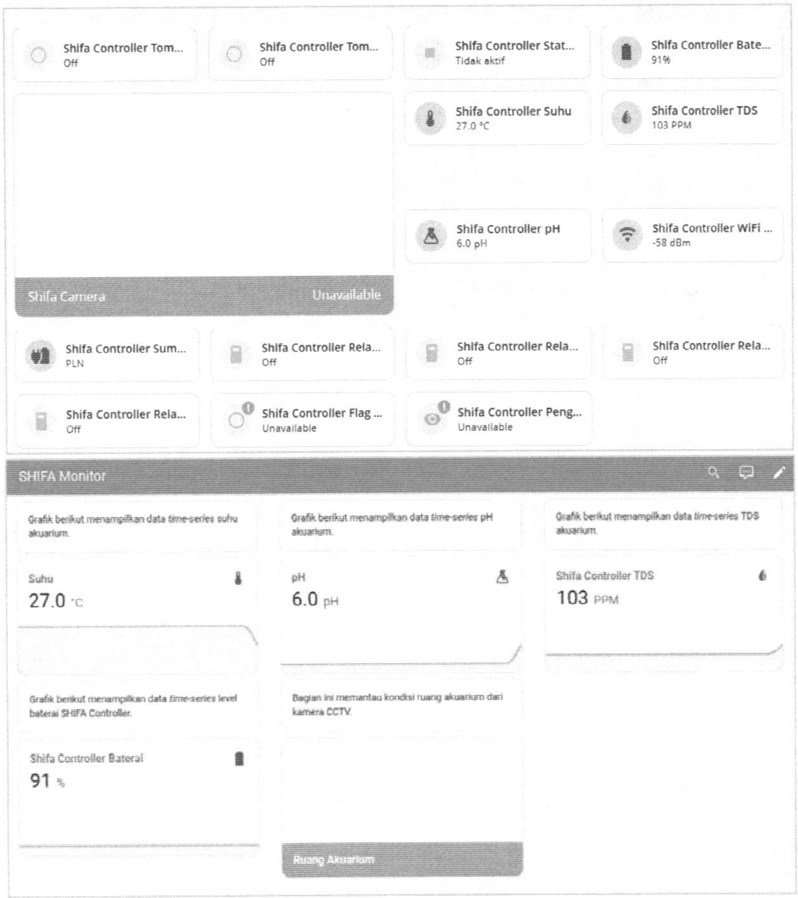

Figure 4. SHIFA web interface for real-time monitoring and control of the aquaculture system.

4.2 pH sensor measurement results

Before taking water pH measurements, the pH sensor must be calibrated to ensure accurate results. The calibration was carried out with buffer solutions with pH values of 4.01 and 6.86; a calibrated pH meter served as the reference device. When the buffer solution was tested at pH 4.01, the pH meter read 4.0, while the uncalibrated pH sensor recorded an output voltage of 2.8 V as logged by the ESPHome system. Likewise, when using the buffer solution with a pH of 6.86, the pH meter showed a value of 6.8 and the pH sensor recorded an output voltage of 2.4 V, which was also logged by ESPHome. These calibration results are summarized in Table 1 below.

Table 1. Calibration results of the pH sensor using buffer solutions.

No	Buffer solution	pH meter	Voltage (V)
1	pH 6.86	6.8	2.4
2	pH 4.01	4.0	2.8

The voltage and pH values from Table 1 were then used to synchronize the pH sensor with the Home Assistant platform. In the ESPHome configuration, calibration was performed using the calibrate linear filter, calibrating the output voltage of the sensor based on the corresponding pH values measured during the test. After calibration and synchronization with Home Assistant, the pH sensor was able to perform automated water pH measurements. The results of these measurements are shown in Table 2 below.

Table 2. pH measurements results.

No	Time of Measurement	pH Meter	SHIFA
1	30-06-2024, 10:00 AM	6.8	6.6
2	30-06-2024, 11:00 AM	4.0	3.9
3	30-06-2024, 12:00 PM	5.5	5.3
4	30-06-2024, 1:00 PM	6.0	5.8
5	30-06-2024, 2:00 PM	6.3	6.1
6	30-06-2024, 3:00 PM	6.5	6.3
7	30-06-2024, 4:00 PM	6.2	6.0

The data in Table 2 shows that the pH values measured by the pH sensor and displayed via the SHIFA Home Assistant interface are consistent with the readings determined by the pH meter. This indicates that the system is capable of accurately monitoring the pH of the water after the calibration process is complete.

4.3 TDS and temperature sensor measurement results

Before measuring TDS and temperature, the TDS sensor was calibrated to ensure precise and reliable readings, similar to the pH sensor calibration process. The calibration took into account the water temperature, which was combined with the output voltage of the TDS sensor to obtain a more accurate TDS value. The TDS calibration results is shown in Table 3.

Table 3. Calibration results of TDS sensor.

No	TDS meter	TDS meter Temp.	TDS (SHIFA)	TDS Voltage (SHIFA)	Temperature (SHIFA)	Description
1	84	4.9°C	66	0.174	6.4°C	Cold mineral water
2	413	24.3°C	270	0.714	26.2°C	Milk coffee water
3	501	25.8°C	350	0.960	26.2°C	Coffee water

The TDS calibration process used three different types of water with varying levels of turbidity. In the ESPHome configuration, the voltage produced by the TDS sensor was adjusted

IOP Conf. Series: Earth and Environmental Science **1446** (2025) 012049 doi:10.1088/1755-1315/1446/1/012049

based on the water temperature using the temperature compensation feature. Following the calibration, the results of the TDS and temperature measurements are displayed in Table 4.

Table 4. TDS and temperature sensor measurements results.

No	Time of Measurement	TDS Meter	Temp.	TDS (SHIFA)	Temp. (SHIFA)
1	30-06-2024, 10:00 AM	450	26.75	428	25.95
2	30-06-2024, 11:00 AM	448	26.65	426	25.85
3	30-06-2024, 12:00 PM	452	26.58	452	25.80
4	30-06-2024, 1:00 PM	450	26.50	429	25.71
5	30-06-2024, 2:00 PM	451	26.45	431	26.56
6	30-06-2024, 3:00 PM	449	26.32	427	26.50
7	30-06-2024, 4:00 PM	450	26.25	428	26.48

The data in the tables show that the temperature and TDS values measured by the calibrated sensors are largely consistent with the readings obtained from the TDS meter, indicating that the system is capable of measuring TDS and water temperature accurately after the calibration process to monitor.

Conclusion

The developed smart fish farming system using ESP32 microcontroller has successfully demonstrated its ability to accurately monitor and control key water quality parameters including pH, turbidity and temperature. Sensor data was effectively processed and transmitted to the Home Assistant platform, enabling real-time monitoring and automated corrective actions. In addition, the integration of a hybrid energy system that combines grid power and solar energy ensured the continuous operation of the system even during power outages. Implementation of the system has shown promising results in increasing the efficiency of fish farms by significantly reducing the need for manual intervention in routine tasks such as water emptying and feeding. These results suggest that the SHIFA system can contribute to more sustainable and efficient aquaculture practices.

The pH, temperature and turbidity sensors were calibrated and tested. The results showed an accuracy of 98% for pH, 97% for temperature and 95% for TDS. The control mechanisms for water circulation, pump and feeding, controlled via relays and servo motor, worked reliably and responded automatically to changing water conditions based on sensor data. Although minor variations in sensor accuracy were observed, the performance of the system remained within acceptable limits, making it a practical tool for managing aquaculture operations.

IOP Conf. Series: Earth and Environmental Science **1446** (2025) 012049 doi:10.1088/1755-1315/1446/1/012049

References

[1] Dinas Kelautan dan Perikanan Provinsi Kalimantan Barat. Rencana Strategis Perubahan Dinas Kelautan dan Perikanan Provinsi Kalimantan Barat Tahun 2018-2023, https://dislautkan.kalbarprov.go.id/renstra/ (2021, accessed 18 May 2024).

[2] Elfitasari T, Albert A. Challenges Encountered By Small Scale Fish Farmers In Assuring Fish Product Sustainability. *omni.akua*; 13. Epub ahead of print 1 December 2017. DOI: 10.20884/1.oa.2017.13.2.256.

[3] Junaidi, Syandri H, Azrita, et al. Floating cage aquaculture production in Indonesia: Assessment of opportunities and challenges in Lake Maninjau. *AIMSES* 2022; 9: 1–15.

[4] Kementerian Kelautan dan Perikanan. Peraturan Menteri Kelautan dan Perikanan Nomor 1 Tahun 2021, https://jdih.maritim.go.id/id/peraturan-menteri-kelautan-dan-perikanan-no-1-tahun-2021 (2021, accessed 19 May 2024).

[5] Acar U, Kane F, Vlacheas P, et al. Designing An IoT Cloud Solution for Aquaculture. IEEE, 2019, pp. 1–6.

[6] Hawari HF, Hazwan MA. Development of Iot Monitoring System For Aquaculture Application. In: *2022 International Conference on Green Energy, Computing and Sustainable Technology (GECOST)*. 2022, pp. 330–334.

[7] Dilli Kumar V, Eshwar M, Manova Sam M, et al. Blynk based aquaculture monitoring system using IOT. *World J Adv Eng Technol Sci* 2023; 8: 262–269.

[8] Islam MM, Kashem MA, Uddin J. An internet of things framework for real-time aquatic environment monitoring using an Arduino and sensors. *International Journal of Electrical and Computer Engineering (IJECE)* 2022; 12: 826–833.

[9] Coutinho RWL, Boukerche A. Towards a Novel Architectural Design for IoT-Based Smart Marine Aquaculture. *IEEE Internet of Things Magazine* 2022; 5: 174–179.

[10] Putra FPE, Ubaidi U, Saputra RN, et al. Application of Internet of Things Technology in Monitoring Water Quality in Fishponds. *Brilliance: Research of Artificial Intelligence* 2024; 4: 356–361.

[11] Salam MdA, Akib ASMAS, Islam M, et al. Precision Fish Farming to Mitigate Pond Water Quality Through IoT. In: *2024 IEEE 3rd International Conference on Computing and Machine Intelligence (ICMI)*. 2024, pp. 1–6.

[12] Hemal MdM, Rahman A, Nurjahan, et al. An Integrated Smart Pond Water Quality Monitoring and Fish Farming Recommendation Aquabot System. *Sensors* 2024; 24: 3682.

[13] Mohd Jais NA, Abdullah AF, Mohd Kassim MS, et al. Improved accuracy in IoT-Based water quality monitoring for aquaculture tanks using low-cost sensors: Asian seabass fish farming. *Heliyon* 2024; 10: e29022.

[14] Flores-Cortez OO, Castro Elizondo JM, Flores Portillo AC, et al. An IoT-Based System for Water Parameters Monitoring. Case Study: Fish Farming in El Salvador. In: *2024 IEEE International Workshop on Metrology for Industry 4.0 & IoT (MetroInd4.0 & IoT)*. 2024, pp. 123–128.

[15] Schröder CR, Weidgans BM, Klimant I. pH Fluorosensors for use in marine systems. *Analyst* 2005; 130: 907–916.

[16] Bhadra S, Tan DSY, Thomson DJ, et al. A Wireless Passive Sensor for Temperature Compensated Remote pH Monitoring. *IEEE Sensors Journal* 2013; 13: 2428–2436.

[17] Oelßner W, Zosel J, Berthold F, et al. Investigation of the dynamic response behaviour of ISFET pH sensors by means of laser Doppler velocimetry (LDV). *Sensors and Actuators B: Chemical* 1995; 27: 345–348.

[18] Wilhelm D, Voigt H, Treichel W, et al. pH sensor based on differential measurements on one pH-FET chip. *Sensors and Actuators B: Chemical* 1991; 4: 145–149.

[19] Zaibudeen AW, Philip J. Temperature and pH sensor based on functionalized magnetic nanofluid. *Sensors and Actuators B: Chemical* 2018; 268: 338–349.

[20] Mitchell HL, Mersch SH. *Nephelometric turbidity sensor device.* US7659980B1, https://patents.google.com/patent/US7659980B1/fr (2010, accessed 20 May 2024).

[21] Bilro L, Prats SA, Pinto JL, et al. Design and performance assessment of a plastic optical fibre-based sensor for measuring water turbidity. *Meas Sci Technol* 2010; 21: 107001.

[22] Tai H, Li D, Wang C, et al. Design and characterization of a smart turbidity transducer for distributed measurement system. *Sensors and Actuators A: Physical* 2012; 175: 1–8.

Palatability response test of hydroponic fodder (*Oryza sativa*, *Zea mays*, and *Vigna radiata*) in thin-tailed sheep (*Ovis aries*)

S B Kusuma[1], S Wulandari[1], A Bahariawan[2], A N Respati[1], M Adhyatma[1]* and B N Z Ruswanto[1]

[1]Department of Animal Science, Politeknik Negeri Jember, PO BOX 164 Jember, East Java, Indonesia
[2]Department of Agricultural Technology, Politeknik Negeri Jember, PO BOX 164 Jember, East Java, Indonesia

*Email : m_adhyatma@polije.ac.id

Abstract. This study aims to determine the effect of plant type on palatability response in thin-tailed sheep. This study used a completely randomized design. The factor is the type of plant consisting of rice (*Oryza sativa*), corn (*Zea mays*), and green beans (*Vigna radiata*). Each treatment was repeated 4 times. Palatability response variables measured included grasping time (seconds), consumption amount/minute (g) and consumption time/100g (minutes). Data were analyzed using Analysis of Variety (ANOVA) and differences between means were analyzed using Duncan's Multiple Range Test (DMRT). The results showed that different plant species had no significant effect ($P>0.05$) on grasping time and consumption time. However, mung bean hydroponic feed showed a significant difference ($P<0.05$) in the number of consumption/minutes compared to corn and rice. The conclusion of this study is that mung bean hydroponic feed has the best palatability, seen from the amount of consumption/minute which is 83 ± 8.12 g/minute. However, it is still necessary to further study the production capacity and price per kg as feed to achieve feed efficiency.

1. Introduction

Hydroponic Fodder is one of the animal feed technologies in the form of forage fodder that is cultivated in a short time on liquid media and under controlled conditions. [1]. Land limitation is one of the challenges in providing forage for livestock. In addition to being able to be grown on limited land, hydroponic fodder also has high digestibility [2]. Provision of high nutritional feed ingredients with self-reliance is a solution to support feed availability [3]. Lack of feed availability affects livestock productivity which affects the profitability of the farm. Hydroponic systems are used to grow seeds of cereals or other types with a certain humidity and without the use of solid media [4]. Hydroponic cultivation takes 7-14 days depending on the crop commodity being cultivated. Hydroponics has the advantage of providing feed that is environmentally friendly, has an energy balance, and increases digestibility [5].

The availability of forage feed in Indonesia is highly dependent on the season which causes the quantity and quality to vary. Feed is a very important factor in a livestock business, because feed has a direct effect on livestock production. The principle of hydroponic fodder cultivation lies in the conditions of humidity, light, temperature and availability of liquid media. Some hydroponic installations are equipped with automatic systems for regulating water flow, nutrients, lighting and

IOP Conf. Series: Earth and Environmental Science **1446** (2025) 012050 doi:10.1088/1755-1315/1446/1/012050

humidity, depending on the needs of each type of plant. [6]. Hydroponic fodder can be used as forage feed for sheep. Thin-tailed sheep are native to Indonesia, known as local sheep. Thin-tailed sheep have a high tolerance for a variety of forage and good adaptability to various environmental conditions, allowing them to live and breed throughout the year. [7]. Hydroponic fodder with potential as forage feed for sheep include corn (*Zea mays*), rice (*Oryza sativa*) and mung bean (*Vigna radiata*) [4]. Providing hydroponic fodder is expected to increase the production of thin-tailed sheep.

2. Methodology

The planting of hydroponic fodder was carried out in the mini greenhouse of Jember State Polytechnic. The planting time was 10 days. The types of seeds used were maize (*Zea mays*), rice (*Oryza sativa*) and mung bean (*Vigna radiata*) seeds purchased at Tanjung Market, Jember. The palatability test was applied to four male thin-tailed sheep with an average age of nine months.

2.1 Seeds Sorting Process

The sorting process begins with weighing each rice (*Oryza sativa*), maize (*Zea mays*), and mung bean (*Vigna radiata*) seed as much as 1 kg. The seeds that have been weighed are then put into a container for soaking for ± 12 hours. The floating seeds were taken and discarded while the sinking seeds were soaked with an anti-mold solution for 20 minutes [8].

2.2 Germination Process

Seeds that are ready for sowing are placed in a tray, covered with a cloth for ± 24 hours. During the cloth covering, liquid fertilizer (AB mix) was applied three times a day. On the next day, the cloth was opened, and water was sprayed three times a day until day ten [8].

2.3 Observed parameters

The parameters observed in this palatability test were grasping time (seconds), consumption amount/minute (g), and consumption time/100g (minute) [9].

2.4 Data analysis

The data obtained were subjected to analysis of variance (ANOVA), if the treatment had a significant effect, it would be followed by the Duncan Multiple Range Test (DMRT).

3. Result and Discussion

3.1 Grasping time

Gripping time is defined as the time it takes for the first time for the animal to eat its feed. The shorter the time taken to grasp the feed, the better the palatability of the feed. The grasping time of hydroponic fodder corn (*Zea mays*) is the fastest at 1.08 ± 0.10 seconds while rice (*Oryza sativa*) is the longest at 1.25 ± 0.21 seconds. This shows that thin-tailed sheep have good adaptability to the type of forage feed from hydroponic fodder. On the part of hydroponic fodder affected by mold causes a fairly pungent aroma that affects the speed of sheep in grabbing it. Part hydroponic fodder susceptible to mold is the root, as found in mung beans hydroponic fodder

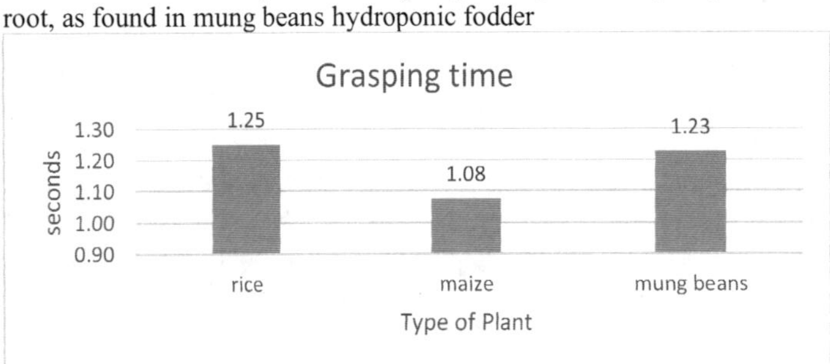

Figure 1. Hydroponic fodder Grasping time among type of Plants

The results of the analysis of variance showed that there were no significant differences (P>0.05) between plant types on grazing time. Livestock palatability is influenced by the physical appearance of feed, texture, color and aroma of livestock feed before the sheep graze the feed [10]. Before grabbing the feed, the sheep will usually smell the aroma of the feed, if it is not used to it, it takes time to adapt.

3.2 Consumption amount/minute

Feed consumption time is defined as the time required for fat-tailed sheep to consume corn fodder. The consumption time in this study was one minute. In one minute, the sheep were given hydroponic fodder and observed the amount of consumption. Hydroponic fodder mung beans showed the highest amount of consumption of 83 ± 8.12g/minutes. This could be due to the fact that mung bean hydroponic fodder contains less cotyledons so that the sheep can eat more in one minute. Hydroponic fodder rice is the least amount of consumption because it still contains rice husks or husks that reduce the palatability of sheep. Whereas in hydroponic fodder corn the seeds have not completely germinated so it is quite hard when grabbed by sheep.

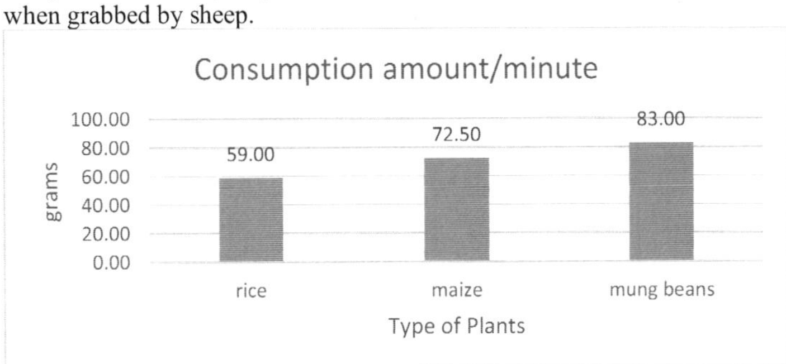

Figure 2. Hydroponic consumption amount/minute among type of plants

The results of variance analysis of consumption amount/minute showed significant differences (P<0.05) in plant types. The amount of feed consumption is influenced by palatability, digestibility, breed, sex, age and health conditions of livestock. The treatment of type of plant affects the amount of consumption of hydroponic fodder.

3.3 Consumption time/100g

Feed consumption time is defined as the time it takes for livestock to consume 100g of feed. Feed consumption time is a palatability parameter which affects the level of feed liking. Good hydroponic fodder has a short consumption time, otherwise if the consumption time is long, the level of palatability is low. mung beans hydroponic fodder showed the fastest consumption time/100g.

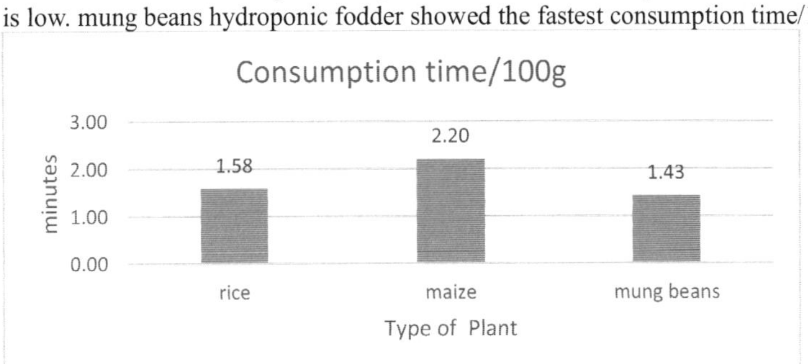

Figure 3. Hydroponic fodder consumption time/100g among type of plants

The results of the analysis of variance consumption time/100g showed results that had no significant effect (P>0.05) on the type of plant. there is a tendency for sheep to consume feed with a softer texture faster than harder ones. This can be seen in the characteristics of hydroponic fodder mung bean which overall has a softer texture compared to the texture of hydroponic fodder rice and corn.

4. Conclusion

From the analysis, it can be concluded that mung bean hydroponic fodder has the best palatability, seen from the consumption amount/minute which is 83±8.12g/minutes. However, it is still necessary to further study the production capacity and price per kg of fresh in order to achieve feed efficiency.

5. Acknowledgements

The authors would like to thank P3M Politeknik Negeri Jember for funding the research through PNBP funds in 2024, number: 827/PL17.4/PG/2024.

References

[1] R. Jolad, S. D. Sivakumar, C. Babu, and N. Srithran, "Performance of Different Crops under Hydroponics Fodder Production System," 2018.

[2] R. Zahera, I. G. Permana, and Despal, "Utilization of mungbean's green house fodder and silage in the ration for lactating dairy cows," *Media Peternakan Fakultas Peternakan Institut Pertanian Bogor*, vol. 38, no. 2, pp. 123–131, 2015, doi: 10.5398/medpet.2015.38.2.123.

[3] A. F. Qohar and E. Hendarto, "Pertumbuhan Rumput Raja (Pennisetum purpureophoides) Defoliasi Kedua Akibat Pemupukan Kompos yang Diperkaya dengan Azolla microphylla," in *Prosiding Seminar Nasional Pembangunan dan Pendidikan Vokasi Pertanian*, Manokwari, 2020.

[4] T. F. Rayani, Y. Resti, and R. K. Dewi, "Kuantitas dan Kualitas Fodder Jagung, Padi dan Kacang Hijau dengan Waktu Panen yang Berbeda Menggunakan Smart hydroponic Fodder," *Jurnal Ilmu Nutrisi dan Teknologi Pakan*, vol. 19, no. 2, pp. 36–41, Aug. 2021, doi: 10.29244/jintp.19.2.36-41.

[5] D. Takanjanji *et al.*, "Pengaruh Umur Panen Terhadap Kecernaan In Vitro dan Nilai Energi Fodder Jagung Sebagai Pakan Pedet (Effect of Harvesting Age on In Vitro Digestibility and Energetic Value of Maize Fodder as Calf Feed)," *Jurnal Peternakan Lahan Kering*, vol. 1, no. 4, pp. 554–561, 2019.

[6] S. Lee and J. Lee, "Beneficial bacteria and fungi in hydroponic systems: Types and characteristics of hydroponic food production methods," Nov. 12, 2015, *Elsevier B.V.* doi: 10.1016/j.scienta.2015.09.011.

[7] M. Najmuddin and M. Nasich, "Produktivitas Induk Domba Ekor Tipis di Desa Sedan Kecamatan Sedan Kabupaten Rembang," *TERNAK TROPIKA Journal of Tropical Animal Production*, vol. 20, no. 1, pp. 76–83, Jun. 2019, doi: 10.21776/ub.jtapro.2019.020.01.10.

[8] S. Wulandari, S. B. Kusuma, B. Utomo, and A. N. Respati, "Characterization and production hydroponic fodder for corn (*Zea mays*) and sorgum (Sorghum bicolor) at different harvest times," in *IOP Conference Series: Earth and Environmental Science*, Institute of Physics, 2024. doi: 10.1088/1755-1315/1338/1/012017.

[9] N. Nuraeni and A. M. Zaen, "Uji palatabilitas fodder jagung (*zea mays*) hidroponik pada ternak domba ekor gemuk," *Jurnal Sains Peternakan Nusantara*, vol. 01, no. 02, pp. 45–52, Dec. 2021.

[10] I. M. A. Sudarma, "Pengujian Konsistensi, Waktu Adaptasi, Palatabilitas dan Persentase Disintegrasi Ransum Blok Khusus Ternak Sapi Potong Antarpulau," *Jurnal Sain Peternakan Indonesia*, vol. 13, no. 3, pp. 265–273, Oct. 2018, doi: 10.31186/jspi.id.13.3.265-273.

IOP Conf. Series: Earth and Environmental Science **1446** (2025) 012051 doi:10.1088/1755-1315/1446/1/012051

The effect of broiler bone nanocalcium citrate fortification on the color and sensory properties of Indonesian traditional beef meatballs '*bakso*'

A H Prayitno[1*], N Asrianto[1], R Rahmasari[1], H T Handayani[2], C Huda[3], S S Abdil[1], I A M Naky[1], L Mafazah[1], M F H Ramadhan[1], M S Umam[1], W M Ramadhana[1], S H Wicaksono[1], and R Yogi[1]

[1] Animal Science Department, Politeknik Negeri Jember, Jl. Mastrip Po Box 164 Jember 68101, Indonesia
[2] Agricultural Production Department, Politeknik Negeri Jember, Jl. Mastrip Po Box 164 Jember 68101, Indonesia
[3] Information Technology Department, Politeknik Negeri Jember, Jl. Mastrip Po Box 164 Jember 68101, Indonesia

*E-mail: agushp@polije.ac.id

Abstract. Green synthesis can produce broiler bone nanocalcium citrate, which could serve as a novel food supplement for Indonesian traditional meatballs known as *'bakso'*. The purpose of this investigation was to ascertain the impact of broiler bone nanocalcium citrate fortification on the sensory properties and color of beef meatballs. Ingredients such as beef, tapioca flour, egg whites, salt, shallots, garlic, monosodium glutamate, pepper, soy protein isolate, sodium tripolyphosphate, carrageenan, ice, lime juice, and aquadest were included in the research materials. As part of the treatments, broiler bone nanocalcium citrate was added to the total meatball dough at the following percentages: P0 (0%), P1 (0.15%), P3 (0.3%), P4 (0.45%), and P5 (0.6%). The color parameters measured included lightness (L*), redness (a*), and yellowness (b*), with each treatment replicated four times. For the evaluation of sensory properties, 40 untrained panelists implemented a hedonic test. The hedonic parameters that were evaluated included color, flavor, taste, texture, chewiness, and overall acceptability. The hedonic test results were analyzed using non-parametric analysis through the Kruskal-Wallis hedonic test, while the data from the color measurements were analyzed using a one-way analysis of variance. Duncan's New Multiple Range Test was used for further analysis if significant differences were found ($P<0.05$). The results showed that different levels of broiler bone nanocalcium citrate fortification affected color measurements (lightness and yellowness) as well as sensory properties (texture, chewiness, and overall acceptability). This study concludes that fortifying beef meatballs with 0.3% broiler bone nanocalcium citrate is the best treatment, resulting in a lighter color, a more preferred texture, and chewiness.

1. Intoduction

Broiler bones are one of the livestock wastes that contain high calcium which is abundant and very cheap. The main components of broiler bones are 57.35% calcium phosphate, 3.85% calcium carbonate, and 33.3% collagen [1]. Broiler bones can be obtained from the poultry industry. The output of broiler meat in 2022 was 3,765,573.09 tonnes, according to BPS [2]. The consumption of broiler meat which continues to increase every year produces bone waste. Broiler bone waste can have its economic value increased by being made into nanocalcium. Broiler bones through the green synthesis method can produce nanometer-sized calcium, which is around 524.1 nm with a calcium content of around 5.63% and calcium oxide 7.88% [3]. The green synthesis method was chosen because it supports more eco-friendly aspects. The green synthesis process uses natural ingredients that are environmentally friendly and safe [4].

The nanoparticles' diminutive size enables the extract to dissolve effortlessly and achieve a high absorption rate in the intestine, in contrast to the microcalcium that is more prevalent [5]. Nano-sized materials exhibit enhanced performance due to their increased surface area [6]. Nanocalcium is a metal oxide extensively used as a catalyst [7], antibacterial agent [8], and drug delivery system enhancing solubility and absorption by the body with about 100% bioavailability [9–11]. Calcium from broiler bones that have been nanosized may be used as a dietary calcium source and as a useful element that has health benefits.

In comparison to men, Indonesian women are at a fourfold increased risk of developing osteoporosis. According to the International Osteoporosis Foundation (IOF), one in four Indonesian women aged 50 to 80 are at risk of developing osteoporosis [12]. The risk of osteoporosis is closely related to low calcium intake in the body [13]. The body's calcium needs can be met from foods such as milk and dairy products, broccoli, soybeans, and spinach. Other ways can be done to meet the body's calcium needs, namely by consuming calcium supplements. Nevertheless, there are alternative calcium sources, such as broiler bone nanocalcium, whose potential as a fortification material for functional food products has not been extensively investigated. One form of food product that is easily accepted by the Indonesian people based on how to consume it is meatballs.

Meatballs are one of the processed meat products and are very popular with the public. Meatballs are becoming increasingly popular thanks to their flexibility in terms of marketing and sales [14]. Quality meatballs must use the right ingredients without causing harm to health [15]. The main reason for choosing meatball products is because the market share of meatball products is still very large. Changes in consumer consumption patterns that are currently shifting to ready to cook (RTC) and ready-to-eat (RTE) products and consumers of meatball products cover all age segments so that the presentation of products in the form of meatballs will make it easier for people to accept meatball products as an alternative food ingredient rich in calcium. The color and sensory properties of meatballs are an important part of consumer assessment of products to be consumed. As a result, this investigation was undertaken to ascertain the impact of nano calcium citrate fortification on the sensory properties and color of traditional Indonesian meatballs known as "*bakso*."

2. Materials and Methods

2.1. Materials

A meat grinder, spoon, gas cooker, pan, knife, plastic container, dicing board, pH meter, scale, and tissue were each employed in this investigation. The following materials were employed in this investigation: beef, broiler bones, tapioca flour, egg white, salt, monosodium glutamate (MSG), sodium triphosphate (STPP), pepper, garlic, shallot, isolate soya protein (ISP), carrageenan, lime, ethanol, ice crystals, and aquadests.

2.2. Methods

This work employed a completely randomized design (CRD) with a one-way pattern to fortify broiler bones with nanocalcium citrate using the green synthesis technique. For the meatball dough, the nanocalcium citrate fortification treatments were P0 (0%), P1 (0.15%), P3 (0.3%), P4 (0.45%), and P5 (0.6%).

2.2.1. Preparation of broiler bone flour. Broiler bones are washed clean and then boiled for 1 hour. Water and broiler bones were combined in a 1:6 (w/v) ratio, and the mixture was then cooked for two hours in a pressure cooker. After a 24-hour oven drying process at 105°C, broiler bones are pulverized using a sample mill.

2.2.2. Preparation of broiler bone calcium oxide. The preparation of broiler chicken bone flour into calcium oxide (CaO) was made as described by Prayitno et al. [16] with modification. CaO powder was obtained by calcining chicken bone flour for four hours at 900°C.

2.2.3. Preparation of nanocalcium citrate of broiler bones. Preparation of nano calcium through green synthesis as described by Umam et al. [3]. Broiler bone calcium oxide was mixed with distilled water. After adding the lime juice and ethanol, the mixture was agitated for 30 minutes at 75°C using a hotplate magnetic stirrer set at 500 rpm/min. The sample was then dried for 5 days in sunlight and ground using a blender to produce nanocalcium citrate powder.

2.2.4. Meatball formulation. The meatball formulation used in this study was made as described by Prayitno & Rahman [17] with modification. The meatball composition comprises beef (60%), tapioca flour (4%), egg white (12.5%), salt (1.4%), shallots (1%), garlic (2.5%), MSG (0.8%), pepper (0.5%), ISP (4%), STPP (0.3%), carrageenan (0.5%), and ice (12.5%).

2.2.5. Meatball processing. The meatball processing is made according to that described by Prayitno & Rahman [17]. The meat is cleaned with water and separated from the fat tissue then cut into small pieces. The pieces of meat are finely ground with a meat grinder. The spices are ground then mixed with salt, pepper, MSG, and broiler bone nano calcium treatment then put into the meat processor. Tapioca flour, ISP, egg white, and ice are added little by little into the meat processor then ground and mixed until smooth. The dough is formed manually into balls according to each treatment. The meatball mixture is immersed in water at 60°C for 2 minutes, and the meatballs are immersed in boiling water for 10 minutes. The meatballs that have been cooked are removed, drained, then cooled at room temperature. The meatballs are then tested for color and sensory evaluation.

2.3. Color properties

The color properties of beef meatballs fortified with nanocalcium citrate from broiler bones were tested using a color reader [18].

2.4. Sensory properties

The cooked meatballs were evaluated for their sensory properties through a hedonic test by 40 untrained panelists [19]. The panelists evaluated the hedonic test characteristics of color, flavor, taste, texture, chewiness, and acceptability [20]. The hedonic scale used a Likert scale, namely 1 (extremely dislike) to 5 (extremely like).

2.5. Statistic analysis

The data obtained from the color properties test were analyzed using analysis of variance, whereas the data obtained from the sensory properties test were analyzed using non-parametric methods, especially the Hedonic Kruskal-Wallis test. If a statistically significant difference ($P<0.05$) was found, it was further investigated by employing Duncan's New Multiple Range Test [21].

3. Results and Discussion

3.1. Color properties

The findings indicated that the fortification of broiler bones with varying quantities of nanocalcium citrate significantly influenced ($P<0.05$) the L* (lightness) value of beef meatballs. The L* number depicts the lightness of the beef meatball's hue. The L* value of beef meatballs enriched with nanocalcium citrate derived from broiler bones varied between 48.45 and 51.90. The L* value of beef meatballs enriched with nanocalcium citrate from broiler bones was higher than that of beef meatballs in research by Widiati et al. [22]. Widiati et al. [22] indicated that the L* value of beef meatballs ranged from around 43.93 to 48.30. Increasing the level of nanocalcium citrate fortification of broiler bones can change the b* value of beef meatballs. The L* value of beef meatballs increased with the increasing level of nanocalcium citrate fortification of broiler bones. Increasing nanocalcium citrate fortification of broiler bones made the resulting beef meatballs have a brighter color. This is indicated by the highest L* value obtained with 0.60% nanocalcium citrate fortification of broiler bones, namely 51.90, while the lowest L* value was obtained in beef meatballs without fortification, namely 48.45.

Table 1. The color properties of beef meatballs fortified with nanocalcium citrate broiler bones with different levels.

Variable	Fortification level of nanocalcium citrate broiler bones				
	0%	0.15%	0.30%	0.45%	0.60%
L*	48.45±0.62^{a}	49.50±0.52^{a}	50.75±0.82^{b}	51.15±0.89^{b}	51.90±0.82^{b}
a*^{ns}	6.05±0.55	6.20±0.83	6.15±0.59	6.10±0.56	6.50±0.43
b*	17.60±0.51^{a}	18.05±0.87^{ab}	19.10±1.17^{bc}	18.35±0.75^{ab}	20.10±0.82^{c}

^{ns} Not significant (P>0.05)
$^{a\text{-}c}$ Distinct superscripts in the same row demonstrate significant differences (P<0.05)

The findings indicated that the fortification of broiler bones with varying doses of nanocalcium citrate did not significantly affect (P>0.05) the a* (redness) value of beef meatballs. The a* value represents the reddish hue of beef meatballs. The a* value of beef meatballs fortified with nanocalcium citrate from broiler bones varied between 6.05 and 6.50. The a* value of beef meatballs enriched with nanocalcium citrate from broiler bones was higher than that of beef meatballs reported in the work by Widiati et al. [22]. Widiati et al. [22] indicated that the a* value of beef meatballs ranged from 16.60 to 20.07. In addition, the concentration of nanocalcium citrate fortification in broiler bones did not impact the a* value of beef meatballs. The a* value of beef meatballs enriched with nanocalcium citrate extracted from broiler bones was almost the same between all treatments.

The findings indicated that the fortification of broiler bones with varying levels of nanocalcium citrate significantly influenced (P<0.05) the b* value (yellowness) of beef meatballs. The b* value demonstrates the yellowish hue of beef meatballs. The b* value of beef meatballs fortified with nanocalcium citrate broiler bones ranged from 17.60-20.10. The b* value of beef meatballs fortified with nanocalcium citrate from broiler bones rose above that of beef meatballs reported in the study by Widiati et al. [22]. The b* value of beef meatballs was found by Widiati et al. [22] to be between 13.53 and 16.00. Increasing the level of nanocalcium citrate fortification of broiler bones can change the b* value of beef meatballs. The b* value of beef meatballs increased with the increasing level of nanocalcium citrate fortification of broiler bones. Increasing nanocalcium citrate fortification of broiler bones made the resulting beef meatballs have a higher b* value. This is indicated by the highest b* value obtained with 0.60% nanocalcium citrate fortification of broiler bones, namely 20.10, while the lowest L* value was obtained in beef meatballs without fortification, namely 17.60.

3.2. Sensory properties

3.2.1. Color

The color formed in processed meat such as meatballs is the result of various processes and reactions that are very diverse [15]. Color significantly influences customer acceptability of a product since it is the first attribute perceived [23]. The study's results indicated that varying levels of nanocalcium citrate fortification in broiler bones did not significantly affect (P>0.05) the color score in the beef meatball hedonic test. The color score of beef meatballs ranged from 3.85-4.05, namely quite like to very like. This is because the color of beef meatballs fortified with nanocalcium citrate broiler bones visually according to the panelists did not show a striking difference so the panelists' level of preference for the color of beef meatballs for each treatment was almost the same. Beef meatballs have a color ranging from pale to brownish [15]. Nanocalcium citrate fortification of broiler bones with different levels is thought to still have the typical color of beef meatballs, namely pale so that they are preferred by the panelists.

Table 2. Sensory properties of beef meatballs fortified with nanocalcium citrate broiler bones with different levels.

Variable	Fortification level of nanocalcium citrate broiler bones				
	0%	0.15%	0.30%	0.45%	0.60%
Color^{ns}	3.98±0.65	4.03±0.65	4.05±0.59	4.00±0.50	3.85±0.76
Flavor^{ns}	3.88±0.75	3.73±0.89	3.95±0.71	3.78±0.96	3.70±0.60
Taste^{ns}	3.68±0.72	3.85±0.73	3.98±0.85	3.65±0.88	3.48±0.77
Texture	3.73±0.74^{bc}	3.88±0.68^{ab}	4.10±0.70^{a}	3.60±0.94^{bc}	3.48±0.87^{c}
Chewiness	3.55±0.80^{a}	4.00±0.77^{b}	4.03±0.72^{b}	3.60±0.89^{a}	3.43±0.95^{a}
Acceptability	3.80±0.68^{a}	4.08±0.65^{ab}	4.25±0.43^{b}	3.85±0.69^{a}	3.80±0.71^{a}

^{ns} Not significant (P>0.05)
$^{a-c}$ Distinct superscripts in the same row demonstrate significant differences (P<0.05)

3.2.2. Flavor
Flavor is one of the important factors in processed meat products such as meatballs that is difficult to define objectively. The flavor is also called long-distance tasting because humans can recognize the deliciousness of food that has not been seen just by smelling its flavor from a distance [15]. The study's findings indicated that the flavor score in the beef meatball hedonic test was not significantly influenced by the fortification of broiler bone nanocalcium citrate at varying levels (P>0.05). The beef meatball flavor score ranged from 3.70-3.95, namely slightly liking to liking. This is because the flavor of beef meatballs fortified with broiler bone nanocalcium citrate by smelling according to the panelists did not show a striking difference so the panelists' level of preference for the beef meatball flavor for each treatment was almost the same. During the preparation of meatballs, the flavour is influenced by the seasonings, flour, and meat used [24]. Meatballs have a distinctive flavor [25]. Fortification of broiler bone nanocalcium citrate with different levels is thought to still have a distinctive flavor of beef meatballs, namely with a meat flavor so that it is liked by the panelists.

3.2.3. Taste
Taste is a sensory component that dominates consumer acceptance of processed meat products such as meatballs. The taste response is received by the sensory on the tongue due to the presence of particles dissolved in water or oil, as well as the solubility of protein binding. Taste ranks first in terms of consumer acceptance of meatballs [24]. The study's findings indicated that the taste score in the beef meatball hedonic test was not significantly influenced by the fortification of broiler bone nanocalcium citrate at varying levels (P>0.05). The beef meatball taste score ranged from 3.48-3.98, namely slightly liking to liking. This is because the taste of beef meatballs fortified with broiler bone nanocalcium citrate by chewing according to the panelists did not show a striking difference so the panelists' level of preference for the taste of beef meatballs for each treatment was almost the same. Meatballs have a distinctive taste [25]. Fortification of broiler bone nanocalcium citrate with different levels is thought to still have a distinctive taste of beef meatballs, namely a meaty taste, so it is preferred by the panelists. Consumers prefer meatballs that have a meaty taste [24].

3.2.4. Texture
Texture in food products such as meatballs can be influenced by the ability to bind water. Changes in texture are influenced by protein coagulation, collagen gelatinisation, water release, expansion, and

starch gelatinisation [15]. The results indicated that the texture score in the beef meatball hedonic test was significantly influenced (P<0.05) by the fortification of broiler bone nanocalcium citrate with varying levels. The beef meatball texture score ranged from 3.48-4.10, namely slightly like to very like. The lowest beef meatball texture score was obtained with broiler bone nanocalcium citrate fortification of 0.60%, namely 3.48, while the highest beef meatball texture score was obtained with broiler bone nanocalcium citrate fortification of 0.30%, namely 4.10. This is because broiler bone nanocalcium citrate fortification up to a level of 0.30% causes the texture of beef meatballs to become more compact. Prayitno et al. [26] reported that nanocalcium fortification up to 0.30% can increase the compactness of meatballs as seen from the microstructure of the meatballs. This is what caused the panelists to prefer beef meatballs fortified with 0.30% broiler bone nanocalcium citrate because the texture is more compact than the other treatments.

3.2.5. Chewiness

Chewiness refers to the capacity of meatball products to return to their original form following compression [15]. The chewiness of meatballs is closely related to the texture of meatballs. The study's findings indicated that the chewiness score in the beef meatball hedonic test was significantly influenced (P<0.05) by the nanocalcium citrate fortification of broiler bones at varying levels. The beef meatball chewiness score ranged from 3.43-4.03, namely slightly liking to very liking. The lowest beef meatball chewiness score was obtained with nanocalcium citrate fortification of broiler bones of 0.60%, namely 3.43, while the highest beef meatball texture score was obtained with nanocalcium citrate fortification of broiler bones of 0.30%, namely 4.03. Nanocalcium citrate fortification of broiler bones up to a level of 0.30% caused the texture of beef meatballs to become more compact. This is what causes beef meatballs fortified with 0.30% broiler bone nanocalcium citrate to be chewy, which is felt when bitten by the panelists. The increased compactness of the meatballs further increases the chewiness of the meatballs because the density of the meatballs is higher. Panelists prefer chewier meatballs [15]. In addition, Prayitno et al. [26] reported that 0.30% nanocalcium fortification can increase the chewiness of meatballs. This is what causes panelists to prefer beef meatballs fortified with 0.30% broiler bone nanocalcium citrate because they are more chewy than other treatments.

3.2.6. Acceptability

Meatball acceptability is the overall level of preference for each sensory parameter assessed by the panelists through a hedonic test. The findings indicated that the fortification of broiler bones with varying levels of nanocalcium citrate significantly influenced (P<0.05) the acceptability score in the hedonic test of beef meatballs. The acceptability score of beef meatballs ranged from 3.80-4.25, namely slightly liking to very liking. The lowest beef meatball acceptability score was obtained with nanocalcium citrate fortification of broiler bones of 0.60% and without fortification, namely 3.80, while the highest beef meatball texture score was obtained with nanocalcium citrate fortification of broiler bones of 0.30%, namely 4.25. Nanocalcium citrate fortification of broiler bones up to a level of 0.30% caused the texture of beef meatballs to become more compact and chewy. This is what causes beef meatballs fortified with nanocalcium citrate fortified with broiler bones of 0.30% to be the most preferred treatment by the panelists. Prayitno et al. [26] reported that nanocalcium fortification of 0.30% was the most preferred meatball compared to meatballs without fortification. This caused the panelists to prefer beef meatballs fortified with nanocalcium citrate broiler bones of 0.30% because they were more compact and chewy than the other treatments.

4. Conclusion

The results showed that different nano calcium citrate fortifications of broiler bones can affect the color properties (lightness and yellowness) and sensory (texture, chewiness, and acceptability). The research concludes that fortifying broiler bones with 0.3% nanocalcium citrate in the whole beef meatball mixture yields the optimal treatment, resulting in a brighter meatball color and a more desirable texture and chewiness.

Acknowledgment
Pusat Penelitian dan Pengabdian Kepada Masyarakat (P3M) of Politeknik Negeri Jember provided the research funds for this paper through DIPA Politeknik Negeri Jember SP. DIPA-023.18.2.677607/2024.

References

[1] Jannah A, Maunatin A, Windayanti A, Findianti Y and Mufidah Z 2013 *Alchemy* **2** 184–9
[2] BPS 2023 *Produksi Daging Ayam Ras Pedaging menurut Provinsi (Ton), 2020-2022* (Jakarta: Badan Pusat Statistik Indonesia)
[3] Umam M S, Ridho M R, Safitrri A R, Roihan N A and Prayitno A H 2023 *Produksi Nano Kalsium dari Limbah Tulang Ayam melalui Green Sythesis menggunakan Ekstrak Jeruk Nipis dan Belimbing Wuluh sebagai Bahan Suplemen Pangan* (Jember: Politeknik Negeri Jember)
[4] Khan M S, Ranjani S and Hemalatha S 2022 *Mater. Chem. Phys.* **282** 125985
[5] Shafey A M El 2020 *Green Process. Synth.* **9** 304–39
[6] Habte L, Shiferaw N, Mulatu D, Thenepalli T, Chilakala R and Ahn J W 2019 *Sustainability* **11** 1–10
[7] Gopalappa H, Yogendra K, Mahadevan K M and Madhusudhana N 2012 *Int. J. Sci. Res.* **1** 91–5
[8] Roy A, Gauri S S, Bhattacharya M and Bhattacharya J 2013 *J. Biomed. Nanotechnol.* **9** 1570–8
[9] Malsch N H 2005 *Biomedical nanotechnology* (Boca Raton (US): CRC Press)
[10] Balaganesh A S, Sengodan R, Ranjithkumar R and Chandarshekar B 2018 *Int. J. Innov. Technol. Explor. Eng.* 2278–3075
[11] Suptijah P 2009 Sumber Nano Kalsium Hewan Perairan *101 Inovasi Indonesia* (Jakarta: Kementerian Riset dan Teknologi)
[12] Kementerian Kesehatan 2005 *Data dan Kondisi Penyakit Osteoporosis di Indonesia* (Jakarta: Pusat Data dan Informasi Kementerian Kesehatan RI)
[13] Jahari A B and Prihatini S 2014 *Gizi Indones.* **30** 1–11
[14] Sari N I, Sari M I and Rahayu J 2022 *J. Ekon. dan Bisnis GROWTH* **20** 204–12
[15] Falahudin A, Somanjaya R, Widianingrum D and Dianto R 2024 *J. Ilmu Pertan. dan Peternak.* **12** 52–9
[16] Prayitno A H, Lorenza F, Suparmi and Naafi'yan M H 2021 *J. Ilmu Ternak dan Vet.* **26** 152–7
[17] Prayitno A H and Rahman T H 2020 *J. Ilmu Ternak dan Vet.* **25** 190–4
[18] Park S-Y, Byeon D-S, Kim G-W and Kim H-Y 2021 *J. Anim. Sci. Technol. J Anim Sci Technol* **63** 180–90
[19] Sujarwanta R O, Jamhari, Suryanto E, Yuliatmo R and Prayitno A H 2019 *IOP Conf. Ser. Earth Environ. Sci.* **387** 012091
[20] O'Neill C M, Cruz-Romero M C, Duffy G and Kerry J P 2019 *Food Packag. Shelf Life* **21** 1–11
[21] Zhang H, He P, Kang H and Li X 2018 *LWT - Food Sci. Technol.* **93** 470–6
[22] Widati A S, Rosyidi D, Radiati L E and Nursyam H 2021 *Int. J. Food Stud.* **10** SI11–20
[23] Saputro M A P and Susanto W H 2016 *J. Pangan dan Agroindustri* **4** 62–71
[24] Rosita F, Hafid H and Aka R 2015 *JITRO* **2** 14–20
[25] SNI 2014 *Bakso Daging* (Jakarta: Badan Standardisasi Nasional)
[26] Prayitno A H, Suryanto E, Rusman, Setiyono, Jamhari and R U 2020 Karakteristik mikrostruktur dan nilai gizi bakso ayam yang difortifikasi kalsium oksida dan nanokalsium laktat kerabang telur ayam *Prosiding Seminar Nasional Teknologi Peternakan dan Veteriner* (Bogor: Pusat Penelitian dan Pengembangan Peternakan) pp 653–63

The Determinant of Excellent Horticulture Commodity in Ngawi Regency

Mohammad Edwinsyah Yanuan Putra[1*], Elis Rahmawati Mar'atus Sholihah[1], Waridad Umais Al Ayyubi[1], Ngani Alwirya Musi[1], Aditya Wardana[1], Anwar Firmansyah[1], Rahmat Dhandy[1]

[1]Politeknik Negeri Jember

[*]Email: edwin_yp@polije.ac.id

Abtract. Ngawi Regency based on statistical data has the potential for horticultural commodities that have the 3 highest rankings, namely red onion, cayenne pepper, and big chilies . Based on these data, further studies need to be carried out to provide recommendations to further increase the optimization of one of the horticultural commodities so that it can become one of the leading commodities for Ngawi Regency. Increasing food security and sustainability of the agricultural sector is a major focus in developing agribusiness in various countries. In this context, horticultural commodities play an important role, considering their strategic role in providing nutrition and improving the local economy. The selection of superior horticultural commodities, that can provide optimal results and meet market needs, is a crucial step to ensure sustainable growth in this sector. The SWOT method helps in evaluating internal and external factors that affect the potential of a horticultural commodity. The AHP method offers a quantitative approach to overcome the complexity in decision-making. The combination of these two methods provides a comprehensive and data-based approach in determining superior horticultural commodities and strategies that can be carried out by Ngawi Regency to improve their quality and quantity. The SWOT method provides a framework for understanding the internal and external context, while AHP offers a tool for making more structured decisions based on comprehensive criteria evaluation. Horticultural commodities from Ngawi Regency that are considered to be superior horticultural commodities are large chili commodities with an AHP value of 0.659, and the strategy that can be carried out to maintain the sustainability of this commodity is to increase production with an AHP value of 0.558. This can be a recommendation for Ngawi Regency to be able to increase attention to large chili commodities so that they can survive as superior horticultural commodities by maintaining and improving both in terms of quality and quantity.

1. Introduction

Increasing food security and sustainability of the agricultural sector is a major focus in developing agribusiness in various countries. In this context, horticultural commodities play an important role, considering their strategic role in providing nutrition and improving the local economy(Berlianantiya, 2017). The selection of superior horticultural commodities, that can provide optimal results and meet market needs, is a crucial step to ensure sustainable growth in this sector (Junari et al., 2020).

IOP Conf. Series: Earth and Environmental Science **1446** (2025) 012052 doi:10.1088/1755-1315/1446/1/012052

Determining superior horticultural commodities in Ngawi Regency that are by market potential and needs requires a systematic and objective approach. The SWOT (Strengths, Weaknesses, Opportunities, Threats) and AHP (Analytic Hierarchy Process) methods are two effective analysis tools in this process (Nurjannah & Mashuri, 2020) .

Ngawi Regency based on statistical data has the potential for horticultural commodities that have the 3 highest rankings, namely shallots, cayenne pepper, and large chilies (BPS Ngawi, 2024). Based on these data, further studies need to be carried out to provide recommendations to further increase the optimization of one of the horticultural commodities so that it can become one of the leading commodities for Ngawi Regency.

The SWOT method helps in evaluating internal and external factors that affect the potential of a horticultural commodity (Sylvia & Hayati, 2023). Through SWOT analysis, We can recognize internal strengths and weaknesses, along with external opportunities and threats, that may impact the success of the commodity. Internal strengths and weaknesses include aspects such as product quality, cultivation technology, and farmer skills, while external opportunities and threats include factors such as market demand, climate conditions, and competition.

On the other hand, the AHP method offers a quantitative approach to overcome the complexity in decision-making. By using AHP, various relevant criteria can be systematically assessed and weighted according to their importance. This process allows decision-makers to determine commodity priorities based on various aspects that have been identified in the SWOT analysis, such as market potential, production costs, and technological capabilities (Rachman, 2019).

The combination of these two methods provides a comprehensive and data-based approach in determining superior horticultural commodities and strategies that can be carried out by Ngawi Regency to improve their quality and quantity. The SWOT method provides a framework for understanding the internal and external context, while AHP offers a tool for making more structured decisions based on comprehensive criteria evaluation (Rachman, 2019; Sylvia & Hayati, 2023).

This article aims to explain the process and application of the SWOT and AHP methods in determining superior horticultural commodities in Ngawi Regency. Thus, it is hoped that readers can understand how these two methods complement each other in producing more precise and effective decisions in horticultural development (Rachman, 2019; Sylvia & Hayati, 2023).

2. Research Method

In this study, the determination of superior horticultural commodities is carried out through a systematic approach by utilizing the SWOT (Strengths, Weaknesses, Opportunities, Threats) and AHP (Analytic Hierarchy Process) methods. This analysis consists of several main steps, which involve data collection, analysis, and evaluation based on relevant criteria (Rachman, 2019; Sylvia & Hayati, 2023).

2.1. Data Analysis Methods

2.1.1 Internal Factor Analysis Matrix/ IFE

Internal Factor Analysis Matrix / IFE

An analysis of the internal environment can be analyzed using the Internal Factors Evaluation (IFE) Matrix to formulate internal factors that have been identified within the framework of the company's strengths and weaknesses (Annisa Nowira et al., 2021).

2.1.2 External Factor Analysis Matrix / EFE

An external environmental analysis can be analyzed using the External Factors Evaluation (EFE) Matrix to formulate external factors that have been identified within the framework of the company's opportunities and threats (Dosen FEBI IAIN Purwokerto, 2019) .

2.1.3 SWOT Matrix

The SWOT matrix is compiled after the IE matrix is compiled. The SWOT matrix is an analysis tool to develop 4 types of alternative strategies. The types of alternative strategies are SO (Strength-Opportunity) strategy, WO (Weakness-Opportunity) strategy, ST (Strength-Threat) strategy, and WT (Weakness-Threat) strategy. Compiling a SWOT matrix requires several stages. (Dosen FEBI IAIN Purwokerto, 2019) including the following:

a. The combination of internal strengths and external opportunities is then documented in the SO cell..
b. The combination of internal weaknesses and external opportunities then documented in the WO cell.
c. The combination of internal strengths with external threats is then recorded in the ST cell.
d. The combination of internal weaknesses with external threats and then record the results in the WT strategy cell.

2.1.4. AHP Analysis

a. Determination of Criteria and Sub-Criteria:
Determine the main criteria relevant to the assessment, such as market potential, production costs, and environmental sustainability. These criteria are then divided into sub-criteria if necessary.
b. Criteria Weight Assessment:
Using the AHP method to assign relative weights to each criterion and sub-criteria. This involves creating a pairwise comparison matrix where each criterion is compared with each other in terms of relative importance.
c. Assessment Data Collection:
Collecting assessment data for each commodity based on predetermined criteria. This assessment can be done through interviews with experts, surveys, or field evaluations.
d. AHP Data Processing:
Calculate the final weight for each commodity using the AHP method to determine the ranking or priority based on predetermined criteria.

2.2 Integration and Evaluation of Results

a. Integration of SWOT and AHP Results:
Integrating the results of the SWOT analysis with the ranking of the AHP method to determine superior horticultural commodities. The results of the SWOT will provide context to the weight values generated by the AHP.
b. Evaluation and Recommendations:
Using the results of integration to provide recommendations on the most superior horticultural commodities. Evaluation is carried out based on the final results and practical considerations of the recommendations

3. Results and Discussion

Based on documents, literature reviews, news published in local media, and conducting a preliminary survey in the research environment, the following internal and external factors were obtained:

1) Internal Factors Are internal conditions within a company or business that can give rise to a company's strengths and weaknesses that the company can fully control or overcome known

IOP Conf. Series: Earth and Environmental Science **1446** (2025) 012052 doi:10.1088/1755-1315/1446/1/012052

weaknesses to be fixed. Some internal factors of Ngawi Regency that influence superior horticultural commodities are as follows:

a. Product quality
b. Cultural superiority
c. Market demand
d. Farmer's experience
e. Seasonal dependency
f. Technology limitations
g. Market limitations
h. Management and Organization

2) External Factors

It is an external condition outside the company that leads to the emergence of a business opportunity or opportunity and can even threaten the company or entrepreneur. Some external factors of Ngawi Regency that influence the superior horticultural commodities are as follows:

a. Market demand is increasing
b. Consumption trends
c. Regulatory support
d. International market access
e. Diseases and pests
f. Climate change and extreme weather
g. Commodity price fluctuations

Additionally, internal factors are categorized as strengths and weaknesses, while external factors are classified as opportunities and threats. The following step is to assign weights IFAS - EFAS and compile a SWOT matrix with the results presented in Table 1. Furthermore, the results of the SWOT strategy priorities are presented in Table 2.

Table 1. IFAS – EFAS matrix

IFAS / EFAS	**Strength** 1. Product Quality 2. Cultural Excellence 3. Farmer Experience 4. Market Demand 5. Natural Resource Potential **Weight: 1,762**	**Weakness** 1. Seasonal Dependence 2. Technology Limitations 3. Market Limitations 4. Management and Organization. **Weight : 0.641**
Opportunity 1. Market Demand Increases 2. Consumption Trends 3. Regulatory Support 4. International Market Access **Weight: 1,941**	**SO Strategy** 1. **Increasing** horticultural production by utilizing farmer experience, regulations, and supporting resources (S1, S2, S3, O2, O4) **Weight: 3,703**	**WO Strategy** 1. **Developing superior varieties** by utilizing efficient technology and marketing (W1,W2,W3,W4,O1,O2,O3) **Weight: 2,582**
Threat 1. Diseases and Pests 2. Climate change and extreme weather 3. Commodity price fluctuations **Weight : 0.679**	**ST Strategy** 1. Maintaining the culture and experience of farmers to **maintain product quality** by utilizing the potential of Natural Resources (S1, S4, S5, T1, T2) **Weight: 2,441**	**WT Strategy** 1. **System improvement** by improving technology, management, and organization at the farmer level. (W2,W3,W4,T1,T2) **Weight: 1,320**

Table 2. SWOT strategy alternative priority order

Priority	Strategy	Weight of Value
I	*Strength – Opportunity (SO)*	3,703
II	*Weakness – Opportunity (WO)*	2,582
III	*Strength – Threat (ST)*	2,441
IV	*Weakness – Threat (WT)*	1,320

The results of the combination of I-E factors show the results with the highest value, namely the combination of Strengths and Opportunities with a strategy to increase horticultural commodity production by utilizing farmer experience, regulations, and various supporting resources. This strategy is considered the most optimal because Ngawi Regency has environmental conditions that support horticultural plant cultivation.

Based on the SWOT analysis, the results show that the combination of SO strategies is a strategy with the highest value, but this strategy is not necessarily the best strategy that can be implemented sustainably. So, it is necessary to conduct further analysis using the AHP analysis method to determine the priority sequence of strategies that can be implemented to increase the superiority of horticultural commodities in Ngawi Regency and determine which commodities have the best potential to be developed in Ngawi Regency.

3) Hierarchical Arrangement

The structure of the hierarchy is the key element of the AHP model, because it will be the basis for respondents to give assessments/opinions more simply. By using the hierarchy in the AHP model, it is expected to simplify and simplify complex problems.

The AHP hierarchy in this study is composed of 3 levels, with the highest peak of the hierarchy being the Superior Horticultural Commodity as the goal of AHP. The AHP hierarchy used as a whole is presented in Figure 3.1 below, and the results of the AHP analysis values are presented in Table 3.1.

Furthermore, at the target level, the 3 strategies with the highest values in the SWOT matrix are used, this is because the most optimal strategies that can be implemented are the three strategies with the highest values. While at the criteria level, the 3 superior horticultural commodities found in Ngawi Regency are used.

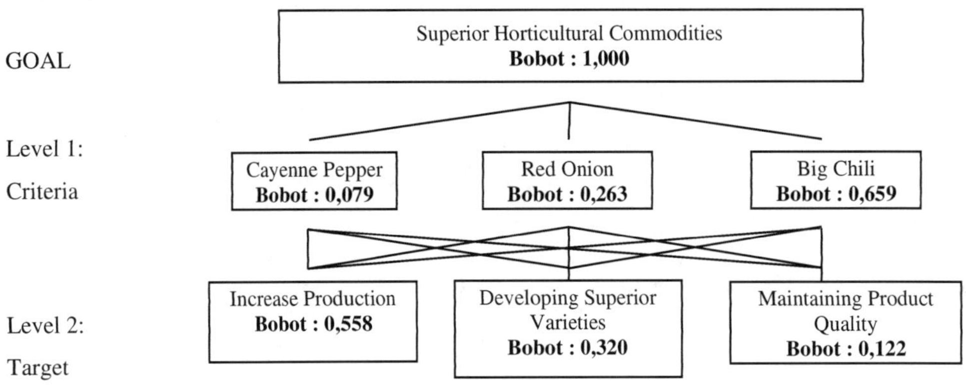

Figure 1. AHP hierarchy and values

Table 3. Target priority with AHP method

Level	DEFINITION	WEIGHT
1	**Criteria**	
	1. Cayenne pepper	**0.079**
	2. Red onion	**0.263**
	3. Big Chili	**0.659**
2	**Target**	
	1. Increase Production	**0.558**
	2. Developing Superior Varieties	**0.320**
	3. Maintaining Product Quality	**0.122**

Based on the results from the AHP analysis, it can be concluded that the target that is the main priority is to increase the production of superior commodities with the commodity that is the main choice to be a priority is large chili. The AHP analysis value can be stated to have met the requirements with an Overall Inconsistency Ratio of 0.3, which means that the analysis and assessment are declared consistent. The Inconsistency Ratio value $\leq$ 0.1 means that the decision taken by the key person respondent in determining the priority scale is acceptable. The smaller the Inconsistency Ratio value, the more consistent the respondent is in determining the priority scale (Saaty, 2000).

Increasing production is the target that has the highest value so respondents assessed that in maintaining the sustainability of commodity superiority, it is expected to maintain and increase production with a value of 0.558, and for horticultural commodities that are considered to have the potential as superior commodities are large chilies with a value given by respondents of 0.659.

4. Conclusion

Horticultural commodities from Ngawi Regency that are considered to be superior horticultural commodities are large chili commodities with an AHP value of 0.659, and the strategy that can be carried out to maintain the sustainability of this commodity is to increase production with an AHP value of 0.558. This can be a recommendation for Ngawi Regency to be able to increase attention to large chili commodities so that they can survive as superior horticultural commodities by maintaining and improving both in terms of quality and quantity.

5. Acknowledgment

Acknowledgments are addressed to the DIPA Politeknik Negeri Jember, SP DIPA-023.18.2.677607/2024 for the funding provided to the researchers through of Affirmation Research scheme for the 2024 Budget Year.

References

[1.] Annisa Nowira, P., Puspa Sari, R., Momon, A. S., Industri, T., Singaperbangsa Karawang Jl Ronggo Waluyo, U. H., Telukjambe Timur, K., & Karawang, K. (2021). Strategi Persaingan Jamu Gunanty Menggunakan Matriks: Internal Eksternal, Bowman Strategy, Grand Strategy Dan Profil Kompetitif. In JISI: JURNAL INTEGRASI SISTEM INDUSTRI (Vol. 3).

[2.] Berlianantiya, M. (2017). Pertumbuhan ekonomi dan ketimpangan pembangunan ekonomi antar wilayah kebijakan pembangunan di Provinsi Jawa Timur. EQUILIBRIUM : Jurnal Ilmiah Ekonomi Dan Pembelajarannya, 5(2), 163. https://doi.org/10.25273/equilibrium.v5i2.1544

[3.] Brojonegoro & Bambang. 1992. AHP (the Analytical Hierarchy Process). Jakarta: Pusat Antar University – Studi Ekonomi Universitas Indonesia.

[4.] Dhandy, R., 2022. Penerapan Metode Location Quetiont (LQ) dalam Penentuan Komoditas Unggulan Tanaman Pangan di Kabupaten Sigi. JMAA, Vol. 2 No. 2.

[5.] Dosen FEBI IAIN Purwokerto, S. (2019). Implementasi Matrik Efe, Matrik Ife, Matrik Swot Dan Qspm Untuk Menentukan Alternatif Strategi Guna Meningkatkan Keunggulan Kompetitif Bagi Usaha Produksi Kelompok Buruh Pembatik Di Keser Notog Patikraja Banyumas. In TAHUN (Vol. 3, Issue 1).

[6.] Junari, T., Rustiadi, E., & Mulatsih, S. (2020). Identifikasi Sektor Industri Pengolahan Unggulan Propinsi Jawa Timur (Analisis Input Output). TATALOKA, 22(3), 308–320. https://doi.org/10.14710/tataloka.22.3.308-320

[7.] Nurjannah, D., & Mashuri. (2020). Analisis Swot Sebagai Strategi Meningkatkan Daya Saing (Studi Pada PT. Bank Riau Kepri Unit Usaha Syariah Pekanbaru). Jurnal Perbankan Syariah, 1(1), 97–112. https://ejournal.stiesyariahbengkalis.ac.id/index.php/jps

[8.] Putra, M. E. Y. 2022. Peningkatan Kinerja Sektor Hulu Kelompok Tani Kopi Robusta Desa Sidomulyo Kabupaten Jember. JMAA, Vol. 2 No. 1.

[9.] Rachman, R. (2019). Penerapan Metode Ahp Untuk Menentukan Kualitas Pakaian Jadi Di Industri Garment. Jurnal Informatika, 6(1), 1–8. https://doi.org/10.31294/ji.v6i1.4389

[10.] Saaty, L. Thomas. 2000. Decision Making For Leaders. University of Pittsburgh

[11.] Sylvia, R., & Hayati, D. (2023). Analisis Swot Dalam Menentukan Strategi Pemasaran Produk Indosat Pada Pt X Stie Nasional Banjarmasin (Vol. 16, Issue 1).

IOP Conf. Series: Earth and Environmental Science **1446** (2025) 012053 doi:10.1088/1755-1315/1446/1/012053

The analysis of consumer perception on quality of canned sardines with chili sauce TeFa Polije used importance performance analysis method

A Brilliantina[1*], I R A Sasmita[1], M Mardiyanto[1], E Kurniawati[1], P T Fadila[1]

[1]Department of Agriculture Technology, Politeknik Negeri Jember, Indonesia

*Email : aulia_b@polije.ac.id

Abstract. Polije had a TEFA Canning unit that produced canned sardines. The TEFA Canning unit only produced without studying and knowing consumer satisfaction with the quality of the products produced. This study aimed to identify key attributes in quality improvement of Canned Sardines With Chili Sauce TeFa Polije based on consumer needs, and to formulate a strategy to improve the quality of canned sardines with chili sauce TeFa Polije products. The methods used in this study were Importance Performance Analysis (IPA) and SWOT analysis. This study elaborated on 100 respondents, who were residents of Jember Regency. The results of the analysis showed that the overall performance level of canned sardines with chili sauce TeFa Polije (3.78) still did not meet the level of importance expected by consumers (4.00). Based on the IPA matrix, priority attributes for improvement were obtained: availability of permits, availability of product information, advertising promotion, and a well-known brand. After weighting and scoring through the IFE (2.83) and EFE (3.25) matrix, the IE matrix showed the position of the canned sardines with chili sauce TeFa Polije in diagram I, which could be managed by aggressive strategies. The SWOT matrix analysis resulted in six suggested main priorities.

1. Introduction

Jember State Polytechnic has TEFA Fish Canning which produces sardines in canned packaging. One of the canned products produced by TEFA Fish Canning is sardine seasoned with chili sauce. So far, TEFA Fish Canning has only produced without studying and knowing consumer satisfaction with the quality of the products produced. The quality of food consumed will determine the level of physical growth, including intelligence. Customers seeking food naturally expect quality products. According to [1], food quality significantly impacts consumer purchasing decisions, meaning that improved food quality can lead to more positive buying choices. High-quality food products play a crucial role in customer satisfaction, so enhancing and sustaining this quality can effectively serve as a foundation for marketing strategies [2].

IOP Conf. Series: Earth and Environmental Science **1446** (2025) 012053 doi:10.1088/1755-1315/1446/1/012053

The marketing mix strategy, known as the 7P framework—product, price, place, promotion, people, process, and physical evidence—serves as a method to analyze the key attributes that food products can offer. This approach helps in identifying the characteristics that influence consumer purchase decisions. The Importance Performance Analysis (IPA) method, due to its advantages over other techniques, is useful for determining which product or service features should be improved or reduced to maintain customer satisfaction. Benefits of IPA include its ease of use, straightforward result interpretation, and low cost. According to [3], the IPA method is a simple tool that assesses both the importance of attributes and their actual performance, making it highly effective for designing targeted marketing programs.

The IPA approach will yield a number of characteristic factors, which will subsequently be examined to produce a strategy for improving the packed chilli sauce seasoned sardines' quality. SWOT analysis is a methodical way to uncover different aspects so that strategies can be developed. Generally speaking, reaching objectives in an effective and efficient manner is the first step in choosing the best strategy for an industry. One of the most fundamental analytical techniques is the SWOT analysis, which is helpful for examining a subject or issue from four angles. The analysis's conclusion is a guideline or advice to maximise opportunities and preserve strengths while reducing weaknesses and averting risks. When applied appropriately, SWOT analysis can reveal hidden or overlooked aspects [4]. The aim of this study is to ascertain the degree of consumer satisfaction, identify critical attributes to enhance the quality of chilli sauce seasoned sardines in cans based on consumer preferences, and develop a plan to enhance the quality of chilli sauce seasoned sardine products in cans using the IPA-based SWOT method.

2. Methods

The research was conducted from May to August 2024. The research location was in Jember Regency, taking all sub-districts in Jember Regency. The methodology in this research includes the following stages:

2.1. Problem identification

During this phase, researchers define the problems to be studied and provide a background of the research. The author provides an explanation of the rationale behind the research's execution in the background section, citing the issues raised by the study's object and its goals.

2.2. Literature review

Using theories, concepts, and material from books or journals to support study is part of this stage for researchers [5].

2.3. Determine and compile instruments.

This step comprises the creation of a questionnaire, the questions of which are derived from the operational variable indicators in this study. Customers' expectations of the marketing mix and perceived performance are gauged through questions. With five possible replies, the Likert scale was selected as the scale to be employed in this investigation through the questionnaire. In this study, the sampling amount was calculated using the Slovin Formula to determine the number of samples. 100 respondents made up the sample, which was selected using non-probabilistic sampling methods.

2.4. Validity and reliability test

According to [3], a validity test measures the extent to which an instrument is accurate and reliable. This test uses the product-moment correlation method, where each item in a variable is compared with the overall score. If the r-value (correlation coefficient) has a probability value of ≤ 0.05, the item is deemed valid. Reliability is defined as steadiness, consistency, correctness or accuracy, predictability, and predictability [6]. Reliability is the degree to which measurement results may be relied upon. The

Cronbach Alpha method is used to test the dependability of measurement devices. It can be concluded that the questionnaire instrument is dependable if the alpha coefficient is less than or equal to 0.6 [7].

2.5. Data collection

Data for this study will be gathered by the distribution of a questionnaire with multiple choice questions. Manual distribution is the method used. The questionnaire includes questions on respondent demographics, consumer satisfaction, expectations, and perceptions related to each variable. Variables and attributes are outlined as shown in Table 1. Expectations are measured through a distributed questionnaire using the Importance Performance Analysis (IPA) method. Responses are rated on a 1 to 5 scale, where 1 represents "strongly disagree" (STS) and 5 represents "strongly agree" (SS).

Table 1. Variables and attributes of the 7P marketing mix of sardines seasoned with chili sauce in cans

Dimension Attributes	Product Attribute Variables
Products	• Taste of sardine seasoned with chili sauce in a can
	• Succulence level of chili sauce seasoned sardines in cans
	• The aroma of sardines seasoned with chili sauce in a can
	• Texture of chili sauce-seasoned sardines in a can
	• Packaging design
	• Permit availability
	• Availability of product information
Price	• Product price of chili sauce seasoned sardines in cans
Place	• Ease of obtaining chilli sauce seasoned sardine products incans
Promotion	• Advertising promotion
	• Recognized brands
People	• Friendliness and courtesy of the producer
	• Manufacturer's appearance
Process	• Speed in service
Physicalevidence	• Completeness of tefa *fish canning* facility
	• Cleanliness and tidiness of tefa *fish canning*
	• Safety tefa *fish canning*

2.6. Data processing

Importance Performance Analysis is the technique used for data analysis. The Importance Performance Analysis approach has the following stages: [8]:

a. By comparing the performance score with the importance score, one may ascertain the degree of agreement between the importance and performance levels for the seasoned chilli sauce-seasoned sardines. The level of conformance formula that is applied is [9]:

IOP Conf. Series: Earth and Environmental Science **1446** (2025) 012053 doi:10.1088/1755-1315/1446/1/012053

$$\text{Tki} = \frac{Xi}{Yi} X\ 100\% \quad \text{.......(1)}$$

Description:
Tki = Level of conformity
Xi = Performance level assessment score
Yi = Importance level assessment score

b. Calculate the average for each attribute perceived by consumers, with the formula:

$$\overline{\text{XI}} = \frac{\Sigma\text{XI}}{n} \qquad \overline{\text{YI}} = \frac{\Sigma\text{YI}}{n} \qquad \text{.......(2)}$$

Description:
$\overline{\text{XI}}$ = Average score of product performance level Xi
$\overline{\text{YI}}$ = The average score of the importance level of the product Yi
n = Number of respondents n

c. Calculate the average of all attributes of the level of importance (Y) and performance (X) which are the boundaries on the Cartesian diagram, with the formula:

$$\overline{\text{X}} = \frac{\Sigma\text{XI}}{k} \qquad \overline{\text{Y}} = \frac{\Sigma\text{YI}}{k} \qquad \text{.......(3)}$$

Keterangan:
$\overline{\text{X}}$ = average score of implementation level or product performance of all factors or attributes X
$\overline{\text{Y}}$= the average level of importance of all attributes that affect product quality Y k = The number of attributes that can affect product quality k

d. Mapping into a Cartesian diagram as shown in Figure 1.

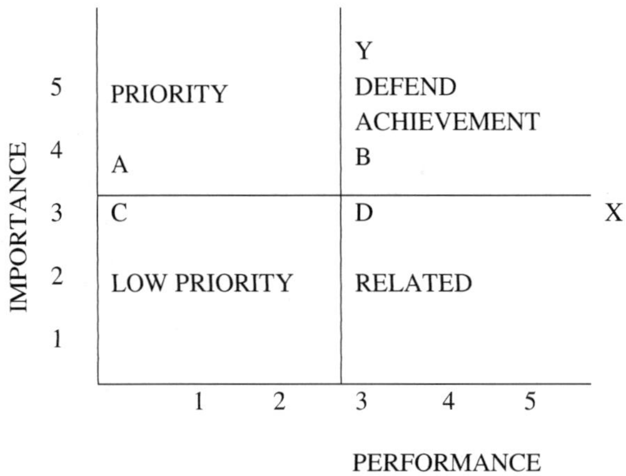

Figure 1. Cartesian diagram

2.7. Internal and external factor analysis

The information gathered will be utilised to calculate the importance of external and internal elements on business development. Analysing internal factor evaluation (IFE) and external factor evaluation (EFE) allows for the detection of internal and external influences.

2.8 Analysis of data processing results and conclusions

In this final step, the author evaluates the outcomes of all data processing and draws conclusions about the findings of the conducted research.

3. Results And Discussion

The test results indicate that the relationship is valid, as the probability value for the r count (p) is ≤ 0.05, confirming a valid correlation between the item score and the overall score. It is possible to establish whether a relationship is dependable by using Cronbach's Alpha to examine the relationship between the variable's items. The correlation coefficient is higher than the predetermined reliability value of 0.6.

3.1. Suitability level

The priority for improving each characteristic within the product quality variable is established by comparing scores for performance level and importance level, referred to as the level of conformance. Table 2 displays the results of the computation of the degree of concordance between the performance and importance levels.

Table 2. Calculation of the level of conformity between the level of performance and the level of importance

Product Attribute Variables	Performance	Importance	GAP
Products			
1. Taste of sardine seasoned withchili sauce in a can	4,6	4,33	-0,27
2. Succulence levlel of chili sauce seasoned sardines in cans	4,43	4,39	-0,04
3. The aroma of sardines seasoned with chili sauce in a can	4,03	4,45	0,43
4. Texture of chili sauce seasoned sardines in a can	4,33	4,33	0
5. Packaging design	3,21	4,53	1,32
6. Permit availability	1,03	5	3,97
7. Availability of product information	2,31	5	2,69
Price			
1. Product price of chili sauceseasoned sardines in cans	3,11	4,36	1,25
Place			
1. Ease of obtaining chili sauce seasoned sardine products in cans	2,63	4,31	1,68
Promotion			
1. Advertising promotion	2,45	4,39	1,93

Product Attribute Variables	Performance	Importance	GAP
2. Recognized brands	2,43	4,31	1,88
People			
1. Friendliness and courtesy of theproducer	4,4	5	0,6
2. Manufacturer's appearance	4,69	4,24	-0,45
Process			
1. Speed in service	3,89	4,63	0,73
Physical Evidence			
1. Completeness of tefa fish canning facility	4,67	5	0,6
2. Tefa fish canning hygiene	4,4	5	0,6
3. Safety tefa fish canning	4,4	5	1,04
Average	3,59	4,60	1,12

Source: Results of data analysis (2024)

The analysis's findings indicate that, overall, performance is greater than importance (3.59 for performance < 4.6 for importance), indicating that, on average, there is still a gap between quality and expectations (with an average gap value of 1.12), indicating that it is still below customer expectations and that more work is needed to bring it closer to what customers want. An Importance Performance Analysis (IPA) cartesian diagram analysis was then performed to determine the priority scale of the items that were already in existence.

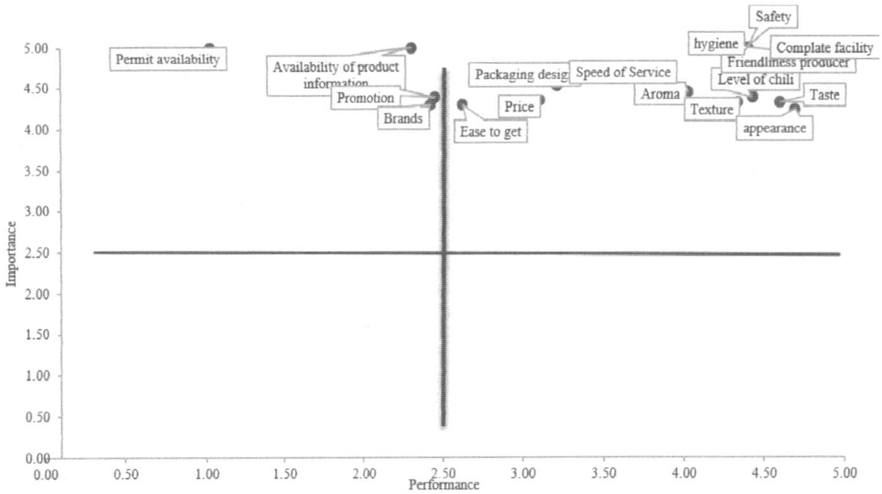

Figure 2. IPA mapping of Tefa Fish Canning's sardine chili sauce

From the IPA mapping image above, it can be analyzed with the following results [10]:

1. *Top Priority (Concentrate Here) quadrant A.* Items that are included in quadrant A are taste, spiciness, aroma, texture, packaging design, price, ease of obtaining products, friendliness of producers, appearance of producers, speed of service, completeness of facilities, cleanliness of production, and safety of tefa fish canning. Factors located in this quadrant are considered important or expected by consumers. According to (1), writing the shelf life on packaging labels affects consumers' buying interest
2. *Keep up the good work quadrant B.* Items located in quadrant B—such as licensed availability, product information availability, advertising promotions, and recognized brands—are deemed important and expected to support customer satisfaction. Therefore, these factors are essential to maintain as they significantly contribute to customer satisfaction.

3.2. IFE (Internal Factor Evaluation) Matrix

Interviews and validation conducted using the IPA method identified 17 indicators as internal factors (strengths and weaknesses) for canned chili sauce seasoned sardines. As shown in Table 3, the total internal matrix score is 2.83. As for these results, it can be seen that chili sauce seasoned sardines have a higherstrength value than their weaknesses.

Table 3. Internal factor matrix of chili sauce seasoned sardines

NO	INTERNAL FACTORS/ STRENGTH	Total	Rating	Weight	WeightxRating
1	Flavor	20	4	0,07	0,27
2	Spiciness level	20	4	0,07	0,27
3	Aroma	17	3	0,06	0,20
4	Texture	18	4	0,06	0,22
5	Packaging Design	10	2	0,03	0,07
6	Price	16	3	0,05	0,18
7	Ease of Getting	15	3	0,05	0,15
8	Manufacturer friendliness	20	4	0,07	0,27
9	manufacturer appearance	15	3	0,05	0,15
10	speed of service	17	3	0,06	0,20
11	Completeness of facilities	12	2	0,04	0,10
12	production hygiene	20	4	0,07	0,27
13	Safety tefa fish canning	20	4	0,07	0,27
NO	WEAKNESSES	Total	Rating	Weight	WeightxRating
1	Permit Availability	20	1	0,07	0,06
2	Availability of Product Information	20	1	0,07	0,06
3	Advertising Promotion	15	1	0,05	0,03
4	Recognized brands	17	1	0,06	0,04
TOTAL		292		1,00	2,83

Source: Result of data analysis, 2024

3.3. EFE (External Factor Evaluation) matrix

Based on the interviews and validation, six indicators were identified, which constituted external indicators (opportunities and threats) for sardines seasoned with chili sauce. According to the calculations in Table 3, the total external matrix is 3.25 which has a positive value, where the value of opportunities is still above the threat. With this value, it means that chili sauce seasoned sardines still have many opportunities compared to threats.

Table 4. External factor matrix of sardines seasoned with chili sauce

NO	EXTERNAL FACTORS OPPORTUNITIES	Total	Rating	Weight	WeightxRating
1	The development of technology isadvancing	8	2	0,09	0,15
2	Consumer trust as an educationalinstitution	17	3	0,19	0,66
3	Open market	9	2	0,10	0,18
4	Population growth is getting higher	20	4	0,23	0,91
NO	THREAT/THREATS	Total	Rating	Weight	WeightxRating
1	Fluctuating production costs	14	3	0,16	0,45
2	Many competitors	20	4	0,23	0,91
TOTAL		88		1,00	3,25

Source: Results of data analysis, 2024

3.4. IE Matrix for Positioning

The values acquired from the IFE and EFE matrices will be entered into the Internal-External matrix to determine the position of chili sauce seasoned sardines. This Internal-External matrix shows the production in a four-cell layout. This IE matrix is built around two important dimensions: the total IFE weight score on the X axis and the EFE weight score on the Y axis [12]. According to the IFE and EFE Matrix, the position on the X axis is 2.83, and the position on the Y axis is 3.25.

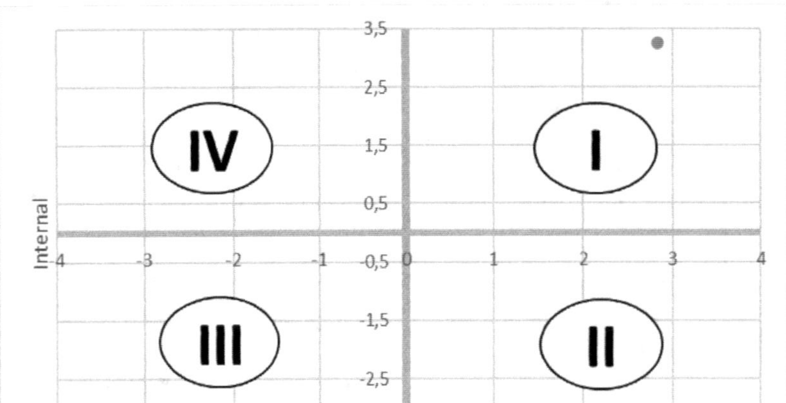

Figure 3. IE Matrix for sardines in chili sauce (Source: [13] and Results of data analysis, 2024)

3.5. SWOT Matrix

The SWOT matrix, which considers strengths, weaknesses, opportunities, and threats, can be used to help build strategies [14]. The SWOT matrix [15] includes methods for Strengths Opportunities (SO), Weakness Opportunities (WO), Strengths Threats (ST), and Weakness Threats (WT). By examining the relationships between the SWOT components, the qualitative method to SWOT analysis might generate alternate strategies for businesses [16]. These alternate tactics each have a unique method for raising an organization's performance [17]. It is possible to determine that the sardines with chilli sauce are in Quadrant I based on the study performed using the IE matrix. Where in quadrant I is the company located? where producers who fit the definition of "aggressive" [4] are among the businesses in the quadrant. It is appropriate to use the following tactics in this cell:

- SO Strategy
  1. Ensure high-quality chili sauce seasoned sardines in cans created with cutting-edge technology and enhance marketing reach.
  2. Immediately apply for a distribution license to expand marketing.
- WO Strategy
  1. Increase cash for the sardine chili sauce business and promote it more effectively by giving training.
- ST Strategy
  1. Develop production with new product innovations, for more variety
  2. Increase production capacity.
- WT Strategy
  1. Create good cooperation with investors.

4. Conclusions and policy recommendations

4.1. Conclusion

According to the findings, the IPA approach can be used to gauge how satisfied customers are with the chilli sauce-seasoned sardine goods found in TEFA Fish Canning products from Jember State Polytechnic. Quadrants A and B contain every item related to a quality attribute. This quadrant's components are seen as significant and are anticipated to support consumer satisfaction. Cans of seasoned sardines with chilli sauce are recognised to be in quadrant I based on SWOT analysis. where the producers in quadrant I are located. In this case, producers in quadrant I are positioned aggressively.

4.2. Policy Recommendations

The policy recommendations that can be given to improve the quality of sardine fish are by increasing production capacity and immediately taking care of BPOM licensing so that it has a wide enough market reach. In addition, the Jember State Polytechnic Cooperative is expected to help monitor product quality so that it can increase promotion and cooperation with investors.

Acknowledgements

The authors would like to thank the Centre for Research and Community Service of Jember State Polytechnic for providing grant funding for the year 2024 funding.

References

[1] Margaretha F, Edwin J. Analisa Pengaruh Food Quality dan Brand Image Terhadap Keputusan Pembelian Roti Kecik Ganep's Di Kota Solo. J Manaj Pemasar [Internet]. 2012;Vol.1(1):hal.1-6. Available from: https://www.mendeley.com/catalogue/9ad89f91-8fe6-33fb-b619-b5077c29ac2b/?utm_source=desktop&utm_medium=1.19.8&utm_campaign=open_catalog&userDocumentId=%7B889ae004-10f5-300d-bcd8-03ae07d9a6ee%7D

[2] Brilliantina A, Novitasari EK, Pratiwi BY, Sasmita IRA. VALUE ADDED ANALYSIS OF SWEET

POTATO CHIPS PROCESSING IN JEMBER REGENCY. 2019;624–8.
[3] Kurnianto MF, Wibowo MJ, Hariono B, Wijaya R, Brilliantina A. The Analysis of Consumer Perception on Quality of Soybean Milk Used Importance Performance Analysis Method. IOP Conf Ser Earth Environ Sci. 2020;411(1).
[4] Brilliantina A, Istiqomah N. Strategi Pengembangan Industri berbasis Ubi Jalar Ungu dengan Metode SWOT dan ANP di Kabupaten Bondowoso. J Ilm Inov. 2021;21(1):13–7.
[5] Hermanuadi D, Brilliantina A, Novitasari EK. Value added analysis of cassava "tape" supply chain in Bondowoso regency. IOP Conf Ser Earth Environ Sci. 2021;672(1).
[6] Hermanuadi D, Brilliantina A, Novitasari EK. Determination of Agro-Industry Area Based on Cassava Commodity in Bondowoso Regency. IOP Conf Ser Earth Environ Sci. 2018;207(1).
[7] Bakri A, Supriyono S, Prasetyo B, Hariono B, Kurnianto MF, Brilliantina A. Performance Analysis and Planning Strategies for Improvement of Banana Chips in Jember Regency. Proc 2nd Int Conf Soc Sci Humanit Public Heal (icosh 2021). 2022;645(Icoship 2021):51–6.
[8] Septiani E, Pujiyanto E, Hisjam M. A Design to Improve the Quality of OVO Electronic Money Payment Services in Tokopedia Using IPA and PGCV. J Tek Ind. 2020;21(2):153–62.
[9] Yang H-C, Kwon W-T, Kim D-H. An IPA-based study on the sextic industry's potential for continued development from producers' viewpoint. J Distrib Sci. 2014;12(11):15–23.
[10] Wong MS, Hideki N, George P. The use of importance-performance analysis (IPA) in evaluating Japan's e-government services. J Theor Appl Electron Commer Res. 2011;6(2):17–30.
[11] Brilliantina A, Wardani DK, Fadhila PT, Hariono B, Wijaya R. Accelerated shelf life test method with arrhenius approach for shelf life estimation of tongkol 'euthynnus affinis' balado in cans. In: IOP Conference Series: Earth and Environmental Science. IOP Publishing; 2022. p. 12038.
[12] Ghazinoory S, Abdi M, Azadegan-Mehr M. SWOT methodology: a state-of-the-art review for the past, a framework for the future. J Bus Econ Manag. 2011;12(1):24–48.
[13] Yaqin, Anfal Ainul. "Analisis Swot Dalam Strategi Pengembangan Usaha Kerupuk Rumahan Di UD. Sumber Abadi Tanggulangin." JISO: Journal of Industrial and Systems Optimization 4.2 (2021): 81-87.
[14] GURL E. SWOT analysis: A theoretical review. 2017;
[15] Kamilia N, Setiyarini T. Strategi Pengembangan Produk Melalui Analisis Strength Weakness Opportunity Threat (SWOT) Pada Industri Rumahan Kerupuk Udang Kwanyar. J Kaji Ilmu Manaj. 2022;1(3):291–8.
[16] Novitasari EK, Hermanuadi D, Brilliantina A. Application of Swot and Anp Methods in Order To Select the Agroindustrial Development Strategy Based on Tapai in Bondowoso. Food Sci J. 2020;2(2):53.
[17] Kasutjianingati K, Wahyono A, Brilliantina A, Novitasari EK. SWOT and Analytical Network Process (ANP) Analysis for Robusta Coffee Bean Development Strategy in Panti District, Jember Regency. IOP Conf Ser Earth Environ Sci. 2020;411(1):0–7.

IOP Conf. Series: Earth and Environmental Science **1446** (2025) 012054 doi:10.1088/1755-1315/1446/1/012054

Development Strategy of Agribusiness Teaching Factory Pilot at PSDKU Sidoarjo using SWOT Analysis and Lean Canvas Business Model

A H H Slamet[1)], D N Mutmainah[2)], S Brillyantina[3)], S A Wulandari[4)], R Dhandy[5)], S E N Julianti[6)]

[1,2,3,4,5,6]Agroindustry Management, Department of Agribusiness Management, Politeknik Negeri Jember, Jalan Mastrip PO BOX 164, Jember 68121, Indonesia
[*]Corresponding author's email: ahmad.haris@polije.ac.id

Abstract. The Teaching Factory is a strategic initiative of Politeknik Negeri Jember for the year 2020-2024, implementing the one department, one teaching factory policy. The TEFA most suitable for establishment at Campus 4 is TEFA Agribusiness. This teaching factory adopts the structure of a minimarket, encompassing all aspects of agribusiness operations from processing to product marketing. The objective of this research was to develop a TEFA Agribusiness design strategy using the lean canvas business model and SWOT analysis. The research findings indicated that PSDKU Sidoarjo's agribusiness teaching factory pilot has significant viability for development. The SWOT analysis classified the teaching factory startup as being in cell I, which represents growth and build. The techniques for implementation include intense operations (market penetration, market development, and product creation) and integrative approaches (backward integration, forward integration, and horizontal integration).

1. Introduction

Labor productivity is still a concern in the context of Indonesian employment. Indonesian labor productivity is still lower than the average labor productivity in ASEAN. The percentage of Indonesian labor productivity is 74.4%, lower than ASEAN (78.2%) [1]. One of the contributing factors is the need for further alignment between the education and industry sectors. Therefore, one of the main objectives of the Directorate General of Vocational Education is to integrate the industry sector into educational activities by establishing a Teaching Factory (TEFA) on campus [2]. *Teaching factory* is an educational concept that combines academic learning with real-world industrial practices. This concept bridges the gap between theoretical knowledge and practical skills by providing students with hands-on experience in a factory-like environment. This model allows students to engage in real-life projects, understand production processes, and develop industry-specific technical competencies. TEFA can also be a practical tool for lecturers to carry out the Tridharma: teaching, researching, and serving the community [3]. PSDKU Sidoarjo is located outside Politeknik Negeri Jember's main campus. PSDKU Sidoarjo has quite an enormous potential for the establishment of a Teaching Factory. It is proven by the many industries in Sidoarjo Regency. In Sidoarjo Regency itself, there are 961 large and medium-sized companies spread across almost all sub-districts [4]. With the establishment of TEFA at PSDKU Sidoarjo, a relationship, and collaboration between the industrial sector and Politeknik Negeri Jember can be established.

The most suitable TEFA to be established in PSDKU Sidoarjo is TEFA Agribusiness. This TEFA adopts a minimarket structure, which covers all aspects of agribusiness operations from processing to product marketing. The establishment of TEFA Agribusiness aligns with the competencies available at PSDKU Sidoarjo, namely the Agro-Industry Management Research Program and the Information Technology Program. TEFA Agribusiness also aligns with the local potential in Sidoarjo Regency, which has superior commodities in fisheries agribusiness, namely milkfish and shrimp. Total milkfish production 2021 reached 34,548,900 kg, while shrimp production was 9,992,098 kg [5]. Other potentials are the many existing industries, as well as the position of Sidoarjo Regency as a buffer city in East Java [6].

The establishment of TEFA agribusiness at PSKDU Sidoarjo is very promising. Several aspects that can be considered are the location of the campus in the center of Sidoarjo, the potential for abundant raw materials, and the availability of human resources that can support the development of TEFA. The development of TEFA agribusiness is expected to be a learning medium according to industry standards, especially in fishery product processing. Furthermore, this TEFA is expected to be a bridge for various parties, both government and private institutions, in developing business cooperation in Sidoarjo Regency.

This study aims to develop a strategy for designing TEFA Agribusiness using the lean canvas business model and SWOT (Strengths, Weaknesses, Opportunities, and Threats) analysis. In assessing the design of TEFA Agribusiness, a business model is needed to evaluate the business conditions that will be developed. The lean canvas business model is very suitable for start-up businesses or businesses in the development stage. This business model follows the problems and needs of the users and consumers that will be targeted. The lean canvas business model can be used to determine the direction of the company and determine the competitive advantages that exist in a business to be able to compete [7]. SWOT analysis is one method that is quite good, effective, and efficient as a tool in formulating business strategies. SWOT analysis will be carried out to formulate a development strategy. This stage aims to determine the strengths, weaknesses, opportunities, and threats in TEFA Agribusiness. Then, formulate the right strategy to overcome weaknesses and challenges while maximizing strengths and opportunities [8].

2. Method

This research was conducted at PSDKU Sidoarjo. This research employs a qualitative descriptive approach, using the lean canvas business model and SWOT analysis to assess strengths, weaknesses, opportunities, and threats. The research utilizes both primary and secondary data. Primary data is obtained directly via surveys and interviews with specialists and pertinent parties based on the research requirements, while secondary data is gathered from numerous relevant materials. The sampling technique used is purposive sampling, with a predetermined number of respondents totaling five, consisting of academics, government, and industry [9]. Respondents from this research were lecturers in Agroindustry Management at PSDKU Sidoarjo, the Sidoarjo Fisheries Service, UPT for Agribusiness Development of Food Crops and Horticulture as representatives of the East Java Province Agriculture and Food Security Service, and PT. ATINA (a company that processes and exports frozen shrimp and seafood).

This study involved a carefully selected group of respondents from diverse sectors, each contributing specialized knowledge relevant to the research topic. Lecturers are experts in the theory and practice of agricultural business management, production processes, and industry trends. Their academic background provides a comprehensive understanding of the strategic frameworks and models, such as the Lean Canvas and SWOT analysis, which are central to this research. The Fisheries Service is in managing and supporting the local fisheries sector. Their expertise is vital for understanding the dynamics of the local fisheries industry, including regulatory challenges, sustainability practices, and market trends, particularly in Sidoarjo. Their input helps link the theoretical aspects of the study with real-world fisheries management practices, enhancing the relevance of the findings for local policy-making. UPT for Agribusiness Development of Food Crops and Horticulture provides a regional perspective on agribusiness development, especially concerning food security and agricultural policies in East Java. As a critical player in the seafood processing and export industry, PT. ATINA brings practical insights from the business side of the agroindustry. PT. ATINA's input provides valuable perspectives on business operations, market access, product development, and international trade dynamics. Limiting the sample size to five respondents allows for a more focused, detailed exploration of the topic, ensuring the study achieves data saturation while maintaining quality and relevance. The insights gained from these experts are crucial for understanding the agroindustry's business side and contribute to the research's academic and practical value.

The research stages begin with an analysis of the issues in the agribusiness teaching factory pilot at PSDKU Sidoarjo. The problem analysis refers to the absence of establishment and related research on TEFA at PSDKU Sidoarjo. The subsequent phase involves the acquisition of primary and secondary data. The primary data comprises interviews, questionnaires, focus group discussions

(FGD), and observations. In contrast, the secondary data comprises supporting data from literature reviews that were acquired through eBooks, journal articles, and other sources of information.

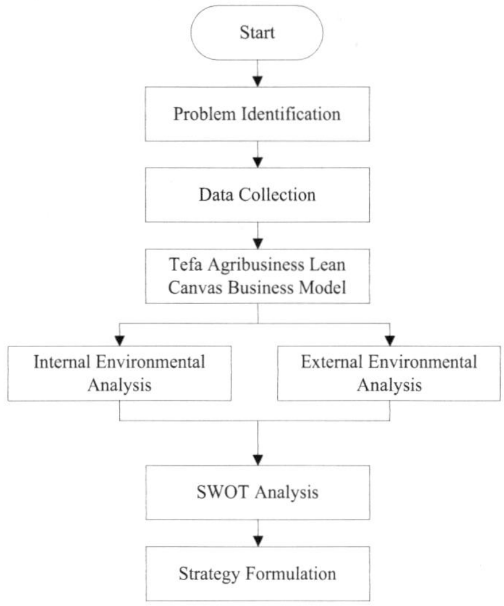

Figure 1. Research Stages

The third stage involves the descriptive statistical analysis of the business model, which examines the data by portraying or illustrating the acquired information as it exists. The business model analysis was conducted using interview and observation methods. The developed business model is the lean canvas, a modification of the business model canvas to facilitate the formation of business models for startups. The lean canvas consists of nine sections, almost the same as the business model canvas. However, several sections and explanations in the lean canvas differ from the business model canvas.

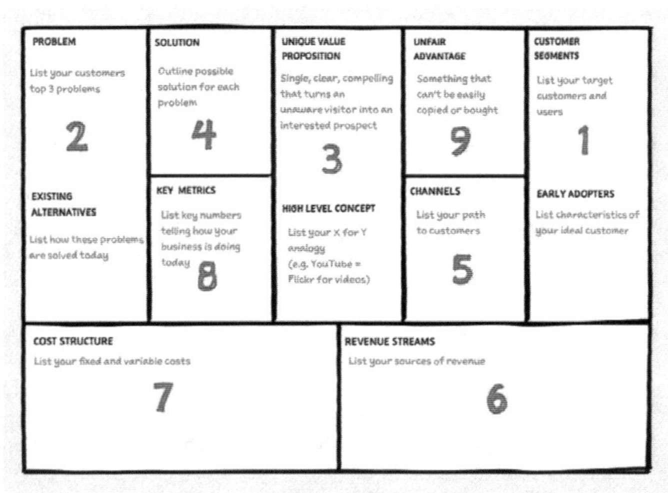

Figure 2. *Lean Canvas Template* [10]

3. Result and Discussion

A. Lean Canvas Business Model of Agribusiness TEFA at PSKDU Sidoarjo

The lean canvas business model was developed to design the agribusiness teaching factory at PSDKU Sidoarjo based on the results of discussions (FGD) with several experts [12]. The experts involved in the FGD included academics, practitioners, and entrepreneurs. Based on the results of the FGD, the problems that emerged were the absence of minimarkets in the PSDKU Sidoarjo area and the limited variety of fishery products currently sold. Other alternatives besides designing the Agribusiness Teaching Factory are to open an expo, do product branding, and optimize online marketing. The consumer segments targeted by TEFA Agribusiness are the academic community of Campus 4 Sidoarjo, the environment around the campus, and the community within and outside Sidoarjo Regency. Early adopters of TEFA Agribusiness are the academic community of PSDKU Sidoarjo from the Agro-Industry Management and Information Technology study programs and the surrounding environment. The uniqueness and superiority of TEFA Agribusiness lie in its minimarket, which emphasizes the sale of superior fishery products in Sidoarjo Regency, which is the result of research by the PSDKU Sidoarjo research team. Figure 3 illustrates the results of the lean canvas business model analysis.

		Designed for:	Designed by:	Date: / Version:
Lean Canvas				
Problem	**Solution**	**Unique Value Proposition**	**Unfair Advantage**	**Customer Segments**
1. There is no Minimarket or canteen on Campus 4 2. There are still few who sell product variants in accordance with local wisdom 3. There is no place or place to market local wisdom products in the area around Campus 4	1. Established an agribusiness minimarket on Campus 4 Sidoarjo 2. Providing product diversification from local wisdom 3. Collaborate with agencies and related parties in marketing products	Providing a variety of local wisdom products typical of Sidoarjo Regency based on the research of the Academic Team	1. Emphasizing local wisdom products of Sidoarjo Regency 2. Providing quality agro-industrial products 3. Teaching Factory-based minimarket	1. Academic community at Campus 4 Sidoarjo 2. The community around campus 4 Sidoarjo 3. Communities outside Sidoarjo
Existing Alternatives	**Key Metrics**	**High-Level Concept**	**Channels**	**Early Adopters**
1. Opening the Expo / bazaar to the public 2. Branding products in accordance with local wisdom 3. Marketed online	1. Consumer preferences and market analysis 2. Marketing mix 3. Market segmentation 4. Target market 5. Positioning 6. Customer satisfaction	Local wisdom products are designed based on **consumer preferences of** the people of Sidoarjo and its surroundings	*1. Social media marketing* 2. Marketing with Polije Campus 4 Sidoarjo partners 3. *Word of mouth* 4. Establishing conconsecution with MSME partners in Sidoarjo district	1. Academic community of the agro-industrial management and informatics engineering study program Polije Campus 4 Sidoarjo 2. Communities in the Sidokare Sidoarjo area
Cost Structure			**Revenue Structure**	
1. Construction costs 2. Equipment investment costs 3. Employee fees 4. Electricity costs 5. Hosting costs 6. Promotion fee			1. Revenue from products sold 2. Income from *teaching factory*	

Figure 3. Lean canvas business model for TEFA Agribusiness PSKDU Sidoarjo.

B. SWOT Analysis of Agribusiness TEFA at PSKDU Sidoarjo

Based on the planning of the Teaching Factory Pilot (TEFA) design, two or more marketing programs can be used simultaneously in the introduction and development of the Teaching Factory (TEFA) because each type of program, including advertising, promotions, sales, personal selling, customer service, and product development—especially in the variety of products sold—will have different influences on the development of the Teaching Factory pilot (TEFA). Therefore, a mechanism is needed to coordinate these programs in line and synergistically integrated. A SWOT analysis is a tool for strategic planning that involves taking stock of a project's or company's internal and external factors, including opportunities and threats [13]. This procedure entails defining the explicit objectives of the Teaching Factory Pilot (TEFA) design plan and recognizing internal and external factors that may facilitate attaining the TEFA Pilot's objectives.

a) Matrix IFE (Internal Factor Evaluation)

The weight value of the internal factors is given on a scale of 0-1, the closer the weight is to 1, the more important the factor is and if the weight value is close to 0, the factor is not important. The stages in obtaining weights start from assigning a value to each factor on a scale of 1-3. A value of 3 indicates a very important value while a value of 1 indicates an unimportant value. Each factor that has been assessed is then totalled to get the total value of the internal factors. The weight value is obtained from dividing the value of the factor by the total value of the internal factors [14]. Identify the development strategies for the Teaching Factory Pilot (TEFA)that have been assigned an average weight value, then proceed to assign average ratings to each internal factor on a scale of 1-4. The final phase involves calculating the weighted score of the internal factors, as illustrated in the subsequent Table 2 of the IFE matrix analysis results.:

Table 2. Matrix IFE (Internal Factor Evaluation)

No	External Strategy Factors	Weight	Rating	Score
Internal Factors Strength				
1	Teaching Factory-based agribusiness minimarket	0,13	3	0,39
2	The available tools are very adequate to support the processing of fishery innovation products	0,15	4	0,6
3	Quality processed products based on research studies by the PSDKU Sidoarjo Research Team	0,17	4	0,68
4	Sidoarjo Regency Government's support related to the development and marketing of processed fishery products	0,14	3	0,52
Total		**0,59**		**2,19**
Internal Factors Weakness				
1	Insufficient diversity of processed fishing products in Sidoarjo Regency	0,13	3	0,39
2	Insufficient availability of marketing media	0,14	3	0,42
3	The packaging of processed fishery products in circulation is less appealing.	0,14	3	0,42
Total		**0,41**		**1,23**
Score Weighted		**1,00**		**3,42**

Source: Processed Data, 2024.

The results of the IFE matrix analysis show that the most prominent internal strength lies in the quality of processed products, as evidenced by research conducted by the PSDKU Sidoarjo Research Team, which obtained a value of 0.68. On the other hand, the most prominent internal weakness is the limited variety of processed fishery products available in the Sidoarjo Regency, reflected in a value of 0.39. The weight value for determining the position of the Teaching Factory Pilot (TEFA)at PSDKU Sidoarjo in the IFE matrix is 3.42.

b) Matrix EFE (External Factor Evaluation)

The external factor weight value is given on a scale of 0-1. The closer the weight is to 1, the more influential the factor is. If it is close to 0, the factor is not important. The stages in obtaining the weight start from giving a value to each factor on a scale of 1-3. A value of 3 indicates a significant value, while 1 indicates an unimportant value. Each factor assessed is then totaled to obtain the total value of the external factor. The weight value is obtained by dividing the factor value by the total value of the external factor. Identify the TEFA pilot development strategy that has been given an average weight, then proceed to the stage of giving an average value to each external factor on a scale of 1-4. The final stage is determining the external factor weight score, with the results of the EFE matrix analysis presented in Table 3 below:

Table 2. Matrix EFE (External Factor Evaluation)

No	**External Strategy Factors**	**Weight**	**Rating**	**Score**
External Factors Opportunities				
1	There is a vocational campus that supports agribusiness activities in Sidoarjo Regency	0,13	3	0,39
2	Utilization of cheap marketing technology	0,16	4	0,64
3	The location of the planning for the establishment of *the Teaching Factory* is in the city center and close to the industrial area	0,14	4	0,56
4	The availability of fishery raw materials is quite abundant	0,15	4	0,6
Total		**0,58**		**2,19**
External Factors Threat				
1	The number of minimarkets that provide typical souvenirs in Sidoarjo Regency	0,15	3	0,45
2	Diverse public tastes related to processed products	0,12	3	0,36
3	Fluctuations in raw material prices	0,15	4	0,6
Total		**0,42**		**1,41**
Weighted Score		**1,00**		**3,6**

Source: Processed Data, 2024.

From the EFE matrix analysis presented above, the most significant external opportunity factor is the effective use of affordable marketing technology, which has a value of 0.64. Meanwhile, the most significant external threat factor is related to the diversity of consumer preferences for processed products, quantified at a value of 0.360. The weight value calculated to assess the Teaching Factory Pilot (TEFA) placement at PSDKU Sidoarjo in the EFE matrix is 3.6.

c) Company Position Analysis

The company's potential is examined through the IE (Internal-External) matrix. The IE (Internal External) matrix employs the parameters of the organization's internal strengths alongside the external influences it encounters. The IFE and EFE matrix, allocated weights and rankings that reflect a weighted average score, will be integrated to create the IE (internal-external) matrix. The Y-axis represents the EFE weight score, while the X-axis represents the IFE weight score [15]. The IE matrix facilitates the establishment of suitable strategies by determining the location of the Teaching Factory Pilot (TEFA) Campus 4 of Politeknik Negeri Jember in Sidoarjo. Teaching Factory Pilot (TEFA) at PSDKU Sidoarjo is located in Cell I, representing growth and build, according to the overall IFE score of 3.42 and the whole EFE score of 3.60. This IE matrix shows the position of the Teaching Factory Pilot (TEFA) at PSDKU Sidoarjo based on the results of the IFE and EFE matrix analysis:

		Total Weighted IFE Value		
		Strong 3,00-4,00	Average 2,00-2,99	Weak 1,00-2,99
Total weighte d EFE Value	High 3,00-4,00	I	II	III
	Medium 2,00-2,99	IV	V	VI
	Low 1,00-1,99	VII	VIII	IX

Source: Primary Data Analysis processed, 2024.

Based on the IE matrix, the Teaching Factory Pilot (TEFA) at PSDKU Sidoarjo is positioned in cell I, indicating a robust internal and external position. The Teaching Factory Pilot (TEFA) at PSDKU Sidoarjo has considerable strengths that can compensate for its weaknesses. Four strengths in the internal factors can be maximized to improve the development of the Teaching Factory Pilot

IOP Conf. Series: Earth and Environmental Science **1446** (2025) 012054 doi:10.1088/1755-1315/1446/1/012054

(TEFA) at PSDKU Sidoarjo. These four strengths are expected to manage the weaknesses of the Teaching Factory Pilot (TEFA) at PSDKU Sidoarjo to maintain and carry out sustainable development. The acute external position shows that the response to opportunities and threats can influence the Teaching Factory Pilot (TEFA) development at PSDKU Sidoarjo.

d) Formulation of Strategy Options

The Teaching Factory Pilot (TEFA) at PSDKU Sidoarjo, with an IFE score of 3.26 and an EFE score of 3.60, is categorized in cell I, designated as "Grow and Build." For growth, the TEFA Pilot at PSDKU Sidoarjo may employ either an intense strategy or an integrative strategy. The approach used by TEFA Pilot Campus 4 of Politeknik Negeri Jember in Sidoarjo is described below [16]:

1) Intensive Strategy
- Market Penetration Strategy
  A market penetration strategy aims to enhance market share for existing items inside the present market by intensifying marketing activities.
- Market Development Strategy
  The market development strategy is employed by firms to launch items in previously untapped regions.
- Product Development Strategy
  A product development strategy aims to enhance sales through the improvement or modification of existing products or services.

2) Integrative Strategy
- Backward Integration Strategy
  Backward integration strategy aims to enhance ownership or control over suppliers of raw materials.
- Forward Integration Strategy
  Forward Integration Strategy pertains to initiatives aimed at acquiring increased ownership or influence of distributors or retailers.
- Horizontal Integration Strategy
  Horizontal integration strategy involves acquiring increased ownership or control over external competitors.

e) SWOT Analysis

The SWOT matrix (Strength, Weakness, Opportunity, and Threat) can generate four columns of alternative strategies. These strategies are SO (strength-opportunity), WO (weakness-opportunity), ST (strength-threat), and WT (weakness-threat). By looking at the company's strengths, weaknesses, opportunities, and threats to developing the Teaching Factory Pilot (TEFA) Campus 4 of Politeknik Negeri Jember in Sidoarjo, 13 different development strategies have been found. These are based on an analysis of the company's position. Table 4 below is the result of the strategic analysis on the SWOT matrix that has been carried out:

Table 4. Matrix SWOT

INTERNAL / EXTERNAL	STRENGTH (S): 1. Teaching Factory-based agribusiness minimarket 2. The available tools are very adequate to support the processing of fishery innovation products 3. Quality processed products based on research studies by the PSDKU Sidoarjo Research Team 4. Sidoarjo Regency Government's support related to the development and marketing of processed fishery products	WEAKNESS (W): 1. Insufficient diversity of processed fishing products in Sidoarjo Regency 2. Insufficient availability of marketing media 3. The packaging of processed fishery products in circulation is less appealing.

OPPORTUNITY (O): 1. There is a vocational campus that supports agribusiness activities in Sidoarjo Regency 2. Utilization of cheap marketing technology 3. The location of the planning for the establishment of the Teaching Factory is in the city center and close to the industrial area 4. The availability of fishery raw materials is quite abundant	STRATEGI (SO): 1. Optimizing the role of the teaching factory so that the community can widely access it in the form of Edu tourism (O1, S1) 2. Optimizing social media for product marketing (O2, S2) 3. Maintaining and continuously improving the quality of marketed products with support from research by the research team (O4, S3) 4. Conducting promotions and participating in events organized by Sidoarjo Regency Government's (O3, S4) 5. Providing education in schools about the potential of processed products from fisheries in Sidoarjo Regency (O1, S4)	STRATEGI (WO): 1. Providing and developing various innovative products according to consumer demand (O4, W1) 2. Creating several creative platforms used as marketing media, such as websites, Instagram, TikTok, and Facebook (O2, W2) 3. Designing trendy product packaging using environmentally friendly materials (O4, W3)
THREAT (T): 1. The number of minimarkets that provide typical souvenirs in Sidoarjo Regency 2. Diverse public tastes related to processed products 3. Fluctuations in raw material prices	STRATEGI (ST): 1. Providing processed products with distinctive characteristics compared to other products (S2, T1) 2. Providing products that align with current market demand trends (S3, T2) 3. Providing products that have obtained PIRT, BPOM, and Halal certifications. (S4, T2)	STRATEGI (WT): 1. Conducting intensive market research to obtain product criteria that align with consumer desires (T1, W1) 2. Providing a barcode on the packaging that contains detailed information about the product being sold (T2, W3)

Source: Primary Data Analysis processed, 2024.

4. Conclusion

The TEFA most suitable for establishment at Campus 4 is TEFA Agribusiness. This teaching factory adopts the structure of a minimarket, encompassing all aspects of agribusiness operations from processing to product marketing. This research aimed to develop a TEFA Agribusiness design strategy using the lean canvas business model and SWOT analysis. The research findings indicated that the agribusiness teaching factory pilot at PSDKU Sidoarjo has significant viability for development. Based on the analysis calculations, the IFE value is 3.26, and the EFE value is 3.60. The SWOT analysis classified the teaching factory startup as being in cell I, which represents growth and build. The available techniques for implementation include intense operations and integrative approaches. Several strategies that can be implemented include optimizing the role of teaching factories to be widely accessible to the community in the form of educational tourism (O1, S1), maximizing social media for product marketing (O2, S2), providing and developing various innovative products according to consumer demand (O4, W1), creating several creative platforms to be used as marketing media such as websites, Instagram, TikTok, and Facebook (O2, W2), as well as providing products that have obtained PIRT, BPOM, and Halal certifications (S4, T2).

Acknowledgments

The author wishes to thank Politeknik Negeri Jember for giving financial assistance in carrying out this research activity. Gratitude is extended to the ICOFA 2024 committee and the State Polytechnic of Jember for successfully organizing the international conference.

References

[1] Armanto J, 2023, Di Bawah Rata-rata ASEAN, Produktivitas Pekerja Indonesia Rendah.

[2] Astuti S and Sidiq M M A, 2021 Analisis Strategi Pengembangan Teaching Farm/Teaching Factory (TEFA) *Jurnal Manajemen, Ekonomi, Keuangan dan Akuntansi (MEKA)* **2**, 2 p. 295–299.

[3] Rina R Ardhiarisca O Andriyani F Triwidiarto C and Govinda B, 2023 Strategi Pengembangan Teaching Factory Tax Center Politeknik Negeri Jember Teaching factory (TEFA) Tax Centre sebagai salah satu media yang digunakan untuk proses belajar mengajar . Pada proses kegiatan belajar mengajar berbasis TEFA, mahasiswa akan me **12**, 1 p. 27–34.

[4] Badan Pusat Statistik Kabupaten Sidoarjo, 2019, Jumlah Perusahaan Industri Besar dan Industri Sedang Menurut Kecamatan 2018. .

[5] Badan Pusat Statistik, 2022, Kabupaten Sidoarjo Dalam Angka 2022.

[6] Yassin M A Rahmadanik D and Widiyanto M K, 2023 IMPLEMENTASI KEBIJAKAN OVER DIMENSION AND OVER LOADING (ODOL) ANGKUTAN BARANG INDUSTRI DI DINAS PERHUBUNGAN KABUPATEN SIDOARJO *PRAJA Observer: Jurnal Penelitian Administrasi Publik,* **3**, 4 p. 32–38.

[7] Sama H and Chandera A, 2021 Perancangan Dan Pengembangan Online Shop Support System Berbasis Lean Canvas *CoMBInES - Conference on Management, Business, Innovation, Education and Social Sciences* **1**, 1 p. 843–850.

[8] Sueb S and Churiyah M, 2023 Strategi Perencanaan Bisnis Teaching Factory Dalam Meningkatkan Kemampuan Entrepreneurship Siswa Melalui SWOT (Studi Kasus SMK Darut Taqwa Purwosari Kab. Pasuruan) *JUBIMA: Jurnal Bintang Manajemen* **1**, 2 p. 33–53.

[9] Pohan S Mavianti M Setiawan H R and Marpaung A H, 2022 Meningkatkan Minat Belajar Siswa dengan Menggunakan Media Bergambar dan Power Point Pada Mata Pelajaran Fiqih *Edukasi Islami: Jurnal Pendidikan Islam* **11**, 03 p. 779.

[10] Khan F, 17-Jul-2023, Lean Canvas Template. [Online]. Available: https://faisalkhan.com/learn/payments-wiki/lean-canvas-template/. [Accessed: 24-Nov-2024].

[11] Naton S Radiansah D and Juniansyah H, 2020 Analisis Nilai Tambah Dan Strategi Pengembangan Usaha Pengolahan Pisang Pada Umkm Keripik Tiga Bujang *Jurnal Sosial Ekonomi Pertanian* **16**, 2 p. 135.

[12] O.Nyumba T Wilson K Derrick C J and Mukherjee N, 2018 The use of focus group discussion methodology: Insights from two decades of application in conservation *Methods Ecol Evol* **9**, 1 p. 20–32.

[13] Wiara Sanchia Grafita Ryana Devi Desty Rara Pringgandinie Henly Yulina and Deni Hadiansah, 2022 SWOT Analysis as a Competitive Strategy at Primkop Kartika Ardagusema Cimahi City, West Java, Indonesia *International Journal of Science, Technology & Management* **3**, 1 p. 134–143.

[14] Rizki M *et al.*, 2021 Determining Marketing Strategy At Lpp Tvri Riau Using Swot Analysis Method *Journal of Applied Engineering and Technological Science* **3**, 1 p. 10–18.

[15] Az Zahra A F Wahyudin W and Nugraha B, 2021 The Implementation of the Strategy of Marketing Management through a SWOT Analysis with the Matrix of IFE, EFE and IE *Jurnal Serambi Engineering* **6**, 2 p. 1721–1729.

[16] Wardhani F K and Dini A, 2020 Strategy Formulation Using SWOT Analysis, SPACE Matrix And QSPM: A Conceptual Framework *Int J Innov Sci Res Technol* **5**, 5 p. 1520–1527.

Determining consumer preferences for orange attributes: a conjoint analysis approach at the teaching factory, politeknik negeri jember

Tia Sofiani Napitupulu[1*], Sumarlina[1], Aulia Nadhirah[1], Amalia Dwi Marseva[2], Lintang Anis Bena Kinanti[2] and Angel Ferinta[1]

[1] Department of Agribusiness Management, Politeknik Negeri Jember, Indonesia
[2] Department of Agro Industry Management, Politeknik Negeri Jember, Indonesia

*tia.sofiani@polije.ac.id

Abstract. Oranges are one of the most widely cultivated and easily accessible horticultural products in the market. At the Teaching Factory (TeFa) Kebun Inovasi, Politeknik Negeri Jember, a variety of horticultural products, including oranges, are produced to support educational and income-generating activities. This study aimed to identify consumer preferences for orange attributes, focusing on appearance, taste, and price, using a conjoint analysis approach. Conducted at the Teaching Factory (TeFa) Kebun Inovasi Politeknik Negeri Jember, this research collected data from 124 respondents to evaluate consumer choices based on peel color, size, taste, and price attributes. The results showed that the most preferred attributes were small-sized oranges with orange-colored peels and a sweet-tart taste, while price was the least important factor influencing preferences. Statistical analysis confirmed that size was the most crucial attribute, followed by peel color, taste, and price. These findings indicate that future product development and marketing strategies should prioritize small-sized oranges with visually appealing orange peels and a balanced sweet-tart flavor to meet consumer demands effectively. The study concludes that a focused product strategy could increase market acceptance and competitiveness, especially when considering potential consumers outside the local community of Politeknik Negeri Jember.

1. Introduction

Politeknik Negeri Jember (Polije), one of vocational higher education institution in Indonesia is known as the campus of teaching factory due to the amount of teaching factory development carried out. The teaching factory that was developed not only acts as a means of vocational learning, but also as an income generating unit for the institution. Kebun Inovasi is one of the teaching factory at Polije that has been running sustainably to produce many kinds of food crop and horticultural products, one of which is orange commodity.

The orange commodities available at teaching factory of Kebun Inovasi Polije are quite diverse, such as Siamese oranges, RGL, pomelo, trigas, santang, seedless lemons, and California lemons, so there are alternatives that can be an option for consumers. The results of a study (1) showed that consumer preferences in purchasing orange fruits vary. Another study showed that orange consumers in California liked all the oranges that were tested, but some types of oranges were preferred over others, the attributes that were considered in consumer preferences were the level of acidity and sweetness (2). Likewise with other similar studies on other commodities such as apples (3,4), raspberries (5), coffee (6), strawberries (7), tomatoes (8), milk (9), and cheese (10).

Consumer preferences can vary significantly depending on factors such as culture, tradition, local tastes, and personal preferences. Research conducted in a location can provide new knowledge about consumer preferences specific to that region. Therefore, it is very important to understand the consumer preference to a product specifically in a region in order to ensure the public acceptance of the product released to the market. In line with this, a study showed that marketers who ignore different and diverse consumer preferences may run into problems (11). Understanding the consumer preference will help the producers to design the most appropriate product attribute in order to meet the consumer's expectation.

Based on interviews with the managers of Kebun Inovasi Polije, 90% of consumers still come from Politeknik Negeri Jember. The determination of consumer preference also needs to consider potential consumers outside the Polije as a consideration for marketing expansion. Thus, the marketing target of the products from Kebun Inovasi Polije can be expanded through an in-depth study first. A study showed that there is positive relationship between the vertical growth of the brand and the increase of customer loyalty (12). Polije that has been known as a vocational higher education focuses on agriculture has the advantage to release many products under the "brand" name of Polije. It is very possible to expanse the marketing of orange fruits from Kebun Inovasi as long as it meets the consumer preference.

Currently, mapping the priority of orange types based on consumer preferences has not been studied in detail, especially for the consumer of orange from Kebun Inovasi Polije.Therefore, a study of consumer preference mapping for various oranges needs to be conducted in order to provide a better understanding of what consumers of Kebun Inovasi Polije oranges want and expect. The results of this study is expected to give comprehensive insight on the most appropriate type of orange for the consumer of Kebun Inovasi Polije in a short term. However, in a long term the results is also expected to be used as consideration to expanse the marketing to the market outside Polije.

2. Method

The research methodology used in this study consists of four main phases: (1) Identification of research problems and objectives followed by the data collection process, (2) Identification of respondents' characteristics, (3) Analysis of consumer behavior, and (4) Evaluation of consumer preferences using a conjoint analysis approach. Each phase is designed to identify respondents' characteristics, analyze consumer purchasing behavior, and evaluate consumer preferences for the orange attributes developed at the Teaching Factory. The methodological framework of the research is illustrated in Figure 1.

2.1. Data Collection

The data collection method used in this study involved a survey of 124 respondents. The selected respondents met the following criteria: (1) they were over 17 years old; (2) they were students, lecturers, or part of the academic community at Politeknik Negeri Jember (Polije); and (3) they had purchased oranges produced by TeFa Kebun Inovasi Polije. The respondents were asked to complete a series of questions using a structured questionnaire.

2.2. Identification of Repondents' Characteristics

Consumers' purchasing behavior was examined through several categories and discussed descriptively. The analysis of consumer behavior covered the following aspects: (1) purchasing behavior based on the reasons for buying, (2) purchasing behavior according to the type of oranges

2.3. Consumer Purchasing Behavior

Consumers' purchasing behavior was examined through several categories and discussed descriptively. The analysis of consumer behavior covered the following aspects: (1) purchasing behavior based on the reasons for buying, (2) purchasing behavior according to the type of oranges most frequently bought, (3) purchasing behavior based on the frequency of orange purchases, (4) purchasing behavior based on the average quantity of oranges purchased, and (5) purchasing behavior according to the location of purchase. Consumer purchasing patterns were systematically analyzed and categorized into multiple dimensions. The discussion includes the following: (1) consumer motives for purchasing, (2) most frequently purchased type of oranges, (3) frequency of purchase, (4) average quantity of oranges bought per transaction, and (5) purchasing locations.

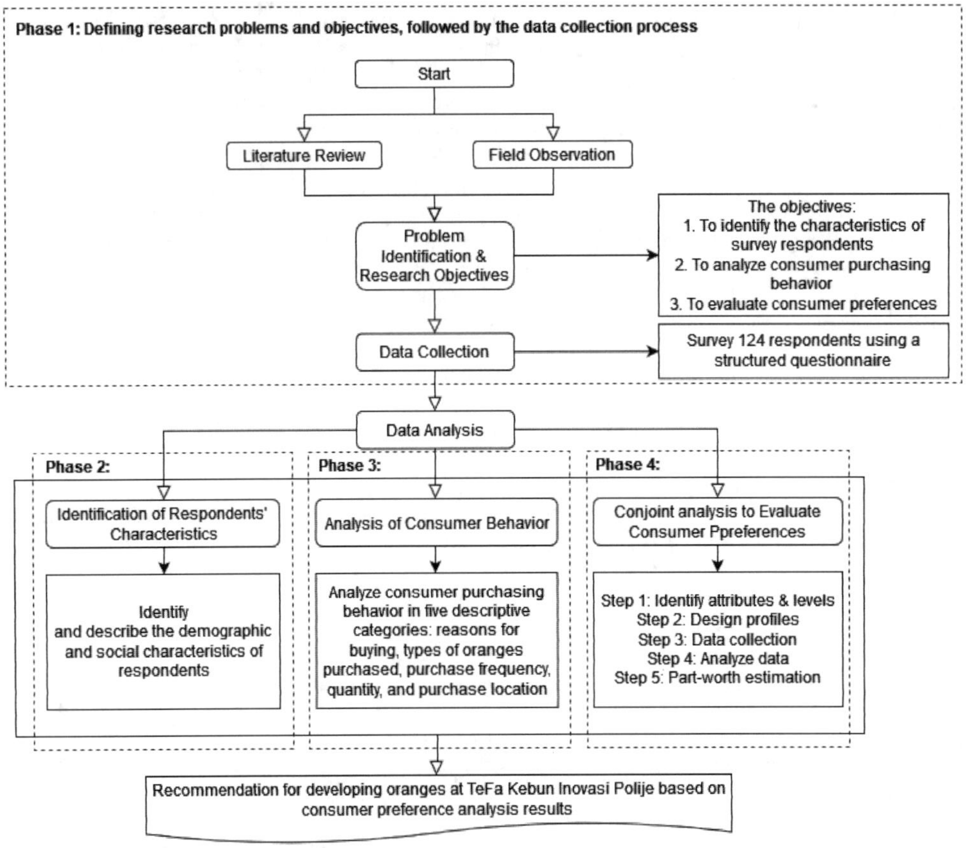

Figure 1. Methodological framework for analyzing consumer behavior and preferences at Tefa Kebun Inovasi Polije.

2.4. Conjoint Analysis

Conjoint analysis is a popular method that is extensively applied in the development of new products, particularly in marketing and related fields (13). It serves as an effective tool for identifying consumer preferences by analyzing their responses to various combinations of attributes presented to them (14). In this study, choice-based conjoint analysis was employed, wherein respondents evaluated sets of attributes that have been combined in specific ways. This method follows a series of structured data analysis steps (15).

1. Identify the attributes to be analyzed and their respective levels.
2. Profiles are designed by compiling combinations of attributes and their levels.
3. Data were collected based on respondents' evaluations.
4. Analyze the data to derive meaningful insights.
5. Determining part-worth values and preferences for the analyzed attributes.

The initial step in this approach is to determine the attributes for the analysis (16). This involves selecting attributes based on prior research with necessary adjustments made to suit the context of this study (2,17,18). In this study, four key attributes were identified: taste, peel color, size, and price. Each attribute consists of two to three distinct levels. These attributes and their sublevels, reflecting consumer preferences for oranges, are outlined in Table 1.

Table 1. Attributes and sub-attributes of consumer preferences for oranges.

Attribute	Sub-attributes	Definition
Taste	1. Sweet	The intensity of sweetness
	2. Sweet-tart	The combination of sweetness and tarness (sourness)
Peel color	1. Green	Low peel color (green)
	2. Yellowish green	Medium peel color (yellowish green)
	3. Orange	High peel color (orange)
Size	1. Small	Approximately 10 oranges per kilogram
	2. Medium	Approximately 7-9 oranges per kilogram
	3. Large	Approximately 5 oranges per kilogram
Price	1. Less than IDR 10,000	The price per kilogram is less than IDR 10,000
	2. IDR 10,000 – 15,000	The price per kilogram is between IDR 10,000 - 15,000
	3. More than IDR 15,000	The price per kilogram is more than IDR 15,000

Figure 2. Illustration of orange peel color attributes: from left to right, green, yellowish green, and orange.

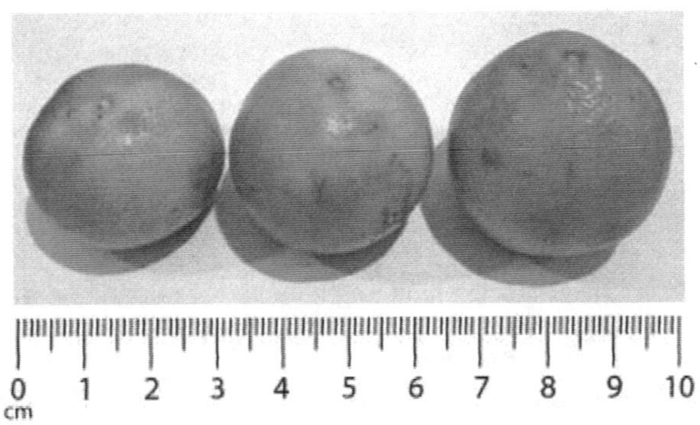

Figure 3. Illustration of orange fruit size attributes: from left to right, small, medium, and large.

Following the attribute selection, the next step involves designing profiles by combining different attribute levels. Given the potential for an overwhelming number of combinations that could complicate the respondents' decision-making process, an orthogonal procedure was employed to narrow the selection to nine specific profiles. The selected profiles are presented in Table 2.

Table 2. Orange profiles: orthogonal procedure results.

Orange ID	Taste	Peel color	Size	Price
Orange 1	Sweet	Yellowish green	Medium	IDR 10,000 – 15,000
Orange 2	Sweet-tart	Green	Large	IDR 10,000 – 15,000
Orange 3	Sweet-tart	Orange	Medium	Less than IDR 10,000
Orange 4	Sweet-tart	Yellowish green	Small	More than IDR 15,000
Orange 5	Sweet	Orange	Small	IDR 10,000 – 15,000
Orange 6	Sweet	Green	Medium	More than IDR 15,000
Orange 7	Sweet	Green	Small	Less than IDR 10,000
Orange 8	Sweet	Orange	Large	More than IDR 15,000
Orange 9	Sweet	Yellowish green	Large	Less than IDR 10,000

Data collection was followed by asking respondents to rank the nine orange profiles based on their preferences, from most preferred (rank 1) to least preferred (rank 9). The collected data were then subjected to conjoint analysis. The result of this analysis is a set of utility values that represents consumer preferences for the different attributes of oranges. Positive utility values indicate a higher level of preference for a given attribute, whereas negative values indicate a lower preference (19–21).

The primary aim of conjoint analysis is to estimate respondents' preference patterns, referred to as Estimate Part-Worth, and to compare these estimates with their actual evaluations during the process (21,22). The accuracy of the resulting model must be assessed to ensure its alignment with real-world conditions. This is typically performed using Pearson's R and Kendall's tau correlation coefficients. A value above 0.5 for these coefficients suggests a strong correlation between the estimated results and actual conditions. Additionally, a significance level below 0.05 (5%) confirms that the results of the analysis are strongly correlated with respondents' evaluations.

Alternative perspectives of consumer preferences can also be seen based on the socio-demographic characteristics of respondents. The results of the analysis that are normally distributed in the Kolmogorov-Smirnov (K-S) test ($p>0.05$) can be applied to parametric tests. The results of the t-test and ANOVA can show consumer preferences for oranges on various attributes of their composition, namely taste, peel color, size and price against the socio-demographic characteristics of respondents including gender, age, and number of family.

3. Result and Discussion

3.1. Characteristics of Survey Respondents

Table 3 presents the respondents' demographic characteristics, showing that the study involved 124 participants from TeFa Kebun Inovasi Polije. The sex distribution included 53 males (42.72%) and 71 females (57.26%). Most respondents fell within the productive age range, with 75% aged between 17 and 30 years, indicating that the majority were young and economically active. In terms of educational background, the majority of respondents had completed high school, accounting for 83.87% of the total, whereas those with higher education degrees represented 16.13%. This finding highlights that most respondents had a secondary education level. Regarding family size, the majority of respondents (57.26%) came from families with three to four members. In contrast, families with more than six members constituted the smallest group with only 3.23% of the respondents. These findings suggest a general trend toward smaller family structures among respondents.

Table 3. Demographic criteria of respondents for consumer preferences of oranges.

No.	Demographic criteria	Number of respondents	Percentage (%)
1	Gender		
	Male	53	42.74
	Female	71	57.26
2	Age		
	17 – 30 years	93	75.00
	31 - 45 years	25	20.16
	46 - 60 years	6	4.84
	>60 years	-	-
3	Education		
	Finished elementary school	-	-
	Finished junior high school	-	-
	Finished high school	104	83.87
	Graduated from college	20	16.13
4	Number of family members		
	1 - 2 people	16	12.90
	3 - 4 people	71	57.26
	5 - 6 people	33	26.61
	>6 people	4	3.23

3.2. Consumer Purchasing Behavior

3.2.1. Consumer Motives for Purchasing Orange

Consumers' decisions to buy oranges are based on various factors that reflect their needs and personal preferences. These reasons can be classified into several main categories, such as health benefits, lifestyle or habits, ease of product availability, affordable prices, and other factors. The analysis revealed that health benefits emerged as the most dominant reason for purchasing oranges (35.48%), indicating that consumers recognize oranges as nutrient-rich fruits, particularly high in vitamin C, which is considered beneficial for maintaining a strong immune system. Another significant factor was affordable price, expressed by 20.97% of respondents, suggesting that the Polije TeFa Kebun Inovasi segment is also highly influenced by economic considerations when choosing to buy oranges. Additionally, lifestyle or habitual factors influenced 13.71% of consumers, indicating that oranges may already be a part of their daily fruit consumption. The ease of obtaining oranges, whether through local markets or easily accessible distribution networks, was cited by 13.71% of the consumers as a reason for their purchase. Other reasons included more specific personal factors such as individual preferences for certain types of oranges or recommendations from family or health professionals.

3.2.2. Most Frequently Purchased Orange Varieties

The fruit market in Jember Regency offers a variety of oranges, each with distinct characteristics in terms of appearance, nutritional content, and taste. Commonly sold orange varieties include Siam oranges, Siam honey oranges, Baby Java oranges, Santang oranges, imported oranges, and pomelos. Each type has unique qualities that may influence consumer choice. An analysis of consumer buying behavior revealed that Siam oranges were the most frequently purchased variety, highlighting their popularity within the local market segment. This is followed by Siam honey oranges, which, while visually similar to Siam oranges, tend to be preferred because of their sweeter taste. Among Polije TeFa Kebun Inovasi, the third most popular variety after Siam oranges and Siam honey oranges is

Santang oranges (8.06%), while imported oranges and other varieties, including baby Java oranges, are purchased by a smaller portion of respondents. This suggests that while there is a variety of oranges available in the market, consumers generally prefer familiar and affordable local varieties.

3.2.3.Frequency of Orange Purchases

The analysis of purchase frequency shows that the majority of consumers (58.06%) buy oranges one to two times per month, indicating that oranges are a regularly consumed product, although not necessarily a daily necessity. This relatively consistent consumption pattern suggests that oranges play an important role in the diet of respondents, even if they are not consumed daily. Furthermore, 27.42% of respondents purchased oranges several times a year, suggesting that this group of consumers may view oranges as a seasonal fruit or as an alternative to other fruits. On the other hand, only a small percentage of consumers (8.06%) buy oranges more frequently, one to two times per week, with an even smaller percentage (6.45%) purchasing oranges almost daily. These varying purchase frequencies indicate that while oranges are popular fruits, factors such as seasonal availability, price, or preferences for other fruits may influence consumer buying patterns.

3.2.4.Average Quantity of Orange Purchases

Based on this analysis, the majority of consumers tend to purchase oranges in relatively small quantities. This buying pattern may reflect consumers' short-term or daily needs, as they prefer to buy oranges for immediate consumption within a short period. Additionally, the tendency to buy in small quantities could be linked to consumers' preferences for keeping the fruit fresh or their desire to purchase a variety of fruits in smaller portions. Another factor that may influence this behavior is the availability of oranges in the market, where buyers feel more flexible in terms of purchasing amounts that align with their daily needs. On the other hand, there is also a segment of consumers who purchase oranges in larger quantities, with 26.61% buying between 2 and 4 kg per transaction. Only 1.61% purchased more than four kilograms per transaction. This indicates that a small percentage of consumers may buy oranges in larger quantities for family consumption or long-term use.

3.2.5.Preferences for Orange Purchase Locations

Oranges are widely available fruit commodities sold in a variety of locations, providing consumers with multiple options to purchase oranges according to their preferences and needs. These locations include traditional markets, supermarkets, fruit shops, and mobile street vendors selling fruit from cars. The findings show that traditional markets are the preferred choice for the majority of consumers (37.90%). The popularity of traditional markets is likely due to their affordable prices and the availability of a wide range of oranges in large quantities. Furthermore, direct interaction with sellers allows consumers to assess the quality and freshness of fruit, thus enhancing their trust in the purchase. The second most popular purchase location is mobile street vendors, chosen by 33.87% of consumers. Convenience of access and flexible timing are often the main reasons for consumers to buy from mobile vendors, as they can easily find oranges at competitive prices and of good quality without the need to visit a market. On the other hand, only a small proportion of consumers prefer to buy oranges at supermarkets, even though supermarkets offer convenience and generally higher quality assurance. However, higher prices and a limited variety of oranges at supermarkets might be factors that reduce consumer interest, leading them to prefer traditional markets or mobile vendors, where they can find more options at lower prices.

3.3. Consumer Preference

Figure 4 presents the results of the conjoint analysis of consumer preferences for the orange attributes studied. This analysis reveals several important findings that provide insight into consumer preferences for orange products marketed by Kebun Inovasi Polije. The utility range resulting from this analysis ranged from -0.23 to 0.22. These utility values reflect the level of consumer preference for various orange attributes, with positive values indicating higher preference and negative values indicating lower preference.

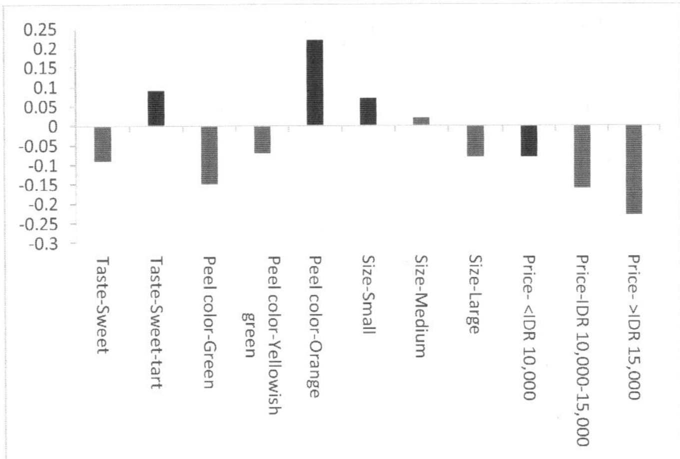

Figure 4. Average utility estimates for selected orange attributes.

The main findings show that consumers prefer oranges with sweet and sour tastes compared to oranges that only have a sweet taste. This indicates that consumers prefer oranges that provide a fresh taste sensation, which is considered more satisfying and refreshing. Orange peel color is also an influential attribute in determining consumer preferences. Oranges with orange skin were preferred, followed by oranges with yellowish-green skins. Oranges with green skin received the lowest preference, which could indicate that a riper skin color (orange) is associated with consumers with better quality or a sweeter taste. Consumer preferences for fruit size tend to favor small oranges. This can be observed from the decrease in utility value as the size of the fruit increases, where small oranges are considered more practical and easier to consume. Price also plays an important role in consumers' decisions. The results show that consumers tend to prefer lower prices for oranges, especially those with the highest preference in the price range below IDR 10,000. Consumer preferences decrease as prices increase, indicating price sensitivity.

Based on the average level of importance of each attribute, size had the highest score (32.69%), followed by peel color (32.03%). The taste attribute ranked third with a score of 18.98%, while price ranked last with a score of 16.30%. These findings indicate that when making purchasing decisions, consumers tend to prioritize the size and color of orange peel over other attributes such as taste and price.

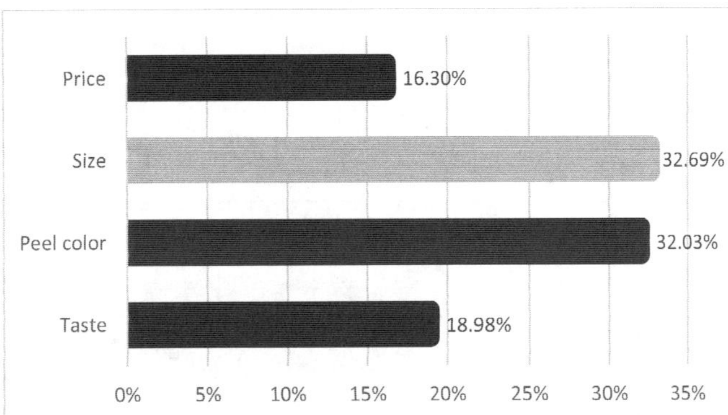

Figure 5. Average level of importance for selected orange attributes.

The ideal estimate closely approximates the actual conditions. Predictive accuracy can be measured through a significant and high correlation between the estimated results and the respondents' actual preferences. The results indicate that the Pearson's R and Kendall's tau values obtained were above 0.5, signifying a strong correlation between the estimates and the real preferences of the respondents. Additionally, a significance level smaller than 0.05 (or 5%) also indicates a significant correlation between the conjoint analysis results and the respondents' opinions. This demonstrates that the estimates generated by conjoint analysis possess a high degree of predictive accuracy. In this study, the Pearson's R value was 0.79, with a significance level of 0.006, whereas Kendall's tau value was 0.56 with a significance level of 0.019. These results confirm that the conjoint analysis method successfully provides estimates that closely align with the respondents' actual preferences, reinforcing the validity of the analysis.

Table 4 shows the results of the ANOVA test showing the significance and probability of differences in the average utility scores of attributes based on the socio-demographic characteristics of the respondents. Respondents aged 46-60 years preferred size as the basis for purchasing oranges compared to respondents aged 17-30 years and 31-45 years (F = 7.300; p <0.05). Respondents aged 46-60 years preferred price as the basis for purchasing oranges. Respondents aged 17-30 years were in second place in choosing price as the basis for purchasing oranges, followed by respondents aged 31-45 years (F = 11.460; p <0.05). The next significant attribute was the number of family members. Respondents who had more than 6 family members preferred color as the main attribute in purchasing oranges. Next, in sequence, respondents who have family members of 1-2 people, 5-6 people and 3-4 people choose skin color as the main attribute in buying oranges (F = 3.957; p <0.05). Three socio-demographic attributes have a response in choosing oranges. However, only the taste attribute has no response. In detail, the attributes that have a response are the attributes of peel color, size and price.

Table 4. Differences in attribute and level utility score averages by socio-demographic variables.

Attribute	Age		Number of Family	
	F	*p*	*F*	*p*
Taste	0.643	0.528	.666	0.574
Peel Color	0.771	0.465	3.957	0.010*
Size	7.300	<0.001*	1.742	0.162
Price	11.460	<0.001*	0.259	0.855

Notes: Age: 1: 17-30 yo; 2: 31-45 yo; 3: 46-60 yo; 4: >60 yo. Number of Family: 1: 1-2 people; 2: 3-4 people; 3: 5-6 people; 4: >6 people. *p < 0.05

The table below shows the mean (M) and standard deviation (SD) of utility scores for four orange attributes (Taste, Peel Color, Size, and Price) categorized by gender (Female and Male). Additionally, it presents the t-test results, including the t-value and p-value, which indicate whether there are statistically significant differences in attitudes toward these attributes between male and female respondents.

Table 5. Utility score average and t-test scores of attitudes and levels by socio-demographic variables.

Attribute	Gender					
	Female (n=72)		*Male (n=52)*		*t*	*p*
	M	SD	M	SD		
Taste	3.373	0.605	3.316	0.762	0.464	0.643
Peel Color	2.877	0.962	3.196	0.845	-1.913	0.580
Size	3.322	0.934	3.124	0.896	1.182	0.240
Price	3.109	0.876	3.030	0.847	0.498	0.619

The t-test results for all four attributes (taste, peel color, size, and price) show no significant differences in preferences between male and female respondents, as all p-values are greater than 0.05. This suggests that gender does not have a substantial impact on the attitudes toward these attributes in this study. Therefore, marketing strategies or product development focusing on these orange attributes can be applied uniformly across genders without the need for gender-specific adjustments.

4. Conclusion

The study revealed that health benefits are the primary motivator for consumers purchasing oranges, reflecting the perceived nutritional value of the fruit, particularly in its vitamin C content. Price also played a significant role, with a substantial portion of respondents being influenced by affordability. The frequency of orange purchases shows that consumers typically buy oranges once or twice a month, indicating that they are regularly consumed, though not necessarily daily. Most consumers tend to buy smaller quantities per transaction, likely to maintain freshness or fulfill immediate needs, with a preference for purchasing from traditional markets and mobile street vendors, which offer competitive pricing and a variety of options. Conjoint analysis revealed that consumers showed a clear preference for oranges with a sweet and sour flavor profile, which is perceived as fresh and refreshing. Peel color significantly affects consumer preferences, with orange or yellowish-green skins being more favored, as these colors are often associated with better quality and sweetness. Smaller oranges are also preferred because they are considered more practical and convenient. Price sensitivity was evident, with a higher preference for oranges priced under IDR 10,000 and a decline in preference as prices increased. Among the evaluated attributes, orange size and skin color were deemed the most important factors in consumers' purchasing decisions, surpassing taste and price considerations. The analysis indicated that there were no significant differences in consumer preferences for orange attributes (taste, peel color, size, and price) between male and female respondents, as indicated by p-values greater than 0.05 for all attributes, such as taste ($p = 0.643$) and peel color ($p = 0.580$). The overall mean preference scores show that size was the most crucial attribute ($M = 3.322$ for females and $M = 3.124$ for males), followed by peel color ($M = 2.877$ for females and $M = 3.196$ for males). In summary, consumers prioritize health benefits, convenience, and affordability when purchasing oranges, with a strong preference for smaller, fresh-tasting oranges with bright peel colors. These insights provide a clear direction for developing marketing strategies and product offerings that better align with consumer expectations.

5. References

1. Dwiastari KY, Susrusa KB, Artini WP. Analisis Preferensi Konsumen terhadap Buah Jeruk Impor dan Buah Jeruk Lokal di Kota Denpasar, Bali. J Agribisnis dan Agrowisata ISSN. 2019;8(3):391–400.
2. Simons T, Sivertsen H, Guinard J-X. Mapping The Preferences of Adult and Child Consumers For California-Grown Navel Oranges. HortScience. 2018;53(5):661–8.
3. Bonany J, Brugger C, Buehler A, Carbó J, Codarin S, Donati F, et al. Preference Mapping of Apple Varieties in Europe. Food Qual Prefer. 2014;32:317–29.
4. Bowen AJ, Blake A, Tureček J, Amyotte B. External Preference Mapping: A Guide for a Consumer-Driven Approach to Apple Breeding. J Sens Stud. 2019;34(1):e12472.
5. Villamor RR, Daniels CH, Moore PP, Ross CF. Preference mapping of frozen and fresh raspberries. J Food Sci. 2013;78(6):S911–9.
6. Harwood WS, McLean KG, Ennis JM, Ennis DM, Drake M. Comparison of Preference Mapping with Projective Mapping for Characterizing Consumer Perception of Brewed Black Coffees. J Sens Stud. 2020;35(3):e12563.
7. Oliver P, Cicerale S, Pang E, Keast R. Identifying Key Flavors in Strawberries Driving Liking Via Internal and External Preference Mapping. J Food Sci. 2018;83(4):1073–83.
8. Oltman AE, Yates MD, Drake MA. Preference Mapping of Fresh Tomatoes Across 3 Stages of Consumption. J Food Sci. 2016;81(6):S1495–505.

IOP Conf. Series: Earth and Environmental Science **1446** (2025) 012055 doi:10.1088/1755-1315/1446/1/012055

9. Mahato DK, Oliver P, Keast R, Liem DG, Russell CG, Mohebbi M, et al. Identifying Ideal Product Composition of Chocolate-Flavored Milk Using Preference Mapping. J Food Sci. 2021;86(7):3205–18.
10. Los PR, Simões DRS, Benvenutti L, Zielinski AAF, Alberti A, Nogueira A. Combining Chemical Analysis, Sensory Profile, CATA, Preference Mapping and Chemometrics to Establish The Consumer Quality Standard of Camembert-Type Cheeses. Int J Dairy Technol. 2021;74(2):371–82.
11. Hasibuan Z, Syahputri N, Hamtheldy RZ, Ramadhan A. Pentingnya Mengembangkan Produk dan Keputusan Merek dalam Pemasaran Global. J Soc Res. 2022;1(4):261–7.
12. Vuković D, Kerum F, Žukina M. Consumer Preferences and Reactions to Brand Expansion. Naše Gospod Econ. 2024;70(2):47–62.
13. Kulshreshtha K, Sharma G, Bajpai N. Conjoint analysis: The assumptions, Applications, Concerns, Remedies and Future Research Direction. Int J Qual Reliab Manag. 2023;40(2):607–27.
14. Jr. GAC. Dasar-dasar Riset Pemasaran. Edisi 4, jilid 1 [Internet]. Erlangga; 2001. Available from: https://books.google.co.id/books?id=bBN9elNsNaYC
15. Rao VR. Applied Conjoint Analysis. New York: Springer; 2014.
16. Shepherd DA, Zacharakis A. Chapter 6 Conjoint Analysis: A Window of Opportunity for Entrepreneurship Research. in 2018. p. 149–83. Available from: https://www.emerald.com/insight/content/doi/10.1108/S1074-754020180000020011/full/html
17. Kurzer AB, Bechtel R, Guinard J-X. Adult and Child Focus Group Views of Oranges and Mandarins. Horttechnology. 2019;29(4):408–16.
18. Ajisola S, Olajide OA, Agulanna FT. Assessing Consumer Perception and Preference for Fresh or Processed Citrus Fruits: Implications for The Sweet Orange Supply Chain in Oyo State Nigeria. 2021;
19. Ghozali I. Aplikasi Analisis Multivariat dengan Program IBM SPSS 21. Semarang: Badan Penerbit Universitas Diponegoro; 2016.
20. Supranto J. Analisis Multivariat: Arti dan Interpretasi. Jakarta: Rineka Cipta; 2004.
21. Jansen SJT, Coolen HCCH, Goetgeluk RW. The measurement and analysis of housing preference and choice. Springer Nature; 2011.
22. Santoso S. Statistik Multivariat dengan SPSS. Elex Media Komputindo; 2017.

Acknowledgments
The authors extend their heartfelt appreciation to Politeknik Negeri Jember for the funding and support provided for this research. We are also grateful to the staff and students of TeFa Kebun Inovasi for their invaluable assistance and collaboration throughout the study. Furthermore, we would like to acknowledge our colleagues and peers for their constructive feedback and encouragement. This research was funded by DIPA PNBP Politeknik Negeri Jember, Year 2024, under grant number 7934/PL17/KP/2024.

Applications of 18650 lithium-ion batteries pack in 3 kw electric vehicles

A T Zain[1]*, D D Suranto[1], A Budiprasojo[1], C N Karimah[1], F A Azhar[1], and M D Haryanto[1]

[1]Automotive Engineering Study Program, Engineering Department, Politeknik Negeri Jember

*e-mail: alextaufiqurrohman@polije.ac.id

Abstract. One of the important components in the development of electric vehicles is battery. Battery conditions must be monitored in real-time. This research begins with assembling a 18650 Lithium-ion battery pack with a maximum capacity of 83.5V. A PZEM-017 is used to calculate the value of voltage and current of the battery. In this study, Arduino Atmega 2560 is used for microcontroller and LCD display to show the measurement results. Before being applied to electric vehicle, the measurement and monitoring system needs to be calibrated first. The callibration results show that the Voltage and Current sensors have maximum error under 2% (0.31% for Voltage and 1.56% for Current). While the minimum precisions are 99.59% for voltage and 98.10% for current. This measurement and monitoring system has been successfully applied to 3KW electric vehicle. Where the measurement results show that the increase in driving distance and driver weight will be proportional to the increase in energy consumption from the battery. The electric motor, as a source of vehicle propulsion, will work longer and harder. The highest decrease in voltage value occurred when the distance was 1.8 km and the driver's weight was 83kg at 3.8V with a power consumption of 136.29Wh.

1. Introduction

The transition of energy resources from conventional to renewable energy is currently being promoted by the Indonesian government. In the automotive industry, the government has provided strategic policies, including legalizing the conversion of conventional vehicles to electric vehicles, downstream battery materials and developing the battery industry, subsidies for purchasing and taxing electric vehicles, and improving supporting infrastructure [1].

Electric vehicles (EVs) have become a major highlight in the global automotive and transportation industry. In contrast to fossil fuel vehicles which rely on burning fuel such as petrol or diesel, electric vehicles use electric motors that are driven by energy stored in the battery. Advances in battery technology and charging infrastructure have made electric vehicles an increasingly attractive option for consumers and policymakers.

Electric vehicles offer potential solutions to some of the biggest environmental challenges facing the world today. With increasing levels of air pollution and global climate change, electric vehicles can help reduce the greenhouse gas emissions, and other pollutants that come from fossil fuel vehicles. According to a report from the International Energy Agency [2], electricity generation is the highest producer of carbon dioxide gas compared to other sectors. However, this sector is also at the top of the renewable energy transition through strategic policies related to the implementation of solar and wind power. The transition to electric vehicles can play an important role in reducing the carbon footprint and promoting sustainability.

Battery electric vehicles (BEVs) are one type of EVs. EVs are renowned for their silent operation and immediate torque [3][4]. Batteries are an important component in EVs because they are used as the main energy source. Several types of battery materials are used. Nickel is one of the materials in making batteries, especially for EVs. It's important to note that Indonesia provides 27% of the world's nickel supply, which is crucial for producing batteries needed for EVs and also other materials such as cobalt and rare earth metals [1], [5]. Lithium-ion batteries have attracted significant interest as supporting devices in electric vehicles than other various batteries. The extraordinary advantages of Lithium-ion batteries are relatively high energy density (up to 200 Wh/kg), high energy efficiency (more than 95 %), and long cycle life (3000 cycles at 80% inner discharge). So far, 77% of US electric power storage systems operating to stabilize the electric grid rely on Lithium-ion batteries, indicating a high-value market for Lithium-ion batteries.. In recent years, lithium-ion batteries have been successfully developed, and their performance has improved greatly [6], [7].

To supply the electrical energy needs of EVs, each cell of the 18650 Lithium-ion battery needs to be assembled. Series circuits are used to increase its nominal voltage, and parallel to increase its capacity. This is because each 18650 Lithium-ion battery cell only has a nominal voltage of 3.7V to 4.2 V and a capacity of around 3500 mAh [8], [9].

2. Method

2.1. Assembly of 18650 Lithium-ion battery pack and battery management system (BMS)

The first stage in this research is assembling an 18650 Lithium-ion battery. The 18650 Lithium-ion battery is a popular type of battery because of its good performance and relatively compact size. The name "18650" refers to the dimensions of the battery: 18 mm in diameter and 65 mm long. These batteries are often used in a variety of applications, from portable electronic devices to electric vehicles. This battery usually has a capacity of between 1800 mAh to 3500 mAh, and the nominal voltage is 3.7 volts. The full voltage when fully charged can reach around 4.2 volts. 18650 batteries should be handled with care. They are sensitive to overcharge, over-discharge, and extreme temperatures. Batteries that are damaged or used incorrectly can leak, catch fire and even cause explosions. These batteries are often used in larger battery packs to provide power to various devices. In applications such as electric vehicles, they are usually arranged in a series-parallel configuration to increase capacity and voltage. To maximize battery life, it is best to store the battery in a dry and cool place. Avoid overcharging and using until the battery is completely discharged.

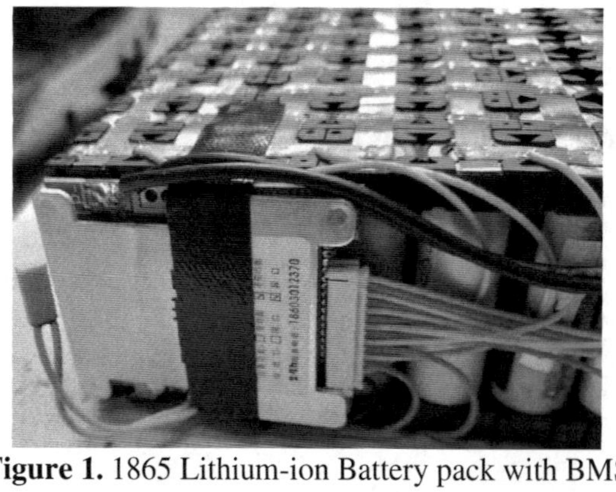

Figure 1. 1865 Lithium-ion Battery pack with BMS

IOP Conf. Series: Earth and Environmental Science **1446** (2025) 012056 doi:10.1088/1755-1315/1446/1/012056

The anode, separator, and cathode of Lithium-ion batteries are rolled up and encased in a steel shell filled with a liquid electrolyte to store electrical energy. Typically, these batteries have a cylindrical shape with flat tops and bottoms to facilitate better arrangement in battery packs. Some 18650 cells are modified by adding a button top to the standard design for a secure connection in specific applications. Commercial 18650 cells that include a protection circuit at the bottom are known as "protected" cells featuring both a button top and a protection circuit [10]. In this study, to generate 3KW Evs and its accessories, We use 20 series circuits and 12 parallel circuits so a total of 240 lithium-ion batteries are needed.

A Battery Management System (BMS) is an essential element designed to oversee and regulate the performance of battery packs. The BMS constantly tracks various metrics, including voltage, current, temperature, and the state of charge (SoC) of each cell or module within the battery. This monitoring ensures that each cell operates within safe parameters. It maintains uniform charge levels across all cells in the battery pack, preventing issues like overcharging or excessive discharging of individual cells, which could degrade battery life or pose safety risks.

The BMS safeguards the battery by avoiding conditions that could cause damage, such as overvoltage, undervoltage, overcurrent, and overheating. It can disconnect the battery or restrict its operation to avoid such conditions. It typically interfaces with external systems to relay information about the battery's status and performance. This includes connecting with the main control systems of a vehicle or device to enhance performance and safety. The BMS assesses and estimates critical battery metrics, including state of charge (SoC), and remaining useful life. This data is vital for effective battery management and performance optimization.

2.2. Assembly of measuring and monitoring battery parameters

To determine the parameters of Lithium-ion battery usage, a monitoring system needs to be assembled [11], [12]. This measuring system is located between the output from the BMS and the input from the electric motor controller. For the monitoring process, the PZEM-017 module is used. The PZEM-017 is a device used for measuring electrical parameters in various applications: 1) The PZEM-017 can measure AC and DC voltage and its current. This makes it useful for monitoring power usage (discharging process) of Lithium-ion batteries and ensuring systems are operating within safe electrical parameters. 2) It provides readings for real power, apparent power, and energy consumption. This allows for a detailed analysis of power usage and efficiency. 3) The PZEM-017 typically includes a digital display that shows the measured parameters, making it easy to monitor real-time data. The PZEM-017 is a versatile tool for electrical measurement, commonly used in applications where monitoring and managing electrical parameters are crucial [13].

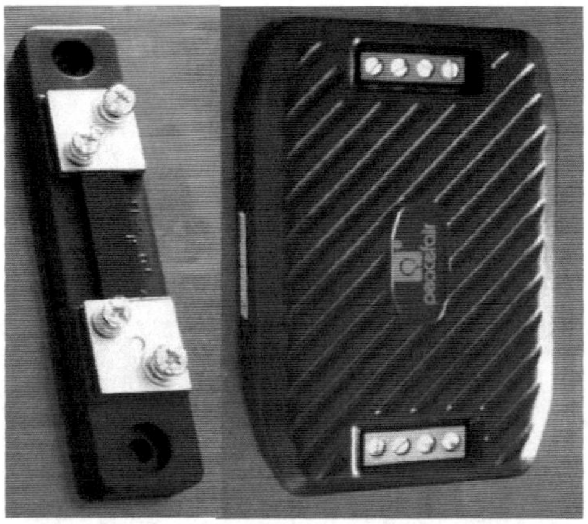

Figure 2. PZEM-017 Module

IOP Conf. Series: Earth and Environmental Science **1446** (2025) 012056 doi:10.1088/1755-1315/1446/1/012056

By knowing the current and voltage values, the power consumption value can be calculated using the equation:

$$P = V.I \quad (1)$$

So, the electric energy generated by the Lithium-ion battery is:

$$E = P.t \text{ or } E = V.I.t \quad (2)$$

Where t is the usage time of EV. In this research, Arduino ATmega 2560 was used as the microcontroller. This board can handle complex projects due to its increased number of input/output pins, PWM outputs, and memory than other commonly Arduino boards [13].

Table 1. Arduino ATmega 2560 Specifications

Parameters	Specifications
Microcontroller	ATmega2560
Digital I/O Pins	54 Pins
Analog Input Pins	16 Pins
PWM Outputs	15 Pins
Power Connection	USB Port & DC Power Jack
Serial Communication	4 hardware serial ports (UARTs)
Flash Memory	256 KB(8 KB used by bootloader)
RAM	8 KB of SRAM
EEPROM	4 KB
Clock speed	16 MHz
Operating Voltage	5 V
Input Voltage (recommended)	7-12 V
Input voltage (limit)	6-20 V

After being processed by the microcontroller, the data is received by the LCD screen to display the measurement results. Here is the block diagram of circuit:

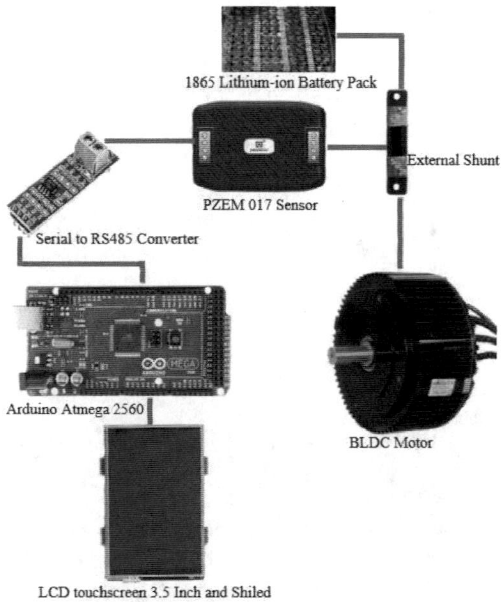

Figure 3. Block Diagram of Sytems

2.3. Sensor calibration

One of the processes that needs to be carried out in the process of assembling a sensor-based measurement and monitoring system is the calibration process. Sensor calibration is the process of adjustments or repairs carried out to ensure that the sensor provides accurate and consistent readings.

This process involves comparing sensor readings to a known standard or reference and making adjustments to correct detected errors. Calibration is carried out periodically to maintain sensor accuracy, especially in applications that require very precise measurements [14], [15].

Accuracy and precision are two important concepts in data measurement and analysis. Although they are often used interchangeably, they have different meanings. Accuracy indicates how close the measurement results or data obtained are to the actual value or known reference value (from standard measuring instruments). Precision describes the consistency or repeatability of measurement results when the same measurement is carried out several times under the same conditions. In the context of data, precision refers to the degree to which measured or generated data is consistent and has low variability [15]. To determine the error, accuracy and precision rate (in %), we use the following equations:

$$error(\%) = \left(\left| \frac{Y - X_n}{Y} \right| \right) \cdot 100\% \quad (3)$$

$$accuracy(\%) = \left(1 - \left| \frac{Y - X_n}{Y} \right| \right) \cdot 100\% \quad (4)$$

$$precision(\%) = \left(1 - \left| \frac{X_n - \overline{X_n}}{X_n} \right| \right) \cdot 100\% \quad (5)$$

Where Y for reference parameter value (measurement results using standard measuring instruments); X_n for n-th measured parameter value using the developed system; and $\overline{X_n}$ = Average value of the measured n parameter using the developed system. For each of the same conditions, data collection was repeated five times. The standard measuring instrument used is the Fluke 101 multimeter.

An adjustable power supply is used in this calibration process. This device is used so that changes in voltage and current values can be regulated. The maximum voltage that can be used is 30V and the maximum current is 10 A. According to Ohm's Law, several resistors are used to measure the current value.

$$V = I.R \quad (6)$$

2.4. *Application of measurement and monitoring battery system to 3 KW electric vehicle*

The final stage of this research is the application of the system that has been developed to a 3 KW electric vehicle. For safety reasons, the output of the 18650 Lithium-ion battery pack is connected to a Miniature Circuit Breaker (MCB).

Figure 4. Assembly of the Measurements and Monitoring Systems to Electric Vehicle

The following is a 3KW electric vehicle used in this research

Figure 5. The 3KW Electric Vehicle

3. Results and Discussions

3.1. Calibration results

The system has good measurement results. This is proven by the error results which are not more than 2% for all conditions.

Table 2. Voltage Calibration Test Results of PZEM-017

Voltage (V)		Error (%)	Accuracy (%)	Precision (%)
Fluke 101 Multimeter	**PZEM-017**			
15.00	15.05	0.31	99.69	99.79
17.50	17.53	0.19	99.81	99.80
20.00	19.97	0.15	99.85	99.80
22.50	22.53	0.12	99.88	99.71
25.00	24.97	0.13	99.87	99.59
27.50	27.53	0.11	99.89	99.76

The accuracy value above 95% indicates that the test results have results that are similar to standard measuring instruments. For the precision results, the test shows that the data consistency is quite good too. This is shown by the precision value above 95% for five repetitions of data in each of the same conditions.

For the voltage test, the error percentage is less than 0.32%. This indicates that in every condition, PZEM-017 has good accuracy (above 99.68%) for voltage measurements. So it can be applied to measuring digital voltage values using a microcontroller. The precision percentage test results also show good values, above 99.50% in each condition. This shows that the level of data consistency is good for measuring voltage values using the developed system.

Table 3. Current Calibration Test Results of PZEM-017

Current (A)		Error (%)	Accuracy (%)	Precision (%)
Fluke 101 Multimeter	**PZEM-017**			
1.00	1.01	1.33	98.67	98.90
1.50	1.52	1.56	98.44	98.10
2.00	2.03	1.50	98.50	99.34
2.50	2.47	1.20	98.80	99.19
3.00	3.02	0.78	99.22	99.71
3.50	3.51	0.19	99.81	99.49

The current test of PZEM-017 shows an error percentage of less than 1.57%. This indicates that in every condition, PZEM-017 still has good accuracy (above 98.44%) for current measurements. So it can be applied to measuring digital current values using a microcontroller. The precision percentage test results show good values too, above 98.10% in each condition. This shows that the level of data consistency is good for measuring current values using the developed system.

If observed, the error value obtained in the current test is greater, and the accuracy and precision values are lower than in the voltage test. This is because the variations in the voltage and current values provided are different. Where the voltage value has a range between 15.00V to 27.50V and the current value ranges from 1.00A to 3.50A. So the divisor value according to Equations (3), (4), and (5) becomes smaller. From the results of sensor calibration testing, error values were still found (although quite low). This is caused by the correction factor of the digital systems and the resolution of PZEM-017.

3.2. Application on 3KW electric vehicle

The track conditions during testing on electric vehicles were straight and relatively flat. By varying the treatment of rider load and distance, the characteristics of the Lithium-ion battery parameters that have been previously assembled will be known. Before testing, the voltage of the Lithium-ion battery was 83.5V in every condition.

Table 4. System Test Results On A 3KW Electric Vehicle

Weight (kg)	Distance (km)	Initial Voltage (V)	Final Voltage (V)	Current (A)	Duration (second)	Capacity (Ah)	Power Consumption (Wh)
50	0.6	83.5	82.1	4.98	245	0.339	27.83
	1.8		81.6	5.21	504	0.729	59.48
63	0.6		82	5.35	259	0.385	31.57
	1.8		81.3	7.13	540	1.07	86.99
83	0.6		81.5	7.64	396	0.84	68.46
	1.8		79.7	8.55	720	1.71	136.29

So based on Table 4, it can be seen that with the same load, but with a greater distance, there will be a greater decrease in the voltage value. The same phenomenon is also found when the distance is the same, but the increasing load will result in a greater decrease in the voltage value (See Figure 5.)

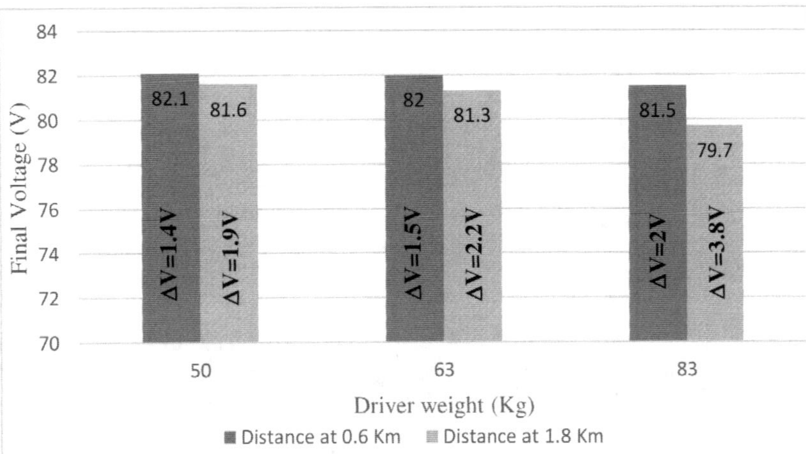

Figure 6. Test Results On A 3KW Electric Vehicle

This decrease is caused by the longer the distance traveled, the longer it takes for the Lithium-ion battery to generate energy to supply electric vehicles, resulting in a greater decrease in voltage values.

Furthermore, with the increase in the driver's load on an electric vehicle, the energy required to drive the electric motor will be even greater, so that the current flowing to the motor will be greater and there will be an even greater decrease in the voltage value [9], [11].

4. Conclusion

The 18650 Lithium-ion battery series has been successfully assembled. This system for measuring and monitoring battery parameters has also been successfully implemented with good measurement accuracy and consistency. The error value obtained in measuring the voltage is below 0.32% and below 1.57% for measuring the current value. Meanwhile, the precision values for voltage measurements are in the range of 99.59% to 99.80, and 98.10 to 99.71. The error occurs due to the accuracy of the tool and the resolution of the microcontroller. This measurement and monitoring system has been successfully applied to 3KW electric vehicles. Where the measurement results show that the increase in driving distance and driver load will be proportional to the increase in energy consumption from the battery. This is because the electric motor, as a source of vehicle propulsion, will work longer and harder.

Acknowledgments

All researchers team would like to thank Politeknik Negeri Jember through the Central Unit of Research and Community Service (P3M) for the financial and technical support in carrying out all research activities.

References

[1] M. Soerijadji, "Indonesia's Electric Vehicle Outlook Supercharging Tomorrow's Mobility," Jakarta, Jul. 2023.

[2] I. - International Energy Agency, "Electricity 2024 - Analysis and forecast to 2026," 2024. [Online]. Available: www.iea.org

[3] Energy Efficiency & Renewable Energy, "Electric Vehicle Basics," Washington DC, Aug. 2023.

[4] F. Alanazi, "Electric Vehicles: Benefits, Challenges, and Potential Solutions for Widespread Adaptation," *Applied Sciences (Switzerland)*, vol. 13, no. 10, May 2023, doi: 10.3390/app13106016.

[5] F. Alamsjah *et al.*, "Potential factors affecting adoption of electric vehicle by indonesia market," *ICIC Express Letters, Part B: Applications*, vol. 12, no. 11, pp. 1081–1090, Nov. 2021, doi: 10.24507/icicelb.12.11.1081.

[6] T. Chen *et al.*, "Applications of Lithium-Ion Batteries in Grid-Scale Energy Storage Systems," Jun. 01, 2020, *Tianjin University*. doi: 10.1007/s12209-020-00236-w.

[7] K. W. Wong and W. K. Chow, "Principle for the Working of the Lithium-Ion Battery," *Journal of Modern Physics*, vol. 11, no. 11, pp. 1743–1750, 2020, doi: 10.4236/jmp.2020.1111107.

[8] M. Rahimi, "Lithium-Ion Batteries: Latest Advances and Prospects," 2021, doi: 10.3390/batteries.

[9] A. T. Zain, D. D. Suranto, C. N. Karimah, F. A. Azhar, and D. A. Tyagita, "Analysis of Lithium-Ion Battery Consumption for Three-Wheeled Electric Vehicle with Variations in Weight and Speed †," *Engineering Proceedings*, vol. 63, no. 1, 2024, doi: 10.3390/engproc2024063013.

[10] B. Xu, L. Kong, G. Wen, and M. G. Pecht, "Protection Devices in Commercial 18650 Lithium-Ion Batteries," *IEEE Access*, vol. 9, pp. 66687–66695, 2021, doi: 10.1109/ACCESS.2021.3075972.

[11] G. Pratap Singh, Y. Lehri, L. Bhatia, and Y. Sehgal, "Monitoring of lithium-ion cells using a microcontroller," *Clean Energy*, vol. 6, no. 1, pp. 853–860, Feb. 2022, doi: 10.1093/ce/zkab058.

[12] P. Harahap, F. I. Pasaribu, and M. Adam, "Prototype Measuring Device for Electric Load in Households Using the PZEM-004T Sensor," *Budapest International Research in Exact Science*, vol. 2, no. 3, pp. 347–361, Jul. 2020, doi: 10.33258/birex.v2i3.1074.

[13] T. T. V. Hong, B. Delinchant, J. Ferrari, and Q. D. Nguyen, "Autonomous Electrical System Monitoring and Control Strategies to Avoid Oversized Storage Capacity," in *IOP Conference Series: Earth and Environmental Science*, Institute of Physics Publishing, Jul. 2020. doi: 10.1088/1755-1315/505/1/012045.

[14] P. M. Ramos, N. B. Bras, and A. C. Serra, "A New Calibration Method for Current and Voltage Sensors Used in Power Quality Measurements," Sorrenti: Institute of Electrical and Electronics Engineers (IEEE), Mar. 2007, pp. 2283–2288. doi: 10.1109/imtc.2006.328579.
[15] J. Lu, Y. Hu, G. Chen, Z. Wang, and J. Liu, "Mutual calibration of multiple current sensors with accuracy uncertainties in IPMSM drives for electric vehicles," *IEEE Transactions on Industrial Electronics*, vol. 67, no. 1, pp. 69–79, Jan. 2020, doi: 10.1109/TIE.2019.2896320.

IOP Conf. Series: Earth and Environmental Science **1446** (2025) 012057 doi:10.1088/1755-1315/1446/1/012057

Optimizing the Process of Extracting Palm Fruit from Politeknik Negeri Jember Plantation into Crude Palm Oil using Taguchi Method

Y Susmiati[1,*], B Rudiyanto[1], Sugiyarto[2], D A Prasetyo[1]

[1]Renewable Energy Engineering Study Program, Engineering Department, Politeknik Negeri Jember
[2]Plantation Crop Cultivation Study Program, Agricultural Production Department, Politeknik Negeri Jember

[*]e-mail: yuana_susmiati@polije.ac.id

Abstract. This research aims to study the effect of steaming time and imbibition water percentage on oil extraction from palm fruit. The independent variables used in this research were steaming time of 60, 90, and 120 minutes and imbibition water percentage of 15%, 20%, and 25%. The steaming process before extraction is carried out to facilitate the decomposition of the fruit flesh and speed up the release of oil from the palm fruit flesh. Research parameters include oil yield and quality of the oil produced, such as water content, free fatty acid (FFA) content, and impurity content. The data obtained were analyzed using the Taguchi method. Research has found that variations in steaming time and percent imbibition water had a significant effect on the yield and quality of the palm oil produced. Steaming for 120 minutes with an imbibition water percentage of 25% provides optimal yield and oil quality results. Steaming Time (A) has the most significant effect on CPO yield, as indicated by both the SNR and Mean values. These findings are expected to increase the efficiency and effectiveness of the palm oil extraction process.

1. Introduction

The global energy crisis is one of the biggest challenges facing the world today, especially due to high dependence on fossil energy sources. Energy sources such as oil, natural gas, and coal are non-renewable energy that will gradually run out. The decline in fossil energy reserves accompanied by a rapid increase in energy demand has caused an energy crisis that has an impact on unstable energy prices, decreased energy security, and increased greenhouse gas emissions. Therefore, the transition towards utilization of renewable energy is an important solution that continues to be pursued in various countries. Renewable energy such as solar power, wind, water, biomass, and biofuels have the potential to replace fossil energy while reducing the environmental impact caused by the use of conventional fuels [1].

One form of renewable energy utilization that is being developed in Indonesia is the processing of palm oil (CPO) into biodiesel. Biodiesel is a fuel made from biological resources such as vegetable oils, including palm oil, and animal waste. Indonesia, as the world's largest producer of palm oil, has great potential in developing palm-based biodiesel. The Indonesian government through the B30 policy (a mixture of 30% biodiesel with 70% conventional diesel fuel) is a concrete step in reducing dependence on fossil energy and maximizing the potential of palm oil for renewable energy [2].

Optimizing the process of extracting palm fruit into crude palm oil (CPO) is very important in the context of modern agriculture and the palm oil industry in Indonesia, one of the largest producers of CPO in the world. The extraction process, which involves a series of stages, such as sterilization, grinding, pressing, and clarification, plays an important role in determining the quality and quantity of oil produced. At the Politeknik Negeri Jember Gardens, this research is very relevant because it not only supports efforts to increase production output but also provides opportunities for technological development in an agro-industry-based educational environment. This optimization research is expected to provide significant results in increasing production efficiency and CPO quality in the plantation [3].

Challenges that face the palm oil industry include low levels of efficiency at several stages of the extraction process. In some cases, the low yield of oil produced from palm fruit can reduce productivity, especially in small and medium-scale industries. Therefore, an effective approach is needed to identify optimal parameters that directly influence the extraction results. Taguchi Method is an experimental design technique that has been proven effective in optimizing processes by minimizing variability and improving product quality [4]. This method provides a systematic approach to determining the process parameter settings that most influence the yield and quality of CPO.

At Politeknik Negeri Jember, oil palm plantations act not only as production units but also as research and education facilities. Thus, improving the extraction process will have a direct impact on student's practical learning and increase research results in the field of agricultural product processing technology. This research will provide new insight into how to maximize the potential of local palm fruit, enrich teaching materials, and contribute to the development of local technology that is applicable in the agro-industry sector [5].

Research conducted, that used the Taguchi method to evaluate various parameters in the CPO extraction process, found that sterilization temperature and pressing pressure were the two factors that most influenced the quality and yield of CPO. By optimizing these two parameters, this research succeeded in increasing palm oil yield by up to 15% compared to conventional processes. This research shows that the Taguchi method is not only able to increase production yields but also reduce energy costs and process waste [3]. Another research focuses on the application of the Taguchi method in the Jember State Polytechnic Garden. They experimented with several key parameters, including sterilization time, pressing pressure, and oil separation speed. The results of this research show that an optimal combination of parameters can improve the quality of palm oil, especially in terms of free fatty acid (FFA) levels. They also noted that the Taguchi method is very effective in reducing process variability, resulting in products that are more consistent in terms of quality [5]. In addition, research found that the Taguchi method can also be used to identify optimal points in the CPO clarification process. This research focuses on the parameters of temperature and stirring speed during the clarification process, which are important for reducing impurities and moisture levels in oil. Through optimization, they were able to reduce dirt levels by up to 25% and increase overall process efficiency. This research strengthens the evidence that the Taguchi method can be widely applied in various stages of the palm oil extraction process [4]. Another study applied the Taguchi method on an industrial scale to increase palm oil extraction yields in palm oil mills. They focused on the pressing and drying parameters in the extraction process. The research results show that this method can identify the optimal combination of parameters, which results in an increase in production efficiency of up to 10% and reduces the energy required in the extraction process. They also emphasize that the Taguchi method allows factories to significantly reduce waste and operational costs [6].

2. Method

2.1. Extraction process

This research is aimed at finding out the most optimal process in the process of extracting CPO from palm fruit seen from the length of steaming time (A), namely 60, 90, and 120 minutes as well as the addition of imbibition water (B) during pressing with the help of a hydraulic press, namely 15, 20, and 25% by weight. The treatment variations and reactor naming can be seen in **Table 1**.

Table 1. Treatment Variations and Reactor Naming

Steaming Time (A)	Imbibition Water (B)		
	B1 (15%)	B2 (20%)	B3 (25%)
A1 (60 min)	A1B1	A1B2	A1B3
A2 (90 min)	A2B1	A2B2	A2B3
A3 (120 min)	A3B1	A3B2	A3B3

The palm fruit used in this research came from oil palm plantations at the Politeknik Negeri Jember taking ripe fruit. Ripe fruit is separated from the bunch manually with the help of tools such as knives, gloves, and several other equipment to make it easier to separate the fruit from the bunch.

Then the separated fruit is immediately processed and each repetition uses 1 kg of palm fruit. So, around 3 kg is needed for each steaming process with 3 repetitions made each time. So, the ripe palm fruit needed in this research is around 30 kg with an additional 3 kg as blank. During the steaming process, 3 kg of ripe palm fruit is added with 6 liters of clean water added which will be used during the 1 hour steaming process for variation A1. After the steaming process, it is immediately continued with the extraction of palm oil without any pause so that the soft fruit does not harden again. Oil extraction was carried out using a hydraulic press which also added imbibition water of 15% by weight or 150 mL for every 1 kg of sample (1 replication) in variation B1.

This steaming and pressing process is carried out repeatedly until all variations and repetitions have been carried out. Where variations A2 and A3 steam the palm fruit for 1.5 and 2 hours respectively. Meanwhile, variations B2 and B3 are treatments with the addition of imbibition water of 20 and 25% of the weight of the palm fruit or 200 and 250 ml per 1 kg of palm fruit (1 sample replication). The complete process of extraction process can be seen in **Figure 1.**

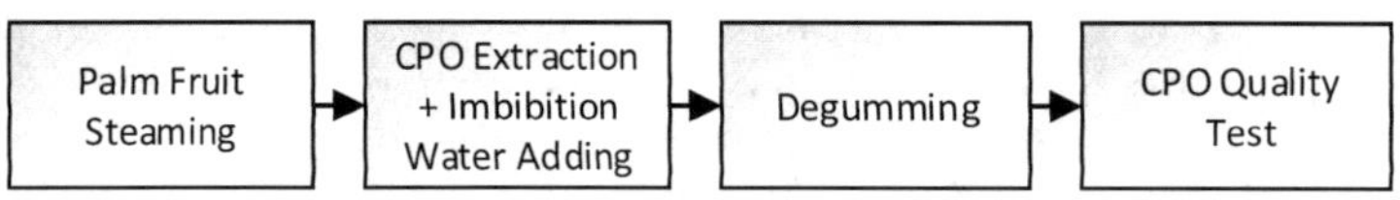

Figure 1. Crude Palm Oil Extraction Process

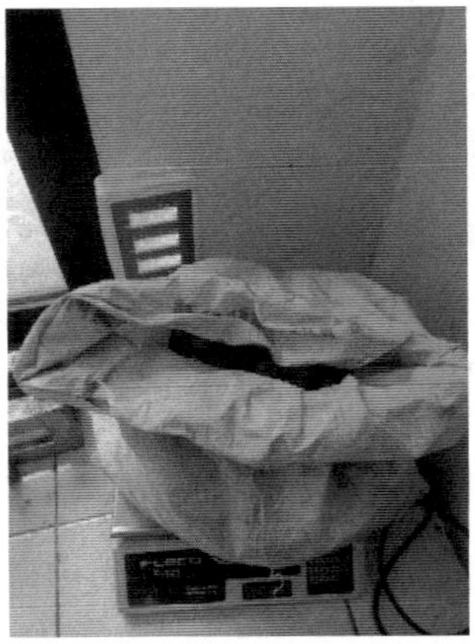

Figure 2. Weighing palm fruit to be steamed

Figure 3. Palm fruit steaming process

Figure 4. Final weighing after steaming

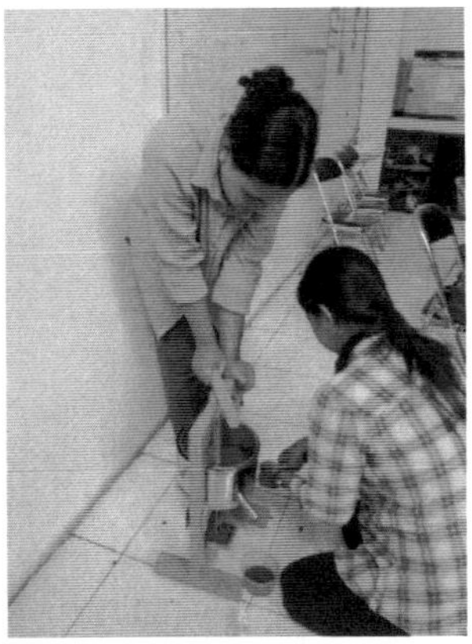

Figure 5. Pressing process with a hydraulic press.

Figure 6. Pressing process with the addition of imbibition water

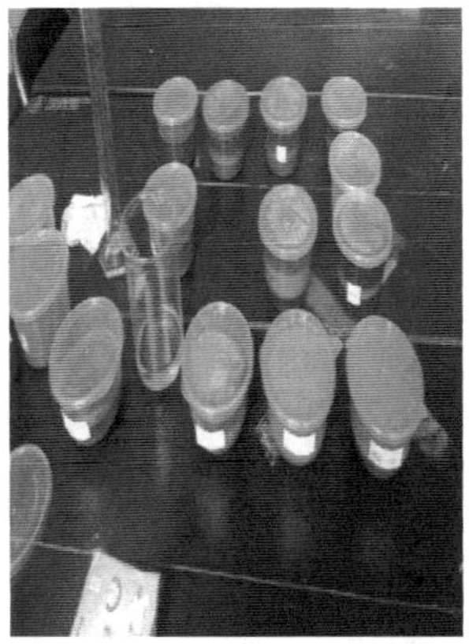

Figure 7. CPO resulting from the extraction process

2.2. Degumming process

Degumming is an important stage in the Crude Palm Oil (CPO) refining process, which aims to remove phospholipids, proteins, and other impurities that can cause problems during further processing. Several methods can be used for degumming CPO, namely water degumming, acid degumming, and enzymatic degumming. However, in this research, the acid degumming process was chosen. The advantage of acid degumming is that it can remove non-hydrate phospholipids which cannot be removed by water degumming [7]. These phospholipids are converted into hydrated phospholipids by adding acid (usually phosphoric acid or citric acid, in this study phosphoric acid was used) before the hydration process. Phosphoric acid is added to CPO at a temperature of around 80°C in an amount of 0.1% by weight of the oil. After that, 100 mL of hot water was added to hydrate the phospholipids. The mixture is heated to a temperature of around 80°C and stirred for a short time using a hot plate and magnetic stirrer. After that, the gum is separated using filter paper.

Figure 8. Add 1 mL (1% mass) of phosphoric acid to the CPO

Figure 9. Heated and stirred on a hot plate at a temperature of 80°C

Figure 10. Gum is separated with the help of filter paper.

2.3. *Quality testing process*

Based on the Indonesian national standard SNI 3741:2013 concerning cooking oil, several parameters must be tested including odor, color, water and volatile matter content, acid value, peroxide value, and several metal contents [8]. So, in this research, we tested several parameters that are important for knowing the quality of the oil that will be used as raw material in making biodiesel. The parameters tested in this research include free fatty acid content, water and volatile matter content, insoluble impurities, iodine value, and oil density.

Figure 11. Iodine value test using AOCS Tg 1-64 standard method.

Figure 12. The FFA level test using the AOCS Ca-5a-40 standard method.

Figure 13. Insoluble impurities test using AOCS Ca 3a-46 standard method.

Figure 14. Density test using AOAC 920.212 and 920.213 standard method

Figure 15. Moisture and volatile matter test using AOCS Ca 2c-25

3. Result

3.1. Optimization of crude palm oil extraction

The results of crude palm oil (CPO) extraction were analyzed using the Taguchi method with the independent variables being steaming time and the addition of imbibition water. The steaming times used were 60, 90, and 120 minutes, while the addition of imbibition water was carried out with variations of 15%, 20%, and 25%. The Taguchi method is used to determine the optimal combination of these two factors to increase CPO extraction efficiency. Data from Matlab in analyzing the extraction of CPO with the Taguchi method can be seen in **Table 2** and **Table 3**.

Table 2. Operating parameter for steaming time and imbibition water

No	Operation parameters	Level 1	Level 2	Level 3
A	Steaming time (min)	60	90	120
B	Imbibition water (%)	15	20	25

Table 3 Taguchi experimental design and Orthogonal Array (codes)

Run of experiment	Steaming time/A (min)		Imbibition water/B (%)	
1	60	1	15	1
2	60	1	20	2
3	60	1	25	3
4	90	2	15	1
5	90	2	20	2
6	90	2	25	3
7	120	3	15	1
8	120	3	20	2
9	120	3	25	3

3.2. Degumming result

The results of the CPO degumming process show that there are changes in the compounds contained in each oil by analyzing using the GC-MS method. Before degumming is carried out, it can be seen in **Figure 16** and the list of compounds is in **Table 4**. Meanwhile, after the degumming process is carried out, the compound content in the oil can be seen in **Figure 17** and **Table 5**. Likewise, the compounds contained in the filtered gum can be seen in **Figure 18** and **Table 6**.

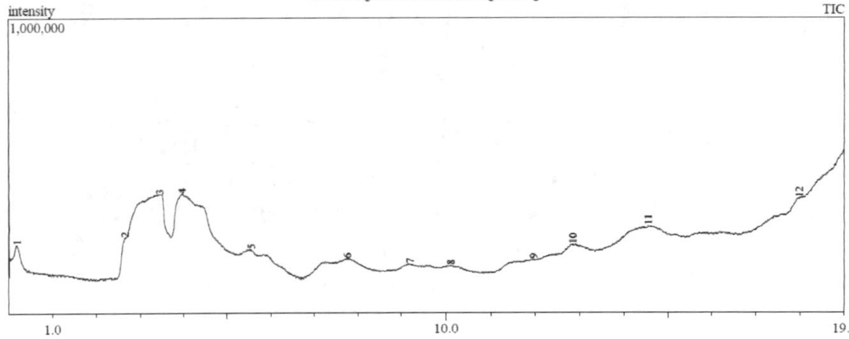

Figure 16. Chromatogram CPO before degumming

Table 7. Compounds contained in CPO before degumming

Peak#	R.Time	Area%	Name
1	0.193	0.72	3-(Cis-2'-Hydroxy-Cyclohexyl) Propanol
2	2.647	0.27	Methane, tetranitro- (CAS)
3	3.433	13.85	Cyclopropyl carbinol
4	3.968	8.04	Cyclotrisiloxane, hexamethyl- (CAS)
5	5.538	0.22	Benzene, methyl- (CAS)
6	7.762	3.71	1-Pentalenol, 1,2,3,3a,4,6a-Hexahydro-
7	9.199	0.66	Furan, 2-pentyl- (CAS)
8	10.107	0.67	1,3-di(1-propenyl)cyclopentane
9	11.954	1.46	Tricyclo[2.2.1.0(2,6)]heptan-3-ol, acetate (CAS)
10	12.837	0.98	Nonanal (CAS)
11	14.542	4.4	.beta.-d-Lyxofuranoside, phenyl
12	17.987	0.48	2-Decenal, (E)- (CAS)
13	20.333	24.01	Borinic Acid, Diethyl-, 5-Hexynyl Ester
14	20.975	26.15	2-Hexyl-1-decanol
15	23.575	14.38	Tricosanoic acid, methyl ester (CAS)
		100	

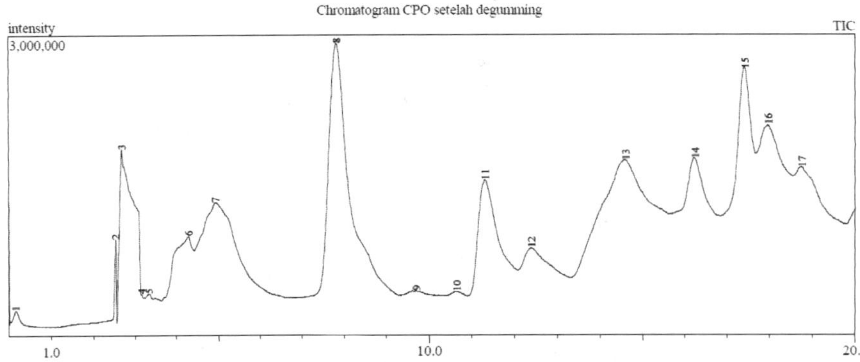

Figure 17. Chromatogram CPO after degumming

IOP Conf. Series: Earth and Environmental Science **1446** (2025) 012057 doi:10.1088/1755-1315/1446/1/012057

Table 8. Compounds contained in CPO after degumming

Peak#	R.Time	Area%	Name
1	0.165	0.20	9-Octadecenal, (Z)- (CAS)
2	2.532	0.42	(S)-(+)-1-Cyclohexylethylamine
3	2.678	7.42	Carbamic acid, monoammonium salt (CAS)
4	3.152	0.15	Methane, chloro- (CAS)
5	3.333	0.55	Oxetane, 2,2-Dimethyl-
6	4.298	4.70	Acetamide, 2,2-dichloro- (CAS)
7	4.933	13.76	Acetic acid (CAS)
8	7.822	18.05	2-Furancarboxaldehyde
9	9.710	0.65	Ethanone, 1-(2-furanyl)- (CAS)
10	10.642	0.19	1,3-Dioxolan-4-one, 2-(1,1-dimethyl ethyl)-5-(1-methyl ethyl)-, (2s-cis)- (CAS)
11	11.312	6.64	5 Methyl Furfural
12	12.396	2.54	2,5-Furandione, dihydro-3-methylene-
13	14.605	15.81	2-Octanol (CAS)
14	16.247	7.39	2-Furancarboxylic acid, methyl ester (CAS)
15	17.430	8.77	2-Furanmethanol (CAS)
16	17.981	8.30	Pentanoic acid, 4-oxo- (CAS)
17	18.766	4.47	Pyrimidine, 4-chloro-5-ethoxy-2-methyl- (CAS)
		100	

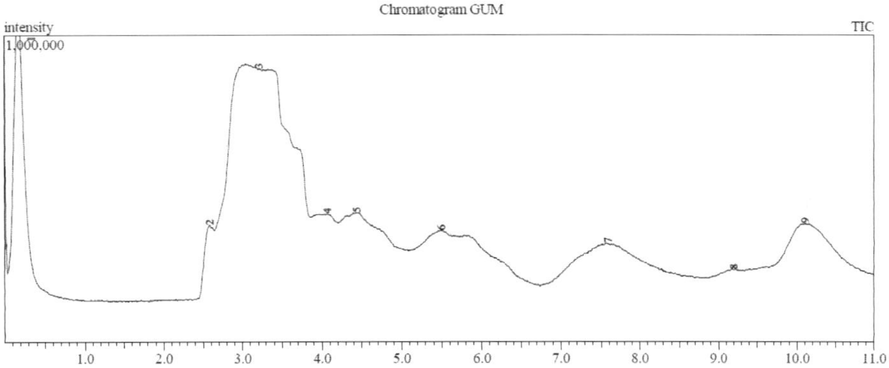

Figure 18. Chromatogram gum

Table 9. Compounds contained in gum

Peak#	R.Time	Area%	Name
1	0.172	14.58	9-Octadecenal, (Z)- (CAS)
2	2.583	4.11	Methane, tetranitro- (CAS)
3	3.212	4.71	3-(3,3-Bis-[1,3]Dioxolan-2-Yl-1-Methyl-Propylcarbamoyl)-Bicyclo[2.2.1]Heptane-2-Carboxide
4	4.066	10.69	1-Heptan-1,1-D2-Ol
5	4.434	15.43	1-Pentanol, 2-methyl- (CAS)
6	5.495	13.49	Benzene, methyl- (CAS)
7	7.577	17.62	7-Butyl-1,3,5-Cycloheptatriene
8	9.195	1.95	Furan, 2-pentyl- (CAS)
9	10.101	17.62	Octanal (CAS)
		100	

3.3. Quality of crude palm oil

The crude palm oil used in this research was obtained from the Politeknik Negeri Jember palm fruit plantation. This CPO is extracted by using manual tools where there is a change in the color of the oil so that you first know the characteristics of the used cooking oil starting from the FFA content, density, and acid number. **Table 10** is test data on the characterization of crude palm oil.

Table 10. The Quality of Crude Palm Oil

Quality parameters	Units	Value	Standard test method
Free Fatty Acid	%	9	AOCS Ca 5a-40
Density	g/cm^3	0.89	AOAC 920.212, AOAC 920.213
Moisture and volatile matter	%	8	AOCS Ca 2c-25
Insoluble impurities	%	12	AOCS Ca 3a-46
Iodine value	g I_2/100g	35	AOCS Tg 1-64

4. Discussion

The results of data processing using the Taguchi method and with the help of the Mathlab application obtained analysis results which can be seen in **Table 11**, **Table 12,** and **Table 13**.

Table 11. Data and Taguchi Analysis of CPO Yield for Larger is better

Run of experiment	Steaming time (A)	Imbibition water (B)	R1	R 2	R 3	SNR A	MEAN
1	1	1	131	151	132	42,7430	138,000
2	1	2	135	139	119	42,2855	131,000
3	1	3	95	167	112	41,2228	124,667
4	2	1	122	121	122	41,7032	121,667
5	2	2	170	129	130	42,8998	143,000
6	2	3	110	111	138	41,4203	119,667
7	3	1	154	128	103	41,8164	128,333
8	3	2	145	112	151	42,4377	136,000
9	3	3	184	198	159	45,0123	180,333

Table 12 Response for Signal to Noise Ratios (Larger is better)

Level	Steaming time	Imbibition water
1	42,08	42,09
2	42,01	42,54
3	43,09	42,55
Delta	1,08	0,46
Rank	1	2

The SNR values for Steaming Time (A) and Imbibition Water (B) show how consistent and high-performing each factor level is under varying conditions. Steaming Time (A): The levels show an increasing trend from Level 1 (42.08) to Level 3 (43.09), indicating that the third level of Steaming Time (longest steaming) produces the most stable and higher yields with less variability. This implies that increasing the steaming time may help extract more oil. Imbibition Water (B): The SNR values are relatively close between levels, with Level 3 (42.55) being marginally higher than the others. This suggests that while more water enhances the yield, its effect is less significant compared to the steaming time. In similar studies on oil extraction processes, the influence of temperature or time-related factors is often more pronounced than that of water addition. This aligns with the findings here. For instance, emphasized that temperature and steaming time had more influence on palm oil yield than other factors like moisture content.

Table 13 Response for Means (Larger is better)

Level	Steaming time	Imbibition water
1	131,2	129,3
2	128,1	136,7
3	148,2	141,6
Delta	20,1	12,2
Rank	1	2

From the mean response table, the average yield increases from Level 1 to Level 3 for both factors, but more drastically for Steaming Time (A): Steaming Time (A): The means increase from 131.2 (Level 1) to 148.2 (Level 3). This supports the conclusion that longer steaming enhances oil recovery, possibly by loosening the oil cells within the palm fruit more effectively. Imbibition Water (B): The means increase from 129.3 (Level 1) to 141.6 (Level 3). This suggests that while higher water usage aids in extracting more oil, it's a secondary factor compared to steaming time.

Delta (Δ) values: Steaming Time (A): Δ = 1.08 for SNR, 20.1 for Means. Imbibition Water (B): Δ = 0.46 for SNR, 12.2 for Means. The larger Delta values for Steaming Time indicate that it has a more substantial influence on the CPO yield than Imbibition Water. This ranking is consistent with findings from other optimization studies in agricultural and food engineering, where process time (steaming, drying, etc.) is often the dominant factor influencing yields.

5. Conclusion

The results of the research showed that the best optional setting in extracting crude palm oil is level 3 at steaming time and imbibition water which is the longest time and the highest number of imbibition water adding that reach the highest SNR and mean.

Acknowledgments

The author would like to thank the PNBP Politeknik Negeri Jember for giving us the funding for this research and also Workshop Energi dan Mekanik laboratory, Politeknik Negeri Jember for allowing us to carry out this research analysis.

Bibliography

[1] D. Hartono and B. Rudiyanto, "Analisis Potensi Energi Terbarukan untuk Mengurangi Ketergantungan Energi Fosil," *Jurnal Energi dan Lingkungan,* vol. 15, no. 2, pp. 112-125, 2020.

[2] H. Purwanto, "Implementasi Kebijakan B30 dalam Mendorong Penggunaan Biodiesel berbasis Minyak Kelapa Sawit di Indonesia," *Jurnal Teknologi Agroindustri,* vol. 8, no. 3, pp. 145-158, 2019.

[3] A. M. Siregar and P. Marbun, "Pengaruh Parameter Proses terhadap Rendemen dan Kualitas Minyak Sawit menggunakan Metode Taguchi," *Jurnal Teknologi Pertanian,* vol. 22, no. 1, pp. 35-45, 2021.

[4] M. H. Yusuf and R. Fahmi, "Penggunaan Metode Taguchi untuk Mengoptimalkan Proses Ekstraksi CPO di Industri Minyak Sawit," *Jurnal Rekayasa Proses,* vol. 18, no. 3, pp. 102-115, 2020.

[5] E. R. Putra and M. Sudrajat, "Optimalisasi Proses Produksi Sawit dengan Pendekatan Taguchi di Politeknik Negeri Jember," *Jurnal Agroindustri,* vol. 11, no. 2, pp. 78-86, 2021.

[6] Y. S. Hasibuan and D. Simatupang, "Optimalisasi Pengepresan dan Pengeringan pada Proses Ekstraksi Minyak Sawit menggunakan Metode Taguchi," *Jurnal Teknologi Agroindustri,* vol. 9, no. 4, pp. 210-220, 2020.

[7] P. Bondioli, L. Folegatti and S. Galli, "Degumming Process in the Oil Industry," *Journal of the American Oil Chemists' Society,* vol. 74, no. 9, pp. 1061-1065, 1997.

[8] Badan Standarisasi Nasional Indonesia, "Standar Nasional Indonesia (SNI) 3741:2013," Badan Standarisasi Nasional Indonesia (BSNI), Jakarta, 2013.

[9] Sekretariat Jendral Dewan Energi Nasional, Outlook Energi Indonesia, Jakarta: Sekretariat Jendral Dewan Energi Nasional, 2019.

[10] M. Nurhadi, "Petani Kelapa Sawit: Kebijakan Biodiesel Pemerintah Habiskan Bahan Baku Minyak Goreng," suara.com, 21 Maret 2022. [Online]. Available: https://www.suara.com/bisnis/2022/03/21/130042/petani-kelapa-sawit-kebijakan-biodiesel-pemerintah-habiskan-bahan-baku-minyak-goreng. [Accessed 7 Mei 2023].

[11] American Oil Chemists Society, "AOCS Official Method Ca 5a-40 Reapproved 1997 'Free Fatty Acid'," AOCS, Urbana, 1997.

[12] Association of Official Agricultural Chemists (AOAC), "Specific Gravity of Oils, Temperature Correction," AOAC Inc, Arlington, 1990.

[13] Association of Official Agricultural Chemists (AOAC), "Specific Gravity (Apparent) of Oils, Pycnometer Method," AOAC Inc, Arlington, 1990.

[14] American Oil Chemists Society (AOCS), "Moisture and Volatile Matter Air Oven Method," AOCS, Urbana, 1997.

[15] American Oil Chemists Society (AOCS), "Insoluble Impurities," AOCS, Urbana, 1997.

[16] American Oil Chemists Society (AOCS), "Iodine Value, Wijs Method," AOCS, Urbana, 2009.

Bioethanol Production with Semi Continue Fermentation Distillation Equipment using Sacharomyces Cerevisiae

Siti Diah Ayu Febriani[1], Yuana Susmiati[1], Hariadi Subagja[2], Qanitah[1], Nuzula Afianah[1], Abdilah Faqih[1]

[1]Engineering Department, Politeknik Negeri Jember, JL. Mastrip 164 Jember
[2]Animal Science Department, Politeknik Negeri Jember, JL. Mastrip 164 Jember

*Email:siti_diah@polije.ac.id

Abstract. Ethanol produced from fermentation of biomass that consist of carbohydrates, cellulose or glucose with the aid of microorganisms namely Bioethanol. This research was to developing a bioethanol production tool with a fermentation and semi continue distillation using a stirred bioreactor and evaluate the productivity of bioethanol fermentation with saccharomyces cerevisiae that capable of concentration ethanol to support development activities at Teaching Factory (TEFA) Renewable Energy as the Bioenergy competency testing center. The method of the research using an experimental method use system of semi continue with the research stages included the design, fabrication, installation of sensors and final functional testing of the equipment. The research was carried out ethanol production were observed every 12-hours interval during 72 hours fermentation time. The results showed that process of semi continue fermentation done only for 72 hours when the brix concentration is 20 (start). Based on the research it is known that reducing sugar concentration (brix) and pH will decrease as fermentation time increase. The fermentation process carried out produced 66,49% bioethanol.

Keyword: bioethanol, fermentation, semi-continuous

1. Introduction

Ethanol is a chemical used as a solvent and ethanol used as a mixture of gasoline fuel with concentration approaching 100% [1] Bioethanol (C_2H_5OH) is liquid biochemistry of the sugar fermentation process from carbohydrate, cellulose or glucose source using microorganisms. Biomass conversion to bioethanol largely depends on type of biomass and biochemical composition. Bioethanol can be used for various purpose depend on concentration; bioethanol at 95-99% used as a substitute for gasoline, 80% concentration for sterilizing medical devices [2]. Bioethanol can be used as vehicle fuel by mixing with gasoline and have high octane number than gasoline. The factors influencing the production of ethanol are temperature (28^0C-32^0C), pH (for saccharomyces cerevisiae 4-5) and incubation time (42-72 hours) [3]

Bioethanol processing generally involves three main stages: saccharification, fermentation and distillation. The production of bioethanol from biomass derived from forest waste begins with pre-treatments of the material using sulphuric acid and sodium hydroxide by hydrolysis with cellulose enzymes and fermentation with saccharomyces cerevisiae for 12-72 hours to produced ethanol. The next process is distillation with ethanol concentrations of 35-45% [4] . Ethanol production in the fermentation process is influenced by the type of microbes used as well as aeration and agitation. The microorganism used is Saccharomyces cerevisiae because it is widely found in the nature, has high survival rates and is capable of producing high amounts of alcohol. Under conditions of 100 rpm agitation and an aeration rate of 0,3 ppm, it can produce 5,43% ethanol with a concentration of 4.07 g/L [5] . The process of separating a mixture of liquid or vapour into pure component fractions is distillation. An ethanol distillation apparatus consist of several main components; an evaporator tank to hold and heat the material, a dehydratorcolumn containing Zeolite to absorb water vapour from the ethanol and distillation column for condensation and cooling, which is converts alcohol vapour into liquid [6] .

The fermentation tools developed is designed to be semi-continuous with a distillation tool, when bioethanol starts to form during the fermentation process, the distillation process can be carried out and the fermentation substrate is slowly reduced while the substrate in the fermenter is still continuing fermentation. According to [7] in the batch ethanol fermentation process, accumulation of toxic dissolved products will slowly decrease so that it can stop the growth and production of ethanol-forming microorganisms. The distillation process for a lot of direct substrate (by batch) also requires quite heating time and energy. The semi-continuous fermentation and distillation systems can make the ethanol production process more efficient. NPK fertilizer is a type of fertilizer that contains the essential nutrients nitrogen, phosphorus, and potassium [8] . NPK is an inorganic fertilizer that is highly soluble in water and easily absorbs moisture. Urea fertilizer has high nitrogen (N) content, amounting to 46%. This means that for every 100 kilograms of urea fertilizer, there are 46 kilograms of nitrogen [9] .

The production of bioethanol has been widely developed through research and education. For example [10] that enhancement of bioethanol productivity by Saccharomyces Cerevisiae in a non-conventional bioreactor under ELF magnetic fields and [5] explored sustainable bioethanol production from office paper waste and its purification through mixed polymer. This research aims to construct semi continue fermentation and distillation equipment for bioethanol production which can be used to produced ethanol to support the Renewable Energy engineering Teaching Factory (TEFA) as well as a bioenergy competency testing center. The fermentation equipment developed is designed to operate semi-continuously with the distillation unit. It is hoped that the development of this equipment design can produce bioethanol in a faster time because when bioethanol is formed (48 hours of fermentation), the distillation process carried out as well. The fermentation substrate is reduced gradually through a tap connected using a hose to the distillation apparatus, while fermentation continues on the remaining substrate within fermenter.

2. Materials and Methods

2.1 Location, equipment, materials

Research conducted at energy and mechanics workshop, department of engineering Politeknik Negeri Jember over period of 8 months. The equipment required for this research includes workshop tools such grinders, welders, hammers and others as well as laboratory experiment tools (measuring cylinders, analytical balances, beakers. Testing tools include hand refractometer for sugar content, oven, alcohol meter, pycnometer, thermocouple/thermometer and pH meter.

2.2 Fermenter and distillatory design

According to [11] The design of fermenter with a stirrer consist of tube design and stirrer design as in figure 2.1.

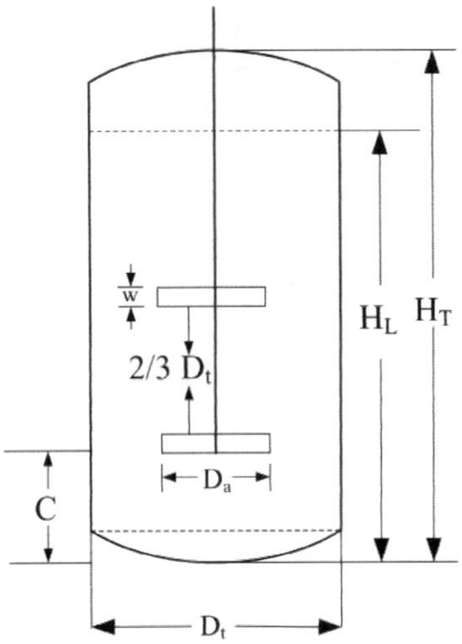

Figure 1. Fermenter Design

The fermenter is a cylindrical vessel with a height and diameter ratio:

$$\frac{H}{D_t} = 2 \qquad (1)$$

The working volume is set to a maximum of 70-80% and a minimum of 20%. The volume fermenter is determined using the formula for a cylinder:

$$V = \frac{\pi}{4} x\, D_t^2 \text{ x H} \qquad (2)$$

Where:
V= fermenter volume (cm^3)
H= fermenter height (cm)
Dt= fermenter diameter (cm)

The stirrer inside the fermenter is designed with the following specifications:

$$\frac{D_a}{D_t} = 0.6 \; to \; 0.8 \qquad (3)$$

$$\frac{W}{D_a} = \frac{1}{6} \; to \; \frac{1}{10} \qquad (4)$$

$$\frac{C}{D_a} = \frac{1}{3} \qquad (5)$$

Where:
Dt = tank diameter (cm)
Da = impeller diameter (cm)
W = impeller width (cm)
C = height of stirrer from the bottom (cm)

Destillaation design is a simple distiller consist of three main components: evaporator tank, distillation column and condenser.

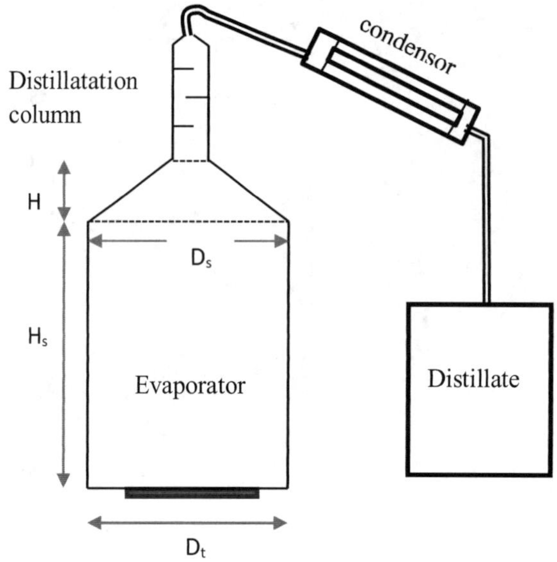

Figure 2. Distillation Design

Design of the evaporator tank using the formula :

$$H_h = \{\frac{D-d}{2}\} \times \tan\theta \quad (6)$$

$$V_s = \pi/4 \, . D_t^2 \, . H_s \quad (7)$$

$$H_T = H_s + H_h \quad (8)$$

$$D_t = D_a \quad (9)$$

Where:
Vs = Tank Volume (cm^3)
Dt = Tank Diameter (cm)
Da = Cover Diameter (cm)
Hs = Height Tank (cm)
Hh = Cover Height (cm)
Ht = Total Height (cm)

The condenser of the distillation apparatus is design as a shell and tube type, where the distillation vapour pass through tubes located inside a shell filled with cooling water. The cooling water circulates from a cooling water tank. In analysing heat transfer in the condenser, the best approach is to determine average temperature difference between the hot fluid and cold fluid using the formula:

$$Q = U.A.\Delta T_m \quad (10)$$

Where:
Q = rate of heat transfer (W)
U = heat transfer coefficient ($W/m^2.^oC$)

A = heat transfer surface area (m^2)
ΔT_m: Log Mean Temperature Difference (°C)
LMTD are temperature difference of heat exchange using natural logarithm with two temperature differences. LMTD is the logarithmic average of temperature difference between hot and cold side. LMTD method is approach to finding value of the temperature difference between two fluids in heat exchange as a totally.

2.3 Research Stages

The research will be conduct in stages, the design phase of the fermenter and distillation, the construction of the fermenter and distillation according the design, functional testing of the equipment, installation of the sensors and performance test. The performance test result include the ethanol yield and ethanol content produced. The fermentation performance is observed through the success of the fermentation process, including temperature, pH, and sugar content.

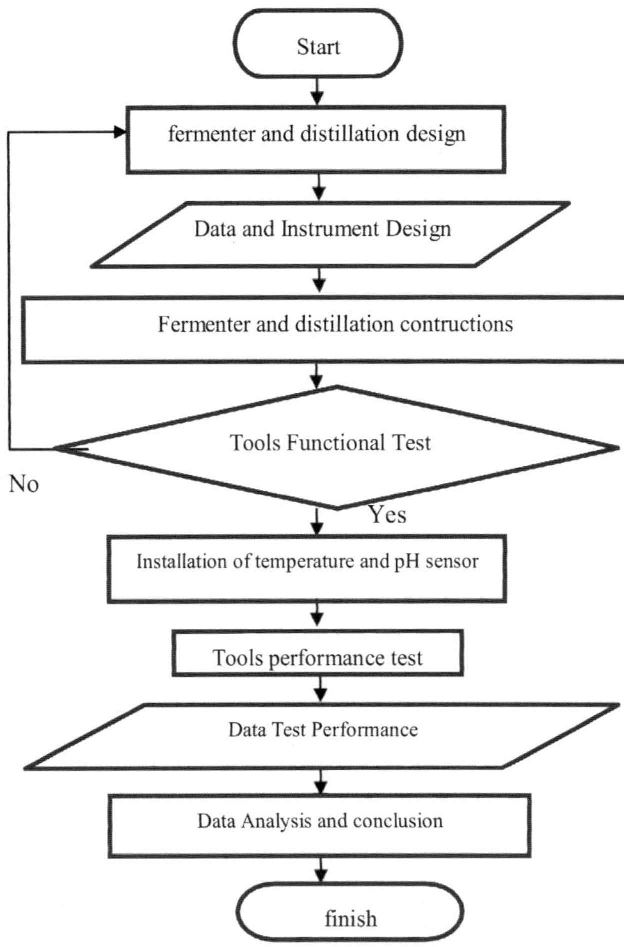

Figure 3. Research Diagram

IOP Conf. Series: Earth and Environmental Science **1446** (2025) 012058 doi:10.1088/1755-1315/1446/1/012058

2.4 Fermentation Process

The medium or substrate used for fermentation is molasses. The fermentation process begin with the preparation of the medium, which is involves diluting the molasses and pH. The sugar content of the molasses is measured, and it is diluted until the sugar concentration reaches 15%. Sulfuric acid (H_2SO_4) is added to lower pH (2-3). The molasses is heated to 90°C to expedite the precipitation of insoluble solids. The molasses solution is filtered to separate dissolved solids. The filtered molasses is then treated with 0.5 M NaOH until the pH reaches 4.5, which is the optimum pH for yeast growth. The next step involves adding nutrients to the fermentation medium, including urea at a concentration of 0.4285 g/L and 10% phosphoric acid at 0.07 mL/L. The prepared molasses is then heated to 121°C for 15 minutes to achieve sterilization. The microbial starter *Saccharomyces cerevisiae* made from bread yeast with volume of 10% into molasses with a sugar concentration of 15% and incubated for 12 hours at 30°C. The bioethanol production process is conducted in the previously prepared medium, with agitation performed at varying speeds. The parameters observed during fermentation include microbial cell density, reducing sugar concentration, pH, and temperature. Fermentation samples are taken every 12 hours, when fermentation process indicates the formation of ethanol, the substrate is gradually transferred to the distillation apparatus (evaporator) for a stepwise distillation process. The remaining substrate in the reactor continues to produce ethanol while the distillation process also progresses. The parameters observed during distillation include the distillation rate, ethanol concentration in the distillate, and distillation yield. Throughout the distillation process, the temperature is maintained consistently over time.

3. Results and Discussion

Organic biofuel that has been obtained from lignocellulose biomass, a renewable substrate is bioethanol that through a process in numerous unit operations converted an ethanol solution with high purity grade. Foods that contain sugar, produced through fermentation, can be used to make bioethanol. Bioethanol can be produced from feedstocks such as sucrose, lignocellulosic, startch, algal and biomass through fermentation process by microorganism [12]. The chemical change in organic substrates caused by microorganisms through enzyme activity is called fermentation. Bacteria, yeast and fungi are the microorganisms involved in the fermentation process. The main principle of fermentation is to activate certain microbial activities to alter the properties of the material by producing fermentation products. During fermentation, sugar is broken down into alcohol and carbon dioxide, primarily by the yeast *Saccharomyces cerevisiae.* According to [13] the ethanol or alcohol reaction produces H_2O and CO_2 in smaller amounts through using reaction:

$$C_6H_{12}O_6 + \textit{Saccharomyces cerevisiae} \rightarrow 2C_2H_5OH + 2CO_2$$

Several factors that can influence fermentation include microbes, pH (acidity), temperature, oxygen, and nutrients for the microbes [6] . In addition to microbes, nutrients, temperature, pH, and oxygen, fermentation is also influenced by time. The variable related to the development stage of microbes during the fermentation process is fermentation time, which affects the fermentation yield. If fermentation is conducted for a short time, the amount of bioethanol produced will be low. According to [14] , if the fermentation time is extended, more bioethanol will be produced. This is influenced by the condition of *Saccharomyces cerevisiae*, which is sensitive to pH changes during fermentation, as pH plays a crucial role.

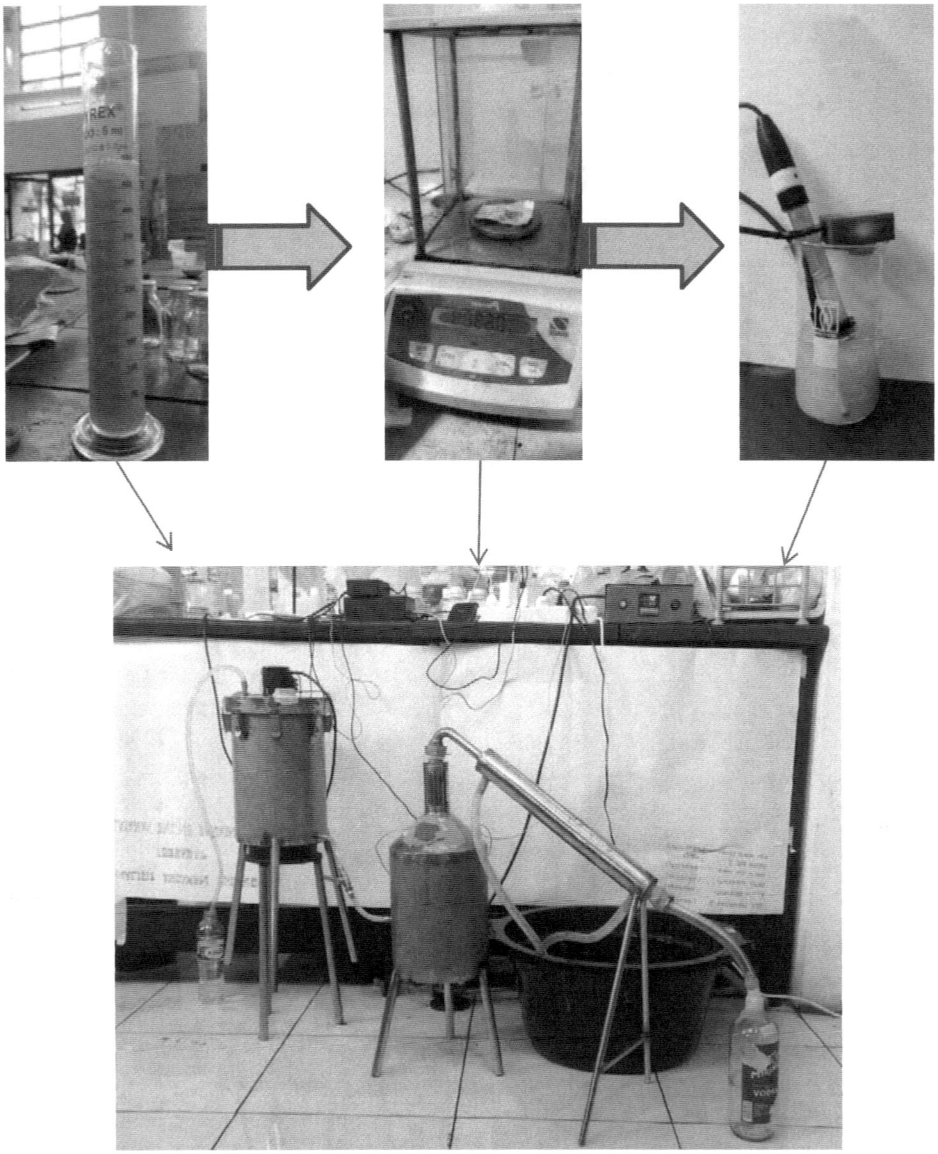

Figure 4. Bioethanol Production Process

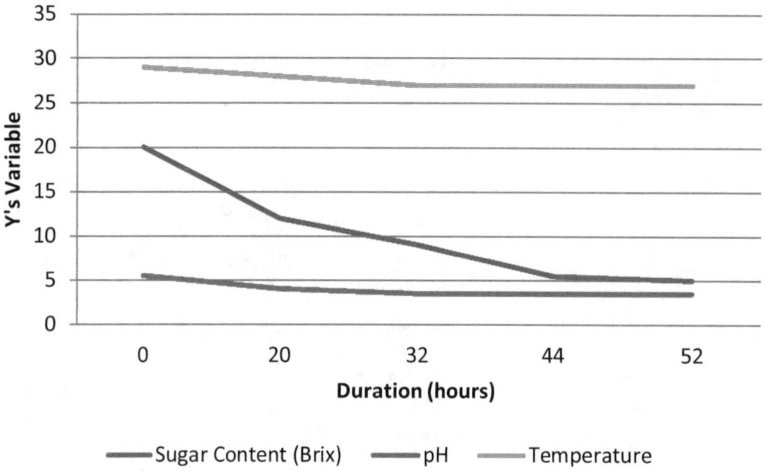

Figure 5. Bioethanol Process

According to Figure 5. Bioethnol process using microba Saccharomyces cerevisiae 1,4352 gram, NPK 0,3744 gram and Urea 3,12 gram produced pH, sugar content and temperature for three days of research with same time duration (12 hours). The optimal pH range for *Saccharomyces cerevisiae* is between 3 and 5. At this pH level, *Saccharomyces cerevisiae* can effectively break down sugars and prevent the growth of other types of bacteria. If the pH falls below 3, the fermentation process will slow down due to enzyme activity [15] .Fermentation by *Saccharomyces cerevisiae* is more effective in producing ethanol between 5.1% and 91.8% [16] . The addition of NPK elements and urea fertilizer is one way to support growth of *Saccharomyces cerevisiae* during the fermentation process [17] . The addition of NPK and urea fertilizer serves the same function in the bioethanol fermentation process, acting as nutrients for microbial growth. This enhances the microbes' metabolism in oxidizing sugar into ethanol more efficiently [18] .

The demand for ethanol show an increase over the years, there will be an immediate need for high yield production strains and the realization of economical viable process for the production of ethanol.Fermentation system operated on semi-continous basis offer several advantages compared to other fermentation process. These benefits include increased volumetric productivity and the ability to use smaller bioreactor volume. The immobilization of cells and ethanol productivity in immobilized yeast cells depend on the surface characteristics such as hydrophilic properties, water content and magnetism. Semi-continous ethanol production from molasses with 20% sugar concentration in the first days (t1) by free cells of Saccharomyces cerevisiae was investigated.

In this research as presented in figure 5, the result of the reduce sugar analysis were obtained. Reducing sugar concentration can serve as an indicator of the fermentation process. As shown in figure 5, there is a tendency for reducing sugar concentration in the substrate to decrease during fermentation. This occurs beacause the reducing sugars are utilized by Saccharomyces cerevisiae as a carbon source to generate energy.

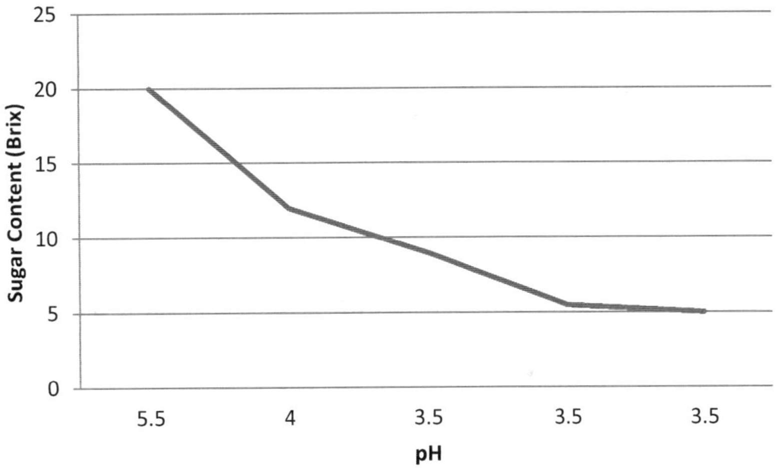

Figure 6. Fermentation process

Based on figure 6. The decrease in reduce sugar concentration corresponds with a decline in Brix value. This is influenced by the fermentation time and the addition of nutrients, specifically urea (3.12 grams) and NPK (0.37 grams), which is affect the number of *Saccharomyces cerevisiae* that survive, growth, and actively consume the sugars present in fermentation medium. The amount of nutrients added during fermentation also needs to be considered. During the fermentation process, the initial reduce sugars gradually decrease as fermentation time progresses (in hours). The reduction in reducing sugar levels with increased fermentation time indicates that simple carbohydrates are being utilized by *Saccharomyces cerevisiae* for growth and product formation in fermentation process.This is consistent with the others studies, which suggested that urea $(NH_2)_2CO$ and $NH_4H_2PO_4$ can function as nutrients during fermentation process. The availability of nutrient sources in the fermentation medium can enhance the growth of microorganisms, increase the concentration of bioethanol produced. Nitrogen is one of the most important nutrient sources for the growth of microorganisms, as it plays a key role in the formation of nucleic acids and amino acids. The decrease in bioethanol levels can be influenced by several factors, one of which is pH. This is realated to the fermentation medium's pH, which tends to decrease as the concentration of urea increases. A drop in pH can hinder the fermentation process. Using excessively high concentrations of urea can inhibit the growth of *Saccharomyces cerevisiae*, leading to a decrease in bioethanol concentration, which tends to be lower as a result. From 430 ml fermentation produced 66,49% ethanol.

4. Conlusions

The addition of urea result in higher bioethanol levels compared to the addition of NPK. The highest bioethanol concentration achieved was 66.49% with the addition of 3,12 gram urea and 0,37 NPK. As fermentation time increases, the bioethanol concentration tends to rise; however, once fermentation reaches optimal time, the ethanol concentration for subsequent fermentation periods tends to decline. The longer the fermentation process continues, the higher the substrate efficiency becomes. Reducing sugar concentration (brix) and pH will decrease as fermentation time increases.

5. Acknowledgements

Thank you for Politeknik Negeri Jember and Pusat Penelitian dan Pengabdian Masyarakat (P3M) for the PNBP 2024 to support with funding this research. This script would give positive contribution to the educational development further research.

Reference

[1] Fathorrahman, A. Swastika, N. Yuliana, and Abrina. S. P. 2012. Proses Pembuatan Ethanol Semikontinyu Melalui Rekayasa Alat Fermentor Fluidisasi. *Buana Sains* 12(2): 1–8.

[2] Wulandari, R., R., A. and Utami., B. 2016. "Pembuatan Bioetanol Dari Air Kelapa Tua Menggunakan Proses Fermentasi." *Prosiding Seminar Nasional Kimia, JURDIK Kimia FMIPA-UNY* 1(1): 147–52.

[3] Susmiati, Y. 2018. Prospek Produksi Bioetanol Dari Limbah Pertanian Dan Sampah Organik. *Industria: Jurnal Teknologi dan Manajemen Agroindustri* 7(2): 67–80.

[4] Hossain, N. 2021. Journal of Environmental Chemical Engineering Bioethanol Production from Forest Residues and Life Cycle Cost Analysis of Bioethanol-Gasoline Blend on Transportation Sector. *Journal of Environmental Chemical Engineering* 9(4): 105542. https://doi.org/10.1016/j.jece.2021.105542

[5] Mansy, Ahmed E. 2024. Sustainable Production of Bioethanol from Office Paper Waste and Its Purification via Blended Polymeric Membrane. *Energy Conversion and Management* 299 (November 2023): 117855. https://doi.org/10.1016/j.enconman.2023.117855

[6] Adi. W. K., Gusti, P., Komang, A. N., and I Desak, P. K. P. 2020. Pengaruh Lama Fermentasi Terhadap Karakteristik Fermented Rice Drink Sebagai Minuman Probiotik dengan Isolat Lactobacillus Sp. F213.*Jurnal Ilmu dan Teknologi Pangan (ITEPA)* 9(2): 181. https://doi.org/10.24843/itepa.2020.v09.i02.p08

[7] Rosyadi, Firda, A, Kurnia, P. P., and Tri, W. 2013. Optimasi Proses Produksi Etanol Dari Molases Menggunakan Fermentasi-Ekstraktif. *Jurnal Teknik POMITS* 3(2): 2–3.

[8] Kaya, E. 2020. Pengaruh Pupuk Hayati dan Pupuk NPK Untuk Meningkatkan Pertumbuhan Tanaman Tomat (Solanum Lycopersicum) yang Di Tanam Pada Tanah Terinfeksi Fusarium Oxysporum Effects of Biofertilizer and NPK Fertilizer to Increase the Growth of Tomato Solanum Lycop. *Agrologia* 9(2): 81–94.

[9] Hidayati, Miftahul, David S.S, Ida F, and Yohandri, B. 2022. Pengaruh pH dan Waktu Fermentasi Molase Menjadi Bioetanol Menggunakan Bakteri EM4. *Publikasi Penelitian Terapan dan Kebijakan* 5(1): 33–40. https://doi.org/10.46774/pptk.v5i1.394

[10] De Andrade, C. M., Cogo, A. J. D., Perez, V. H., dos Santos, N. F., Okorokova-Façanha, A. L., Justo, O. R., and Façanha, A. R. 2021. Increases of bioethanol productivity by S. cerevisiae in unconventional bioreactor under ELF-magnetic field: New advances in the biophysical mechanism elucidation on yeasts. *Renewable Energy*, *169*, 836–842. https://doi.org/10.1016/j.renene.2021.01.074

[11] Irvan, Bambang, T., Luri, A., and Reviana, R. 2015. Rancangan Alat Pembuatan Bioetanol Dari Bahan Baku Kulit Durian. *Jurnal Teknik Kimia USU*, *4*(1), 53–59. https://doi.org/10.32734/jtk.v4i1.1461

[12] Mohdazhar, S. H., Abdulla, R., Jambo, S. A., Marbawi, H., Gansau, J. A., Mohdfaik, A. A., and Rodrigues, K. F. 2017. Yeasts in sustainable bioethanol production: A review. *Biochemistry and Biophysics Reports*, *10*(March), 52–61. https://doi.org/10.1016/j.bbrep.2017.03.003

[13] Khairiah, H., and Ridwan, M. 2021. Pengembangan Proses Pembuatan Bioetanol Generasi Ii Dari Limbah Tandan Kosong Kelapa Sawit. *Jurnal Pangan Dan Agroindustri*, *9*(4), 233–240. https://doi.org/10.21776/ub.jpa.2021.009.04.5

[14] Putri, Z. S., Ruyani, A., Uliyandari, M., Wardana, R. W., and Sukarso, A. A. 2024. Pengaruh Penambahan Nutrient (NPK dan Pupuk Urea) terhadap Bioethanol Hasil Fermentasi Biji Alpukat (Persea americana Mill.). *Bioscientist : Jurnal Ilmiah Biologi*, *12*(1), 662. https://doi.org/10.33394/bioscientist.v12i1.11283

[15] Herliati, Sefaniyah, and Indri, A. 2018. Pemanfaatan limbah kulit pisang sebagai. *Teknologi*, *6*(1), 1–10.

[16] Qomariyah, L., and Sindhuwati, C. 2023. Pengaruh Penambahan Npk Dan Urea Pada Pembuatan Etanol Dari Air Tebu Melalui Proses Fermentasi. *DISTILAT: Jurnal Teknologi Separasi*, *7*(2), 82–

88. https://doi.org/10.33795/distilat.v7i2.186

[17] Susmanto, P., Yandriani, Y., Dania, B., and Ellen, E. 2020. Pengaruh Jenis Nutrient Dan Waktu Terhadap Efisiensi Substrat Dan Kinetika Reaksi Fermentasi Dalam Produksi Bioetanol Berbahan Baku Biji Durian. *Jurnal Integrasi Proses*, *9*(2), 1. https://doi.org/10.36055/jip.v9i2.8056

[18] Artini, R., Gusti, I., Agung, N., and Wartana, W. 2023. The Effect Of Fermentation Time On Bioethanol Production From Molasses. *Surabaya : The Journal of Muhamadiyah Medical Laboratory Technologist*, *6*(6), 87–96.

IOP Conf. Series: Earth and Environmental Science **1446** (2025) 012059 doi:10.1088/1755-1315/1446/1/012059

Fuzzy logic model on web-based citrus mapping and marketing geographic information system

Nugroho Setyo Wibowo[1*], Sri Sundari[2], Wenny Dhamayanthi[3], Endro Sugiartono[4], Suparto[5]

[1] Information Technology Department, Politeknik Negeri Jember, Indonesia
[2,3] Agribusiness Management Department, Politeknik Negeri Jember, Indonesia
[4] Business Department, Politeknik Negeri Jember, Indonesia
[5] Department of Agricultural Technology, Politeknik Negeri Jember, Indonesia

*nugroho@polije.ac.id

Abstract. *Jember Regency is the second largest producer of citrus in East Java Province with a production of 131,855 tons or 27.45% of the total citrus production. Siamese orange production continues to increase, from 852,359 in 2011 to 1,370,422 in 2015. In 2016, the production of siamese oranges was the largest production compared to other types of fruit in Jember Regency, which amounted to 1,348,767 tons of 4,222,455 tons of fruit crop production in Jember Regency (BPS Jember Regency, 2017). In terms of sales, these siam oranges are not only in the local market and the nearest inter-regional market but have also been marketed outside the city such as Yogyakarta, Jakarta, Semarang, Surabaya and Bali (pertanian.jatimprov.go.id). In order to improve and optimize the agricultural land of citrus plants and to find out the marketing of citrus fruit in Jember Regency, a fuzzy logic model was developed as an artificial intelligent base that can not only visualize the distribution of citrus agricultural land but also can provide an overview of citrus fruit marketing data in Jember Regency so that later it can be used as a basis for decision making by interested parties in Jember Regency, especially those related to citrus fruit. This research aims to apply fuzzy logic models as artificial intelligence about the value and parameters of land data and citrus fruit marketing in Jember Regency so that the model can be applied to a web-based geographic information system application of land mapping and citrus marketing that can help farmers and related agencies in Jember Regency.*

1. Introduction

The Economist Intelligence Unit (EIU) released the Global Food Security Index (GFSI), an instrument that measures a country's food security performance. The GFSI index measurement is carried out by comparing the food security conditions of a country with other countries based on aspects of affordability, availability, food safety and quality. In 2020, Indonesia ranked 65th with a score of 59.5. This position increased compared to the position in 2015 which was in 74th position. The increased value of Indonesia's food security index is due to the improved position of the three pillars that make it up, namely the pillars of affordability and availability as well as quality and safety. For the affordability aspect of food, Indonesia's position rose from 68 in 2017 to 63 in 2018. Indonesia's ranking for the aspect of food quality and safety in 2018 also improved and was at position 84. In 2019, Indonesia is in 86th position. Furthermore, for the aspect of food availability, not only the rank, but the score has also increased. In 2017 Indonesia was ranked 64th, rising to 58th in 2018. Likewise, the food availability score has increased by 13.93 points, from a score of 50.77 in 2015 to a score of 64.7 in 2020 [1][2][4].

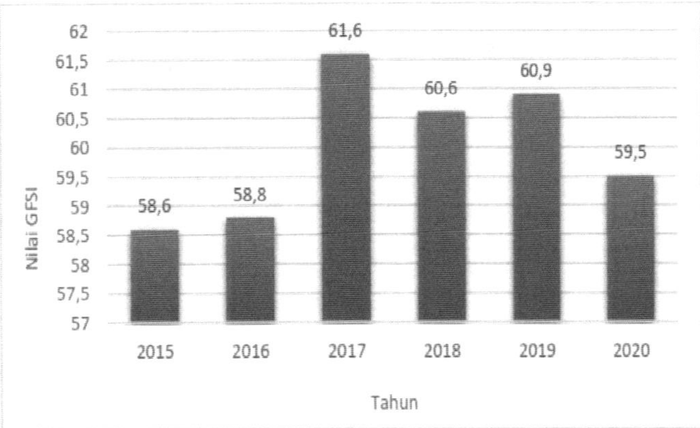

Figure 1. Indonesia's global food security index value 2015-2020

Based on data from the Center for Research and Development of Land Resources (2016), it shows that the land area of Indonesia is around 191.1 million ha. This land area consists of wetlands covering 43.6 million ha and dry land covering 144.5 million ha. Of the total land area, 15.9 million ha is very potential land for agriculture, namely for APL (Other Use Areas) covering 3.4 million ha, HP (Production Forest) covering 3.7 million ha, and HPK (Convertible Production Forest) covering 8.9 million ha. The potential availability of land for developing paddy fields is 7.5 million ha, then for food crops, shallots, chilies, and sugar cane is 7.3 million ha, and for chili plants in the highlands and shallots is 154.1 thousand ha [5].

In 2015, East Java Province was the province with the largest production of siamese oranges in Indonesia (Ministry of Agriculture, 2016). In addition, from 2012 to 2016 the amount of production of siamese oranges continued to increase from 362,680 tons in 2011 to 837,369 tons in 2016 (BPS East Java Province, 2017). This shows that in terms of production of Siamese oranges in East Java has considerable potential. Based on the Regional Spatial Plan (RTRW) of East Java Province in 2011-2031, it has been determined in the cultivation area that Jember Regency is one of the centers of chayote production. In 2015 Jember Regency was the second largest producer of siamese oranges in East Java Province with a production of 131,855 tons or 27.45% of the total production of siamese oranges [1][2][6].

In addition, in Jember Regency, the production of siamese oranges also continues to increase, which amounted to 852,359 in 2011 to 1,370,422 in 2015 (BPS Jember Regency, 2017). In 2016, the production of siamese oranges was the largest production compared to other types of fruit in Jember Regency, which amounted to 1,348,767 tons out of 4,222,455 tons of fruit crop production in Jember Regency (BPS Kabupaten Jember, 2017). In terms of sales, these siam oranges are not only in the local market and the nearest inter-regional market but have also been marketed outside the city such as Yogyakarta, Jakarta, Semarang, Surabaya and Bali (pertanian.jatimprov.go.id). This condition shows the potential of siamese orange production in Jember Regency [2][6].

In order to improve and optimize the agricultural land of citrus plants and to know the marketing of citrus fruits in Jember Regency through the use of information technology, it is necessary to develop an artificial intelligent model that can be applied to a web-based geographic information system. This Geographic Information System was developed by applying a fuzzy logic model as the basis for artificial intelligence so that it can visualize the distribution of citrus agricultural land and can provide an

overview of citrus fruit marketing data in Jember Regency so that this data can be used as decision making by interested parties in Jember Regency related to citrus fruit plants. The application of the Fuzzy Logic Model to the Geographic Information System is the best choice because in addition to having the ability to visualize the distribution well, it can also provide accurate information in the decision-making process. Fuzzy Logic Model is one of the Artificial Intelligent functions that has excellent capabilities in decision making. This research aims to apply fuzzy logic models as artificial intelligence about the value and parameters of land data and citrus fruit marketing in Jember Regency so that the model can be applied to a web-based geographic information system application of land mapping and citrus marketing that can help farmers and related agencies in Jember Regency.
The Geographic Information System developed in this study makes it easier for farmers and related agencies in Jember Regency to obtain more in-depth information related to land data and citrus fruit marketing. With this application, it is expected to help the local government and citrus farmers in Jember Regency in an effort to improve the people's economy, especially in Jember Regency.

2. Related Work

Geographic Information Systems (GIS) is a computer system used for data processing to integrate, manipulate, analyze, and display location data on the earth's surface [2]. The system to be developed applies a fuzzy logic model as an artificial intelligent base that will can visualize the distribution of citrus agricultural land but also can provide marketing data for citrus fruits in Jember Regency so that later it can be used as a basis for decision making by interested parties in Jember Regency, especially those related to citrus fruit. The application of the Fuzzy Logic Model to the Geographic Information System is the best choice because in addition to having the ability to visualize the distribution well, it can also provide accurate information in the decision-making process. Fuzzy Logic Model is one of the Artificial Intelligent functions that has excellent capabilities in decision making. This research aims to apply fuzzy logic models as artificial intelligence about the value and parameters of land data and citrus fruit marketing in Jember Regency so that the model can be applied to a web-based geographic information system application of land mapping and citrus marketing that can help farmers and related agencies in Jember Regency. Fuzzy logic applied in Geographic Information Systems (GIS) can visualize the distribution of agricultural land and citrus fruit marketing data so that it can provide important and accurate information in the decision-making process.

3. Research Methods

The method used in this research refers to the Waterfall Methodology [11][12][13][14][15], as shown in Figure 2. Field study is the initial stage of the research, in which data and information from the Jember Regency Agriculture Office are extracted. The next stage is to formulate the problem. The main problem in this research is how a fuzzy logic model can solve problems about the distribution of citrus crop land data and citrus marketing in Jember Regency through a web-based geographic information system application based on existing parameters. The next stage is to determine the purpose of this research activity, which is to apply a fuzzy logic model to a web based Jember Regency Citrus Mapping and Marketing Geographic Information System application. This fuzzy logic model is used in an application development through the calculation of land data parameters and citrus marketing in Jember Regency. With the hope that the resulting application can help farmers and related institutions in Jember Regency in making decisions related to land data and existing citrus marketing through a web-based Geographic Information System application. Furthermore, literature studies and formulation of objectives are carried out to can solve the problems. The solution to this research activity is to develop a Geographic Information System by applying a fuzzy logic model to the mapping and marketing application of oranges in Jember Regency.

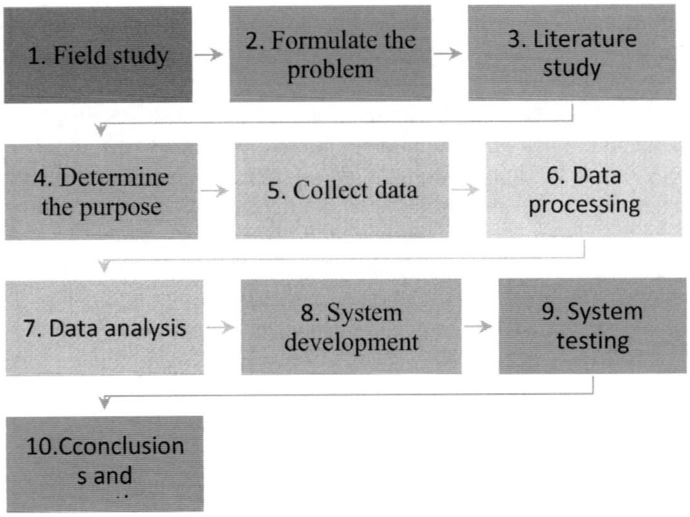

Figure 2. Research methods

Next is collecting data. In order to create and apply a fuzzy logic model into a web-based geographic information system, it is necessary to collect geographical data of the Jember Regency area, citrus crop data, land data in Jember Regency, and citrus crop marketing data. The next stage is to process the data that has been successfully collected. This data processing is done to find a fuzzy logic model as the basis for making artificial intelligence in a geographic information system application that will be developed. The next step is to do data analysis.Analysis of the results of data processing is done to provide valid and measurable data in order to create a fuzzy logic model of citrus crop land data system and citrus marketing in Jember Regency. Furthermore, the application development stage is carried out. This stage was carried out by involving the Jember Regency Agriculture Office. This stage is the application of fuzzy logic models as artificial intelligence into a web programming language program code in the development of a geographic information system for mapping and marketing of oranges in Jember Regency. With this fuzzy logic model, the performance of the system will increase in order to help farmers and related agencies in Jember Regency in decision making. Involvement with partners in the process of developing this system aims to produce research results that can have novelty and can solve problems faced by partners. Then after the system is created and developed, the next stage is to test this geographic information system. Black box and white box testing methods are used for testing this geographic system. This system testing is carried out to ensure that all parts have been tested so that they produce the expected output. The last is to make conclusions and suggestions so that they can provide more useful input in the future.

4. Discussion

To improve and optimize citrus agricultural land and to know the marketing of citrus fruits in Jember Regency through the use of information technology, it is necessary to develop an artificial intelligent model that can be applied to a web-based geographic information system. The system that will be developed applies a fuzzy logic model as an artificial intelligent base that will not only be able to visualize the distribution of citrus crop agricultural land but also be able to provide an overview of citrus fruit marketing data in Jember Regency so that later it can be used as a basis for decision making by interested parties in Jember Regency, especially those related to citrus fruit.

The application of the Fuzzy Logic Model to the Geographic Information System is the best choice because in addition to having the ability to visualize the distribution well, it can also provide accurate information in the decision-making process. Fuzzy Logic Model is one of the Artificial Intelligent functions that has excellent capabilities in decision making. This research aims to apply fuzzy logic models as artificial intelligence about the value and parameters of land data and citrus fruit marketing in Jember Regency so that the model can be applied to a web-based geographic information system application of land mapping and citrus marketing that can help farmers and related agencies in Jember Regency. This system has the following features: login menu for user registration, admin menu is a dashboard menu to manage user data, regional data, citrus crop data, marketing transaction data, and monitor sales and purchase transactions of oranges. It is hoped that this geographic information system can make it easier for farmers and related parties to monitor the condition of agricultural land, especially citrus plants, and conduct citrus fruit marketing transactions between farmers and buyers so that it can achieve independent, quality, competitive and resilient trade in order to improve the people's economy in Jember Regency.

The Fuzzy Logic model generated in this study can be represented with the Fuzzy membership function as follows:

1. Soil Drainage membership function

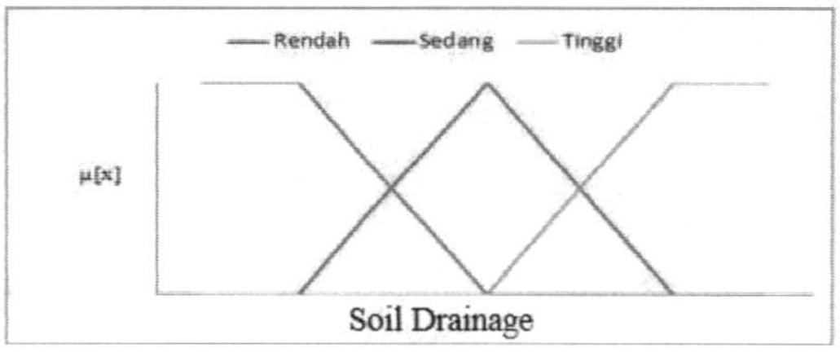

Figure 3. Soil drainage membership function

2. Soil Texture membership function

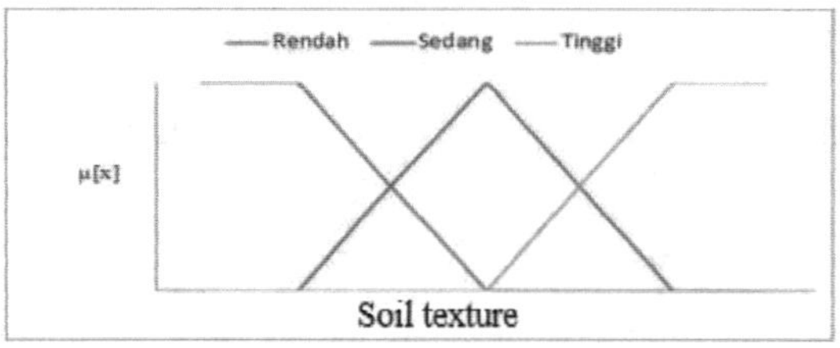

Figure 4. Soil texture membership function

3. Soil Depth membership function

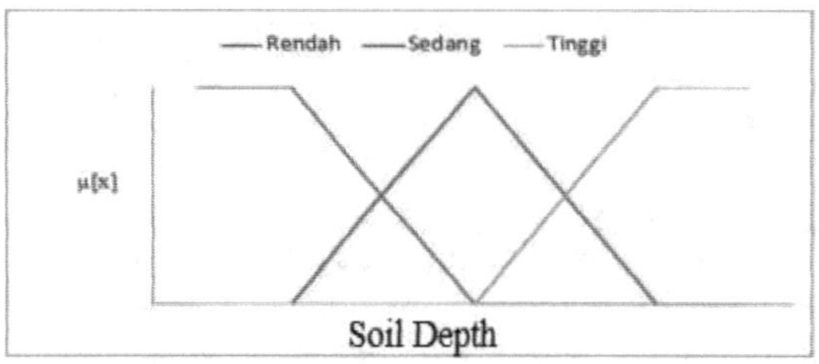

Figure 5. Soil depth membership function

After the fuzzy logic model is obtained, the model is applied in a geographic information system of land mapping and citrus marketing in Jember Regency. The results of the development of this model obtained a Web-based Geographic Information System of Land Mapping and Citrus Marketing in Jember Regency as in the following figure.

IOP Conf. Series: Earth and Environmental Science **1446** (2025) 012059 doi:10.1088/1755-1315/1446/1/012059

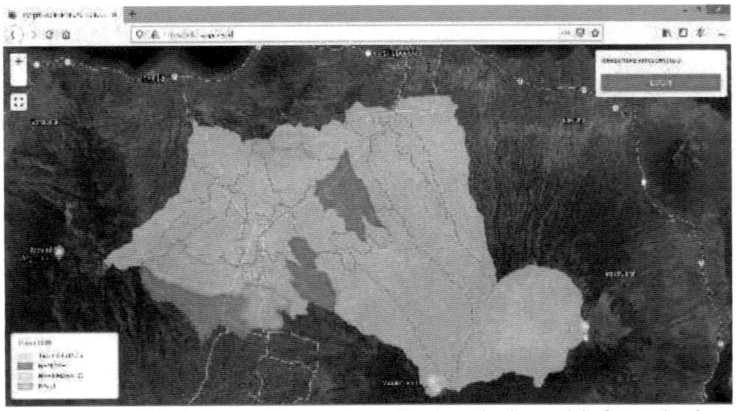

Figure 6. Geographic information system display design with fuzzy logic model

To test the reliability of this Geographic Information System, application testing was conducted using three parameters, namely Information Quality, System Quality, and User Satisfaction. The test results using the black box method obtained values as in Table 1. The average value of 100 respondents for Information Quality, System Quality, and User Satisfaction was 8.87, 8.98 and 8.89.

Table 1. SIG quality testing

Aspecs	Result
Information Quality	8.87
System Quality	8.98
User Satisfaction	8.89

5. Conclusion

This research obtained a fuzzy logic model of the distribution of citrus crop agricultural land in Jember Regency which can then be used as the basis of Artificial Intelligent to be applied in a web-based geographic information system application of land mapping and citrus marketing in Jember Regency. So that this geographic information system application has better performance in helping citrus farmers and related agencies in Jember Regency. This developed Geographic Information System application can help citrus farmers and related agencies in Jember Regency in making a decision in order to mapping and marketing of citrus crops in Jember Regency. To test the reliability of this Geographic Information System, application testing was conducted using three parameters, namely Information Quality, System Quality, and User Satisfaction. The average value of 100 respondents for Information Quality, System Quality, and User Satisfaction was 8.87, 8.98 and 8.89.

Acknowledgment

Thank you to Politeknik Negeri Jember and P3M Politeknik Negeri Jember for supporting this research fund through the Politeknik Negeri Jember PNBP grant.

References

[1] Department of Economic and Social Affairs Population Dynamics, World Population Prospects 2019, United Nations, https://population.un.org/wpp2019/, 2019.
[2] Kementerian Pertanian Republik Indonesia, Rencana Strategis Kementerian Pertanian Tahun 2020-2024, Jakarta, 2020.
[3] Kementerian Pertanian Republik Indonesia, Rencana Strategis Balai Penelitian Tanaman Jeruk dan Buah Subtropika Tahun 2015-2019, Jakarta, 2017
[4] Sekretariat Kabinet RI, Peraturan Presiden Republik Indonesia Nomor 18 Tahun 2020 tentang Rencana Pembangunan Jangka Menengah Nasional 2020-2024, Jakarta, 2020.

[5] Balai Penelitian Tanaman Jeruk dan Buah Subtropika, Laporan Kinerja Balai Penelitian Tanaman Jeruk Dan Buah Subtropika Tahun 2022, Pusat Penelitian dan Pengembangan Hortikultura, Badan Penelitian dan Pengembangan Pertanian, Kementerian Pertanian, 2022.

[6] Annisa Denar Oktaviana, Arahan Pengembangan Agrowisata Berbasis Komoditas Jeruk Di Kecamatan Semboro Kabupaten Jember, Tugas Akhir, Institus Teknologi Sepuluh Nopember Surabaya, 2018.

[7] S. H. P. Kusumadewi, "Aplikasi Logika Fuzzy Untuk Pendukung Keputusan," Yogyakarta, Graha Ilmu, 2010.

[8] A. V. C. D. &. C. A. Vitianingsih, Analysis and Design of Web-Geographic Information System for Tropical Diseases-Prone Areas: A Case Study of East Java Province, Indonesia., In 2017 4th International Conference on Information Techn, 2017.

[9] Q. &. Z. S. Cheng, Fuzzy Weights of Evidence Method Implemented in GeoDAS GIS for Information Extraction and Integration for Prediction of Point Events., Symposium, 2002. IGARSS'02. 2002 IEEE, 00(C), 2933–2935. Retrieved from http://ieee, 2002.

[10] Z. A. &. M. M. H. Latif, Mapping of Dengue Outbreak Distribution Using Spatial Statistics and Geographical Information System., In 2nd International Conf on Information Science and Security (pp. 1–5)., 2015.

[11] C. L. G. Z. S. &. C. J. Cao, Research On The Environmental Impact Factors of Hand-Foot-Mouth Disease in Shenzhen, China using RS and GIS Technologies, 2012.

[12] Aronoff, Geographic Information Sistem : A Management Perpective., Ottawa, Canada: WDL Publication, 1989.

[13] W. L. X. C. X. C. H. &. W. P. Zeng, Remote Sensing and GIS for Identifying and Monitoring The Environmental Factors Associated with Vector-Borne Disease: An verview., In International Geoscience and Remote Sensing Symposium (IGARSS) (pp, 2006.

[14] L. S. Z. D. L. &. L. L. Guo, Spatial Distribution and Variation Analysis of Lyme Disease in The Northeastern United States., https://doi.org/10.1109/Agro-Geoinformatics.2016.7577627, 2016.

[15] A. V. C. D. &. C. A. Vitianingsih, Analysis and Design of Web-Geographic Information System for Tropical Diseases-Prone Areas: A Case Study of East Java Province, Indonesia., In 2017 4th International Conference on Information Techn, 2017.

IOP Conf. Series: Earth and Environmental Science **1446** (2025) 012060 doi:10.1088/1755-1315/1446/1/012060

Determination Of Processed Products From Pomelo Orange Outer Peel Using Analytic Hierarchy Process (AHP)

Andi Ridwan Makkulawu[1]*, Soemarno[2]*, Imam Santoso[3], Siti Asmaul Mustaniroh[4]

[1]Agroindustry Study Program, Pangkep State Polytechnic of Agriculture. Pangkep 90655, Indonesia
[2]Department of Agroecotechnology, Faculty of Agriculture, University of Brawijaya. Malang 65145, Indonesia
[3,4] Department of Agroindustrial Technology, Faculty of Agricultural Technology, University of Brawijaya, Malang 65145, Indonesia

*Email: andiridwanm@polipangkep.ac.id, smno@ub.ac.id

Abstract. The increasing production of Pomelo oranges and the need to diversify their use to enhance economic value have prompted research into developing processed products from Pomelo's outer peel. This study aims to rank potential processed products derived from the outer peel of Pomelo using the Analytic Hierarchy Process (AHP). Data were collected by distributing questionnaires to five experts, who conducted pairwise comparisons of the selected products. The AHP method was applied to rank these products based on their priority weights. The results show that candy ranked first with a priority weight of 0.35, followed by snack (0.25), aromatherapy (0.20), scrubs (0.10), and sauce (0.09). The consistency of expert evaluations was validated through the Consistency Ratio (CR), calculated at 0.059, indicating a high level of consistency. The findings highlight candy as having the highest potential for further development, with consistent results supporting its prioritisation for product development from Pomelo Peel.

1. Introduction

As Pomelo production continues to increase over time, and farmers strive to maintain the quality of Pomelo during storage when market demand decreases, there is a need to diversify Pomelo into processed products [1]. This diversification is expected to extend the shelf life and maintain the quality of Pomelo in a different form, as well as to increase the economic value of Pomelo, which has only been sold in whole fruit form [2].

Pomelo Citrus, often referred to as large citrus, is a type of fruit that has been known in Indonesia for a long time [3], [4]. The name Pomelo was proposed by the Ministry of Agriculture because this citrus has no relation to Bali. The large citrus (Citrus maxima Merr), known as Pomelo, originates from Southeast Asia, including Indonesia, Vietnam, Singapore, Thailand, the Philippines, and Malaysia [5]. In Indonesia, the primary development areas for Large Citrus (Citrus Maxima) include Magetan, Pangkep, and Aceh [6].

The Pomelo fruit is round with a flat bottom. It is large, and its outer skin is green when young, turning yellowish as it matures. The skin is thinner compared to other citrus fruits [7]. The flesh is very dense. It has few seeds, and some varieties are even seedless. The tree height ranges between 5 to

15 meters. The tree crown is relatively low with irregular branches. The tips of the branches often droop. The trunk diameter ranges from 10 to 30 cm. The bark is quite thick and yellowish-brown in colour. Like other citrus species, the branches and twigs of the Pomelo tree are angular when young and become rounded as they mature. Pomelo has been cultivated for generations, making it a significant source of livelihood for many locals. The success of Pomelo cultivation heavily depends on the quality of the seeds planted and the care given after planting. Thriving in lowland areas, this plant requires a fairly open space or direct sunlight. Pomelo contains a substantial amount of vitamin C [8]. Here are some types of processed skin products in Pomelo Oranges:

1.1. Candied

Candied food is preserved with a high sugar content. Besides adding sweetness, the purpose is also to prevent mould growth, thus extending the shelf life. Candied Pomelo peel has cholesterol-lowering benefits; in addition, this candy also contains protein (2.6%) and vitamin C (4 mg/100g). Pomelo peel can be recycled from waste and further processed into food that has naturally undergone certain processes and treatments [9].

1.2. Toffee

Toffee One of the processed products made from Pomelo is toffee. Pomelo toffee is made from ripe fruit pulp that is mashed and then cooked with the addition of sugar and other food ingredients, or without the addition of other ingredients such as glutinous rice flour, tapioca flour, fugue flour, colourings, and preservatives [10].

1.3. Crackers

The process begins with thinly slicing the pomelo peel and carefully drying it to preserve the natural texture and flavour of the fruit [11]. Once dried, the slices are soaked in a mixture of salt and sugar water to enhance the flavour and add a touch of sweetness. These crackers have a dry and crunchy texture with a sweet taste accompanied by a slight tang from the pomelo fruit.

1.4. Pomelo Noodles

Noodles made from pomelo peel have unique characteristics, such as a strong and distinctive aroma that comes from the pomelo peel [12]. These pomelo noodles typically have a fresh and slightly tangy taste, adding a unique touch to dishes that use them. They usually have a darker and richer colour compared to regular noodles, due to the pigments and oils from the pomelo peel that impart colour and flavour.

1.5. Pomelo Flour

Flour made from the inner peel of pomelo has several distinctive characteristics, including a strong and fresh aroma derived from the essential oils within. Typically, this pomelo flour has a slightly bitter yet refreshing taste, often used as a spice or flavouring in dishes, cakes, or beverages. Its texture can be fine or slightly coarse depending on the processing method [9].

The AHP method is a decision-making method that optimizes factors such as logic, intuition, experience, knowledge, emotions, and feelings in a systematic process that allows for pairwise comparisons of intangible values and solid quantitative data in a qualitative manner [13], [14].

The topic to be discussed in this book chapter is how to rank processed Pomelo products that can be developed based on the region or village that produces Pomelo using the AHP method. The purpose of this book chapter is to design a decision support system that can assist in determining processed products based on village regions using the AHP method. The use of the AHP method is quite widespread across various industries.

IOP Conf. Series: Earth and Environmental Science **1446** (2025) 012060 doi:10.1088/1755-1315/1446/1/012060

2. Method

The location of this research was in Segeri Sub-district, Pangkep Regency, South Sulawesi. The research stages are shown in Figure 1, starting with problem identification, literature review, determination of the research sample, questionnaire development, data processing using AHP, data analysis, conclusion, and recommendations.

The data collection method involved distributing questionnaires to 5 experts, including 2 experts from the government, 1 expert from SMEs, and 2 lecturers competent in the field of processing. The questionnaires used a rating scale, as shown in Table 1, where the experts were asked to evaluate different criteria using a scale from 9 to 1 for positive judgments and from 1 to 1/9 for negative judgments. A value of 9 indicates the highest level of importance or preference, while a value of 1/9 indicates the lowest. This scale helps in systematically and consistently comparing the relative importance of each criterion.

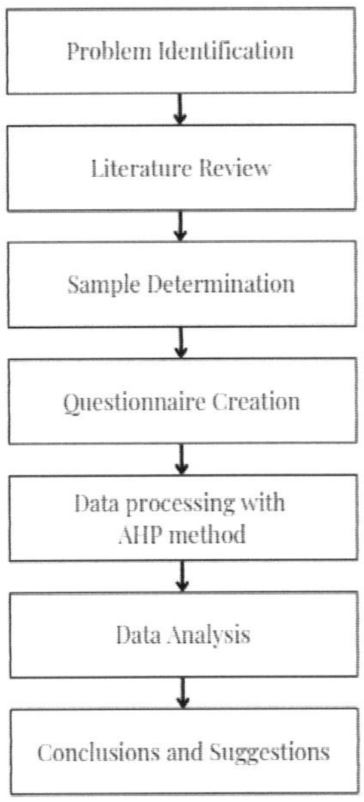

Figure 1. Flowchart of the research

In Figure 1, the research stages are explained, starting with problem identification, which focuses on the need to develop processed products from the outer peel of Pomelo to enhance its economic value and reduce waste. This was followed by a literature review to gather information related to processed Pomelo peel products and the Analytic Hierarchy Process (AHP) method. The next step was the determination of the research sample, where five experts were selected, consisting of 2 from the government, 1 from SMEs, and 2 academics with expertise in processing. After that, a questionnaire was developed using a specific rating scale, followed by data processing using AHP to conduct pairwise comparisons of the established criteria. Once the data was processed, data analysis was carried out to rank the most potentially processed products, with the consistency of responses checked through the Consistency Ratio (CR) calculation. The research concluded with recommendations and conclusions, summarising the ranking of processed Pomelo peel products and providing suggestions for further development.

Table 1. Rating scale

SCALE	9	8	7	6	5	4	3	2	1	1	2	3	4	5	6	7	8	9
VALUE	9	8	7	6	5	4	3	2	1	1	½	⅓	¼	⅕	⅙	⅐	⅛	⅑

Ranking processed products from the outer peel of pomelo using the Analytic Hierarchy Process (AHP) method. AHP involves several stages, namely defining the problem and compiling a hierarchy, creating a pairwise comparison matrix to assess the relative importance of each criterion, calculating priority weights, and checking the consistency of the assessment with the Consistency Ratio (CR). AHP is known for its ability to break down complex problems into more structured and consistent judgements, making it reliable in objective decision-making.

3. Result

AHP is a decision support method developed by Thomas L. Saaty. This decision-support model represents highly complex problems with many elements in a hierarchy [15]. A hierarchy is defined as the representation of a complex problem in a multilevel structure, where the first level is the goal, followed by the last level of alternatives. With a hierarchy, complex problems can be divided into groups and then organised into a hierarchical form, so that they become more structured and systematic. In this research, AHP was used to compare five potential products from the outer peel of Pomelo, namely Candy, Snack, Aromatherapy, Sauce, and Body Scrub, based on the evaluation of five experts.

The pairwise comparison matrix of the respondents for the outer peel of Pomelo orange is described in Table 2, the geometric mean matrix of the respondents' matrix is described in Table 3, the normalisation matrix and priority weights are described in Table 4, and the ranking results of processed products from the outer peel of Pomelo orange are described in Table 5.

Table 2. Pairwise comparison matrix from respondents.

K (Outer peel)	Expert 1	Expert 2	Expert 3	Expert 4	Expert 5
Snack (Ka) >< Candy (Kb)	6	0.125	9	0.14	0.14
Snack (Ka) >< Sauce (Kc)	6	0.125	6	6	7
Snack (Ka) >< Scrub (Kd)	0.16	8	5	6	7
Snack (Ka) >< Aromatherapy (Ke)	0.14	0.1	6	6	7
Candy (Kb) >< Sauce (Kc)	6	9	5	6	8
Candy (Kb) >< Scrub (Kd)	0.14	8	4	9	8
Candy (Kb) >< Aromatherapy (Ke)	0.125	9	3	5	0.14
Sauce (Kc) >< Scrub (Kd)	0.14	0.14	0.125	3	7
Sauce (Kc) >< Aromatherapy (Ke)	0.125	0.1	2	3	7
Scrub (Kd)>< Aromatherapy (Ke)	0.14	0.14	2	2	0.16

Table 2 shows the pairwise comparison results of five experts (EXPERT 1 to EXPERT 5) who evaluated various processed products from the outer skin of Pomelo Orange, namely Snack (Ka),

Candy (Kb), Sauce (Kc), Scrub (Kd), and Aromatherapy (Ke). In this table, each expert is asked to compare two products at a time to determine which one is more important or more prioritised, and how much importance it has.

Table 3. Geometric mean of expert matrix

Criteria	Ka	Kb	Kc	Kd	Ke
Snack (Ka)	1	0.67	2.85	3.06	1.29
Candy (Kb)	1.50	1	6.65	3.17	1.19
Sauce (Kc)	0.35	0.15	1	0.55	0.88
Scrub (Kd)	0.33	0.31	1.81	1	0.42
Aromatherapy (Ke)	0.78	0.84	1.14	2.40	1
Total	**3.95**	**2.97**	**13.45**	**10.19**	**4.77**

Table 3 is the calculation result of the geometric mean value derived from the pairwise comparison matrix given by the experts. This matrix shows how each processed product of Pomelo Orange (Snack, Candy, Sauce, Scrub, and Aromatherapy) is compared to each other based on the assessment of the experts.

Table 4. Normalization matrix and priority weights

Criteria	Ka	Kb	Kc	Kd	Ka	Avg.
Snack (Ka)	0.25	0.22	0.21	0.30	0.27	0.25
Candy (Kb)	0.38	0.34	0.49	0.31	0.25	0.35
Sauce (Kc)	0.09	0.05	0.07	0.05	0.18	0.09
Scrub (Kd)	0.08	0.11	0.13	0.10	0.09	0.10
Aromatherapy (Ke)	0.20	0.28	0.08	0.24	0.21	0.20
Total	**1**	**1**	**1**	**1**	**1**	**1**

Table 4 shows the normalisation results from Table 3 (Geometric Mean of Expert Matrix). This normalisation is done by dividing each value in Table 3 by the total value in the corresponding column. The normalisation results are then used to calculate the Priority Weights for each criterion.

Table 5. Ranking results of processed products from the outer peel of pomelo

Code	Description	Priority weight	Ranking
(K1a)	Snack	0.25	II
(K1b)	Candy	0.35	I
(K1c)	Sauce	0.09	V
(K1d)	Scrub	0.10	IV
(K1e)	Aromatherapy	0.20	III

Table 5 shows the final result of the analysis, i.e. the ranking of Pomelo Citrus processed products based on the priority weights calculated in Table 4. This table presents the ranking order of the products from the most priority to the least, based on the assessment of the experts. Candy gets the first rank because it has the highest priority weight (0.35). This indicates that Candy is considered the most important or potential to be developed among all the evaluated products. Table 5 shows that Candy (Kb) has the highest priority weight and is ranked first. The Consistency Ratio (CR) is 0.059, which indicates that the comparisons in the matrix are consistent and the AHP analysis results can be considered valid.

4. Conclusion

Based on the results of the analysis using the Analytic Hierarchy Process (AHP) method, it can be concluded that Candy is the most potential Citrus Pemelo processed product to be developed in Segeri District, Pangkep Regency, South Sulawesi, Indonesia, followed by Snacks and Aromatherapy. Scrubs and Sauce products have lower potential and therefore require further study before being developed. The consistency of the experts' evaluation proved to be valid with a Consistency Ratio (CR) value of 0.059, which is well below the threshold of 0.1, indicating that the analysis is reliable for decision-making. Therefore, the focus of product development should be on Candy and Snacks, while Aromatherapy, Scrubs, and Sauces can be considered as alternatives.

5. References

[1] S. Sharma, B. Singh, G. Kaur, Y. Srivastava, and R. S. Sandhu, "Nutritional, Bioactive, and Health Potential of Pomelo (Citrus maxima): An Exotic Underutilized Fruit," *Curr. Res. Nutr. Food Sci. J.*, vol. 12, no. 2, 2024.

[2] W. Makkumrai, Y. Huang, and Q. Xu, "Comparison of pomelo (Citrus maxima) grown in China and Thailand," *Front. Agric. Sci. Eng*, vol. 8, no. 2, pp. 335–352, 2021, doi: https://doi.org/10.15302/J-FASE-2021391.

[3] M. El-Otmani, A. Ait-Oubahou, and L. Zacarías, "Citrus spp.: Orange, mandarin, tangerine, clementine, grapefruit, pomelo, lemon and lime," in *Postharvest biology and technology of tropical and subtropical fruits*, Elsevier, 2011, pp. 437--516e. doi: https://doi.org/10.1533/9780857092762.437.

[4] G. Liu *et al.*, "Comprehensive utilization of immature honey pomelo fruit for the production of value-added compounds using novel continuous phase transition extraction technology," *Biology (Basel).*, vol. 10, no. 8, p. 815, 2021.

[5] S. S. Leong, S. C. T. Leong, and G. A. C. Beattie, "Integrated Pest Management Strategies for Asian Citrus Psyllid Diaphorina citri Kuwayama (Hemiptera: Psyllidae) and Huanglongbing in Citrus for Sarawak, East Malaysia, Borneo," *Insects*, vol. 13, no. 10, p. 960, 2022.

[6] S. Susanto *et al.*, "Indonesian pummelo (Citrus maxima (Burm.) Merr.): its cultivars and characteristics," in *IV Asian Horticultural Congress-AHC2023 1404*, 2023, pp. 129–134.

[7] A. Green, *Field Guide to Produce: How to Identify, Select, and Prepare Virtually Every Fruit and Vegetable at the Market.* Quirk Books, 2004.

[8] X. Yan *et al.*, "Establishing quality evaluation standards for pomelo fruit: The role of harvesting time and appearance characteristic," *Postharvest Biol. Technol.*, vol. 212, p. 112863, 2024.

[9] L. Xiao, F. Ye, Y. Zhou, and G. Zhao, "Utilization of pomelo peels to manufacture value-added products: A review," *Food Chem.*, vol. 351, p. 129247, 2021.

[10] I. Ahmad, A. Adililham, K. Karma, and A. Andryanto, "PIPK agrowisata jeruk Keprok Selayar," in *Prosiding Seminar Nasional Politeknik Pertanian Negeri Pangkajene Kepulauan*, 2022, pp. 738–746.

[11] I. Ahmad, A. Adilham, K. Karma, and A. Aman, "Application of Science and Technology for Regional Development: Tangerine Agrotourism Selayar," *Mattawang J. Pengabdi. Masy.*, vol. 3, no. 4, pp. 436–441, 2022.

[12] S. Bhatt and M. Gupta, "Formulation of instant noodles incorporated with insoluble dietary fiber from fruit peel: in vitro starch digestibility, biophysical, structural and textural characteristics," *Appl. Food Res.*, vol. 3, no. 2, p. 100326, 2023, doi: https://doi.org/10.1016/j.afres.2023.100326.

[13] Y. Wang and J. Zhang, "Research on cultural diversity and sustainable land-use management

assessment model," *Front. Environ. Sci.*, vol. 12, p. 1359521, 2024.

[14] G. Rivera and A. Yunusa-Kaltungo, "A Reliability-Based Appraisal of the Cultural Challenges of Developing Asset Management Systems in the Energy Industry: A Case of a Leading Player in Argentina," in *Key Themes in Energy Management: A Compilation of Current Practices, Research Advances, and Future Opportunities*, Springer, 2024, pp. 531–575.

[15] H. A. Nefeslioglu, E. A. Sezer, C. Gokceoglu, and Z. Ayas, "A modified analytical hierarchy process (M-AHP) approach for decision support systems in natural hazard assessments," *Comput. \& Geosci.*, vol. 59, pp. 1–8, 2013.

Preprocessing vannamei skeleton image for feature classification

Mariam[1], Ismail Gaffar[1]*, Andi Ridwan Makkulawu[1], Ardiansyah[1]

[1]Politeknik Pertanian Negeri Pangkep, South Sulawesi 90655, Indonesia

E-mail: ismail.gaffar@polipangkep.ac.id

Abstract. The use of visual imaging technology on underwater objects is still being developed, one of which is vaname shrimp (Litopenaeus vannamei). It is one type of shrimp that is popularly consumed by the world community. Super intensive shrimp pond farming can produce high amounts of production, but sometimes farmers also find very low production or in this case, harvest failure. It is influenced by many factors, one of which is the supervision of post-molting shrimp that require stable calcium levels in water to form new shells. This is done to keep soft shrimp without skin (post-molting) from dying due to cannibalism. Handling this requires monitoring using underwater imaging. The purpose of this research is to interpret the image of vaname shrimp molting underwater. The method used is image processing which consists of: Resize image, grayscale, thresholding, erosion and skeleton techniques. Skeleton features are well used for prediction of molting and not molting vanamei shrimp, where the prediction accuracy value using Support Vector Machine (SVM) classification Linear model shows an accuracy value of 71.43%.

1. Introduction

Today's aquaculture is developing in line with the advancement of information technology. Smart farming has made many advances in the field of fisheries [1]. One of them is a monitoring system using surveillance cameras to monitor the growth of living things. The use of sensors is widely applied to aquaculture to help farmers in cultivation, especially vaname shrimp.

Image processing technology is no longer new knowledge at this time. The technology has been applied in various human lives in the 21st century [2]–[4]. The use of visual imaging technology on underwater objects is currently still being developed, one of which is vaname shrimp (Litopenaeus vannamei). Vannamei is one type of shrimp that is popularly consumed by the world community. The largest importer of vanamei shrimp comes from China, which is supplied mostly from Ecuador, which is then followed by Southeast Asian countries such as Indonesia, Malaysia, Thailand & Vietnam [5].

Super intensive shrimp pond farming can produce high amounts of production [6], but sometimes farmers also find very low production or in this case, harvest failure. It is influenced by many factors, one of which is the supervision of post-molting shrimp that require stable calcium levels in water to form new shells [7][8]. This is done to keep soft shrimp without skin (post-molting) from dying due to cannibalism.

Handling this requires surveillance using underwater imaging, monitoring the activities of vaname shrimp. This approach was taken to address the problem of high molting failure in vaname shrimp. Several surveillance methods were developed to detect underwater objects including CNN method with size and shape features [9], Multilayer Perceptron multispectral features [10], and SVM with Histogram Oriented Gradient (HOG) features [11]. Using image processing techniques with skeleton features and recognition with Support Vector Machine (SVM) classification is considered capable of providing good accuracy [12]. Although image processing technology has developed rapidly, the application of skeletonization in vaname shrimp for post-molting detection is still a new approach. Using skeleton features to identify changes in the physical shape of shrimp after molting is an innovation that is rarely applied in shrimp farming. As a first step in this research, linear and RBF models of SVM classification were tested to get the best model in detecting molting shrimp.

2. Research method

The research was carried out with respect to the flow that had been designed, useful for connecting the running process. As shown in figure 1.

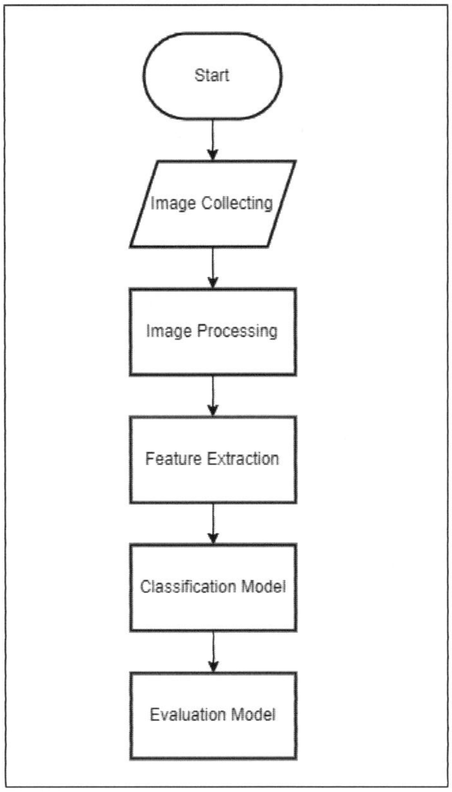

Figure 1. Research method flowchart

Each process that runs on the flowchart method is computed using python programming language with libraries including: numpy, pandas, matplotlib and sklearn..

2.1. Image collecting

Building a prediction model requires good data, as a first step the image of vaname shrimp is simulated in an aquarium pond adapted to living conditions in shrimp farming ponds. Vaname shrimp that became the sample of the experiment amounted to 10 tails with the age of 60 days where the average weight of 10 grams / tail. Imaging using an underwater camera to get clear results.

2.2. Image processing

2.2.1 Resize image

In order to get good performance at the classification stage, the image size needs to be minimized. Using a smaller image size will provide lightweight computational results [6]. Where the original image size of 1280 x 960 pixels is then reduced to a size of 30 x 20 pixels with the ratio = W / H equation. From this equation, we have obtained a 30 x 20 matrix which will be used 600 pixels in an image data.

2.2.2 Grayscale image

Before processing into a black and white image, the image is first converted to grayscale. The processed grayscale image is derived from RGB color frequencies. RGB information has three dimensional properties that can impact its signal processing to be very complex and heavy [13]. Conversion of red, green, blue (RGB) images to grayscale can be applied using the equation (1).

$$0.30 * R + 0.59 * G + 0.11 * B = val \tag{1}$$

In general, the conversion of RGB images into grayscale form is by multiplying the red color value by 0.30, the green number value by 0.59 and 0.11 for blue values.

2.2.3 Binary image

Binary images are representations of objects that are made significantly different from the background color [13]. The use of black and white colors is used to represent objects and backgrounds. This technique is used for the image segmentation process. Some studies use this technique to obtain features on images or can be developed into further techniques, such as Local Binary Pattern (LBP) or skeleton features. For segmentation into binary images, we can use the equation (2) below.

$$b_m = \begin{cases} 1 & if\ a_m \geq t_m, \\ 0 & otherwise, \end{cases} \tag{2}$$

Where, a_m is the value at the pixel coordinate position and t_m is the threshold value for separating grayscale colors. If the first condition is met, pixel will be assigned a value 1, and if not fulfilled will be replaced with a value 0.

2.2.4 Skeleton image

Applying the skeletonization function requires morphological operations. Image skeletonization works by not removing the shape, but keeping the topological shape of the object. [14] [15]. The image erosion operation is used to remove the edges of the image based on the median point and leaves only the center line that is worth one pixel. Running the erosion operation uses the equation (3).

$$dst(x, y) = min(x', y'): element(x', y') \neq 0\ src(x + x', y + y') \tag{3}$$

Where, $dst(x, y)$ is the minimum distance from the point (x, y) to the nearest zero point. $min(x', y')$ berfungsi untuk menemukan nilai terkecil dari semua kemungkinan titik (x, y). Untuk *element(x', y') ≠ 0* merupakan syarat bahwa kita hanya mencari titik yang memiliki nilai bukan yang bernilai 0 dan $src(x + x', y + y')$ merupakan sumber pixel/kordinat pada sebuah citra.

2.3. Evaluation

Support Vector Machine (SVM) merupakan satu dari banyak metode dalam melakukan prediksi by classification or regression. SVM works by finding the hyperlane that has the best margins [10]. Hyperlane is useful for separating between each class [16]. SVM classification has several kernel functions that can be used, namely linear, polynomial, radial base function (RBF) etc. To determine the performance of skeleton features, the accuracy of linear & RBF kernel functions is compared. Measuring the accuracy of the model can be done using the equation (4).

$$Accuracy = \frac{correct\ prediction}{All\ Prediction}\ x\ 100 \tag{4}$$

Where, $correct\ prediction$ is the number of correct predictions and all predictions the number of all data sets.

3. Result and discussion

3.1. Image collecting

The image data obtained are 11 molting shrimp and 9 shrimp that have not molted. The shape of the difference between the two can be distinguished by human vision. Molting vaname shrimp leave the exoskeleton, therefore each molting shrimp leaves the shell in an open cephalothorax condition. Shown in figure 1 and figure 2.

Figure 2. Image of shrimp not molting

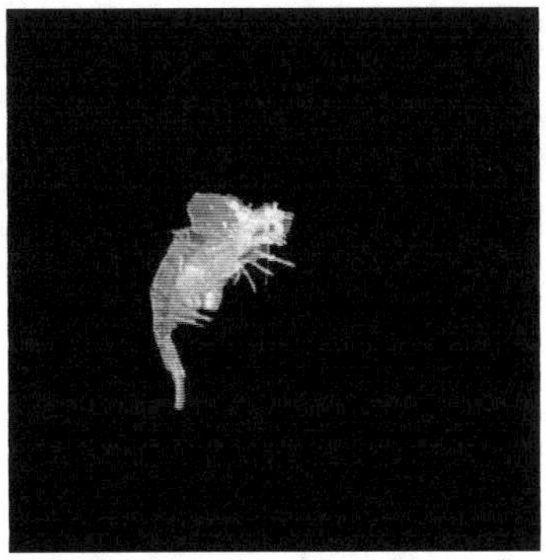

Figure 3. Image of molting shrimp

3.2. Image processing

3.2.1 Grayscale image

The RGB color of the vaname shrimp object is converted to a grayscale color object, in order to merge the core of the object. The gray scale color is interpreted with a value of 0 to 255. Where the value of 0 is a dark color until it scales to white color at the value of 255. The grayscale image can be seen in figure 4 and figure 5.

Figure 4. Grayscale image vannamei

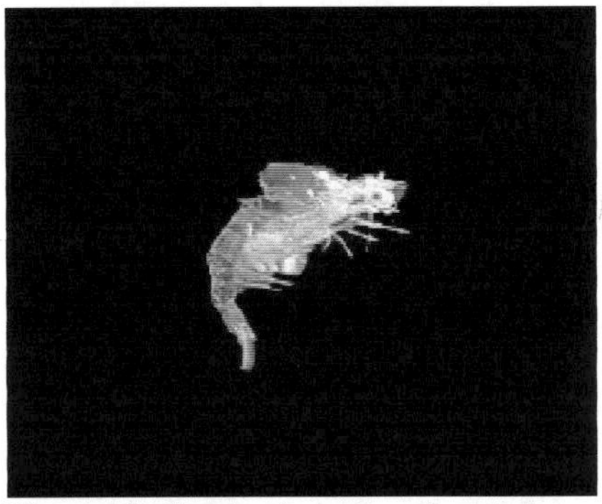

Figure 5. Grayscale image eksoskeleton vannamei

3.2.2 Binary image

Separate the object from the background with the threshold function. The threshold value used is 75 from a scale of 0-255. So as to produce a binary image as in figure 5 and figure 6.

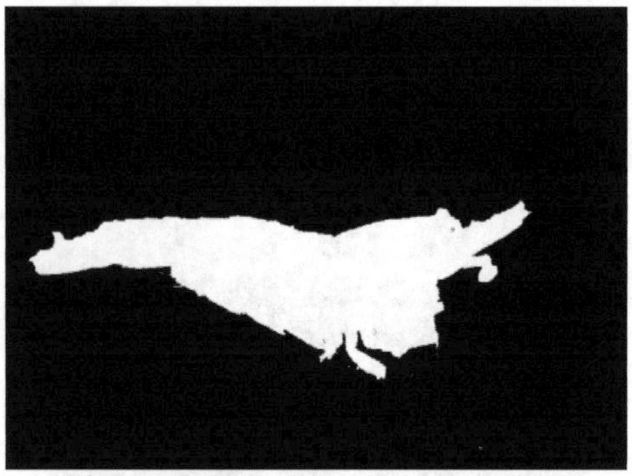

Figure 5. Image threshold vannamei

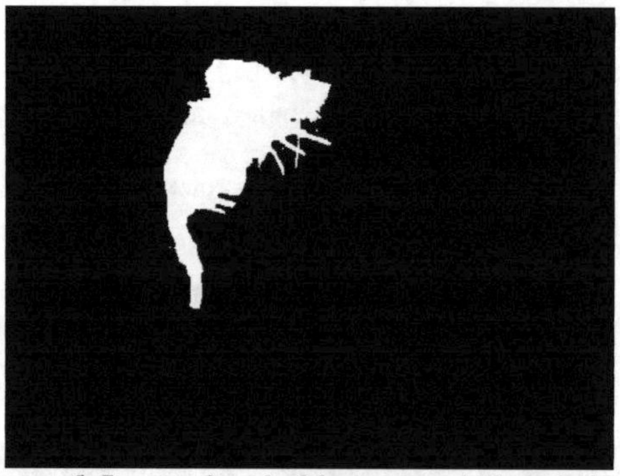

Figure 6. Image threshold vannamei eksoskeleton

3.2.3 Skeleton image

After getting the binary image, then run the image erosion operation to get the skeleton features. Figure 7 shows the skeleton features of a crab that is not molting, and figure 8 the skeleton features of a shrimp that has molted.

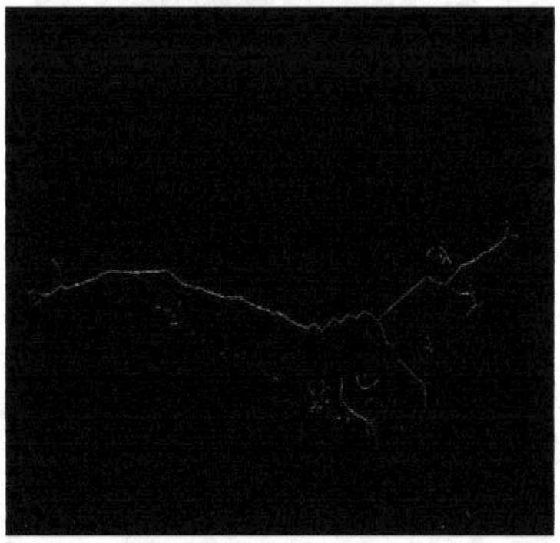

Figure 7. Skeleton feature vannamei

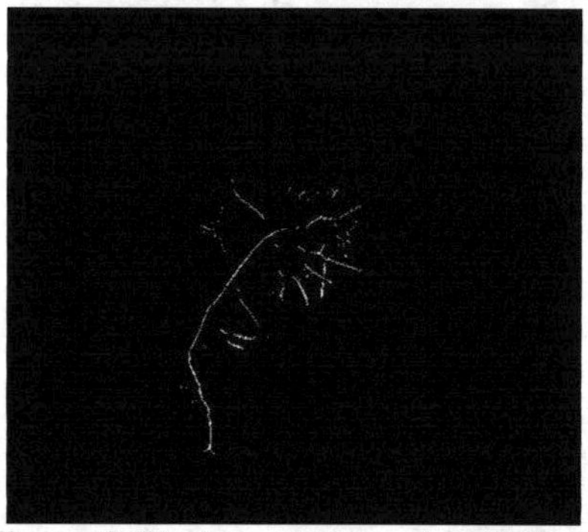

Figure 8. Skeleton feature eksoskeleton vannamei

The skeleton feature contains the number 1 at one pixel coordinate point in the form of a 30x20 matrix. To enter the classification process, the data needs to be made into a 1-dimensional array using the flatten() function. So as to produce data as in table 1 below.

Table 1. Dataset transpose feature image

Index	Label	Pixel 1	Pixel 2	Pixel …	Pixel 599	Pixel 600
1	0	0	0	1	0	0
2	0	0	0	0	0	0
…		…	…	…	…	…
21	1	0	1	0	0	0

3.3. Evaluation result

This stage is conducted to test how well the skeleton features predict vaname shrimp objects. SVM classification using linear kernel and radial base function (RBF) was compared. Using 21 datasets where 10 data are images of molting shrimp and 11 data of non-molting shrimp. The dataset was divided into 0.7 training data and 0.3 testing data.

The results show that the linear kernel is better at classification than the RBF kernel. Where the Linear kernel obtained an accuracy of 71.43% and for RBF of 42.86%.

4. Conclusion

In this paper, vaname shrimp object recognition using skeleton features can be used in the future. SVM successfully predicts molting and non-molting shrimp using linear SVM kernel with 71.43% accuracy. Furthermore, shrimp image features need to be optimized again to get better performance and test more datasets because molting and non-molting shrimp objects have close similarities.

5. Acknowledgments

The authors would like to express their gratitude to Kementerian Riset dan Teknologi for supporting this research in terms of funding for lecturers Politeknik Pertanian Negeri Pangkajene Kepulauan.

6. Reference

[1] P. Mohanty, V. Valagadri, and S. Ramya, "Smart Farming enabled by IoT and Spectral Imaging," *J. Phys. Conf. Ser.*, vol. 2161, no. 1, 2022, doi: 10.1088/1742-6596/2161/1/012044.

[2] J. K. Lakshmi and M. Punithavalli, "A survey on skeletons in digital image processing," *Proc. - 2009 Int. Conf. Digit. Image Process. ICDIP 2009*, pp. 260–269, 2009, doi: 10.1109/ICDIP.2009.21.

[3] P. M. Prihatini, I. K. Suryawan, and I. N. Mandia, "Feature extraction for document text using Latent Dirichlet Allocation," *J. Phys. Conf. Ser.*, vol. 953, no. 1, 2018, doi: 10.1088/1742-6596/953/1/012047.

[4] Y. P. Pasrun, M. Muchtar, A. N. Basyarah, and Noorhasanah, "Indonesian License Plate Detection Using Morphological Operation," *IOP Conf. Ser. Mater. Sci. Eng.*, vol. 797, no. 1, 2020, doi: 10.1088/1757-899X/797/1/012037.

[5] F. Organization, "US shrimp imports declined in 2023," *FAO*, 2024. https://www.fao.org/in-action/globefish/news-events/trade-and-market-news/q3-2024-jul-sep/en/ (accessed Mar. 09, 2024).

[6] M. Junda, "Development of Intensive Shrimp Farming, Litopenaeus vannamei in Land-Based Ponds: Production and Management," *J. Phys. Conf. Ser.*, vol. 1028, no. 1, 2018, doi: 10.1088/1742-6596/1028/1/012020.

[7] J. R. de Oliveira Cesar, B. Zhao, S. Malecha, H. Ako, and J. Yang, "Morphological and biochemical changes in the muscle of the marine shrimp Litopenaeus vannamei during the molt cycle," *Aquaculture*, vol. 261, no. 2, pp. 688–694, 2006, doi: 10.1016/j.aquaculture.2006.08.003.

[8] P. M. Ochoa-Pereira and P. C. Velásquez-López, "Effect of fasting on molting and survival rate in post-larvae of the shrimp Litopenaeus vannamei," *Rev. Biol. Mar. Oceanogr.*, vol. 58, no. 1, pp. 10–18, 2023, doi: 10.22370/rbmo.2023.58.1.4133.

[9] J. H. Christensen, L. V. Mogensen, R. Galeazzi, and J. C. Andersen, "Detection, Localization and Classification of Fish and Fish Species in Poor Conditions using Convolutional Neural Networks," *AUV 2018 - 2018 IEEE/OES Auton. Underw. Veh. Work. Proc.*, pp. 1–6, 2018, doi: 10.1109/AUV.2018.8729798.

[10] F. Monteiro *et al.*, "Classification of Fish Species Using Multispectral Data from a Low-Cost Camera and Machine Learning," *Remote Sens.*, vol. 15, no. 16, 2023, doi: 10.3390/rs15163952.

[11] R. R. Baharuddin, M. Niswar, A. A. Ilham, and S. Kashihara, "Crab Molting Identification

using Machine Learning Classifiers," *2021 Int. Semin. Mach. Learn. Optim. Data Sci. ISMODE 2021*, pp. 295–300, 2022, doi: 10.1109/ISMODE53584.2022.9743136.

[12] S. Vijayprasath, "A simple feature extraction method for analysis of hand written characters," *J. Phys. Conf. Ser.*, vol. 1717, no. 1, 2021, doi: 10.1088/1742-6596/1717/1/012066.

[13] V. M. Mohan, R. Kanaka Durga, S. Devathi, and K. Srujan Raju, "Image processing representation using binary image; grayscale, color image, and histogram," *Adv. Intell. Syst. Comput.*, vol. 381, pp. 353–361, 2016, doi: 10.1007/978-81-322-2526-3_37.

[14] M. G. A. Komang, M. N. Surya, and A. N. Ratna, "Human activity recognition using skeleton data and support vector machine," *J. Phys. Conf. Ser.*, vol. 1192, no. 1, 2019, doi: 10.1088/1742-6596/1192/1/012044.

[15] Y. Chen, K. Drechsler, W. Zhao, and C. O. Laura, "A thinning-based liver vessel skeletonization method," *Proc. - 2011 Int. Conf. Internet Comput. Inf. Serv. ICICIS 2011*, pp. 152–155, 2011, doi: 10.1109/ICICIS.2011.44.

[16] M. A. Ahmed, M. S. Hossain, W. Rahman, A. H. Uddin, and M. T. Islam, "An advanced Bangladeshi local fish classification system based on the combination of deep learning and the internet of things (IoT)," *J. Agric. Food Res.*, vol. 14, no. June, p. 100663, 2023, doi: 10.1016/j.jafr.2023.100663.

Feature Extraction of Digital Image Processing for Optimization Chicken Egg Quality Classification

L Perdanasari[1], B Etikasari[1*], R Ayuninghemi[1], M D Widianta[1] and B Utomo[2]

[1]Department of Information Technology, Politeknik Negeri Jember, Indonesia
[2]Department of Animal Science, Politeknik Negeri Jember, Indonesia

[*]Email : bety.etikasari@polije.ac.id

Abstract. Manual external egg quality testing often faces various problems, one of which is the low level of accuracy. Testing processes that rely on human ability to identify eggshell defects can produce inconsistent results. So more efficient and accurate quality testing method is needed, namely imaging technology. Using digital imaging to evaluate egg quality offers advantages, including the ability to automatically and consistently detect shell defects. In applying digital image classification, good training data is needed to get accurate classification results, so the aim of this research is to extract chicken egg color features as training data to optimize the classification system. The method used on classifying chicken eggs based on shell color is feature extraction based on color in HSV (Hue Saturation Value) and RGB (Red Green Blue). Research results obtained 95 image data of chicken eggs consisting of 61 Grade A eggs with average RGB value of 75.39 and average HSV value of 0.36, 28 Grade B eggs with average RGB value of 87.92 and average HSV value is 0.35, and 6 Grade C eggs have average RGB value of 92.36 and average HSV value of 0.33, with color extraction in HSV and RGB spaces.

1. Introduction

Eggs are a cheap source of animal protein that can meet nutritional needs [1]. The growth in the number of laying hens is closely correlated with this increase in consumption. Between 2018 and 2022, the Badan Pusat Statistik (BPS) reported that the number of chickens increased to 7.32% of the total population of 261 million [2]. It has been proven that the growth of laying hen breeders began as a result of society's increasing need for egg protein [3]. The nutritional content of eggs consists of: 73.7% water, 12.9% protein, 11.2% fat and 0.9% carbohydrates. The fat in eggs is found in the egg yolk, which reaches 32%, while the fat content in the egg white is very small [4].

Egg quality control can be carried out on the physical condition, freshness of the egg contents, checking for damage and measuring physical composition. The physical condition of the egg includes size (weight, length and width), color (white, slightly brownish, brown), condition of the egg shell (thin and thick), shape (round and oval) and cleanliness of the egg shell. Subjectively, the quality of whole eggs can be assessed by candling, namely by placing the egg in the path of a strong beam of light (sun or electric light) so as to allow examination of the shell and inside of the egg [5][6].

Examination of eggs is carried out by means of a telescope with a sunlight source or incandescent lamp carried out by a worker manually so that cracks in the egg shell can be found, as well as the position of the yolk, the size and position of the air sacs, blood spots, damage by microorganisms and fungal growth [6][7]. Manual external quality testing of eggs often faces various problems, one of which is a low level of accuracy. Testing processes that rely on humans' ability to identify defects in egg shells can result in inconsistent results [8]. Apart from that, the time required to carry out manual testing is also relatively long, considering that each egg must be examined one by one [6]. The durability of manual test results is also often questioned due to the subjectivity of assessment and operator fatigue which can affect accuracy [9][10]. The Indonesian government through the National Standardization Agency (BSN) tries to ensure that eggs circulating on the market meet the expected quality criteria according to quality standards [11]. However, variability in egg quality remains a challenge that must be overcome by farmers and egg producers [12].

It is important to be able to classify egg quality as a research urgency, so that farmers can group egg types with the quality that buyers or suppliers are interested in. Farmers can sort egg colors according to the SNI standard for Egg Consumption in Indonesia [11]. Therefore, more efficient and accurate quality testing methods are needed, one of which is using imaging technology. The use of digital imaging to evaluate egg quality offers various advantages, including the ability to detect shell defects automatically and consistently [13][14][15]. In applying digital image classification technology, good training data is needed to get accurate classification results [16][17].

Digital imaging technology allows testing to be performed more quickly, significantly reducing the time required for the inspection process. Additionally, imaging can produce more durable and reliable data for further analysis [18]. So it is effective to apply in the problem of choosing the quality of chicken eggs, one of which is based on shell color. The science of creating, processing, and analyzing images to provide information that people can understand is known as digital image processing [19]. Digital image processing has been widely applied in monitoring laying hens for a variety of purposes, including individual identification, health monitoring, and egg production estimation.

Color segmentation is known as color filtering. Segmentation is an image processing method used to change images based on certain colors. This method works by comparing the color component of each image pixel with a certain color, so that if the color matches a certain color then the pixel color component will be left alone, but if the color does not match a certain color then the pixel color component will be left alone. Known color spaces include HSV (Hue, Saturation, Value), YCbCr, and RGB (Red, Green, Blue) [20]. RGB is an abbreviation of Red, Green, and Blue, the three main colors that are usually used as a reference for other colors [19][21]. RGB color coding can be written in hexadecimal numbers (base 16) for each R, G or B component [18]. The segmentation method with HSV color detection uses pixel samples as a color reference to create the desired segment. Because digital images use the RGB color model as a standard color reference, the initial process of this method requires converting the RGB color model to HSV [22].

The way to taken an image can determine the quality of the image and subsequent image processing. Several studies have carried out several steps in image preprocessing such as image resizing, image cropping, and background removal [23]. In this research, we only carried out the process of resetting the images, because the image taking process had conditioned the box where the images were taken with expert validation. The image results are then processed using RGB and HSV because based on several studies using this segmentation method can support the optimization of classification system results when compared with other color spaces such as YCbCr and HSL [24] [21] [25]. So the aim of this research is to extract RGB and HSV color features of chicken eggs as training data to optimize the classification system that will be developed.

2. Method

This research uses digital image processing techniques. The technique in this study to categorize chicken eggs according to shell color is feature extraction based on color extraction in the RGB (Red Green Blue) and HSV (Hue Saturation Value) color spaces to yield the parameter values H, S, V, R, G, and B. The stages in creating the dataset show in Figure 1, that are include:

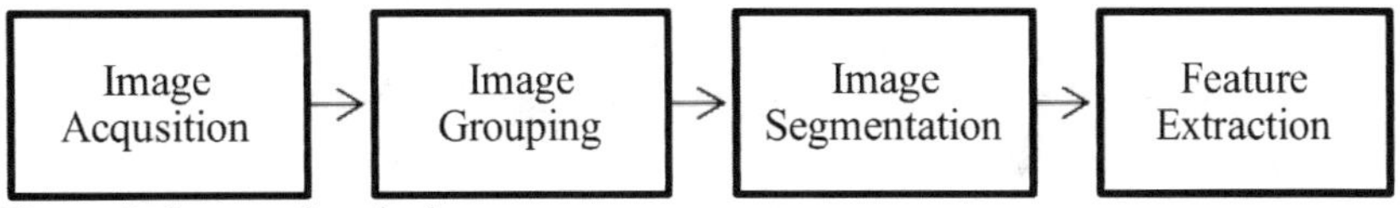

Figure 1. Stages of method

2.1. Image Acquisition
At this stage, analog images are taken, the process of converting them into digital form which can be processed further [18]. The process of taking pictures of eggs requires standardization of shooting techniques such as light conditioning, background, position of chicken eggs and camera.

2.2. Image Grouping
At this stage, the images are grouped into three quality classes based on the SNI standards for Egg Consumption in Indonesia together with validation from experts in the livestock sector. SNI for Consumable Eggs explains that there are three groupings, namely Grade A, B and C [11].

2.3. Image Segmentation
At this stage, image processing is carried out, including: resizing the image, changing the image to Grayscale, changing it to a binary image.

2.4. Feature Extraction
At this stage the aim is to retrieve the RGB and HSV values for the Egg object. At this stage there are two processes, namely calculating the average RGB value and the average HSV value and the values are saved into Excel as a dataset which will be used as training data in the chicken egg quality classification system.

3. Experiment and Result

The research began by collecting egg data for feature extraction. Eggs that undergo feature extraction are taken randomly to be able to group grade A, grade B and grade C. The next step before the feature extraction process is to prepare a precise place for taking images with the next process of image taking and preprocessing.

3.1. Image Acqusition
The image acquisition process is shown in Figure 2. The Figure 2 show that in the process of taking pictures of eggs, an image taking box is made by conditioning the box space to get more precise image results, thus making the data preprocessing process easier.

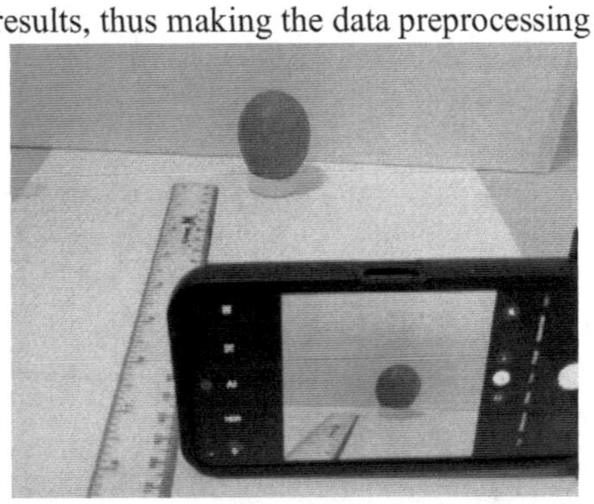

Figure 2. Acquisition of the image of consumable chicken eggs

At this stage what is done is to take images of the eggs manually using a POCO M5S cellphone with a distance of 20 cm between the camera and the egg object without using the camera flash. The eggs are placed on a white background with the egg holder placed using a white bottle cap which aims to make it easier to remove the background.

3.2. Image Segmentation

a. Variable Initialization and Preparation

Create a new variable to retrieve object files from the folder that has been created and determine the resolution of the new image.

```
clc; clear; close all; warning off all;

ukuran_baru = [400 400];
nama_folder = 'Telur Baru';
nama_file = dir(fullfile(nama_folder, '*.jpg'));
jumlah_file = numel(nama_file);
```

b. Displaying Original Image (Before Resize)

In this process,the imread() function is used to display the original image by calling the image storage directory. The original image before the resizing process in Figure 3.

Figure 3. Original of image before resize

The Figure 3 show that the egg image in original size that is 3472 x 3472 pixel and the background of egg image still in original white with shadow from light.

c. Image Resize

The egg image is resized to a size of 400x400 pixels and saves the resized image shown in Figures 5a and 5b. The results of the image that has undergone the resizing process are shown in Figure 4.

```
% Melakukan resize gambar
Img_resized = imresize(Img, ukuran_baru);

% Menyimpan gambar hasil resize
[~, nama_file_noext, ~] = fileparts(nama_file(n).name);
nama_file_resized = fullfile(nama_folder, [nama_file_noext, '_resized.jpg']);
imwrite(Img_resized, nama_file_resized);

% Menampilkan gambar setelah resize
figure, imshow(Img_resized)
title('Gambar Setelah Resize');

ukuran_baru = [400 400];
```

Figure 4. Image result after resize

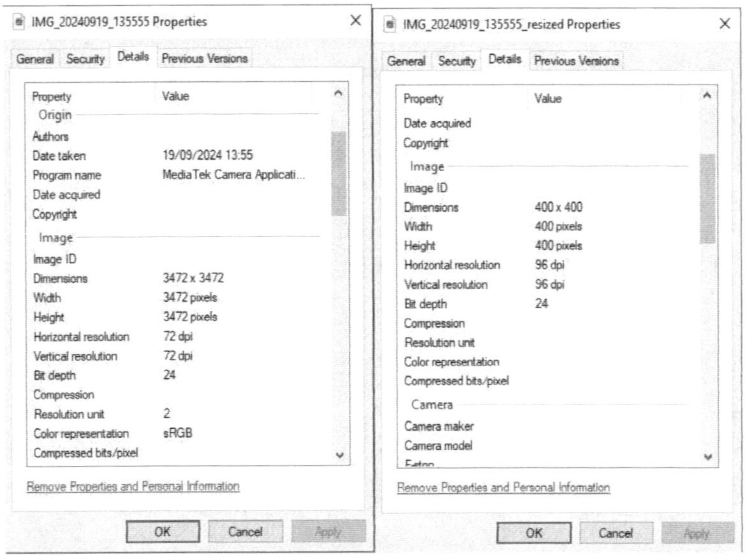

(a) (b)

Figure 5. (a) Egg image resolution before resizing; (b) Egg image resolution after resizing

The Figure 4 show image after rezise using imrezise() function. Resizing image from 3472 x 3472 pixel into 400 x 400 pixel in order to when processing images by a computer it will be faster and not use up much storage memory in temporary memory [23].

d. Hanging the Image Resize Results to Grayscale

After the image resizing process is successful, the image will be converted to Grayscale with source code. The results of changing to grayscale are shown in Figure 6.

IOP Conf. Series: Earth and Environmental Science **1446** (2025) 012062 doi:10.1088/1755-1315/1446/1/012062

```
% Mengubah gambar ke grayscale
Img_gray = rgb2gray(Img_resized);
figure, imshow(Img_gray)
title('Gambar Grayscale Setelah Resize');
```

Figure 6. Grayscale Image Result

Figure 6 show that grayscale process used in this step of feature color extraction. This process aims to facilitate segmentation and creation of binary images. This process also to reduce the information required to process each image element which is can increase the image quality.

e. Grayscale Image become Binary Image
In this process, the imbinarize() function is used with a threshold value of 0.5. Creating binary images aims to segment objects (separating objects from background). Binary image result in Figure 7.

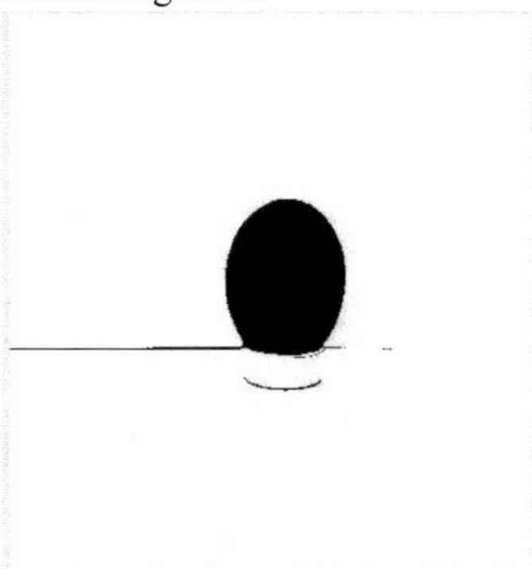

Figure 7. Results of Binary Images

Figure 7 is process of image segmentation aims to separate areas of an object from other areas based on similar characteristics in the image so that it is easy to analyze and can be used to recognize objects. In this research, MATLAB was used to create an easy GUI toolbox application as well as a source of information for further research. The most basic method for the segmentation process, for example through reducing image boundaries or image floating [26].

f. Applying Complement to the resulting Binary Image
This process aims to reverse the value of a binary image. Pixels that were originally white become black and vice versa. The following is the source code and implementation results in Figure 8.

IOP Conf. Series: Earth and Environmental Science **1446** (2025) 012062
doi:10.1088/1755-1315/1446/1/012062

```
% Melakukan komplement gambar
bw = imcomplement(bw);
figure, imshow(bw)
title('Gambar Biner Setelah Komplement');
```

Figure 8. Binary Image Results After Complementation

Figure 8 show segmentation process with complementation which result a binary egg image that is complemented by the image so that the object has a value of one (white) and the background value is zero (black), so that the geometric shape of the egg becomes clear binary image [27].

g. Filling Holes in Egg Objects

The imfill() function aims to fill the hole area in the binary egg object and to ensure the egg object becomes a solid area. The following is the source code and the implementation results are shown in Figure 9.

```
% Melakukan pengisian lubang pada objek
bw = imfill(bw, 'holes');
figure, imshow(bw)
title('Gambar Biner Setelah Pengisian Lubang');
```

Figure 9. Binary Image After Hole Filling

Figure 9 illustrates the process of dilatation and erosion that results in the solidification of the egg object. Closing seeks to join adjacent things and fill in minor gaps in objects..

h. Removing Small Areas

The bwareaopen() function aims to remove noise or very small objects less than 100 pixels in size and irrelevant. The following is the source code and implementation results in Figure 10.

IOP Conf. Series: Earth and Environmental Science **1446** (2025) 012062 doi:10.1088/1755-1315/1446/1/012062

```
% Menghilangkan area kecil
bw = bwareaopen(bw, 100);
figure, imshow(bw)
title('Gambar Biner Setelah Menghilangkan Area Kecil');
```

Figure 10. Binary Image After Removing Small Areas

The result of removing small areas leaves image 10 clean of unwanted small components, especially those originating from noise or objects that are too small to be considered relevant.

i. Chanel RGB processing
Chanel RGB processing is separated in channels: red (R), green (G), and blue (B). Object pixels outside (based on the bw binary image) are set to black (0) in all three channels. Then these channels are combined again using paint (3, R, G, B) to form a masked RGB image (only the object displayed, background removed). The results are shown in Figure 11.

```
% Mengisolasi channel R, G, B
R = Img_resized(:,:,1);
G = Img_resized(:,:,2);
B = Img_resized(:,:,3);

% Set piksel di luar objek menjadi 0 (hitam)
R(~bw) = 0;
G(~bw) = 0;
B(~bw) = 0;

% Membuat gambar RGB yang baru setelah filter biner
RGB = cat(3, R, G, B);
figure, imshow(RGB)
title('Gambar RGB Setelah Penerapan Masker');
```

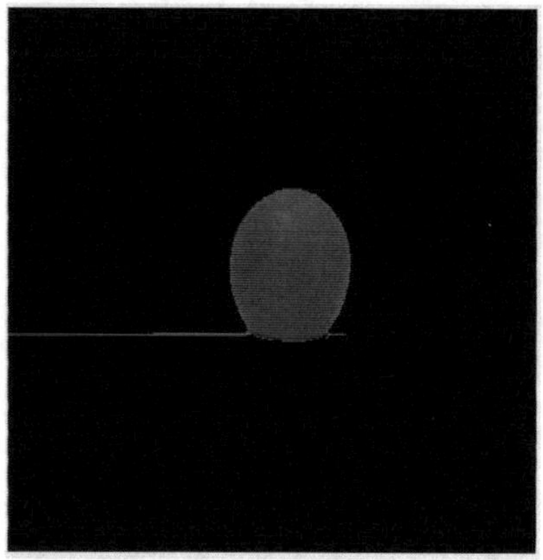

Figure 11. RGB Image implementation results

Figure 11 represents the process of obtaining color intensity information. The Red channel displays the intensity of red in the image, pixels that have a lot of red will appear brighter. The Green channel displays the intensity of green, areas that contain a lot of green will appear brighter in the image. The Blue channel displays the intensity of blue, areas that are rich in blue will appear brighter.

j. HSV Converting
Converting RGB images into HSV (Hue, Saturation, Value) color space using the rgb2hsv() function. The HSV color space is often used in image processing because it is more in line with human color perception. In particular, Hue (color), Saturation (saturation), and Value

(brightness) can be easier to process than RGB in some image processing applications. The results in Figure 12.

```
% Mengubah ke HSV dan menampilkan
HSV = rgb2hsv(RGB);
figure, imshow(HSV);
title('Gambar dalam Ruang Warna HSV');
```

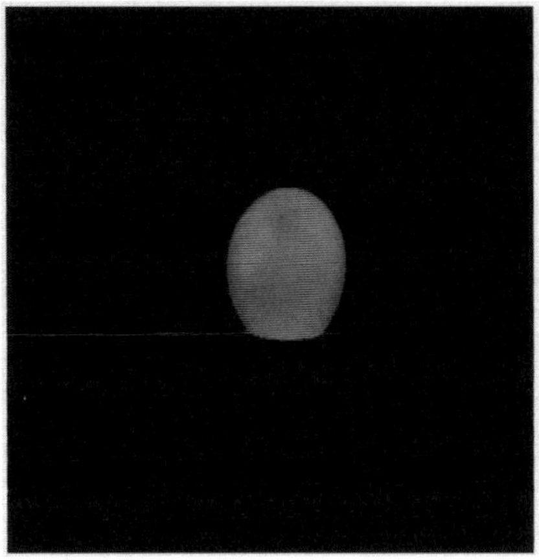

Figure 12. Results implementation RGB images become HSV

Figure 12 is the result of converting RGB to HSV, where the Hue image displays color variations in the form of hue intensity that shows the distribution of colors without regard to brightness or clarity. The Saturation image displays the intensity of the color where the pale or gray areas will appear darker in the image. The Value image shows the brightness of the image with the brightest areas in the RGB image will appear white, while the dark parts will appear black or dark gray.

3.3. *Feature Extraction*

Feature extraction aims to extract RGB and HSV values for the Egg object. At this stage there are two processes, namely calculating the average RGB value and the average HSV value and the values are saved in Excel. The feature extraction results in Table 1.

```
% Menghitung rata-rata nilai RGB
Red = sum(sum(R))/sum(sum(bw));
Green = sum(sum(G))/sum(sum(bw));
Blue = sum(sum(B))/sum(sum(bw));
```

```
% Menghitung rata-rata nilai HSV
Hue = sum(sum(HSV(:,:,1)))/sum(sum(bw));
Saturation = sum(sum(HSV(:,:,2)))/sum(sum(bw));
Value = sum(sum(HSV(:,:,3)))/sum(sum(bw));
```

Table 1. RGB and HSV Color Space Feature Extraction Results

No	Sample Egg Image	Process in MATLAB Classification Description						Classification Description
		RGB Value			HSV Value			
		R	G	B	H	S	V	
1.	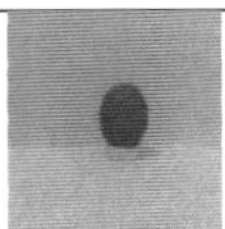	107,9799813	77,01720908	59,28740342	0,06170732971	0,4521408704	0,423500948	Egg A
2.	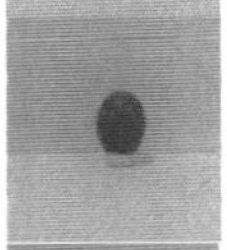	107,72754	76,45849012	56,9971778	0,06495378197	0,4724034494	0,4225135115	Egg A
3.	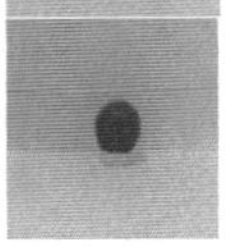	106,9994076	66,55938981	40,44757109	0,0662208853	0,6271041765	0,419631656	Egg A

IOP Conf. Series: Earth and Environmental Science **1446** (2025) 012062 doi:10.1088/1755-1315/1446/1/012062

No	Sample Egg Image	Process in MATLAB Classification Description						
		RGB Value			HSV Value			Classification Description
		R	G	B	H	S	V	
4.	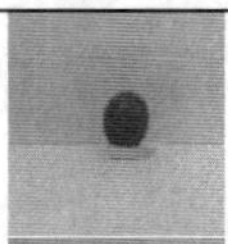	109,2356763	68,20023627	41,22445363	0,06726998143	0,6278188767	0,428393732	Egg A
5.	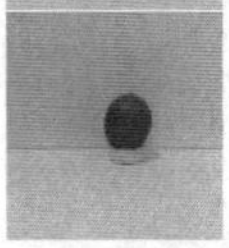	106,2822973	65,69986486	40,36743243	0,0649779751	0,6275519843	0,4168161102	Egg A
...	...	...	...	...	...	...	...	...
62.	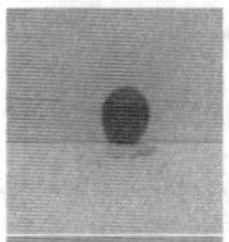	127,319086	92,13494624	56,37755376	0,08473349164	0,5569267001	0,4993042378	Egg B
63.	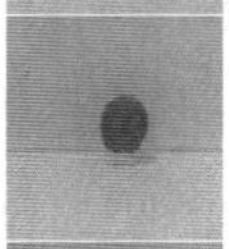	130,2616835	96,0644985	59,31525007	0,08670194663	0,5437801838	0,5108435284	Egg B
64.	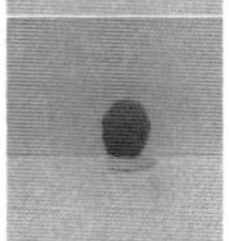	116,9102394	88,74348404	73,90398936	0,05874730211	0,370565614	0,4584918648	Egg B
65.	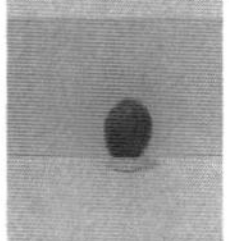	115,4141307	80,93702038	57,47486107	0,06873450437	0,5040050431	0,4526189624	Egg B
66.	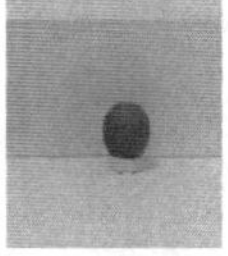	123,2057947	84,14419171	54,30801516	0,0726592936	0,5588728636	0,4831812172	Egg B
...	...	...	...	...	...	...	...	...
90.	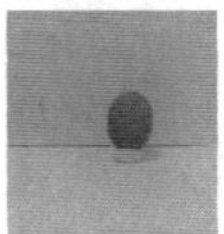	125,7411372	97,25066036	70,90365633	0,0879273111	0,4338138058	0,4933647364	Egg C
91.	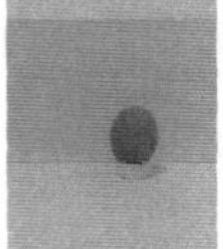	126,2049066	98,24254251	71,07973237	0,08806722295	0,4336197496	0,495126191	Egg C
92.	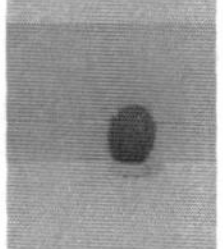	117,7170857	90,1116518	71,83806434	0,07505847754	0,3877217639	0,4619165752	Egg C
93.	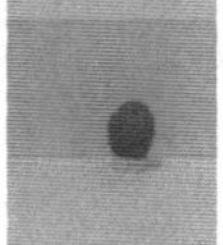	119,116229	90,55164983	72,67057239	0,06947381439	0,3871844219	0,4672840827	Egg C
94.	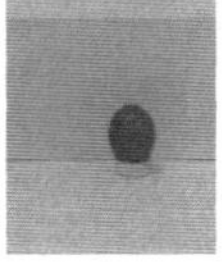	117,6809383	86,61918892	56,60039565	0,08938498358	0,5172380461	0,4617382462	Egg C

No	Sample Egg Image	Process in MATLAB Classification Description						Classification Description
		RGB Value			HSV Value			
		R	G	B	H	S	V	
95.	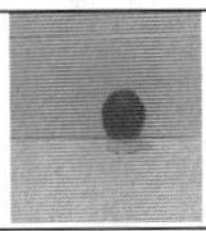	114,2192394	82,54837808	53,30313199	0,08650773984	0,5331026288	0,4481258499	Egg C

Table 1 presents the results of feature extraction from 95 images of eggs. The eggs are classified into grades A, B, or C based on their RGB (Red, Green, Blue) and HSV (Hue, Saturation, Value) values, which are obtained using MATLAB. These color values are derived from the images of the eggs and are used for classification. The differences in RGB and HSV values are correlated with the classification. An increase in the R-value from grade A to grade C indicates that grade B and C eggs tend to have a redder color. The G and B values also increase from grade A to grade C, but the increase is not as pronounced as with the R value. The H value increases gradually from grade A to grade C, indicating a shift in the color nuances of the eggs. Grade A eggs have higher saturation, showing a more saturated color. Saturation decreases in grade B and slightly increases again in grade C. Brightness increases from grade A to grade C, indicating that grade C eggs are brighter than grade A. This suggests that grade A eggs tend to have a darker and more saturated color, grade B eggs have increased brightness and decreased color saturation, and grade C eggs show the highest brightness with varying saturation.

4. Conclusion

Feature extraction for egg quality classification has been grouped into grades A, B and C. The results from 95 egg image data have varying average RGB and HSV values for each grade. Grade A has an average RGB value of 75.39 and an average HSV value of 0.36. Grade B has an average RGB value of 87.92 and an average HSV value of 0.35. Grade C has an average RGB value of 92.36 and an average HSV value of 0.33. Of the 95 grade A eggs, the number was greater with the smallest average RGB value. This can happen because the better the quality of the egg, the less the egg shell will have a mixture of colors. The results of the training data which have been validated by animal husbandry experts and through feature extraction processing in RGB and HSV space are expected to improve the optimization of the chicken egg quality classification system that will be developed.

5. Acknowledgments

The head of the P3M State Polytechnic of Jember, the head of the information technology department and its employees, the head of the software engineering laboratory, the head of the engineering and information systems laboratory, and Rizqie Laying Hens Farm, who has viewed it as a research case study location, are all acknowledged by the authors. The Jember State Polytechnic, which supplied funding for the PNBP program to carry out this study, is also acknowledged by the authors..

References

[1] B. Sakti P, S. D. Andriyani, and R. A. Rifandi, "Perlunya Implementasi Konsep Pembangunan Berkelanjutan pada Aktivitas Peternakan, Perkebunan dan Pertambangan," *Envoist Journal: Environmental Sustainability Journal*, vol. 3, no. 1, pp. 100–105, 2022, doi: https://doi.org/10.31331/envoist.v3i1.2248.

[2] Badan Pusat Statistik, "Populasi Ayam Ras Petelur." Accessed: Mar. 03, 2022. [Online]. Available: https://www.bps.go.id/indicator/24/477/1/populasi-ayam-ras-petelurmenurut-provinsi.html

[3] F. Nurdiansyah and A. Y. Rahman, "Penerapan Teknologi Untuk Mengontrol Suhu dan Kelembaban pada Budidaya Ayam Jenis Petelur," *Jurnal Aplikasi dan Inovasi Ipteks*, vol. 2, no. 2, pp. 43–49, Oct. 2019, doi: https://doi.org/10.31328/js.v2i2.1340.

[4] W. Bilyaro, D. Lestari, and A. S. Endayani, "Identifikasi Kualitas Internal Telur Dan Faktor Penurunan Kualitas Selama Penyimpanan," *Journal of Agriculture and Animal Science (Agrimals)*, vol. 1, no. 2, Oct. 2021, doi: https://doi.org/10.47637/agrimals.v1i2.418.

[5] A. Rizqita, N. Haryuni, and Lestariningsih, "Pengaruh Umur dan Tipe Kandang (Close House dan Open House) terhadap Kualitas Fisik Telur Ayam," *BRILIANT: Jurnal Riset dan Konseptual*, vol. 8, no. 2, pp. 433–440, May 2023, doi: https://doi.org/10.28926/briliant.v8i2.989.

[6] M. C. Simanjuntak and P. Robinson, "Kualitas Fisik Telur Ayam Ras Petelur Di Beberapa Peternakan Ayam Petelur Di Kabupaten Nabire," *Jurnal PARA PARA, Jurnal Ilmu Peternakan*, vol. 4, no. 2, Dec. 2023.

[7] R. E. Mudawaroch and R. Rinawidiastuti, "Kualitas fisik Telur Ayam yang di Kemas dalam Plastic Warp," *Jurnal Ilmiah Ilmu-Ilmu Peternakan*, vol. 26, no. 1, pp. 53–59, Jun. 2023, doi: 10.22437/jiiip.v26i1.24796.

[8] E. Loffredi, S. Grassi, and C. Alamprese, "Spectroscopic approaches for non-destructive shell egg quality and freshness evaluation: Opportunities and challenges," *Food Control*, vol. 129, Nov. 2021, doi: 10.1016/j.foodcont.2021.108255.

[9] A. Qurniawan, S. Ananda, K. Asgaf, H. I. Susanti, and N. Baharuddin, "Evaluasi Kualitas Eksterior dan Interior Telur Ayam Ras di Pasar Moderen Kota Makassar Sulawesi Selatan," *JITRO: Jurnal Ilmu dan Teknologi Peternakan Tropis*, vol. 9, no. 1, pp. 252–258, Dec. 2020, doi: 10.33772/jitro.v9i1.17745.

[10] A. Anwar and N. Hudaningsih, "Perancangan Alat Pengangkut Telur yang Ergonomis Menggunakan Pendekatan Ergonomi dan Value Engineering," *Jurnal Manajemen Teknologi dan Teknik Industri*, vol. 6, no. 2, pp. 146–159, 2024, doi: 10.30737/jurmatis.v6i2.5776.

[11] S. Nasion and B. Standard, "Standar Nasion al Indonesia Telur ayam konsumsi," pp. 1–16, 2018, [Online]. Available: http://slidepdf.com/reader/full/sni-39262008-telur-konsumsi

[12] R. Hanifa and D. Pitoyo, "Pengendalian Kualitas Produk Telur Ayam Di Peternak Kampung Neglarasa Menggunakan Metode Seven Tools Kombinasi Dengan Metode Six Sigma," *Prosiding Seminar Sosial Politik, Bisnis, Akuntansi dan Teknik*, vol. 5, no. 5, Dec. 2023, doi: https://doi.org/10.32897/sobat.2023.5.0.3128.

[13] F. Nurdiyansyah, S. Fatriana Kadir, I. Akbar, and L. Ursaputra, "Penerapan Convolutional Neural Network Untuk Deteksi Kualitas Telur Ayam Ras Berdasarkan Warna Cangkang," *Jurnal Mnemonic*, vol. 7, no. 1, Feb. 2024, doi: https://doi.org/10.36040/mnemonic.v7i1.8767.

[14] D. Miftahul, J. Musdar, S. Azis, A. B. Kaswar, and S. Sasmita, "Implementasi Jaringan Saraf Tiruan Backpropagation Untuk Klasifikasi Tingkat Kesegaran Wortel Berbasis Pengolahan Citra Digital Disertai Operasi Morfologi," *JIPI: Jurnal Ilmiah Penelitian dan Pembelajaran Informatika*, vol. 9, no. 3, pp. 1518–1533, 2024, doi: 10.29100/jipi.v9i3.5672.

[15] A. Kurniawan Saputro and D. N. Purnamasari, "Identifikasi Jenis Plastik Berdasarkan Reflektansi Cahaya Menggunakan Transformasi Ruang Warna HSV," *Jurnal Simantec*, vol. 11, no. 1, 2022, doi: https://doi.org/10.21107/simantec.v11i1.19732.

[16] E. Apriliyani and Y. Salim, "Analisis performa metode klasifikasi Naïve Bayes Classifier pada Unbalanced Dataset," *Indonesian Journal of Data and Science (IJODAS)*, vol. 3, no. 2, pp. 47– 54, 2022, doi: https://doi.org/10.56705/ijodas.v3i2.45.

[17] F. Baharuddin and A. Tjahyanto, "Peningkatan Performa Klasifikasi Machine Learning Melalui Perbandingan Metode Machine Learning dan Peningkatan Dataset," *Jurnal Sisfokom (Sistem Informasi dan Komputer)*, vol. 11, no. 1, pp. 25–31, Mar. 2022, doi: 10.32736/sisfokom.v11i1.1337.

[18] Y. M. F, B. Yuwono, and D. B. P, *Dasar Pengolahan Digital Edisi 2022*, 2022nd ed. Yogyakarta: Lembaga Penelitian dan Pengabdian kepada Masyarakat UPN Veteran Yogyakarta, 2022.

[19] S. Ratna, "Pengolahan Citra Digital dan Histogram dengan Python dan Text Editor Phycharm," *Technologia*, vol. 11, no. 3, pp. 181–186, May 2020, doi: http://dx.doi.org/10.31602/tji.v11i3.3294.

[20] D. Hardiyanto and D. Anggun Sartika, "Ekstraksi Fitur Citra Api Berbasis Ekstraksi Warna pada Ruang Warna HSV dan RGB," *FAHMA : Jurnal Informatika Komputer, Bisnis dan Manajemen*, vol. 16, no. 3, Sep. 2018, doi: https://doi.org/10.61805/fahma.v16i3.85.

[21] A. Firlansyah, A. B. Kaswar, and A. A. N. Risal, "Klasifikasi Tingkat Kematangan Buah Pepaya Berdasarkan Fitur Warna Menggunakan Jaringan Syaraf Tiruan," *Techno Xplore: Jurnal Ilmu Komputer dan Teknologi Informasi*, vol. 6, no. 2, pp. 55–60, Oct. 2021.

[22] Ellif, S. H. Sitorus, and R. Hidayati, "Klasifikasi Kematangan Pepaya Menggunakan Ruang Warna HSV dan Metode Naive Bayes Classifier," *Coding : Jurnal Komputer dan Aplikasi*, vol. 09, no. 01, pp. 66–75, 2021.

[23] J. E. Widyaya and S. Budi, "Pengaruh Preprocessing Terhadap Klasifikasi Diabetic Retinopathy dengan Pendekatan Transfer Learning Convolutional Neural Network," *Jurnal Teknik Informatika dan Sistem Informasi*, vol. 7, no. 1, Apr. 2021, doi: 10.28932/jutisi.v7i1.3327.

[24] Yohannes, M. E. Al Rivan, S. Devella, and Meiriyama, "Ekstraksi Fitur Warna Dengan Histogram HSV Untuk Klasifikasi Motif Songket Palembang," *Jurnal Teknik Informatika dan Sistem Informasi*, vol. 11, no. 2, pp. 386–396, Jun. 2024.

[25] W. Aliansa, H. N. Ifayatin, and R. A. Saputra, "Segmentasi Kematangan Pisang Raja Berbasis Fitur Warna HSV Menggunakan Metode KNN," *Jurnal Sains Komputer & Informatika (J- SAKTI*, vol. 7, no. 2, pp. 595–608, Nov. 2023, doi: http://dx.doi.org/10.30645/j-sakti.v7i2.669.

[26] S. Bhahri and Rachmat, "Transformasi Citra Biner Menggunakan Metode Thresholding Dan Otsu Thresholding," *Jurnal SIstem Informasi dan Teknologi Informasi*, vol. 7, no. 2, pp. 195– 203, Oct. 2018.

[27] J. Hidayat and Syafriwel, "Implementasi Metode Otsu Thresholding Pada Binerisasi Citra Wajah," vol. 1, no. 1, pp. 1–6, Apr. 2022.

AUTHOR INDEX